Fundamentals of Fracture Mechanics

Pietro Paolo Milella

Fundamentals of Fracture Mechanics

Theory, Test Methods and Applications

Springer

Pietro Paolo Milella
Mechanical Analysis and Technologies
University of Cassino, ENEA
Rome, Italy

ISBN 978-3-032-23080-5 ISBN 978-3-032-23081-2 (eBook)
https://doi.org/10.1007/978-3-032-23081-2

This Springer imprint is published by the registered company Springer Nature Switzerland AG
The registered company address is: Gewerbestrasse 11, 6330 Cham, Switzerland

To the memory of my mother Zaira and my father Michele, who read this book well before I wrote it.

Preface

This book celebrates 55 years of professional activity. There is a long story behind. A story of work and commitment to assure the structural integrity of the primary circuit of nuclear power plants in any possible operating condition. «Safety first», we used to say and that statement required a continuous R&D effort in three main fields, in particular: Fatigue, Corrosion and Fracture Mechanics that 50 years ago were not so advanced as they are today. I have treated the first two subject matters, Fatigue and Corrosion, in the previous book I wrote: Fatigue and Corrosion in Metals, whose second edition was published 2 years ago by Springer. The third argument, i.e., Fracture Mechanics, is the subject of this new book: *Fundamentals of Fracture Mechanics, Theory, Test Methods and Applications*. This book is the translation into English of a previous book I wrote in Italian: Meccanica della Frattura Lineare Elastica ed Elastoplastica, sponsored and edited by Ansaldo Nucleare in 1999, completely revised to account for advances occurred in this quarter of century and for the experience gained teaching Fracture Mechanics to practicing engineers during the last 25 years. It is also a history of Fracture Mechanics written by researchers, scientists, and experts who were initially challenged to prevent brittle fracture of welded steel structures without adequate tools using an empirical approach based on the Nihil-Ductility Transition Temperature, NDT. This temperature was measured with an old forgotten specimen introduced at the beginning of the last century by S. B. Russell, a young American engineer, today known as Charpy V-notch specimen or CV-notch specimen. That empirical approach was also used in the design of pressure vessels of the first generation of nuclear power plants entered into operation at the beginning of the 60s of the last century in the US and Italy. Later, George Irwin, who can be considered the father of modern Fracture Mechanics, introduced the stress intensity factor, K, that under the general name of fracture mechanics soon became the most powerful analytical tool that for the first time provided the long-searched relation between crack size, applied stress and toughness of metals. Parallel to this, in the UK a new approach was followed based on General Yield Fracture Mechanics that made use of material selection by measurement of its critical crack-opening displacement or COD. Today, those approaches apparently so different from each other to refuse any integration, merged into the last and most promising fracture assessment method, the R-6 Fad. The intention of this book is to offer to practicing engineers, fabrication and inspection engineers, material specialists as well as researchers and students the state-of-the-art of fracture mechanics theory, applications, and test methods as they relate to material selection techniques for metal structures. At variance with most tests on Fracture Mechanics, this textbook dedicates a particular attention to material science and metallurgy since it is my personal opinion, gained through decades of research and practice, that Fracture Mechanics cannot be fully understood and used in practice without the knowledge of material behavior and mechanical responses of different metallurgical structures. The mathematical framework, though very important, is not sufficient for using Fracture Mechanics in practical cases. Therefore, Chap. 1 examines the different macroscopic mechanical responses of metals in relation to their tensile stress-strain response under quasistatic and high strain rate conditions. Chapter 2 links these responses to the different crystal structures of metals introducing dislocations and their motion and the two main mechanism of plastic deformation: slip and twinning. Chapter 3 introduces the important concept of transition temperature and the first empirical tentative to measure this temperature with the CV-notch specimen and the drop weight test and other dynamic tear tests that led to the introduction of FAD diagram. Chapters 4 and 5 deal with the important issue of size effect, triaxiality, and temperature on the mechanical response and apparent toughness of materials. The morphological aspects of brittle fracture are treated in Chap. 6 where it is shown how brittle fracture is always preceded by ductile tearing whose extension makes the transition from a cleavage fracture to a ductile failure. After these introductory chapters, Chap. 7 enters into the mathematical treatment of linear elastic fracture mechanics (LEFM) introducing Griffith's theory and the stress analysis of cracks with the Irwin stress intensity factor, K, evidencing their limits of validity. Chapter 8 covers the issue of the crack tip plastic zone which is the basis for understanding the limits of LEFM and the transition to elastic-plastic fracture mechanics (EPFM) evidencing the effect of temperature on materials toughness. Chapter 9 treats the subject of experimental LEFM and material testing introducing all different specimens and methods used to measure material toughness and relative standards including fracture testing of welds. Chapter 10 describes the transition from LEFM to EPFM from the use of the strip-yield model and CTOD to the Rice J-integral. Experimental EPFM is the subject of Chap. 11 in which the limits of validity of the J-integral are introduced. Chapter 12 cover the issue of crack arrest and relative methods of dynamic fracture toughness measurement. Worm prestressing is treated in Chap. 13. This is a subject almost forgotten in fracture mechanics textbook which is of rather great importance since it avoids brittle fractures under conditions where conventional fracture mechanics would predict it to occur. Applications of LEFM and EPFM to pressure vessels and piping are presented in Chap. 14. This is a new experimental and analytical presentation of how to design against brittle fracture and general yield of relatively thin structure under internal pressure. Metallurgical variability is treated in Chap. 15 where the use of the Weibull analysis is presented to cover data scatter. The reader will certainly notice the absence of a chapter on both fatigue and corrosion. In the first writing of the book these important issues were present. However, I finally accepted the suggestion of the Publisher to skip over these subjects since they were already fully treated in the previous book I wrote: "Fatigue and Corrosion in Metals" using fracture mechanics, in particular, in five different chapters that were: Chapter 14 "Fracture Mechanics Approach to Fatigue Crack Propagation", Chapter 15 "Crack Tip Plastic Zone Effect on Fatigue Crack Propagation", Chapter 18 "Hidrogen Embrittlement and Sensitization Cracking", Chapter 19 "Fracture Mechanics Approach to Stress Corrosion" and Chapter 20 "Corrosion Fatigue". The interested reader, then, is suggested to consult those Chapters in the book mentioned. Finally, almost each chapter of this textbook has a Chapter Concluding Remarks section that summarizes in a short format the fundamental points of what has been discussed in the chapter.

Rome, Italy Pietro Paolo Milella

Contents

Acronyms

ASTM	American Society for Testing and Materials
BCC	Body Centered Cubic
BCL	Batelle Columbus Laboratory
BCT	Body Cubic Tetragonal
BMI	Battelle Memorial Institute
C(T)	Compact Specimen in Traction
CC(T)	Central Crack Panel (specimen) in Traction
CEGB	Central Electricity Generating Board
CMOD	Crack Mouth Opening Displacement
COD	Crack Opening Displacement
CTOD	Crack Tip Opening Displacement
DCPD	Direct Current Potential Drop
DEC(T)	Double Edge Crack Panel (specimen) in Traction
DT	Dynamic Tear Test
DWT	Drop Weight Test
ENEA	Ente Nazionale per l'Energia Atomica
EPFM	Elastic-Plastic Fracture Mechanics
EPRI	Electric Power Research Institute
FAD	Failure Analysis Diagram
FAD	Failure Assessment Diagram (R-6 Method)
FCC	Face Central Cubic
FIM	Field Ions Microscope
FTE	Fracture Transition Elastic
FTP	Fracture Transition Plastic
GY	General Yielding
HAZ	Heat-Affected Zone
HRTEM	High-Resolution Transition Electron Microscope
HSST	Heavy Section Steel Technology
ICR	Isolated Cleavage region
LEFM	Linear Elastic Fracture Mechanics
LGC	Large Geometry Change

LSY	Large-Scale Yielding
MRL	Material Research Laboratory
NDT	Nihil Ductility Temperature
NIST	National Institute of Standards and Technology
NRC	Nuclear Regulatory Commission
NRL	Naval Research Laboratory
NSCL	Net Section Collapse Load
ORNL	Oak Ridge National Laboratory
PVRC	Pressure Vessel Research Committee
PWHT	Post Weld Heat Treatment
PZSC	Plastic Zone Screening Criterion
RNB(T)	Round Notch Bar in Traction
RT_{NDT}	Reference Transition Temperature
SEC(T)	Single Edge Crack panel (specimen) in Traction
SEM	Scanning Electron Microscope
SEN(B)	Single Edge Notch specimen in Bending
SFF	Shear Fracture Factor
SGC	Small Geometry Change
SSY	Small-Scale Yielding
STM	Scanning Tunneling Microscopy
TDCB	Tapered Double Cantilever Bend Specimen
TEM	Transition Electron Microscope
TSE	Thermal Shock Experiment
USAEC	United States Atomic Energy Commission
USE	Upper Shelf Energy
WOL-T	Wedge Opening Load specimen Type T
WOL-X	Wedge Opening Load specimen Type X
WPT	Wide Plate Test

Chapter 1
Brittle Fracture of Metals

1.1 Introduction and Historical Sketch

I will here give a brief sketch of the progress of knowledge on the behavior of materials affected by structural discontinuities such as flaws or cracks. All metals subjected to external loads deform and eventually fail. Not all metals, when they fail, also fracture. Fracture is a word that has always been used in engineering to indicate brittle cleavage failures. Cleavage defines any severance that occurs along crystallographic planes. However, before World War two the term fracture had never been associated with the presence of a crack. The crack-fracture link was not really known before the Forties of the last century. Today, we shall use the term fracture to defines those particular failures caused by the presence of a crack, however small. A crack is a very sharp notch whose tip has a radius of curvature, ρ, extremely small, theoretically equal zero. In practice, a structural discontinuity can be considered a crack when its radius of curvature is equal or smaller than 0.005 mm, i.e., 5 micron. The crack is the most dangerous macroscopic defect that can jeopardize the strength of materials and structures. Therefore, no design code or rule allows its presence in a component before entering service life. If a crack is preset in a work piece, cleavage brittle fracture may occur under a stress level lower than the yield strength of the material. Brittle fracture is characterized by the absence of any plastic deformation, at least on a macroscopic level. Cleavage is the mechanism of sudden separation along crystallographic planes. Without crack failure would occur under stress higher than the yield strength and is said to be a ductile rupture. Rupture, not fracture. Figure 1.1a presents an example of ductile rupture occurred in a conventional traction test in a smooth prismatic bar of low carbon steel used for piping. Failure is characterized by necking like an hourglass indicative of large plastic deformations (permanent deformations), as would be expected in this kind of steel at room temperature. At variance, Fig. 1.1b shows a specimen of the same steel that failed by brittle fracture because a tiny corner crack was accidentally introduced by grinding during specimen

P. P. Milella, *Fundamentals of Fracture Mechanics*,
https://doi.org/10.1007/978-3-032-23081-2_1

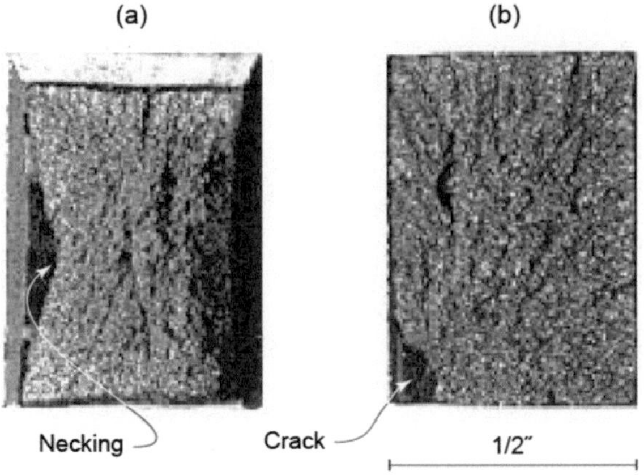

Fig. 1.1 **a** Ductile rupture of a prismatic smooth specimen of low carbon steel in a traction test, **b** brittle fracture of a specimen of the same steel, containing a tiny corner crack resulting from specimen machining

preparation. Now, the fracture is clean, flat and occurred without any apparent plastic deformation.

The tiny crack has completely changed the type and appearance of failure, turning a ductile rupture into a brittle cleavage fracture, though the material is the same. The phenomenon of brittle fracture was known to engineers since steel has been used as a structural material. It seriously worried engineers who, however, had never really understood its cause.

A gruesome illustration, appeared on the British magazine Punch in 1891, depicted this feeling of fear to cross steel bridges (Fig. 1.2). Note the long cracks drawn on the steel plates. However, they are the effect not the cause. The list of steel bridges that failed by brittle fracture is exceptionally long. However, it was not before 1941 that the phenomenon of sudden fracture of large steel structures became a real problem no longer tolerable. Historically, in fact, it has been the massive use of welding in the technique of naval constructions to turn into a rather frequent and even catastrophic event a phenomenon otherwise kept almost silent. Before the introduction of welding, in fact, hulls and steel structures, in general, were assembled by riveting so that fracture initiated somewhere in a steel plate could not cross the plate border, arresting without propagating into the adjacent plate. On the contrary, the monolithic-type structure of the new welded vessels provided a continuous path to a crack being initiated somewhere with the consequence to even demolishing the entire ship. Over a total of about 4,000 Liberty ships, T-2 tankers and Victory vessels built from 1941 to 1945, at least 1,000 experienced a brittle fracture event during World War II, developing cracks up to one meter long. In two hundred cases, brittle fracture was rather severe. Nineteen ships broke catastrophically in two. Figures 1.3 and 1.4 show two spectacular cases of complete severance of T-2 tankers that have become historical references in fracture mechanics. At that time, the cause of those disasters was completely unknown and remained unknown for more than a decade. It was not before 1957 that it was finally understood that the fracture of Liberty ships was due to three concurring events:

Fig. 1.2 "ON THE BRIDGE". Illustration appeared on the British magazine Punch in 1891 warning people not to trust steel bridges on which the Great Reaper was patiently waiting for clients. Note cracks in steel plates

1. crack formation, generally at the square corner of a welded hatch;
2. ambient temperature too low for the steel used, which resulted in a too low toughness of the metal (dependence of toughness from temperature);
3. stress levels not acceptable in relation to points 1 and 2 above, though always lower than the yield strength and within the code limits.

In general, none of those three factors individually considered or combined with only one of the other two, provides the basis for establishing whether the structure will undergo a fracture event. Figure 1.5 is a schematic interpretation of this concept. Brittle fracture is possible only in the area A of superposition of the three circles of crack dimension, a, temperature, T, hence toughness, and stress state, σ.

For example, the inspection of many Liberty ships that developed cracks but survived brittle fracture, revealed the presence of cracks even one meter long that arrested. The apparent incongruence, unexplainable at that time, can be understood to-day on the base of the cited interdependency of the three factors of Fig. 1.5. The large flaw that did not become critical, in fact, started from a very small crack and propagated through steel plates because the stress state and the material toughness

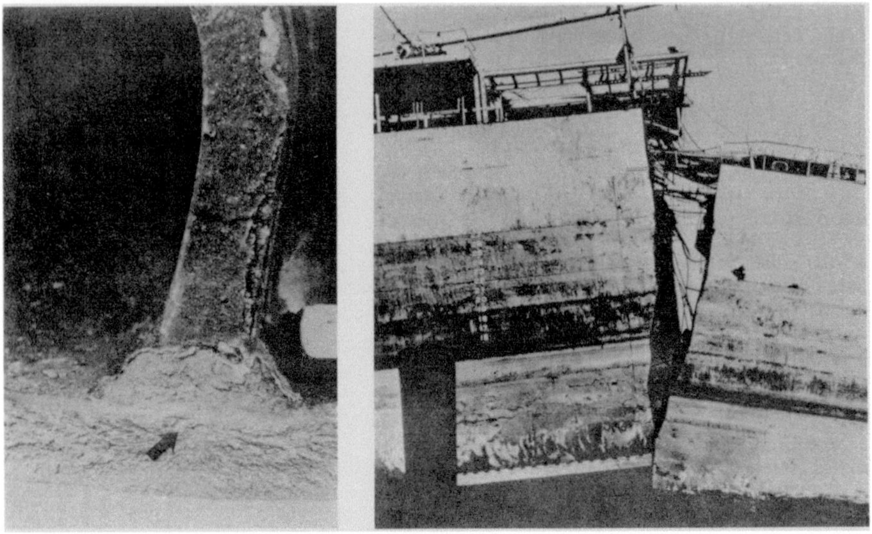

Fig. 1.3 Ship fracture at 2 °C (35 °F) resulting from tiny cracks within an arc strike (see arrow) located at the toe of a chock-bracket fillet weld [1]

Fig. 1.4 SS Schenectady T-2 tanker that broke in two amidships on 16 January 1943 [2]

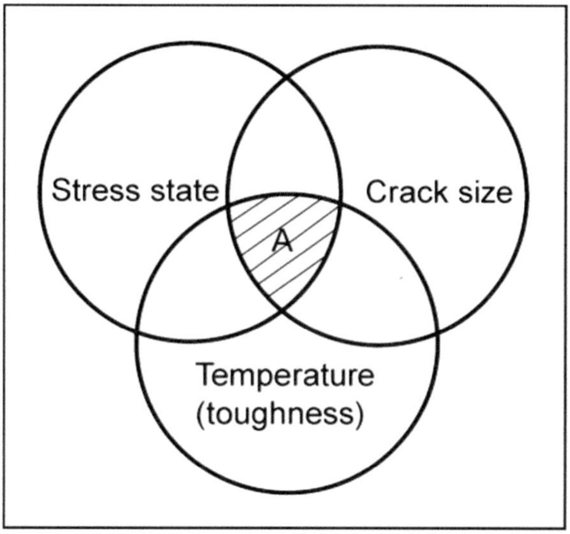

Fig. 1.5 Brittle fracture is possible only in the area A of the three concurring factors: stress state, crack size and temperature (toughness)

provided the conditions for propagation up to a distance where a new combination of crack size, stress state and toughness arrested it. Incidentally, the same tragic end of the famous Titanic liner in April 1912 might have been caused by the properties of steel used too rich in phosphorus and sulfur with low toughness at those polar temperatures.

As it will be shown in the following chapters, the purpose of Fracture Mechanics is to provide the relationship linking those three factors: crack size, a, stress state, σ and toughness, K_c. Temperature, T, has been introduced in Fig. 1.5 as relevant factor because, as it will be shown in the following sections, in ferritic steels and, in general, in all metals having Body Centered Cubic structure (BCC) it represents the parameter that, more than any other, affects the metal capability to withstand brittle fracture, known as toughness and indicated by the symbol K_{Ic}. At the time when those failures occurred, toughness was simply a concept indicating the property of materials to resist brittle fracture and was not a measurable physical property. In most cases, cracks would originate in welds because of the poor quality of the consumables or the non-consolidated weld practice and, as in the case of Liberty ships, the hurry exerted by war needs that would not allow any thermal treatment of the steel plates that were used as they would come from the mill. Moreover, electrodes not well kept in a dry environment would absorb hydrogen from ambient moisture. Hydrogen would then enter the steel during welding, embrittling the alloy. Striking the arc is a critical step to initiating the welding process. In many cases the electric arc was struck directly on the weld groove rather than away from the area to be welded and this was responsible for the formation of a tiny crater right into the weld. Moreover, the spot where the electric arc strikes the steel is heated above the steel Critical Transformation Temperature and then suddenly quenched by the large heat sink provided by the surrounding metal.

This leads to the formation of martensite, which is a very hard and brittle phase that breaks resulting in small cracks inside the crater. A tremendous combination that resulted, in most cases, in a sudden local fracture that propagated through steel plates that had very poor toughness properties further reduced by the outside temperature too low. Figure 1.6 is an example of arc struck by a negligent welder on a 30 mm diameter rebar coming out of a concrete block, improperly used as sacrificial arc starter. He saved the work piece, but not the rebar. It is possible to recognize the ark strike in the form of a tiny nail-shaped surface crack, about 3 mm depth, on the lower right side of the rebar. The concrete block was later lifted to be set on place during a rigid winter night with temperature close to zero. During the positioning just the rebar with the ark strike hit, by chance, another concrete block and the high strength steel rebar broke in two like a grissino. Crack, temperature and stress combination (impact loads are tremendous) were responsible for that failure. Note the flat fracture surface (absence of plastic deformation) with fan-shaped radial marks leading back to the initiation point, typical of brittle fracture. Liberty ships were built in some 15–40 days, with a record of just 5 days, Fig. 1.7. This fast construction, due to war time needs, would leave no time for accurate welding procedures to follow. But the real problem, as it emerged only some 10 years later, was the rather high brittle to ductile transition temperature of steel plates used in the construction of the ships. In those times the concept of transition temperature or NDT, *nihil ductility temperature*, was completely unknown. Steel plates used in the construction of Liberty ships were welded as they came out of the rolling mill without any heat treatment that could restore their mechanical properties, beside that toughness was just a concept and not a material property that could be measured, same with the transition temperature. As a consequence of that lack of any thermal treatment, the NDT of steel plates was rather high compared to some working temperatures very low during winter months at northern latitudes. In the case of the Liberty ship of Fig. 1.3, for example, the working temperature at fracture was only 2 °C much lower than the NDT of the steel plates that was about 10 °C. All ships launched in warmer months or that were initially sent to warmer seas never underwent a brittle fracture event, though they were successively sent in arctic seas. The steel plates, exposed first to temperatures higher than the NDT of the material, were protected by the beneficial effect of a warm prestressing treatment, see Chap. 13. However, at that time the concept of warm prestressing was completely unknown. It is not a case that the other two ocean liners twin to the Titanic, the Olympic and the Britannic, did not experience any brittle fracture event. The first ship was launched in summer and the second in winter, but she made her maiden journey to Naples, Sicily and other warm harbors of the Mediterranean Sea.

It is of fundamental importance to understand the role of temperature on the brittle fracture process. Brittle fracture occurs only below or near the NDT temperature of the material. However not all metals have a brittle-to-ductile transition temperature or NDT. Those having a BCC crystal lattice (α-iron or ferritic steels, tungsten, vanadium, molybdenum, chrome and their alloys) have

Fig. 1.6 Arc strike on a 30 mm rebar producing a nail-shaped tiny crack that triggered brittle fracture. Note the marks on the flat fracture surface heading back to the point of fracture initiation (the crack)

an NDT. Already at 30–40 °C over the NDT, the fracture toughness, K_{Ic}, becomes so high ($K_{Ic} \gg 150$ MPa$\sqrt{m}$) to prevent any real brittle fracture event even in very thick members (see Sect. 3.7). At those temperatures, Linear Elastic Fracture Mechanics or LEFM cannot be applied or, if used, it is conservative and it may predict brittle fracture that cannot be possible at all (see Sect. 9.10). Metals having a Face Centered Cubic lattice (FCC), such as γ-iron or austenite or austenitic steels, aluminum, copper, nickel and their alloys do not have, in general, an NDT. As such, their toughness is almost independent of temperature that does not play any role in the fracture process. Martensitic steels do not show, in general, a transition temperature. However, their toughness and that of aluminum alloys may be so low to make them susceptible to brittle fracture being $K_{Ic} < 50$ MPa$\sqrt{m}$.

1.2 The Safety Factor

Brittle fracture represents one of the major threats a designer can face. Brittle fracture happens suddenly, without any warning and propagates through a structure thousands of meters per second with consequences often catastrophic. The first step historically

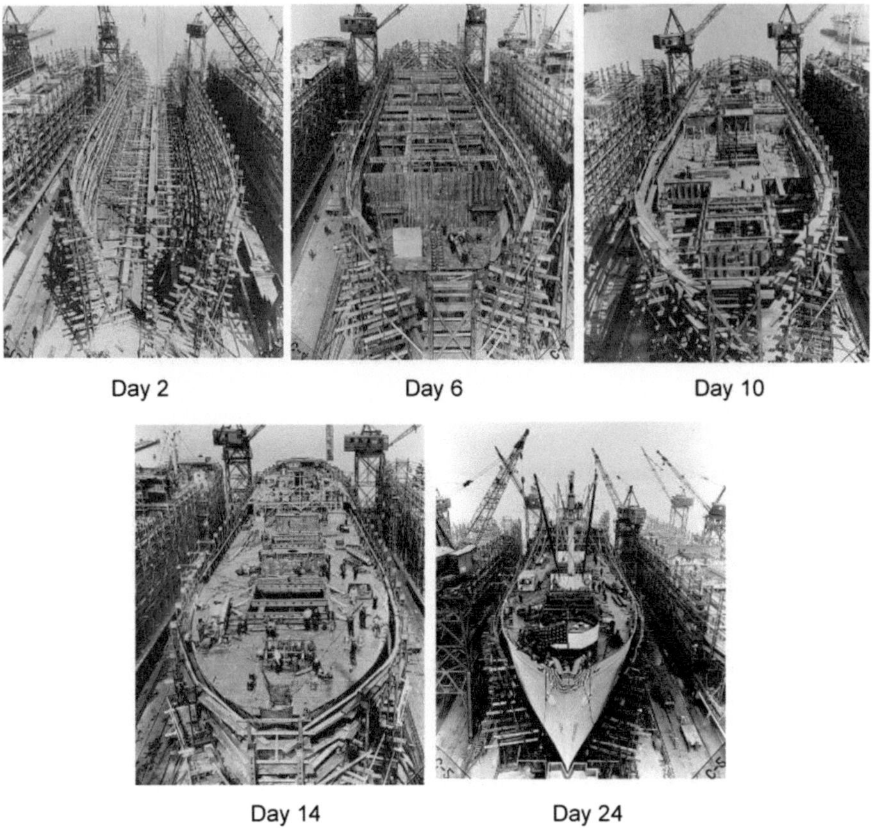

Fig. 1.7 The construction of a Liberty ship at the Bethlehem-Fairfield shipyards, Baltimore, Maryland in March–April 1943

undertook to prevent such a failure followed the approach of traditional engineering revealing the total lack of a theory, though approximate, about the initiation and propagation of brittle fracture and its relation to metallurgical and mechanical properties of materials. It was thought, in fact, that the problem of brittle fracture could be solved by simply reducing the working stress to a fraction of the yield strength through the introduction of a safety factor, well known in the engineering practice. However, to constrain a stress below a given limit value having no idea at all of how high or low that limit had to be, could not represent a valid design criterion, proof is it the great number of brittle fracture of large section of medium–high strength steel used in the construction of bridges, tanks and forged turbine generators occurred all over the world around mid-Fifties of last century [3]. The disasters of steel bridges, evocated in the nineteen century with great suggestiveness, Fig. 1.2, continued in the last century. Figure 1.8 presents the dramatic scene of the brittle fracture event of Duplessis Bridge at Three Rivers, Quebec, Canada on January 31, 1951 at −31 °C.

Four of the eight spans kept by steel welded beams collapsed. The steel composition was 0.4% C and 0.12 S. In those years, the construction of steel bridges seemed not to pay any attention to the problem of brittle fracture until the tragic event of Point Pleasant suspended bridge, in West Virginia, took place on December 15, 1967, Fig. 1.9. The bridge collapsed without any warning taking the lives of 46 people. The disaster was due to the failure of the lower limb of one of the high strength eyebars (more than twice the tensile strength of mild steel!) that was affected by a 2.5 mm deep crack. Also in this case, the winter temperature played its role. Figure 1.9b shows an eyebar of the suspension rods [4]. The electric power generation industry was also struck by this type of catastrophic failures. Figures 1.10 and 1.11 show two examples of catastrophic brittle fracture of the 1346 mm diameter steam generator rotor retaining ring made of SAE 3335 forging, of 690 MPa yield strength and 854 MPa ultimate and that of a turbine spindle, 210 cm maximum diameter made of ASTM A293 forging 595 MPa yield and 777 MPa ultimate, respectively [5]. The accidents continued during the Sixties and even in the Seventies always in welded structures under stresses lower than the yield strength. Practically, no field of engineering applications survived the shame of brittle fracture. Two accidents became notorious in 1954, two years after the introduction of the first commercial jet plane in the aviation history in 1952.

Two de Havilland Comet passenger jet planes, Fig. 1.12, crashed within a few months of each other. The common cause was metal fatigue. The crack originated at the corner of a far too sharp window, curiously not circular or elliptical, but rather square shaped. The cracks once originated could not arrest since, even more singular, the fuselage was a one-piece shell chemically bonded of ultra-high strength

Fig. 1.8 Duplessis Bridge, Three Rivers, Quebec, Canada, collapsed on January 31, 1951

(b)

(a)

Fig. 1.9 **a** Point Pleasant Silver Bridge collapsed on 15 December, 1967, **b** an eyebar used in the construction of the bridge [5]

aluminium alloy of the 7000 series, too brittle for that application. It comes to mind the failures of the Liberty Ships and T2 tankers during World War II whose hull was monolithic due to introduction of welding in the ship construction technique and made of brittle steel. If a crack would develop at a square hatch corner on the welded deck it would not arrest. A more recent brittle fracture of a ship occurred in 1972 is shown in Fig. 1.13, again recalling the Liberty ship disasters of thirty years before. Failures caused by brittle fracture continued during the 50s. Medium–low strength materials were very much penalized by the use of a safety factor that required large sections to cover higher loads. The overall weight would also increase. Just think that the safe-service record for brittle cast iron would suggest a safety factor of 10 or larger, even 20. A safety factor that high cannot be accepted in the engineering practice. It clearly indicated the limits of design practice of that time and represented a toll much too high to pay to the lack of knowledge. New high-strength alloys were developed to cope with the stringent requirements of a safety factor well too high. Only increasing the strength, in fact, both the thickness and the weight of a structure could be reduced. Unfortunately, this last misplaced confidence would reveal tragically unsuccessful with the failure of large rocket cases built with high-strength steel. Figure 1.14 is an example of a high-strength steel rocket case shed into pieces by brittle fracture during a proof test [1]. It would be definitely understood that, in general, high strength alloys are more prone to brittle fracture than low-strength mild steel that can count on a large reservoir of ductility to face brittleness.

Fig. 1.10 Turbine generator failure resulting from the fracture of 53-in-diameter retaining ring [5]

1.3 Ductility, Resilience and Toughness

The old saying of Italian metallurgists that iron is a gentleman revealed definitely. Let's consider, for example, a family of carbon steels as shown in Fig. 1.16. For carbon contents lower than 0.2%, curves 1 and 2, the yield strength is rather low and the metal is ductile. The more the yield strength increases, the more the ductility reduces to disappear completely for the steel having a carbon content of about 1.5%, curve 7 of Fig. 1.16.

Ductility is the property by which a material can absorb energy through plastic deformations (Fig. 1.16). The area A under the stress–strain curve of Fig. 1.16 represents the deformation energy density W (energy per unit volume) absorbed by the material

Fig. 1.11 Remains of a turbine spindle following fracture and ejection of the 7-ft-diameter center section [5]

Fig. 1.12 De Havilland Comet passenger jet plane that crashed in 1954 because of a fatigue crack initiated at the corner of a square shape window

$$W = \int_0^{\varepsilon_{\max}} \sigma \cdot d\varepsilon. \tag{1.1}$$

In ductile materials this energy is almost completely fed by plastic deformation, while elastic strain provides a negligible contribution. As we will see, the only way to survive a brittle fracture is through plastic deformation, in particular ahead of the crack tip where very large strains need to be accommodated by the material. If the material cannot respond to this request of deformation it fractures. Then, any increase of the yield strength, to which a decrease of ductility is generally associated, will be followed also by a loss of toughness that represents the material property to

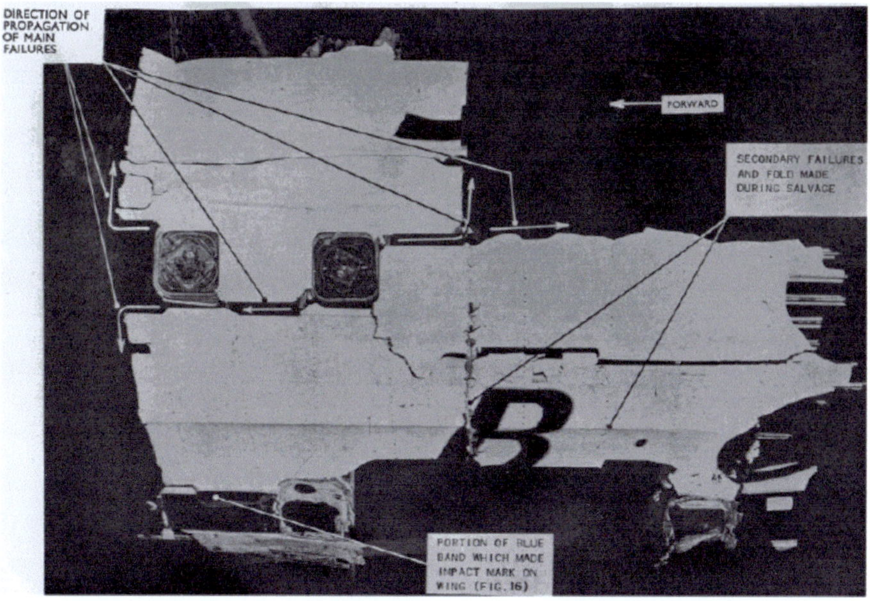

Fig. 1.13 One piece shell chemically bonded of ultra-high strength aluminium alloy of the 7000 series. The crack originated by fatigue at the corner of a window and propagated flat on the brittle shell

Fig. 1.14 Oil barge, Martha R. Ingram, which failed on January 10, 1972 (from The New York Times)

Fig. 1.15 Brittle fracture of a high-strength steel rocket case during a proof test [1]

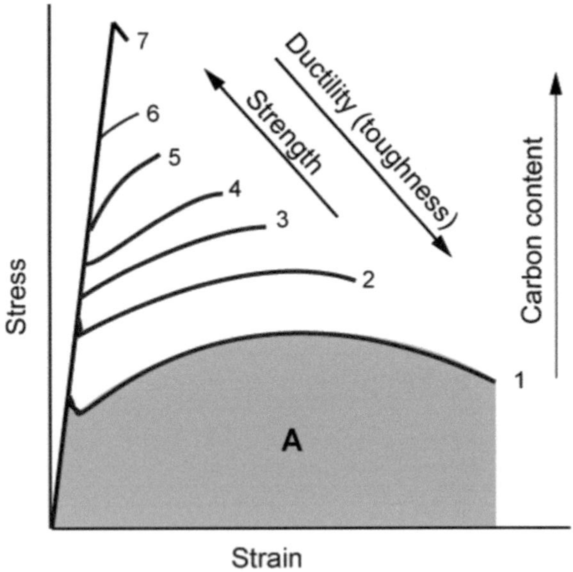

Fig. 1.16 Stress–strain characteristics of a family of carbon steels having different carbon content

withstand brittle fracture, i.e., failure caused by the presence of a crack, and indeed, the loss of toughness is by far larger than the increase of strength that caused it.

This is true independently of the factors that are responsible for the ductility change that can be of metallurgical nature, such as the carbon content already seen, or a heat treatment or the presence of non-metallic inclusions or a temperature variation since we know that both the yield and ultimate strength of BCC materials, in particular, depend on temperature (see Sects. 2.11 through 2.14 and 9.10). Under dynamic loads the plastic energy absorbed by the material, that is referred to as impact energy or *resilience*, can sharply decrease because the characteristic stress–strain curve of the material changes and the metal initially ductile, may become brittle (see Sect. 3.4). Toughness is generally higher in steels like those having low yield strength and high resilience with a stress–strain characteristic like that of curves 1, 2 or 3 of Fig. 1.16. At variance, steels having a stress–strain characteristic like that of curves 6 or 7 of Fig. 1.16 have high strength but low resilience and toughness with modest when almost inexistent plastic deformations.

We will see that ductility, resilience and toughness have a strong, intimate connection. The stress state, whether uniaxial or multiaxial, may also change the ductility of a material and, therefore, its toughness and even some environmental conditions can do it. We will be discussing in the next chapters all those factors that can have an impact on the ductility and toughness of a material.

1.4 Experimental Stress–Strain Curves

The tension test is not only the most common, but also the first test introduced in engineering practice to infer material properties from relatively small and simple specimens. As matter of fact, it dates back some 600 years when Leonardo da Vinci first decided to measure the traction strength of iron wires using a test rig like that shown in Fig. 1.17a that he drew between 1484 and 1490 and accurately described in what is today known as Codex Atlanticus (sheet 222). An iron wire was loaded in traction by the weight P (*Peso* = Pound) of a basket continuously fed by sand flowing through a hopper. As the wire broke, the basket fell closing with its hook the opening of the funnel so that no more sand could flow in and the exact weight of the basket could be measured. The traction test was later reintroduced by Galileo Galilei in 1638 [6] who presented the first serious mathematical treatment of the elastic strength of a material subjected to bending. Figure 1.18a, b are two amazing drawings made by Galilei showing the test rig for measuring the strength of materials in traction and bending, as well. It should be noted how the sketches seem to be inspired by the Italian Renaissance: the traction specimen, Fig. 1.18a is a column with its capitol which, by the way, a today engineer would ironically say it was introduced to prevent the stress concentration effect at the fix end. The bending frame of Fig. 1.18b is the romantic wall of a ruin, so abundant in the Italian countryside, where even details about weeds are drawn. Centuries have gone since those pioneering days and the traction test has come to maturity, remaining one of the basic tests that engineers use the most. The entire strength of materials is based on the stress–strain curves of materials. However, a tensile test is not that simple rehearsal one may so naïvely think. Orowan [7] in 1944 put it very clearly saying that «..*it is not possible to do much with its [traction test] results, because one does not know what they really mean*». The Orowan doubts were sheared by an even more skeptical J. J. Gilman [8] who in 1969 wrote: «*Of all mechanical tests for structural materials, the tension test is the most common. This is true primarily because it is a relatively rapid test and requires simple apparatus*».

«*It is not as simple to interpret the data it gives, however, as might appear at first sight*».The test encrypts information about the real nature of deformations and failure of the material that can be unveiled by just having the key to do that. Therefore, a complete knowledge of traction tests and the information they can actually yield are fundamental in engineering. When a polycrystalline specimen is subjected to a traction test up to its final failure one of the two basic types of stress–strain responses will be observed, as shown in Fig. 1.19a, b, known as engineering curves. In both cases, an initial phase of proportionality, 0-A, can be identified in which strains are linearly related to stresses through the Hooke's law named after Robert Hooke who first noticed this linear relation in 1676.

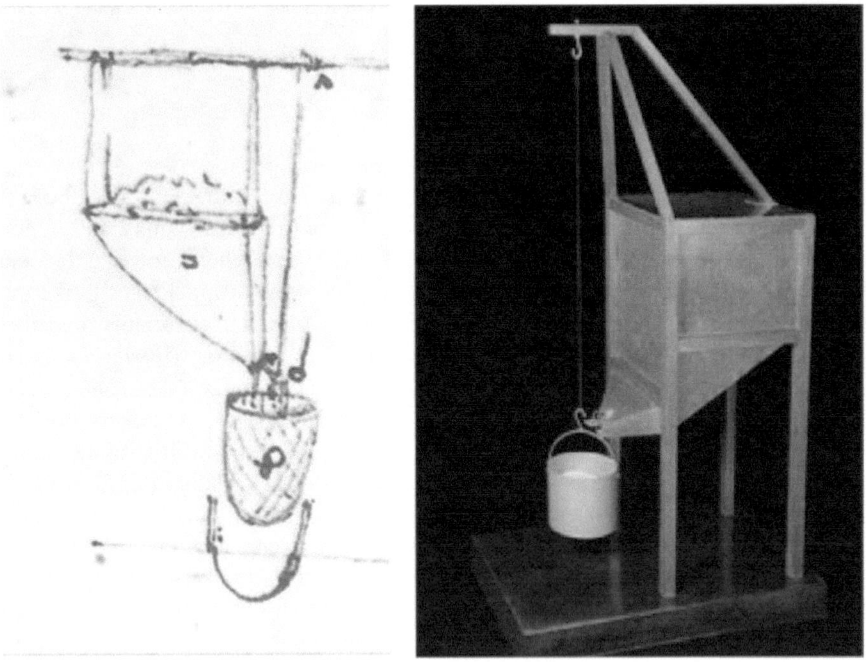

(a) (b)

Fig. 1.17 **a** Drawings by Leonardo da Vinci of a traction test rig and **b** its reconstruction at the Museum of Science and Technique in Milan, Italy

Fig. 1.18 Galileo Galilei drawings showing traction and bending test rigs for material strength evaluation [6]

$$\sigma = E \cdot \varepsilon \tag{1.2}$$

with E being the Young's modulus of the material. Curious enough, Robert Hooke published his finding in the form of an anagram in Latin "*c e iii n sss tt uu*" whose solution is "*ut tension sic vis*" which translated into English gives "as strain (grows) so (does) force". This is the elastic part or state of the diagram where strains are fully reversible. Strain is a rather generic term to describe both elastic and plastic deformations, which are actually caused by two completely different mechanisms and are, therefore, totally different. During the elastic phase, strains are the result of the stretching of the atomic bonds. Atoms of the crystalline lattice are displaced from their equilibrium position by the action of a traction stress. The crystal lattice, then, elongates in the load direction y according to Eq. (1.2) and contracts in the other two x and z directions by a quantity

$$\begin{aligned} \varepsilon_y &= \frac{\sigma}{E} \\ \varepsilon_x &= \varepsilon_z = -\nu \cdot \varepsilon_y \end{aligned} \tag{1.3}$$

where ν is Poisson's coefficient. This mechanism of elongation that changes the volume of the material is schematically shown in Fig. 1.20a. From the sub-microscopic scale up to the macroscopic one, the elastic elongation is always uniform,

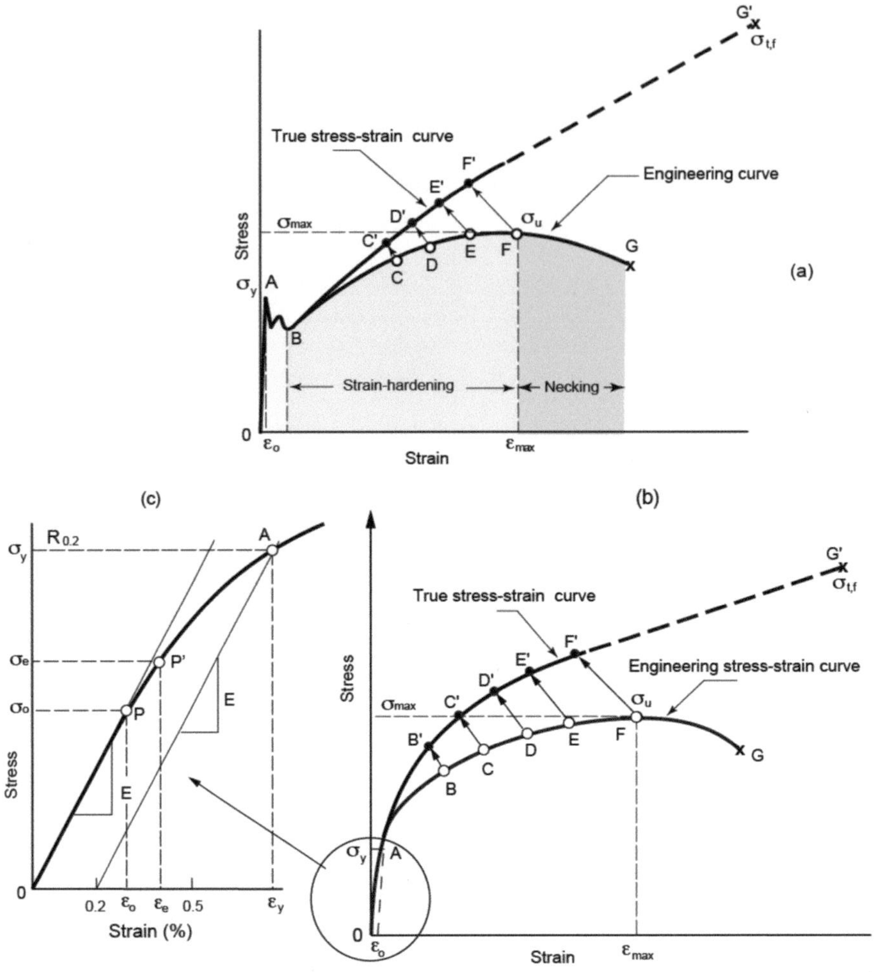

Fig. 1.19 **a** Type I and **b** type II stress–strain curves. **c** Detail of **b**

well-ordered and mostly affine. The work needed to cause elastic elongation is stored as elastic strain energy in the crystal and can be quickly recovered almost completely upon removing the applied load. Because of contractions in x and z directions, elastic elongation is always resulting in a ΔV variation of the overall volume.

Even though the elastic dilatation can theoretically continue up to the failure of the specimen and pass 100%, it cannot go over a mere percent. This is because as traction increases, a second component of stress starts to play a role becoming more and more competitive. This is the shear stress. In the competition between normal stress, σ, and the shear stress, τ, acting at 45° (see detail of Fig. 1.20a) it is this last that at a certain moment prevails. Slip becomes easier than dilatation. When

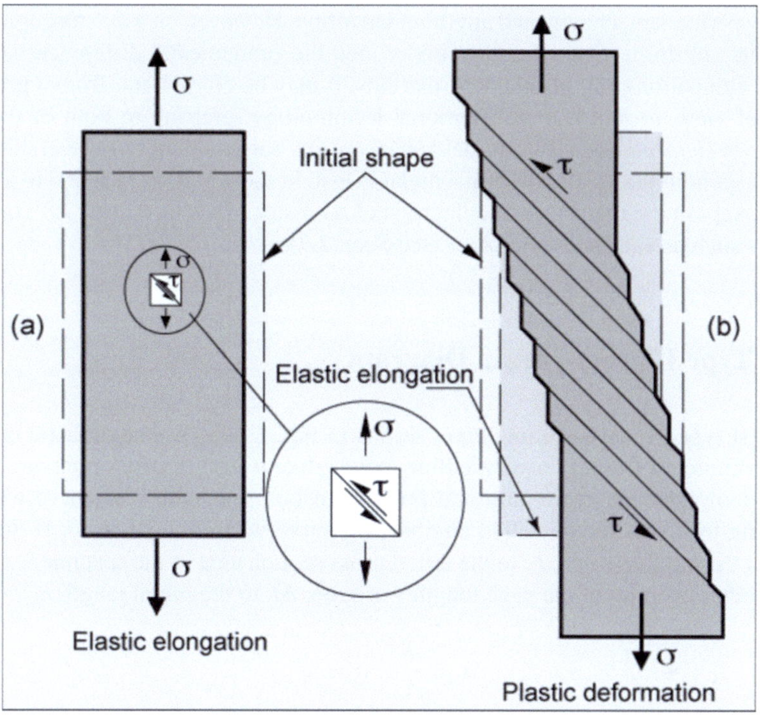

Fig. 1.20 **a** Elastic elongation of a specimen under traction occurs with volume change and **b** plastic deformation of a specimen under traction occurs without volume change

this happens, the stress is said to have reached the elastic limit, σ_o, and the material to enter the plastic domain characterized by large deformations. The point at which this elastic limit, σ_o, occurs is indicated by the letter *A* in Fig. 1.19a and *P* in the detail of Fig. 1.19c. This limit, σ_o, is also referred to as the yield strength, σ_y, of the material even though they are not precisely the same thing. Plastic deformation by shear stress is a completely different process that occurs primarily by slips along crystalline planes. The metal is said to flow. This produces a block-like structure, at microscopic level, with crystalline planes moving relative to each other, as schematically shown in Fig. 1.20b. Deformation, then, is localized into shear or slip bands producing a kind of composite texture made of alternate layers of non-deformed and plastically deformed material. This fragmentation continues beyond microscopic dimensions for 3–4 orders of magnitude further, down to atomic dimensions. Therefore, plastic deformation is inhomogeneous because, at variance with the elastic dilatation, it occurs by the shearing of all blocks of crystals in a polycrystalline aggregate. This process bears very important consequences since it produces local, tiny notches where slip bands cut the metal surface. The local stress concentration, at these microscopic notches, aids the slip process in the bands to continue. At variance with the elastic elongation of the crystalline solid, the plastic deformation or flow is usually permanent, chaotic and anything but affine. However, on a macroscopic scale it appears uniform. Once we have underlined the fundamental difference between elastic elongations and plastic deformations, it may be convenient, from a practical point of view, to return to conventional terminology referring to both of them as *strains* or *deformations*. Plastic deformations are not completely recoverable. The energy spent to plastically deform a metal is lost, in most part, as heat and to a lesser extent, kept in the metal as line defects, such as dislocations (see Sect. 2.8), or point defects such as vacancies and other (see Sect. 2.7).

1.5 Type I Stress–Strain Diagram

The first type of stress–strain curve, shown in Fig. 1.19a, is characteristic of BCC (Body-Centered Cubic) polycrystalline metals, like α-iron (ferrite), tungsten, vanadium, molybdenum, chromium and ferritic steels, in general. The curve we'll be referring to first is the so called engineering curve where the stress, σ, is given as ratio of the traction force, F, to the initial cross section area of the specimen, A_o, and the strain, ε, as ratio of the gage length variation, Δl, to the initial length l_o

$$\sigma = \frac{F}{A_o}; \; \varepsilon = \tfrac{\Delta l}{l_o} \,. \tag{1.4}$$

Carbon steels exhibit what is known as *sharp yield point* (point A in Fig. 1.19a) or *upper yield point* followed by a *lower yield point* (point B in Fig. 1.19a. See also Sect. 2.13). Afterwards, before entering the field of plastic deformations, the material shows a kind of plateau where a series of serrations appears of very small amplitude. In this plateau, strain increments seem to happen spontaneously almost without any stress increment, though they remain confined within just some tenth of percent. This behavior is typical of metals having narrow dislocations (see Sect. 2.10) with a concentrated Cottrell's atmosphere made of small interstitial atoms, such as carbon, nitrogen and oxygen that exert a drag force when dislocations are forced to move. The complete significance of the yield point is explained in Sect. 2.13. Beyond this plateau, there is little difference between type I and type II behavior of materials, and the stress–strain curves appear practically the same. Therefore, all what will be said about this type of stress–strain behavior will also hold for type II diagram. Continuing to increase the applied load, the material is said to strain harden (see Sect. 1.6) regaining its capability to withstand external loads, though at lower rate with respect to the linear elastic behavior. It opens, now, the field of large plastic deformations, at least in medium–low strength materials, and the material is said to flow. At first, deformations are uniform, at least on a macroscopic scale since, as previously said, they are not on a microscopic level. It is within this phase and because deformations are non-homogeneous on a microscopic scale that damage appears in the form of pores and cavities around dislocations, inclusions and foreign phases, in general. Now, deformations occur without variations of the plastic volume since they are the result of slips along planes at 45° to the stress direction (see Fig. 1.20b)

$$V_p = A \cdot l = \text{constant} \tag{1.5}$$

where V_p is the plastic volume, A is the cross section area and l the instantaneous specimen length. Because the relative change of volume $\Delta V/V$ is related to the three strain components by the relation

$$\frac{\Delta V}{V} = \varepsilon_x + \varepsilon_y + \varepsilon_z. \tag{1.6}$$

and the three deformation components are linked together by Eq. (1.3), Eq. (1.6) yields

$$\frac{\Delta V}{V} = \frac{\sigma}{E}(1 - 2\nu). \tag{1.7}$$

This actually means that as the material enters the plastic domain where deformations occur without volume variation, $\Delta V_p = 0$, the only way to comply with Eq. (1.7) is that the Poisson's ratio, ν, be equal exactly to 0.5. Actually, the variation (increase) of Poisson's ratio develops gradually as the deformation regime changes from linear elastic to plastic, as shown in Fig. 1.21 for carbon steel. The fact that Poisson's ratio or coefficient is 0.5 in the plastic domain has a dramatic consequence on the ductility of materials, as will be shown in Sect. 4.2. As more and more deformation advances in the slip bands and the gage length l increases, the cross-section area of the specimen becomes smaller and smaller, as predicted by Eq. (1.5). This raises the effective cross section stress, thereby increasing the material tendency to flow. What generally happens in the plastic domain is that at a certain moment some part of the gage length deforms at a slightly higher rate than the rest. Deformations begin to concentrate at that point that are no longer uniform. The continuous slip in this reduced section will occur more easily than in the remainder of the specimen. In this manner neck can eventually form (Fig. 1.22).

The overall continuous process of elastic uniform strains followed by plastic uniform deformations, neck initiation and progression and eventually necking failure is schematized in Fig. 1.22. The letters A, B, C, D, E, F and G associated with each stage of deformation are the same used to indicate stages in Fig. 1.15a, b. After a certain plastic deformation, prior to neck formation, damage starts to develop and accumulate. It generally takes the form of void formation around inclusions and foreign phases (see Sect. 6.3). Initially, this process of inclusion/matrix decohesion is modest, but it heavily proceeds during continuous deformation. Pores so created grow more and more and eventually coalesce contributing to the crisis of the material. The increasing number and size of these voids determines a loss of net cross section area and, therefore, of strength. This loss of material strength can be seen through the parallel reduction of the Young's modulus E. Therefore, if during the plastic deformation phase the specimen is unloaded and then reloaded at different plastic

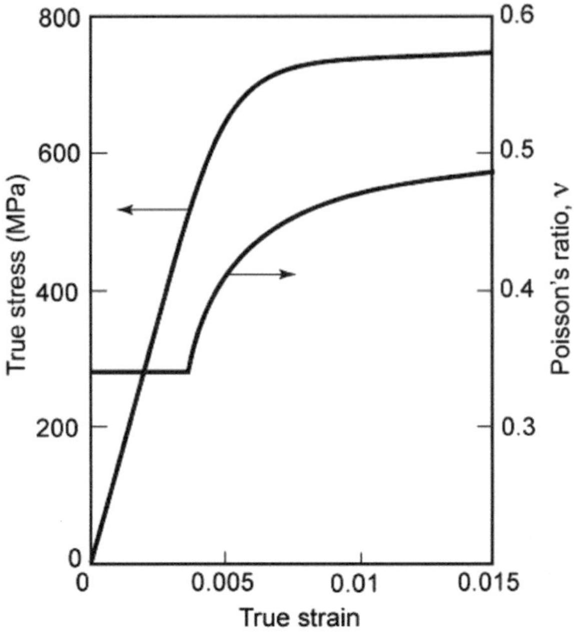

Fig. 1.21 Variation of the Poisson's ratio as a function of true stress–strain for carbon steel

strains it can be noticed that during each unloading the material would follow always a linear path, but with a different slope lower than the initial one that represents the Young's modulus that continuously reduces. This is schematized in Fig. 1.23. This slope variation or the ratio of the ith E_i unloading to the initial one E is a measure of the damage introduced in the material. What actually happens entering the plastic domain and, in particular, during neck formation and progression is that the cross-area reduction experienced by the specimen, because of necking and internal void formation, is opposed by the continuous strain hardening of the material which prevails so that the resulting stress–strain curve is always increasing, tough at lower and lower rate. At necking, instead, this unbalance is lost. Now, the cross-section reduction $-dA$ equals the strain hardening increment $+d\sigma$

$$\frac{d\sigma}{\sigma} = -\frac{dA}{A}. \tag{1.8}$$

Note that when this happens the contribution of voids formation to the net cross area reduction is still modest. From this point on, the material capability to withstand the applied load decreases more and more and the specimen collapses. This is shown by the F-G downtrend of the engineering curve (see Fig. 1.15). At necking, then, the material exerts its maximum strength, and the corresponding stress is named maximum stress σ_{max} or ultimate strength σ_u. In the engineering stress–strain curve the corresponding strain is the maximum strain ε_{max} apparently reached by the material. Deformations beyond this point will be considered in Sect. 1.6.1.

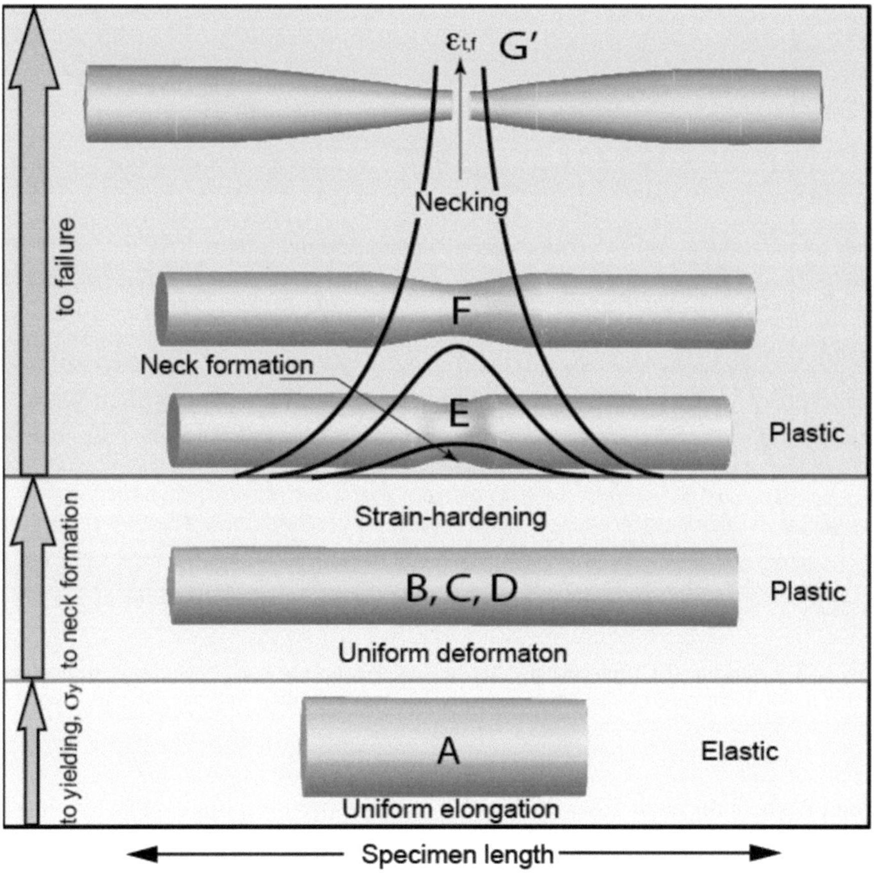

Fig. 1.22 Schematic of deformation stages in a specimen under traction. Letters A, B, C, D, E, F and G refer to Fig. 1.11a, b

1.5.2 Type II stress–strain diagram. The second typical stress–strain curve is shown in Fig. 1.19b. This is characteristic of FCC (Face-Centered-Cubic) materials like copper, aluminum, nickel and γ-iron (austenite) (see Sect. 2.1). Austenitic stainless steels show this behavior. There is no longer a sharp yield point. A continuous curve now appears on which it is very difficult or even impossible to accurately establish where the linear part, where Hooke's law (1.2) applies, terminates and the non-linear plastic behavior of the material initiates. Yield strength may be very arduous to find, and its value becomes arbitrarily defined. For this very reason material testing standards introduce a conventional yield strength by tracing a 0.2% strain offset line parallel to the initial slope of the stress–strain curve (tangent to the curve by the origin). The conventional yield strength, or 0.2% proof stress $\sigma_{0.2}$, is taken where this line intersects the stress–strain curve of the material, point A in Fig. 1.19c. This offset yield point or yield stress is also indicated as $R_{p0.2}$ or simply $R_{0.2}$ or even and σ_y.

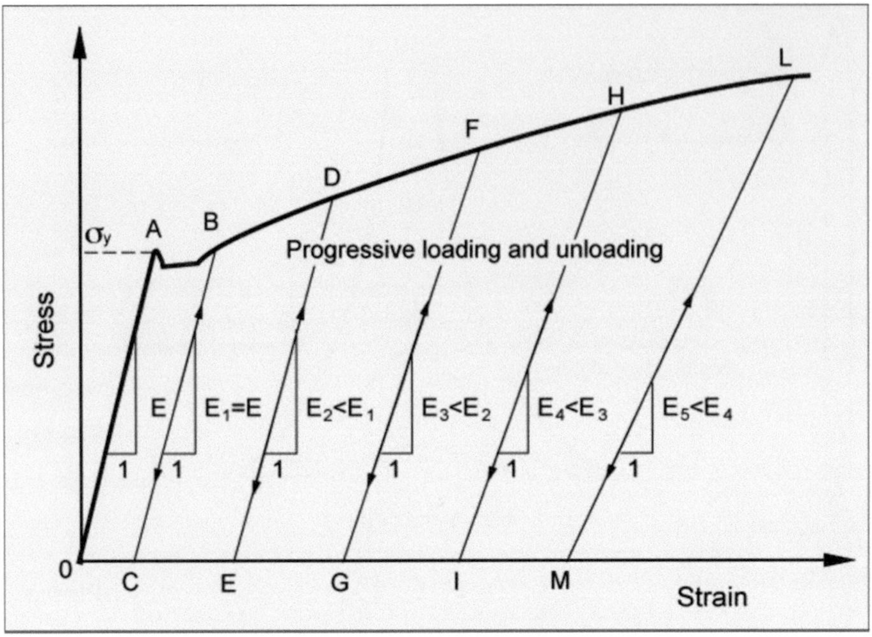

Fig. 1.23 Unloading (BC) immediately after yielding does not show any slope change with respect to the initial linear elastic behavior. Conversely, loading–unloading well into the plastic domain shows a continuously decreasing slope indicating damage

Point P where the curve loses its linearity is the proportional limit ε_o that terminates the Hooke's law validity. The corresponding stress is σ_o. The elastic behavior, where strains are completely recoverable, terminates with the elastic limit ε_e just a little bit beyond ε_o in P′, as shown in Fig. 1.19c. Beyond the $R_{0.2}$ yield point the stress–strain curve becomes similar to the BCC curve of Fig. 1.19a and all what has been said for that curve about necking, in particular, also applies.

Worth noting is the linear relationship between the $\sigma_{0.2}/\sigma_u$ ratio and the nondimensional 0.2 proof stress, $\sigma_{0.2}/E$, found by Rasmussen [9] for a large variety of austenitic stainless steels and duplex alloys.

This is shown in Fig. 1.24 [9]. The steels, in particular, were type 304 and 316 stainless steels, in various annealed or cold worked conditions, having 0.2 proof stress varying from 286 to 612 MPa and ultimate strength σ_u from 611 to 780 MPa and, type UNS31803, with 0.2 proof stress from 575 to 699 MPa and ultimate strength from 805 to 878 MPa. Some ferritic steels can show this behavior. The linear trend is defined by the equation

$$\frac{\sigma_{0.2}}{\sigma_u} = 0.2 + 185\frac{\sigma_{0.2}}{E}. \tag{1.9}$$

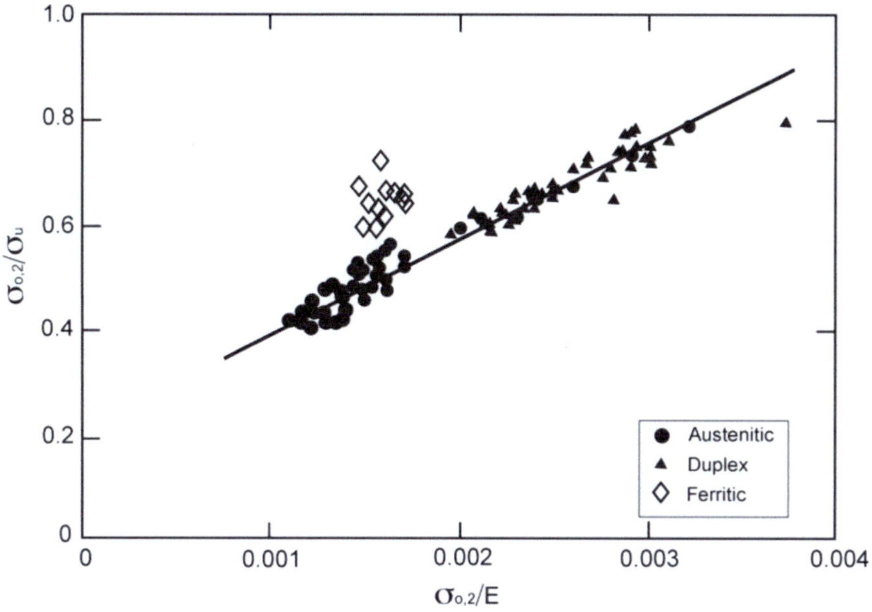

Fig. 1.24 Plot of the ratio $\sigma_{0.2}/\sigma_u$ against the nondimensional proof stress $\sigma_{0.2}/E$ (modified from [9])

The characteristic shape of type II stress–strain curve is attributed to the existence of wide dislocations (see Fig. 2.33) with a diluted Cottrell's atmosphere made of small interstitial atoms, such as carbon, nitrogen and oxygen that cannot exert a large resistance when forced to move. Therefore, the passage from linear elastic behavior to elastic–plastic flow is made easier and continuous. However, many high strength steels also present type II behavior.

1.6 True Stress–Strain Curve

Independently of the particular stress–strain response of the material, whether that of Fig. 1.19a or that of Fig. 1.19b, both stresses σ and strains ε that appear in the engineering diagrams are not the true ones. During the traction test the force F applied to the specimen of initial cross section area A_o is recorded while a strain gage measures the displacement Δl between two selected points on the shank, at a known initial distance l_o. At any time during the test, stresses and strains are, then, non-directly measured but are derived quantities by using Eqs. (1.4). In real life, at each load increment dF the initial area A_o is continuously reducing while the instantaneous elongation dl is applied to a gage that is no longer l_o but $l_o + \Delta l$, Δl being the sum of all prior dl dilatations. It follows, then, that at any instant during the test the *true* stress σ_t acting on the specimen and the *true* strain ε_t are respectively larger and smaller than the corresponding engineering stress σ and strain ε. Each point of the engineering stress–strain curve σ-ε, such as points B, C, D, E and F of Fig. 1.19a, b, will move a little bit back and up to points B′, C′, D′, E′, and F′, respectively, giving rise to the so called true stress–strain curve. The true and engineering stress–strain diagrams present a basic difference, as shown in Fig. 1.19. While the engineering curve reaches a maximum at maximum load or ultimate strength at necking (point F of Fig. 1.19) and decreases thereafter to failure (point G), the true curve rises continuously to failure (point G′). Note that true stress and true strain at final failure are completely different from the corresponding engineering values. Before the onset of necking, the true values σ_t and ε_t can be obtained from the engineering ones. At each elementary load step dF, in fact, the corresponding true strain $d\varepsilon_t$ will be given by the ratio of the incremental strain dl to the instantaneous length l of the gage

$$d\varepsilon_t = \frac{dl}{l}. \tag{1.10}$$

The total true strain ε_t experienced by the gage from its initial length l_o to the final one l will be, then

$$\begin{aligned} \varepsilon_t &= \int_{l_o}^{l} \frac{dl}{l} = \ln \frac{l}{l_o} \\ &= \ln\left(\tfrac{l_o+\Delta l}{l_o}\right) = \ln(1+\varepsilon). \end{aligned} \tag{1.11}$$

Equation (1.11) clearly indicates that the difference between engineering strain ε and true strain ε_t starts to be appreciated for values larger than some percent, that is to say well beyond the elastic field, and becomes larger and larger as the plastic deformation increases, as schematically shown in Fig. 1.19a, b. In terms of area reduction ΔA, differentiating Eq. (1.5) that express the volume constancy during plastic deformation, it yields

$$dA \cdot l + A \cdot dl = 0 \tag{1.12}$$

and therefore

$$-\frac{dA}{A} = \frac{dl}{l} = d\varepsilon_t \tag{1.13}$$

and integrating between the initial area A_o and the final A it is obtained

$$\varepsilon_t = -\int_{A_o}^{A} \frac{dA}{A} = \ln\left(\frac{A_o}{A}\right). \tag{1.14}$$

By introducing the engineering reduction in area RA defined as

$$RA = \frac{A_o - A}{A_o} \tag{1.15}$$

Equation (1.14) becomes

$$\begin{aligned} \varepsilon_t &= \ln\left(\frac{A_o}{A}\right) = \ln\left(\frac{1}{1 - RA}\right) \\ &= -\ln(1 - RA)\,. \end{aligned} \tag{1.16}$$

Equation (1.16) can be solved for A and recalling Eq. (1.11) we get

$$A = A_o e^{-\varepsilon_t} = \frac{A_o}{(1 + \varepsilon)}. \tag{1.17}$$

Now, the true stress can be inferred as

$$\sigma_t = \frac{F}{A} = \frac{F \cdot (1 + \varepsilon)}{A_o} = \sigma \cdot (1 + \varepsilon). \tag{1.18}$$

It must be well understood that Eq. (1.10) can be integrated to yield the true strain given by Eq. (1.11) as long as deformations are uniform over the entire gage length, that is to say up to necking. Beyond necking Eqs. (1.11) and (1.18) are not valid any longer. Hardening lows under uniaxial monotonic loading can be postulated as an explicit dependence of the true plastic stress $\sigma_{t,p}$ on the true plastic strain $\varepsilon_{t,p}$

$$\sigma_{t,p} = K \cdot \varepsilon_{t,p}^m \tag{1.19}$$

where K is a material constant also known as modulus of plasticity or strength coefficient and m is the strain hardening exponent or strain hardening coefficient. This power law used to represent the whole flow curve, that is to say the true stress–strain curve from yielding to the onset of neck formation (point F′ in Fig. 1.19), is generally attributed to Hollomon [10]. Actually, already in 1909 Lüdwik introduced a similar expression including the yield strength σ_y

$$\sigma_t = \sigma_s + K \cdot \varepsilon_t^m \tag{1.19a}$$

The magnitude of the strain hardening exponent reflects the ability of the material to resist further plastic deformation. It is always lower than unity, this last value corresponding to an ideally elastic behavior while a zero value would indicate an ideally plastic material, as schematized in Fig. 1.25.

Some selected values of moduli of plasticity and strain hardening exponents for some metal alloys are listed in Table 1.1 [11]. From Eq. (1.19) it derives

$$\log \sigma_t = \log K + m \log \varepsilon_t \tag{1.20}$$

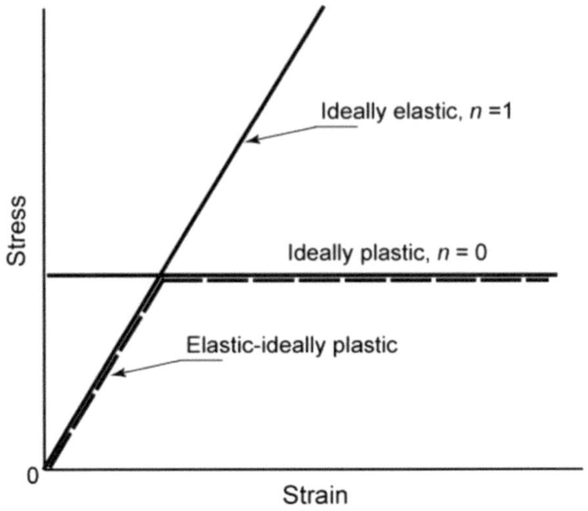

Fig. 1.25 Schematic of ideally elastic, ideally plastic and elastic-ideally plastic behavior

Table 1.1 Typical values of K and m at room temperature for selected metal alloys [11]

Steel	K (MPa)	m
A 533 B (93 °C)[a]	800	0.10
Fe 510[a]	958	0.26
4135 annealed	1015	0.17
4135 cold-worked	1100	0.14
4340 annealed	640	0.15
304 austenitic SS	1275	0.45
410 SS	960	0.10
Ductile cast iron	619	0.14
Aluminum alloys		
1100-O	180	0.20
2024-T4	690	0.16
6061-O	200	0.20
6061-T6	410	0.05
7075-O	400	0.17
Brass		
70–30 annealed	900	0.49
85–15	580	0.34
Copper annealed	315	0.54

[a] Milella

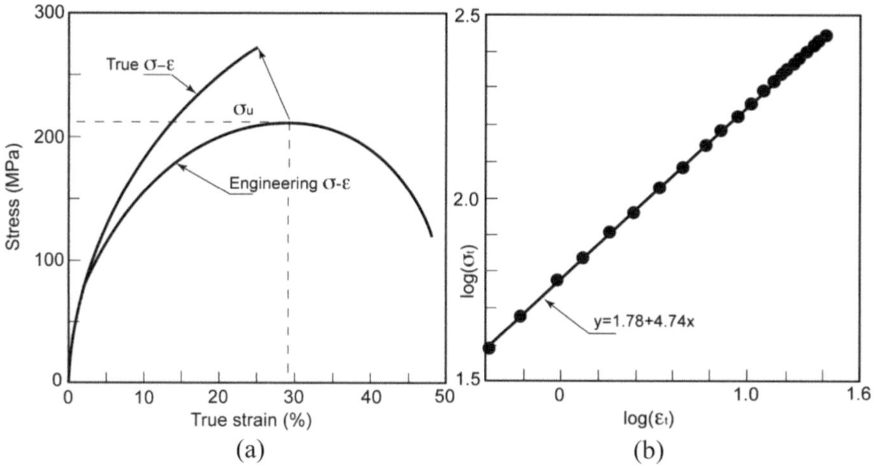

Fig. 1.26 **a** Engineering versus true stress–strain curves for copper and **b** power law of the plastic stress–strain of Fig. 1.23a (modified from [12])

Equation (1.20) represents a straight line in a log–log plot of true stress–strain data. Figure 1.26 [11] demonstrates this relation for copper. Recalling that true stress is given by

$$\sigma_t = \frac{F}{A} \tag{1.21}$$

where F is the applied load and A the actual cross section area, by differentiating it is

$$dF = \sigma_t dA + A d\sigma_t. \tag{1.22}$$

At necking (maximum load) it is

$$dF = 0 \tag{1.23}$$

so that

$$\frac{d\sigma_t}{\sigma_t} = -\frac{dA}{A} \tag{1.24}$$

and from Eq. (1.13) we find

$$\sigma_t = \frac{d\sigma_t}{d\varepsilon_t}. \tag{1.25}$$

Equation (1.25) expresses the so called Considère's criterion [13] that anticipates necking formation at the moment when the slope of the true stress–strain curve numerically equals the value of the true stress. By using the Hollomon expression (1.19)

$$K\varepsilon_t^m = Km\varepsilon_t^{m-1} \tag{1.26}$$

Therefore

$$m = e_t \tag{1.27}$$

which means that the true plastic strain at the onset of necking is numerically equal to the strain hardening coefficient *m*.

1.6.1 *True Stress–Strain Curve After Necking*

After the onset of necking there is no relation between engineering and true stress–strain quantities. Engineering stress rapidly goes down to final failure while true stress continues to grow till failure occurs. The simplest way to assess true stress at failure $\sigma_{t,f}$ is to take the ratio between the load at failure F_f and the final area of the specimen cross section A_f that may be much smaller than the initial one A_o because of the occurrence of necking, depending on the ductility of the material

$$\sigma_{t,f} = \frac{F_f}{A_f}. \tag{1.28}$$

Unfortunately, the true strain at failure remains unknown. A simple method may be used to infer the true strain at failure. The region where necking develops, in fact, may be thought as a small or reduced specimen within the main specimen, as schematized in Fig. 1.27a, b. Its initial dimensions are: length l_{red} (*reduced*) and area A_{bn} (*before necking*). The final dimensions at failure are:$l_{red,f}$ and A_f.

The cross section area A_{bn} of the specimen right before necking does not change till the final rupture because deformation and cross section reduction concentrate in the necking region. During the phase of uniform plastic elongation, as we know, the volume of the specimen does not change, see Eq. (1.5), so that between two loading steps 1 and 2 it is always

$$\begin{gathered} l_1 A_1 = l_2 A_2 \\ \frac{l_1}{l_2} = \frac{A_2}{A_1}. \end{gathered} \tag{1.29}$$

In terms of reduction of area RA, see Eq. (1.15), it is

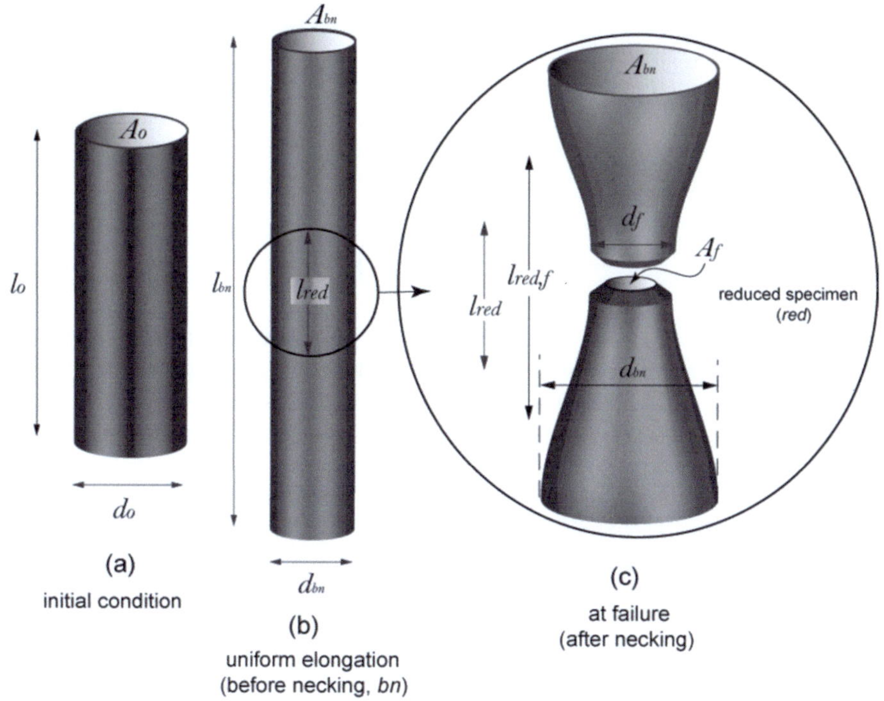

Fig. 1.27 **a** Specimen before testing, **b** specimen during testing right before necking, **c** enlargement and details of the area interested by necking

$$
\begin{aligned}
RA &= \frac{A_1 - A_2}{A_1} = 1 - \frac{l_1}{l_2} \\
(1+\varepsilon) &= 1 + \frac{l_2 - l_1}{l_1} = \left(\frac{1}{1 - RA}\right) \\
&= \left(\frac{A_1}{A_2}\right).
\end{aligned}
\tag{1.30}
$$

Note that RA that appear in Eq. (1.30) is not the engineering RA defined by Eq. (1.15) that refers to the initial cross section area A_o and not to the actual one. The same shall be supposed to hold also for the reduced specimen, Fig. 1.27c and, as first approximation, it can be written

$$
l_{red} A_{bn} = l_{red,f} A_f . \tag{1.31}
$$

Writing Eq. (1.29) for the reduced specimen and recalling Eqs. (1.11) and (1.18)

$$
\varepsilon_{t,red,f} = \ln\left(1 + \varepsilon_{red,f}\right) = \ln\left(\frac{1}{1 - RA}\right) = \ln\left(\frac{A_{bn}}{A_f}\right). \tag{1.32}
$$

It will be enough, then, to measure the final cross section A_{bn} of the specimen far from necking area and that A_f at necking to infer the true deformation $\varepsilon_{t,f,red}$ of the reduced specimen at failure, see Fig. 1.27. As a first approximation, the total deformation $\varepsilon_{t,f}$ of the entire specimen will be that $\varepsilon_{t,f,red}$ of the reduced specimen given by Eq. (1.32) plus the other one that has interested the specimen right before necking

$$\varepsilon_{t,f} = \ln\left(\frac{A_{bn}}{A_f}\right) + \ln\left(\frac{A_o}{A_{bn}}\right) = \ln\left(\frac{A_o}{A_f}\right) \tag{1.33}$$

which is precisely Eq. (1.14) used with $A = A_f$. From an operative point of view the procedure is the following: knowing the engineering stress–strain curve of the material the true stress-true strain curve can be drawn, as schematically shown in Fig. 1.15. This construction terminates at point F′ corresponding to the engineering ultimate stress or maximum stress $\sigma_u = \sigma_{\max}$ where necking occurs. From point A′ draw a line that terminates at the point of coordinates $\varepsilon_{t,f}$ given by Eq. (1.33) and $\sigma_{t,f}$ given by Eq. (1.28). There are arguments that would indicate this line be the tangent in F′ to the true stress–strain curve.

1.7 Bridgman Correction

A basic error is affecting the true stress value at failure as calculated by Eq. (1.28). At necking, in fact, the specimen, initially cylindrical, assumes the shape of a hourglass. This introduces two effects. The first is that the stress is no longer uniform over the entire cross section. The second is that in the area of the neck the stress state is no longer uniaxial. Now, three stress components exist that are: axial σ_a, radial σ_r and circumferential σ_c, as schematized in Fig. 1.28. This triaxial stress state becomes higher as necking gets sharper. Triaxiality allows the metal section to react better than in the uniaxial case because it constrains the material from deforming in the neck region (see Sect. 4.3). This actually reduces the equivalent stress σ_e. Bridgman [14, 15] was the first to address the issue of the stress state beyond necking in 1944. He proposed a solution for the equivalent stress σ_e on the reduced section at rupture that depended on the shape of necking

$$\sigma_e = \frac{\sigma_{t,f}}{\left(1 + \frac{2R}{r}\right)\ln\left(1 + \frac{r}{2R}\right)} = k \cdot \sigma_{t,f} \tag{1.34}$$

where R represents the radius of curvature of the neck contour, r the specimen radius of the minimum cross-sectional area, as schematized in Fig. 1.26, and $\sigma_{t,f}$ is the true stress at failure given by (1.28). The Bridgman correction factor k is always lower than one, which means that the equivalent stress σ_e is always smaller than $\sigma_{t,f}$ obtained from Eq. (1.28) and tends asymptotically to one for $R \to \infty$, i.e., in

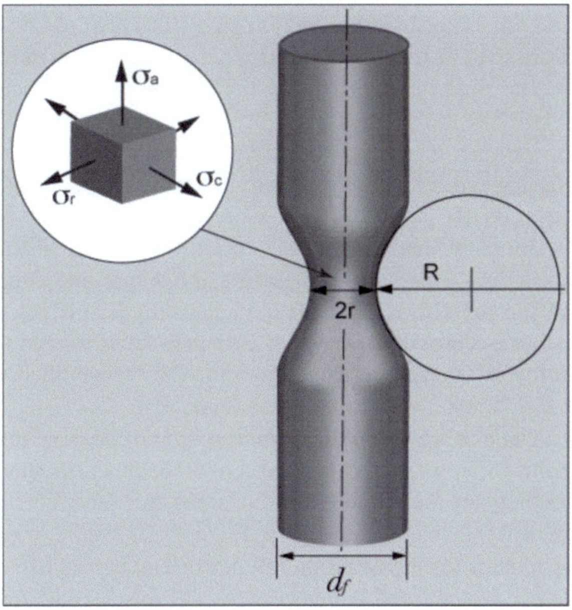

Fig. 1.28 Triaxial tension stress state in the necked region. Symbols shown are those used in the Bridgman equation

complete absence of necking and, therefore, of triaxiality. However, the Bridgman solution is based upon a series of working hypotheses

1. axial deformation on necking section is uniform;
2. the stress state on necking section is proportional;
3. the material is ideally elastic plastic (without strain hardening).

The first hypothesis is likely to be true. The second one is not completely true. It would be true if the radius of curvature were not varying during the loading phase, which is not the case since the specimen is continuously reducing its section as the load increases. The third hypothesis raises the question of whether or not strain hardening may have an effect on the effective response of the specimen to loading and how it may be taken into consideration. In highly strain hardening materials having also very high ductility, as austenitic stainless steel, Eq. (1.34) may underestimate the real equivalent stress at rupture. In high strength materials and, in particular, in metals showing a brittle behavior necking may not take place almost at all and its effect can be neglected.

1.8 Plastic Work and Heat Generation

The issue of the conversion into heat of the work spent to plastically deform a metal has long been matter of interest among researchers. Probably, the first to run experiments to investigate this conversion was Henri Tresca. His results on platinum and other metals were always above 70% [16]. As matter of fact, most of the plastic work spent in the deformation of a metal specimen during a traction test, see Eq. (1.1), is converted into heat. When this plastic work is done at temperatures that are low relative to the melting point of the metal, it is said to be cold worked. At the onset of necking, also heat generation concentrates in the reduced section increasing the local metal temperature. Traction tests are normally run at a strain rate $\dot{\varepsilon} = d\varepsilon/dt$ between 10^{-4} and 10^{-3} s^{-1}. Considering, as a first approximation, the tensile test machine as ideally rigid, the strain rate $\dot{\varepsilon}$ is equal to V/l where V is the crosshead speed and l the specimen length (the actual rate depends on the overall testing rig stiffness). With an effective specimen length of 10 cm to reach a strain rate of 10^{-4} s^{-1} requires a crosshead speed of 10^{-2} mm/s, which means that to reach a, let's say, 30% deformation requires about 50 min. This is why the loading condition for a standard traction test is said to be quasi-static.

Under such a condition the test cannot be considered ideally adiabatic from the point of view of heat evolution and specimen temperature increase. Increasing the loading rate the phenomenon becomes more and more adiabatic since no time is given to heat to leave the specimen. Figure 1.29 is an infrared picture of a low alloy steel specimen during a traction test performed at $\dot{\varepsilon} = 10^{-2}$ s^{-1} [17]. This strain rate is not really dynamic and heat generation not perfectly adiabatic, yet it can be seen the local, though modest, temperature increase in the central area of the specimen. Necking is not yet occurred. The specimen ends are better cooled by heat conduction through the adjacent metal, therefore they appear cooler. The effect of strain rate on local temperature increase can be seen in Fig. 1.30a, b for a polycrystalline nickel [18]. At a strain rate of 10^{-1} s^{-1}, Fig. 1.30a, the maximum temperature increase, at necking, is about 45 and 70 °C at failure, which means a final temperature of about 95 °C. This local temperature increase may change the material characteristic reducing its strength. If the strain rate is lowered to 10^{-2} s^{-1}, Fig. 1.30b, the temperature increase is much smaller, about 17 °C at necking and 25 °C at failure. It is worth noting that the true stress–strain curves for the two loading conditions are practically the same. This means that the specific work spent in deformation is about the same, but the final maximum temperature is not. What is actually happening is that more deformation energy is converted into heat at higher strain rate and that the strength decrease due to the higher temperature of the metal in the failure zone is balanced by the contemporaneous strength increase due to the higher strain rate (see Sect. 1.13). The same behavior has been observed in α-titanium [19]. This time, experiments were run at 1 and 3000 s^{-1}, using a Hopkinson bar. Results are shown in Fig. 1.31 as function of strain rate and amount of plastic deformation [19]. Again, as for nickel more plastic work is converted into heat at higher strain rate, but not all. A large amount of plastic work seems to remain in the material microstructure as stored energy. Same results have been obtained on 2024-T3 aluminum alloy [19] and by Manson et al. on Ti–Al–4V, aluminum and steel where the percentage of plastic work converted into heat was found to vary from 60 to 80–90% [20]. The issue seems to be still open. Nemat-Nasser et al. [21] were arguing that experiments made using infrared detectors to measure the amount of heat generated during plastic work underestimate the rise in temperature of the specimen, probably due to the difference between calibration

Fig. 1.29 Temperature increase in a low alloy steel specimen in traction at a strain rate $\dot{\varepsilon} = 10^{-2}\ \mathrm{s}^{-1}$ [17]

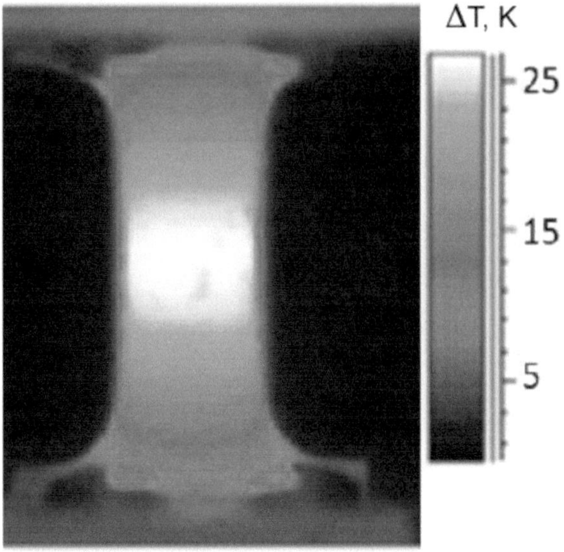

and testing that are not done at the same strain rate. Using indirect technique to compensate for that, they found that 100% of the work done was converted into heat. However, many metallurgical transformations, such as strain aging, recrystallization and polygonization that are normally observed at annealing temperatures can also occur at room temperature when metals are plastically deformed, though requiring a much longer time (months). The most reasonable explanation for that is that a kind of stored energy must be present in the crystal matrix that requires a much longer time to be released and to operate metallurgical transformations. A strain rate of $10^{-1}\ \mathrm{s}^{-1}$, though 100–1000 folds higher than a quasi-static loading, is not precisely a very high strain rate traction test but just a rapidly varying load test.

Real dynamic loading conditions start from strain rates of the order of 10–100 s^{-1} and grow to 10^3–$10^5\ \mathrm{s}^{-1}$ in explosion type tests and to more than $10^6\ \mathrm{s}^{-1}$ in detonation tests (see Sect. 1.13 and Chap. 4). In performing high strain rate tests, this temperature increase must be taken into consideration.

1.9 Ramberg–Osgood Equation

From Eq. (1.19) valid for the true stress–strain curve, it derives

$$K = \frac{\sigma}{\varepsilon^m} \tag{1.35}$$

At the yield point, σ_s, it is

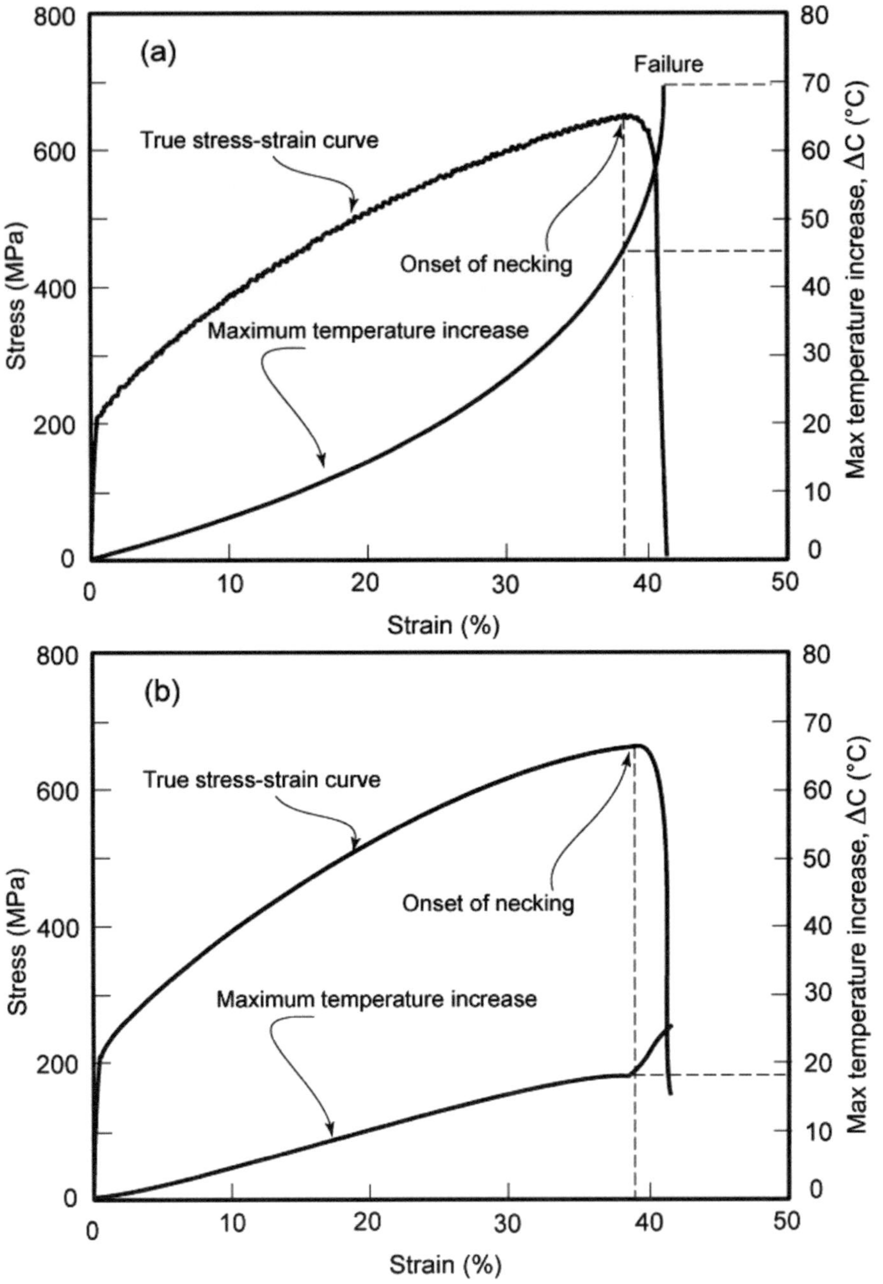

Fig. 1.30 Maximum temperature increase in a nickel specimen put in traction at **a** 10^{-1} s^{-1} and **b** at 10^{-2} s^{-2} (modified from [18])

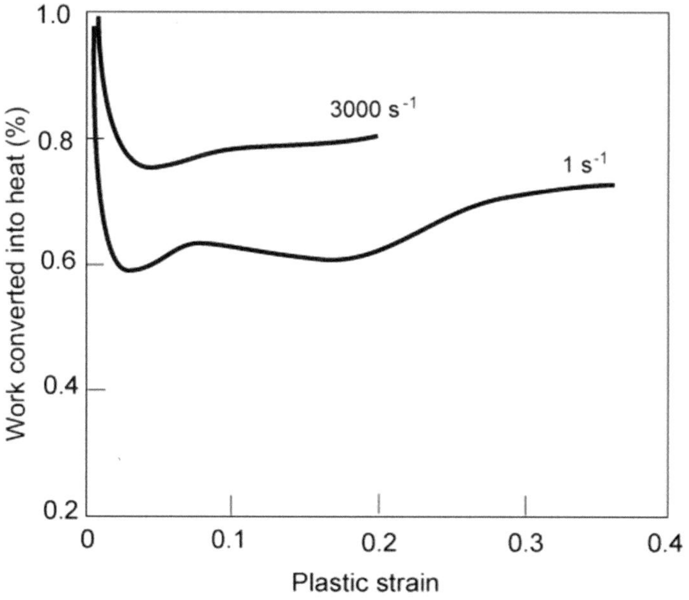

Fig. 1.31 Conversion of plastic work into heat in α-titanium at two different strain rates (modified from [18])

$$K = \frac{\sigma_s}{\varepsilon_s^m} \tag{1.36}$$

Note that the yield stress σ_y has been indicated as σ_s (the subscript s stands for slip) just for convenience to avoid confusion, since, as it will be shown, in fracture mechanics σ_y is the stress in the y direction. Accordingly, ε_s is the strain at yield. Substituting Eq. (1.36) into Eq. (1.19)

$$\frac{\sigma}{\sigma_s} = \left(\frac{\varepsilon}{\varepsilon_s}\right)^m \tag{1.37}$$

$$\begin{aligned}\frac{\varepsilon}{\varepsilon_s} &= \left(\frac{\sigma}{\sigma_s}\right)^{\frac{1}{m}} \\ &= \left(\frac{\sigma}{\sigma_s}\right)^n \end{aligned} \tag{1.38}$$

with $n = 1/m$. Note that conventional symbols σ and ε have been used instead of σ_t and ε_t just for convenience, but it must be kept in mind that they are true quantities. Also note that, since m varies from 0 to 1, n varies from 1 to ∞. In the elastic domain, where both n and m are equal one, Eq. (1.38) is written as

$$\frac{\varepsilon}{\varepsilon_s} = \left(\frac{\sigma}{\sigma_s}\right) \tag{1.39}$$

which is the Hooke's law. The entire true stress–strain curve up to the offset of necking can be, then, represented by the equation

$$\frac{\varepsilon}{\varepsilon_s} = \left(\frac{\sigma}{\sigma_s}\right) + \left(\frac{\sigma}{\sigma_s}\right)^n \tag{1.40}$$

Often, instead of σ_s and ε_s the proportional limits σ_o and ε_o are used. However, it is often assumed $\sigma_o = \sigma_s$ and $\varepsilon_o = \varepsilon_s$.To better represent the entire plastic curve a new equation is used

$$\frac{\varepsilon}{\varepsilon_o} = \left(\frac{\sigma}{\sigma_o}\right) + \alpha\left(\frac{\sigma}{\sigma_o}\right)^n \tag{1.41}$$

with α characteristic constant describing, together with *n*, the hardening behavior of the material. Equation (1.40) or (1.41) are referred to as the Ramberg–Osgood relationship or ROR. ROR is not always able to represent equally well the experimental curve both in the linear field and in the low-medium and high deformation domain. Trying to fit also the large plastic deformation field ROR becomes GROR or generalized Ramberg–Osgood relationship

$$\frac{\varepsilon}{\varepsilon_s} = \left(\frac{\sigma}{\sigma_s}\right) + \alpha\left(\frac{\sigma}{\sigma_s}\right)^n + \beta\left(\frac{\sigma}{\sigma_s}\right)^p \tag{1.42}$$

Or

$$\frac{\varepsilon}{\varepsilon_o} = \left(\frac{\sigma}{\sigma_o}\right) + \alpha\left(\frac{\sigma}{\sigma_o}\right)^n + \beta\left(\frac{\sigma}{\sigma_m}\right)^p \tag{1.42a}$$

with the β constant to be determined by a trial-and-error procedure and σ_m chosen to make $p/m \gg 1$. The last term of Eq. (1.42) becomes dominant in the large plastic deformation field. ROR without the linear term is generally used to represent the true stress–strain curve in the plastic domain.

Figure 1.32 is an example of ROR for austenitic stainless steel type A 304 at room temperature. Only plastic deformations were considered. The RO equation is

$$\frac{\varepsilon}{0.001} = 1.691\left(\frac{\sigma}{207}\right)^{5.42} \tag{1.43}$$

with $\sigma_o = 207$ MPa, $\alpha = 1{,}691$ and $m = 5.42$ ($n = 0.184$). The Young's modulus was equal to 207,000 MPa. Closed circles represent experimental points. Another example, this time for a carbon steel type A 533 B at 93 °C is shown in Fig. 1.33. The corresponding ROR is

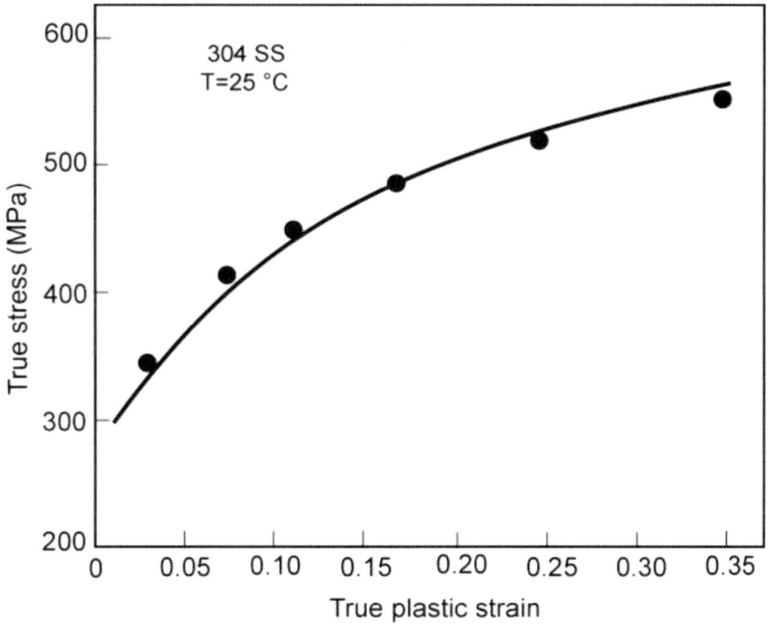

Fig. 1.32 Ramberg–Osgood representation of the true stress-plastic strain characteristic of A 304 SS at RT

$$\frac{\varepsilon}{0.002} = 1.12\left(\frac{\sigma}{414}\right)^{9.71} \tag{1.44}$$

with $\sigma_o = 414$ MPa, $\alpha = 1{,}12$ and $m = 9.71$ ($n = 0.10$). Closed circles represent experimental points.

1.10 Strain Hardening Exponent

From Eqs. (1.27) and (1.11) it can be inferred

$$m = \ln(1 + \varepsilon_{\max}) \tag{1.45}$$

or, as first approximation

$$m = 1 - \frac{\sigma_s}{\sigma_u} \tag{1.46}$$

Table 1.2 lists some values of the $m = 1/n$ exponent for several steels of common use with the respective yield and ultimate strengths, maximum strain and reduction

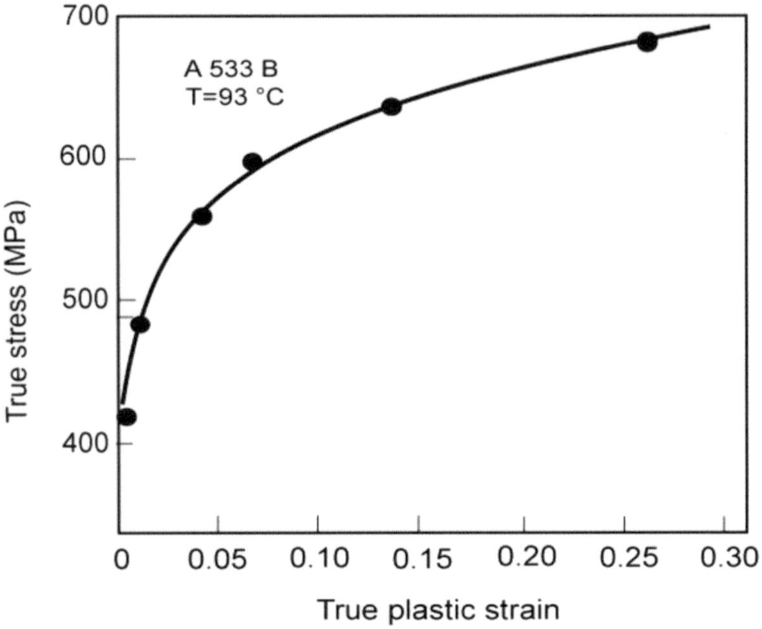

Fig. 1.33 Ramberg–Osgood representation of the true stress-plastic strain characteristic of A 533 B at 93 °C

of area attained in a traction test. Other values of the n exponent can be found in Table 1.1.

It is worth noting that the strain hardening exponent $m = 1/n$ decreases as the yield strength of the material increases while both maximum deformation and reduction of area RA, and therefore the material ductility, decrease. Also note the very low value of strain hardening, $m = 0.1$, i.e., $n = 10$, for high strength 250 martensitic steel (1747 MPa). With such a high strain hardening coefficient a twofold increase in deformation can be obtained with a rather modest 1% increase of stress. Figure 1.34

Table 1.2 Strain hardening coefficients

Steel	m	σ_y (MPa)	σ_u (MPa)	ε_{max}	RA (%)
304 SS	0.5	230	560	0.50	75
ABS-C	0.24	275	440	0.30	65
A 537 B	0.16	420	580	0.19	71
A 533 B	0.14	420	550	0.18	68
A 517 F	0.11	650	810	0.09	50
250 Mart	0.1	1747	1820	0.04	55

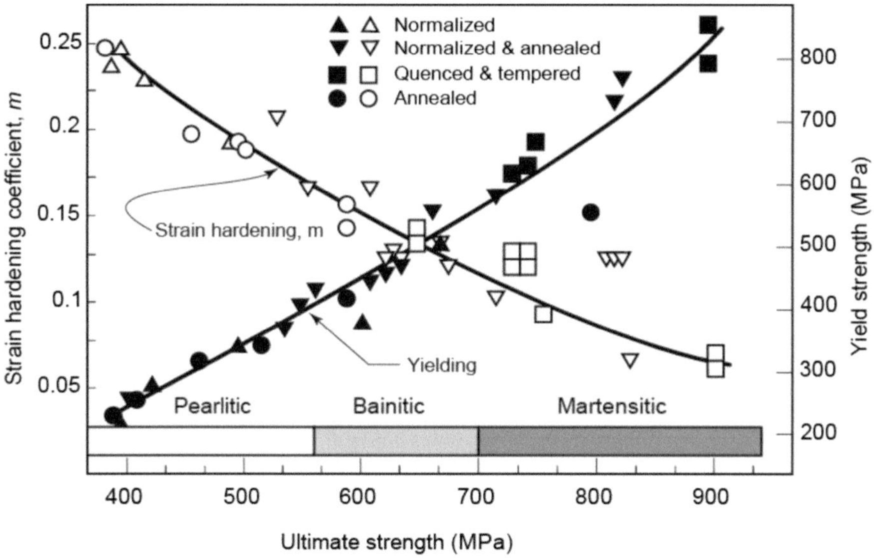

Fig. 1.34 Variation of the strain hardening coefficient $m = 1/n$ with material strength and metallurgical structure (modified from [22])

[22] shows the variation of strain hardening coefficient n with material strength and metallurgical structure.

1.11 Compression Test

The results of a simple compression test on specimens of ductile materials do not differ, generally, from those obtained in tension. The problem is that compression increases the net section area of the specimen so that the actual stress is reduced and material appears to behave better than in traction. This apparent strength increase becomes more and more evident as the yield point is passed. Things get worst since the section area increase is not uniform, but makes the specimen look like a barrel, being larger at center and smaller at both end, as schematized in Fig. 1.35. However, this tendency to harden is affecting the engineering curve and disappears when the true stress–strain diagram is used since the latter refers to the actual section of the specimen and not to the nominal one, as the former does (see Sect. 1.5). An example of such an apparent uneven behavior is shown in Fig. 1.36 for mild steel (low carbon steel). The metal is very ductile and its apparent strength in compression is three times that in traction. But when the tests results are plotted in terms of true stress-true strain values it sudden becomes clear that the material behaves exactly the same in traction as in compression. Another effect may alter the result of a compression

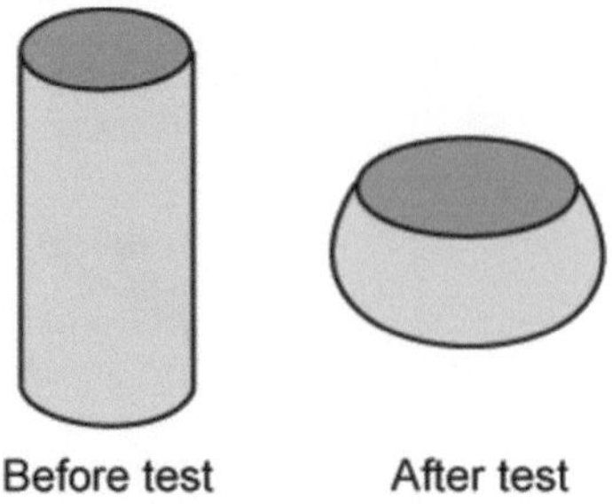

Fig. 1.35 Barrelling of a cylindrical specimen after compression

test with respect to the traction one, which is always due to section enlargement. It involves the friction between the specimen ends and the pressure plates (Fig. 1.37).

Because of this the specimen shows uneven heights and the state of stress is no longer that of a simple compression, as schematically shown in Fig. 1.35. Lubricating the interfaces helps to get rid of the problem. Unlike ductile metals, brittle solids behave quite differently in tension and compression, the latter being generally many times that attained in traction. This is typical, for example, of two ideally brittle

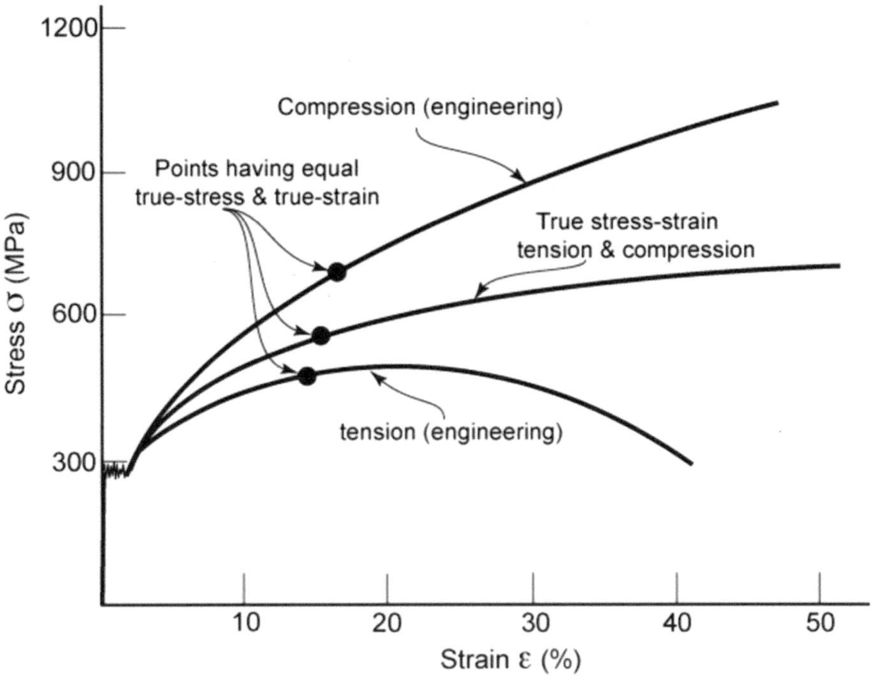

Fig. 1.36 Engineering uneven behavior of mild steel in traction and compression. The uneven behavior disappears when data are plotted as true stress-true strain

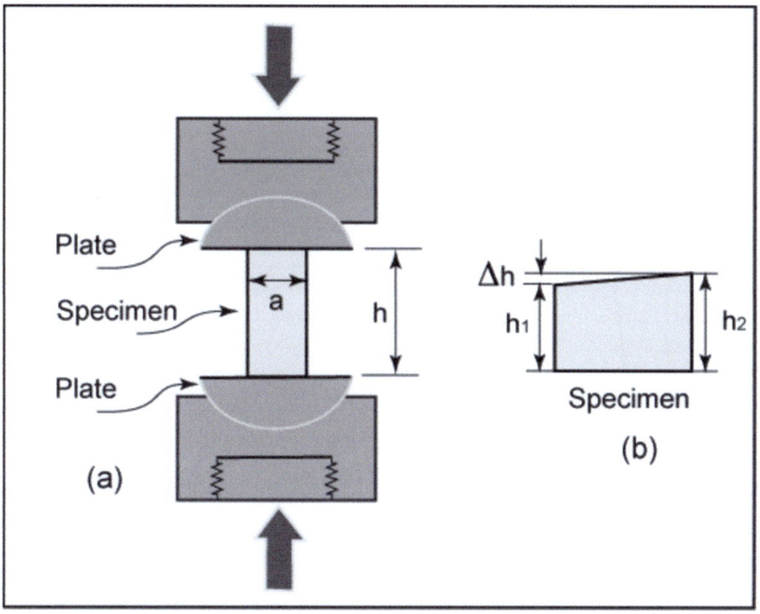

Fig. 1.37 Uneven size possibly shown by a specimen in compression

materials as glass or cast iron. Figure 1.38 is an example of uneven behavior offered by aluminum alloy, glass and gray cast iron in tension and compression, respectively. Note the rather negligible values of the strain at fracture. The fact that strength in compression in cast iron is larger than in traction is due to the composite structure of the material in which graphite is dispersed in the iron matrix. During solidification a high traction residual stress state remains at the graphite-iron interface due to the very different thermal expansion coefficients. This residual stress state is reducing the material strength in traction, but has a large beneficial effect in compression.

1.12 Bauschinger Effect

Suppose to load a specimen in traction to some plastic strain, as shown in Fig. 1.39 by point B. This first loading curve will be called the virgin curve of the material or flow curve. If the specimen is unloaded and then subjected to increasing compression, it is generally found that the material starts to yield before it did in traction, which actually means that the yield stress in compression is lower than that in traction. This is schematically shown in Fig. 1.39 by point D that does not correspond in compression to point A relative to the yield strength $\sigma_{y,t}$ in traction. Yielding in compression (point D) starts earlier than expected (point E) so that $|\sigma_{y,c}|<\sigma_{y,t}$. The phenomenon is called Bauschinger effect from the German engineer Johann Bauschinger [23] who first

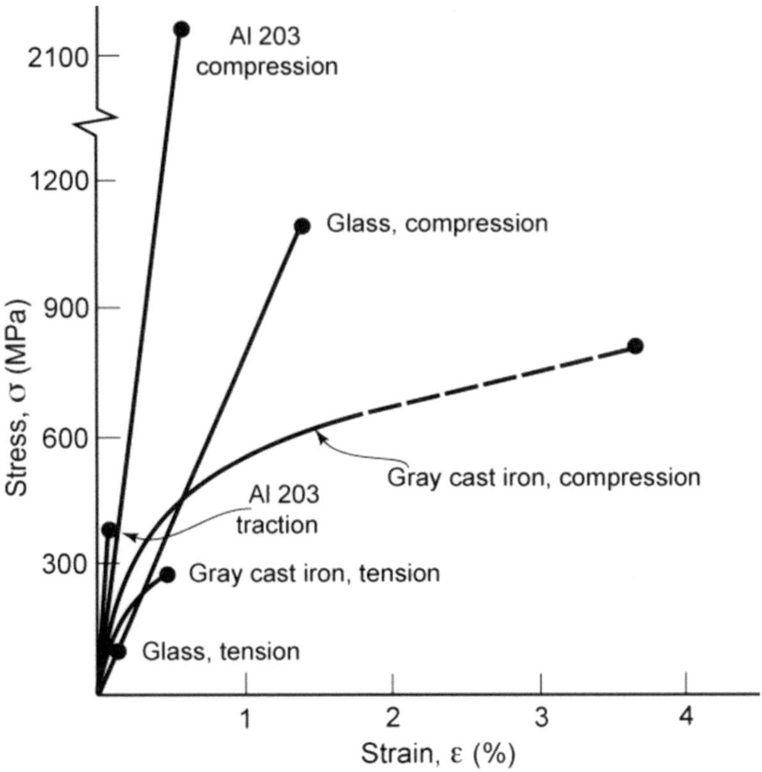

Fig. 1.38 Uneven behavior offered by Al 203, glass and gray cast iron in tension and compression

reported it in 1886. The skepticism of both Gilman and Orowan, we were talking about in Sect. 1.4 referring to traction tests: "..*it is not possible to do much with its [traction test] results, because one does not know what they really mean*" reappears. It is now clear, in fact, that the Bauschinger effect is more complex than originally thought and far from being fully understood and explained. There are two main schools trying to explain the effect. One is based on internal stresses [24–26], the other on dislocation theories. As to the first school, it assumes that the material is made of different elemental pieces (crystals) or, which is equivalent, the material is not homogenous, but is a kind of composite material made by two phases; the matrix and a second phase with two different mechanical properties. Therefore, during plastic deformation the stress distribution in the specimen section is not uniform. Because of this non-uniformity or non-homogeneity, after unloading residual stresses set up which are responsible for the yield-lowering effect upon load reversal. The internal stress theory has a major drawback in that numerous experiments have shown the Bauschinger effect to take place also in single crystal metals.

As to the dislocation theory, there are two main approaches. The first, originally proposed by Mott [26] and later developed by Seeger [27], suggests that during

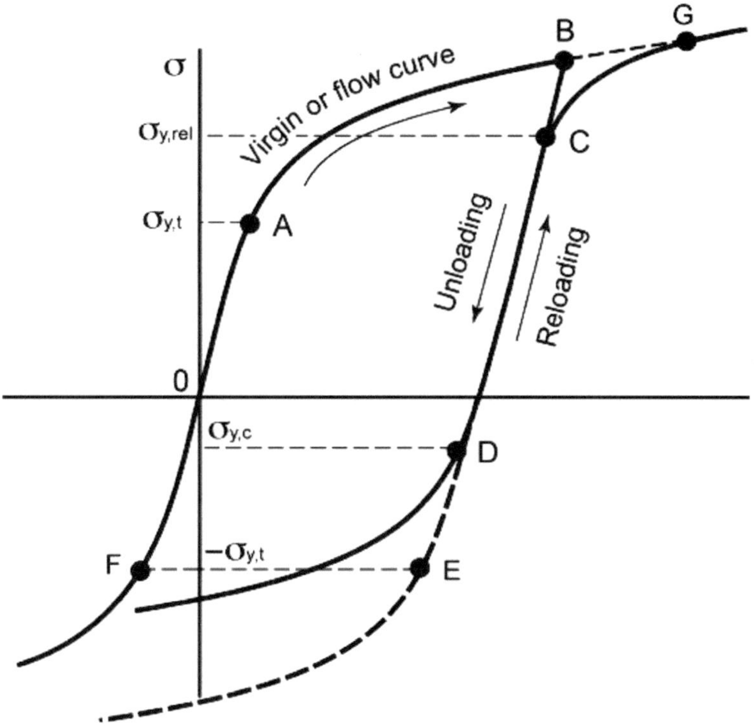

Fig. 1.39 Upon unloading in the plastic domain (point B) the material yields in compression earlier than in traction (point D). Reloading, it starts to yield earlier than before (point C vs. point B) thereafter continuing to follow the virgin curve (point G)

prestraining when dislocations are blocked and pile up at a barrier, such as grain boundary or inclusion, residual stresses build up that were called back stresses by Prager [28]. It is precisely these back stresses that aid dislocation motion in the reverse direction reducing the shear stress τ needed to move then back

$$\tau = \tau_F + \tau_{BS} \tag{1.47}$$

where τ_F is the friction stress on glide plane and τ_{BS} the back stress developed by dislocations piled up. Equation (1.47) indicates that the larger τ_{BS} the lower the reverse flow stress since τ_{BS} opposes the flow in the forward direction, but assists it the reverse direction. At the time it was thought that that these residual stresses were the consequence of anisotropic properties of individual grains. Those grains more favorably oriented yield first. The result is an internal residual stress state due to the mismatch of grains. Embury [29] added a third term σ_{for} to Eq. (1.47) to take into consideration the hardening effect due to dislocation forest generation

$$\begin{aligned} \sigma_F &= \sigma_o + \sigma_{for} + \sigma_{BS} \\ \sigma_R &= \sigma_o + \sigma_{for} - \sigma_{BS} \end{aligned} \tag{1.48}$$

where σ_F and σ_R are the forward and reverse flow stress, respectively. The validity of the theory implies that dislocations piled up and blocked at a barrier must remain in the same position before unloading. In other words, load removal must not make dislocations run back. In such a condition, dislocations are said to be frozen. The second approach is due to Orowan [30] who noticed that there is a permanent softening of the material rather than a real Bauschinger effect. A permanent softening has been, in fact, observed and its magnitude indicates that the back stress effect is relatively small. As matter of fact, if residual stresses were responsible for the Bauschinger effect, heating the specimen to above the recrystallization temperature all traces of the effect would disappear. But Orowan students Wu [31] and Canal [32] showed that this relaxation did not happen. Orowan figured out that the Bauschinger effect was due to dislocation pile-up at barriers, such as foreign particles. This would not generate residual stresses due to the mismatch of grains, but highly localized stresses that would make reverse glide easier than forward glide. After decades of conflicting arguments about the real cause at the base of the Bauschinger effect, it is recent common understanding that the softening observed in reverse loading may be due to a partial recovery of the isotropic hardening following a reverse strain. Classical isotropic hardening cannot predict this behavior. A kind of dynamic recovery during reverse flow occurs that erases some of the previous hardening. This can be envisioned by a mechanism of dislocation annihilation. However, the Bauschinger effect continues to be quantified by the amount of the reverse softening. Consider Fig. 1.40 where the compressive portion of the unloading curve has been reversed into the positive sector. According to the Orowan theory the softening effect is expressed by the term $\Delta\sigma_p$. In most recent years, most of the estimates of σ_{BS} have been made from measurements of $\Delta\sigma_p$. Atkinson [33] found that σ_{BS} was about half the softening $\Delta\sigma_p$

$$\sigma_{BS} \approx 0.5\Delta\sigma_p. \tag{1.49}$$

Rewriting Eq. (1.48) and substituting Eq. (1.49) for σ_{BS} it is

$$\sigma_R \approx -\left(\sigma_F - \Delta\sigma_p\right). \tag{1.50}$$

As general tendency, the greater the tensile cold working the lower the compressive yield strength. Any time the specimen is unloaded a new material emerges which is different from the virgin one. Upon reloading, in fact, the new yield strength is higher than that of the virgin material $\sigma_{y,rel} > \sigma_y$, as shown in Fig. 1.39 (point C vs. point A). However, it is lower than the previously reached stress at the moment when unloading started (point B). Yielding, in fact, starts somewhat earlier at point C and this is also considered pertaining to the Bauschinger effect and is known as generalized Bauschinger effect. After that the reloading curve continues over what would have

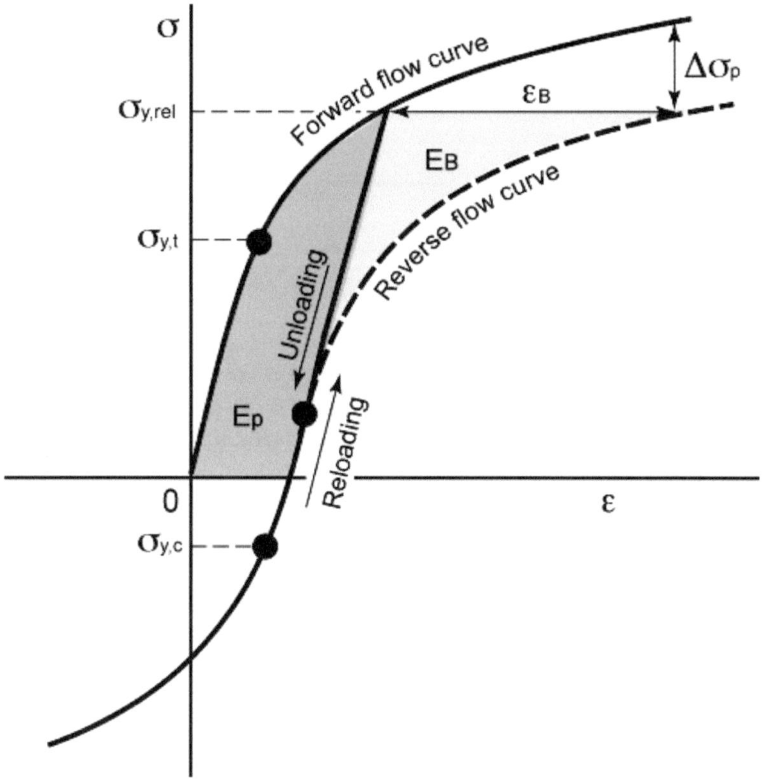

Fig. 1.40 Softening introduced by Bauschinger effect

been the virgin one (point G) as if the material had a memory of the original course. Figure 1.41 is an example of generalized Bauschinger effect in copper [34]. Not all metals obey the Bauschinger effect or not in the same manner. Figure 1.42 [35] shows the Bauschinger effect measured in different steels at different prestrain values. It can be seen how higher strength steels respond with larger decrease of the yield strength in compression with respect to traction. In all cases the percent reduction of the yield strength increases with increasing prestraining. Typically exposed to Bauschinger effect are stainless steel and aluminum alloys. Figure 1.43 presents two sets of experimental curves obtained on 316 stainless steel and 1100-0 aluminum alloy, respectively, tested under both direct traction loading and compression loading conditions [36]. The compression curves have been reversed and shown in the first quadrant to be compared to the traction curves to evidence the softening. The reversed loading tests were all conducted by first loading in tension followed by reversing into compression.

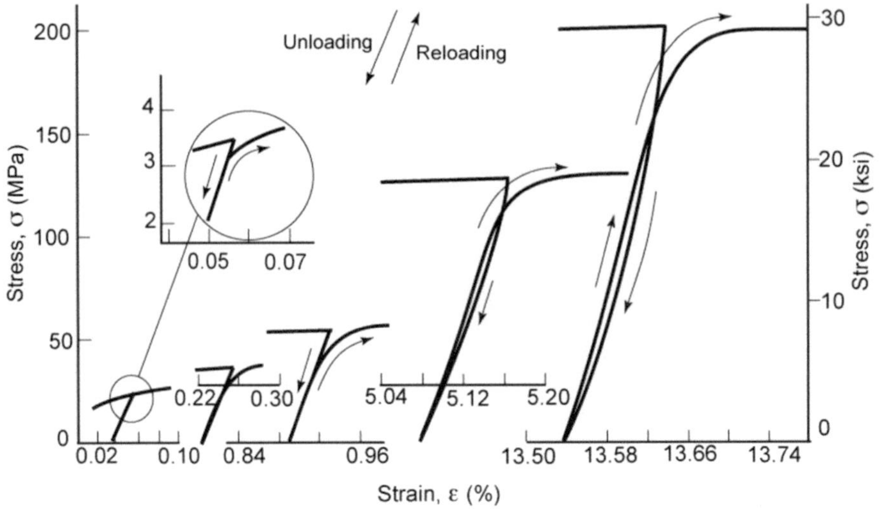

Fig. 1.41 Generalized Bauschinger effect in copper (modified from Ref. [34])

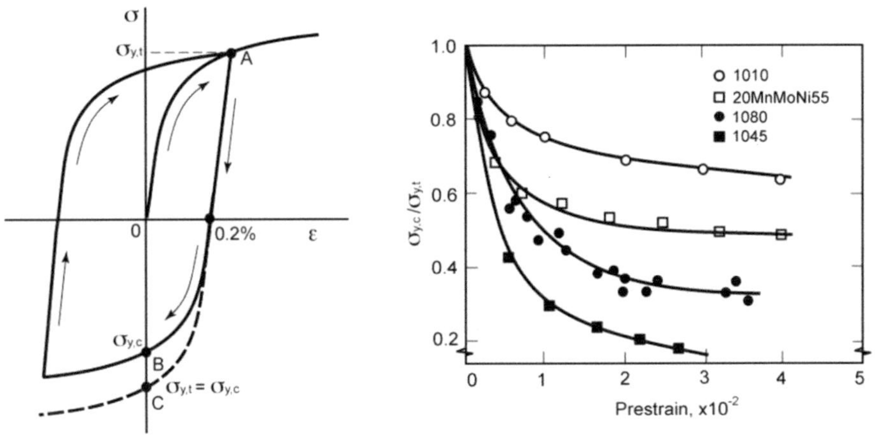

Fig. 1.42 Decrease of the yield strength in compression versus prestrain for different steels the left graph is a schematic stress strain curve showing the decrease of the yield strength in compression the right graph shows the amount of that reduction for several steels normalizes to the corresponding true stress (modified from Ref. [35])

1.13 Strain Rate Conditions in Traction Tests

It has been shown in the preceding sections that the material response depends on the strain rate $\dot{\varepsilon} = d\varepsilon/dt$ at which the traction test is performed. The strain rate can, in fact, completely change the mechanical behavior of materials, particularly as far as

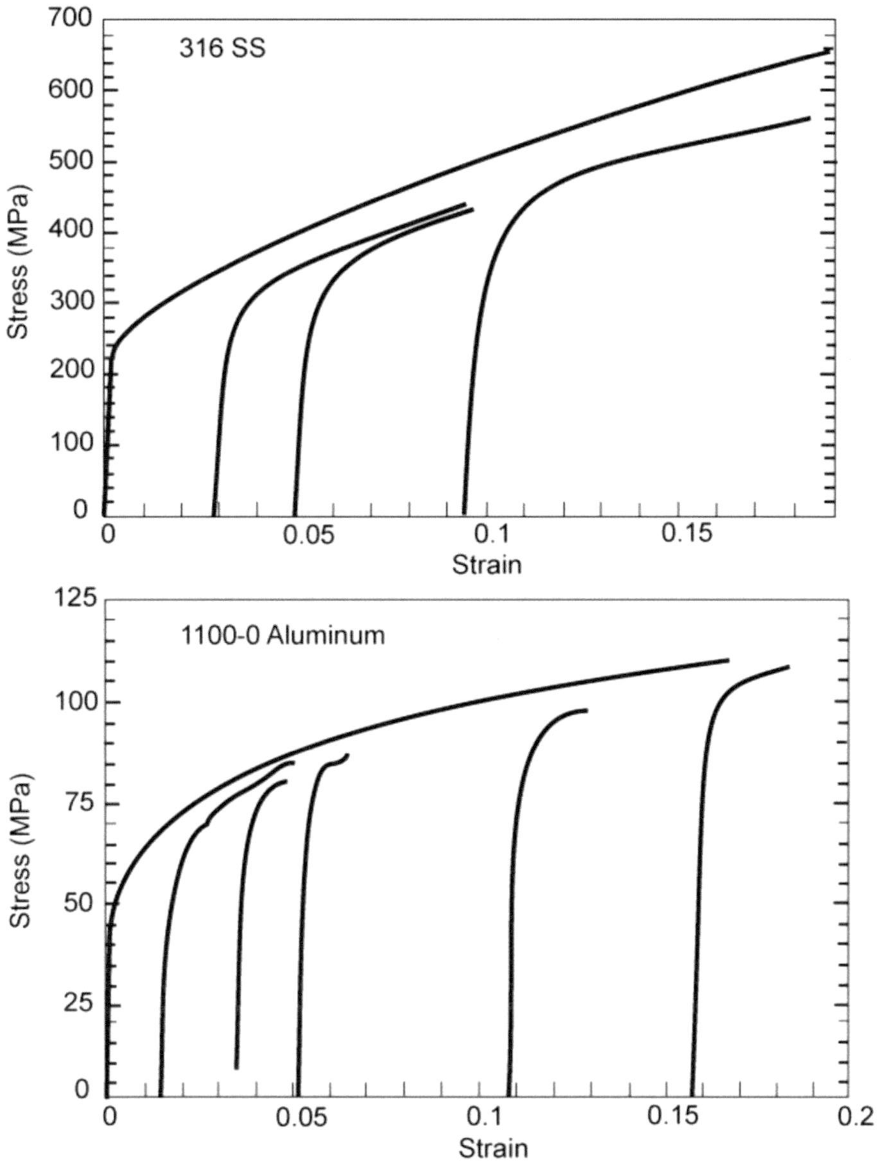

Fig. 1.43 Composite figure showing the tensile curve and the reversed compression curves for 316 SS and 1100-0 aluminum alloy, respectively (modified from [36])

Table 1.3 Classification of loading conditions based on strain rate

Loading conditions	Strain rate (s^{-1})
Very slow rate	10^{-8} to 10^{-5}
Standard or "quasi-static" traction tests	10^{-4} to 10^{-3}
Dynamic tension or compression tests	10^{-2} to 10^{2}
High speed testing (wave propagation effects)	10^{3} to 10^{4}
Hypervelocity impact tests (shock wave propagation)	10^{5} to 10^{8}

its toughness is concerned. The following Table 1.3 reports, indicatively, how strain rates can be grouped and classified from an operative point of view.

Very slow rate tests cannot be done with normal servo-hydraulic loading cells. They require an electro-mechanical, screw-driven machine capable of moving the crosshead of the loading machine at an extremely low speed, of the order of 60–100 μm/day. Indicatively, the strain rate in a traction test is given by the crosshead speed divided by the specimen length, $\dot{\varepsilon} \approx v/l$. As matter of fact, a 10^{-8} s^{-1} strain rate test requires something as a full year to be completed. At such a low strain rate, 10^{-8} s^{-1} or below, the yield strength of ferritic steels approaches the athermal value (see Sect. 5.3) becoming practically independent of temperature. The range 10^{-4} to 10^{-3} is that recommended by standards for traction tests. Dynamic tests in the range of 10^{2} s^{-1} require loading machines with speed crosshead of the order of 10 m/s. Also an instrumented Charpy pendulum can be used with Charpy-V notch type specimens (see Sect. 3.2). A typical strain rate of a Charpy-VN test is of the order of 300 s^{-1}. The author has shown that during the fracture of a Charpy VN specimen the strain rate remains constant through the entire section containing the notch, which is necessary condition to relate the material response to a fix strain rate. Beyond 10^{2} s^{-1}, between 10^{3} to 10^{4}, high speed testing can be done using impact bars or the Hopkinson bar test rig. In the split Hopkinson bar, a small test piece, typically 63 × 9.5 mm, is placed between two straight bars, called incident bar and transmitted bar, as schematically shown in Fig. 1.44. the two bars are precision aligned using a series of bar supports. At one end of the incident bar, a stress wave is generated by the impact of a striker bar fired by a gas gun.

The initial compression wave would propagate trough the incident bar and bypass the specimen without loading it by placing a collar over the specimen. Once arrived at the end of the transmitted bar, the tress wave would reverse and enter the specimen as traction wave, loading it dynamically. Another way of applying a dynamic loading to the specimen is to clamp the midsection of the incident rod while a torque is applied to the free end. By sudden releasing the clamp, using for instance an explosive bolt, a torsion wave is created that enters the specimen dynamically. It can be shown that the specimen undergoes a constant strain rate loading condition.

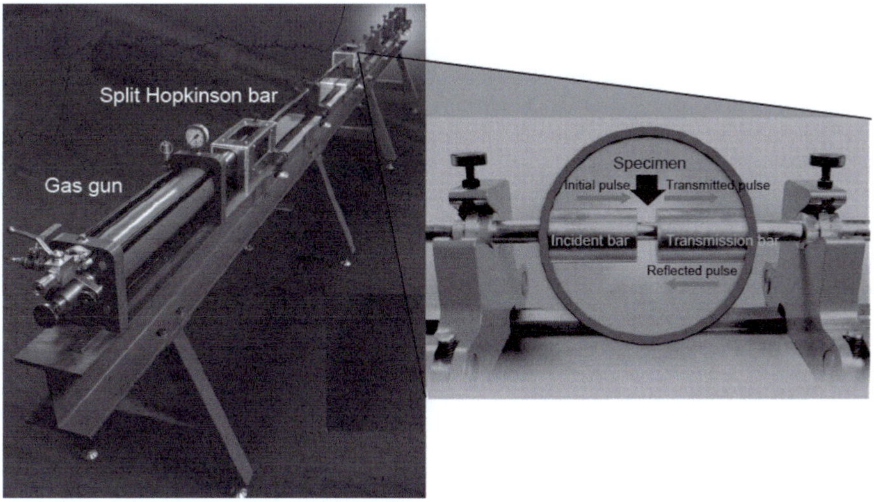

Fig. 1.44 Split Hopkinson bar on its supports with detail of the specimen holder (modified from REL, Inc. High Strain Rate Material Testing, Calumet, Michigan, USA)

References

1. Pellini, W.S.: Integration of analytical procedures for fracture-safe design of metal structures. NRL Rep. **7251**, 2 (1971)
2. Parker, E.R.: Brittle Behavior of Engineering Structures, p. 1. Wiley, New York (1957)
3. Shank, M.E.: A Critical Survey of Brittle Failure in Carbon Plate Steel Structures Other Than Ships, Welding Research Council Bull., Ser. 17 (1954)
4. Highway Accident Report, Collapse of the U.S. 35 Highway Bridge, Point Pleasant, West Virginia (1967)
5. Pellini, W.S., Puzak, P.P.: Fracture Analysis Diagram procedures for the Fracture-Safe Engineering Design of steel structures. NRL Report 5920 (1963)
6. Galilei, G.: Discorsi e dimostrazioni sopra due nuove scienze attenenti alla mecanica & i movimenti locali (1638)
7. Orowan, E.: Proceedings of the Institution of Mechanical Engineers, vol. 151, p. 133 (1944)
8. Gilman, J.J.: Micromechanics of Flow in Solids, p. 246. McGraw-Hill Book Co., New York, USA (1969)
9. Rasmussen, K.J.R.: Full-Range Stress-Strain Curves for Stainless Steel Alloys. University of Sydney, Dep. Civil Engn., Res. Report N. R811 (2001)
10. Hollomon, J.H.: Tensile deformation. Tans. AIME **162**, 268–290 (1945)
11. Kalpakjian, S., Schmid, S.R.: Manufacturing, Engineering and Technology, 5th edn. Pearson Education Inc., Upper Saddle River, NJ (2006)
12. Roylance, D.: Stress-Strain Curves. MIT, Cambridge, MA (2001)
13. Considère, A.: Mémoire sur l'emploi du fer et de l'acier dans le contructions. Annales de Pont et Chaussees, 9, Ser. 6, p. 574 (1885)
14. Bridgman, P.W.: The stress distribution at the neck of a tension specimen. Trans. Am. Soc. Met. **32**, 553–574 (1944)
15. Bridgman, P.W.: Studies in Large Plastic Flow and Fracture. McGraw-Hill Book Co., New York, USA (1952)
16. Tresca, H.: On further applications of flows of solids. Proc. Inst. Mech. Eng. **30**, 301 (1878)
17. Rusinek, A., Klepaczko, J.R.: J. Mat. Des. **30**, 25 (2009)

18. Ghan, C.: Stress-Induced Generation and Strain Localization in Polycrystalline and Nanocrystalline Nickel. Dep. of Mat. Sci. and Eng., University of Toronto, Canada (2011)
19. Ravichandran, R., Rosakis, A.J., Hodowany, J., Rosakis, S.P.: On the conversion of plastic work into heat during high-strain-rate deformation. In: CP 620, Shock Compression of Condensed Matter, 2001, pp. 557–562. American Institute of Physics (2002)
20. Manson, J.J., Rosakis, A.J., Ravichandran, G.: On the strain and strain rate dependence of the fraction of plastic work converted to heat; an experimental study using infrared detectors and Kolsky bar. Mech. Mater. **17**, 135–145 (1994)
21. Kapoor, R., Nemat-Nasser, S.: Determination of temperature rise during high strain rate deformation. Mech. Mater. **27**, 1–12 (1998)
22. Bauschinger, J.: Mechanisch-Technischen laboratorium. Mater. Forum (1886)
23. Heyn, E.: Metal and Erz, pp. 441-2 and 436-41, Heft (1918)
24. Masing, G.: ibid. p. 135 (1926)
25. Schimid, E., Boas, W.: Plasticity of Crystals. Hughes & Co., London (1950)
26. Mott, N.F.: Philos. Mag. 1151 (1952)
27. Seeger, A.: Dislocations and Mechanical Properties of Crystals, p. 243. Wiley, New York (1957)
28. Prager, W.: A new metod of analysing stresses and strains on work-hardening plastic solids. ASME J. Appl. Mech. **23**, 493–496 (1956)
29. Embury, J.D.: Plastic flow in dispersion hardened materials. Metall. Trans. **16A**, 2191–2200 (1985)
30. Orowan, E.: Cause and effects of internal stresses. In: Internal Stresses and Fatigue in Metals, Symposium, Detroit (1958)
31. Wu, T.M.: Master of Science Thesis, Dep. of Mech. Eng., Massachusetts Institute of Technology, Cambridge, MA, USA (1958)
32. Canal, J.R.: Master of Science Thesis, Dep. of Mech. Eng., Massachusetts Institute of Technology, Cambridge, MA, USA (1960)
33. Atkinson, J.D., Brown, L.M., Stobbs, W.M.: Philos. Mag. **30**, 1247 (1974)
34. Lubahn, J.D., Felgar, R.P: Plasticity and Creep of Metals. Wiley, New York (1961)
35. Scholtes, B., Vöhringen, O., Macherauch, E.: Proc. ICMA6, New York Pergamon, vol. 1, p. 255 (1982)
36. White, C.S.: A Combined Kinematic-Isotropic Hardening Model for Large Deformation Metal Plasticity. U.S. Army Material Technology Laboratory, MTL TR 88-46, Watertown, Massachusetts (1988)

Chapter 2
Slip in Crystals

2.1 Introduction

In the previous chapter we have discussed standard tensile tests run on small specimens trying to understand their true meaning. Traction tests represent a macroscopic scale of examination of material behavior. We can well imagine how this macroscopic behavior depends on many microscopic factors. Therefore, it cannot be fully understood without entering this smaller scale, up to the extremely small: the lattice structure with its atoms and its defects, as schematized in Fig. 2.1.

In this work we will descend into the domain of grains and phases (from 150 to 5 μm), continuing in the much smaller one of plastic slips and will touch, though briefly, also that of dislocations where the optical microscopy yields to scanning electron microscopy (SEM) and transmission (TEM) which descends into details of the fraction of microns and arrive, finally, to *touch*, the world of the extremely small, that of the lattice structure of materials that can only be "*seen*" with an X-ray diffraction tool or very advanced high-resolution transmission electron microscopy (HRTEM) or *field ions microscopy* (FIM) or tunnel effect (STM). The last two images of Fig. 2.1 should be observed almost in a state of ascetic meditation because they are *images* of atoms, in this case silicon and titanium atoms, respectively. Is what has always been men's dream of *seeing* atoms about to come true? Maybe yes, who knows. However, this is out of the scope of the book. In this work we will not go beyond plastic slip scale as the most important phenomenon which material behavior depends on. Plastic slip represents the only real barrier against the brittle fracture of materials. The plastic flow is activated by the motion of dislocations or, though in a more particular fashion, by twining. We will deal in some detail with dislocations and twinning, which in many respects is unusual in a text of Fracture Mechanics. However, we will do this because plastic slips made it possible to explain plastic deformation, yield strength and hardening, in particular, which are indispensable premise to understand the different nature and aspects of fracture. Many factors affect the fracture resistance of materials. Starting from the microscopic scale, it is

P. P. Milella, *Fundamentals of Fracture Mechanics*,
https://doi.org/10.1007/978-3-032-23081-2_2

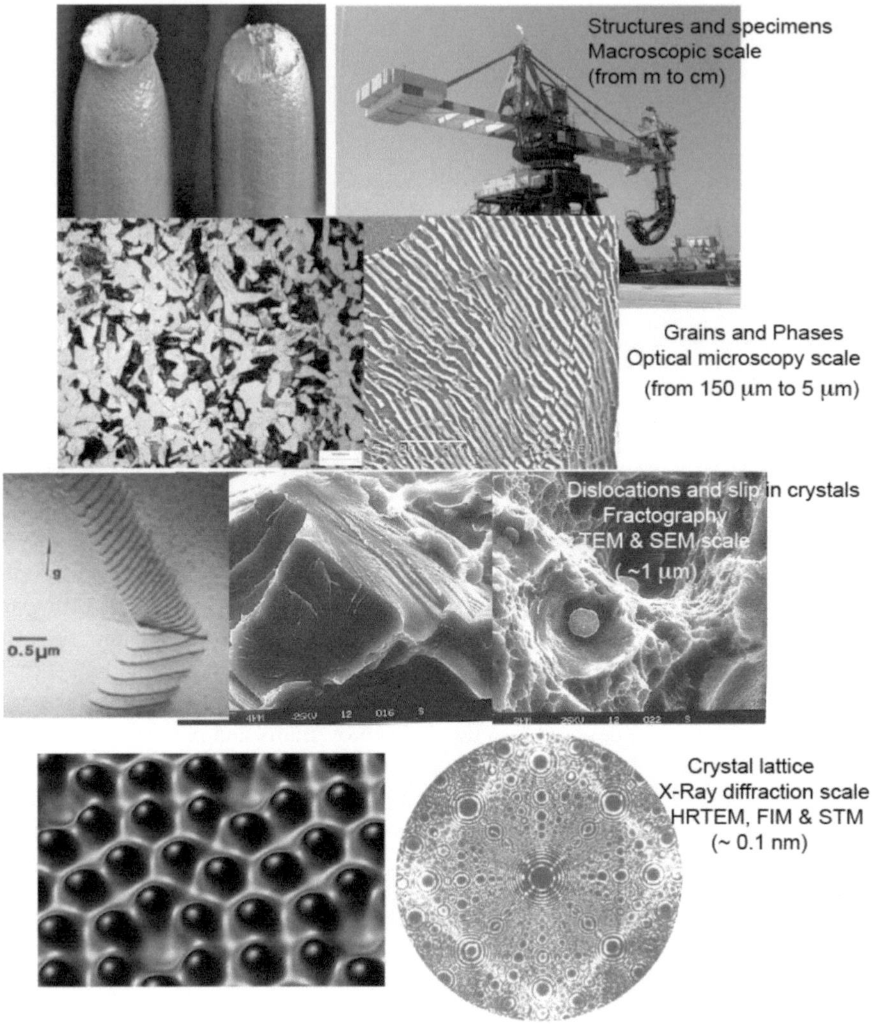

Fig. 2.1 The knowledge of material behavior necessarily requires entering a smaller and smaller scale, till the lattice structure. The last two images in the crystal lattice scale show the silicon atoms in their hexagonal position and the titanium atoms on the very sharp tip of a sample, respectively

the lattice structure and the microstructure of alloys that play a fundamental role. These two factors, in particular, enable the plastic deformation of metals by means of a process known as slip. Plastic deformation through slip is the only real barrier against brittle fracture. Slip is activated by dislocations motion. In a single metal crystal or alloy, atoms are arranged mostly in one of the three major lattice structures that are BCC (Body Centered Cubic), FCC (Face Centered Cubic) and HCP (Hexagonal Close-Packed) crystal lattices. A schematic of these lattice structures is shown in

Figs. 2.2, 2.3 and 2.4. Their mechanical response to a traction test has been already treated in Sects. 1.4 and 1.5. Table 2.1 is a short list of metals and alloys subdivided by their lattice structure. Iron can be found in different forms, known as allotropes.

Among iron allotropic forms there is α-iron or ferrite that has a BCC crystal structure, γ-iron or austenite that has a FCC lattice and martensite, named after the German metallurgist Adolf Martens (1850–1914). This very hard phase has a body-centered tetragonal crystal lattice (BCT) that can be assumed to be an intermediate structure between phases of ferrite and austenite. When high-carbon steel is quickly

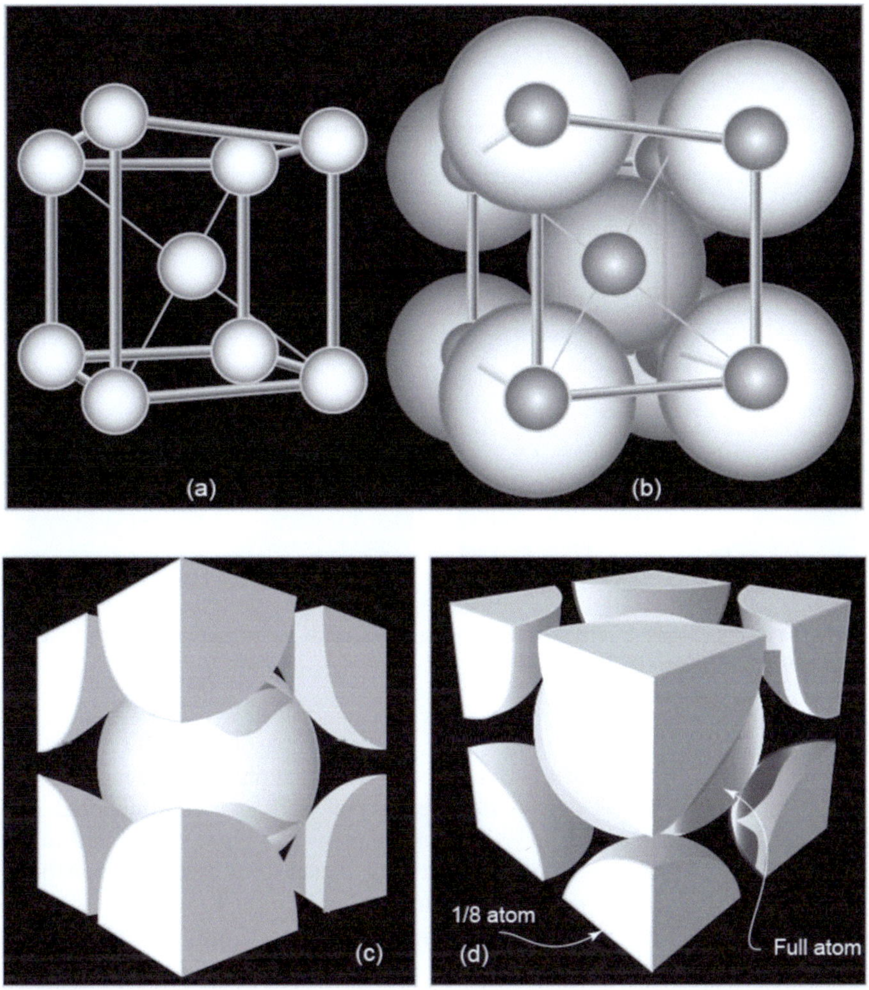

Fig. 2.2 **a** Reduced-sphere model of BCC lattice structure; **b** Hard-sphere model; **c**, **d** 3D views of the unit cell

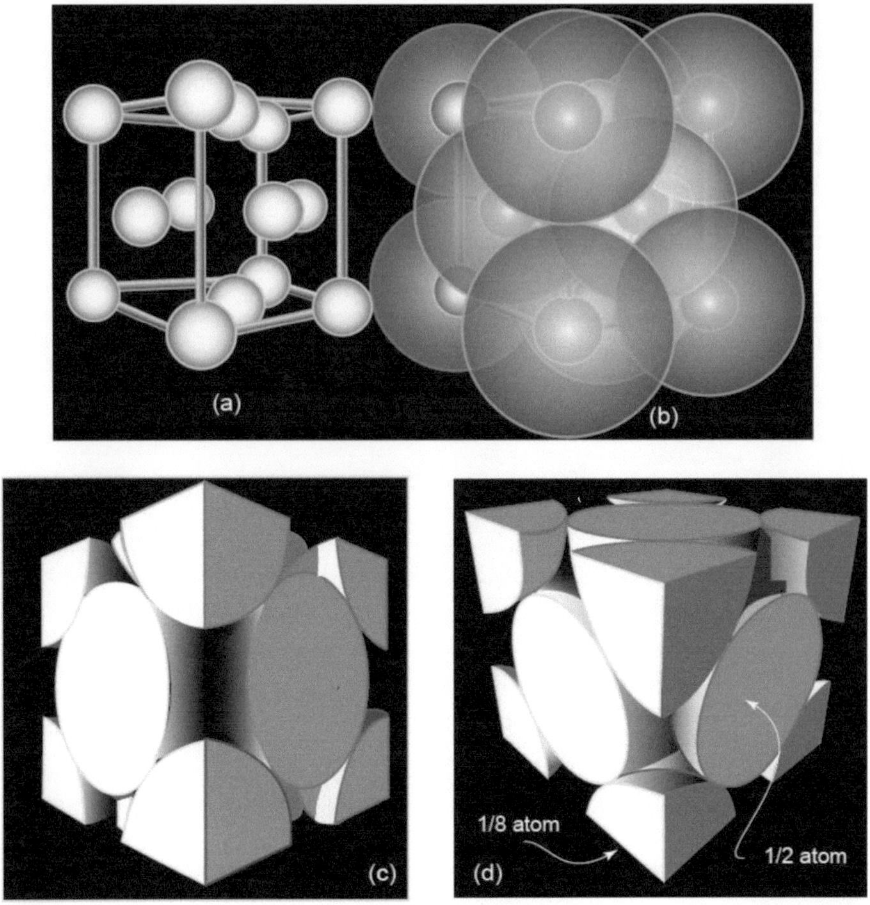

Fig. 2.3 **a** Reduced-sphere model of FCC lattice structure; **b** hard-sphere model; **c**, **d** unit cell separation, **e**, **f** 3D views of the unit cell

cooled to room temperature the FCC phase, which is stable over 910 °C, is converted into small crystalline units of a body-centered tetragonal phase called martensite.

This transformation is due to the presence of carbon atoms inherited from austenite. In practice, below 0.3% carbon, martensite cannot be formed in any significant amount. Figure 2.5 is an interesting sequence of austenite transformation into martensite in ferromagnetic Ni_2MnGa alloy taken at Tom Shield University of Minnesota. In Fig. 2.5a the austenite grains can be seen almost completely free of martensite. In Fig. 2.5d it can be seen the final transformation that still retains the original austenitic grain, filled with martensite plates. Without carbon atoms or with carbon content below 0.1–0.2% the transformation of austenite to martensite during cooling would not occur and cooling would normally proceed to an eventual BCC crystal structure at room temperature. Now consider the FCC crystal lattice of

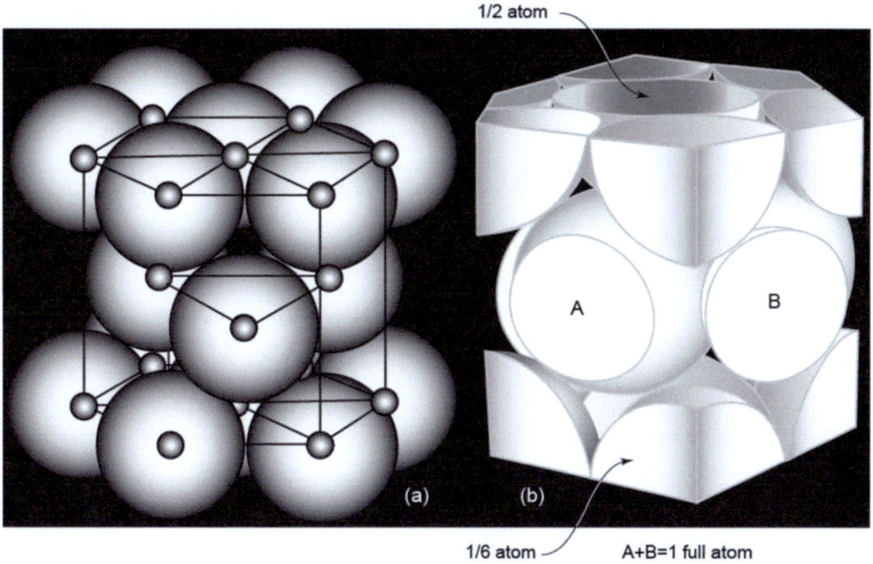

Fig. 2.4 **a** Reduced-sphere and hard-sphere model of HCP lattice structure, **b** isolated unit cell

Table 2.1 Lattice structure of some metals and alloys

BCC Lattice structure	FCC Lattice structure	HCP Lattice structure
Chromium	Aluminum and alloys	Cadmium
α-Iron and ferritic steels	Copper and alloys	Cobalt
Molybdenum	Gold	Titanium (α)
Tantalum	γ-Iron and austenitic steels	Zinc
Tungsten	Lead	
Vanadium	Nickel and alloys	
	Platinum	
	Silver	

austenite, as shown in Fig. 2.6a. In this structure we can identify a tetragonal crystal lattice of the type BCT evidenced in Fig. 2.6a, b as ABCDEFGHI. This is another mode to consider a FCC crystal structure. In a tetragonal structure, the crystallographic axes lie at right angles to each other, as in a cubic structure, but one lattice constant c differs in magnitude from the other two. When considered as BCT structure, the normal FCC lattice has a c/a ratio of about $1/\sqrt{2}$. In 1924, Bain suggested that a BCC crystal structure could be obtained from a FCC lattice by exerting a compression along the c axis, which would yield an intermediate martensite structure, and expanding along the other axis to yield a final cubic structure.

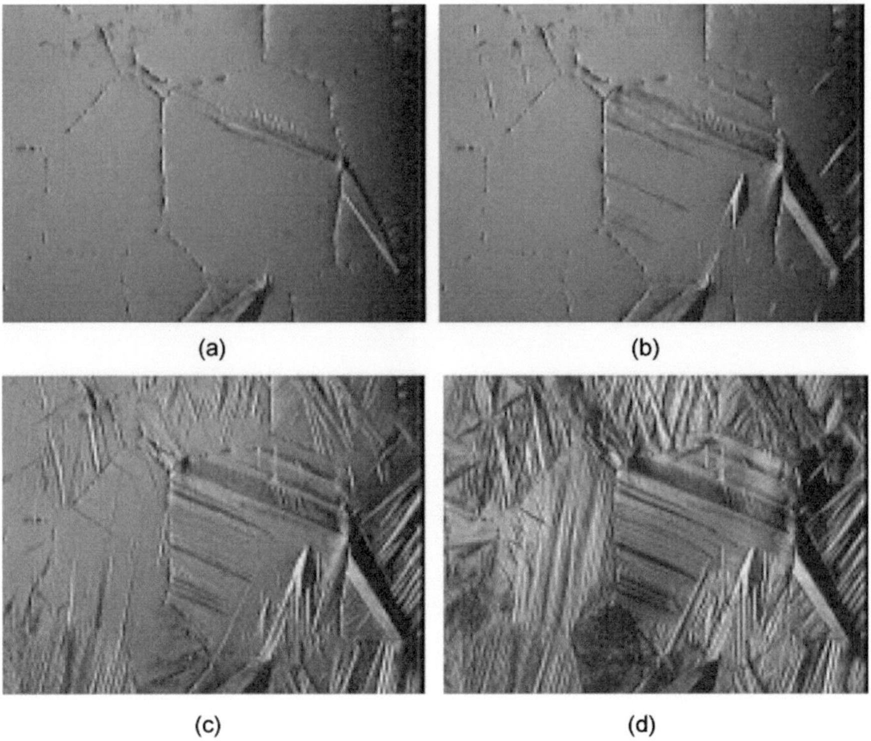

Fig. 2.5 Progression of austenite transformation into martensite in ferromagnetic Ni_2MnGa alloy. **a** Austenite is the only phase present; **b**, **c** martensite needle-like platelets grow in the retained austenite grains and **d** austenite transformation into martensite is completed (from Tom Shield, University of Minnesota)

This is referred to as Bain distortion. What is worth noting is that the unit volume of martensite lattice is different from both the FCC crystal structure volume and BCC lattice volume. This actually means that any metallurgical transformation that changes austenite or ferrite partially into martensite will introduce necessarily a residual stress state at the interface of the two phases as a consequence of the local adaptation strain field. Material response to traction tests of BCC and FCC polycrystalline metals has been treated in Sect. 1.5. We shall be now discussing the slip process that occurs as the linear elastic response of metals ends and the plastic deformation starts. Most metals and alloys are polycrystalline materials. When a polycrystalline material is subjected to an increasing external load, such as during a traction test, individual grains can deform plastically. In this new scale grains that are aggregate of equally oriented crystals, will also be called crystals. Strain is transmitted from grain to grain and the overall assembly responds without any void or defect or crack formation, at least to some extent. In fact, after a certain amount of plastic deformation, damage starts to nucleate and accumulates, see Fig. 1.19 and, eventually, the material fails. Since the plastic deformation of a polycrystalline metal is definitely the result

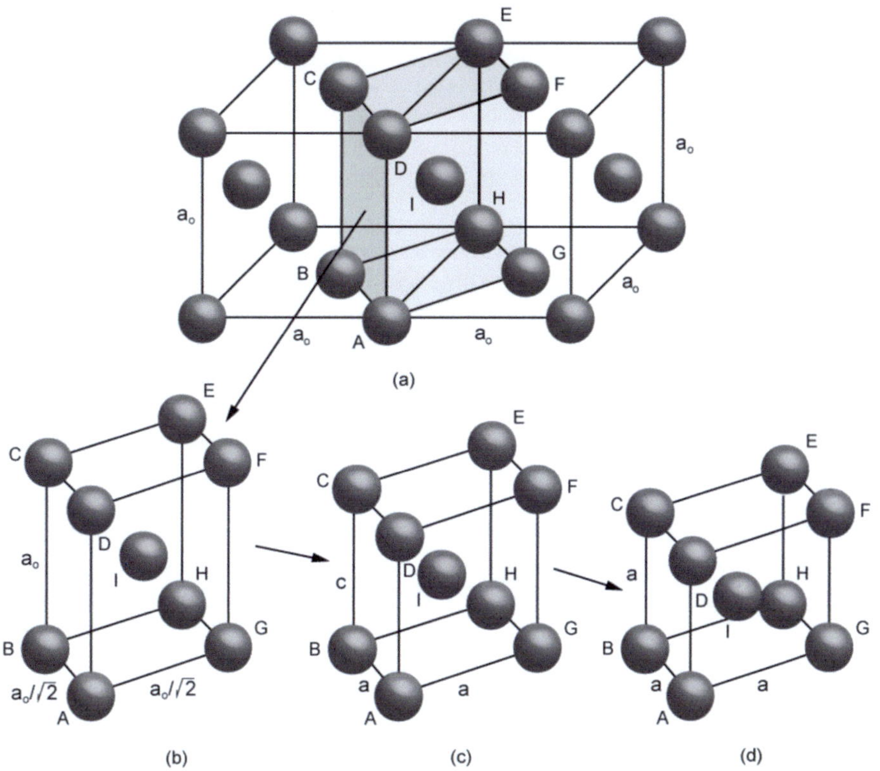

Fig. 2.6 **a** FCC lattice; **b** tetragonal representation of austenite; **c** tetragonal martensite; **d** BCC crystal lattice

of each individual grain deformation, it is convenient to start studying the plastic deformation of single crystals. Observation of single crystal specimens subjected to traction tests indicate that plastic deformation occurs by slip on crystallographic planes or close-packed planes, along crystallographic or close-packed directions that constitute a slip system, see Sect. 2.10. Slip, represents the displacement of one part of the crystal relative to the other, under the action of a shearing stress.

The plastic or permanent deformation of materials is precisely the result of slip within individual grains. Worth noting is the fact that the theoretical strength to plastic deformation is by far higher than that really observed in any material. This discrepancy is due, in particular, to a defect of the crystal lattice called dislocation, see Sect. 2.8, and to a lower extent, to another defect called twinning, see Sect. 2.12.

Slip by dislocations and twinning represents the only mechanism that can deform crystals in solids permanently. It must be understood that slip doesn't occur as result of a massive simultaneous slide of the entire half crystal with respect to the second half. This kind of deformation would require a tremendous amount of energy that would practically vanish any realistic tentative to deform solids. The atoms, instead, move one row at a time into the next position along the slip plane, as described in Sect. 2.8. Slip, then, involves the passage of a large number of dislocations on nearby slip planes within the majority of individual grains.

2.2 Lüders Lines

During a traction test of a polycrystalline specimen, whatever the behavior of the material may be following one of the two modes of deformation that were presented in Sect. 1.5, it may be observed that just beyond the elastic limit some sort of parallel, close-packed and elongated markings appear on the surface of the specimen, grouped in bands, as schematized in Fig. 2.6. These lines are known as Lüders lines after the German scientist who first observed them in 1854. Figure 2.7 presents an original drawing made by Lüders showing these lines on spring steel [1].

These lines appear to be oriented at about 50° to the specimen axis. They represent the macroscopic evidence of some plastic deformation originated by the slip of thousands of crystals on a microscopic scale. Lüders lines and bands are nucleated at points of stress concentration where the yield strength is first reached. Being the result of a slip process, Lüders lines are activated by shear stresses that have a maximum at 45° to the normal stresses (Fig. 2.8).

As matter of fact, Lüders lines are observed at 45°–55° or, some times, 60° to the tensile axis of the specimen rather than at exactly 45°.This is because, though activated by shear stresses, at moderate temperature and strain rate slip occurs on crystallographic planes, also called natural slip planes, as already discussed in Chap. 1. Therefore, actual slip angle depends on the slip planes-to-load direction combination or, better, on the single crystal orientation with respect to the load applied, as it will be explained in Sect. 2.3. At low temperatures and/or high strain rates, mechanical twinning dominates, see Sect. 2.14. Increasing the applied load, Lüders bands multiply and propagate along the specimen. Figure 2.9 is an example of Lüders lines, grouped in bands, evidenced by Crussard in a flat traction specimen using photoelastic lacquer coating activated by polarized light.

As it can be seen, Lüders lines developed at about 52° to the specimen axis. Another example of Lüders lines is visible in Fig. 2.10 [3]. This time it was used a brittle-lacquer technique. Brittle lacquer technique was introduced after the observation of naturally formed brittle coatings, such as thin oxides layers on heated surfaces. The flaking and cracking of these coatings indicate yielding or cracking of the base metal. The propagation of Lüders lines and bands soon after yielding does not require, in general, any load increment because each band, once activated, extends and multiplies under the action of lower stresses leaving the sum practically constant. This may explain the appearance in BCC metals of the plateau described in Sect. 1.5 and shown enlarged in Fig. 2.11, which shall be thought to be independent of the specimen geometry, see Sect. 2.12. Only after Lüders bands have spread to cover the entire gage section, the material starts to strain harden and the stress–strain curve to rise again. This is schematically shown in Fig. 2.10.

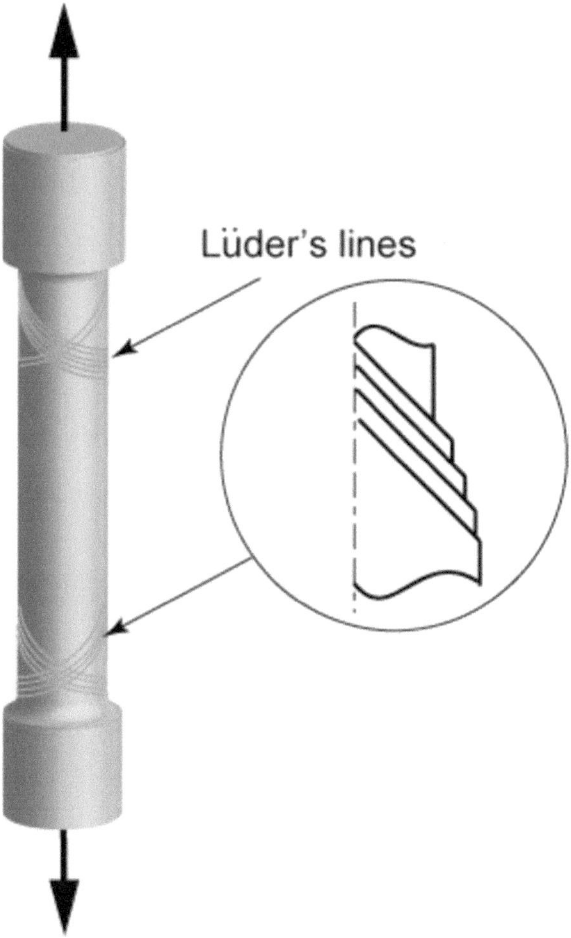

Fig. 2.7 First appearance of Lüders lines in a traction specimen and their detail on a microscopic scale

2.3 Slip Lines

Lüders lines and bands should not be confused with slip lines since they are a macroscopic evidence of slip occurring in thousands of crystals having different orientation. It may be thought that between Lüders bands and slip lines there is the same difference that exists between beach marks and striations in fatigue. The former are visible, the latter are not. It is quite possible, in fact, to have hundreds or thousands of crystals participating to form a Lüders band, each slipping in a complicated fashion on its own slip planes or on a single slip system. Slips in crystals occur, as we have already said, on crystallographic planes along crystallographic directions that are plane and directions where atoms are most densely packed. Slips that are responsible of plastic deformations, on a micro scale are not homogeneous, as schematized in Fig. 2.12. In polycrystalline metals, grains are oriented at random with respect to the loading

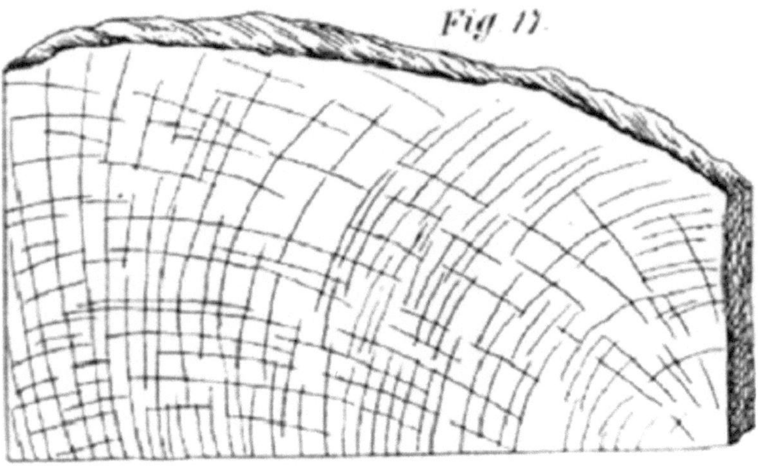

Fig. 2.8 Original drawing by Lüders showing lines on spring steel [1]

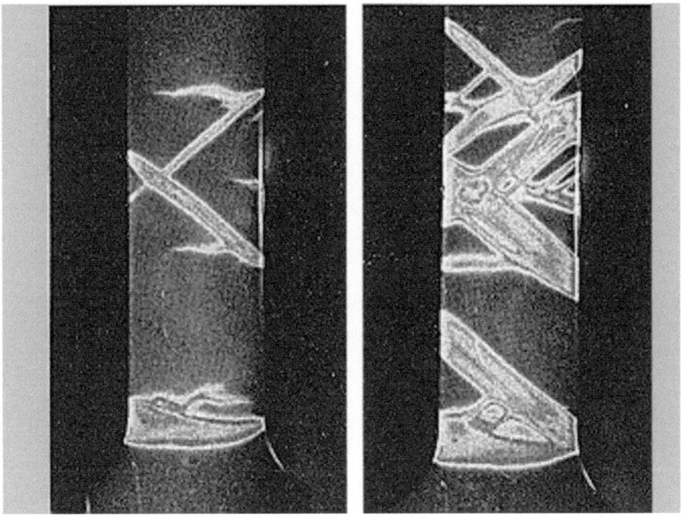

Fig. 2.9 Appearance, propagation and multiplication of Lüders lines and bands in a flat specimen loaded in traction (from Crussard)

direction and so are crystallographic planes. Therefore, in polycrystalline metals slips appear at the microscope as clusters of parallel lines whose direction changes from grain to grain. This is visible in Fig. 2.13 [4] that refers to polycrystalline copper plastically deformed. Note how slips run parallel within each individual grain but have different directions in different crystals. Also note that slips can be seen on the surface of the crystal because, as said, they are not homogeneous and emerge as slip steps from the crystal surface. However, slips can be activated on different directions

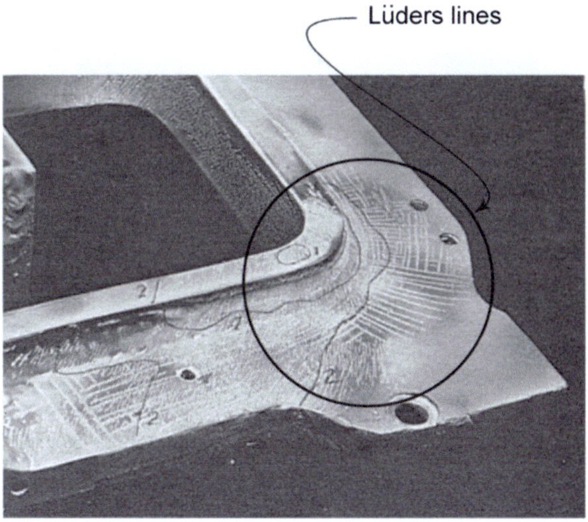

Fig. 2.10 Typical brittle-lacquer Lüders lines pattern [2]

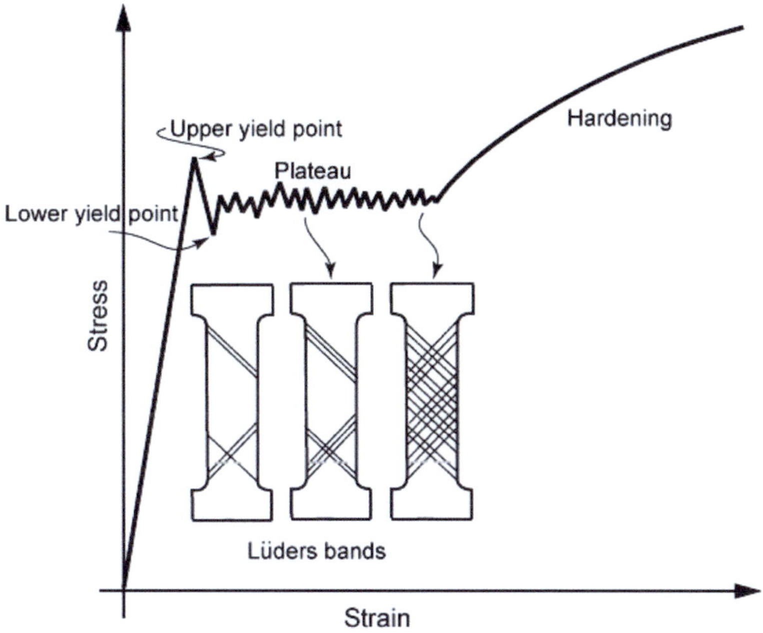

Fig. 2.11 Schematic of the appearance of Lüders lines and bands in the plateau of BCC metals

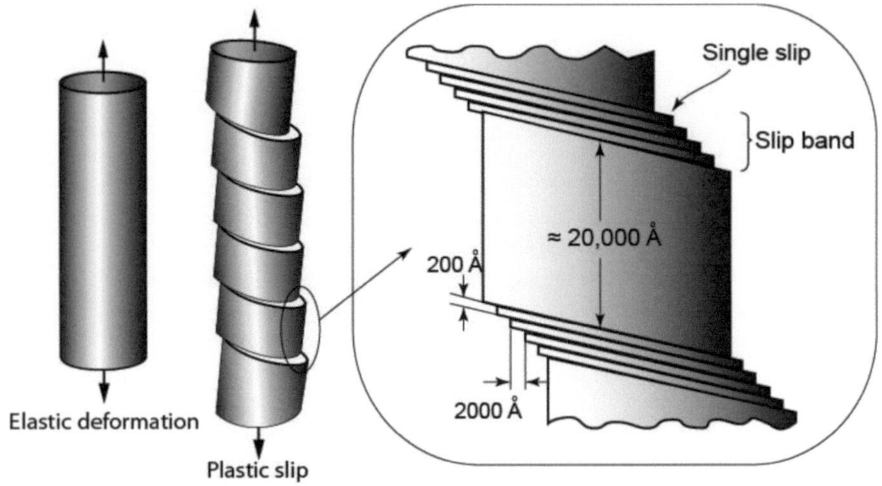

Fig. 2.12 Schematic of non-homogeneous plastic deformation by slip

also within the same crystal. This is shown in Fig. 2.14 for aluminum. Aluminum has a face centered cubic crystal structure, FCC. Characteristic is their appearance in magnesium single crystals oriented with the basal plane, the only crystallographic plane of the crystal, at 45° to the stress axis, as shown in Fig. 2.15 [5]. This orientation allows slips to occur under the lowest possible applied stress.

2.4 Slip in Single Crystals

It has been said that slip occurs on crystallographic planes along crystallographic directions. Therefore, in a single crystal the orientation of both the crystallographic plane and the crystallographic direction with respect to that of the applied load plays a fundamental role. Slip, in fact, occurs when the shear stress on the slip plane in the slip direction reaches a critical value.

Consider, then, a single crystal cylindrical specimen of cross section area A, as shown in Fig. 2.16a with one of its crystallographic plane and direction evidenced. The normal $\vec{n}$ to the slip plane makes an angle ϕ to the axis of the specimen. The corresponding area A_{slip} intercepted by the specimen is

$$A_{slip} = \frac{A}{\cos\phi}. \tag{2.1}$$

The component of the force on the slip direction is

$$F_{slip} = F \cdot \cos\lambda. \tag{2.2}$$

Fig. 2.13 Slip lines developed in polycrystalline copper [4]

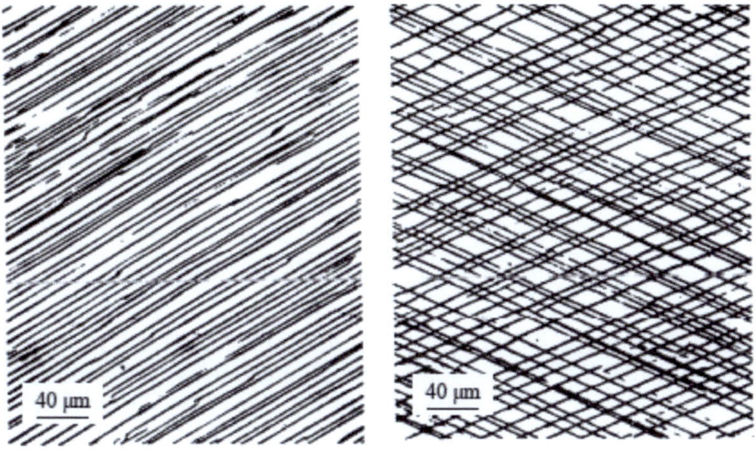

Fig. 2.14 Single and double slip in aluminum on slip plane (111) and direction [011] and [$\bar{1}$ 01]

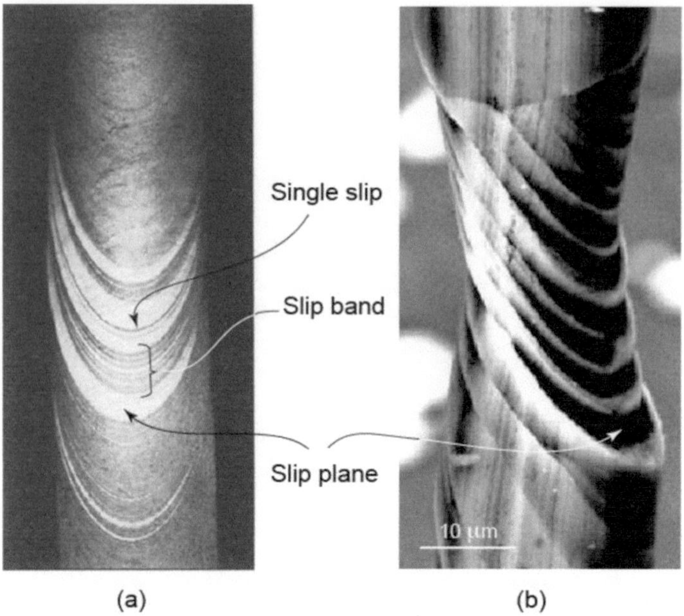

Fig. 2.15 Slips in single crystals of **a** zinc specimen along basal plane oriented at 45° to the traction axis [5] and **b** aluminum specimen [6]

On the slip plane one of the possible crystallographic directions makes an angle λ to the specimen axis, Fig. 2.16a. Assume to apply an external traction force F along the specimen axis as shown in Fig. 2.16b. The shear stress on the slip plane along the slip direction that will be called *resolved shear stress*, τ_{rss}, is, therefore

$$\tau_{rss} = \frac{F_{slip}}{A_{slip}} = \frac{F}{A} \cdot \cos\phi \cos\lambda. \tag{2.3}$$

The term $\cos\phi\cdot\cos\lambda$ is known as Schmid factor, M [7]. M can vary from 0 to 1. Slip will occur when the resolved shear stress reaches a critical value, $\tau_{c,rss}$, called critical resolved shear stress. Equation (2.3) indicates that slip normally occurs on that plane and along that direction having the highest Schmid factor possible, known as primary slip system. From Eq. (2.3) it turns out that the yield stress, σ_y, is given by

$$\sigma_y = \frac{\tau_{c,rss}}{(\cos\phi \cos\lambda)_{\max}}. \tag{2.4}$$

If a crystallographic plane is oriented at 90° to the specimen axis ($\phi = 90°$) and/or the slip direction is normal to the applied force F ($\lambda = 90°$) the resolved shear stress, Eq. (2.3), is null, $\tau_{rss} = 0$. On the contrary, slip will proceed under the lowest critical resolved shear stress when the Schmid factor reaches its maximum

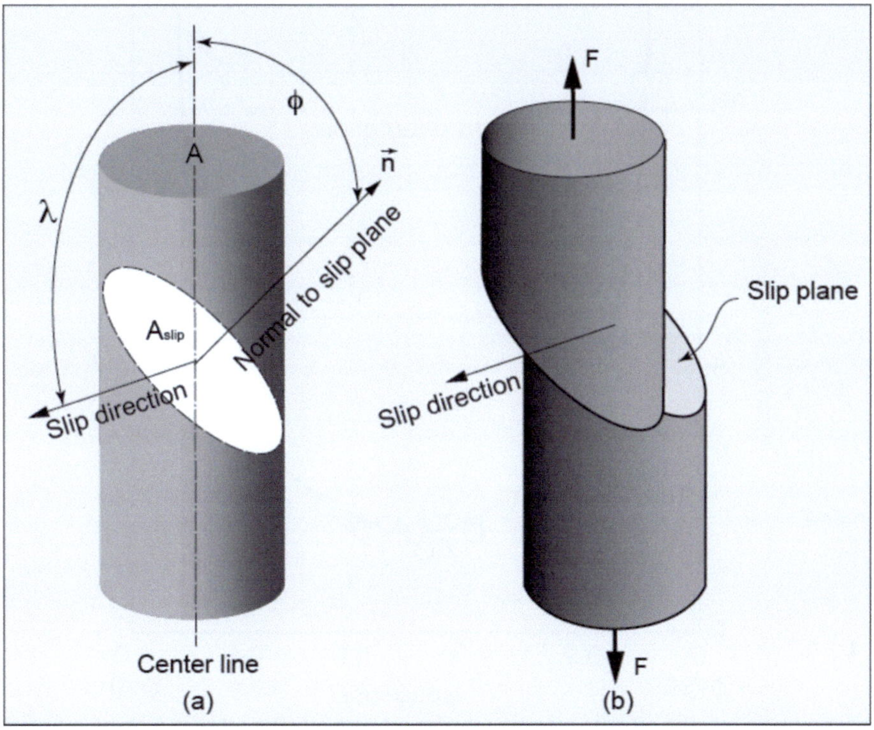

Fig. 2.16 Schematic of a single crystal cilindrical specimen indicating **a** slip plane and slip direction, **b** upon load application slip may occur on the chosen slip plane and direction

value, which is equal to one. This actually means that the flow stress of a single crystal is highly anisotropic. For instance, in the zinc crystal previously considered and shown in Fig. 2.15 the yield stress under uniaxial tension varies by a factor of at least six. Figure 2.17 presents the results obtained with specimens made of a single crystal of magnesium under axial tension. It can be seen the large variation of the yield strength as a function of the crystal orientation [8] with respect to the direction of the outside traction indicated by the Schmid factor M. Between the hard orientation corresponding to $M = 0$ and the softer one obtained with $M = 0.5$ there is a tenfolds reduction of the yield strength. FCC crystals have a rather large number of slip systems and plastic deformation by slip is relatively easy. Conversely, HCP materials do not possess such a large number of slip systems and plastic deformation occurs preferentially by twinning. Strictly speaking, the close-packed planes and direction that can be found in BCC crystals are not the most close-packed planes and direction possible, as in FCC crystal lattice materials. This is why BCC crystal metals need temperature to favor slip. The critical resolved shear stress, $\tau_{c,rss}$, is strongly dependent on temperature, which means that the yield strength of BCC crystals is temperature dependent (see Sect. 5.2.1). At low temperatures BCC crystals are very hard and brittle. This is also why BCC metals, at variance with FCC and HCP crystal

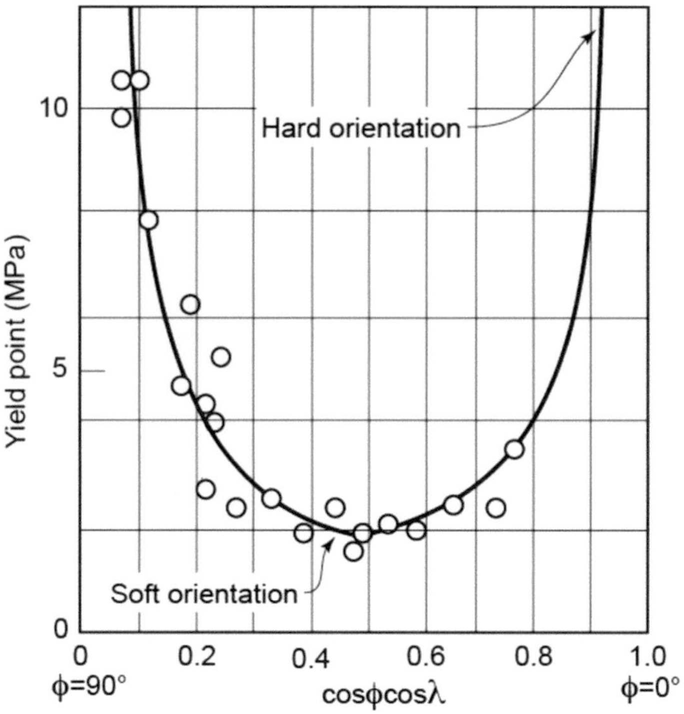

Fig. 2.17 Trend of the yield point in magnesium single crystal subjected to axial traction as function of the Schmid factor M (modified from [8])

lattice metals, present a transition temperature from brittle to ductile (see Chap. 3). In general, it is $\cos\phi{\cdot}\cos\lambda \cong 0.5$ so that yielding occurs when $\sigma = \sigma_y = 2\tau_y$. This also holds for policrystals and the material is assumed to obey the Tresca criterion for yielding. Table 2.2 [9] lists the critical resolved shear stress for some selected single crystal pure metals at room temperature. It can be seen the enormous difference between $\tau_{c,rss}$ in FCC and BCC crystal structure metals. The largest difference is of the order of 30. HCP metals behave like FCC metals with the exception of Ti that seems to follow the BCC behavior.

2.5 Slip in Policrystals

Consider, now, a polycrystalline material, as most metals and alloys are. As first, consider a bi-crystalline sample, as schematized in Fig. 2.18a with the two blocks A and B. We can define six strain components for each of the two crystals: 3 tensile and 3 shear components, ε_{ii} and γ_{ij} with $i = 1, 2, 3$ and $j = 1, 2, 3$. As clearly pointed out by Rice [10], strains cannot vary in an arbitrary manner from point to

Table 2.2 Room temperature slip systems and critical resolved shear stress for metal single crystal [9]

Metal	Crystal structure	Purity	Slip plane	Slip direction	$\tau_{c\ rss}$ (MPa)
Zn	HPC	99.999	(0001)	$\left[11\bar{2}0\right]$	0.18
Mg	HPC	99.996	(0001)	$\left[11\bar{2}0\right]$	0.77
Cd	HPC	99.996	(0001)	$\left[11\bar{2}0\right]$	0.58
Ti	HPC	99.99 99.9	$(10\bar{1}0)$ (1010)	$\left[11\bar{2}0\right]$ [1120]	13.7 90.1
Ag	FCC	99.99 99.97 99.93	(111) (111) (111)	[110] [110] [110]	0.48 0.73 1.3
Cu	FCC	99.999 99.98	(111) (111)	$\left[10\bar{1}\right]$ $\left[10\bar{1}\right]$	0.65 0.94
Ni	FCC	99.8	(111)	[110]	5.7
Fe	BCC	99.96	(110)	[111]	27.5
Mo	BCC		(110)	[111]	49.0

point within a deformable body. That is because the six strain components are all derivable with respect to the three displacement components u_i, with $i = 1, 2, 3$. Restriction on strain resulting from such considerations, are commonly known as compatibility relations; upon deformation the body would not fit together unless they are satisfied. This condition is what is also called cohesiveness (no void or crack formation). Therefore, at interface between A and B it must be

$$\varepsilon_{11}^{A} = \varepsilon_{11}^{B},\ \varepsilon_{33}^{A} = \varepsilon_{33}^{B},\ \gamma_{13}^{A} = \gamma_{13}^{B}. \tag{2.5}$$

Assuming that the two grains are differently orientated to the external load, one grain has a larger Schmid factor $M = \cos\phi \cdot \cos\lambda$. Therefore, the constraint introduced by the compatibility relations restricts the deformation of the most favorably oriented crystal and result in higher yield strength of the bi-crystal. Nevertheless, slips occur in both crystals that are called primary slips, as shown in Fig. 2.18b. If the plastic incompatibilities between two crystals under different deformations are large enough to generate high stress concentration at grain boundary, the local stress field can be relieved by the activation of a secondary slip system. This multiple slip is schematically shown in Fig. 2.18c.

Figure 2.19 [11] is an example of FCC bicrystal system in which it has been activated a secondary slip along the grain boundary in addition to the primary slip systems. If the crystals were three, as shown in Fig. 2.20 for zinc at 5% strain [12], the same secondary slips would occur. Note that a kind of intragranular boundary appears between primary and secondary slips, indicated by arrows A and B. The extension of these intragranular boundaries grows with increasing strain. Moving to a polycrystalline material, more restrictive constraint conditions arise. It is clear that

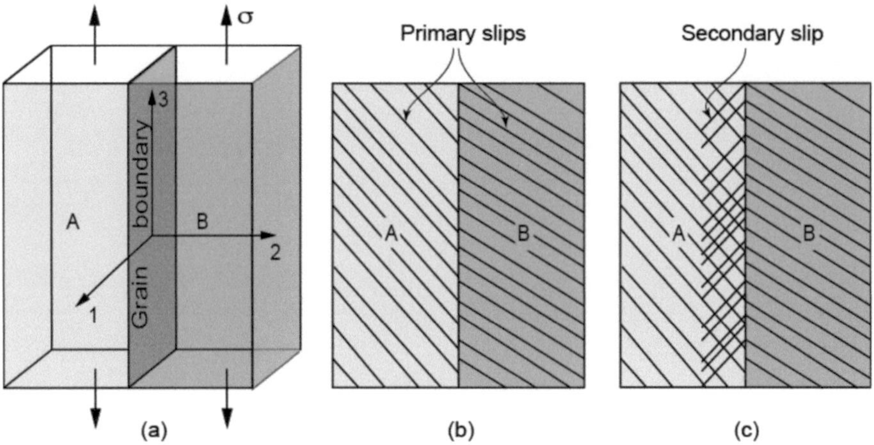

Fig. 2.18 **a** Bicrystal system having different slip directions. **b** Primary slips + **c** secondary slip

the more the number of crystals that are forming the multicrystal structure, the higher the interference among them and, therefore, the higher the resulting yield strength and hardening with respect to the single or bicrystal or tricrystal grouping. This effect is schematically shown in Fig. 2.21 for niobium [13] that has a BCC structure.

Von Mises [14] first observed that for a crystal to undergo a general change of shape by slip would require the operation of a minimum number of active independent slip systems in the vicinity of a grain boundary and that minimum number was five. This

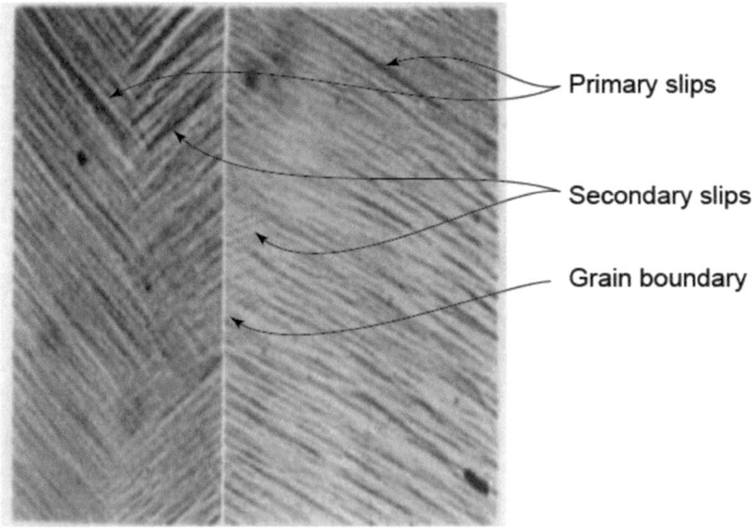

Fig. 2.19 Photomicrograph showing double glide in the vicinity of grain boundary in FCC crystals with a disorientation of about 85° (modified from [11])

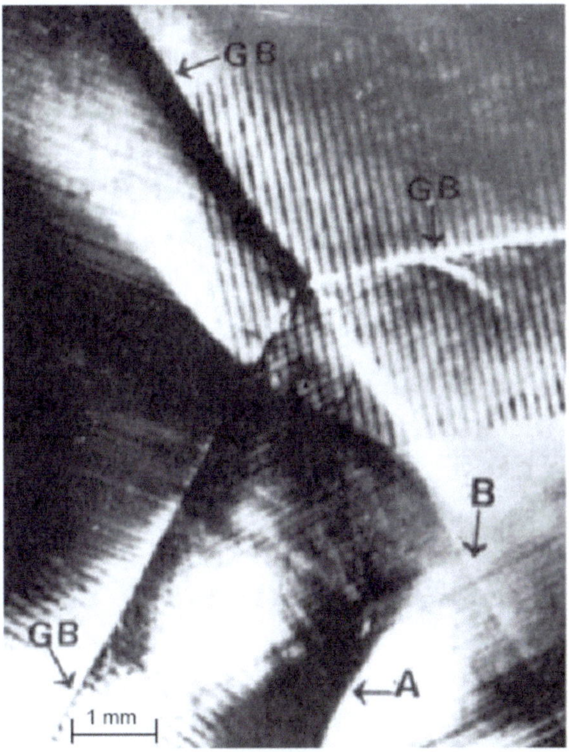

Fig. 2.20 Zinc tricrystal; GB indicates grain boundary; arrows A and B indicate intragranular boundaries [12]

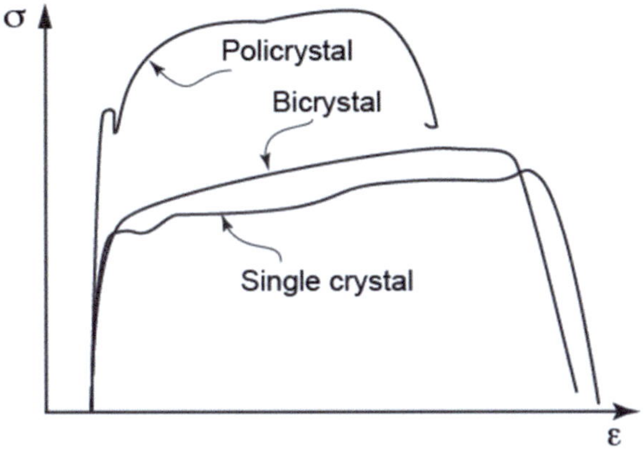

Fig. 2.21 Stress strain response in niobium single and multiple crystals (modified from [13])

is because the six components of strain, Eq. (2.5), are not completely independent. In fact, since deformations are occurring in the plastic domain they must satisfy also the requirement of constant volume, i.e., $\Delta V = 0 = \varepsilon_{11} + \varepsilon_{22} + \varepsilon_{33} = 0$, see Eq. (1.6), which reduces the number of independent variables to five. Therefore, crystals that do not possess five independent slip systems cannot be ductile in polycrystalline form unless twinning is providing the deformation mechanisms, as will be shown in Sect. 2.14. This is the case of HCP metals that have poor ductility at room temperature in polycrystalline form. However, they became ductile at some elevated temperature at which slip planes, other than basal, become operative and increase the number of slip systems to at least five. Face Centered Cubic lattice metals easily satisfy these conditions, which account for their generally high ductility. At variance, Body Centered Cubic lattice metals that do have a large number of slip systems do not possess real close-packed planes and directions so that they also need temperature to activate slip and show the characteristic transition temperature from brittle to ductile behavior. Taylor assumed that the actual five systems that became active were those requiring the least amount of work and established the following relation for the yield strength of policrystals

$$\sigma_y = \tau_{c,rss} \frac{\sum_i d\gamma_i}{d\varepsilon} = M_s \tau_{c,rss} \tag{2.6}$$

where $\tau_{c,rss}$ is the critical resolved shear stress, see Eq. (2.3), $\sum_i d\gamma_i$ is the sum of the shear strain increments of all active slip systems, $d\varepsilon$ the strain increment in an uniaxial traction test and $M_s = 1/M$. Taylor, then, calculated the average value of M_s for all orientation and found it to be equal to 3.06 thus concluding that

$$\sigma_y = 3.06\tau_{c,rss}. \tag{2.7}$$

This average value of the inverse Schmid factor is referred to as the Taylor factor. In practice, the yield stress of polycrystalline metals is often much higher than this due to the effect of grain boundaries. Equation (2.7) is in good agreement with experimental results obtained with aluminum specimens.

2.6 Yield Strength of Single Crystals

It has been said, in the previous sections and in Chap. 1 that slip occurs on close-packed planes along slip directions. Let's try to calculate the theoretical stress required to activate the slip process. Consider two layers of atoms on slip planes of a single crystal in their equilibrium position, as schematized in Fig. 2.22a. No mutual force is acting between any two atoms. The spacing between the two adjacent lattice planes is h and the distance between atoms in the slip direction is d_o. Suppose to apply a shearing stress τ that forces the plains to slide over one another by a distance u. Each atom of the upper plane rises from the initial stable position to a saddle point, Fig. 2.22b, afterwards it falls into the next stable well position, Fig. 2.22c. It is clear that the shear stress τ has to be a periodic function of the displacement u that can be assumed, as a first approximation, sinusoidal with period d_o

$$\tau = \tau_{max} \sin\left(\frac{2\pi u}{d_o}\right). \tag{2.8}$$

The shear stress τ is initially zero when $u = 0$, Fig. 2.21a and is zero for $u = d_o$ when the initial configuration has been restored, Fig. 2.21c. The shearing stress is also zero in between this two positions for $u = d_o/2$, etc. Since elemental displacements are very small the shear strain γ may be assumed to be equal to $\tan\gamma = u/h$ so that, approximating the function sin with its argument

$$\tau = \tau_{max} 2\pi\gamma \frac{h}{d_o}. \tag{2.9}$$

Recalling that the shear strain is related to the shear modulus G by the relationship $\tau = G\gamma$, Eq. (2.9) yields

$$\tau_{max} = \frac{G}{2\pi}\frac{d_o}{h}. \tag{2.10}$$

As a first approximation $d_o \sim h$ so that what Eq. (2.10) is actually saying is that the theoretical shear stress that activates slip in a single crystal is approximately equal to the shear modulus divided by 2π. This is an incredibly high value that is by far larger than that observed in real materials. Table 2.3 [15] indicates, for some selected metals, the corresponding shear modulus G, the experimental shear stress at yielding τ_y and the ratio τ_y/G. It can be seen how the real τ_y/G ratio is ranging from 1/1,000 to 1/100,000 of the shear modulus.

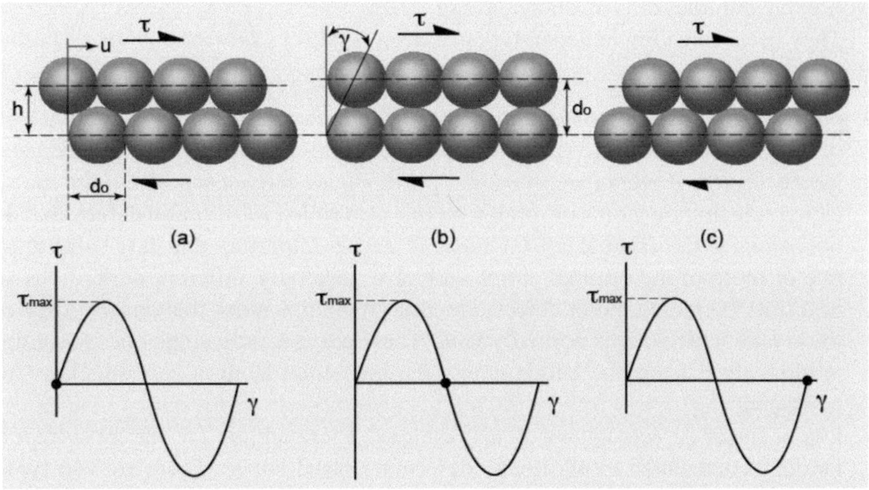

Fig. 2.22 **a** Initial position of the atoms on a slip plane; **b** elemental displacement with saddle point; **c** final position of the atoms moved by one atomic distance. The shear stress is periodic

Table 2.3 Shear modulus G, shear stress at yielding τ_y and τ_y/G ratio for some selected metals [15]

Metal	Shear modulus G (GPa)	τ_y (MPa)	τ_y/G
Alluminum	27	0.78	$\sim 2.6 \cdot 10^{-5}$
Silver	30.4	0.37	$\sim 1.2 \cdot 10^{-5}$
Berillium	119	1.4–52	$1.3 \cdot 10^{-5} - 4.2 \cdot 10^{-4}$
Cadmium	24.2	0.57	$2.34 \cdot 10^{-5}$
Iron	80.8	27.5	$3.4 \cdot 10^{-4}$
Magnesium	16.9	0.39–39	$2.3 \cdot 10^{-5}–2.3 \cdot 10^{-3}$
Molibdenum	131	72	$5.5 \cdot 10^{-4}$
Nickel	77.3	3.2–7.35	$4.1 \cdot 10^{-5}–9.5 \cdot 10^{-5}$
Niobium	40	33.3	$8.3 \cdot 10^{-4}$
Titanium	40.8	13.7	$3.3 \cdot 10^{-4}$

2.7 Defects in Crystals

Though derived simplifying a very complex reality using very simple arguments, Eq. (2.10) establish a theoretically acceptable order of magnitude for the shear stress τ_{max} necessary to plastically deform a metal, which is, let's say, 1/10 of the shear modulus G of the material. Real life indeed rejects such an incredibly high value. If this happens it is because crystals structures cannot be considered defect free. Speaking of imperfections or defects in a crystalline solid, it is convenient to introduce the schematic representation of lattice defects, according to Shockley, Seitz, Hollomon and Maurer [16], shown in Fig. 2.23.

They listed three fundamental types of defects, always present in polycrystalline structures that are

- point defects, also known as atomic size defects, characterized by the virtual lack of dimensions. These are vacancies and interstitials that are present in the crystal lattice anywhere one or more atomic positions are left vacant. These are the so called substitution atoms or foreign atoms, also called substitution defect, that do not belong to the original crystal lattice. They are impurities that have substituted one or more of the original atoms and have, generally, different dimensions so to distort the lattice. Point defects are also interstitial atoms that occupy a site in the crystal lattice that is normally void. These are, generally, high energy configurations that distort the lattice unless the interstitial atom is very small, as for hydrogen;
- line or linear defects known as dislocations. These are, by far, the most important defects deputed to plastically deform a crystal lattice. There are two types of dislocations: the edge dislocations and the screw dislocation. They can also combine to form mixed dislocations;
- bidimensional defects, which are small along one single dimension. Grain boundaries are such a defect that result where the crystallographic direction of the lattice

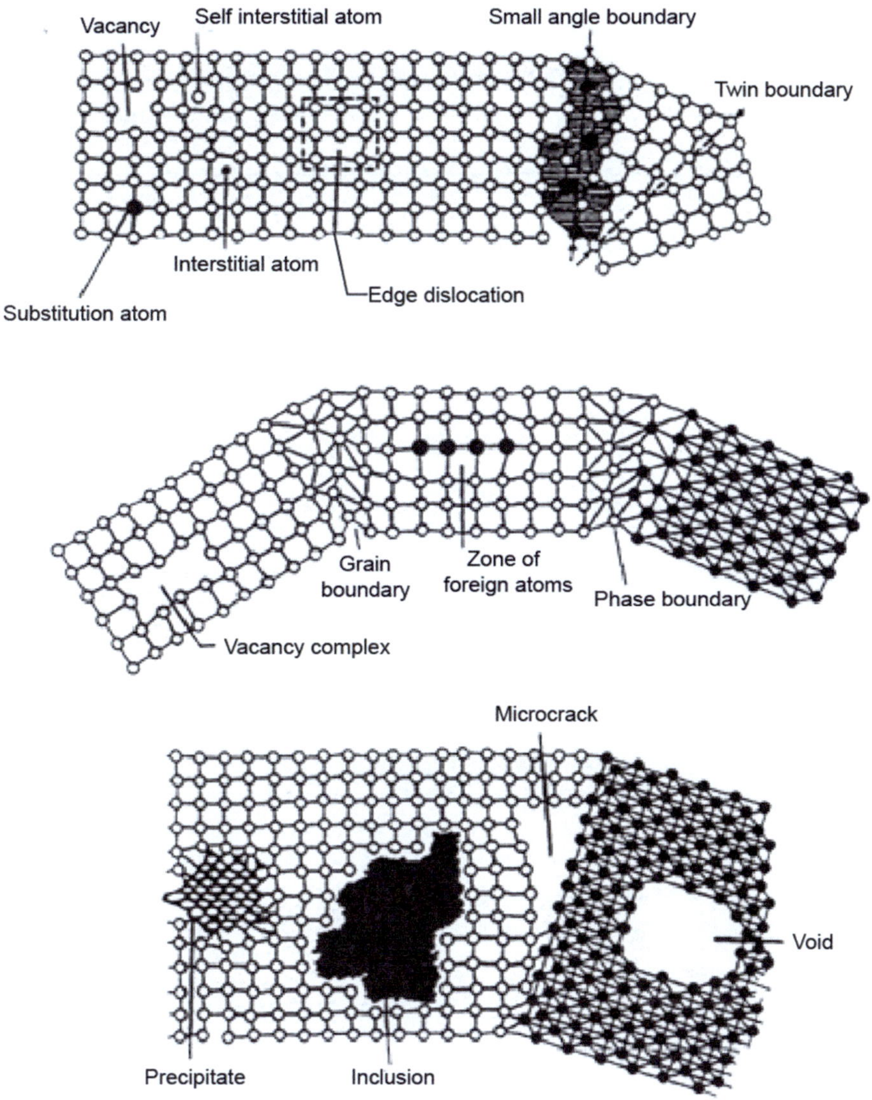

Fig. 2.23 Schematic representation of lattice defects [16]

changes. They originates when crystals form in the molten pool growing from an initial site or germ to meet each other. Stacking faults, precipitates, usually involving impurity atoms, voids as agglomerate of vacancies, inclusions are also such type of defects.

Table 2.4 provides a bar-chart showing size ranges for several structural features that characterize and can be found in materials [17]. It can be said that the most

Table 2.4 Bar-chart showing size ranges for several microstructural features found in materials [17]

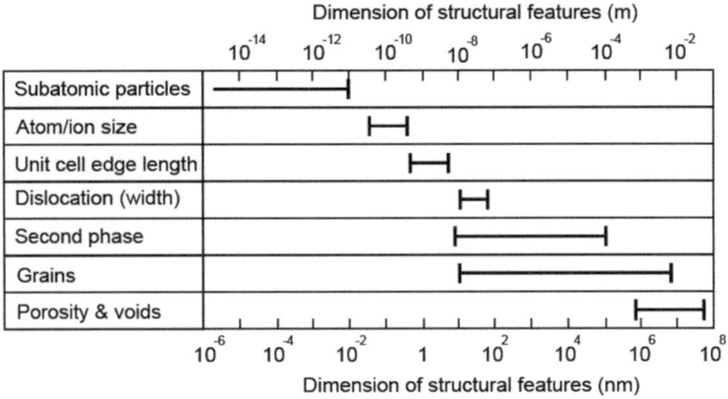

important contribution to interaction between a point defect and a dislocation is due to the distortion the point defect produces in the surrounding crystal.

The crystal lattice either contracts or expands locally, depending on the size of the imperfection with respect to the metal atoms. The distortion field may interact with that of the dislocation to raise or lower the elastic strain energy of the crystal. When this interaction results in a strong locking effect to dislocation motion the material yield strength increases, followed by subsequent hardening. This actually means that imperfections of the crystal lattice may well be desirable. When this is the case, we shall not really speak of defects but, rather, of deliberate addition of foreign atoms or other elements or phases to reach a wanted target of material hardening. The ability to change the response of a pure metal by adding *defects* to the crystalline structure has been engaging researchers for centuries no less than that to eliminate as much as possible all the unwanted defects. Having this in mind, we will continue to call a *defect* any imperfection of the crystal lattice even if it is wanted and deliberately introduced to produce a useful change of the material properties. However, the real effectiveness of defects deliberately introduced to strengthen any alloy is not just connected to the mere presence of the imperfections but, rather, to what those defects can do after the alloy has been heat treated or aged. Metallurgy, then, becomes a science and, perhaps, also an art.

2.8 Dislocations

Among all possible defects, dislocations are the most important of them all. It is this kind of defect that accounts for the discrepancy between the theoretical shear yield strength τ_{max} given by Eq. (2.10) and the shear yielding actually measured on metallic materials. Dislocations were first conceived in 1907 just as a theoretical geometrical discontinuity in continuum mechanics by the Italian mathematician Vito Volterra [18] and no connection was even conceived with the crystal structure of metals. That was because the crystal structure of metals was proved only thanks to the discovery of X-ray diffraction technique in 1912. But we have to wait till 1934 when the cause-effect relationship between dislocations and plastic deformation was first proposed by the Hungarian physicist Egon Orowan [19], the British physicist Geoffrey I. Taylor [20], who coined the word dislocation and the Hungarian physician-polymath Michael Polanyi [21] who, independently of each other, published papers on dislocation plasticity. Nevertheless, it was not until late '50s that the existence of defects such dislocations was finally accepted by metallurgists thanks to their direct observation with the development of transmission electron microscope (TEM). Dislocations can be of two types: edge dislocations and screw dislocations. Edge dislocation is a two-dimensional or line defect that has the configuration of an extra half plane of atoms inserted mid-way through the crystal matrix like a wedge, as schematized in Fig. 2.24. Under the action of a shear stress the upper half space slides over the bottom one resulting in a net displacement b, Fig. 2.24b. Therefore, the complete movement of a single dislocation through a crystal produces a step on the surface, the depth of which is one atomic distance, $b = d_o$. Figure 2.24c is a tridimensional view of an edge dislocation showing the extra half plane of atoms, the line of the dislocation that marks the boundary between the slipped and the unslipped part of the crystal, the slip plane over which the dislocation can move and its center, which is the trace of the extra plane on the slip plane. The slip plane is defined by the dislocation line, also called line vector $\vec{t}$ and by the slip vector$\vec{b}$ also called Burger vector. The one shown in Fig. 2.24 is conventionally defined as positive edge dislocation and indicated by the symbol ⊥. If the extra half plane of atoms lies below the slip plane the dislocation is said negative edge dislocation and its symbol is T. It is clear that two dislocations having opposite sign but same glide plane can move and annihilate each other, as schematized in Fig. 2.25. If the glide plane is not the same, then they can equally annihilate, but they would leave a vacancy in the crystal structure, as shown in Fig. 2.26.

A pure edge dislocation can slip or glide in a direction parallel to its slip vector $\vec{b}$ and perpendicular to its line vector $\vec{t}$. As to the screw dislocation, it appears as schematically shown in Fig. 2.27. The definition of screw dislocation given to this lattice defect comes from the fact that any continuous circuit from A to B turns around the dislocation line like a helical spiral, as sketched in Fig. 2.27. The axis of the helix defines the dislocation line vector, $\vec{t}$. In both edge dislocation and screw dislocation the motion is perpendicular to the dislocation line. At variance with the edge dislocation, in which the slip vector $\vec{b}$ and the line vector $\vec{t}$ are perpendicular, in the screw dislocation the two vectors, $\vec{b}$ and $\vec{t}$, are parallel to one another.

Sometimes, the symbol ↻is used to indicate a crew dislocation. Looking down the dislocation line, if the helix advances one plane when a clockwise circuit is made round it, it is referred to as a right-handed screw dislocation. When the reverse is true it is referred to as a left-handed screw dislocation. Note that when the screw dislocation line arrives to the free surface opposite to the starting one, it leaves the crystal deformed in the same manner as the edge dislocation, as schematically shown

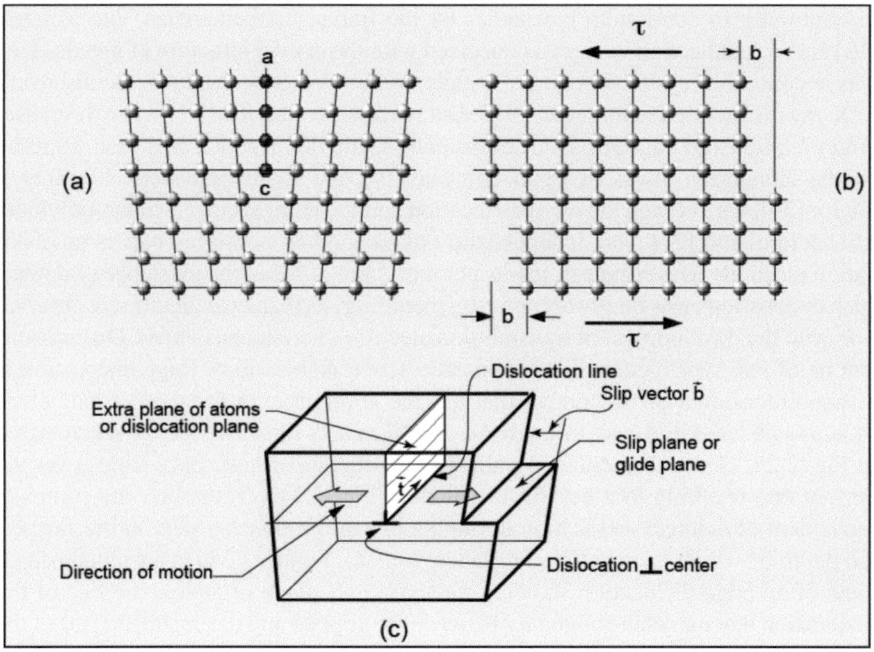

Fig. 2.24 **a** Schematic of an edge dislocation, **b** slip activated by shear produces a net displacement *b* of the upper part of crystal respect to the lower one and **c** tridimensional view of an edge dislocation

in Fig. 2.28. When the dislocation, either edge or screw, leaves the crystal the stress field associated with it is completely relieved.

2.9 Slip by Dislocation Motion

It has been shown, in Sect. 2.6, how the theoretical shear stress at yielding is in the range of 1/10 of the shear modulus G. In real life this value happens to be by large, much higher than that actually measured with traction specimens, from 10^3 to 10^5 times higher, as listed in Table 2.1. Responsible for that discrepancy are dislocations. The reason for that can be understood trough the following considerations. Consider a perfect crystal lattice without defects. Atoms in two adjacent layers on a generic slip plane are aligned as shown schematically in Fig. 2.29a. The top curve indicates the energy of each atom in the lower plane as a function of its position relative to the atoms of the upper plane. It can be seen how these equilibrium or stable positions correspond to a minimum of potential energy, as already indicated in Fig. 2.21. This is why, it was explained in Sect. 2.6, the theoretical yield shear stress is so high. Now consider the situation where the crystal contains a dislocation, schematized in Fig. 2.29b.

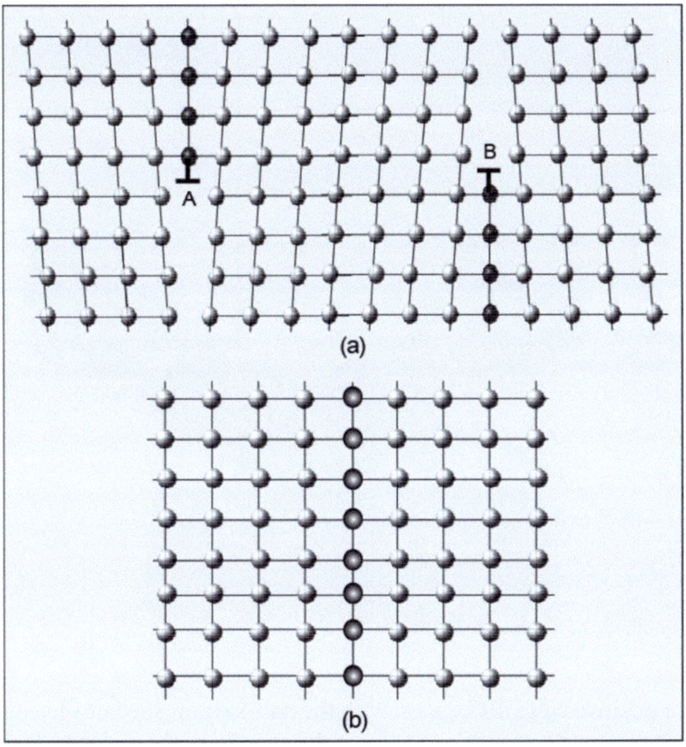

Fig. 2.25 **a** Two dislocations, A and B, having opposite sign can **b** annihilate each other by moving on the same glide plane

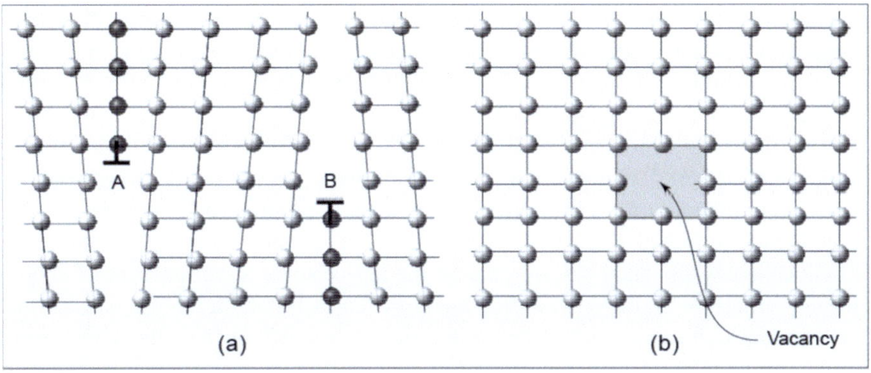

Fig. 2.26 **a** Two dislocations, A and B, having opposite sign, but different slip plane can **b** annihilate each other leaving a vacancy

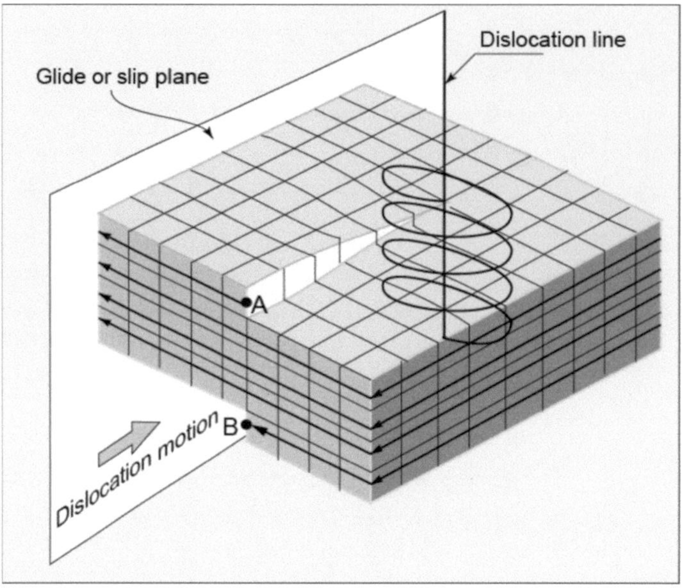

Fig. 2.27 Schematic of a screw dislocation

This is a positive edge dislocation with the half-extra plane between atoms 3 and 5 in the upper layer. Atoms located at large distance from the dislocation line, let say three or four atomic distances, would not *feel* its presence. This is because, according to the Saint Venant principle, the perturbation introduced in the crystal lattice by the dislocation decays within two or three times its characteristic dimension, as described in Sect. 2.11. Now, one or two atomic spaces can be assumed as characteristic dimension of a dislocation. Therefore, from three to five atom positions from the dislocation line its perturbation of the crystal lattice will not be perceived any more. Consider, then, the atoms of the layers below the dislocation center and, in particular, those lying on the first plane of trace A–B. The couple of atoms 3–4, located symmetrically on opposite sides of the dislocation center are both in a well of energy whose depth is smaller than that of atoms that were not displaced, as atoms 1–6. The energy barrier, also called Peierls-Nabarro barrier (see Sect. 2.10) is even lower towards the center of dislocation where they can move with the aid of some small amount of energy. Thermal vibration could provide this energy even at low temperature. Under such circumstances, atoms on both sides of the dislocation center have the same probability to move in opposite direction so that there will not be a definite displacement of the dislocation in a single, preferred direction. However, applying a shear stress will help atoms to decide where to go and the dislocation to move in the direction of the shear force. It will not be necessary to apply a very large shear stress since the energy barrier to overcome is not so large and it is not the entire half crystal to slip over the other, but only a row of atoms, the dislocation line, at a time. Of course, the resulting net displacement is just one atomic distance or the Burgers vector b,

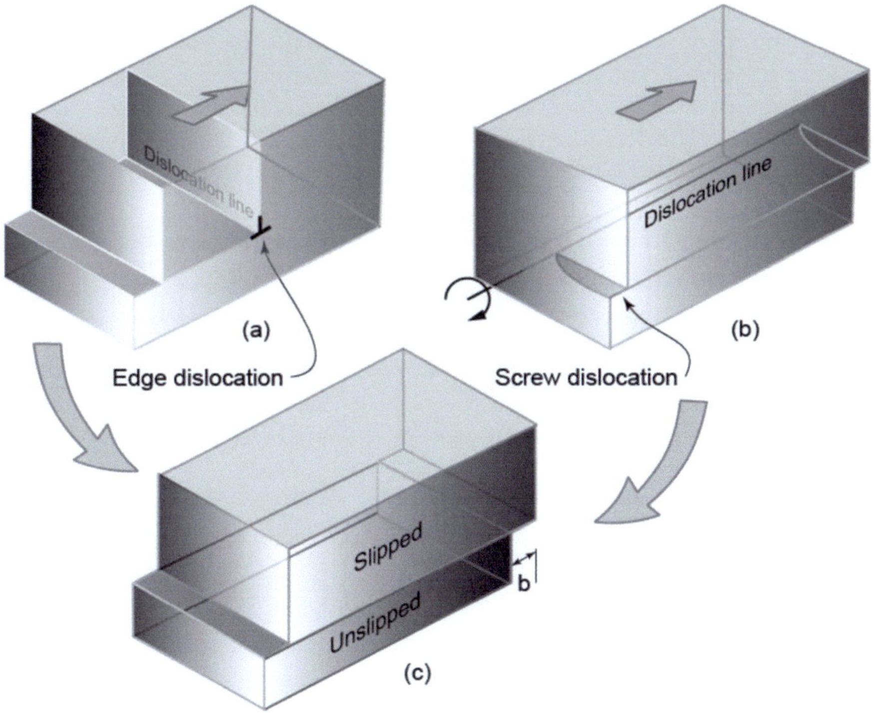

Fig. 2.28 The movement of **a** an edge dislocation and that of **b** a screw dislocation leads to **c** the same step *b* in the crystal lattice

but the shear stress applied will be much smaller than that required to split a defect free, perfect crystal. However, when a dislocation is continuously displaced through a step-by-step sequence or many thousands of dislocations operate on the same slip plane then the resulting plastic slip becomes relevant and can be seen with an optical microscope or even at naked eye, as shown in Fig. 2.14. To plastically deform a crystal lattice operating on each single dislocation rather than on an entire block of atoms is a lot like to move a heavy carpet by giving a relatively small impulse that would create a wrinkle that propagates like a wave. This would require a relatively low force compared to that needed to pull the carpet overcoming the friction with the pavement. This is schematically shown in Fig. 2.30a. The analogy with the carpet is rather effective to visualize the reason why the Peierls-Nabarro stress or, simply, Peierls stress is also called lattice friction stress. But it also recalls the motion of a caterpillar that advances by moving forward just a small section at a time, like a wave, as schematized in Fig. 2.30b [22]. Plastic slip, then, is made possible by the existence of defects like dislocations that overcome the potential barrier just locally in a step-like fashion. Actually, this step-like move can be even more fractioned. This is schematically shown in Fig. 2.31a, b. There is a driving force for the kink to move from a high energy level to a low energy state. This is a configurational force

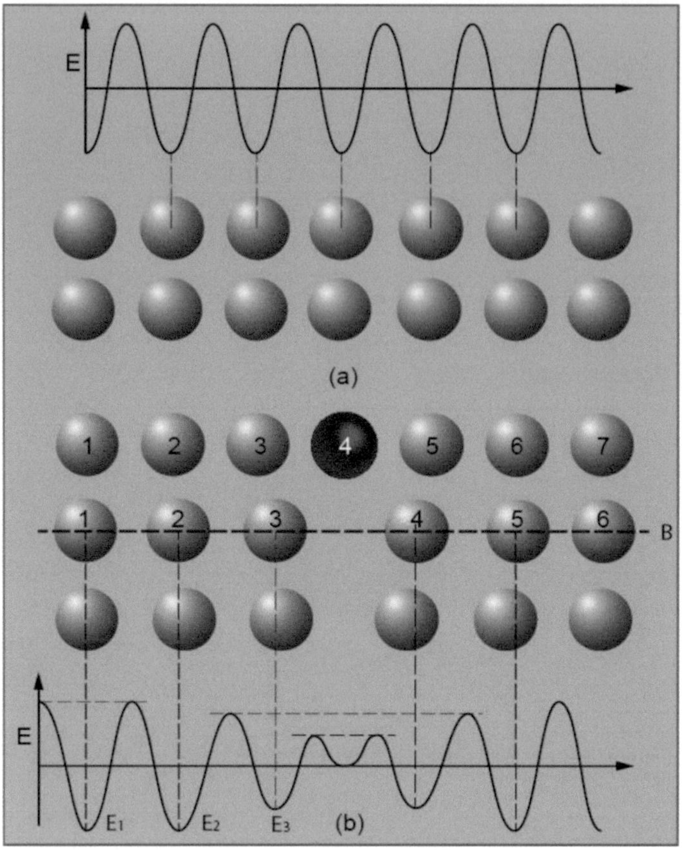

Fig. 2.29 Schematic of **a** perfect crystal lattice with its energy field, **b** lattice containing an edge dislocation that alters the energy field

since it originates by the change in the configuration of the system, which results in a lowering of the energy of the system. Once kinks overcome the barrier, they can spread laterally along the energy valley, Fig. 2.31c, resulting, eventually, in a forward motion of the total dislocation line, Fig. 2.31d. Therefore, the dislocation velocity, $v_{D,}$ becomes a function of the kink velocity, v_k

$$v_D = v_k \frac{b}{l} \tag{2.11}$$

where l is the length of the dislocation segment and b the Burgers vector. Kink velocity, in turn, is a function of the kink formation energy, W_k

$$v_k = \frac{2\sigma b^2 D_k}{kT} e^{\left(-\frac{Q}{kt}\right)} \tag{2.12}$$

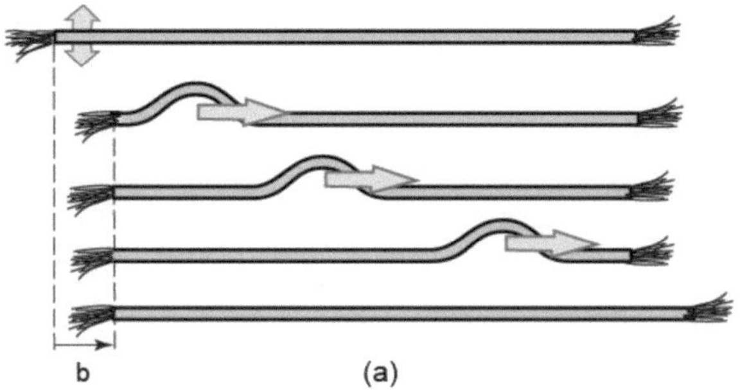

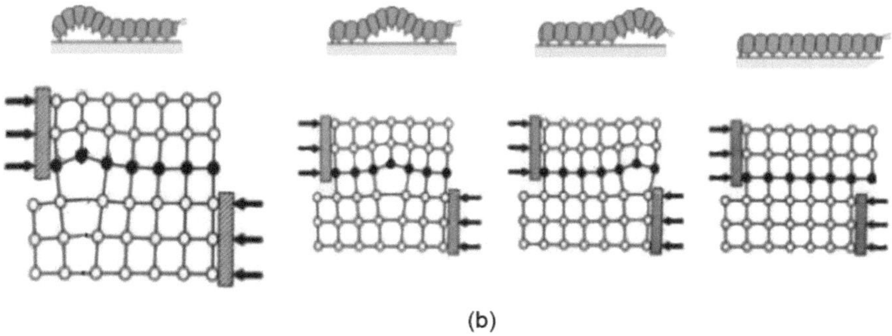

Fig. 2.30 Similitude between the motion of a dislocation and that of **a** a heavy carpet or **b** a caterpillar (modified from [17])

where the activation energy $Q = W_k$ for a single kink and $2W_k$ for a double kink, D_k is a diffusion coefficient and k is the Boltzmann's constant. Since it is $v_k \gg v_D$ the process allows dislocations to move faster than they could do by regular forward motion of the entire line over the Peierls-Nabarro barrier. In BCC crystal structures, thermally assisted formation of kink pairs can produce slip at stresses below the Peierls-Nabarro stress.

2.10 Peierls-Nabarro Barrier

In 1940 the German-born British physicist Rudolph Ernst Peierls calculated the minimum shear stress needed to move a dislocation. In his original model Peierls recognized the periodicity of the shear stress as the dislocation moves through the crystal lattice. Later, in 1947 also the English-born South African Physicist Frank Nabarro arrived to the same conclusions. The magnitude of this shear stress, known as Peierls stress or Peierls-Nabarro stress, is a sensitive function of the width of the dislocation. The width of a dislocation is a measure of the distance over which the crystal lattice is distorted because of the presence of the dislocation. Unit strength

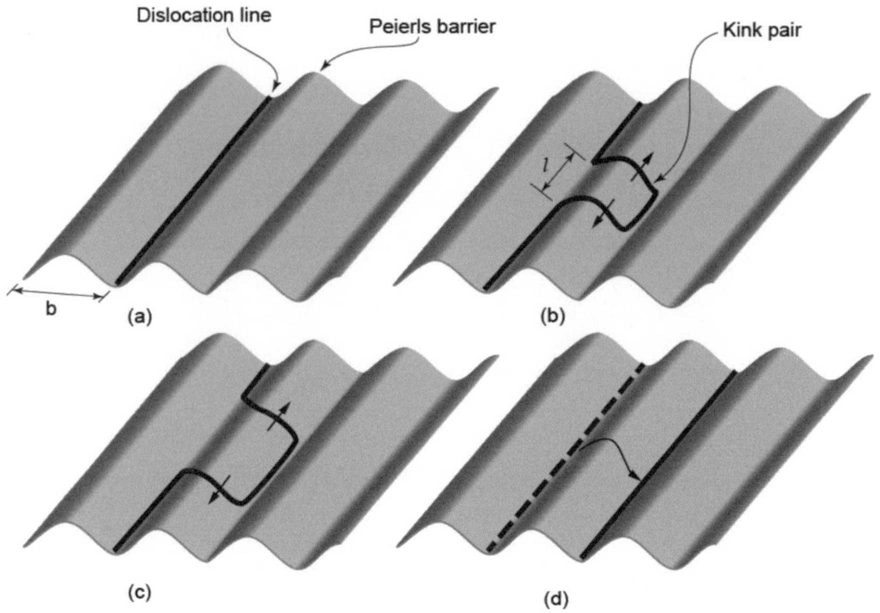

Fig. 2.31 Movement of a dislocation by kink pair formation; **a** dislocation line sitting in a potential valley; **b** kink pair formation overcoming the barrier; **c** kink lateral propagation and **d** total dislocation movement

is that of a dislocation in which the Burgers vector b is equal to one lattice spacing. The Peierls or Peierls-Nabarro stress takes the form of an exponential function of both the width W and the Burgers vector b

$$\tau_{PN} = \frac{2G}{(1-\nu)} e^{\left(\frac{-2\pi W}{b}\right)} \tag{2.13}$$

where G is the shear modulus of the crystal and the width W is given by

$$W = \frac{h}{1-\nu} \tag{2.14}$$

h being the planar spacing, see Fig. 2.32 and ν is the Poisson coefficient. Equation (2.13) indicates that the Peierls stress for a given slip plane decreases with increasing distance h between like planes and increasing atomic separation b. Since the planar spacing h varies inversely with the atomic density, slip occurs preferentially on closely packed planes, as already said. Equation (2.13) is plotted in Fig. 2.32a for four different h values and shown for two cases in Fig. 2.32b. In both cases the slip plane is a close-packed plane. However, in the first case, Fig. 2.32b, the slip direction is a close-packed direction of high atomic density; whereas in the second case, Fig. 2.32c, the slip direction is not a close-packed one, but a generic one having a low atomic density. In this latter case the planar spacing h is much lower, but the slip distance b is by far larger and this will result in a much higher Peierls-Nabarro stress, τ_{PN}. Formally, the width of an edge dislocation is defined as the distance over which atoms offset between slipped and unslipped planes is greater than one quarter of the magnitude of the Burgers vector, b, as schematically shown in Fig. 2.33. At the center of the dislocation the offset is always $b/2$. When W is several atomic distances the dislocation is wide, Fig. 2.33a. At variance, when W is of the order of one or two atomic spacing, the dislocation is narrow, Fig. 2.33b. Metals with BCC crystals structures have narrow dislocations, while FCC and some HCP metals, such as zinc, cadmium and magnesium, have wide dislocations. As such, the Peierls-Nabarro stress, τ_{PN}, is higher in BCC crystal lattice materials than in FCC ones and dislocation motion is much easier in austenitic steels than in carbon steels. A dislocation of unit strength has a minimum energy when its Burgers vector is parallel to a direction of closest atomic packing. Therefore, crystals always slip in the close-packed directions. Foreman, Jaswon and Wood [23] have shown that assuming a periodic sinusoidal dependence of the Peierls stress on displacement, by reducing the amplitude of the Peierls stress by half, the width of the dislocation would increase four-fold. The magnitude of the Peierls-Nabarro stress is extremely small and difficult to measure. It varies from 10^{-5} to 10^{-6} MPa. Note that it is much smaller than the critical resolved shear stress, $\tau_{c,rss}$, needed to activate slip (see Sect. 2.4) and smaller than the yield stress. This actually means that the critical resolved shear stress and the yield stress represent the condition where a large number of dislocations are put into motion.

2.11 Activation Energy

The effective height of the Peierls-Nabarro barrier depends on temperature. This is schematically shown in Fig. 2.34. The area under the curve represents the energy needed to overcome the barrier. It is the so-called activation energy or Gibbs free energy ΔG. Its derivative is the corresponding force. At zero temperature, $T_o = 0$ K, the activation energy is ΔG_o and the Peierls-Nabarro force is given by the full eight of the barrier. Increasing the temperature the effective eight of the barrier reduces since a certain amount ΔG of the thermal energy will be made available to dislocation to overcome the barrier. The more the temperature increases, the more the barrier reduces its effective eight. At T_2 the energy barrier will be

$$\Delta G = \Delta G_o - \Delta G_2. \tag{2.15}$$

By increasing the temperature, eventually there will be a temperature T_3 at which thermal energy will completely fill the barrier. At this point dislocations will move freely without any impediment; there will not be any barrier at all. In terms of probability, the probability P that at the equilibrium temperature T there is an energy fluctuation equal or larger than ΔG is given by

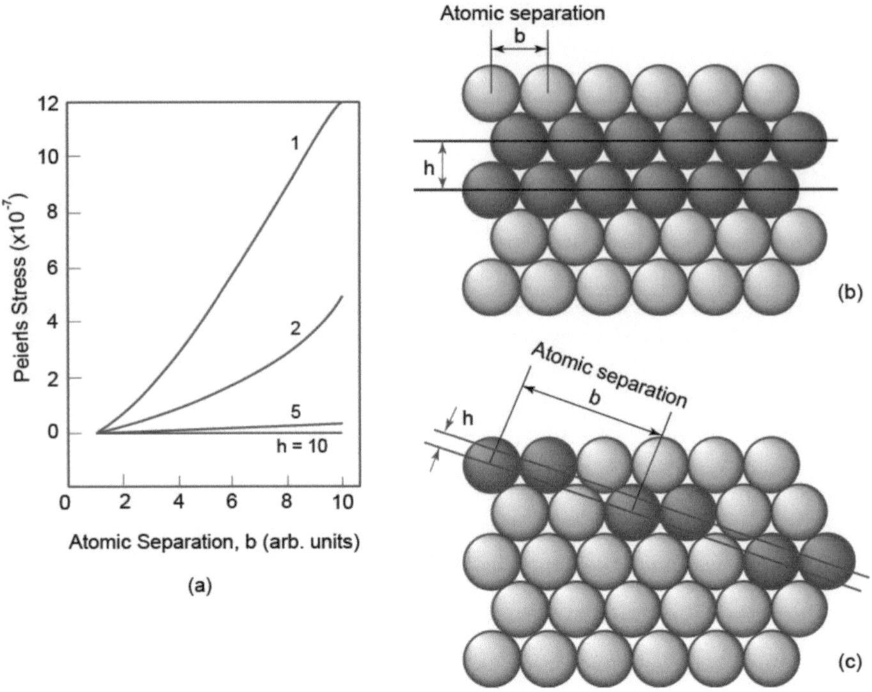

Fig. 2.32 **a** Plot of the Peierls-Nabarro stress versus the atomic separation. **b** Slip along a close-packed direction; **c** slip along a generic non-close-packed direction

$$P = \exp\left(-\frac{\Delta G}{kT}\right) \tag{2.16}$$

where k is the Boltzmann's constant equal to $8.62\cdot 10^{-5}$ eV/K = $3.30\cdot 10^{-24}$ cal/K. This is also the probability that a dislocation will overcome the Peierls-Nabarro barrier. That is to say, given a barrier of height ΔG the probability that a dislocation will jump over the barrier is P. As the temperature goes up, this probability increases because thermal energy will help dislocation to jump over the barrier. At a certain temperature the activation energy for the kink pair mechanism, see Fig. 2.31, is entirely provided by thermal activation. The frequency ν of successful jumps is given by

$$P = \frac{\nu}{\nu_o} \rightarrow \nu = \nu_o \exp\left(-\frac{\Delta G}{kT}\right) \tag{2.17}$$

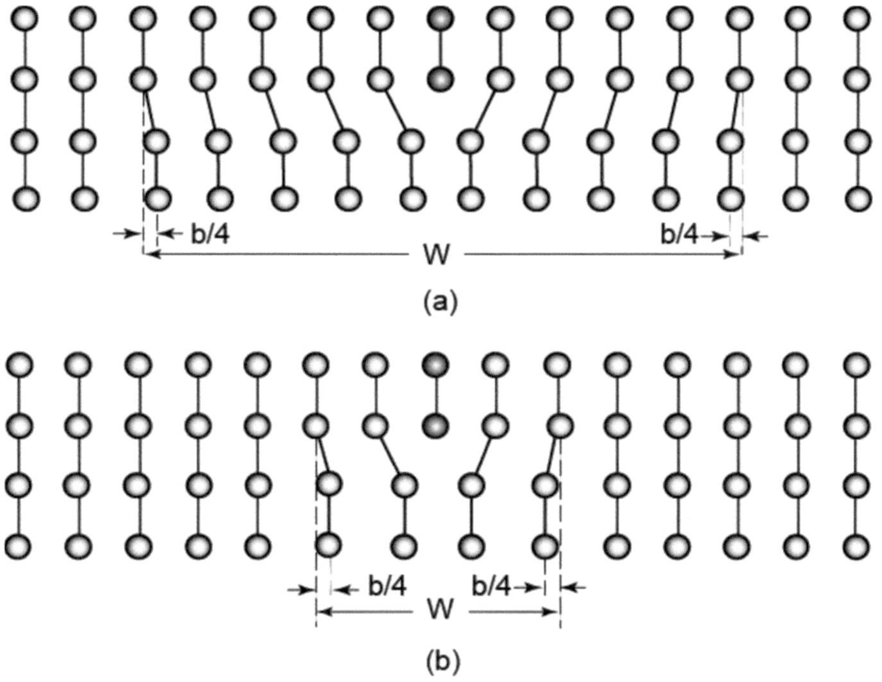

Fig. 2.33 **a** Wide dislocation, **b** narrow dislocation

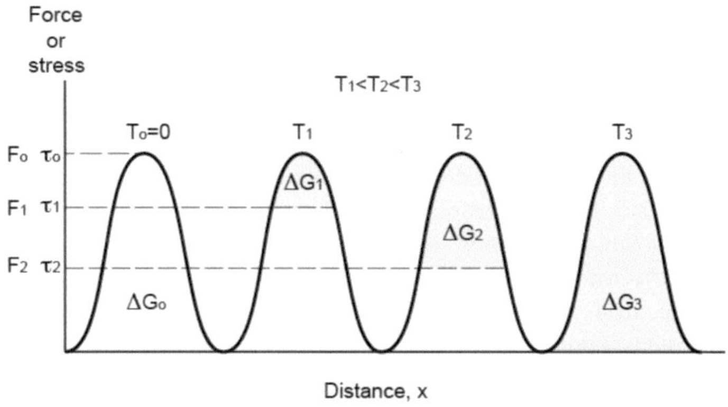

Fig. 2.34 Schematic of the Peierls-Nabarro barrier showing the thermal energy avalability

2.12 Dislocation Atmosphere

Solute atoms or interstitials may have a strong influence on dislocation motion. They change the so called interaction energy, E_I, of the crystal. The equilibrium concentration, c, of interstitials at a position close to the dislocation is following an expression similar to the Arrhenius formula valid for vacancies and carbon solubility

$$c = c_o \exp\left(-\frac{E_I}{kT}\right) \tag{2.18}$$

where c_o is the concentration far from the dislocation, given by the ratio of the number of defects to the number of non-lattice site. Note the similitude between Eqs. (2.17) and (2.16). Interstitials, therefore, tend to concentrate where E_I is large and negative. This is precisely the situation near the dislocation core. Accordingly, interstitials tend to concentrate in the core region forming a relatively large cloud or atmosphere known as the *Cottrell atmosphere*. The condition for that atmosphere to form is that temperature must be sufficiently high for small foreign solute atoms to migrate and remain in the core region of the dislocation but not so high to escape the atmosphere and go back into the solvent matrix. Equation (2.15) indicates that even in dilute solutions with, say, $c_o = 0.001$, dense atmospheres with $c \geq 0.5$ can be formed at $T = 0.5\ T_m$, where T_m is the melting temperature, in regions where $-E_I$ is about $3kT_m$. This corresponds to a solute atom-dislocation binding energy of typically 0.2–0.5 eV for metals. The fact that solute atoms tend to migrate towards the dislocation center can be well understood also considering that the dislocation generates a stress field. An edge dislocation acts like a wedge into the crystal lattice, generating a stress field that is compressive above the dislocation line and traction underneath that line, around the dislocation core. This determines a redistribution of small solute atoms, like carbon, nitrogen or oxygen, in particular, that are squeezed off the compression area and pushed into the traction zone where they found more room to stay. With the exception of hydrogen that is detrimental to material strength, these small atoms are always present in the solid solution. It must be understood, however, that these small solute interstitials have not such a comfortable accommodation in the interstices or voids between larger solvent atoms that form the crystal lattice. Void defines the space extending as a continuous network across the whole solid between atoms of the crystal lattice, as schematized in Fig. 2.35b for BCC structure, typical of iron, carbon steel, chromium, vanadium and molybdenum and Fig. 2.35c for FCC crystals lattice, common to austenitic steel, aluminum, copper and nickel. The dimensions of most common interstitials are also shown against that of the iron atom. It is clear, from the figure, that the crystal matrix does not leave interstitials enough room at all. Though small, these atoms are not that small to find an adequate site to stay. In FCC crystal lattice the largest void available has a size of about 1 Å against 1.54 Å of a carbon atom. As with BCC crystal lattice, Fig. 2.35, the situation is even more dramatic. The largest space available is 0.73 Å wide. Therefore, carbon, nitrogen and oxygen atoms cannot find a site without deforming the crystal lattice. Deformation means internal strains and stresses and also internal energy that claims to be released, somehow. Only hydrogen atoms, being the smallest of them all, with a diameter less than 1/3 that of carbon, can be dwelled, as shown in Fig. 2.35b, c. Hydrogen atoms are shown as ions H^+ since hydrogen can enter the metal only as ion. Because of their small size, these atoms can move more freely that any other in the crystal lattice. This is also the cause of the great damage and toughness reduction caused by hydrogen atoms. Larger impurity atoms, then, may be better sited in a vacancy rather than in interstitial positions. It can be understood how the presence of an edge dislocation in the crystal lattice with the enlarged traction area below its center must be seen as a bonanza to interstitials. They are squeezed off the compression area above the dislocation core towards the expanded region in tension below the core. By doing that a lower energy state is reached. The greatest distortion of the crystal lattice is achieved at about one atomic distance below the center of an edge dislocation. It is there that the Cottrell atmosphere concentrates. Wilde and co-workers mapped a tridimensional distribution of carbon atoms around a dislocation in low carbon martensite with a field ion microscopy [24]. The atoms of carbon formed a cloud extending about 7 nm outwards from the dislocation core, with about 105 carbon atoms per 10^{-3} μm of dislocation length. Assuming a grain 40 μm wide, a single edge dislocation having a line running from one border to the other will contain ~$105 \cdot 40 \cdot 10^3$–$4 \cdot 10^6$ atoms of carbon. The atmosphere of a dislocation, therefore, consists of an excess concentration of interstitial atoms below the edge and a deficiency of these atoms above the edge. A pictorial view of carbon atoms distribution in the core region along an edge dislocation line is given in Fig. 2.36. Just out of the core region the concentration of carbon and other interstitials drops to that existing in the crystal as a whole. More recently it has been shown that screw dislocations that do not possess a traction stress field also have a Cottrell atmosphere. This is because there is a strong interaction between interstitial atoms and screw dislocations when lattice is not symmetrically deformed by solute atoms so that a tensile component of stress is generated.

Cochardt et al. [25] suggested that the core of a screw dislocation can draw even more carbon atoms, about twice as many as the core of an edge dislocation, though the attraction force is lower.

2.13 Dislocation Microstructure

Dislocations are activated by shear stresses. Once activated, they produce plastic deformation by slip. As the shears stress increases also plastic deformation progresses by a phenomenon known as dislocations multiplication. Each dislocation can produce more dislocations by cutting as they encounter barriers on their motion (see Figs. 3.26 and 3.27) through a mechanism known as Frank-Reed source. Therefore, during plastic deformation the dislocation density, ρ, increases. Dislocations begin to interact with each other. It is this interaction, as we know, that produces hardening. This is because in their motion dislocations garble and their motion becomes more and more difficult to sustain. We can easily understand how these interactions become more and more severe as the dislocation density increases leading to a continuous increase of the flow stress. Moreover, tangles or *forest dislocations* form when dislocation motion leads to the formation of immobile dislocation configurations. Examples include jogs and sessile dislocation segments. These segments will keep other dislocations from moving freely. The general observation is that hardening depends on edge dislocation

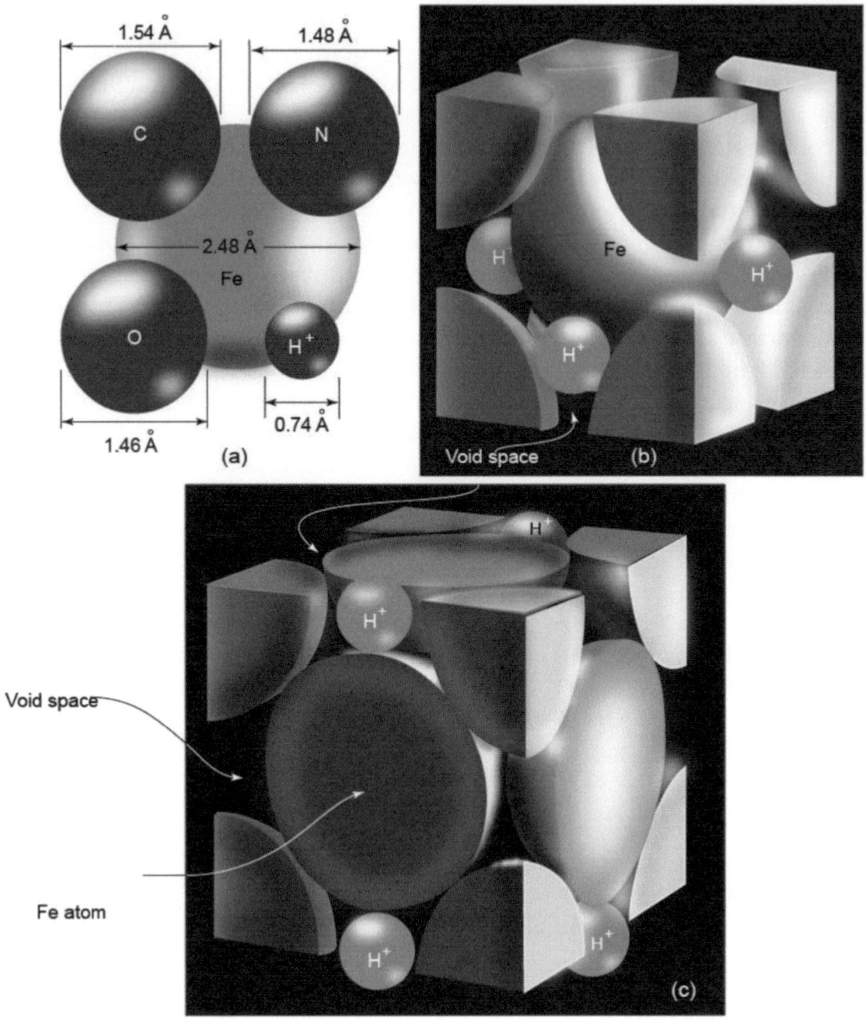

Fig. 2.35 **a** Schematic of dimensions of most common interstitial atoms vs the iron atom size. Void space in **b** the unit BCC and **c** FCC cell. Only hydrogen ions can be accommodated without deforming the crystal lattice (**b**, **c**). Note the cell size that for FCC lattice is larger than that of BCC

density ρ

$$\tau = \tau_o + \alpha Gb\sqrt{\rho} \tag{2.19}$$

where τ_o is an intrinsic flow strength for materials free of dislocations and α is a constant (0.2 for FCC and 0.4 for BCC crystal lattice materials). The shear stress τ that appears in Eq. (2.16) may be taken as the resolved shear stress (see Sect. 2.4). The strengthening increment $\alpha Gb\sqrt{\rho}$ that derives from hardening due to edge dislocation

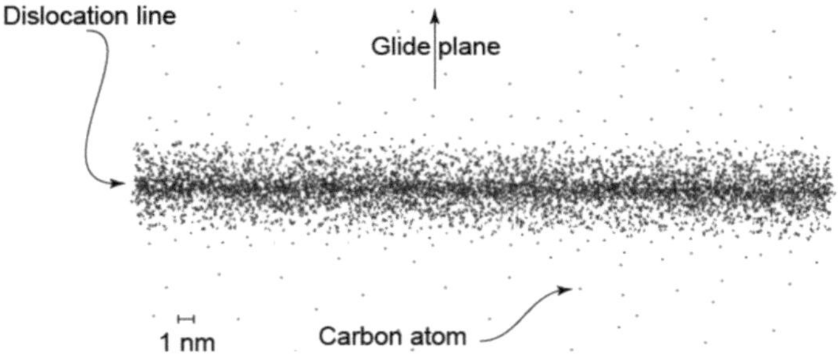

Fig. 2.36 Pictorial view of carbon atoms concentration around the line of an edge dislocation in the traction area of the core region. Just out of the core region the concentration of carbon atoms and other interstitials drops to that existing in the rest of the crystal. Scale is only indicative

multiplication and interactions can be approximated as [26]

$$\Delta\sigma = \frac{M\alpha Eb\sqrt{\rho}}{(1+\nu)} \tag{2.20}$$

where M is the Taylor factor. This dependence on $\sqrt{\rho}$ is shown in Fig. 2.37 [27] for copper poly-crystals and single crystals, with one, two and six slip systems, respectively. The same dependence has been observed in nickel [28]. During the plastic deformation phase when hardening is continuously growing, walls of high dislocation density form.

These are dense dislocation tangles that form a sort of cellular structure within the grain. The boundaries of these cells have high dislocations density while the interior has low dislocations density. These cells are also called subgrains. They look like small grains, but they are not really grains as we normally think of. However, they also produce a grain boundary effect but of lesser strength. It is easier to move dislocations across cells rather than through grain boundaries. Figure 2.38 [29] shows the dislocation microstructure in a polycrystalline stainless steel type 304 as it develops during the plastic deformation and hardening phase. It can be seen the cellular structure created by dislocation tangles and forests at high strain, in particular.

2.14 Upper and Lower Yield Point

The effect of the Cottrell atmosphere is that of anchoring the dislocation. If we want to plastically deform a metal, we have to displace dislocations. To displace a dislocation it is necessary to move also its atmosphere. Interstitial atoms of the Cottrell atmosphere, in fact, cannot be left behind the dislocation otherwise a wake of densely populated interstitial atoms would be formed hosted in regions no longer dilated by the stress field of the dislocation, which is not possible. Those interstitials

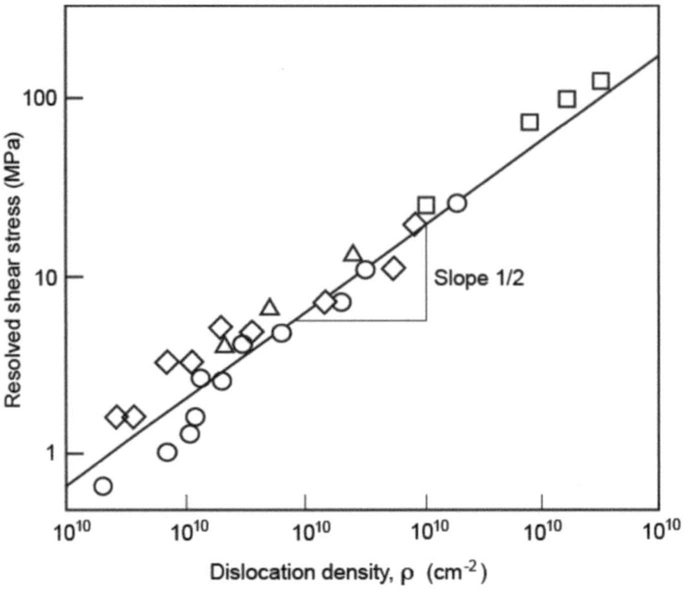

Fig. 2.37 Dependence of the resolved shear stress on dislocation density for poly-crystal copper and single crystals, with one, two and six slip systems, respectively (adapted from [27])

must, therefore, re-enter into solid solution or move into the newly formed expanded region of the dislocation freed from the atmosphere. However, they will find this new dilated area already occupied by other interstitials pushed off the compression area of the dislocation stress field, or at least partially occupied by them. It is, then, clear that at each elemental act of dislocation motion a net number of interstitial atoms equal to the population of the Cottrell atmosphere must reenter into solid solution to allow dislocation to move. However, while dislocation motion is activated by shear stresses, the movement of interstitials responds to the law of atomic diffusion. Atomic diffusion, in turn, depends on temperature so that the two movements may be not compatible, especially at very low temperatures, with the result that the atmosphere will exert a drag force to its motion, before being left and temporarily lag behind the dislocation. The more the temperature decreases the more atoms mobility diminishes, the more the drag force increases creating a serious obstacle to dislocation motion. At this point the width of a dislocation comes to play a role, see Sect. 2.11. Metals with BCC crystal structure, like ferritic steels, have narrow dislocations, while FCC metals, like austenitic steels and aluminum and some HCP metals, like zinc and cadmium, have wide dislocations. A narrow dislocation means that the stress field around the core and underneath its line is concentrated and so is the dilated area and the dwelling Cottrell atmosphere. On the contrary, in a wide dislocation the Cottrell atmosphere is very much diluted, almost nonexistent. This is the reason why the dislocation width enters into Eq. (2.14) affecting the Peierls stress. When the atmosphere of interstitials is concentrated the drag force to dislocation motion is

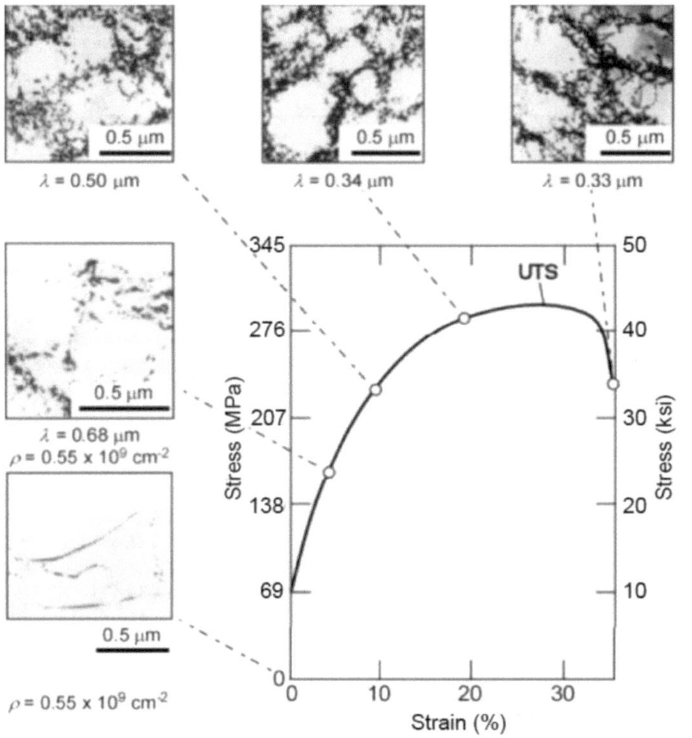

Fig. 2.38 Dislocation cells formation during plastic deformation in polycrystalline 304 stainless steel (adapted from [29])

high and temperature dependent; when it is diluted this resistance force may not even exist and doesn't depend, almost at all, on the temperature. The Cottrell atmosphere is also responsible for the appearance of the upper-lower yield strength followed by a plateau with serrations in BCC crystal lattice metals, see Fig. 2.10. In fact, when the elastic limit is approached during the traction test of a BCC metal, to further deform the material by slip and enter the plastic domain it is necessary to overcome the Peierls stress and move dislocations. Initially, the anchoring action exerted by the Cottrell atmosphere is high. This is precisely when the upper yield point is reached. But continuing to increase the applied load, the first dislocations start to glide getting rid of the respective Cottrell atmospheres that are left behind. Just a sudden jerk followed by a microscopic step ahead. This first glide, due to the breaking away of dislocations from their atmospheres, is seen by the crosshead of the testing cell as a reduction of the specimen stiffness and results in the abrupt drop of the applied load, with the appearance of the so called lower yield point, see Fig. 2.10. But the load drop slows down dislocations motion that may eventually come to a complete halt. This temporary dislocation slow down and rest gives the atmosphere the time and opportunity, depending on the test temperature, to take again the dislocation core and block its movement again. This process of dislocations lock-unlock continues with the appearance of serrations in an almost constant stress regime, the so-called plateau where Lüders lines appear, see Sect. 2.2. During this phase, deformation increases, but stress does not. This is because the continuous activation and multiplication of dislocations from the one hand facilitates the slip process, but from the other hand also increases the interaction with crystal defects and, hence, the resistance to slip. The balance is lost when dislocations multiplication becomes massive and so large to create forests and junctions. Forests and junctions make dislocations movement more and more difficult. They may also turn glissile dislocation into sessile and immobile dislocations. This is when the hardening phase appears. The plateau, then, is a stick and slip phenomenon.

2.15 Deformation by Twinning

We shall now speak of another important form of plastic deformation whose effects on the fracture process are of great importance: twinning. A twin is a region of a crystal that has been displaced by an atomic distance so that the reorientation of the lattice is a mirror image of the parent crystal, as schematically shown in Fig. 2.39. Note that it is just a limited region of the crystal or grain that displaces. Twins do not extend throughout the entire grain, but tapper to a point within it and terminate abruptly. Therefore, they tend to be lenticular, as shown in Fig. 2.40 that refers to polycrystalline zinc [30]. Note that in Fig. 2.40a the heavy etching has revealed both twinning and slips that appears as light parallel thin markings transverse to the large twinning lens.

By repolishing the sample and etching again in a lighter fashion the slip lines do not appear anymore and only the twin lenses emerge, Fig. 2.40b. As for the plastic slip activated by dislocations movement, shear stresses are responsible for twinning. In both cases, then, the crystal lattice is sheared, but while in the former mechanism slip occurs in discrete multiples of the atomic spacing and the macroscopic deformation is the result of the accumulation of microscopic slips along crystallographic planes and directions, in the latter process deformation is due to a solid block that rotates and the atoms movement are much less than an atomic distance.

The basic difference between slip and twinning is, then, schematized in Fig. 2.41. We may say that twinning is actually helping slip by dislocation movement to occur.

In fact, this crystallographic reorientation may place new slip systems in orientations that are more favorable relative to the stress axis so that slip can occur at lower stress. Therefore after twinning, extensive plastic deformation by slip usually takes place. It is not known whether or not there is a precise resolved shear stress for twinning as for slip. However, the stress field accompanying a twin formation is much higher than that of a dislocation. Hence the ratio between actual stress and theoretical stress that in the case of slip was exceeding 1000, as said in Sect. 2.6, is lowered. The critical event in twinning is, for most cases, nucleation. Growth, in fact, can occur at stresses that are a fraction of the nucleating stress, therefore, once created,

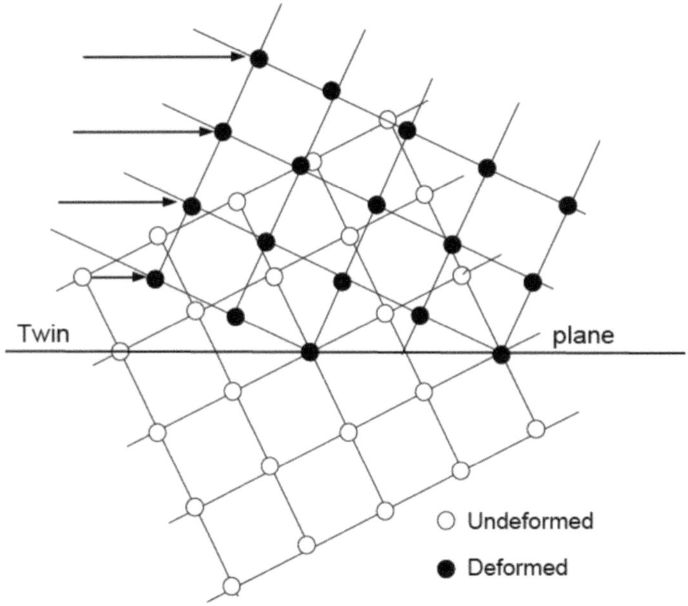

Fig. 2.39 Schematic of deformation by twinning in a cubic crystal

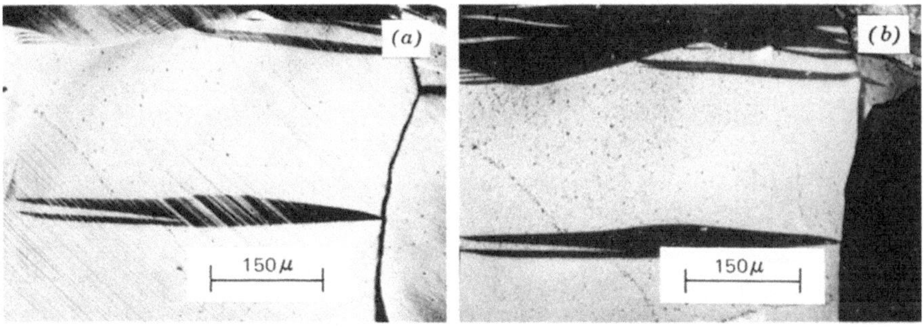

Fig. 2.40 Surface markings resulting from plastic deformation on zinc crystals [30]

twin propagates under lower stresses. This circumstance produces two characteristic features. First, the onset of twinning during a traction test is immediately revealed by the presence of serrations in the stress–strain curve, as shown in Fig. 2.42 for a single crystal of cadmium [31]. Serrations due to twinning have the same cause of those produced in the plateau of ferritic steels where dislocations are continuously locked and unlocked, see Fig. 2.10. Secondly, the excess of stress needed to start a serration, with respect to its progress, accelerates twins that reach immediately a very high velocity.

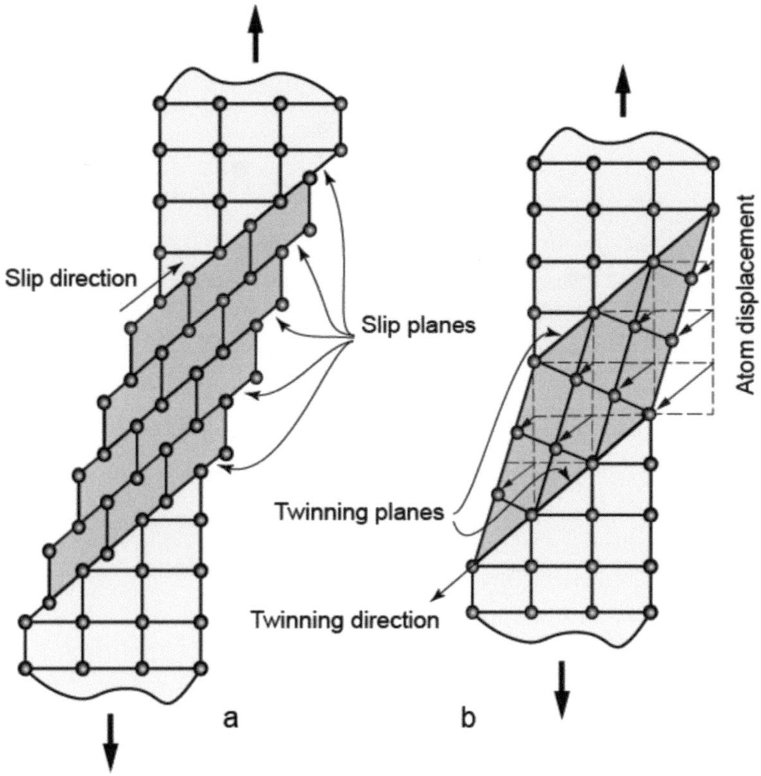

Fig. 2.41 Schematic of plastic deformation mechanisms **a** by slip and **b** by twinning

Fig. 2.42 Serrated stress–strain curve due to twinning in a cadmium single crystal (adapted from [30])

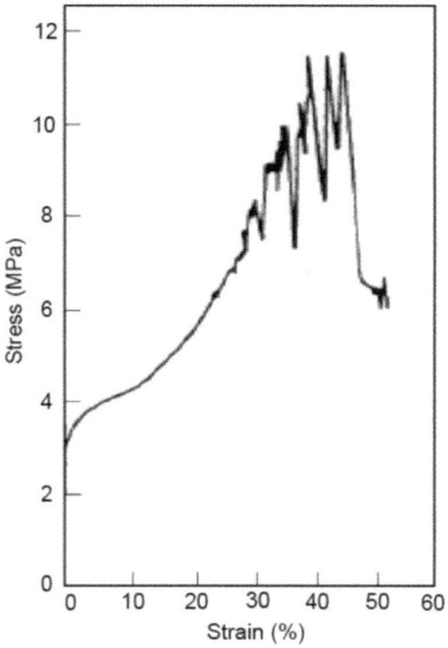

Twins can form in a time as short as a few milliseconds, while for slip bands to form there is a delay time of several milliseconds. The result is that a shock wave is produced that can be heard as in the case of tin, known as tin cry. In general, as said, localized stress fields associated with a twin are higher than those relative to slip. In BCC crystal lattice metals and alloys as in HCP ones, micro twinning can easily reach 1 μm, which actually means that, if blocked, it is equivalent to a pileup of ~10^4 dislocations. This explains why the relative stress fields are so high and why twinning is so an important cause of crack nucleation in coarse grain BCC metals at low temperatures, in particular. Twinning is also important in HCP metals because they do not possess enough easily activated slip systems to produce deformations. The twinning shear stress is practically independent of temperature and strain rate, as well.

However, the issue has been debated among researchers and there are diverging results. Anyhow, Fig. 2.43 shows this independence of the twinning shear stress of temperature for several metals and alloys [31, 32]. It is worth noting the behavior of iron deformed at 10^{-3} and 10^3 s^{-1}. The increase in the strain rate by a six order of magnitude resulted in only a mere 20% increase in the yield strength. This denounces the very limited sensitivity of twinning to strain rate. However, often it seems not be a precise value of the shear stress that activate twinning. Significant are the experiences of Bell and Cahan [32] who found that zinc crystals had an activation shear stress, $\tau_{y,twin}$, varying from 5 to 35 MPa. This may be due to the fact that highly localized stress situations, of the most different nature, can favor the onset of twinning.

Dislocations piling up or dislocation forests and tangling can trigger a twin. Therefore, the shear stress at which twinning occurs is influenced by prior deformation. If crystals are prestrained at room temperature, deformation by twinning is completely suppressed and the crystals deform by slip activated by dislocation motion even at −196 °C [33]. At variance with twinning, the slip shear stress, $\tau_{y,slip}$, in BCC crystal lattice metals, as ferritic steels, has a large thermally activated component. At normal temperatures $\tau_{y,slip}$ is generally much lower than $\tau_{y,twin}$ and in the competition between slip and twinning as primary cause of deformation, slip prevails over twinning. However, as the temperature is reduced the slip shear stress, $\tau_{y,slip}$, sharply increases so that there will be a point at which it will exceed the corresponding shear stress that activates twinning. At this point, the situation reverses and twinning prevails over slip as main mechanism of plastic deformation. This situation is schematized in Fig. 2.44. Here, the crossing temperature T_{qs} is marked, where the suffix indicates the quasi-static condition of strain rate in a normal traction test (10^{-4} to 10^{-3} s^{-1}). Note that an increase of the strain rate moves the $\tau_{y,slip}$ toward higher temperatures shifting the crossing point with twinning from T_{qs} (quasi-static) to T_{hsr} (high strain rate). Figure 2.45 shows the slip curves computed by Meyers [34] for strain rates of 10^{-3}, 10^3 and 10^6 s^{-1} and the experimental results obtained by Halding [35] for iron with 100 μm grain size.

Since twinning has a very low sensitivity to strain rate whereas slip is particularly sensitive, there will be an extremely high strain rate ($\gg 10^7$ s^{-1}) at which slip can be activated only in a very narrow temperature range below which twinning will always come first. On the base of Fig. 2.45 at a strain rate of 10^8–10^9 s^{-1} the crossing is

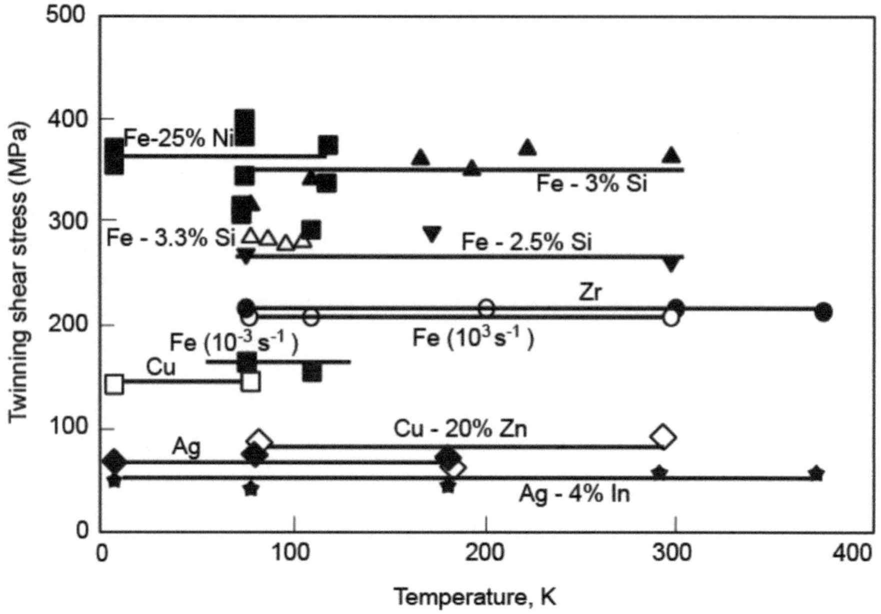

Fig. 2.43 Effect of temperature on twinning stress for several metals and alloys (adapted from [31])

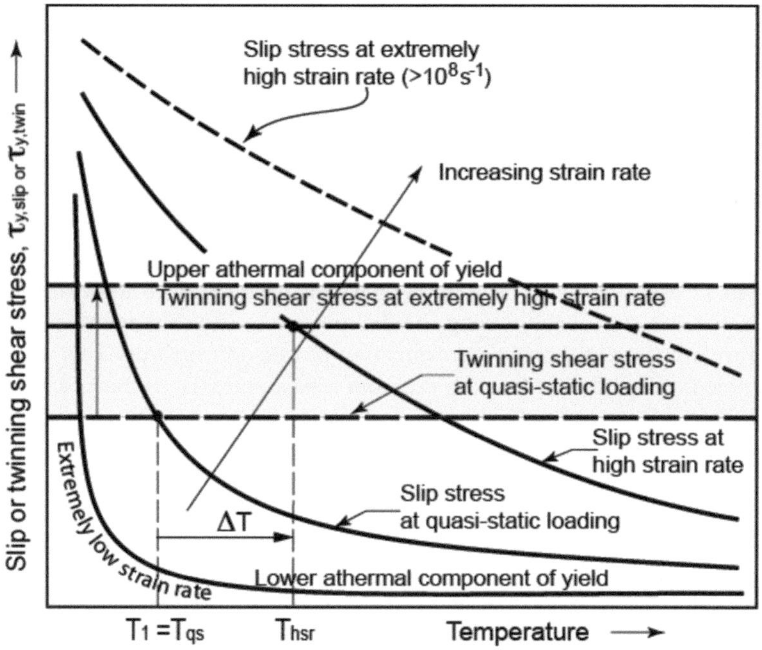

Fig. 2.44 Temperature and strain rate dependence of shear stress for slip and twinning

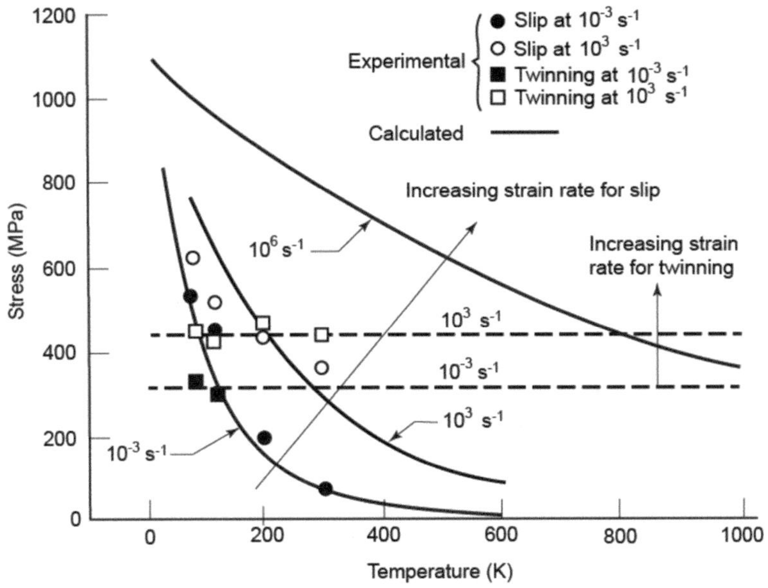

Fig. 2.45 Experimental and calculated slip and twinning stresses for single crystal iron (with data from [34, 35])

likely to occur over 600 K. Slip by dislocation motion will be necessary, anyhow, to activate twinning. Therefore, its upper value at such a very high strain rate sets the upper athermal component of the yield strength; at no temperature and strain rate this value can be exceeded. On the opposite side, at extremely low strain rate ($\ll 10^{-8}$ s^{-1}) dislocation motion becomes so slow that it can overcome any barrier under a low applied stress. It is just a matter of time. A strain rate of 10^{-8} s^{-1} means that to run a traction test it will take more than 200 days! Thermal energy is no longer necessary to aid dislocation motion. A kind of easy glide occurs that becomes independent of temperature. The stress needed to activate dislocation motion is so low that twinning will never occur. This very low stress level sets the lower athermal component of the yield strength. Deformation twinning is found less frequently in FCC crystal structure metals except under extreme conditions as cryogenic temperatures and high strain rates.

This is because these materials have a large number of active slip planes and directions and do not show any temperature activated component of the slip stress, being ductile almost at any temperature. Slip prevails as deformation mechanism. This may not be the case of high Ni austenitic stainless steel that is particularly prone to annealing twins as result of a welding procedure that heats the base metal in the temperature range of 600–650 °C. HCP crystal structure metals, despite their close-packed lattice, have just one active slip plane (basal plane) at normal temperatures. Therefore, twinning can provide an excellent alternative to slip deformation. Twinning is fairly easily obtained in larger grain metals at low temperature. Decreasing

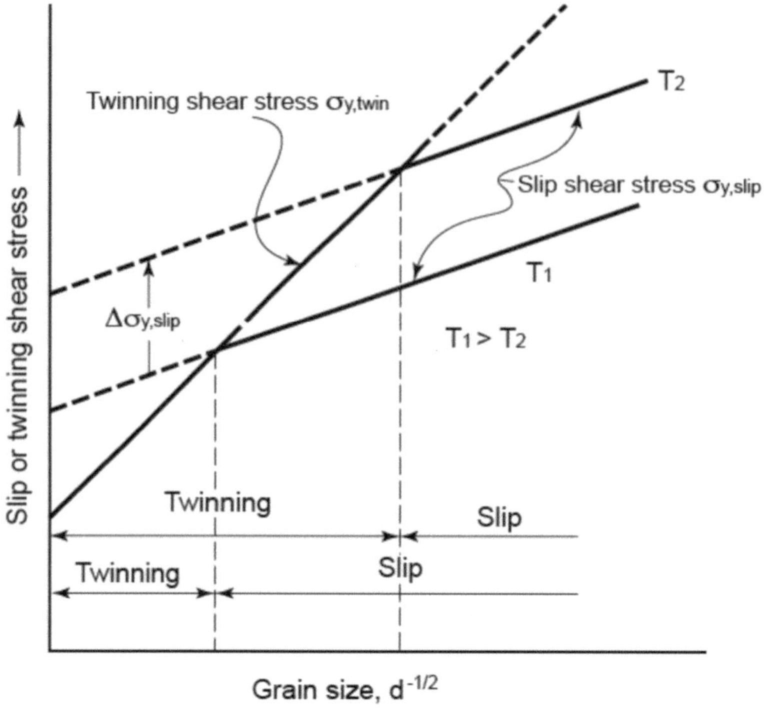

Fig. 2.46 Dependence of slip and twin shear stress on grain size

the grain size would require a temperature reduction to further sustain twin against slip as plastic deformation mechanism. This is schematized in Fig. 2.46.

2.16 Grain Size

Among the microscopic factors that influence the fracture resistance of materials, of particular importance is the grain size. It affects the yield strength and the ductility of a material. Grain boundaries act as pinning point impeding further dislocation motion. Moreover, grains in polycrystalline materials have different orientations, which require more energy for a dislocation to change direction and move into the adjacent grain on a new slip plane. The result is that by lowering the grain size the yield strength of a material increases and ductility decreases. This gives rise to the so called grain-boundary strengthening. Cottrell [36] in his monumental work has shown that dislocations that pile up against a barrier cannot move until the first start to move. When this first dislocation moves by a quantity Δx the following too will move by the same amount. If the dislocations that pile up are *n*, the total external work, E_{est}, per unit © of dislocation is

$$E_{est} = n\tau b\Delta x. \tag{2.21}$$

This work will balance that of the internal reactions, E_{int}

$$E_{\text{int}} = \tau_i b\Delta x. \tag{2.22}$$

By equating (2.21) to (2.22) it yields

$$\tau_i = n \cdot \tau. \tag{2.23}$$

What Eq. (2.23) says is that the pile up effect exerted by blocked dislocations is to generate an action n fold the external shear stress that has activated their movement. As the grain size decreases, the number n of edge dislocations that can pileup against a grain boundary also decreases and the yield strength increases. At variance, large size grains allow more dislocations to pile up and break through the grain boundary barriers. Note that, being of the same sign, dislocations piling-up do not combine. They interact elastically and their spacing decreases towards the front of the pile-up. Assuming that the dislocation source is at the center of the grain of size d, the number of dislocations that can pileup is

$$n = \frac{k\pi\tau d}{2Gb} \tag{2.24}$$

where τ is the shear stress that feeds dislocation movement, which can be assumed to be the resolved shear stress τ_{res}, $k = 1$ for screw dislocations or $(1-\nu)$ for edge dislocations, G is the shear modulus and b the Burgers vector. Dislocations emitted from the center crystal point source, for example a Frank-Read source, encounter a lattice friction stress as they move on a slip plane towards a grain boundary. The lattice friction stress, which is the Peierls-Nabarro stress τ_{PN} (see Sect. 2.10), opposes the applied shear stress $\tau = \tau_{res}$. The effective shear stress that makes dislocations move and produces the plastic deformation is then given by

$$\tau_{eff} = \tau - \tau_{PN}. \tag{2.25}$$

The effective stress τ_{eff} is also known as back stress caused by dislocation pileup. Since dislocations are blocked by the grain boundary, they will pileup until the stress is large enough to break through the barrier. To break through the grain boundary does not necessarily mean to break it. It can be possible to overcome the barrier at some stress concentration large enough to initiate slip in the neighboring grain. Once activated, the new source begins to emit new dislocations, as schematized in Fig. 2.47. Imagine, then, a dislocation pile up at a grain boundary exerting a pressure on the barrier so that the stress τ_r at a point in the next grain at distance r from the grain boundary is given by

$$\tau_r = \tau_{eff}\sqrt{\frac{d}{2r}} = (\tau - \tau_{PN})\sqrt{\frac{d}{2r}} \tag{2.26}$$

where d is the grain size. Let τ_o be the shear stress required to activate slip in the neighboring grain, which is always a resolved shear stress, but different because of the different orientation. This means that yielding will occur when the shear stress τ_r equals τ_o then Eq. (2.26) yields

$$\tau_o = \tau_y - \tau_{PN}\sqrt{\frac{d}{2r}} = \tau_y - kd^{-1/2} \tag{2.27}$$

where τ_y is the yield shear stress. Equation (2.27) can be written in terms of tensile stresses

$$\sigma_y = \sigma_o + kd^{-1/2} \tag{2.28}$$

which is the well famous equation established in the early 50's by Hall-Sylwestrowicz [37] and Petch [33], known as the Hall–Petch equation, relating the yield strength to the grain size d through the Hall–Petch parameter k which depends on material but is independent of temperature and grain size. Table 2.5 is a list of k values for some selected materials [38]. It is worth noting that for very large d values the term expressing the dependency on grain size in Eq. (2.28) becomes increasingly smaller. Very large grain size approaches the condition of single grain metal. Therefore, σ_o can be considered as the yield strength of a single crystal. However, since the

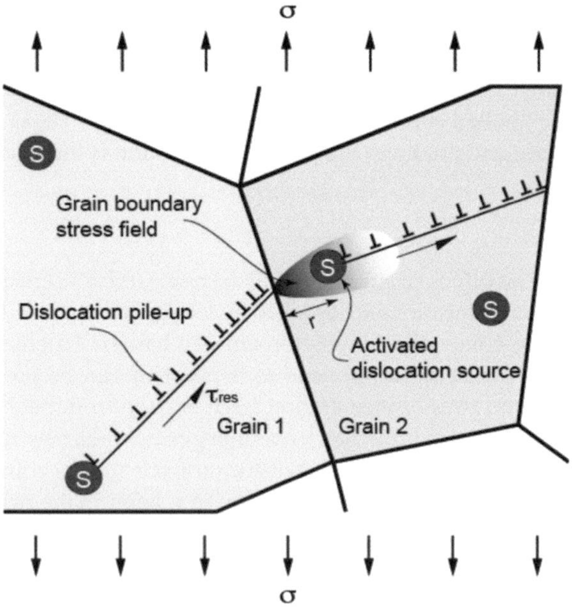

Fig. 2.47 Activation of a dislocation source in grain 2 by the high stress field generated at grain boundary by dislocation pile-up in grain 1

Table 2.5 Values of k for several materials (adapted from [38])

Material	Crystal structure	k (MPa$\sqrt{}$m)
Molybdenum	BCC	1.768
Armco iron	BCC	0.583
Titanium	HPC	0.403
Low-carbon steel	BCC	0.307
Magnesium	HPC	0.279
Zinc	HPC	0.220
Copper	FCC	0.112
Aluminum	FCC	0.068

term $kd^{-1/2}$ is independent on temperature, σ_o may be also regarded as the thermal component of the yield strength, whereas $kd^{-1/2}$ is the athermal component. For mild steel, $\sigma_o \approx 70$ MPa and $k \approx 0.74$ MPa$\sqrt{}$m. A compilation of experimental results obtained by Hall and Petch [33, 37], Armstrong [39], Morrison and Leslie et al. [40] on iron and steel materials is shown in Fig. 2.48.

Note that the data by Hall and Petch were obtained at very low temperature, −196 °C, with mild iron specimens containing 0.036% carbon or with ingot steel with 0.155% carbon. Therefore, the yield strength was effectively depending on grain diameter without any influence of carbon content. The testing temperature was −

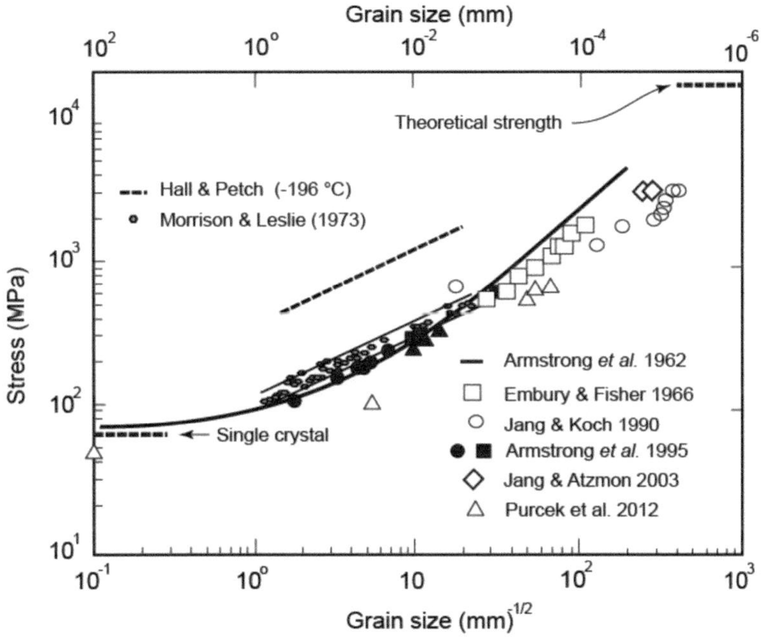

Fig. 2.48 Compilation of experimental results obtained on iron and steel (data from [33–39])

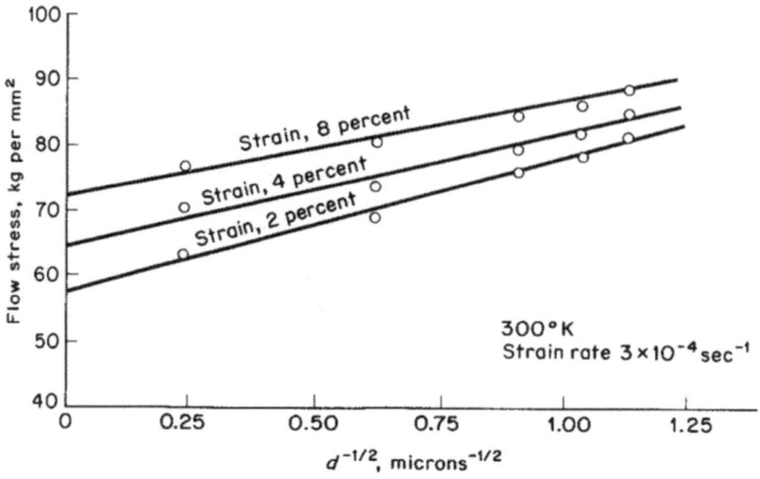

Fig. 2.49 Flow stress versus grain size for titanium at different strain levels [41]

196 °C. At that temperature slip in ferrite is likely to occur by twinning (see Sect. 2.14) rather than by dislocation movement. Figure 2.49 [41] shows the trend of the flow stress versus the inverse square root of the grain size at different deformation values for titanium deformed at 2%, 4% and 8%, respectively.

This means that the Hall–Petch relation applies to the entire stress–strain curve. The same trend is seen as far as temperature is concerned, as shown in Fig. 2.50 [42] for mild steel. Note that in both cases the intercept on the ordinate axis depends on strain level or temperature. Slope doesn't. The Hall–Petch constant, k, in Eq. (2.24) or (2.25) is a material characteristic. It is higher for metals having a BCC crystal structure and lower for FCC ones.

2.17 Grain Refinement and Ultrafine Grain Size Materials

What the Hall–Petch Eq. (2.25) is predicting has long been noted. Already in eighteenth century Grignon [43], a French foundry man and manufacturer of cannons, noted that fine-grained iron was better in its properties. Réaumur [44] also observed this difference stating that excellent steels distinguished from poor ones by the size of their grains.

As the grain size decreases, the term $d^{-1/2}$ in the Hall–Petch equation becomes progressively dominant. All this induce to think that remarkably strong materials can be produced if very small grain sizes can be achieved. This is the reason for the introduction of what is known as nanocrystalline materials. Nanocrystalline materials are multiphase crystals of nano-scale grain size ranging from 250 to 1000 nm (from 0.25 to 1 μm). They are also called ultrafine grain size materials. However, the trend

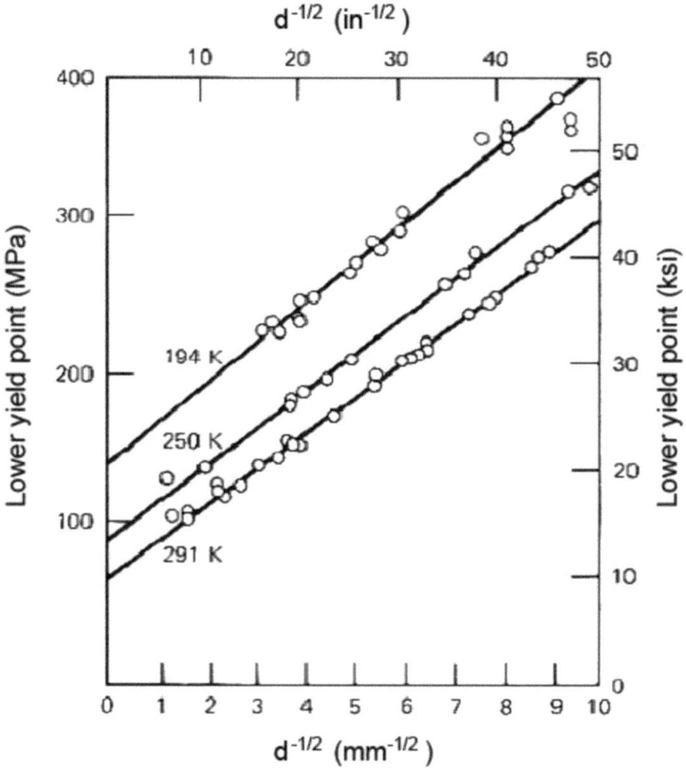

Fig. 2.50 Lower yield point versus grain size for different temperatures in carbon steel [42]

predicted by the Hall–Petch equation does not continue indefinitely. It has long been theorized that there is a lower limit to the grain size for the validity of Eq. (2.26). As matter of fact, the σ_y versus $d^{-1/2}$ curve goes through a maximum, though with lower slope. Afterwards the flow stress is assumed to reach a plateau or to decrease, as in copper. This schematized in Fig. 2.51. Beyond this limiting value grains are so small that there is not that much room left for dislocations to pileup and exert enough pressure over the grain boundaries. Moreover, dislocations cannot bow out at obstacles and multiply because they would touch the grain boundaries. The overall result would be a much higher strength. Figure 2.52 [45] shows the stress–strain characteristic of nanocrystalline copper versus grain size.

It can be seen that the flow stress after reaching a maximum at a grain size of about 15 nm, precipitates. This is because the plastic deformation is no longer controlled by dislocation motion inside grains. As the grain size is decreased, the number of dislocations piled up against a grain boundary decreases, at a fixed stress level. Since this number is function of the distance to the source, but also on the applied stress, an increased stress level is then needed to generate the same number of dislocations

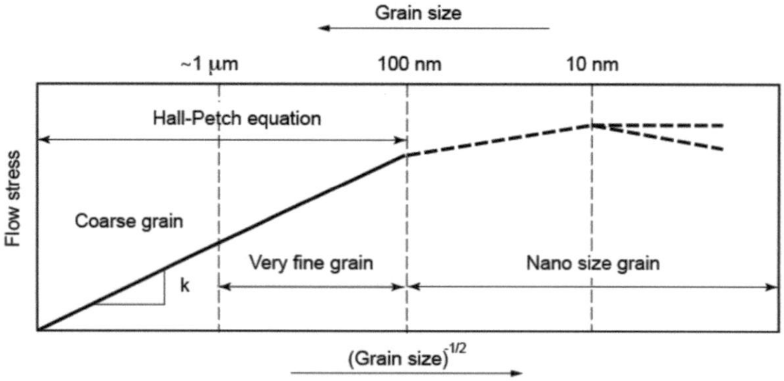

Fig. 2.51 Schematic view of the flow stress variation with grain size

Fig. 2.52 Flow stress versus grain size of nanocrystalline copper (adapted from [45])

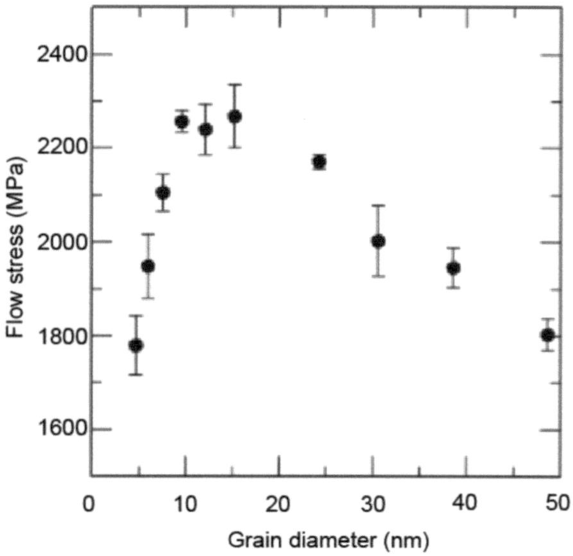

at the pile-up to explain the plastic flow. However, at a critical grain size, this model of dislocation pile-ups fails to explain the plastic flow.

In nanocrystal materials grains are so small that their size becomes almost similar to grain boundaries thickness, while the fraction of grain boundaries increases. Nanocrystals are characterized by a large volume fraction of grain boundaries. Therefore, grain boundaries start to take the lead in controlling plastic slip. It is, now, within grain boundaries that atomic sliding takes place, leading to the observed softening. Meyers and Ashworth [46] proposed a model different from the grain boundary dislocations pileup proposed by Cottrell and shown in Fig. 2.53c. In their model of dislocation mantel larger than grain, Fig. 2.53b, dislocations generated at grain boundary create a hardened reinforcing second-phase network. In order to insure a

$d^{-1/2}$ dependence of the yield stress σ_y at larger grain sizes, Meyers and Ashworth set the thickness, t, of the hardened layer also proportional to $d^{-1/2}$. The resulting expression for σ_y contains the usual $d^{-1/2}$, but also a second term proportional to d^{-1}. This second term becomes more and more important as the grain size reduces and acts to lower the slope of the Hall–Petch trend in the ultrafine region of grain size. Now, in fact, deformations occur by activation of dislocations within grain boundaries, Fig. 2.53d, which lowers the hardening. It must be considered that what could be affecting strength the most is the drastic volume reduction of grains. Since defects distribution probability depends on the process volume, i.e., the larger the process volume the higher the probability to find a larger defect, by reducing the volume to almost zero also the probability to have any defect goes to zero. Under such conditions the strength of a material reaches the theoretical value which is about one tenth of the Young's modulus E (see Sect. 7.2). This is precisely the reason why *whiskers*, better known as fibers, have such an incredibly high strength: their cross-section is on the order of just few microns so that the process volume of each single fiber is practically zero and can be considered flaw free. However, below a certain grain size, grain boundaries become the dominant factor, as schematized in Fig. 2.52d. Since they are, by definition, crystal defects it is practically impossible to reduce to zero the probability to have any defect. The full advantage would be achieved if the grain boundary ledge could shrink back to few atoms thick. Note that for the very same reason ductile fracture in ultrafine-grained metals presents the same characteristic observed in coarsened-grain materials. It is a dimple fracture generated by inclusions, but this time they are not located inside the grain, but in the mantle around grains. Figure 2.54 presents the stress–strain characteristic of ultrafine-grained dual phase C–Mn steel [47].

Grain sizes were: 12.4 μm for the coarse grained steel (CG), 2.4 μm for the fine grained steel (FG) and only 1.2 μm for the ultrafine grained steel (UFG). The figure also shows the microstructure of the three steels considered that consist of a soft α-ferrite matrix (~70%) and hard martensite particles.

2.18 Stacking Faults

The stacking sequence of atoms in the crystalline matrix is determined by the position they occupy in the crystal. The atomic arrangement of a FCC crystal lattice structure and that of a HCP one are shown schematically in Fig. 2.55. As it can be seen from this 2D and 3D view of the stacking sequence, the difference between FCC and HCP crystal structure lies in the third layer of atoms. The ABC sequence of the FCC crystal structure becomes the AB–AB sequence in the HCP atomic arrangement. It has been observed, using X-ray diffraction measurements that faults in the stacking sequence can be introduced in metals by a plastic deformation activated through dislocation motion. A slip in an FCC structure between an A and a B layer that moves each atom layer above the slip plane one atomic distance produces a stacking

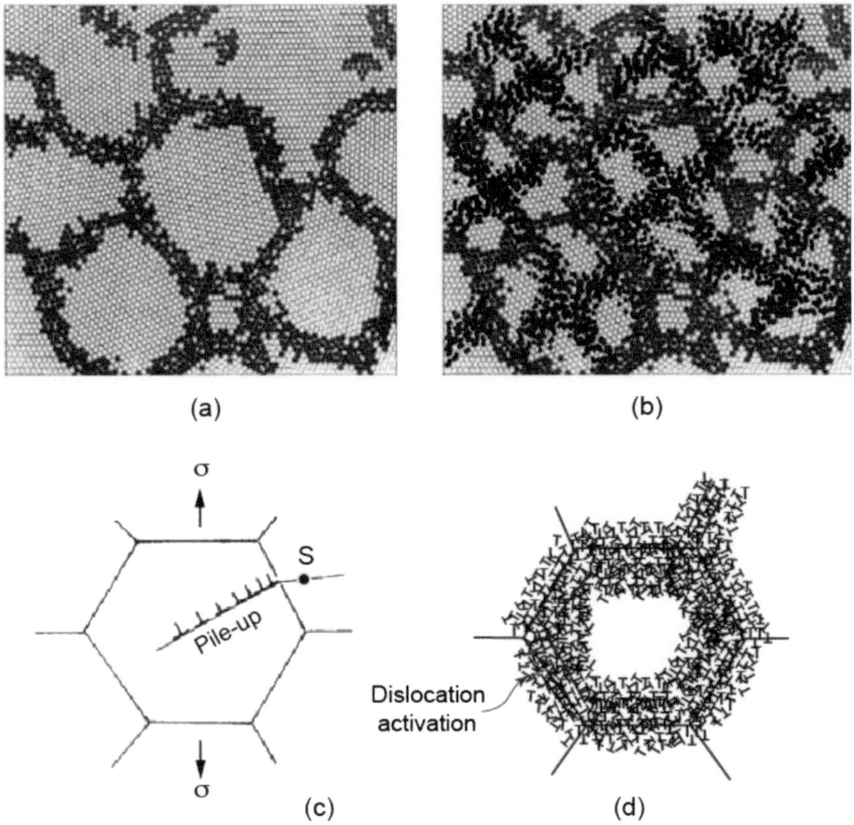

Fig. 2.53 Basic difference between: **a** coarse-grained metal and **b** nanocrystalline materials; the grain boundary becomes as wide as the grain itself; **c** the Cottrell model of dislocation pile-up that activates a source S in the neighboring grain fails in nanocrystalline metals where **d** slip occurs in the grain boundary

sequence of the type ABCA $\vdots$ CAB. Comparison of this faulted stacking sequence with the regular stacking sequence of a HCP structure shows that the deformation stacking fault contains four layers of a HCP sequence. Figure 2.55 schematizes the way this fault can be introduced in a crystal lattice. Therefore, the formation of a stacking fault in an FCC crystal structure metal is equivalent to the formation of a thin HCP region. Also, a twin can change the stacking sequence. Actually, a stacking fault in FCC metals can be considered as a submicroscopic twin of nearly atomic distance. From the point of view of dislocation theory, a stacking fault in an FCC metal can be considered to be an extended dislocation consisting of a thin hexagonal region bounded by partial dislocations, as schematically shown in Fig. 2.56. The wider the separation between the partial dislocations, the lower the stacking fault energy. Stacking faults enter into the plastic deformation of metals in many ways.

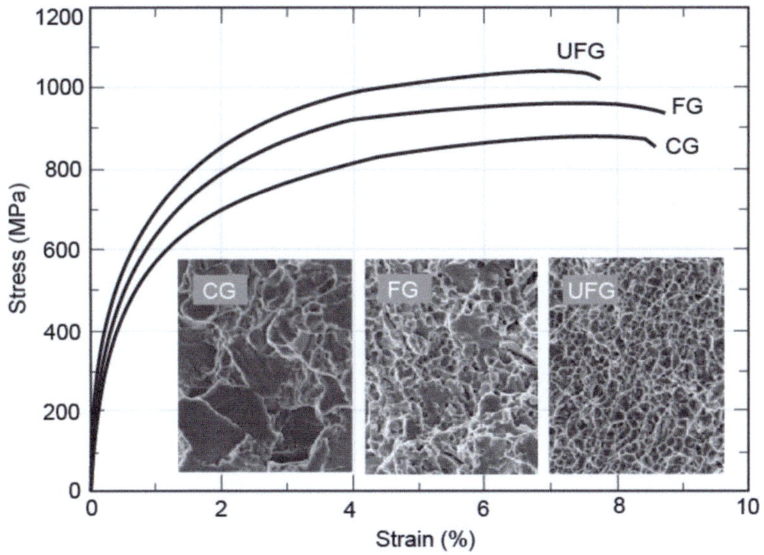

Fig. 2.54 Stress–strain behavior of nanocrystalline dual phase C–Mn steel (modified from [47])

Metals with low stacking fault energy strain-harden more rapidly and show different temperature dependence of flow stress from metals with high stacking-fault energy.

Lower stacking fault materials have more difficulties to flow because the mobility of dislocations decreases. Aluminum has a very high stacking fault energy of about 160–200 mJ/m^2 while silver shows one of the lowest stacking fault energy of

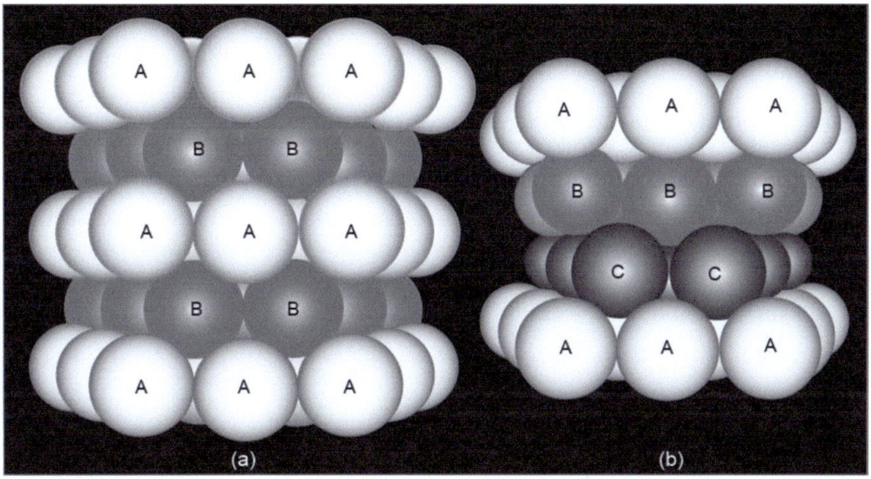

Fig. 2.55 3D view of the stacking sequence of **a** FCC crystal lattice and **b** of HCP crystal lattice

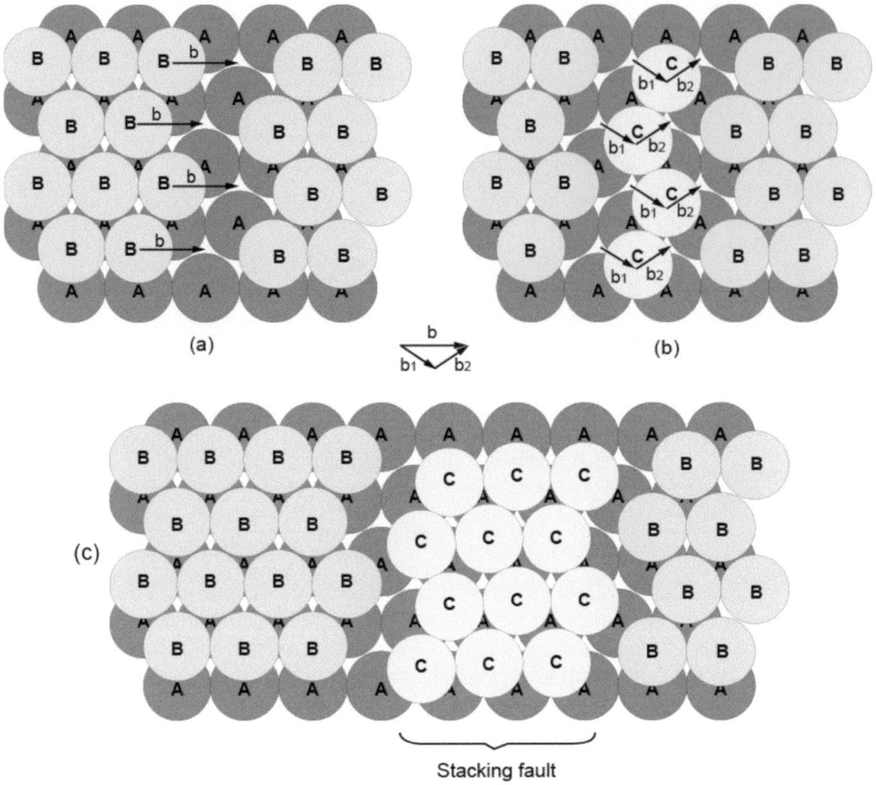

Fig. 2.56 Dislocation motion in an FCC crystal lattice metal; **a** a total dislocation, **b** partial dislocation and **c** an extended dislocation

about 20–30 mJ/m^2. Nickel and copper are in between with 90–80 mJ/m^2. Experimental evidence strongly suggests that a twin boundary can frequently act as a grain boundary. As a consequence of that, twins can often be considered to reduce the grain size by supplying additional grain boundaries. In general, localized stress fields associated with a twin are higher than those relative to slip. In BCC crystal lattice metals and alloys as in HCP ones, micro twinning can easily reach 1 μm, which actually means that, if blocked, it is equivalent to a pileup of ~10^4 dislocations. This explains why the relative stress fields are so high and why twinning is so an important cause of crack nucleation in coarse grain BCC metals at low temperatures.

2.19 Adiabatic Shear Bands

We have been discussing of Lüders lines and shear bands that develop in the crystal lattice under quasi-static loading conditions. When high speed deformations are involved, it happens that the plastic deformations develop in localized areas more and more, as higher and higher is the applied strain rate, which, in turn, amplifies the adiabatic nature of the phenomenon. The distance λ_{dh}, also called thermal diffusivity distance, that the heat produced by the plastic deformation (see Sect. 1.8) travels in a given interval of time can be assessed as

$$\lambda_{dh} = 2\sqrt{\alpha t} \tag{2.29}$$

where α is the thermal diffusivity coefficient of the material and t is the time. For example, if we have a specimen of copper with α equal to 1.14 cm^2s^{-1}, subjected to a plastic deformation of 50% under a strain rate of 10^{-4} s^{-1}, the time interval would be $t = \varepsilon/\dot{\varepsilon} = 0.5/10^{-4} = 5{,}000$ s and the distance λ_{dh} will be

$$\lambda_{dh} = 2\sqrt{\alpha t} = 2\sqrt{1.14 \cdot 5,000} = 151 \text{ cm}. \tag{2.30}$$

This actually means that the heat has given enough time to propagate throughout the entire specimen, without localizing in a single slip band. But if the applied strain rate goes up to 10^3 s^{-1}, then the time reduces to just 0.5 ms and the distance run by the heat reduces to 0.5 mm, which effectively allows us talking of adiabatic bands heat confinement. 90% of the plastic deformation energy turns into heat so that we can say that the temperature increment ΔT produced per unit volume at a given deformation final ε_f will be

$$\Delta T = 0.216 \cdot \frac{0.9}{\rho C_s} \int_0^{\varepsilon_f} \sigma \, d\varepsilon \tag{2.31}$$

where ρ is the material density, C_s is the specific heat, 0.216 the heat-to-work conversion factor (calories per Joule of energy) and the integral represents the plastic deformation energy per unit volume (area under the stress–strain curve). It is clear the temperature increment in the adiabatic shear band will be given by ΔT of Eq. (2.31) times the ratio of the volume V of the specimen to that adiabatic shear band V_{band}.

$$\Delta T_{band} = \Delta T \frac{V}{V_{band}}. \tag{2.32}$$

Consider, now, the curve of Fig. 4.33 of aluminum under a strain rate of 2,800 s^{-1}. The area under the curve is equal to about 29 MPa/m^3. Given a density of 3.76 gr/cm^3 for aluminum and a specific heat C_s of 0.216 cal/gr °C, if the shear band volume is 500 times smaller than that of the specimen under dynamic loading, we will be surprise to discover that the local temperature in the shear band may reach 1,000 °C. Much higher adiabatic conditions will be achieved under military applications such as explosions, detonations and high-speed impact, as schematized in Fig. 2.57 [48].

Also, metal working and forming processes can cause the formation of dynamic shear bands, as schematized in Fig. 2.58 [48]. Note that while dynamic shear bands can be detrimental to the integrity of the work piece in cases (a) and (b) because they are expected to be sites of crack formation, they are instead favoring the break up and separation of chips formed by the cutting tool. This enables the application of higher machining rates. There are two types of adiabatic shear bands

- deformed bands,
- transformed bands.

Deformed bands consist of severely distorted grains, while transformed bands develop within the grain. This difference can be caught at the optical microscope, as shown in Fig. 2.59 for an AISI 1015 type carbon steel hat shaped specimen deformed at $5{\cdot}10^4\ s^{-1}$ and $2{\cdot}10^5\ s^{-1}$ [49], respectively, and Fig. 2.60 that refers to an AISI 4340 steel cylindrical specimen impacted at 19 m/s [50].

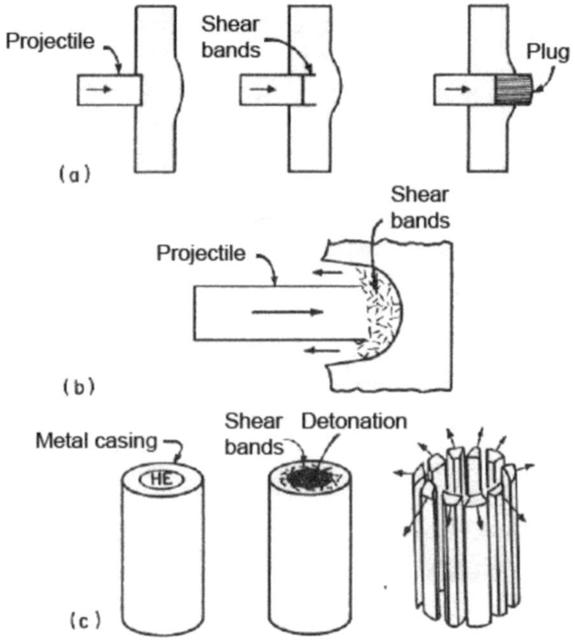

Fig. 2.57 Formation of shear bands in dynamic deformation causing: **a** perforation of armor, **b** breaking up of a projectile, **c** fracture of exploding cylinder (adapted from [48])

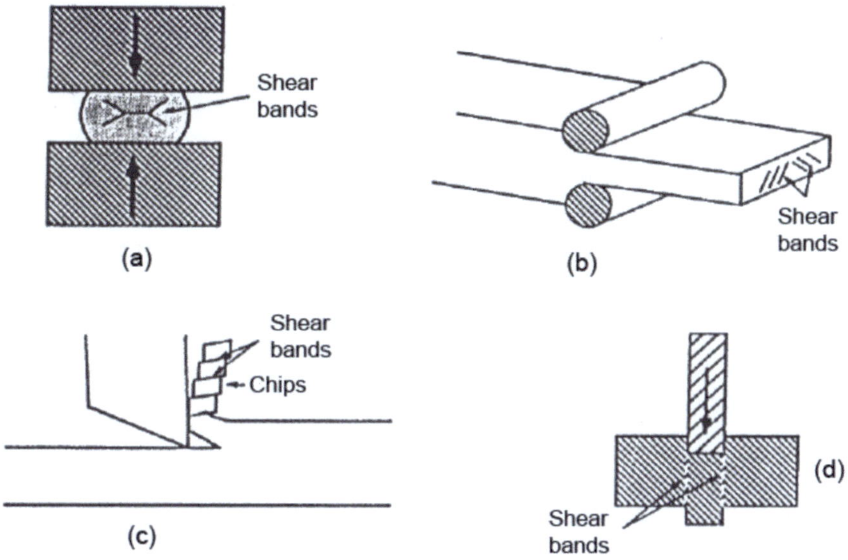

Fig. 2.58 Shear localization in plastic forming processes: **a** upset forging, **b** rolling, **c** machining and **d** punching and shearing (adapted from [48])

Note that cracks initiated at the extremities of the shear bands in Fig. 2.59 and all along the band in Fig. 2.60. Also note the white appearance of the shear band that makes it referred to as white band. As early as 1943, Zener and Hollomon identified the reasons for the formation of an adiabatic shear band in the competition between two opposing factors: thermal softening and work hardening. They wrote: "*When the material undergoes an element of strain, dE, adiabatically, the stress is raised by the strain hardening and lowered by the associated rise in temperature*". This actually means that they attributed the formation of shear bands to the prevailing softening by highly localized temperature rise over the strain hardening. The mechanism of shear band formation at macroscopic level is schematized in Fig. 2.61 [48]. In a homogeneous material subjected adiabatically to a shear action, once a critical shear strain, γ_c, is attained, the original uniform deformation concentrates in a limited region, as in the case of necking (see Fig. 1.19).

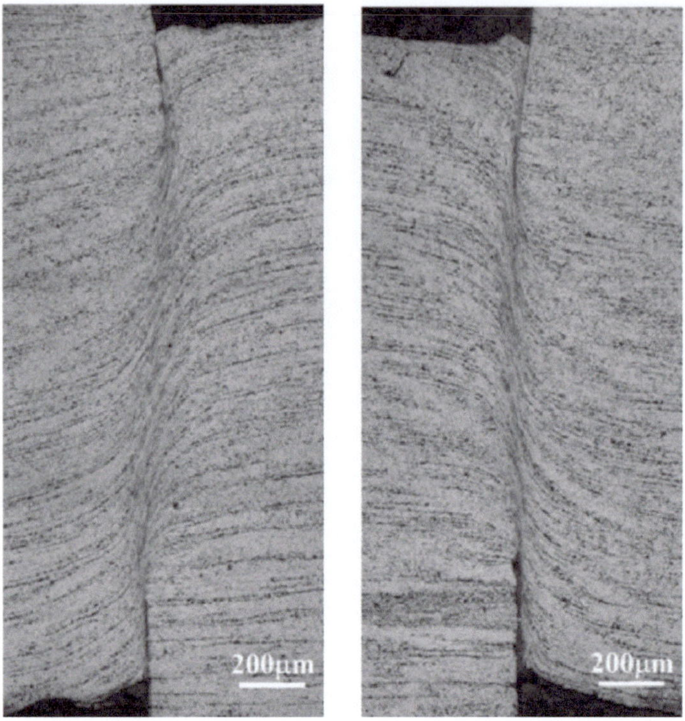

Fig. 2.59 Optical micrograph of shear zone in S15C (AISI 1015) steel specimen under strain rate of $5{\cdot}10^4\ s^{-1}$ and $2{\cdot}10^5\ s^{-1}$, respectively. Cracks initiated at the extremities of the shear bands [49]

Fig. 2.60 Optical micrograph showing a white etching dynamic shear band for martensitic AISI 4340 steel impacted at 19 m/s. Note the shear band has developed a crack [50]

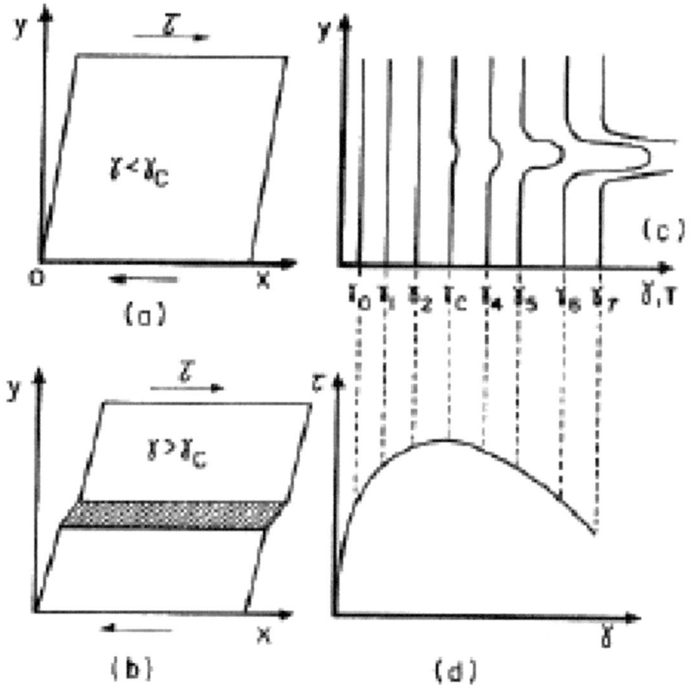

Fig. 2.61 Dynamic (adiabatic) shear band formation: **a** initial homogeneous deformation, **b** instability by sudden localized shear band formation, **c** strain profile in the localized region and **d** adiabatic stress–strain curve [48]

The reasons of such an unpredictable choice of the region are geometrical and metallurgical. The stress wave concentrates its action where the local conditions foster the non-homogenous deformation. From this point on the adiabatic increase of temperature favors its prosecution, since the local heat softening prevails over the strain hardening and metal loses its strength. Note that original metal has become a kind of composite material in which the matrix is much harder than the bands in which deformations concentrate. At the microscopic level the mechanism is schematized in Fig. 2.62 [51]. During the deformation phase at high strain rate, tensile stresses are generated inside the shear bands due to residual stresses or tensile reflection. Since the flow stress of the material is lower in the shear band compared to the surrounding matrix micro voids or pores will be generated, Fig. 2.61a. Continuing to load, these micro-pores aggregate and coalesce to form voids that elongate to become elliptical with the major axis along the shear band, Fig. 2.62b, c. These elliptical voids that are wide as the shear band connect into a micro-crack (d). Eventually, the micro-crack grows and propagates, Fig. 2.62e. This process of pores formation and coalescence is well visible in Fig. 2.63 for a titanium sample [48]. As it can be expected, the process of formation and growth of transformed shear bands is not that simple because it is usually accompanied by a series of events of various nature. At variance with the

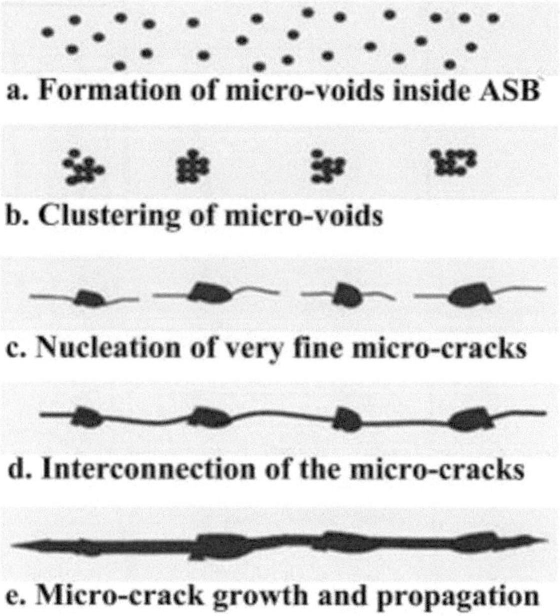

Fig. 2.62 Microstructural generation and evolution of a transformed adiabatic shear band [51]

deformation bands, the heat generated and localized in a narrow band, allows to reach the transformation temperature of the metal. Several options are then opened. In steels, two fundamental metallurgical transformations can take place. In both cases the temperature has to reach 723 °C for the phase transformation to occur. The first type is observed in eutectoid steels, type AISI 1080, that pass from the α-phase, typical of ferrite, to the γ-phase, typical of austenite. When the load is removed, if the cooling of the shear band is sufficiently rapid, austenite may be retained also at room temperature. However, it is more likely that the γ-phase is transformed into brittle and hard martensite. TEM observation has revealed that in quenched and tempered steels, the bands consist of extremely small equiaxed grains, even 0.2–0.4 micron, in which carbides are completely dissolved. The thickness of these bands is 10–50 microns with very high hardness so that they represent the real hard element of the matrix that remains softer, though still hard. In titanium alloys, such as Ti–6% Al–4% V, grains are very small, from 0.05 to 0.3 microns. Also in AZ5G aluminum alloy grains are very small, some 0.3 μm.

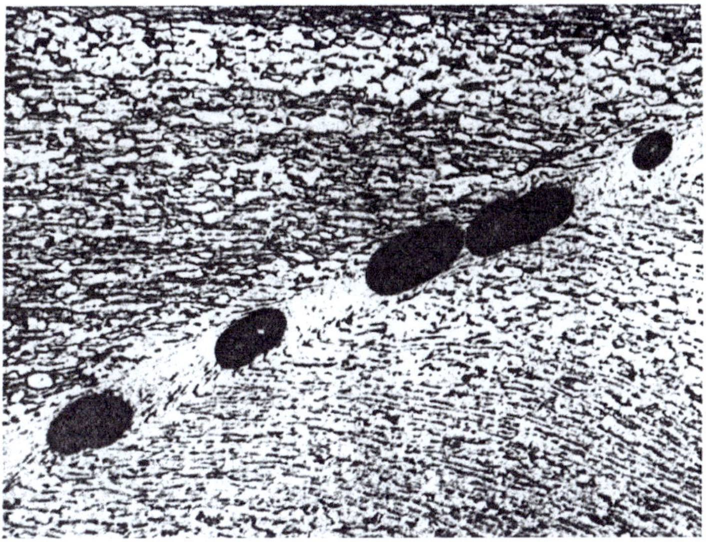

(a)

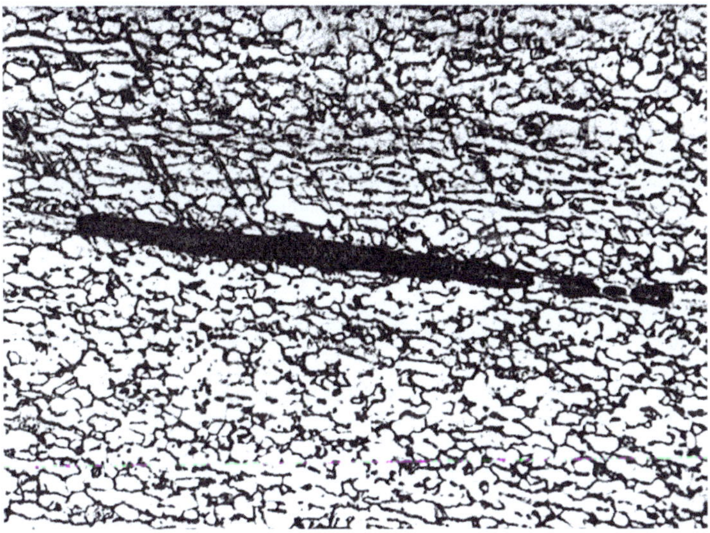

(b)

Fig. 2.63 **a** Voids and **b** coalescence with crack forming along a transformed shear band due to tension in titanium [48]

References

1. Barisic, B.: Lüders' Bands Phenomenon Analysis in the Process of Thin-Walled Sheet Products Manufacture. Faculty of Engineering, Rijeka (2005)
2. Statts, H.N.: Principles of Stress Coat. Magnaflux Corp, Chicago (1955)

3. Ewing, J.A., Humphrey, J.C.: The fracture of metals under repeated alternation of stress. Philos. Trans. R. Soc. A **200**, 241–250 (1903)
4. Callister, W.D.: Material Science and Engineering. Wiley
5. Parker, E., Washburn, J.: Deformation of single crystals. In: ASM Seminar on Modern Research Techniques in Physical metallurgy, pp. 186–204 (1952)
6. Voorhees, P.W., Shull, K. R.: Microstructural Dynamics 1. Department of Materials Science and Engineering, Northwestern University, Evaston, Illinois
7. Schmid, E.: Z. Elektrochem. **37**, 447 (1931)
8. Burke, E.C., Hibbard, W.R., Jr.: Trans. AIME **194**, 295 (1952)
9. Dieter, G.E., Jr.: Mechanical Metallurgy. McGraw-Hill, Book Co., 100 (1961)
10. Rice, J.M.: Solid Mechanics. Harward University, Cambridge, MA, February (2010)
11. Hauser, J.J., Chalmers, B.: The plastic deformation of bicrystals of f.c.c. metals. Acta Metall. **9**, 802 (1961)
12. Rey, C., Mussot, P., Vroux, A.M., Zaoui, A.: Effects of interfaces on the plastic behaviour of metallic aggregates. J. Phys. **46**, 645–650 (1985)
13. Johnson, D.D.: MSE406, Thermal and Mechanical Properties of Materials. Illinois (2005)
14. Von Mises, R.: Angew. Z.: Math. Mech., **8**, 161 (1928)
15. Tegart, W.J. McG.: Elements of Mechanical Metallurgy. Macmillan Company, New York (1966)
16. Shockley, W., Seitz, F., Hollomon, J.H., Maurer, R.J.: Imperfections in Nearly Perfect Crystals. Wiley, New York (1952)
17. Callister, W.D. Jr.: Materials Science and Engineering. An Introduction, 7th edn. Wiley (2007)
18. Volterra, V.: Teoria fisico-matematica delle dislocazioni nei materiali metallici. Annali della Scuola Normale Superiore di Pisa **24**, 400 (1905)
19. Orowan, E.: Uber den Mechanismus des Gleitvorganges. Z. Physik **89**, 634 (1934)
20. Taylor, G.I.: The mechanism of plastic deformation of crystal. Proc. R. Soc. Lond. A **145**, 362 (1934)
21. Polanyi, M.: Z. Physik **89**, 660 (1934)
22. Callister, W.D., Jr.: Fundamentals of Material Science and Engineering, 200. Wiley (2001)
23. Wilde, J., Cerezo, A., Smith, G.D.W.: Three dimensional atomic-scale mapping of the Cottrell atmosphere around a dislocation in iron. Scr. Mater. **43**, 39–48 (2000)
24. Cochardt, A.W., Shoek, G., Wiedersich, H.: Interactions between dislocations and interstitial atoms in body-centered-cubic metals. Acta Metall. **3**(6), 519–609 (1955)
25. Taylor, G.I.: Proc. R. Soc. A **145**, 855, 362–388 (1934)
26. Wiedersich, H.J.: Work hardening in polycrystalline Cu. Metals **16**, 425 (1964)
27. Jones, R.L., Conrad, H.: Trans. AIME **245**, 779–789 (1969)
28. Foulds, J.R., Ermi, A.M., Moteff, J.: Mater. Sci. Eng. **45**, 137–141 (1980)
29. Hertzberg, R.W.: Deformation and Fracture Mechanics of Engineering Materials, 103. Wiley (1976)
30. Boas, W., Schmid, E.: Z. Phys. **54**, 16 (1929)
31. Meyers, M.A.: Dynamic Behaviour of Materials, 527. Wiley (1994)
32. Bell, L.R., Cahan, R.W.: The initiation of cleavage fracture at the intersection ofdeformation twins in zinc single cristals. Procur. R. Soc. Lond. 239–494 (1957)
33. Petch, N.J.: J. Iron Steel Inst. **174**, 25. Progress in Metal Physics 5, Chap. 1, 1, Pergamon Press (1953)
34. Meyers, M.A.: A constitutive description of the slip-twinning transition in metals. In: The Minerals, Metals & Materials Society, Advances in Twinning (1999)
35. Harding, J.: Yield and fracture of high-purity iron single crystal under repeated tensile impact loading. Mem. Sci. Rev. Met. **65**, 245 (1968)
36. Cottrell, A.H.: Dislocations and Plastic Flow in Cristals, vol. 140. Oxford University Press, London (1953)
37. Sylwestrowicz, W., Hall, E.O.: The deformation and aging of mild steel. Proc. Phys. Soc. Lond. B **64**, 495 (1951); Hall, E.O.: Ibid. B 64, 742–747 (1951)

38. Embury, J.D.: Strengthening methods in crystals. In: Kelly, A., Nicholson, R.B. (eds.) Wiley, New York (1971)
39. Armstrong, R.W.: Engineering science aspects of the Hall-Petch relation. In: 50th Annual Meeting of Society of Eng. Science, Inv. Paper, Providence, RI (2013)
40. Morrison, W.B., Leslie, W.C.: The yield strength-grain size relation in iron substitutional alloys. Metall. Trans. **4**, 379–381 (1973)
41. Jones, R.L., Conrad, H.: The effect of grain size on the strength of alpha titanium. Trans. AIME **245**, 779–789 (1969)
42. Petch, N.J.: Fracture, Proceedings of Swampscott Conference, 54 (1959)
43. Grignon, P.C.: Le metamorphoses de l'acier, Memoires de Physique sur l'Art de Fabriquer de Fer, 56–90, Paris (1775)
44. De Réaumur, R.A.F.: Que le fer est de tout le métaux celui qui se moule le plus perfaitment; et quelle en est la cause, 273–287, L'Art de Convertir le Fer Forge and Acier, 63–106, Paris (1722)
45. Sciøtz, J., Jacobsen, K.W.: A maximum in the strength of nanocrystalline copper. Science **301**, Sept. (2003)
46. Meyers, M.A., Ashworth, E.: A model for the effect of grain size on the yield stress of metals. Phil. Mag. **46**, 737–759 (1982)
47. Song., R., Ponge, D., Raabe, D.: Mechanical properties of an ultrafine grained C-MN steel processed by warm deformation and annealing. Elsevier, Acta Mater. **53**, 4881–4892 (2005)
48. Meyers, M.A.: Dynamic Behavior of Materials. Wiley (1994)
49. Lee, W.S., Liu, C.Y., Chen, T.H.: Adiabatic shearing behavior of different steels under extreme high shear loading. J. Nucl. Mater. **374**, 313–319 (2008)
50. Owolabi, G., Odoh, D., Odeshi, A., Withworth, H.: Occurrence of dynamic shear bands in AISI 4340 steel under impact loads.World J. Mech. **3**(2) (2013)
51. Basiim, M.N., Odeshi, A.G.: Shear strain localization and fracture in high strength structural materials. Mater. Sci. Eng. **31**(2), 69–74 (2008)

Chapter 3
Transition Temperature

3.1 Introduction. The 15 Ft-Lb Criterion

The finding that brittle fracture can be triggered by a stress lower than the yield strength of the material is certainly the most worrying aspect of this kind of failure. Such an event cannot be prevented by just keeping the stress below the elastic limit, without even knowing how much below that limit the stress must be kept. Since its first appearance up to the large-scale fracturing of the Liberty vessels during WW II, engineers had no idea whatsoever about the cause of the brittle fracture and how to prevent it. It took a long and extensive investigation to realize that responsible for those disasters were the three factors already outlined in Fig. 1.5, i.e. the presence of a crack of length, a, the stress state, $\sigma\sigma$, and the temperature, T. Once understood that point, the next step was to sort out the kind of relationship that could possibly link those three factors. In this respect, the classical theories of strength such as the Tresca theory of maximum shear stress, or that of von Mises, based on the distortion energy, however advanced, could not be considered as theories of "fracture", in the sense just given to this word. They could predict plastic occurrence but not brittle fracture. It was necessary, then, to develop a new fracture theory and to identify new testing techniques other than the conventional quasi-static traction test, the only available at the time, that we have been discussing in Chap. 1. In this search for new approaches to the brittle fracture problem, historically two independent paths were basically followed. The first was analytical and based, in the US, on the Irwin's Stress Intensity Factor K_I concept and its limit K_{Ic}, known as the toughness of the material, and in the UK on the Wells Crack Opening Displacement (COD) concept. The second was a semi-empirical approach aimed at developing procedures for Fracture-Safe Design leading to a final Fracture Analysis Diagram (FAD) to be used for design purposes based on data coming from an incredible variety of new specimens continuously introduced and developed to that purpose. In this Chapter we will be discussing, as first, this second semi-empirical approach that was developed at the Naval Research Laboratory (NRL) of the Navy, in Washington,

P. P. Milella, *Fundamentals of Fracture Mechanics*,
https://doi.org/10.1007/978-3-032-23081-2_3

D.C., and at Battelle Memorial Institute in Columbus, Ohio. This second approach was lacking the sophistication of mathematical theory and calculation, considered basically unrealistic, and pushed very much on the granitic solidity of experimental evidence. It was understood that temperature was a key factor in the process of brittle fracture of steel but there was no way to identify the value of the temperature that would change a ductile behavior into a brittle one, a safe situation into a fatal one, the so called Transition Temperature known as *NDT* (Nihil Ductility Temperature) or *Zero Ductility Temperature*. As it often happens in the development of a science, the basis for a theory and the most proper test to detect that transition temperature already existed at the time when the phenomenon of brittle fracture became a real problem. In 1849, in fact, a Commission was formed in the UK to report to the parliament on the causes of failures in the railway industry. After examining thousands of axles failures, the commission came to the conclusion that impact or dynamic loads were playing a major role in the fracture process [1]. However no specific test was recommended by the commission but just further investigation. Later, in 1892 it was observed and reported by LeChatelier [2], better known for his discoveries in Chemistry rather than in Mechanics, that some steels were presenting a ductile behavior if tested in the form of a smooth prismatic bar, and a brittle behavior when a notch was introduced in the specimen. Five years later, in 1897, a young American engineer, called Russell [3], recalling the recommendations of the British railway commission on impact loads and LeChatalier observation about the notch effect on material behavior, had the brilliant idea to combine these two effects into a new specimen having an edge V-shaped notch loaded by a striking hammer, Fig. 3.3. The specimen was called Charpy V-notch specimen, or CharpyV specimen, or simply CV-N, rather than Russell-V specimen, from the name of the researcher who first presented it to the scientific community during an international congress in 1901 [4] and spread its knowledge afterwards. However, Russel introduced this new specimen having no idea about its possible use in brittle fracture design, therefore it remained confined to research centers and an ASTM subcommittee began to prepare a standard test method for pendulum impact testing in 1923, though there was no rational basis for using this test in predicting structural performance and there was no requirement in purchase specifications and construction standards. The US Navy had been hit hard by a long series of disasters during WW II, therefore researchers, at the Naval Research Laboratory (NRL) in Washington, D.C., started a massive investigation to solve the puzzle of those brittle fractures. As Pellini put it out very clearly: at the beginning investigators were running tests on naval steels without any clear idea of what they were really looking at». The steel used in the construction of the hull was carbon steel, 0.18 C 0.54 Mn 0.07 Si, used, for economic reasons, as it came out of the rolling mills. However, the more they went on, the more they would develop the idea that a crack, already existing, temperature and, probably, also dynamic loads were the real cause of those disasters. In all cases, in fact, they could identify on ships that had developed a brittle fracture a preexisting crack, generally associated with a sharp discontinuity or a weld not well done, as the starter point of the fracture. It was thought, then, that smooth specimens, free as much as possible from surface defects, were not really suitable to address the issue. Somebody at NRL recalled the Russel-Charpy V-Notch specimen suggesting to use it and in 1945 the first testing program with the Charpy V-Notch specimen started. CV-N specimens were machined from steel plates taken from those vessels that experienced a brittle fracture event in service, tested at different temperatures. In 1951, it was discovered that brittle fracture had never occurred at temperatures higher than that corresponding to 15 ft-lb of energy (about 20.5 Joule) absorbed by the Charpy V-Notch specimens, as depicted inFig. 3.1 [5], which was indicated as the critical temperature, T_c. It was the so-called Charpy V-Notch criterion of the 15 ft-lb. This value of Charpy V energy also called resilience was assessed because the steel plates collected at NRL were classified according to the three different events of initiation, propagation and arrest of the running crack, because some cracks did arrest after running, even after meters of propagation. It was also known the temperature at which those three phases occurred as to establish that no crack ever started at temperatures higher than that at which 10 ft.lb were absorbed by the Charpy V specimen (13.6 J), and none had ever propagated catastrophically at temperatures above 20 ft-lb (27.2 J), as shown in Fig. 3.2 [5].

It was assumed, conservatively, that the value of 15 ft-lb (20.5 J) would represent that upper energy limit beyond which no crack could ever start and then propagate catastrophically under an elastic stress field, since the plates were working under this condition, and it was thought that this criterion should be true for any type of steel. By 1952 the 15 ft-lb Transition Temperature was accepted as a definitive criterion. It was believed that this energy level was an invariant characteristic of any steel. Consequently, to determine the NDT of a steel it would be sufficient to develop the relative resilience curve making use of CV specimens and identify the relative *NDT* as the temperature corresponding to the energy of 15 ft-lb. At the time researchers believed to have found a low of "universal" type, but in 1953, just one year later, to the great surprise and disappointment of everybody, it was discovered that the criterion

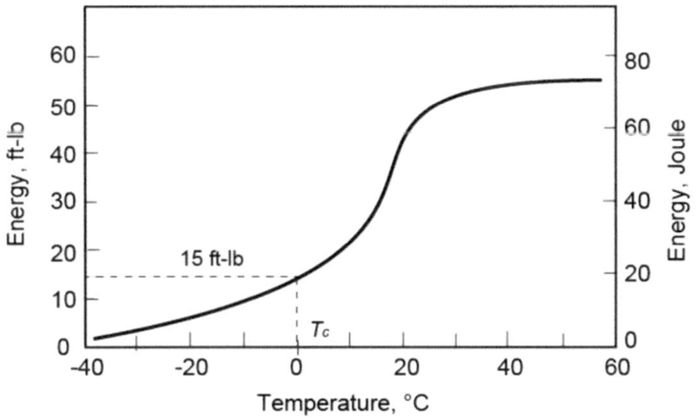

Fig. 3.1 CV energy versus temperature obtained with CharpyV specimens made of steel used by the US Navy for the construction of Liberty and T2 ships during WW II

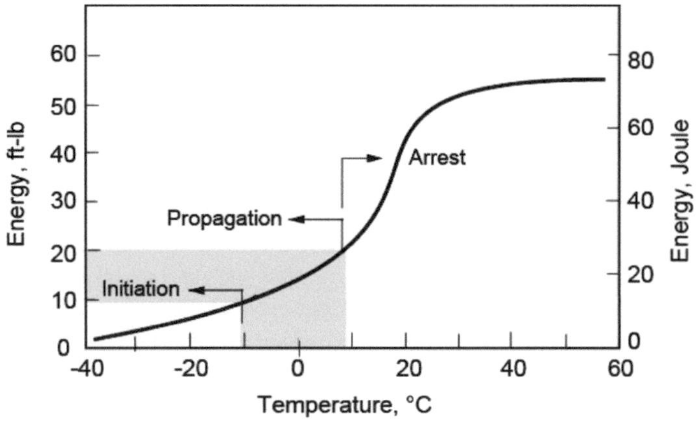

Fig. 3.2 On the CV energy versus temperature curve of naval steel it was possible to identify the energy levels at which the three events of initiation, propagation and arrest of brittle fracture occurred in service

not only failed on other types of steels (in particular with a C/Mn ratio different from that, rather high, of the naval steel), but also on the same naval steel if, instead of being used as it came out of the rolling mills, it was given a simple heat treatment of homogenization and refinement of the grain. Therefore, the 15-ft-lb criterion was definitely abandoned though researchers continue to trust that kind of test. If the 15-ft-lb criterion was abandoned, the Charpy V test did not. It was improved and new procedures were introduced to finally arrive to catch the chimera of the Transition Temperature.

3.2 The Charpy V-Notch Test

As mentioned in the previous paragraph, the Charpy V test marked a fundamental step in the study and prevention of brittle fracture, despite its clear limits. The test is performed on a prismatic specimen, 55 × 10 × 10 mm, in which a surface "V" notch is introduced, two millimeter deep, with a radius of curvature of just 0.25 mm, very small but not very sharp as it should be to simulate a crack. Figure 3.3 shows the standard CV specimen and the testing apparatus. This consists of a pendulum carrying a mass with a striking edge, or hammer, that is released from a height more than sufficient to break the specimen in one blow. The specimen is placed in its anvil, Fig. 3.4, so that the striking hammer puts in traction the fibers containing the notch that will be forced to open, fracturing the specimen. At each test temperature it is recorded the energy absorbed, or CV energy or *resilience*, by the test piece, read as difference between the initial height, h_o, and the final one, h, after the blow, the percentage of ductile rupture and the dilation of the fibers on the face opposite to the one carrying the notch, as schematically shown in Fig. 3.5. There are two types

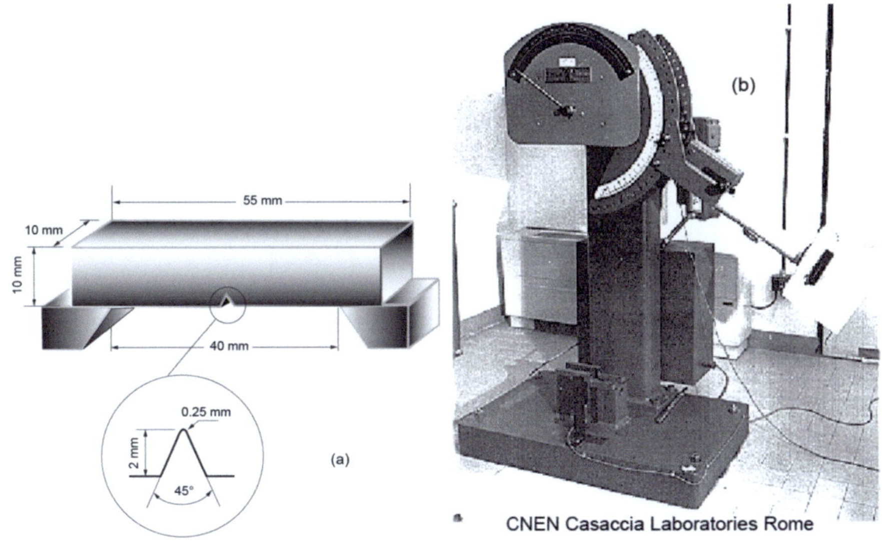

Fig. 3.3 **a** Charpy V-Notch specimen and **b** testing pendulum

of strikers: the American and the European. The first, also referred to as the 4 mm striker since its flat head is 4 mm wide, is shown in Fig. 3.4c, the second, shown in Fig. 3.4d has an almost round head of 2 mm radius.

The percentage of ductile failure is evaluated on the basis of a simple visual examination of the broken section, as the ratio of the ductile area, A_d, that appears dark to the brittle one, A_b, that appears bright. Figure 3.6 shows a series of 10 CV specimens of carbon steel, type A 508 Cl 3, used in the construction of a forged pressure vessel 4 m in diameter and 30 cm thick, having a longitudinal orientation, Fig. 3.30, broken at various temperatures [5].

At very low temperature −80 °C, the section of the specimen remains square, without any permanent deformations. It appears bright, crystalline and shows those characteristic marks that point back towards the initiation site, at the center of the notch. The fracture is produced by a cleavage mechanism. Already at –30 °C it appears a contraction, though very small, on the notch side with a tiny darker area below the notch tip, almost a nail (plastic area) that indicates some permanent deformation and, therefore, the onset of a slip process, with ductile behavior. At –5 °C there are two little shear lips on both sides and the plastic zone reaching a 20–25% of the total area. At 40 °C the fracture surface is now completely fibrous, with extensive shear lips that will grow further with temperature. From this point on, any further increase in temperature will have almost no effect on the resilience of the material that will remain almost constant (*maximum resilience* or *USE upper shelf energy*) and the failure will occur only through large plastic slips activated by stresses higher than yield strength. The relative resilience, ductility and lateral expansion curves are shown in Fig. 3.7 [5]. The trend of instrumental resilience confirms what has

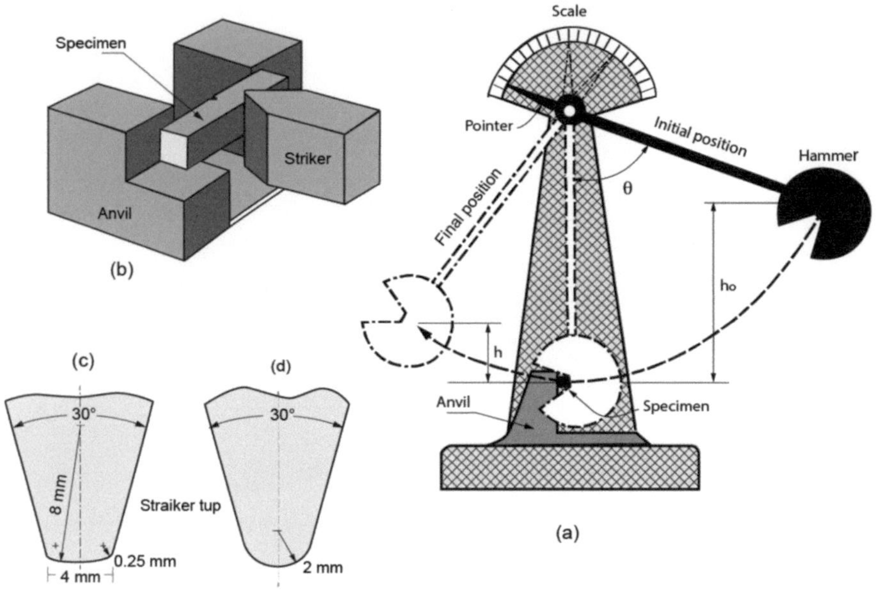

Fig. 3.4 Schematic of Charpy V-Notch testing equipment: **a** pendulum, **b** anvil holding the specimen, **c** American striker tup and **d** European striker tup

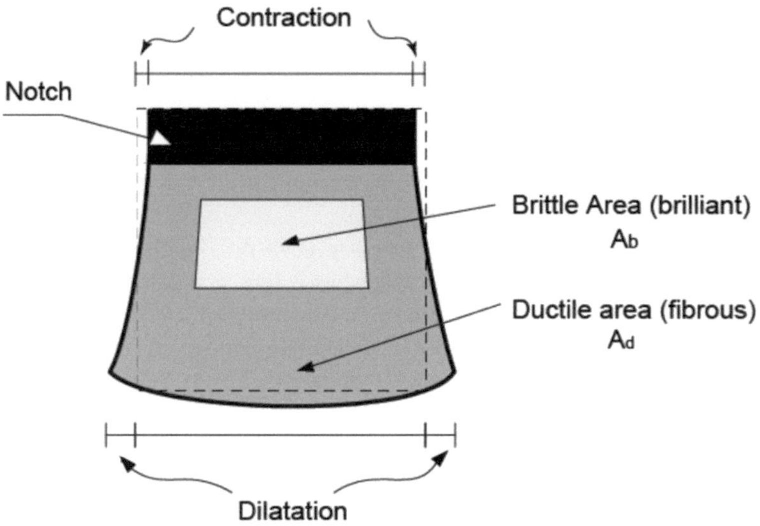

Fig. 3.5 Section of a broken Charpy V-Notch specimen showing lateral expansion and contraction, the brittle fracture and shear fracture (ductile fracture)

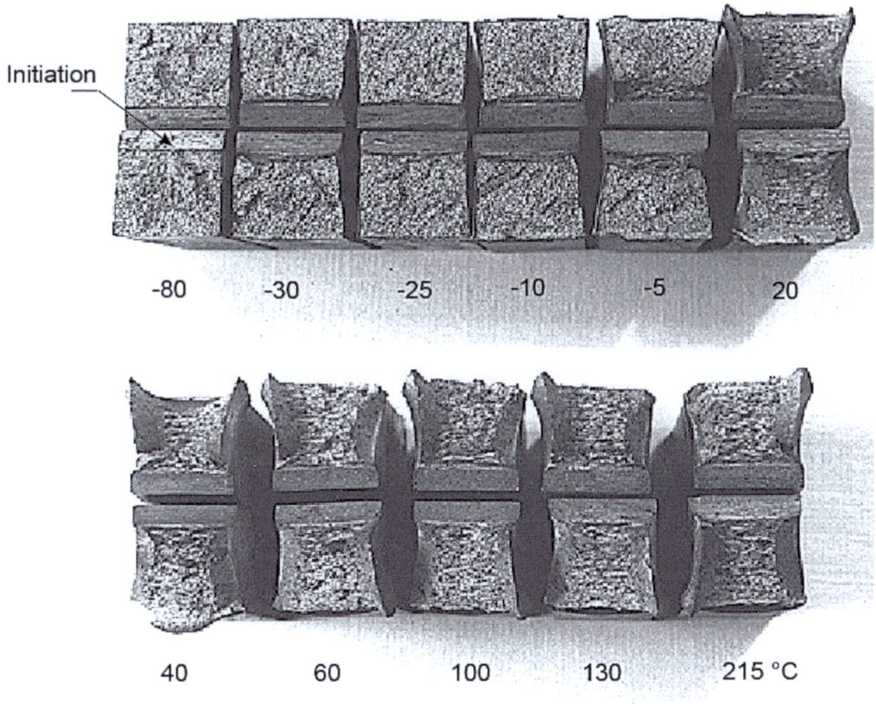

Fig. 3.6 Series of ten CV specimens of carbon steel type A508 Cl. 3 forging, broken at various temperatures showing the progressive transition from a completely brittle cleavage fracture at -80 °C to a totally ductile plastic fracture by shear already at 40 °C. The arrow indicates the point where the brittle fracture was triggered at −80 °C with the characteristic marks irradiating from that point

been found by the simple observation of the fracture surface. Resilience, shown in Fig. 3.7a, presents a temperature range, between –50 °C and 50 °C, in which there is a transition from brittle behavior, characterized by a very low energy absorption, to ductile fracture, with high energy absorption. The ductility curve, Fig. 3.7b, has been traced by reporting, in percent, the fraction of ductile area over the total area, obtained by visual examination. Graphs such as those shown in Fig. 3.7 cannot be used to detect the NDT by simply placing that temperature at the beginning of the rising ramp, for instance at –50 °C, Fig. 3.7a, or at the middle point of the ramp, about 0 °C or at the middle point of Fig. 3.7b where the fracture is 50% brittle. It needs other tests to assess the actual transition temperature and once it is done the Charpy curves can be used, but only because they have been validated. Once the Charpy V 15 ft-lb criterion failed to detect the NDT temperature of any steel, the next step was to search for a kind of improvement of that test that could reach the target. Few years later a new test was found that reached the purpose, it was the Pellini Test, (see Sect. 3.7). However, the CV test was not abandoned but further improved because of its simplicity. A monumental correlation work between CV tests and Pellini tests was carried out for a very large number of steels, leading to the formulation of more precise and specific criteria for the identification of NDT through the use of Pellini Test validated by a set of Charpy V-Notch tests. To-day the ASME design code, Sec. III or VIII [6] uses CV tests to validate the NDT obtained with a Pellini test. It identifies for the large family of carbon steels, a so-called *Reference Transition Temperature* or RT_{NDT}, as that temperature corresponding to the Pellini NDT plus 33°C (60 °F) when at this temperature the Charpy V-Notch resilience is at least 67.8 J (50 ft-lb) and the lateral expansion of Charpy V-Notch specimens is at least 0.889 mm ($35 \cdot 10^{-3}$ in). If these two CV conditions, i.e., CV energy and lateral expansion, are not met, the temperature is increased until both CV requirements are satisfied, and this last temperature becomes the valid RT_{NDT} of the material while its NDT becomes this final RT_{NDT} minus those 33°C previously added.

3.3 Deformation Fields in the Charpy V-Notch Specimen

To fully understand a CV test and why this specimen cannot be used to measure directly the NDT temperature of a material, it is useful to look at the deformation fields that develop under the notch tip during the test. The maximum stress is obtained on the bottom of the notch, because of stress concentration and grows as the applied M moment increases from zero to its maximum. Once the elastic limit is reached at the notch tip, the first Lüders lines (see Sect. 2.2) or slip lines start to appear in the form evidenced in Fig. 3.8a in a FE dynamic analysis [7]. Actually, on the real specimen slip lines are not directly visible, but can be highlighted with a staining technique that makes use of reagents, such as the Fry's Reagent[8] for mild steels, especially those with a high nitrogen content, or the Morris Reagent for Si steels [9]. If fracture starts at this early stage, we can claim that it is of a brittle type, generated by a pure cleavage mechanism, since it is supported by an almost entirely elastic stress field. However, the fracture rarely develops when the slip lines just appear or before, except at extremely low temperatures, below −100 °C for steels, or in extremely brittle materials, such as non-tempered martensite or some very hard aluminum alloys.

As the applied moment M increases the slip lines widen and extend becoming bands, Figs. 3.8b, c, until a point is reached where the upper slip bands, in traction, join the lower bands, in compression, to form the so-called plastic hinge, Fig. 3.8d [7] and [10, 11]. The theory of slip lines [12] states that during plastic hinge formation these two plastic fields consist of two families of mutually orthogonal curves of equal maximum shear stress $\tau = \tau_s$ where τ_s is the shear yield stress or slip stress [10–14].

When the hinge is formed in pure bending, the maximum bending moment or slip moment or general yield moment M_{GY} is [16]:

$$M_{GY} = 1.23\sigma_s \times (W - a)^2 W/4 \tag{3.1}$$

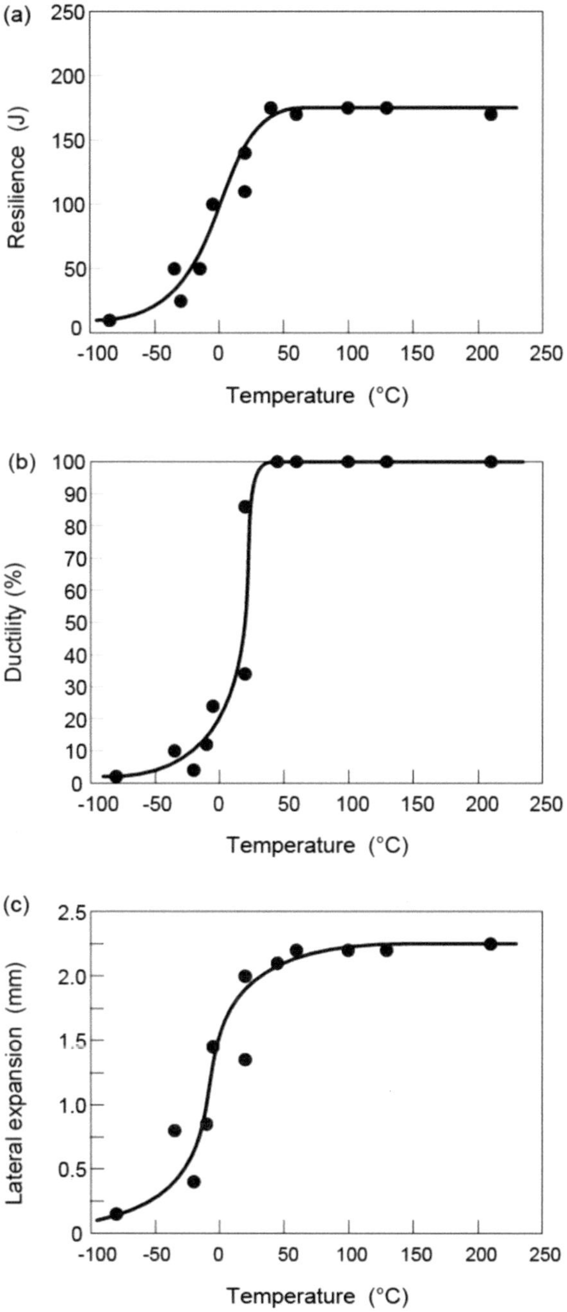

Fig. 3.7 Trend of resilience, ductility and lateral expansion obtained from CV specimens for a carbon steel type A 508 Cl 3 vs temperature [5]

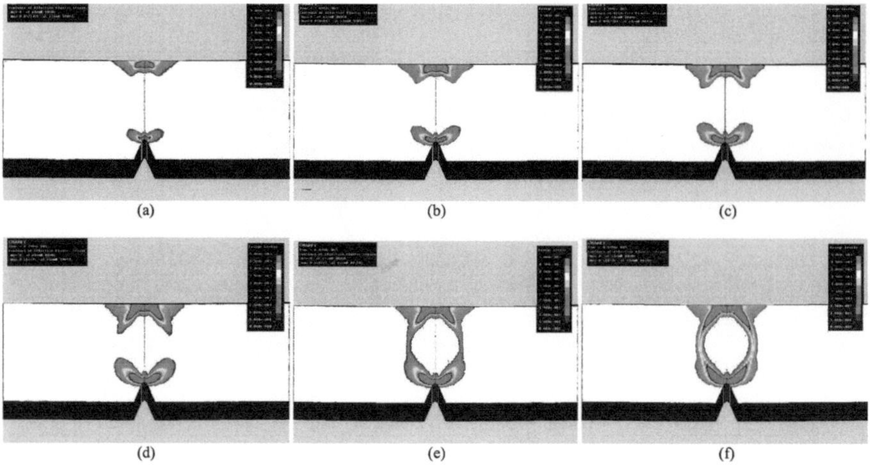

Fig. 3.8 FE analysis showing the development of plastic deformations in a CV specimen in pure bending as the applied moment increases. In (e) and (f) the upper (traction) and lower (compression) bands joined creating a plastic hinge within a rigid elastic region. This stage always precedes the fracture of the specimen at general yield at temperature over the RT_{NDT} of the material. The hammer had an impact velocity of 5 m/s [7]

where *W-a*-is the height of the reduced section under the notch. If we consider that in the same bar of equal section *W-a* but without a notch, the plastic failure is reached when *M* is worth

$$M_{slp} = \sigma_s \times (W - a)^2 W/4 \tag{3.2}$$

we have that the notch has increased *M* by a factor 1.23. This value is defined as *the plastic constraint factor, C*. However, in the Charpy V-notch test the loading condition is not exactly that of pure bending since the specimen is loaded in three-point bending, as shown in Fig. 3.4. Ewing[10] found that the value of *C* is influenced by the width of the tip of the hammer. For a hammer with a 4 mm wide percussion tup, European hammer, Ewing found that *k* was worth 1.279, while for one with a 8 mm wide tip, American hammer, it is worth 1.336. Using this values and Eq. (3.1), the load that produces the plastic hinge, P_{GY}, is:

$$\begin{aligned} P_{GY} &= 2.046 \cdot 10^{-5} \sigma_s \\ P_{GY} &= 2.137 \cdot 10^{-5} \sigma_s \end{aligned} \tag{3.3}$$

for the European and the American hammer, respectively. This issue will be treated in more details in Sect. 3.6. It is interesting to compare the dynamic loading condition of the Charpy V test with a quasi-static test that enhances the slip band and the plastic hinge formation. This is shown in Fig. 3.9 for six CV specimens of a low carbon steel tested at various temperatures under three-point quasi-static bending [15].

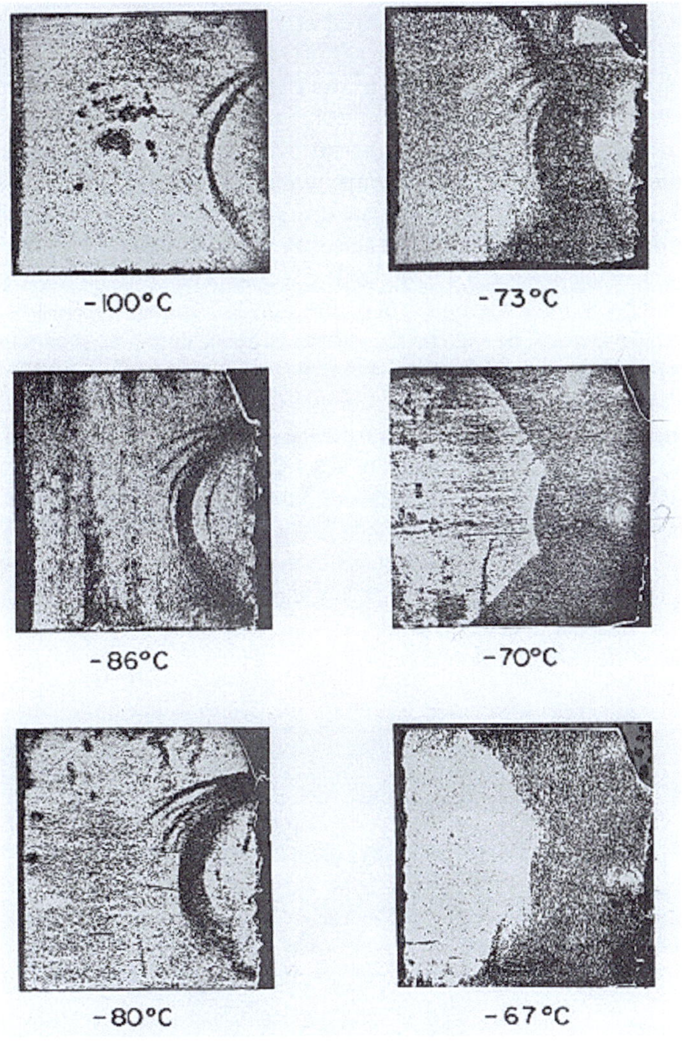

Fig. 3.9 Development of the plastic hinge at the notch tip of Charpy-V specimens of mild steel at different temperatures, loaded under quasi-static three-point bending conditions. Even at very low temperature, −100 °C, the plastic hinge is formed and widens by increasing the temperature. At −73 °C it is already so wide that the plastic stress field prevails over the elastic one resalting a fully ductile fracture [15]

Under quasi-static condition the plastic hinge forms even at a temperature as low as −100 °C and at −73 °C and is so wide that the upper shelf resilience is already reached, indicating a full ductile fracture generated by a lack of triaxiality or plain strain condition (see Chap. 4). To just have another term of comparison, Fig. 3.10 shows the plastic hinge for A508 Cl. 3 at 0 °C under dynamic loading [17]. It is as large as the one of Fig. 3.9 developed at −70 °C. The high loading rate has shifted the fully plastic condition from −70 °C to 0 °C. Specimen surface had been treated with Fry's reagent and the image transferred from the optical microscope to a screen where, with a tone elimination technique, it was possible to preserve only the image of the plastic area, brighter than the rest. The absorbed energy (resilience) was 96 Joules, rather high at this low temperature. Between quasi-static and dynamic conditions (besides that the two steels are different) the relative NDT is likely to be 70 °C that does not come unexpected since, as we have seen in Sect. 1.13 and will be further discussed in Sect. 5.4.3, dynamic conditions increase the yield strength of materials and favor the development of plain strain conditions. This loss of triaxiality in a quasi-static test will be seen on the entire resilience curve, as shown in Fig. 3.11 for four different steels type ABS-C, A 302 B, HY 130 and 18 Ni of 273 MPa, 394 MPa, 900 MPa and 1750 MPa yield strength, respectively. It is very interesting to note the differences between the low resilience region and the upper shelf behavior. For the two low strength steels, namely ABS-C and A 302 B of Fig. 3.11a, b, the change from a quasi-static to a dynamic condition penalizes very much the material response at low temperature, the opposite happens in the upper shelf region of high temperatures. This is because an increase in the yield strength due to the dynamic conditions, promotes brittle fracture at low temperatures while in the upper shelf region where fracture is always ductile, any increase in the strength of the material also increases the energy absorption.

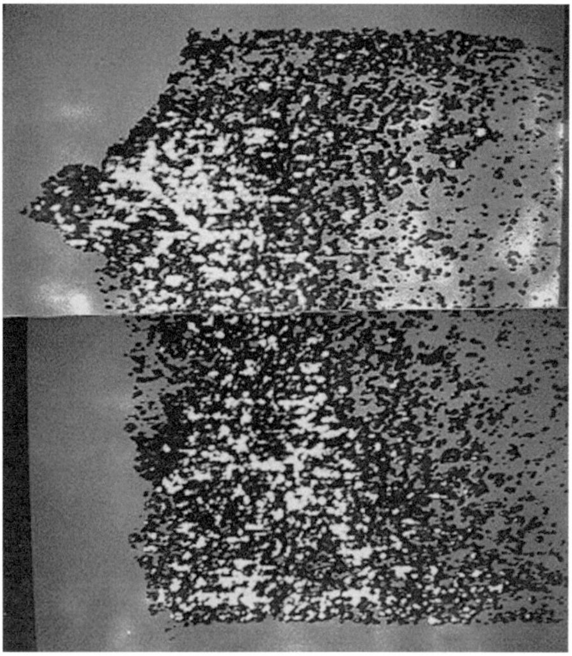

Fig. 3.10 Optical microscope, photo treatment of the image. Evidence of the plastic zone developed in a Charpy V-Notch carbon steel specimen of type A 508 Cl 3 carbon steel, broken at 0 °C. The notch is on top left. The fracture is fully plastic [17]

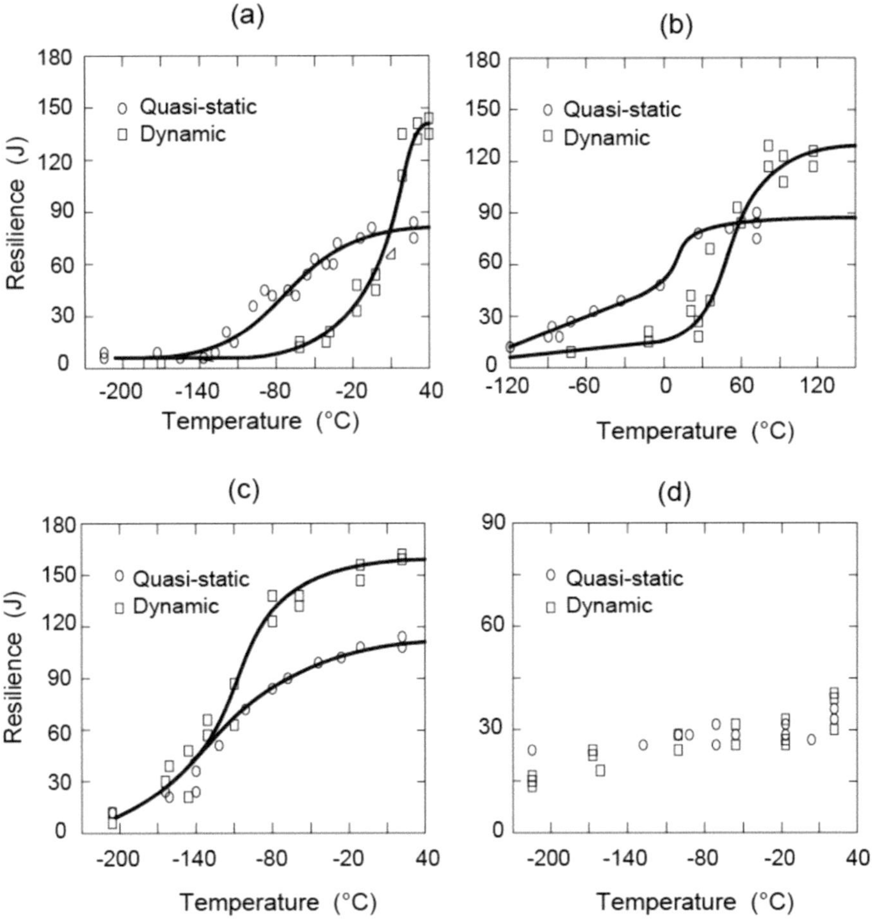

Fig. 3.11 Trend of CV resilience in carbon steel type **a** ABS-C, **b** A 302 B, **c** HY 130 and **d** 18 Ni for the two quasi-static and dynamic conditions

The HY 130 steel of 900 yield strength, Fig. 3.11c, shows no difference at low temperature up to about −120 °C but where the two curves start to separate. In the upper shelf region dynamic conditions always result in a better behavior. It is very interesting the response of the very high strength, 1750 MPa,18 Ni steel, Fig. 3.11d. The resilience is almost flat with a very low gain from −200 °C to 40 °C, just about 20 Joule, revealing a tendency to brittle fracture at all temperature. Here, the change from a quasi-static condition to a dynamic one reveals practically no difference at all.

It is clear that the gain or loss of resilience, moving from a quasi-static to a dynamic condition, will be all the more marked and significant the more ductile the metal, in other words as more space is available for a change in resilience. For a very high strength alloy, this space will be very small if not zero at all and will not allow any significant change in the characteristics of resilience passing from a static to a dynamic condition, as they are already very poor. The embrittling effects of various factors, such as temperature, thickness and dynamic or static load, cannot be combined as a simple individual sum of effects, but as a complement to the maximum possible damage. If this maximum has already been achieved, no space will be left to other factors to operate. Another evidence of the plastic hinge formation is shown in Fig. 3.12 relative to a CV specimen taken from the Heat Affected Zone of a weld on a carbon steel type A516-70 of 350 MPa yield strength broken at−90 °C [18].

Images taken at the scanning electron microscope (SEM) show precisely the very shallow area of the notch tip where the plastic hinge starts to form, positions B and D. At that low temperature the fracture was completely brittle, as revealed by SEM at locations A and C where fracture is by cleavage and shows the typical appearance of flat facets or lettuce leaves, as in Figs. 6.6, 6.23 and 6.24 while at locations B and D, just below the notch, the fracture appearance is ductile, showing the typical dimple formation, as in Fig. 6.13.

3.4 Tanh Regression Analysis

Charpy-V energy or resilience measured at different temperatures can be conveniently interpolated using a non-linear four parameter regression of the form

$$CV = A + B\tanh\left(\frac{T - T_0}{C}\right) \tag{3.4}$$

where T is the test temperature and$A$, B, C and To are four parameters defined in Fig. 3.13.

An example of the *tanh* regression analysis is shown in Fig. 3.14 for A508 CV specimens tested at different temperatures. The *tanh* equation is

$$CV(J) = 113.05 + 142.22\tanh\left(\frac{T + 29}{67.53}\right) \tag{3.5}$$

The same regression can be used for the ductility and the lateral expansion measurements.

3.5 Factors that Affect Resilience

The resilience measured in a Charpy V-Notch test is an index of the material capability to absorb energy through plastic deformations. The more slip is prevented, the lower the energy absorbed by the material at failure. Temperature, it has been seen, is one of those factors, perhaps the most important, that favor slip developing permanent

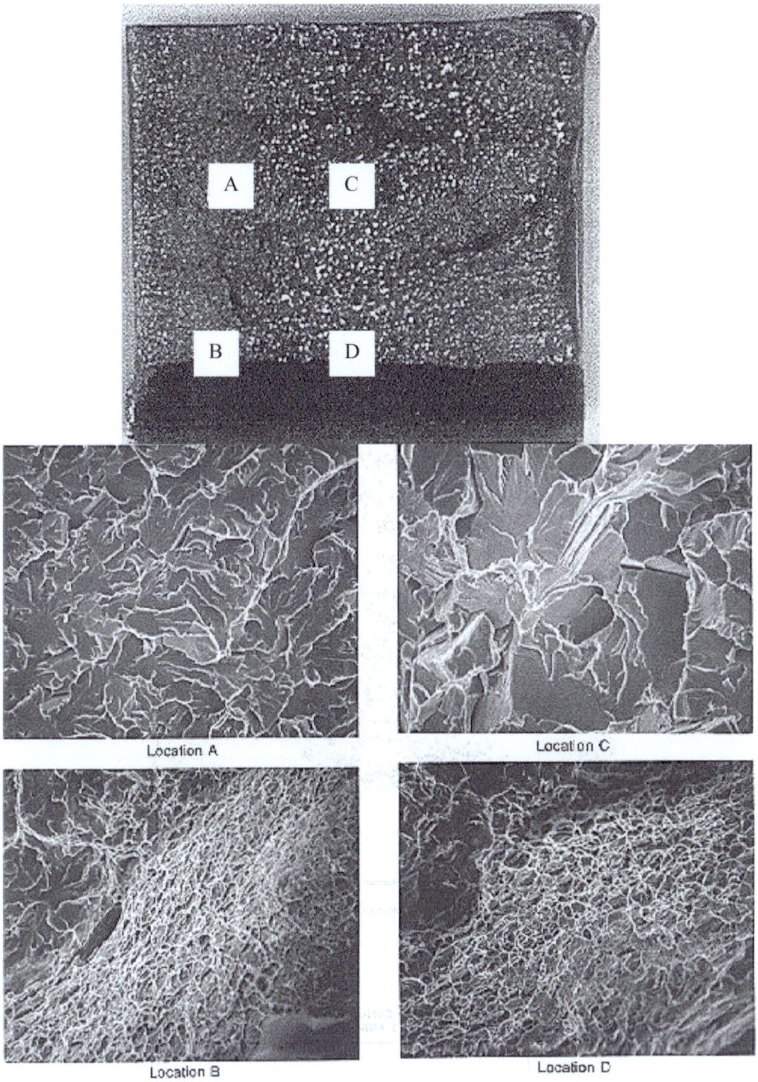

Fig. 3.12 Fracture surface of a Charpy V specimen broken at −90 °C. Four locations are evidenced with the letters A, B, C and D. The SEM examination of Locations A and C in the brittle region, photo below, shows the typical cleavage appearance. Location B and D, at the notch front where the tiny plastic hinge develops, shows the typical ductile fracture like a porous sponge tissue [18]

deformations. However, it is not the only factor that affects resilience. It has been seen in Sect. 2.2, Fig. 1.13, how the carbon content drastically reduces plastic deformations at failure. Therefore, carbon content must have an impact also on resilience of steels. This is clearly visible in Fig. 3.15, which shows the trend of the resilience of a carbon steel versus the C content [19].

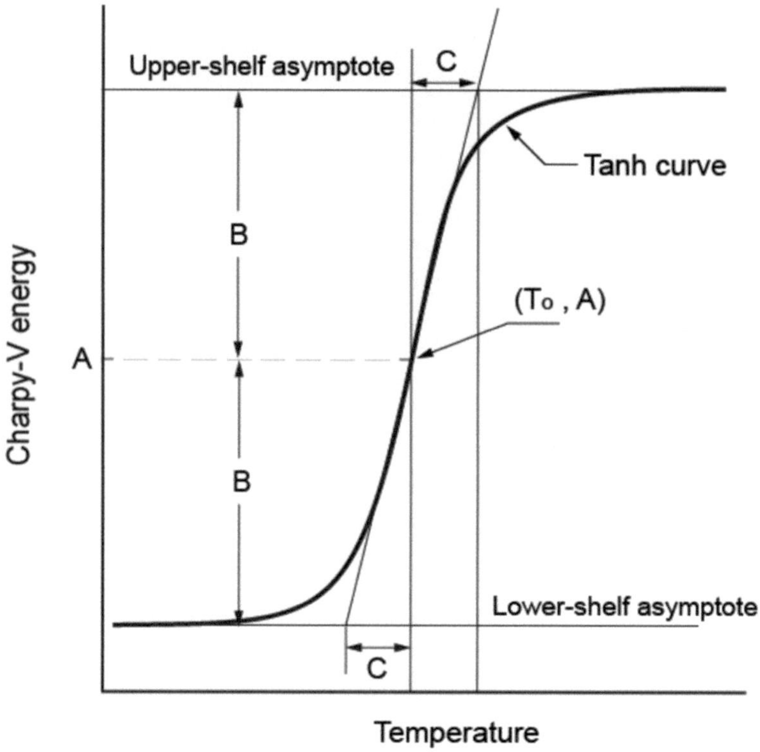

Fig. 3.13 Graphical representation of tanh coefficients

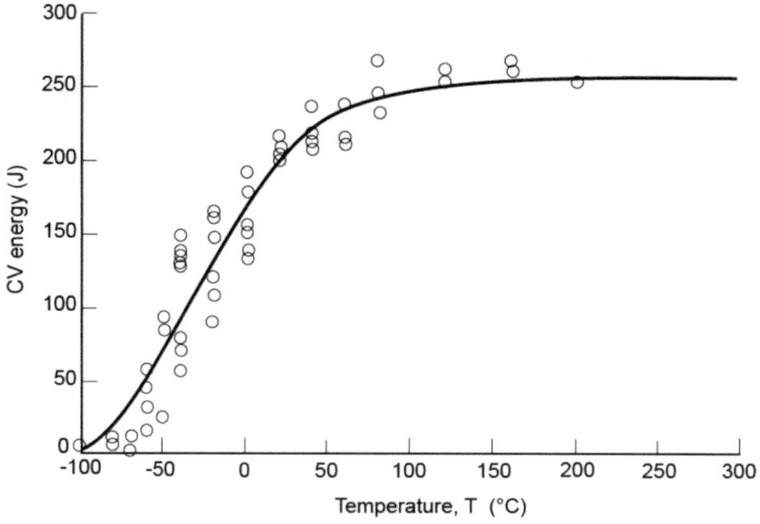

Fig. 3.14 Regression analysis of CV energies measured with A508 steel specimens using tanh method

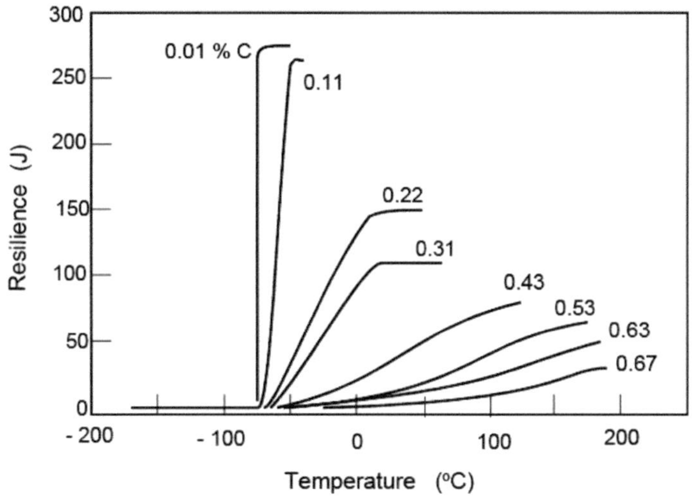

Fig. 3.15 Effect of carbon content on the CV resilience of a carbon steel (Adpt. from [19])

As it can be seen in the figure, the increase in the C content not only decreases the upper shelf, but also shift the entire curve to the wright, i.e., towards higher *NDT*. Same as carbon, also any increase in the yield strength, σ_y, will have the effect to reduce the upper shelf energy and increase the NDT, embrittling the steel. This is shown in Fig. 3.16 for a CrMoV steel with 0.45% C hardened to five different yield strength [20]. There is a precise relationship between the hardness and tensile strength of metal alloys. As the first increases, the second also increases: hard metals are the most resistant. Figure 3.17 shows the empirical correlation [21] between tensile strength and hardness measured in Vickers (HV) and Rockwell C (HRC) degrees. If the increase in tensile strength has, as we have seen, an effect on resilience through the decrease in the capability of the metal to undergo plastic deformations, the same effect must also be observed in relation to hardness [22]. This is shown in Fig. 3.18 [23] for carbon steels with a C content varying between 0.18 and 0.3%. The curves of Figs. 3.15, 3.16, 3.17 and 3.18 are of experimental derivation and, therefore, constitute a solid but empirical basis that must necessarily be used with due caution for practical purposes. This is because the interdependence between C content, yield strength, hardness and resilience are much more complex than may appear at first glance. For example, a steel with a yield strength of 590 MPa shows a maximum resilience of no more than 80 J, Fig. 3.16.

On the basis of Fig. 3.17, the same steel would have a hardness of about 170–180 HV which gives it, instead, a more than double resilience, equal to about 170 J, if the curves of Fig. 3.18 were used. The apparent inconsistency lies in the fact that the curves of Fig. 3.16 are valid for a steel with 0.45% C that goes far beyond the normal carbon contents an that gives the alloy a metallurgical structure very different from that relating to a hardness of 170 HV obtained with 0.15–0.25% of C. They must be compared with the relationship of Fig. 3.15 which show that with a content of C =

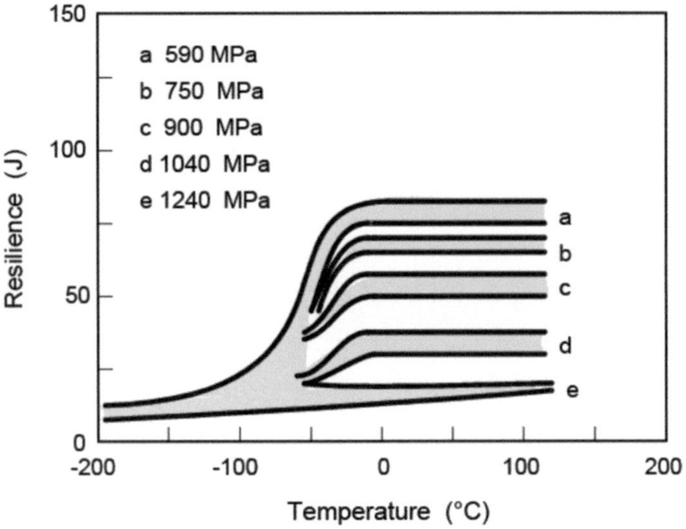

Fig. 3.16 Trend of CV resilience with the variation of the yield strength of a CrMoV steel 0.45% C (Adpt. from [20])

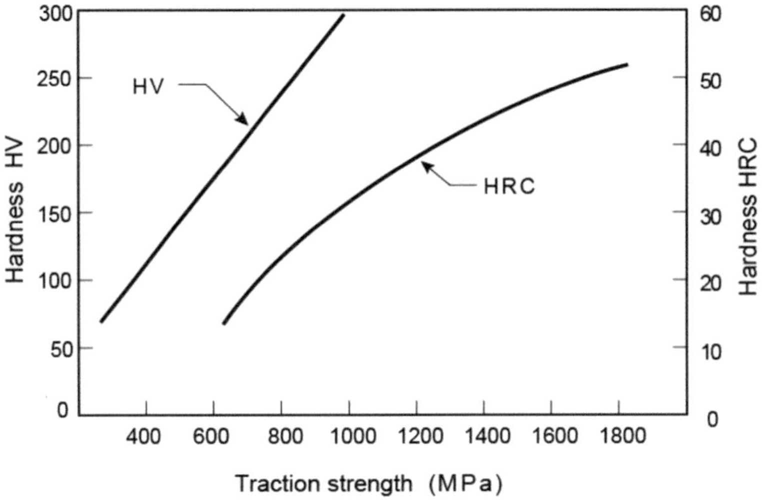

Fig. 3.17 Relationship between traction strength and hardness (from table of [21])

0.45% the steel reaches at most just 70–80 J of resilience. It has already been shown in Sect. 2.15 that any increase in grain size leads to the simultaneous reduction of the yield stress and, therefore, to an increase of the NDT but also to a loss of ductility in the upper shelf of the alloy because the alloy becomes less hard. This effect is

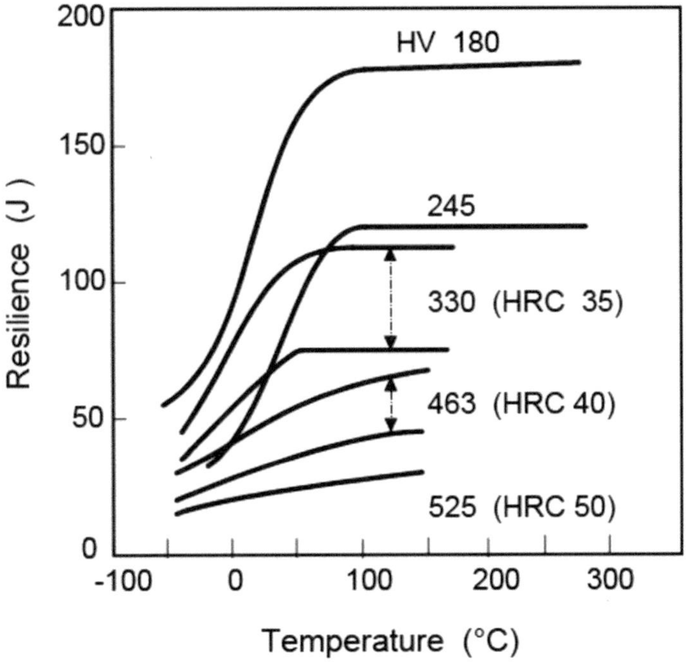

Fig. 3.18 Trend of CV resilience of a steel of medium–high carbon content (C 0.18–0.3%) as a function of hardness. In brackets is the hardness expressed in Rockwell C degrees (Der. from [23])

similar to the static versus dynamic effect seen in Fig. 3.11. Therefore, we expect that resilience will also have a dependence on the grain size.

This is evident in Fig. 3.19 which refers to a series of batches obtained from the same plate of carbon steel type A 533 B indicated in the figure as "Reference", but subjected to annealing treatment at different temperatures for different times, in order to increase the size of ferritic grain [24]. If we take as a reference NDT the temperature T corresponding to the resilience of 68 J we can quantify the temperature rise Δ(NDT) as a function of the size of the ferritic grain. This is being shown in Fig. 3.20. Even now, the cautions mentioned above regarding relationships of empirical nature apply. In fact, if the specimens heated to 1000° C were given a subsequent hardening process through a rapid cooling to about 50° C/minute a grain no longer ferritic but bainitic with a size of about 80 μm instead of 10 μm would be achieved leading to an increase in ΔT of only 70° C [23] instead of the >>140 °C as for Fig. 3.19. If the cooling rate is above 50 °C/min, a fragile martensitic structure can be obtained and the response in terms of resilience and transition temperature will still be different (Fig. 3.21).

Therefore, comparison must be made *ceteris paribus* (all other things being equal). In all the cases examined, the Charpy V-Notch curves show the presence of a decisive fragile-to-ductile transition. Such behavior is characteristic of all metals and alloys with A BCC structure that have a thermally activated component of the yield strength. The steels so far considered are low carbon steels with a ferrite structure,

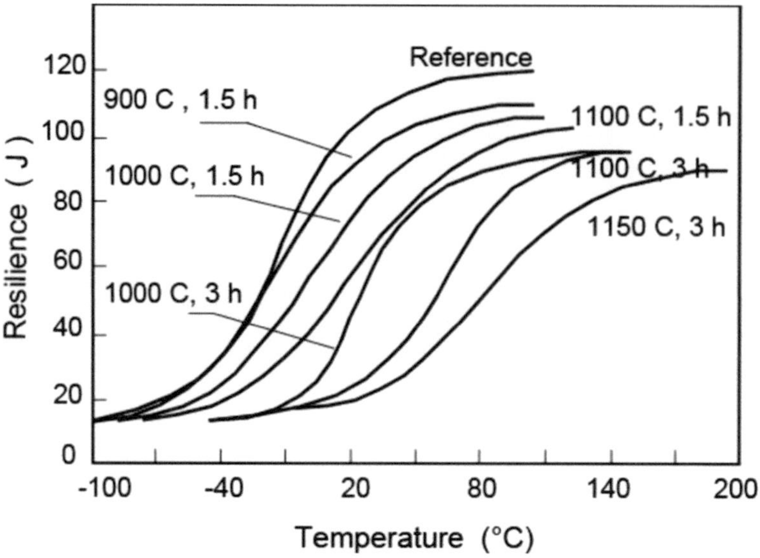

Fig. 3.19 CV resilience of a carbon steel, type A 533 B, versus temperature as a function of the annealing temperature and time, which increases the size of the ferritic grain [24]

Fig. 3.20 Trend of the increase ΔT of the temperature at 68 J of resilience in a carbon steel, type A 533 B, as function of the size of the ferritic grain

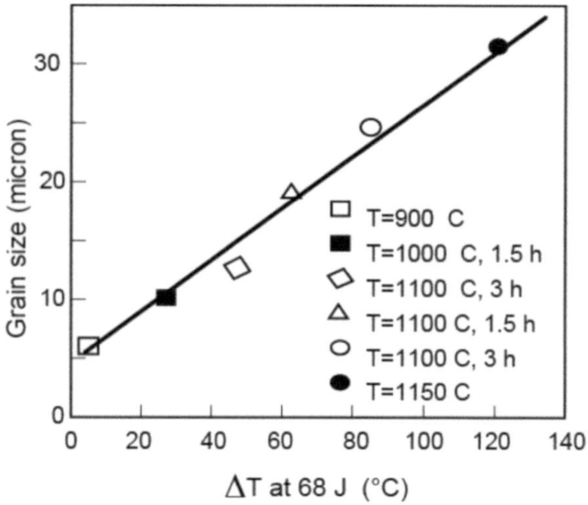

with the exception of those of Fig. 3.18 having a hardness over 35 RC, which have a martensitic structure. Ferrite has BCC lattice and is characterized by a sharp transition temperature effect. The behavior may change with steel or alloy structure as shown in Fig. 3.21 that refers to HY-80 and 1380 quenched and tempered steel, respectively [25]. The first steel is a member of the low carbon, low alloy family of steels with Ni–Cr-Mo of alloying elements. It is a medium strength steel of about 550

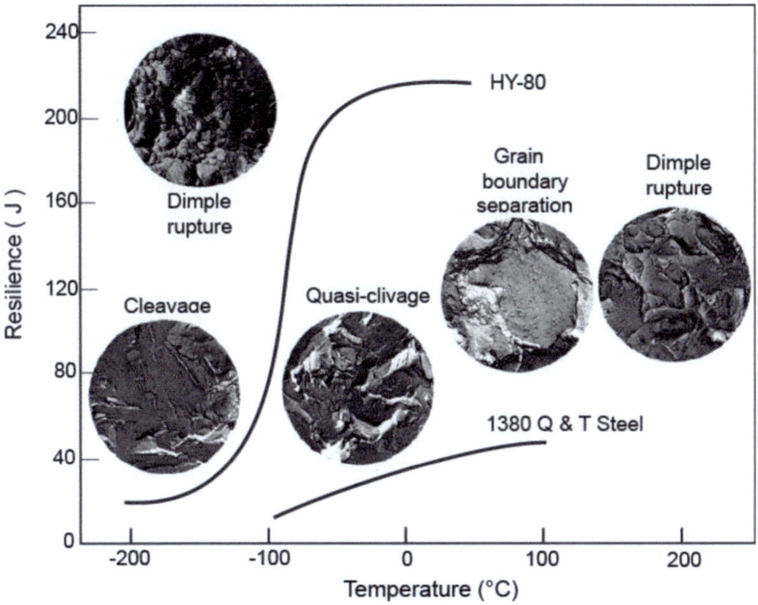

Fig. 3.21 TEM fractographs illustrating various micromechanisms of fracture in two Q&T steels in relation to their CV resilience. 4000 X (Adapt. to SI units from [25])

MPa. Though it has a martensitic structure, its low carbon content provides an upper shelf energy (USE resilience) rather high (over 200 J) indicating high toughness. Its fracture appearance at low temperature is brittle, characterized by flat facets typical of cleavage, while in the USE region of the Charpy V-Notch curve is ductile, with large dimples. At variance, the high strength steel has an almost flat CV resilience without any real transition temperature effect. Its USE is as low as 50 J. At low temperature it shows a quasi-cleavage fracture (flat facets with shear ridges) and a dimple separation at high temperature but dimple are rather large compared to those of the HY-80 steel. The same behavior is seen in titanium alloys of various strength, Fig. 3.22 [25]. Ti-alloys have a HPE lattice. As the strength of the alloy increases its resilience decreases and the transition temperature effect disappears completely. High strength Ti-alloys have low CV USE and do not have an NDT as high strength martensitic steels, though their fracture mode at all temperature remains ductile. The same behavior is observed in aluminum alloys that have a FCC lattice, as shown in Fig. 3.23 [25]. These light metals and alloys with low USE show fracture of dimple rupture type, but with fine details of dimples changing with decreasing CV energy. However, in all cases naked eye fracture appears of brittle type since the fracture surfaces remain flat without any apparent plastic deformation. The failure by cleavage is the result of second phase particles fracture which serve as crack initiation sites at a microscopic level. The same behavior will be observed on austenitic steels whose crystal structure, austenite or γ-iron, is of FCC type. Unlike aluminum, austenitic steels, being heavy metals, always have very high resilience, in general higher than

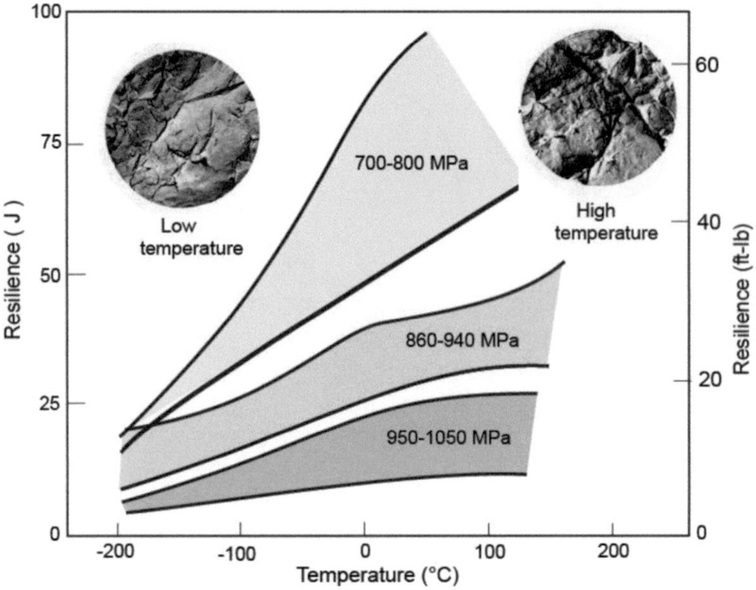

Fig. 3.22 Charpy V-Notch energy and fracture appearance with temperature for several Ti-alloys. The TEM micrographs show a dimple (ductile) fracture at all temperatures though at very low temperatures dimple are rough. 4000 X (Adapt. to SI units from [25])

200 J, which reveals high ductility, see Sect. 3.11 and Fig. 3.60. Figure 3.24 presents the resilience with temperature for three different austenitic stainless steels [26] from −280 °C to 40 °C.

These FCC steels have not got an NDT with the exception of type 301 which, between −200 and −150 °C, has a very clear fragile-to-ductile transition due to a transformation of metastable austenite into ferrite or brittle martensite, probably caused by the low content of Ni, ~7% against >11% of the other two, which is not sufficient to keep austenite at very low temperature under the effect of notch (triaxiality) and impulsive loads.

It must be recalled that austenite (γ phase) converts to ferrite (α phase) at about 720 °C. Its existence below that temperature and, in particular, at room temperature is made possible by a massive addition of nickel (over 7%) nevertheless it always remains in a metastable conditions. Due to the lack of an NDT, both aluminum and austenitic stainless steels are widely used in cryogenics, at temperatures where most alloys are dangerously brittle. However, caution must be used with aluminum alloys since their USE may be too low, see Fig. 3.61.

In ferritic steels of BCC lattice having an NDT, Ni is able to increase the maximum resilience and decrease the transition temperature. This is visible in Fig. 3.25 which refers [22] to two low-alloy carbon steels with equal Mn and C content, 1,3% and 0,21% respectively, but different Ni concentration. The beneficial effect of Ni is mainly due to a refinement of the ferritic grain, Fig. 3.20, which results in the USE

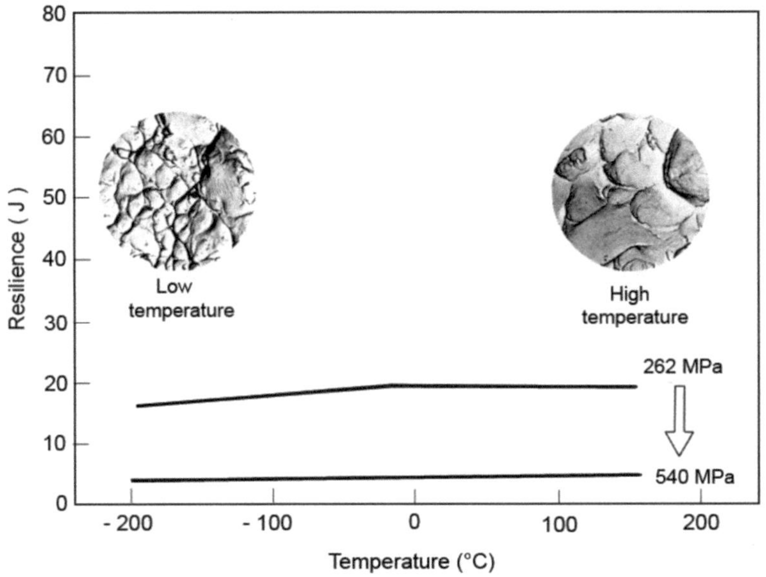

Fig. 3.23 Charpy V-Notch energy versus temperature and TEM observation of fracture surface of aluminum alloys of different strength (4000 X)

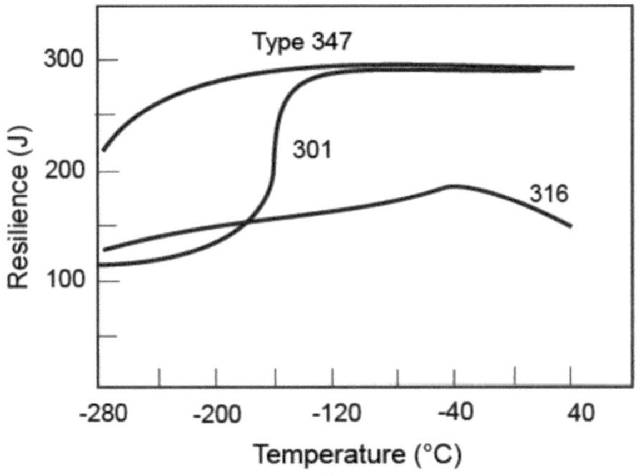

Fig. 3.24 CV Resilience versus temperature for three austenitic stainless steels [26]

of the materials. Another very important parameter that has a marked influence on resilience is the orientation of the Charpy V-Notch specimen with respect to the rolling direction of the plate from which it was been taken. This is particularly evident in steels that underwent a heavy mechanical cold work, such as extrusion or cold rolling. This is because during the mechanical working process grains elongate in the form of fibers in the direction of cold work, making the material strongly anisotropic.

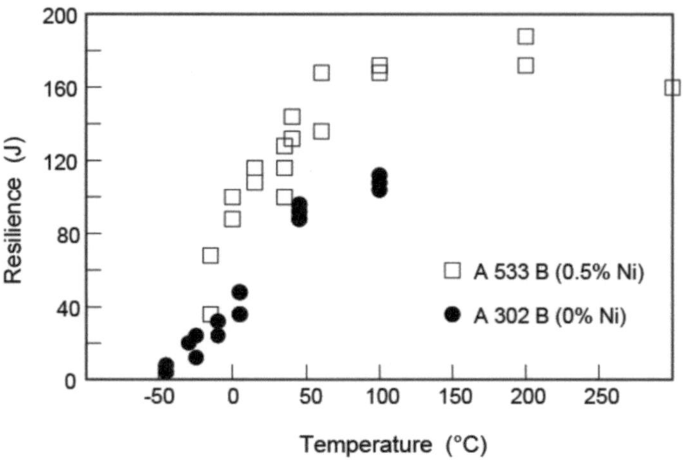

Fig. 3.25 Effect of Nickel on the CV resilience of a carbon steel (C ~ 0.2) [22]

An example of this anisotropy is shown in Fig. 3.26 which refers to a sample of carbon steel, type A 106 B, taken from the inner surface of a pipe [27] and examined in light microscopy at various magnifications. It can be clearly seen the disposition in alternate bands of ferrite (withe grains) and pearlite (dark carbides) in the direction of rolling. The effect on resilience is dramatic and clearly visible in Fig. 3.27, for a pipe of A106 steel, 15 cmin diameter [28]. Charpy-V energy or resilience is maximum when the specimen is oriented along the axis of the pipe, with the notch on a circumferential plane, and reaches its minimum value when oriented in a transverse direction with respect to the rolling direction and fibers.Fig 3.28 [29] is an example of Charpy-V energy dependence on the specimen direction observed in A302 B carbon steel heavily affected by MnS inclusions at various temperatures. Longitudinal specimens yielded much better resilience in the upper shelf region. Those taken in the longitudinal direction show an appreciable resilience of about 110 J, though not very high, while in the transverse direction the maximum resilience drops to just 50 J. Particularly interesting is the external appearance of the fracture surface of a transverse specimen, Fig. 3.29 which appears woody, almost flat, with marginal shear lips even at 180 °C and with an impressive density of large cavities indicating the large amounts of MnS inclusion that were responsible for the low resilience in the upper shelf region [29]. The direction of the Charpy V-Notch specimen taken from a plate is identified by two letters. The first refers to the orientation of the specimen with respect to the direction of rolling of the plate, which can be longitudinal, *L*, transverse, *T*, or along the thickness, *S* or *ST* (Short Transverse), the second refers to the notch orientation, as schematized in Fig. 3.30.

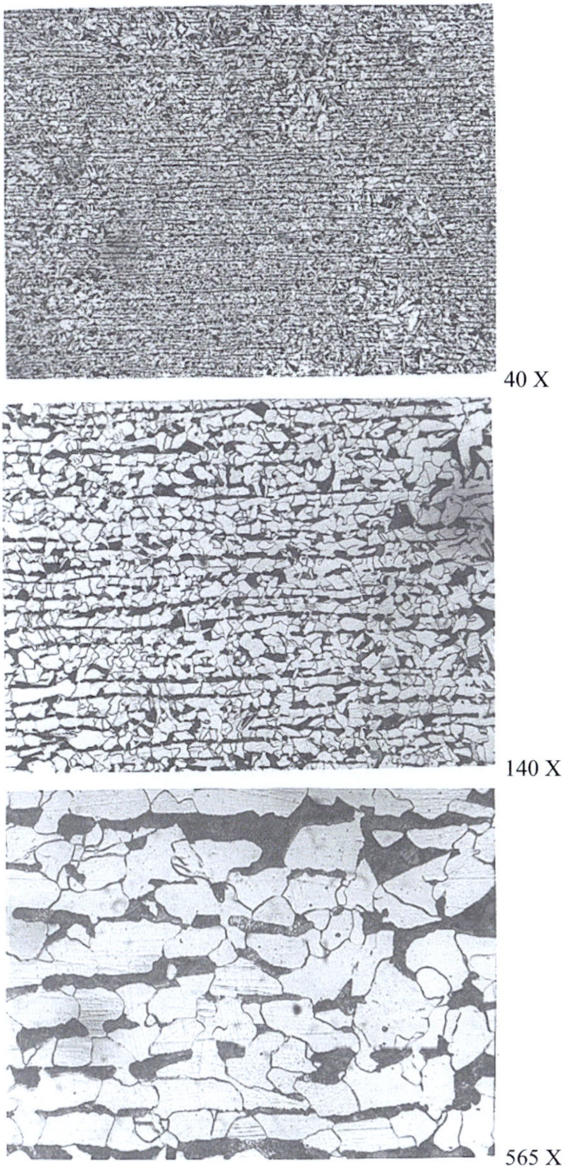

Fig. 3.26 Arrangement in alternate bands of ferritic grains (light) and carbides (dark) elongated along the direction of cold rolling, observed at the light microscope at various magnifications, on a sample taken on the inner surface of a carbon steel pipe, type A106 [27]

3.6 Instrumented Charpy V-Notch Test

The failure of the Charpy V-Notch specimen to measure the NDT of the material did not lead to its abandonment. It possessed all the elements necessary to trigger a brittle fracture, i.e., the notch and dynamic loading, except for the V-Notch itself

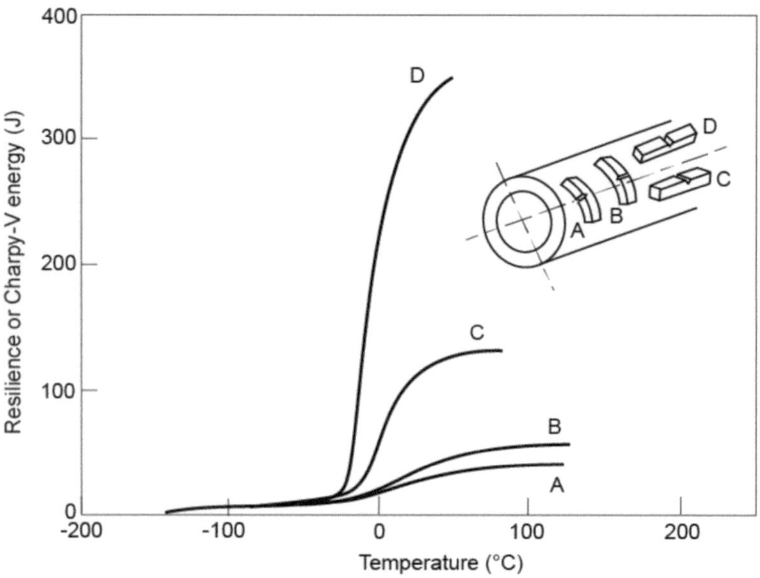

Fig. 3.27 Effect of anisotropy on the CV resilience of a carbon steel pipe type A 106 (Adapt. to SI units from [28])

Fig. 3.28 CV energy or resilience in a carbon steel type A302 B with a high MnS inclusion content, according to the sampling directions of Charpy V-Notch specimens [29]

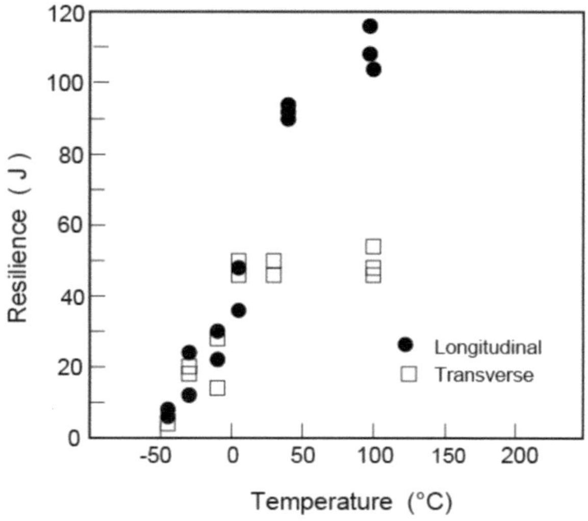

which, though sharp, was not precisely a crack. On the contrary, the CV specimen received a powerful impulse by inserting two strain gauges on both sides of the tup of the hammer.

This simple upgrading of the testing machine allows to gain additional data from the test such as the time within which the event develops, the amplitude of the force

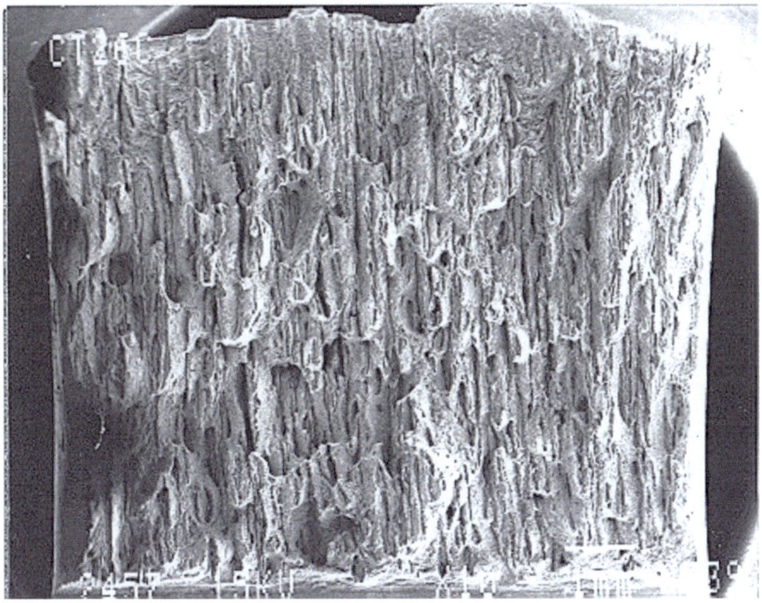

Fig. 3.29 Macrography of the fracture surface of a Charpy V-Notch specimen of A302 B carbon steel with high MnS content broken at 180 °C. Caverns produced by inclusion are clearly identifiable [29]

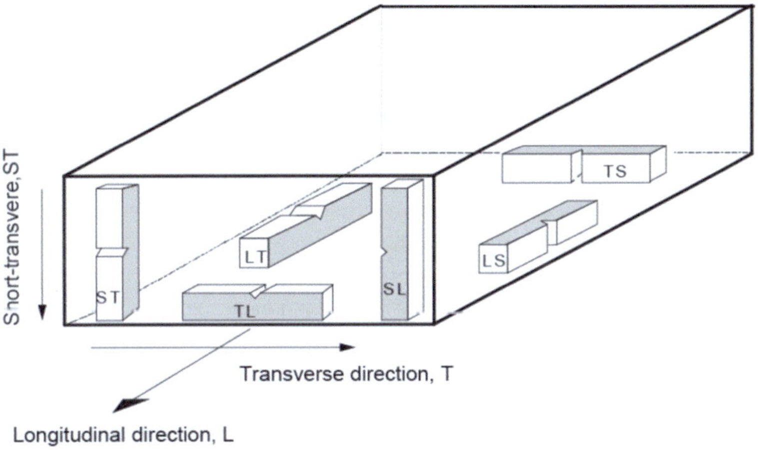

Fig. 3.30 Identification of CV specimens according to their picking position from a plate

applied during that time, the displacement and the energy absorbed from the moment of impact to the complete failure of the specimen. The diagrams in Fig. 3.31 are an example of this information obtained in an instrumented Charpy V-Notch test [30]. They show the trend of the force$F$ versus time from the moment of impact to the complete fracture, at six different test temperatures, equal to −80 °C, −30 °C, 0 °C, 20 °C, 40 °C and 60 °C, respectively [8]. The material used was a carbon steel forging, type A508 Cl.3. They also show the CV energy absorbed during the fracture versus time. It is to note how at the lowest temperatures, −80 °C and −30 °C, in which the specimen absorbs less than 5 J and the behavior is completely brittle, the applied load,F, falls to zero in just 0.1 to 0.6 ms since fracture occurs through elastic stress fields and propagates at velocities of several hundred meter per second. In the upper shelf region, where CV energy is over 150 J, fracture is ductile and develops through the plastic stress fields that characterize the plastic hinge formation, Figs. 3.8 and 3.9, therefore it takes more than 4 ms before the load drops to zero after reaching its maximum value. The peaks that appear on the loading curve, particularly at very low temperatures, are due to specimen vibrations during the transient. From the distance between two consecutive peaks, it turns out that the natural frequency of the Charpy specimen on a three-point configuration, Fig. 3.8, is about 25 kHz. The test also provides the trend of load and resilience versus the displacement of the Charpy V-Notch specimen, Fig. 3.32.

3.7 Dynamic Yield Stress

Charpy V-Notch test can also be used for calculating the yield stress of materials under dynamic conditions. The first attempt was done by Server already in 1974 [31]. Let's start from the static case. For a Charpy V-Notch specimen statically loaded in for-point bending (pure moment) for an ideally plastic material with strain hardening $n = 0$, we can assume that the plastic hinge on the section containing the notch is formed when the external moment M_{GY} (general yield) equals the internal reaction system as shown schematically in Fig. 3.33. It must be considered that if plain strain conditions prevail, the uniaxial yield stress,σ_s, increases by the triaxiality introduced by the notch, see Sect. 4.3 and Fig. 4.8, so that we can write

$$M_{GY} = C\frac{B(W-a)^2}{4}\sigma_s \tag{3.6}$$

where C is a constraint factor that takes into account the triaxiality effect. In case of three-point bending, at general yield it is

$$M_{GY} = \frac{P_{GY} \cdot S}{4} = P_{GY} \cdot W \tag{3.7}$$

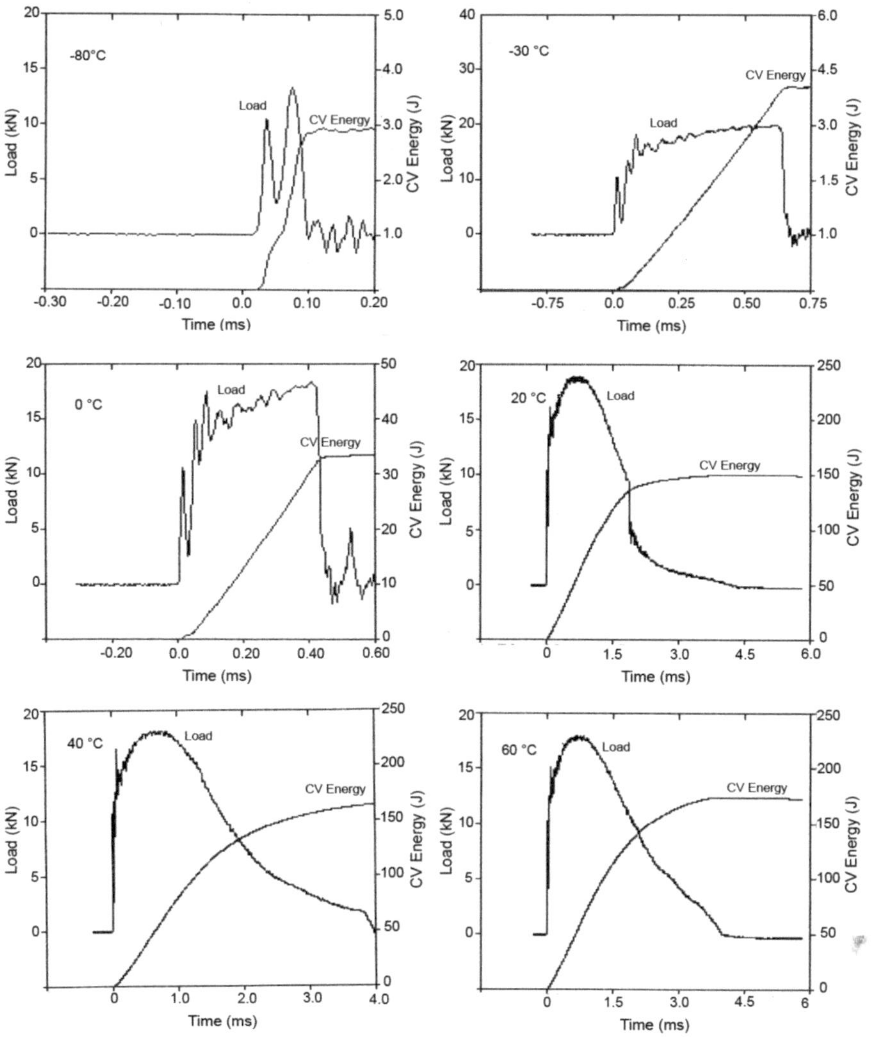

Fig. 3.31 Instrumented Charpy V-Notch test. Load and CV energy versus time for type A508 Cl.3 steel at various temperature. At low temperature where the CV energy is very low, less than 4 J, the fracture is completely brittle and takes less than 0.1 ms while in the upper shelf where CV energy is over 150 J, fracture occurs by ductile tearing and takes about 4 ms to occur, 40 times longer [30]

where P_{GY} is the load that produces the plastic hinge and S is the specimen span length, $S = 4 \cdot W$. Wilshaw [15] has shown that the load P_{GY} is that corresponding to the first departure from linearity of the load displacement diagram, as shown in Fig. 3.34. Equation (3.6) then becomes

$$\sigma_s = \frac{4P_{GY}W}{CB(W-a)^2} \tag{3.8}$$

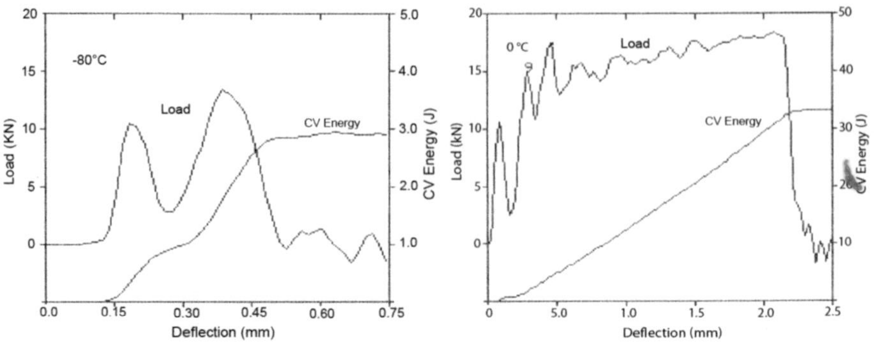

Fig. 3.32 Load and CV energy versud the deflection of the impact point at −80 °C and 0 °C, respectively, for two of the six sample presented inFig. 3.31 [30]

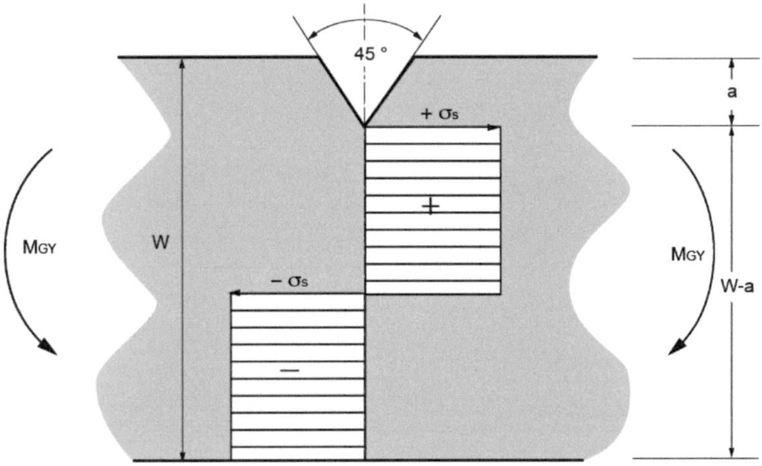

Fig. 3.33 Schematic of internal reactions balancing the external moment M_{GY} at general yield for ideally plastic material

which provides the value of the yield stress σ_s under static conditions. Green [10, 13, 16], Green and Hundy[11] and Ewing [10] provide a full list of C values for different loading configurations and geometries. For a Charpy V–N specimen loaded with a flat striking edge 4 mm wide under three-point condition, $C = 1.279$ and the yield stress given by Eq. (3.8) is

$$\sigma_s = 3.127 \frac{P_{GY} W}{B(W-a)^2} = 4.678 \cdot 10^4 P_{GY} (MPa) \tag{3.9}$$

while for the American indenter 8 mm wide $C = 1.336$ and Eq. (3.8) yields

$$\sigma_s = 2.99 \frac{P_{GY} W}{B(W-a)^2} = 4.678 \cdot 10^4 P_{GY} (MPa) \tag{3.10}$$

Milella et al. [7] has verified Eq. (3.8) and the impact of strain hardening, n, in real materials through FE calculation. The material was A508B Cl.3 steel with a uniaxial yield stress $\sigma_s = 492$ MPa and Young's modulus E = 196.000 MPa for an ideally elastic–plastic condition, $n = 0$. The Charpy V–N specimen was schematized using a 2D model under plain strain conditions. The tup indenter was 4 mm wide. The general yield load was found to be $P_{GY} = 11{,}193.4$ N. Entering this value in Eq. (3.8) gives $C = 1.4219$ against the theoretical $C = 1.279$ which is about 10% higher. However, Eq. (3.6) applies to four-point loading conditions, or pure moment. In three-point conditions for a Charpy V–N specimen having ideal rigid-plastic behavior the plastic hinge is formed when the moment M_{GY} reaches the value

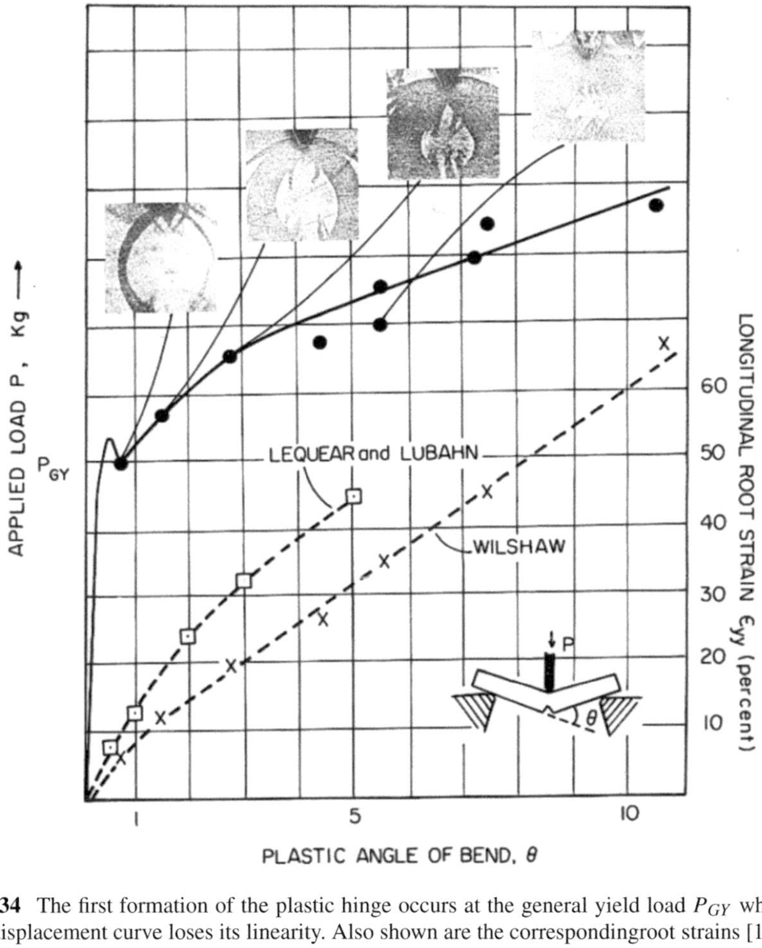

Fig. 3.34 The first formation of the plastic hinge occurs at the general yield load P_{GY} when the load–displacement curve loses its linearity. Also shown are the correspondingroot strains [15]

$$M_{GY} = \frac{B(W-a)^2}{3.2}\sigma_s \tag{3.11}$$

This is 1.25 times higher than that pertaining to pure moment. Therefore, the new relation between σ_s and P_{GY} for three-point bending and ideal rigid-plastic behavior is

$$\sigma_s = \frac{3.2}{1.4219}\frac{P_{GY}W}{B(W-a)^2} = 2.2505 \cdot \frac{P_{GY}W}{B(W-a)^2} \tag{3.12}$$

Three more FE calculations were also run for the same yield stress $\sigma_s = 492$ MPa but with different strain hardening n equal to 0.1, 0.2 and 0.3, respectively [7]. Results in terms of constraint factor C are listed in Table 3.1 and shown in Fig. 3.35.

The variation of the constraint factor with n is due to the fact that the moment of plastic hinge formation is not always the same, as indicated by the load at general yield.

The constraint factor plotted in Fig. 3.35 has a linear trend of equation

$$C = 1.429 + 1.370 \cdot n \tag{3.13}$$

In all cases considered, FE calculations were run using a 2D specimen mesh under plain strain conditions. Therefore, the moment of plastic hinge formation was the same for any section throughout the thickness. However, a 3D FE analysis of the same Charpy-VN specimen simulating the real behavior of all sections has shown that, as expected, the plastic hinge doesn't form at the same time all over the specimen thickness.

This is due to the fact that constraint is higher on the center of the specimen with respect to the external surface where the plastic hinge forms first because the external surface is necessarily in a plain stress condition, as shown in Fig. 3.36 for two different values of the strain hardening $n = 0$ and $n = 0.3$, respectively [7]. The respective plastic constraint factors, C, change with respect to the 2D analysis and are reported in Fig. 3.37 [7]. This raises a question on the actual constraint factor, C, to use in Eq. (3.12).

It is evident that the Charpy V–N test can also provide the value of the yield stress, $\sigma_{s,d}$, under dynamic conditions. Equation (3.8) can be always used by substituting 4 with 3.2 for the three-point bending, Eq. (3.11), using the proper constraint factor, C.

Table 3.1 Time, displacement, force and constraint factor C at plastic hinge formation for different strain hardening, n. ($\sigma_S = 492$ MPa and E = 196.000 MPa)

n	Displacement at GY (mm)	Load at GY (N)	C
0.0	0.200	11,193.4	1.4219
0.1	0.205	12,080.4	1.5346
0.2	0.215	13,017.9	1.6546
0.3	0.220	13,915.6	1.7677

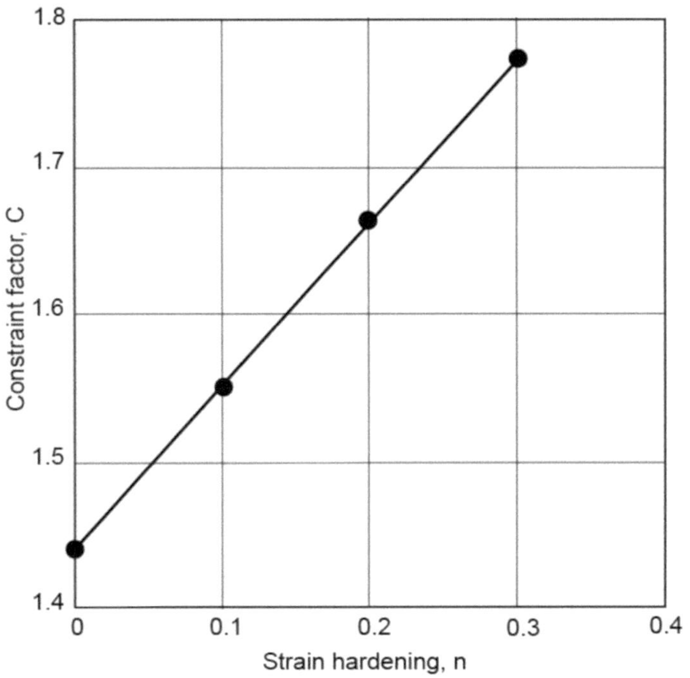

Fig. 3.35 Trend of the constraint factor *C* versus strain hardening, *n*, for a steel with $\sigma_s = 492$ MPa and E = 196,000 MPa [7]

To this purpose, two conditions need to be verified. It must be demonstrated that during the plastic hinge formation the strain rate, $\dot{\varepsilon}$, remains constant and is the same all over the section. This has been checked by 3D FE analysis as shown in Fig. 3.38 [7]. Effectively, the strains measured in 12 points along the plastic hinge during its formation are almost linear with time. The value of the strain rate, $\dot{\varepsilon}$, was found to vary between 200 and 280 s^{-1}. Question remains as to the value of the load at general yielding, P_{GY}, to be used in Eq. (3.12). At very low temperature, for example −80 °C in Fig. 3.31, where fracture occurs without plastic deformation, the peaks that precede fracture, two in Fig. 3.31, refer to free vibrations of the specimen on the first mode. Therefore, the second peak may be overestimating the actual P_{GY} value. Probably, the best estimation of P_{GY} is at half the difference between the last two peaks. In case of ductile behavior, the load at general yielding, P_{GY}, to be used in Eq. (3.12) is that at the formation of the plastic hinge, as shown in Fig. 3.39.

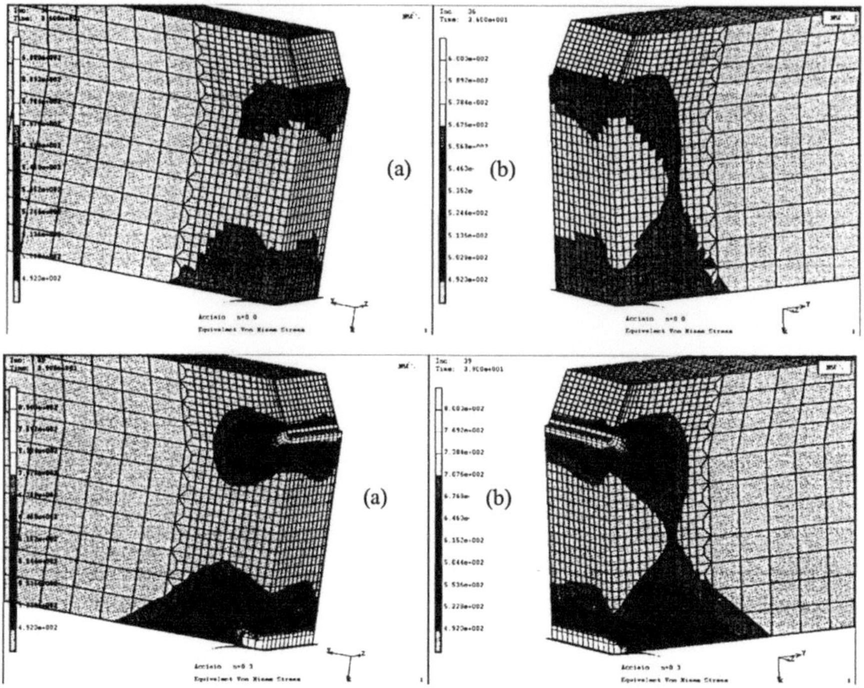

Fig. 3.36 Plastic hinge formation at **a** mid-thickness and **b** on the external surface of a Charpy-VN specimen in a 3D FE analysis for $n = 0$ (upper picture) and $n = 0.3$ (lower picture), respectively, at the same time into the transient. In a real situation, as simulated by the 3D analysis, the external surface is not under a plain strain condition and yields first [7]

3.8 Energy and Displacements in the Charpy Test

With reference to Fig. 3.4, the *energy* E_o available to break the CV specimen is

$$E_0 = \frac{1}{2} I \dot{\theta}_0^2 \tag{3.14}$$

where I is the moment of inertia of the pendulum and $\dot{\theta}_o$ the angular velocity at the moment of impact. The energy available E_o can also be written as a function of the impacting mass m and the height h_o from which the hammer falls

$$E_o = mgh_o \tag{3.15}$$

g being the acceleration of gravity. Expressing the height h_o as a function of the angular position θ and the length l of the hammer, Eq. (3.15) is written as

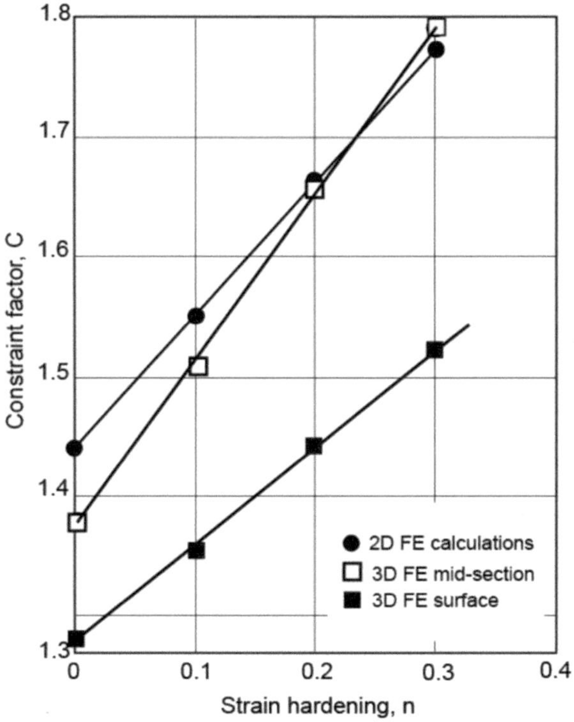

Fig. 3.37 Constraint factor, *C*, versus strain hardening, *n*, obtained under 2D FE calculations in plain strain condition and 3D analysis at mid-section and on the surface [7]

$$E_o = 2mgl \cdot \sin^2 \frac{\theta}{2} \tag{3.16}$$

At a generic instant t after the impact, the angular velocity of the hammer is $\dot{\theta}(t)$ and the energy $\Delta E(t)$ given to the specimen can be expressed as:

$$\Delta E(t) = \frac{1}{2} I\left[\dot{\theta}_0^2 - \dot{\theta}^2(t)\right] \tag{3.17}$$

The momentum conservation is

$$I\left[\dot{\theta}_0 - \dot{\theta}(t)\right] = l \int_o^t P dt \tag{3.18}$$

with P being the load acting on the striking point. Equation (3.9) can be written as

$$\Delta E(t) = \dot{\theta}_o I \cdot \left[\dot{\theta}_o - \dot{\theta}(t)\right] - \frac{I}{2} \cdot \left[\dot{\theta}_o - \dot{\theta}(t)\right]^2 \tag{3.19}$$

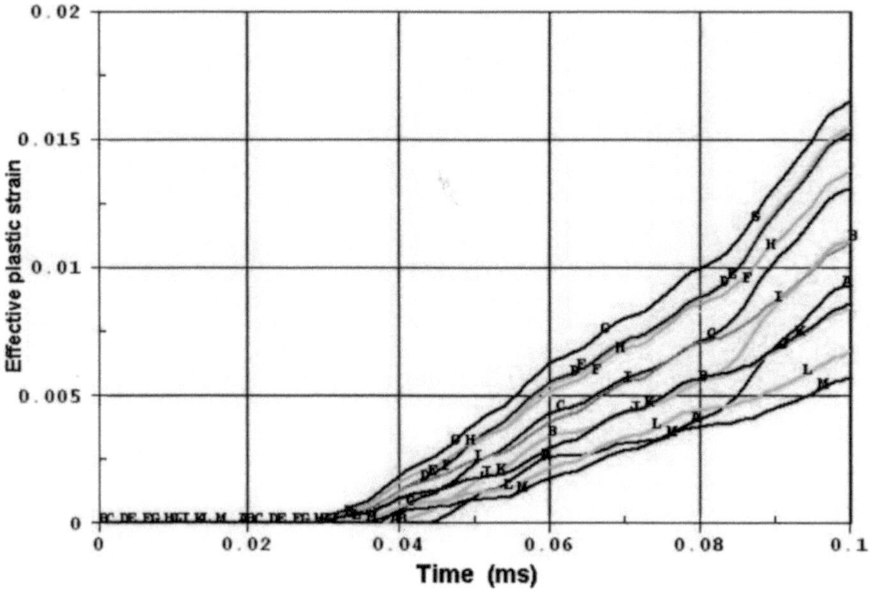

Fig. 3.38 Effective plastic strain in the traction part of the plastic hinge versus time [7]

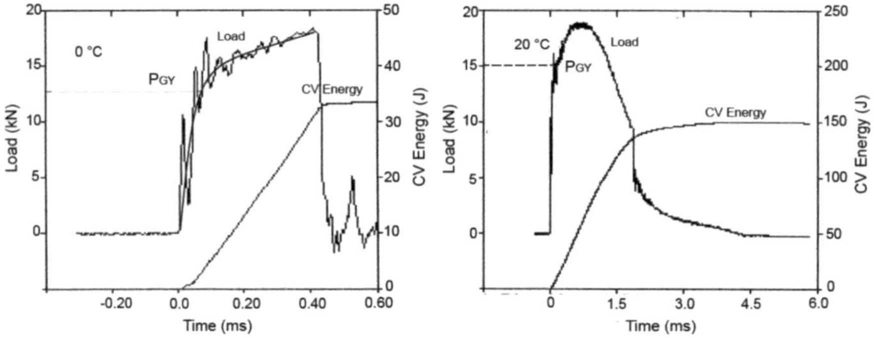

Fig. 3.39 Determination of the General yield load point, P_{GY} at different temperatures at which the plastic hinge forms

$$\Delta E(t) = v_o \int_0^t P dt - \frac{\left(v_0 \int_0^t P dt\right)^2}{4E_0} \tag{3.20}$$

being v_o the speed of the hammer at the time of impact, equal to, Fig. 3.4:

$$v_0 = \sqrt{2gh_o} \tag{3.21}$$

calling E_a the product

$$E_a = v_0 \int_0^t P\mathrm{d}t \tag{3.22}$$

the integral being the area under load time curve, it yields

$$\Delta E(t) = E_a - \frac{E_a^2}{4E_0} \tag{3.23}$$

Equation (3.22) is used for the calculation of the energy absorbed by the specimen at any instant into the transient, note the characteristics of the pendulum. Generally, the length of the pendulum l is equal to 0.9 m and its mass m is equal to 27.24 kg, and the moment of inertia is $I = m \cdot l^2$ or 22 kg·m^2. The displacement on the load line, $\Delta(t)$, at the contact point specimen-striking tup can also be achieved through an integration:

$$\Delta(t) = l \int_o^t \dot{\theta}(t) dt \tag{3.24}$$

Recalling Eq. (3.18)

$$\Delta(t) = l \int_o^t \dot{\theta}_o(t) dt - \frac{l^2}{I} \iint_{oo}^{t\tau} P d\tau dt \tag{3.25}$$

Or

$$\begin{aligned} \Delta(t) &= l \int_o^t \dot{\theta}_o(t) dt - \frac{l^2}{I} \int_o^t \int_o^\tau P d\tau dt \\ &= v_o t - \frac{l^2}{I} \int_o^t \int_o^\tau P d\tau dt \end{aligned} \tag{3.26}$$

The assumption that the striker velocity linearly decreases with time during impact yields

$$\Delta(t) \approx v_0 t - \frac{l^2}{2I} \overline{P} t^2 \tag{3.27}$$

where $\overline{P}$ is the load assumed to remain constant during the impact until time t. The load can be calculated from Eq. (3.18) as

$$l\overline{P}t = I\left[\dot{\theta}_o - \dot{\theta}(t)\right] \tag{3.28}$$

From Eqs. (3.18) and (3.22) it follows that

$$E_a = \frac{Iv_0}{l}\left[\dot{\theta}_o - \dot{\theta}(t)\right] \tag{3.29}$$

Equations (3.20) and (3.21) yield

$$\overline{P} = \frac{E_a}{v_0 t} \tag{3.30}$$

Replacing $\overline{P}$ in Eq. (3.27) as per Eq. (3.30) yields

$$\Delta(t) \approx v_0 t\left[1 - \frac{E_a}{4E_0}\right] \tag{3.31}$$

The last step is correcting Eq. (3.31) to consider the machine displacement $\Delta_M(t)$

$$\Delta(t) \approx v_0 t\left[1 - \frac{E_a}{4E_0}\right] - \Delta_M(t) \tag{3.32}$$

with $\Delta_M(t)$ given by the machine compliance C_M

$$\Delta_M(t) = P(t) \cdot C_M \tag{3.33}$$

3.9 Explosion Crack Starter Test

The inability of the CV specimen to measure the Transition Temperature of a metal or alloy and to establish the safe operating temperature to exclude a brittle fracture event under a stress state lower than the yield strength, reopened the race to identify new test methods. By the end of the 40s of the last century, it was not yet fully understood or widely understood the difference between a sharp notch and a crack in the initiation of a brittle fracture event, though it was already clear at investigators that brittle fracture in the Liberty ships was triggered in some areas, even microscopic, of metallurgical embrittlement such as, as already said, that generated by an arc strike, schematized in Fig. 3.40, see also Fig. 1.3, and then propagated from the microscale to the macroscale of the entire structure.

Once the cleavage instability was developed within relatively few grains in the embrittled region, the adjacent base metal grains were then subjected to a dynamic load transfer due to the very fast propagation of a sharp natural crack. A brittle fracture by cleavage propagates in the metal at the speed of sound, over 600 m/s. The

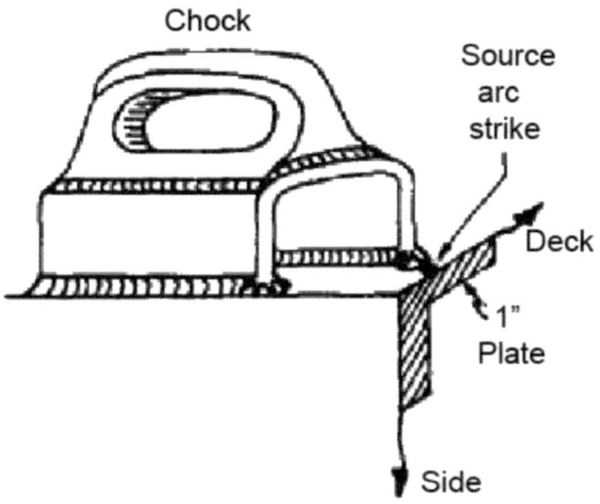

Fig. 3.40 Schematic of the fillet weld of a chock of the T-2 tanker Ponaganset embrittled by an arc strike [32]

basic premise was the recognition of the role of cleavage microstructure processes, which are inherent to metallurgical considerations of the problem, suggesting that cleavage fracture instability is first developed in embrittled metal adjacent to a weld, i.e., in the Heat Affected Zone (HAZ).

Thickness was thought to be another key factor since it would provide the plain strain conditions necessary to prevent plastic slip. Thus, began the study of a new test that reproduced precisely those worst conditions of use of the material and that allowed, finally, the direct and precise measurement of the Transition Temperature NDT. The experiments were conducted on square plates in which an embrittled zone with a possible natural microcrack had been deliberately practiced through a weld *arc strike.*

The plate was simply resting, along the edges, on a circular die and was dynamically charged by the fall of a weight on its center. Subsequently, the test was modified by replacing the falling mass with an offset explosion and the arc strike with a weld made brittle by impurities deliberately introduced. Figure 3.41 schematically shows the arrangement of the test. In these tests the energy to break the specimen was not measured, but only the temperature at which the fracture occurred without plastic deformations, therefore in a brittle fashion. Obviously, the first experiments took place on plates taken from ships that underwent a brittle fracture in service, and it was found that brittle fracture initiated only at temperatures no higher than 15 °C, precisely those of service fracture that led to the criterion of 15 ft-lb.

The same plates then showed a fully ductile behavior, with strong plastic deformations, when the very small, but sharp initial crack was not introduced.

The result was really exceptional because it demonstrated, without any doubt, that brittle fracture was always to be linked to the presence of cracks. Of particular significance is the analysis of the appearance of the fracture in the plates at the various test temperatures, presented in Fig. 3.42. As it can be seen, at a temperature of −8 °C (20 °F) the plate fractures remaining almost completely flat with very limited

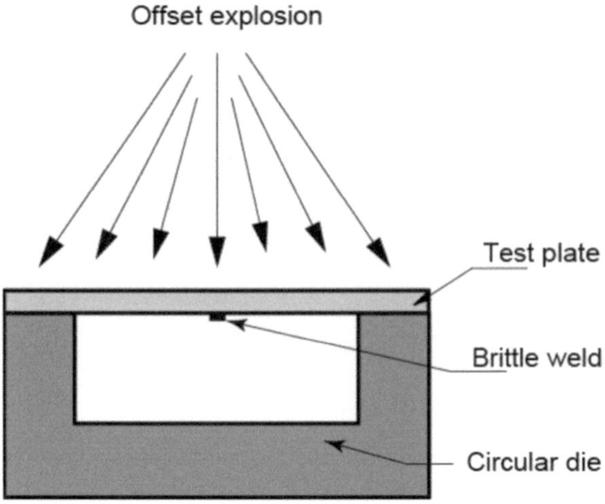

Fig. 3.41 Schematic of the NRL Explosion Crack Starter Test. The test plate was simply resting on a circular die with free ends

permanent plastic deformation, indicating a cleavage brittle fracture that propagated under the prevalent action of elastic stress fields. It is really interesting to note that fracture has developed even within the resting extremities that are not subjected to any traction, but virtually unloaded, if not even in compression.

Already at 5 °C (40 °F) there is a certain, albeit small, central curvature of the test plate, indicative of stresses above the elastic limit that have left some permanent deformation. At 5 °C, therefore, the fracture is no longer totally brittle. It is intuitive how between −8 °C and 5 °C it should be placed the temperature at which the transition from a totally brittle fracture to a predominantly, but not totally brittle fracture, must be placed. The temperature at which this fundamental moment of brittle-to-ductile transition occurs is the Nil Ductility Temperature (NDT) or *Transition Temperature with Zero Ductility*. Returning to Fig. 3.42, it can be seen that between 15 °C and 25 °C (60–80 °F) the fracture is no longer able to propagate along the edges of the plate, while it affects only the central area, quite bulged, therefore plastically deformed. This temperature interval can be thought of as the passage from a partially ductile fracture to a predominantly ductile one supported by a stress field entirely above the yield strength. This phase is referred to as Fracture Transition Elastic or FTE and represents the highest temperature at which a fracture can still propagate through a field of stresses predominantly, but not entirely elastic. Above this temperature, only stresses higher than yield strength can cause the material to fracture. Another significant moment is the transition from 60 °C (140 °F) to 70 °C (160 °F). To produce the enormous swelling of the plate it was necessary to use an explosive charge much higher than the one used before. We are, now, in the presence of stresses well above the yield strength that caused a general plastic flow. The transition from 60 °C to 70 °C is, therefore, referred to as FTP or *Fracture Transition Plastic*. Since the NDT of the material is between −8 °C and 5 °C, the FTP is located at about 65 °C to 78 °C above the *NDT*. To fully understand the importance of the crack on the fracture process, it is interesting to observe the result of an explosive test conducted at −40

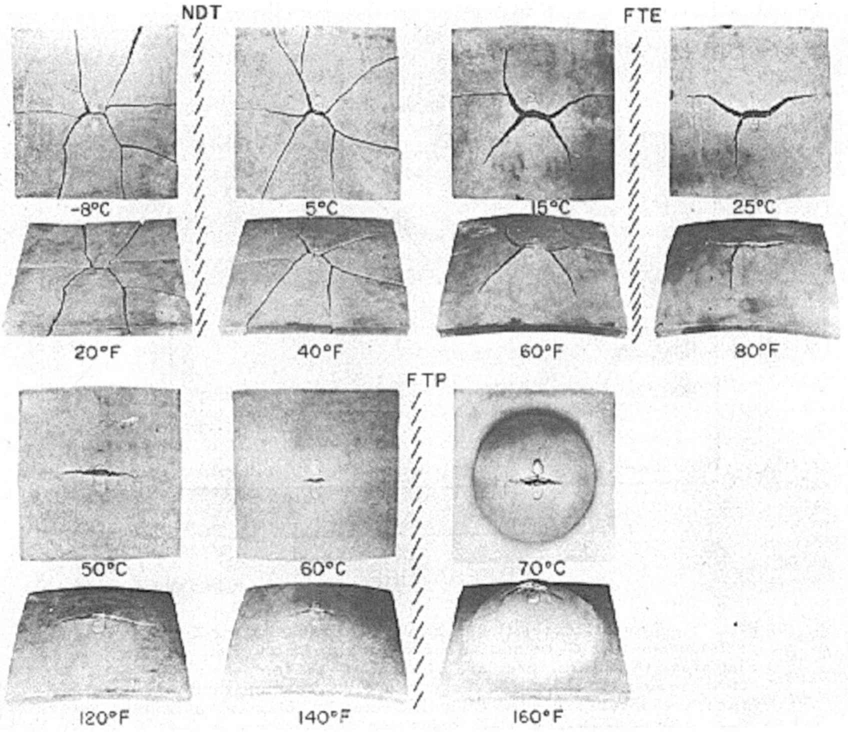

Fig. 3.42 Features of Explosion Crack Starter Tests of ship plates steels [31]

°C on the same steel used in the tests of Fig. 3.42, but without the brittle weld deposit, Fig. 3.43. We can see how despite the very low temperature, much lower than the NDT, the plate did not fracture at all, but even developed a mammillary shape that in tests with brittle weld deposit could only be observed between 60 °C and 70° C.

Interesting enough is the correlation between the appearance of the fracture in the explosive tests and the corresponding Charpy V-Notch energy, as shown in Fig. 3.44 [31]. We can see how the flat appearance of the brittle fracture obtained at −8 °C in Fig. 3.42 is located on the Charpy V–N curve at the very low resilience of 20 ft-lb (27 J). Probably, the best guess of the NDT would have been between 60 °F and 80 °F (15 °C to 27 °C) while the explosive test locates that temperature between −8 °C and 5 °C, right where the Charpy energy curve starts its upswing trend. Finally, the FTE temperature that occurs between 60 °F and 80 °F is positioned much over the lower shelf. Such a correlation is valid, in general, for all alloys with high USE, above 100 J.

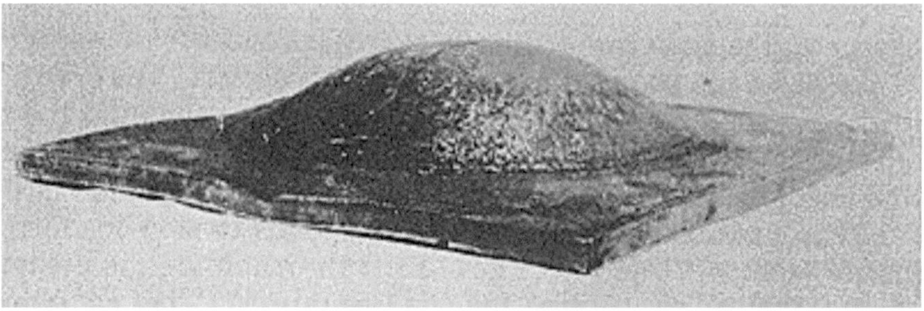

Fig. 3.43 Permanent deformation obtained at –40 °C in an explosive test on a plate of the type used in Fig. 3.42 without any brittle starter. The steel did not fracture. Its behavior is totally ductile developing a large permanent bulge that is completely absent even at much higher temperatures when the brittle weld starter is introduced [31]

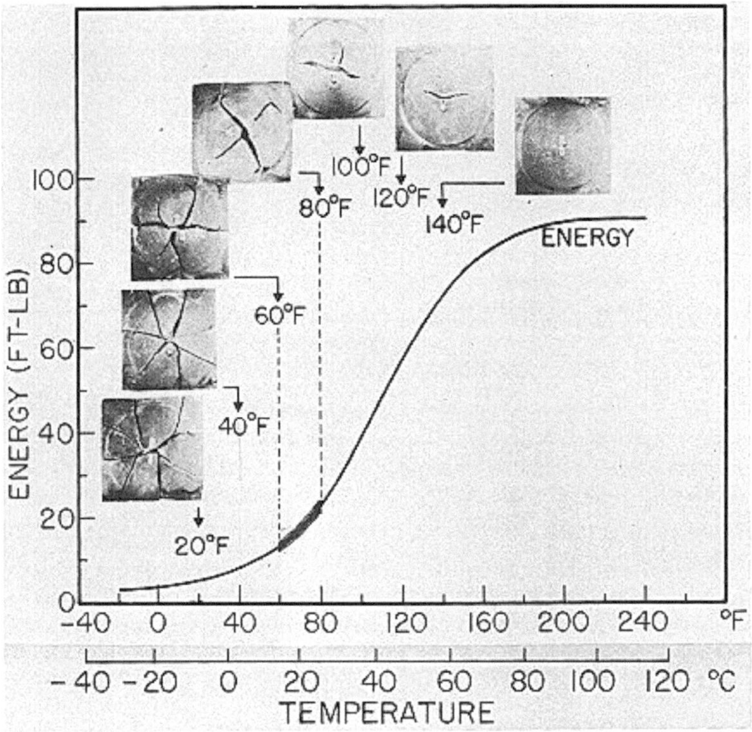

Fig. 3.44 Typical correlation of Explosion Crack Starter Test results with CV energy curve for Liberty ship steels. Flat cleavage fracture (brittle) is obtained at temperatures il the lower shelf region of the curve. Crack arrest starts to occur in the rising portion of the curve over 20 ft-lb and completely ductile fracture is developed over 80 ft-lb CV energy while in USE region the plate does not break [31]

3.10 Drop-Weight Test (DWT) or Pellini Test

The explosion crack starter test was certainly not easy to use, particularly outside a Navy laboratory. Therefore, a new, simpler test was needed. Much more practical will be the test that Pellini introduced in 1953, known as "Drop-Weight Test" or DWT or simply Pellini Test. This test uses a specimen, typically 50 × 130 × 16 mm, loaded by a vertical-dropped-weight, typically 20 kg, from different heights to have a potential energy from 300 to 1,300 J. The usual brittle weld seam is deposited along the specimen in the form of a bead 60 mm long and 12 mm wide and a notch of about 12 × 1.5 mm is introduced in the center of the bead length. The specimen is placed on an anvil, resting on two supports, where it remains free to bend. The specimen and its supports are schematized in Fig. 3.45.

The bending of the specimen, however, is limited by a stop block, at a distance such as to guarantee the absence of plastic deformations; therefore, if the specimen breaks during the test, it will be due to an elastic stress field and the fracture will be totally brittle. A sufficiently low temperature is selected to start the test.

At the impact, the notch in the brittle weld bead, placed in the region of the specimen loaded in traction, will act as a crack starter. The crack will propagate in the base metal, giving fracture and breaking the specimen in two halves. The temperature is then increased by 5 °C proceeding to the next test. The highest temperature at which the specimen no longer breaks is the NDT of the material. Figure 3.46 shows a series of three Pellini specimens broken at the NDT and at temperatures of 10 °C and 17 °C above the NDT of the steel, respectively [31].

While at the NDT the specimen breaks without any apparent permanent deformation (totally brittle fracture) already at 10 °C over that temperature it develops permanent plastic deformations that reveal the ductile nature of the failure. At 17

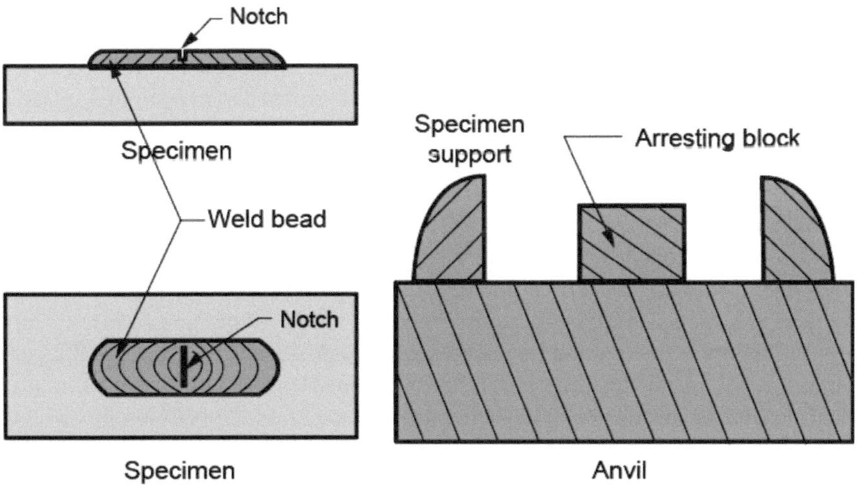

Fig. 3.45 Schematic of Pellini DWT specimen and its anvil

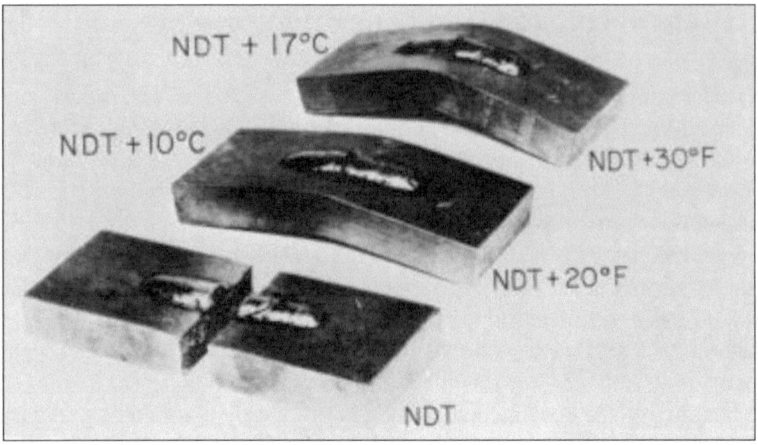

Fig. 3.46 Pellini specimens of ferritic steel broken at three different temperatures. At the NDT (–12 °C) the specimen broke by brittle cleavage without any apparent permanent deformation, while at just 10 °C above the NDT clear permanent deformations occurred without fracture and increased at 17 °C over the NDT. The plastic deformation of the specimen was allowed by the removal of the stop block [31]

°C over the NDT this permanent deformation is still more evident. To allow plastic deformation to develop, the arrest block was removed. The Pellini Test is insensitive to specimen orientation with respect to the rolling or forging direction. The DWT is covered by the ASTM E 208 test method. More recent studies [32–35] have shown, however, that the NDT measured with the Pellini Test can vary considerably, depending on the location of the notch on the weld bead. The reason for such variability must be sought in the way the brittle weld bead is deposited. The entire weld, in fact, is not always laid in a single pass, but in two passes. In the first pass, weld is laid from one side to the center of the specimen. In the second, the weld starts on the opposite side and laid until it joins the first pass. At this point the notch is introduced more or less at the junction point of the two weld seams. There is no real technical reason for doing this, but rather a consolidated tradition. However, the first weld pass overheats the underlying metal causing a surface layer, a few millimeters thick, to reach the transformation temperature of ferrite into austenite, about 727 °C. This shallow zone, or HAZ, of metal thermally treated by the welding process, is rapidly quenched by the colder metal determining the transformation of austenite into hard and brittle martensite that would thus be right under the notch, helping it to fracture. However, the second weld bead causes some martensite of the first pass to undergo a second thermal process and be tempered, as schematized in Fig. 3.47. Tempered martensite has higher toughness than quenched martensite with the result that the NDT measured with the Pellini specimen will depend on the position of the notch relative to this area of second thermal treatment. Khondo et al.[36] found that the elevation of the NDT as a function of the distance d of the notch from the deposition interface can reach even 40 °C.

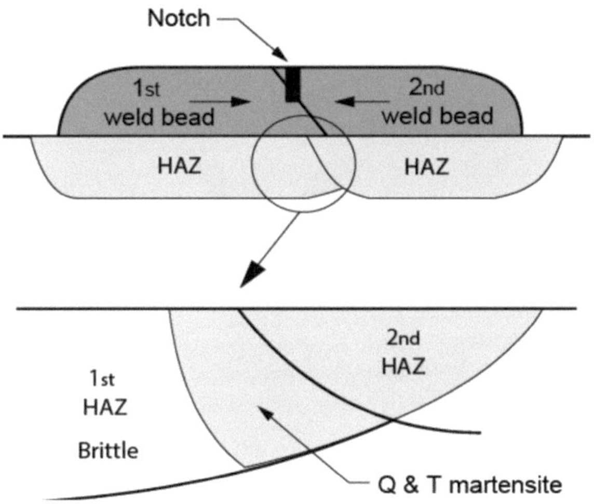

Fig. 3.47 The second weld bead tempers the quenched martensite produced by the first weld deposit, in a shallow surface layer of the specimen,

3.11 The Drop-Weight Tear Test (DWTT)

The Pellini Test was finally yielding the true transition temperature from brittle to ductile behavior of materials, NDT. However, researchers were not completely satisfied with that test since it did not provide any indication of the energy absorbed at fracture. Therefore, new specimens were introduced at NRL and BMI, independently of each other. At the NRL, early in the 60s of last century, Pellini et al. improved the DWT by using a much bigger specimen to which a brittle bar was welded having the role of crack starter. On drop-weight loading, the brittle bar fractures and transmits a running crack to the welded bar to test the material.

For a metal having high toughness, this results in high energy absorption while for a low fracture toughness material a low value is recorded. Figure 3.48 illustrates such a bimetal specimen after fracture, for a steel of high fracture toughness [24]. Contemporarily, at Battelle Memorial Institute, BMI, to-day known as Battelle Research Laboratory, in Columbus, Ohio, in 1962 McClure, Eiber, Hahn et al. developed a new DWTT (Drop Weight Tear Test) specimen to overcome some limitations of the NRL specimen [37–39]. The test specimen was a rectangular bar of 305 mm length and 76 mm width, having a thickness up to at least 19 mm (full thickness).

The specimen has a shallow pressed notch introduced by a sharp indenter with a 45° included angle and a tip radius of 0.0127 mm to 0.0254 mm and is subjected to a three-point bending, as the Charpy V-Notch specimen. The test machine is either a pendulum type or a vertical-dropped-weight type. During the notch indentation process, the specimen material is pressed beyond the yield strength; upon removal of the indenter, a notch is left on the specimen surface that develops a state of residual traction stresses, close to the yield strength. These very high residual traction stresses favor the generation of a crack at the tip of the notch during the impact loading. The DWTT is covered by the ASTM Standard Test Method for Drop-Weight Tear Test

Fig. 3.48 Fracture appearance of the NRL Drop-Weight Tear Test (DWTT) specimen 1-in-thick. The brittle crack initiated in the brittle bar on the bottom, welded to the test specimen, entered the tough steel and converted into a ductile fracture producing shear lips [24]

E 436. NRL investigators evolved the test using two types of notch design as crack starters by either machining a deep notch filled by a brittle TIG weld or using a pressed knife edge sharpening process. The ultimate design of the DWTT specimen was to replace the pressed knife edge notching process with an electron beam weld metallurgically embrittled for crack starter purposes, using a pendulum machine of 5000 ft-lb capacity as loading apparatus. In 1964 the new test was redefined as the Dynamic Tear (DT) test. Figure 3.49 shows the pendulum machine and two specimens 1-in.-thick of high and low toughness steel, respectively [24].

3.12 The Robinson Crack Arrest Test

We have already said, in paragraphs 1.2 and 3.1, how the extensive examination of all ships that underwent brittle fracture during World War II, had led to the surprising discovery that some cracks had run through the hull and eventually stopped, even after some meters run. It was amazing that some of these cracks had relevant sizes which, however, had not caused the catastrophic failure of the vessel. Responsible for arresting the crack, it was said, were the level of stress, not necessarily the same everywhere or, even more probable, the different toughness of the steels used. Starting

Fig. 3.49 Fracture appearance of DT 1-in.-thich specimen of ductile (up) and brittle (down) steel and pendulum machine test. The ductile specimen presents large shear lips indicative of FTP regime while the brittle one has a flat fracture surface [24]

from these observations, in the 40s of the last century Robertson [40] introduced a test in which it was possible to stop a crack propagating through a material having an increasing toughness. The increase in toughness was achieved by maintaining a temperature gradient throughout the test specimen. Further testing led to the development of the specimen of Fig. 3.50a. In this specimen the fracture was triggered at a protruding nub of the test plate, containing a hole with a sharp notch, embrittled by a liquid nitrogen bath. In such conditions of extreme fragility, a simple hammer blow, given to the protruding nub, triggered a crack at the tip of the notch that penetrated the test section. The test section was put in traction by a remote load as to generate a stress state below or equal, at most, to the yield strength of the material. On the other side of the plate, opposite to the notch, a gas torch was used to heat the metal creating a temperature gradient throughout the test plate ranging from −60 °C to 60 °C.

The crack produced by the hammer blow entered the test section facing a continuously growing temperature field eventually stopping when the combination of stress and temperature allowed it. Since the variables are only the stress and the temperature, the test itself provides a stress versus arrest temperature curve, known as CAT, *Crack-Arrest Temperature*. An example of such a curve is shown in Fig. 3.51, that refers to tests carried out in the Esso laboratories on 16″ deep carbon steel sheets for piping [40]. The experimental CAT transition curve shows that at the operating stress of 125 MPa the crack starts growing at 15 °C but arrests at 20 °C. The same steel develops a crack at 5 °C under a dynamic stress of 80 MPa that arrests when the stress

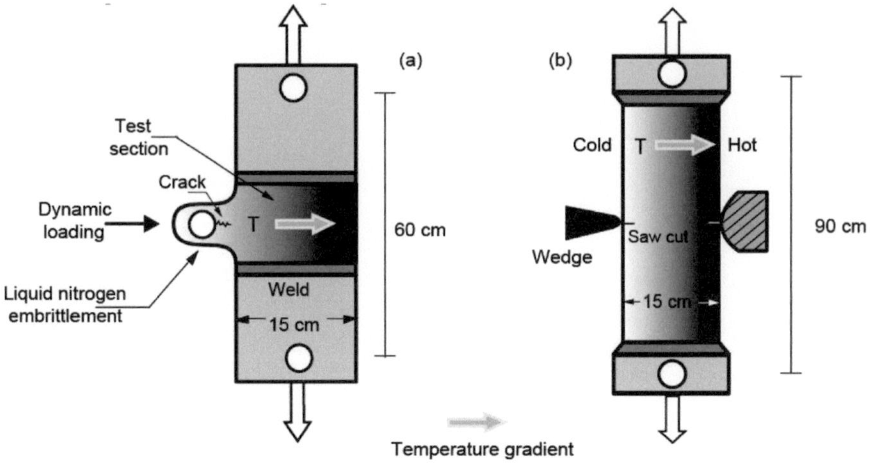

Fig. 3.50 **a** Original Robertson test specimen and **b** modified version in which fracture is initiated by wedge impact over a saw cut. The second saw cut on the opposite face is done to balance the test section

drops down to 60 MPa. Figure 3.52 shows a series of experimental data obtained by Mosborg et al. at the University of Illinois [41] with a similar test specimen, in which fracture very easily started in a brittle steel plate welded to the test plate.

The figure shows that the NDT, measured using DWT specimens, is located at the toe of the CAT curve, as expected. The intersection of the *CAT* curve with the yield strength marks the *FTE* point, Fig. 3.52 which is located at about 2 °C. Figure 3.53 shows a set of arrest data and explosion test FTE data (see Sect. 3.6) for a wide range of steels [33]. The important feature of Fig. 3.53 is that all data are indexed to the relative NDT, easily measured with a Pellini Test. The diagram also includes data shown in Figs. 3.51 and 3.52. This procedure allows the use of data coming from different steels and different types of tests that will be positioned on the temperature scale using the corresponding ΔT.

Three curves are shown in the figure: the upper bound, the average and the CAT curve.All curves bear a fix relationship to the NDT and provide the corresponding stress value. Below the NDT the upper bound curve runs flat at the value of the yield strength.

It may appear in contrast with the idea that below the NDT a brittle fracture can be triggered by such a high stress level, but the curve represents the behavior of specimens free of any crack or sharp defect. Therefore, on the vertical axis the length of the critical crack can also be inserted, as it will be done in the FAD diagram that will be introduced in the next section. In this region, the lower bound CAT curve is based on specimens having very long cracks as those of Robinson Tests. It is important to note that at NDT + 30 °F (NDT + 17 °C) the CAT curve is crossing the $1/2\sigma_y$ yield strength and the NDT + 60 °F (NDT + 33 °C) provides a similar index for the yield strength. This actually means that any structure operating at 33 °C over

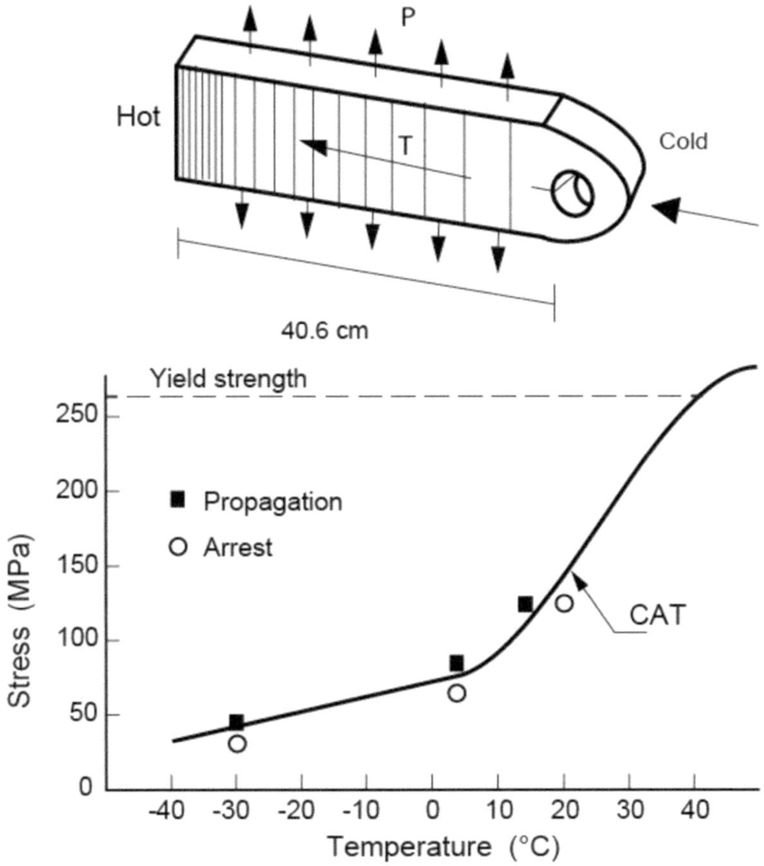

Fig. 3.51 Typical crack arrest temperature (CAT) transition curve obtained in a Robinson test for a piping steel of 255 MPa yield strength (mod. from [40])

the NDT of the material will never fail catastrophically even if it works at a stress close to the yield strength and contains a very long crack.

3.13 The Fracture Analysis Diagram, FAD

The tests and studies presented in the previous paragraphs had the purpose to arrive at a final diagram that would allow to relate with each other the three elements that are at the base of any brittle fracture: the crack length, the stress and the temperature, as shown in Fig. 1.2. The long-lasting study conducted at the Naval Research Laboratory, in Washington, D.C., in particular, led to the closing of the circle with the formulation, in 1960, of the FAD, or *Fracture Analysis Diagram*. This diagram

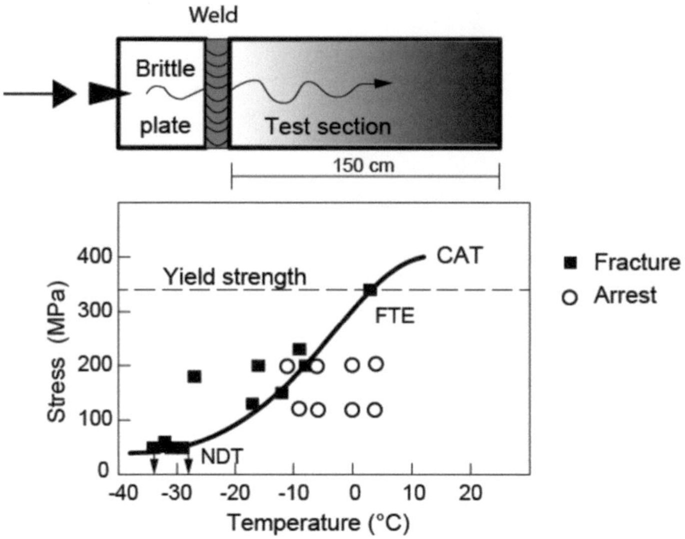

Fig. 3.52 Mosborg wide plate test and relative CAT transition curve for a pipe steel of 331 MPa yield strength. The NDT, determined using DWT specimens, is located at the toe region of the CAT curve. At the operating stress of 125 MPa the temperature must be kept over –15 °C to prevent catastrophic fracture of the pipeline

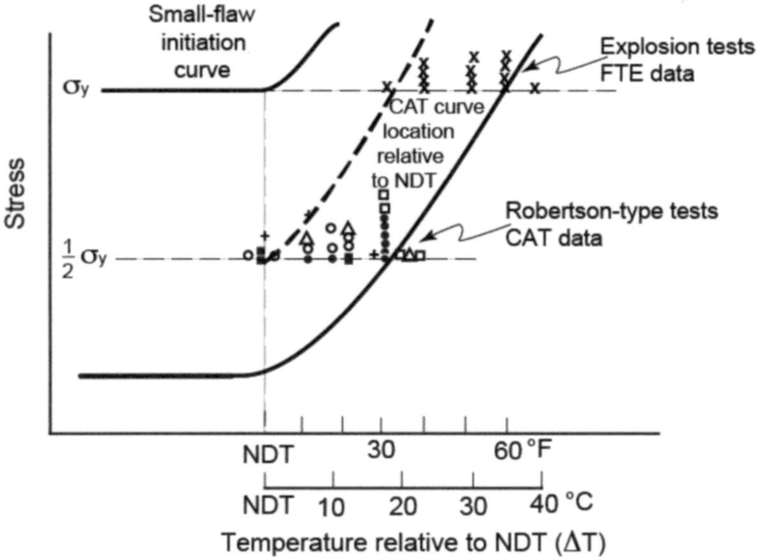

Fig. 3.53 Summary of tests for a wide variety of steels which indicate that the temperature range location of the CAT curve may be established by temperature increment, ΔT, reference to the NDT of the steels. The points represent CAT or FTE determination obtained with different test types [33]

was able, eventually, to realize the seam between the three parameters: crack size -temperature-stress for design purposes, though in a semi-empirical form,. The CAT curve indicated not only the lower bound of the permissible crack size-temperature-stress combination, but also the value of the FTE where it crossed the yield line, Fig. 3.45. That point was really representative of FTE, as required by logic and confirmed by the solidity of explosive experiments on plates (see § 3.6). Moreover, that point was located at 33 °C above the NDT. The same experiments also provided another essential point: the FTP at about 80 °C above the NDT. It was clear that the lower and upper bound CAT curves had to converge on this single point. In other words, at FTP, fracture occurs when the ultimate stress is reached, with or without a crack because at general yielding a crack is simply seen as a lack of material and does not contribute to the stress elevation.

In this way, a sort of closed domain was created, as shown in Fig. 3.54, outside of which any crack size-temperature-stress combination had to be safe, while the allowable relationship among those three factors had to be found inside the diagram. The problem of the explicit introduction of the crack length parameter in the FAD was solved always empirically through hydraulic burst tests expressly carried out with pressure vessels of small, medium and large size, pressurized by air or water or with the interpretation of fractures occurred in service [4, 25, 32, 33, 42, 43]. It was, then, possible to link the size of the defect to the temperature and the applied stress that produced fracture. The result of all this efforts is shown in Fig. 3.54 for five cases of failures in carbon steel cylinders under pressure.

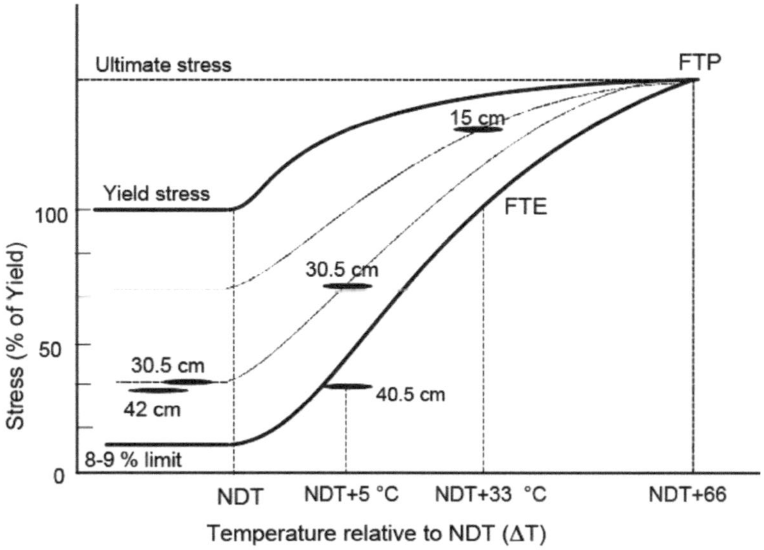

Fig. 3.54 Experimental results, obtained from carbon steel pressure vessels tests in which flaws of various sizes were introduced, positioned on the FAD curve (data coming from [33])

Consider, for example, the 40.5 cm (16″) crack case. The large crack stopped at about 5 °C above the NDT under an applied stress of 91 MPa, equal to about 0.33·σ_s, simply because it was to the right of the *CAT* curve where fracture cannot occur, causing a water flow without catastrophic rupture. A crack of almost the same size, 42 cm, would cause fracture at the same stress at a temperature below the NDT of the material. Equally interesting is the case of the 15 cm (6″) crack, grown by fatigue in a nozzle just over 3″ of maximum thickness, which had propagated without stopping in the cylindrical belt of a pressure vessel 2″ (5 cm) thick and had caused it to burst at a temperature equal to the FTE. At this temperature, we know, it takes a stress higher than the yield strength to give fracture. The vessel had been pressurized to an average stress level equal to the yield σ_s, and, therefore, probably not able to fracture. However, the local stress concentration, due to the geometric discontinuity represented by the nozzle, had caused the unstable propagation. Once again it must be highlighted how indexing the temperature scale to the NDT, allowed the representation of the five cases, relating to three different materials, on a single diagram. The analysis of all the tests available led, finally, to the construction of the FAD diagram of Fig. 3.55. Note that the stress level for plastic fracture is not indexed because of the lack of analytical procedures for its definition. Ultimate stress signifies only that maximum load and strain tolerance is attained at FTP for the specific flaw size cited. It obviously does not indicate the equivalent of the tensile test specimen maximum load or maximum stress limit. According to the FAD diagram of Fig. 3.55 a crack 12 inch long would produce a brittle fracture below the NDT if subjected to a stress equal to half the yield stress or higher, while a 4 inch crack would fail at the yield stress.

The NRL experiments and studies were later confirmed by those of the PVRC (Pressure Vessel Research Council) conducted on carbon steel pressure vessels, type A 201 B and A 302 B [37, 38, 44, 45]. Also, in these cases the same crack size-temperature-stress relationship was obtained, on a temperature scale indexed to the

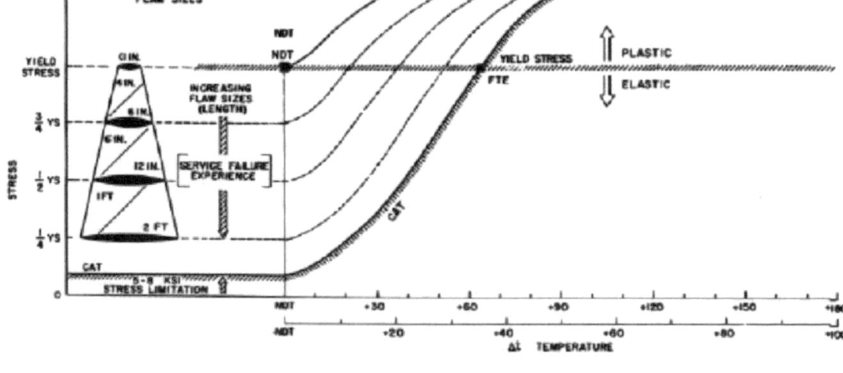

Fig. 3.55 Original drawing of FAD [33]. YS indicates the yield stress

NDT. They also showed that the FTE and FTP temperatures were about 33 °C and 80 °C above the NDT, respectively. Some important observations need to be made about the FAD. It must be emphasized that the area of the diagram above the yield strength, is more indicative than really exact. This is because, over to the yield strength, it is particularly difficult to determine the real value of the stress that caused the rupture and, above all, of the size of the critical crack, since in the plastic field a crack would grow subcritically, that is, in a stable way before triggering fracture (see Sect. 10.1 and Fig. 10.5) and it is not said that this subcritical tearing should be the same in all material and structure, that is, independent of the geometry of the system. Equally indicative is the diagram below the NDT. Here the curves become parallel, but only by convention, leaving us to believe that going down in temperature there will no longer be any worsening of the situation, with the same crack length. In reality under the NDT there will continue to be a further reduction in the fracture resistance capacity of the material. Another important aspect is that the stress at which the FAD diagram must be used is the local one that affects the area of the crack and not the nominal one. This means, for example, that in welds that have not undergone heat treatment for the distension of residual stresses, it must reasonably be assumed that there is a residual local stress equal to the yield strength affecting the crack. For different situations, such as corner cracks, complex geometries or non-uniform loads, the FAD diagram is only indicative, although quite valid conceptually. It will take the Stress Intensity Factor of fracture mechanics, K_I, to assess the true relationship among those three factors crack size-temperature-stress. Another problem that is not considered by the FAD diagram of Fig. 3.55 is the thickness effect. We know that the increase of thickness favors the occurrence of a plain strain condition and therefore of brittle fracture (see Sect. 9.10). The shift of FTE towards higher temperature values was well indicated by DT tests, in particular. Figure 3.56 is an example of such a shift obtained in DT tests at Oak Ridge National Laboratory on low alloyed steel, type A 533 Gr B Cl 1, used in the construction of nuclear reactor pressure vessels [46]. Specimen size ranged from 5/8 inch to 12 in. This effect was introduced in the FAD and led to the so-called expanded version of the FAD, illustrated in Fig. 3.57, in which FTE and FTP were placed at twice the temperature shift of the normal *FAD*.

The transition between small and large thickness is more likely to occur at about 7.58 cm (3″) for a ferritic steel. It can be seen, in Fig. 3.57, how the *FAD* relative to small cracks does not vary as the thickness increases. This is because a small section size is already large with respect to a very small flaw. The last element to be considered in this discussion about FAD is the issue of highly brittle metals having very low level of initial and propagating resistance. Fracture resistance to initiation and crack propagation may be assessed in terms of increased plastic work energy absorbed per unit increment of initial extension or R-curve. Early measurements of the R-curve were obtained in 1970 using DT specimen of increasing path length.

Figure 3.58 illustrates two test series relative to two steels having ductile and brittle behavior, respectively. The ductile steel features large shear lips and a steep R-curve, while the very brittle steel is characterized by flat surface appearance and a flat R-curve. In the brittle domain, the rate of release of elastic deformation energy, associated with an act of fracture progression, is sufficient to feed the energy demand

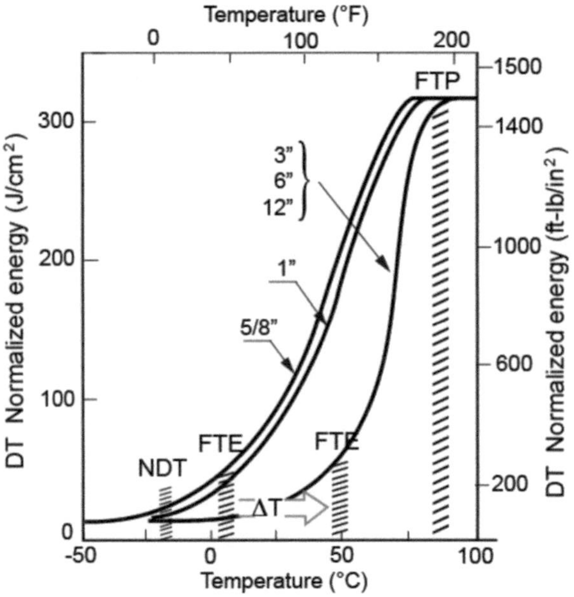

Fig. 3.56 Results obtained at ORNL on A 533 GR B Cl 1 carbon steel. DT energy is normalized to the fracture surface [46]

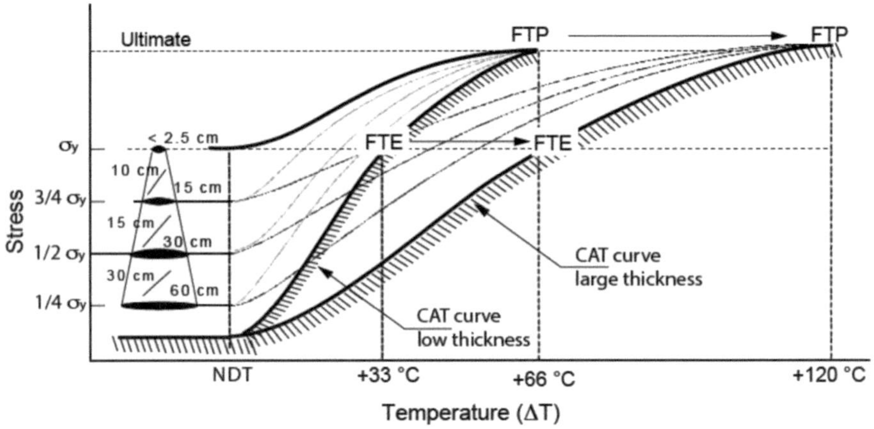

Fig. 3.57 FAD diagram expanded to section sizes ranging from about 15 cm (6″) to over 25 cm (10″). It involves a shift of the FTE point by a temperature increment of about 33 °C to 40 °C. The NDT is not affected by the section size [46]

for the next increase. The fracture is self-sustaining and the R-curve is flat. These steels do not have an NDT transition temperature, they are brittle at any temperature. It should be borne in mind that the concave aspect of the R-curve in the ductile field, which seems to diverge with fracture propagation, is the result of an overestimation of the absorbed energy compared to a similar specimen, but with the same initial crack plus the only measured advancement, as described in Sect. 11.4. The true resistance curve turns its concavity downwards. The final FAD that takes into consideration also low energy fracture brittle steels is then that of Fig. 3.59 [33]. By the time this new Fracture Analysis Diagram or FAD was proposed in 1970 the George Irwin Stress Intensity Factor, K_I, had already appeared and dominated fracture mechanics studies with the strength of its solid theoretical basis. The empirical FAD had no sense any longer. It had made its time and was abandoned. However, it must be underlined how the FAD was used to design the first generation of nuclear power reactor vessels by the turn of the 50s beginning of 60s of last century. The startup of nuclear power plants was done by circulating the primary water using the recirculation pumps. By doing so, the mechanical energy of the pumps was converted into heat up to the moment when water temperature reached the 66 °C over the transition temperature NDT of the vessel steel. Only at that point designers were sure to rule out any possibility of brittle fracture on the base of FAD and the control rods of the reactor were slowly extracted form the core allowing the nuclear reaction to initiate.

3.13.1 Charpy V-Notch Impact Energy of Materials Other Than Carbon Steel

The introduction of the Charpy V-Notch specimen and the Pellini test that finally was able to define the transition temperature, NDT, from brittle-to-ductile fracture was the result of efforts made to understand the brittle fracture of the Liberty ships during World War 2. However, those ships were made of carbon steel that has a BCC lattice structure. Question arises as to the Charpy V specimens behavior with temperature for materials that have a FCC structure like austenitic stainless steels or aluminum and nickel alloys and whether the same transition from brittle-to-ductile fracture can be observed in this class of materials. It must be recalled that FCC materials have not got a thermally activated component of the yield stress which remains insensitive to temperature, as will be shown in Sect. 5.2.2.

Figure 3.60 shows the Charpy V impact energy for two low carbon austenitic stainless steels type 316L and 304L, respectively [47]. As can be seen the energy absorbed is exceptionally high at all test temperatures, even at −200 °C where still maintains over 200 Joule. There is no sign of brittle-to-ductile transition because these steels have got no NDT unless either mechanical or thermal treatment made possible the transformation of some austenite into ferrite which do has an NDT.

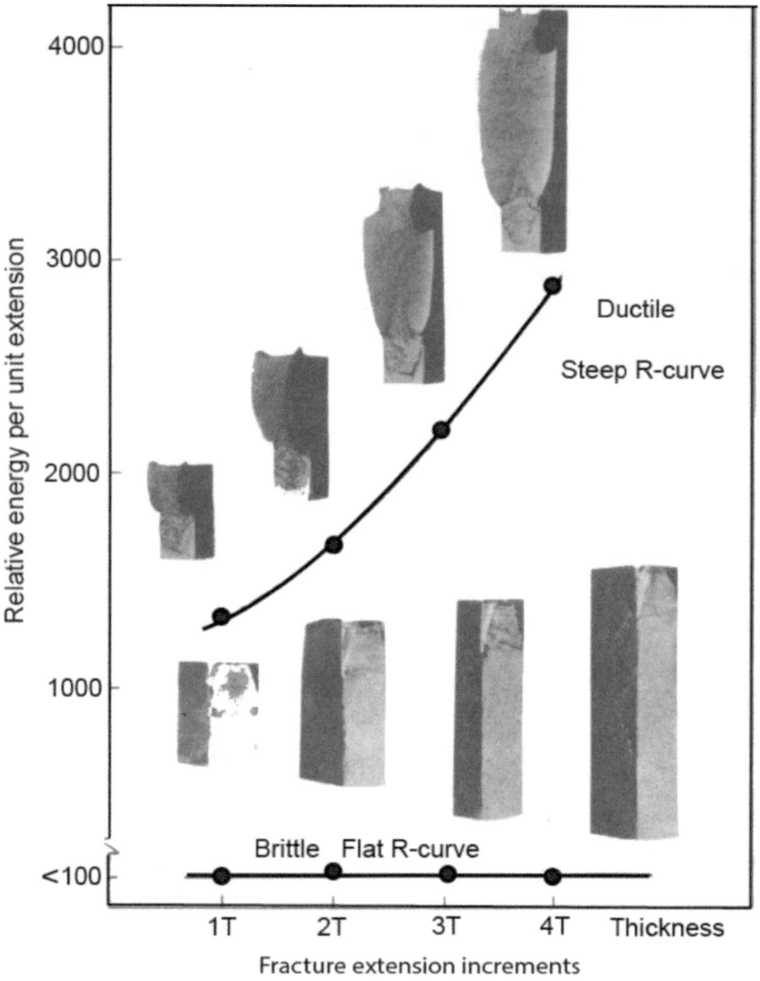

Fig. 3.58 R-curves obtained with DT test specimens 1-in thick of two steels having ductile and brittle behavior. The ductile behavior results in exponential increase of plastic work energy with increasing path length. For the steel with brittle behavior the R-curve is flat [43]

The same behavior is observed in aluminum alloys, as shown in Fig. 3.61 [48]. Note the difference in the energy absorbed with respect to the austenitic steels of Fig. 3.60 because the much lower density of the aluminum alloys. So far, no case of cleavage brittle fracture ha ever be found in FCC materials.

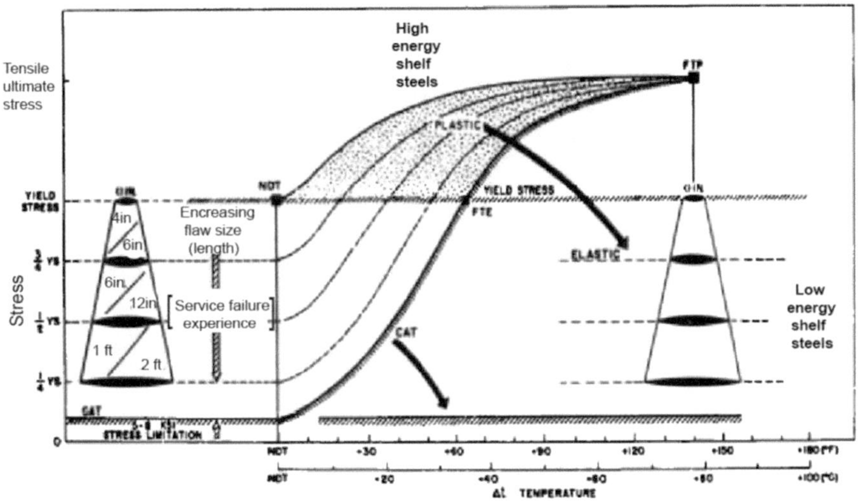

Fig. 3.59 Schematic illustration of the consequence of a drastic reduction in the shelt energy to fracture. The plastic fracture regime of the FAD disappears. The low shelf steels are insensitive to temperature change [33]

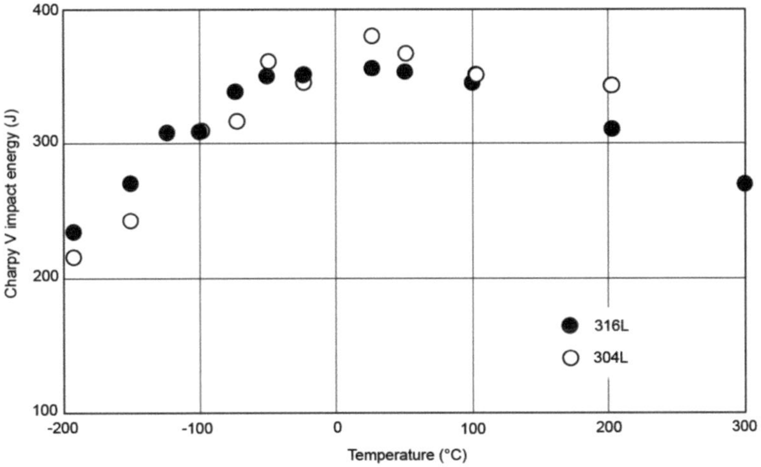

Fig. 3.60 Charpy V impact energy for austenitic stainless steel type 316L and 304L, respectively. The absorbed energy is exceptionally high at all test temperatures, even at −200 °C. There is no sign of brittle-to-ductile transition (data from [47])

Concluding Remarks

The Charpy V-Notch specimen has been the first in the history of fracture mechanics to address the issue of brittle fracture by coupling a notch with a dynamic loading.

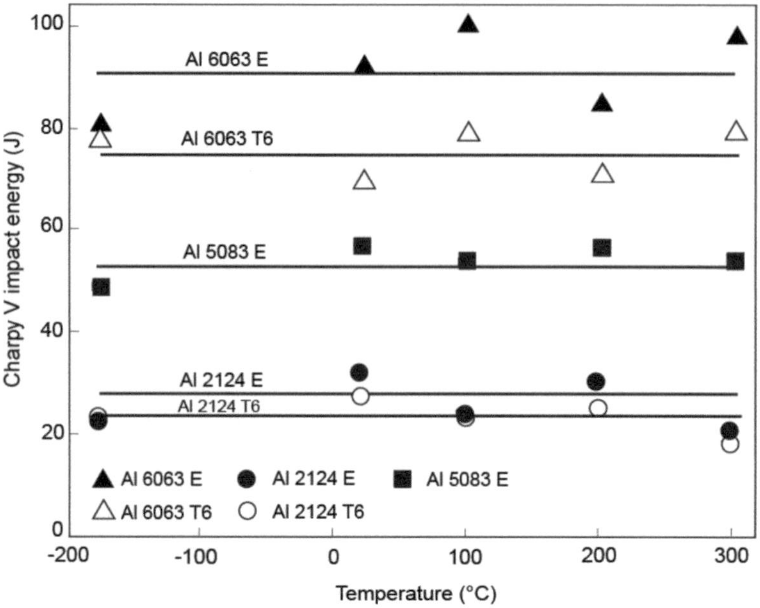

Fig. 3.61 Charpy V impact energy for some aluminum alloys (data from [48]). None of the alloys shows a NDT since all curve are practically flat at all temperatures

However, the notch, though sharp, cannot be considered a real crack. This is why the Charpy V-Notch test failed to measure the Ductile-to-Brittle Transition Temperature or Nihil Ductility Temperature, NDT, of steels and cannot be considered a real fracture mechanic specimen. The Charpy V test, tough still frequent, cannot be used without some other validation test such as, for example, the Pellini test. The Fracture Analysis Diagram, FAD, represented the first empirical tentative to relate crack size, temperature and stress to avoid brittle fracture. Its empirical basis had to yield to the solidity of the Irwin stress intensity factor, K_I, that by 1970 had already reached its maturity. A fundamental result of the Charpy test and the FAD was that ferritic steels having a transition temperature NDT cannot undergo a brittle fracture event when used at 33 °C over the NDT or, in thick pieces over 15–20 cm, at 66 °C over that temperature. This actually means that it has no sense at all to try to measure the toughness, K_{Ic}, of a ferritic steel at those temperatures over the NDT because Linear Elastic Fracture mechanics (LEFM) would no longer be valid and so does the K_I concept. However, during the 70s and even the 80s of the last century, in particular, tremendous efforts and resources were spent trying to obtain toughness K_{Ic} data over the transition temperature NDT that were completely useless.

References

1. White, A.E., Clark, C.L.: Bibliography of Impact Testing, Department of Engineering Research, University of Michigan (1925)
2. LeChatelier, A.: On the Fragility after Immersion in a Cold Fluid, French Testing Commission, Vol. 3 (1892)
3. Russell, S.B.: Experiments with a new machine for testing materials by impact. ASCE Trans. **826**, 237–272 (1898)
4. Pellini, W.S.: Evolution of Engineering Principles for Fracture-Safe Design of Steel Structures, U.S. N*R*L Report 6957, Washington, D.C., September 23, 1969
5. Milella, P.P., Barracaracciolo, M., Bigagli, F.: Caratterizzazione Meccanica di una Virola di A 508 Cl3. Programma di Ricerca ENEA-BATTELLE sul danno da radiazione (1984)
6. ASME Boiler & Pressure Vessel Code, Sec. III, Nuclear Power Plant Comp., Division 1, Sub. Sec. NB 2331, Test *R*equirements & Acceptance Standards and Appendix G, Ed. (1977)
7. Milella, P.P., Lombardi, P.: FE Analysis under EOARD Research Contract N. F61775-00-WE029, Advance Modeling for Ductile Behavior of Materials under High Strain Rates and Temperature Using Continuous Damage Mechanics and Solid State Equation, Authors N. Bonora and P.P. Milella, London (2001)
8. Knott, J.F., Cottrell, A.H.: Notch Brittleness in Mild Steel, Journal of Iron and Steel Institute, pp. 249–260 (1963)
9. Hahn, G.T., Mincer, P.N., Rosenfield, A.R.: The Fe-3Si Steel Etching Technique for Local Strain Measurement, Experimental Mechanics, pp. 248–253 (1971)
10. Green, A.P.: The plastic yielding of notched bars due to bending, quart. J. Mech. Appl. Mathem. **VI**, Pt. 2, 223–239 (1953)
11. Green, A.P., Hundy, B.B.: Initial plastic yielding in notch bend tests. J. Mech. Phys. Solids **4**, 128–144 (1956)
12. Hill, R.: The Mathematical Theory of Plasticity. Oxford University Press (1950)
13. Green, A.P.: The plastic yielding of shallow notched bars due to bending. J. Mech. Phys. Solids **4**, 259–268 (1956)
14. Lianis, G., Ford, H.: Plastic yielding of single notched bars due to bending. J. Mech. Phys. Solids **7**, 1–21 (1958)
15. Wilshow, T.R.: The Deformation and Fracture of Mild Steel Charpy Specimens, Stanford University SU-DMS Report No. 66–6 (1966)
16. Ewing, D.J.F.: Calculation of the bending of rigid/plastic notched bars. J. Mech. Phys. Solids **16**, 205–213 (1968)
17. Milella, P.P.: Plain Strain and Plain Stress Conditions in Charpy V-Notch Specimen. NRC Workshop on Thermal Shock, Washington, D.C. (1990)
18. Lundin, C.D., Zhou, G., Khan, K.K.: Metallurgical Characterization of the HAZ in A516–70 and Evaluation of Fracture Toughness, WRC Bulletin 403 (1995)
19. Metals Handbook: Properties and Selection, American Society for Metals, Vol. 1, p. 227, 8th Edition (1975)
20. Metals Handbook: Properties and Selection, American Society for Metals, Vol. 1, p. 232, 8th Edition (1975)
21. Guzzoni, G.: Gli Acciai Comuni e Speciali, Vol. 4, p. 1335, Hoepli, 6° Ediz. (1966)
22. Milella, P.P., Bigagli, F., Charpy-V Testing of Specimens of the TMI-2 Vessel Lower Head ENEA/DISP Contribution to the TMI Project, OECD/NEA Report TMI V(92)I01 (1992)
23. Metal Handbook, Properties and Selection, American Society for Metals, Vol. 1, p.231,232, 8th Edirion (1975)
24. Beghini, M., Carmignani, C., Vitale, E.: Prove di Thermal Shock su Lamiere di A 533, Contratto ENEA-Università di Pisa RL 358 (88) (1988)
25. Pellini, W.S., Goode, R.J., Puzak, P.P., Lange, E.A., Huber, R.W., Review of Concepts and Status of Procedures for Fracture-Safe Design of Complex Welded Structures Involving Metals of Low to Ultra-High Strength Levels. U.S. NRL Report 6300, p. 58–61, Washington, D.C. (1965)

26. DeSisto, T.S.: Low Temperature Charpy, True Stress Strain and Notched Properties of Base and Weld Deposits of AISI Types 301,310,316 and 347 Stainless steels, 65th Meeting of the Am. Metal. Society, June 24–29, Trans., pp. 756–766 (1962)
27. Lucon, E., Lazzati, A.: Caratterizzazione di Tubazioni in Acciaio Ferritico ed Inossidabile, Contratto ricerca ENEA/DISP, Rapporto CISE 7229 (1992)
28. Reynolds, M.B.: Failure Behavior in ASTM A 106 B Pipes Containing Axial Through-Wall Flaws, General Electric Report GEAP-5620, p. 6, Pleasanton, CA (1968)
29. Borghi, L., Lucon, E.: Prove di Resilienza su Provini CV Ricostruiti. Rapporto CISE 4880 (1989)
30. Milella, P.P., Maricchiolo, C., Pini, A., Marchetti, M.: Analisi Numerica e Sperimentale dei Modi di Rottura di un Provino Charpy V, Gruppo Italiano Frattura IGF8, pp. 205–213, Genova (1992)
31. Server, W.L., Ireland, D.R., Wullaert, R.A.: Strength and Toughness Evaluation from an instrumented Impact test, Dynatup TR 74–29 R, Effects Technology, Santa Barbara, California (1074)
32. Pellini, W.S., Puzak, P.P.: Fracture analysis diagram procedures for the fracture-safe engineering design of steel structures. NRL Report **5926**, 28 (1963)
33. Pellini, W.S.: Evolution of Engineering Principles for Fracture Safe Design of Steel Structures, NRL Report 6957, p. 15 (1969)
34. Onodera, S., Tsukada, H., Suzuki, K., Iwadate, T., Tanaka, Y.: A Study on Pellini Test: Reproducibility and Welding Procedure, 6 MPA Seminar, Stuttgart (1980)
35. Tsukada, H., Suzuki, K., Iwadate, T., Tanaka, Y.: A Study on Drop Weight Test Using A 508 Cl. 2 Steel, The Japan Steel Works, *R*(MS) 81–60 (1981)
36. Khondo, M., Kusida, S., Abe, T., Suzuki, A.: Effect of Crack Starter Bead Welding Conditions of Drop Weight Test Specimen on Nil-Ductility Transition Temperature, 101 IslJ Meeting (1981)
37. McClure, G.M., Eiber, R.J., Hahn, G.T., Boulger, F.W., Masubuchi, K.: Catalogue No. 40/PR. American Gas Association, Inc., New York (1962)
38. McClure, G.M., Duffy, A.R., Eiber, R.J.: Trans. ASME, Ser. B. J. Eng. Ind. 87, pp. 265–278 (1965)
39. Lange, E.A., Puzak, P.P., Cooley, L.A.: Standard method for the 5/8 inch Dynamic tear test, NRL Report 7159 (1970)
40. Robertson, T.S.: Propagation of brittle fracture in steels. J. Iron Steel Inst., 361–374 (1953)
41. Mosborg, R.J.: An investigation of welded crack arresters. Weld. J. **39**(1), 409–483 (1960)
42. Lange, E.A., Klier, E.P.: A Study of fracture development and materials properties in PVRC vessels 1 and 2. Welding J. **41**(2), 53–61 (1962)
43. Pellini, W.S.: Integration of analytical procedures for fracture-safe design of metal structures, NRL Report 7251 (1971)
44. Kooistra, L.F., Lange, E.A., Pickett, A.G.: Full-Size, pressure vessel testing and its applications to design, transactions of ASME. J. Eng. Power **86**, Series A, No. 4 (1964)
45. Lange, E.A., Pickett, A.G., Wiley, R.D.: Failure Analysis for PVRC Vessel No. 5. Part I Study of Materials, Properties and Fracture Development, WRC Bull. No. 98 (1964)
46. Whitman, G.D.: Heavy Section Steel Technology Program, NUREG/CR-4489, ORNL-6259 (1986)
47. Ajeesh, S.V., Kumeresan, V.: Analysis of ductile-to-brittle transition temperature of stainless steels. Proceedings of 4th International Conference on Energy Efficient Technologies for Sustainability–ICEETS'18, TamilNadu, India, April (2018)
48. Ozden, S., Ekici, R., Nair, F.: Investigation of impact behaviour of aluminium based SiC particle reinforced metal–matrix composites. Elsevier, Compos. Part A **38**, 484–494 (2007)

Chapter 4
Plain Strain Fracture

4.1 Introduction

At the end of World War II, an intense activity of research and study started in the United States, in particular, aimed at understanding and, therefore, preventing brittle fracture. Some of this study done at NRL and BMI were presented in Chap. 3 that led to the formulation of the Failure Analysis Diagram (FAD). Parallel to that effort, work began in earnest to develop a plan of action to resolve the question of the size effect on the brittle fracture problem. It became clear that the intrinsic metallurgical ductility of materials could be reduced and even completely eliminated by a full plain strain condition or triaxial stress state. It was also understood that in very ductile steels a plain strain condition could be generated only by a large thickness, while in high and very high strength materials even a small thickness, of the order of 1.5–2 mm, could be enough to trigger a brittle fracture. It was understood that this variation of the intrinsic ductility or metallurgical ductility had to be related to the size of the plastic zone ahead of the crack that would swell enormously with the decrease of the strength of materials. The brittle fracture was, therefore, the result of a sort of *competition* between two antagonistic effects: the local embrittlement by triaxiality of the area just ahead of the crack where fracture would start and the metallurgical ductility of the bulk material. As the studies of the effects of triaxiality on brittle fracture went on, two completely different schools of thought developed. The first argued that the generation of a state of plain strain or triaxial stresses ahead of the crack was necessary and also sufficient condition to trigger brittle fracture, regardless of the ductility of the bulk material and, therefore, of the size of the plastic enclave formed at the crack tip. It was just a question of adequate thickness to confine the apical plastic zone and create triaxiality. It was, therefore, only a matter of scale: small plastic zone-small thickness needed to confine it, large plastic enclave-large thickness of confinement, but with the same effects. This school of thought introduced a worrying consequence since it argued that brittle fracture would always be possible at any temperature in any steel or alloy. By the end of the 60 s of last century, research at NRL and BMI had

P. P. Milella, *Fundamentals of Fracture Mechanics*,
https://doi.org/10.1007/978-3-032-23081-2_4

virtually rule out this possibility in ferritic steels and alloys having a BCC structure that showed a transition temperature NDT, but this final outcome was the result of much too empirical procedures to be widely recognized. On the contrary, the second school of thought asserted that it was not only a question of size and scale, but also of intrinsic metallurgical ductility and, therefore, of the absolute size of the plastic zone and that the plain strain state, although necessary, however was not sufficient to sustain a cleavage brittle fracture when this plastic enclave would grow too much. A plain strain state could also be obtained with an adequate thickness to constrain a large plastic enclave, but its effects would not be the same. To give a definitive answer to the thorny question and shed full light on the thickness effect, it started in the United States, already in the 60 s, a large research activity within a much larger research program called Heavy Section Steel Technology Program (HSST) sponsored by the American Nuclear Regulatory Commission (NRC), following the recommendation of the PVRC (Pressure Vessel Research Council), that involved numerous research centers, laboratories and universities and, in particular, the Oak Ridge National Laboratories (ORNL), in Tennessee, which acted as coordinator, the Westinghouse Nuclear Corporation that built nuclear pressure water reactors also embarked on the Navy submarines, with pressure vessels over 30 cm thick, the Battelle Memorial Institute (BMI) in Columbus, Ohio, and the Electric Power Research Institute (EPRI), in California that was also concerned with the safety of nuclear power reactors for energy production. In the nuclear field, the demand for safety was essential and dominant. The program would last well over four decades.

The introduction of the Stress Intensity Factor, K_I, by George Irwin already in the 60 s gave a sudden, incredible impulse to Fracture Mechanics applications giving a definite answer to old, unresolved questions.

At Westinghouse Research Laboratories in Pittsburgh, Pennsylvania, Wessel, Wilson, Clark et al., were using Compact or CT specimens of variable thickness from one inch (25.4 mm) to 12″ thick, Figure 4.1 [1]. The CT specimen, Fig. 9. 9, was developed by Manjoine in 1964 [2] to simulate biaxial loading conditions existing ahead of a crack tip. DT type specimens (see Sect. 3.8) were also used, and so did Pennell and Whitman et al. at Oak Ridge, as shown in Figure 4.2 [3]. Much later, in the 80 s, at the MPA in Stuttgart, Germany, a huge load cell was set up capable of competing with that of the National Institute of Standards and Technology, NIST, in Gaithersburg, Maryland, to break 600 mm thick (24″) CT type specimens in order to measure the fracture toughness of medium-strength steels at temperatures much over the ambient temperature up to 100 °C that required large thickness to confine the crack tip plastic zone. At those temperature this plastic zone in a ductile steel reaches dimensions of the order of centimeters, it's like a ping pong ball. Figure 4.3 shows one of those very large specimens, 24″ thick, broken open at MPA in Germany [4].

The lesson learned was that it is always possible to fully constrain a large plastic zone in the transverse direction, z, through the proper thickness. However, a large plastic zone changes a sharp crack into a blunt notch that reduces triaxiality, $\sigma_x << \sigma_y$ Fig. 4.18g, and, therefore, the probability of brittle fracture, see Sect. 6.4. It became clear that to have a real cleavage brittle fracture, the plastic zone ahead of the crack

Fig. 4.1 CT specimens of A 533 GR B Cl 1 steel, ranging from 1″ to 12″, used at Westinghouse laboratory by Wessel et al. to study the size effect [1]

tip had to remain very small, below a millimeter, so small that it could even be neglected, from a practical point of view. This is the so-called criterion of SSY or Small-Scale Yielding. The result of all those researches, eventually, granted reason to the second school of thought that gave the metallurgical ductility of materials the full recognition in preventing brittle fracture: even if it were possible to measure fracture toughness values, K_{Ic}, equal to 200 MPa$\sqrt{}$m or higher, valid according to the ASTM E 399-72 standard because sufficiently large, those values, however, would never be representative of a real cleavage brittle fracture and, revealed, once and for all, that the titanic effort to break specimens that large was useless. The following paragraphs are dedicated precisely to the study of the thickness effect on the fracture resistance of materials and explain why the triaxial stress state though necessary to cause brittle fracture may not be sufficient when the metallurgical ductility is so high as to wash out the sharpness of the original real crack.

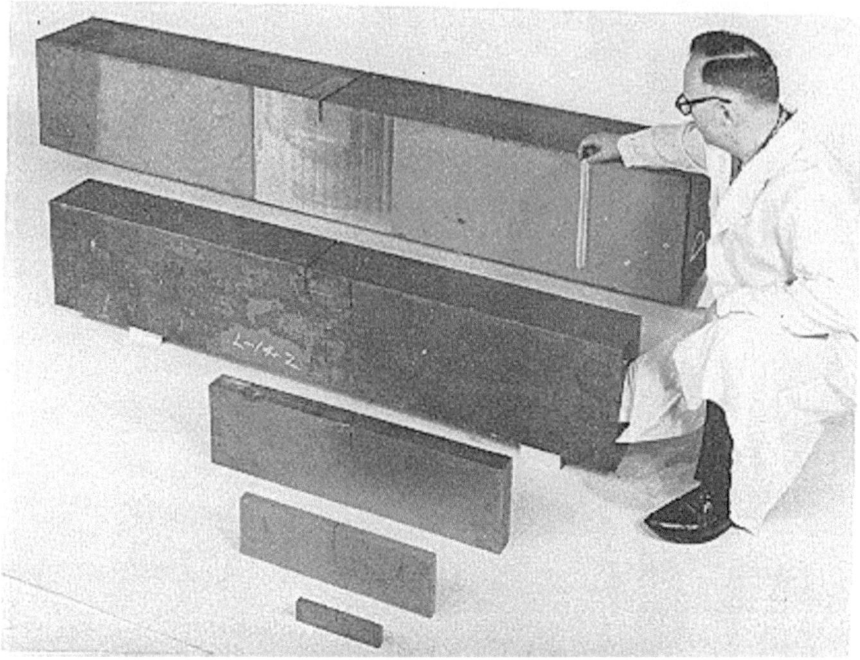

Fig. 4.2 Dynamic tear (DT) specimens ranging from 5/8″ to 12″ used at ORNL [3]

4.2 Effect of Stress State on Ductility and Fracture

The mechanical properties of materials, such as yield and ultimate stress and, even more, the ductility that are so important in the fracture process, can vary depending on the working conditions of the material. After a short possible settling phase, typical of ferritic steels (see Sect. 1.4), the recovery of strength through strain hardening appears in which the material follows a trend of the type expressed by Eq. (1.19), with K and n constants whose value depends on the alloy considered. For simplicity of treatment, without losing the conceptual value, it will be useful to represent the characteristic stress strain curve of the material in the plastic domain with a line AB, as shown inf Fig. 4.4.

Fig. 4.3 Carbon steel specimen, type 20 MnMoNi 55, broken at the MPA in Stuttgart (Germany) in the upper shelf temperature range [4]. The high thickness prevented lateral contraction

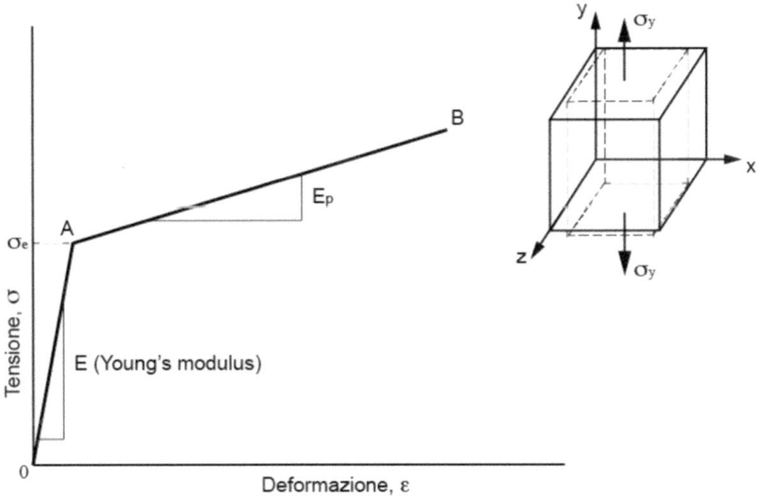

Fig. 4.4 Schematization of the stress–strain characteristic of the material and reference system for the generic stress element

Beyond the yield strength the material still follows a law of the type (1.19) in which $n = 1$ and K is a sort of plastic modulus, E_p, having a value lower than the Young's modulus, $E_p < < E$. Imagine, then, to subject the material to a simple tensile test in which the generic element of material is subjected to a traction stress in the y direction, σ_y, as shown in Fig. 4.4. The state of tension is clearly uniaxial. Under the action of the stress σ_y the element will tend to stretch in the y direction and to contract, by Poisson effect, in the x and z directions. The dilation is

$$\varepsilon_y = \frac{\sigma_y}{E} \tag{4.1}$$

and the lateral contractions are

$$\varepsilon_x = \varepsilon_z = -\nu\varepsilon_y \tag{4.2}$$

being v the Poisson coefficient. Imagine applying to the element of Fig. 4.4 also a stress in the x direction, σ_x, as to prevent the contraction ε_x. For an elastic and isotropic material, deformations are linear functions of the main stresses σ_1, σ_2, σ_3 and are expressed as:

$$\begin{aligned} \varepsilon_1 &= \frac{1}{E} \cdot [\sigma_1 - \nu(\sigma_2 + \sigma_3)] \\ \varepsilon_2 &= \frac{1}{E} \cdot [\sigma_2 - \nu(\sigma_1 + \sigma_3)] \\ \varepsilon_2 &= \frac{1}{E} \cdot [\sigma_3 - \nu(\sigma_1 + \sigma_2)] \end{aligned} \tag{4.3}$$

where E represents the Young's modulus of the material. Equation (4.3) are known as the *Hooke's generalized law*. In the case considered the main stresses σ_1, σ_2 and σ_3 are precisely those acting in the x, y and z directions of the reference system. Since it is $\sigma_3 = \sigma_z = 0$. Equation (4.3) can be written as

$$\begin{aligned} \varepsilon_x &= \frac{1}{E} \cdot \left[\sigma_x - \nu\sigma_y\right] = 0 \\ \varepsilon_y &= \frac{1}{E} \cdot \left[\sigma_y - \nu\sigma_x\right] \neq 0 \\ \varepsilon_z &= \frac{\nu}{E} \cdot \left(\sigma_x + \sigma_y\right) \neq 0 \end{aligned} \tag{4.4}$$

From the first of Eq. (4.4) we obtain

$$\sigma_x = \nu\sigma_y \tag{4.5}$$

and substituting that value in the second of Eq. (4.4) it yields

$$\varepsilon_y = \frac{\sigma_y}{E} . \left(1 - \nu^2\right) \tag{4.6}$$

Defining $E' = E/(1-\nu^2)$, Eq. (4.6) becomes

$$\varepsilon_y = \frac{\sigma_y}{E'} \tag{4.6a}$$

where E' is the Young's modulus for plain strain conditions. Comparing Eqs. (4.6) with (4.1), obtained for the elastic uniaxial case, we realize that moving from the uniaxial to the biaxial stress state a reduction of the strain ε_y occurred by a factor $(1 - \nu^2)$. For most metallic materials the Poisson coefficient is about 0.3. The comparison between (4.6) and (4.1) shows that

$$\varepsilon_{y,biax} = 0.9 \cdot \varepsilon_{y,uniax} \tag{4.7}$$

Much larger will be, however, the reduction of the strain ε_y entering the field of plastic deformations, beyond the elastic limit. Suppose, in fact, to increase the traction σ_y entering the plastic domain. With reference to Fig. 4.4, if we neglect the contribution of the elastic component of strain, ε_e, to the total strain this will be represented by the plastic component, $\varepsilon_{y,p}$, which is

$$\varepsilon_{y,p} = \frac{\left(\sigma_y - \sigma_s\right)}{E_p} \tag{4.8}$$

where σ_s is the yield strength (elastic limit) and E_p the plastic Young's modulus. We have used, as usual, σ_s (slip) to indicate yield stress in order to avoid confusion with the stress in the y direction σ_y. If a stress σ_x is added as to prevent the contraction in the plastic domain, $\varepsilon_{x,p}$, and supposing that Eq. (4.4) are still valid substituting the Young's modulus E with E_p and σ_x, σ_y, σ_z with differences $(\sigma_x - \sigma_s)$, $(\sigma_y - \sigma_s)$ and $(\sigma_z - \sigma_s)$ then, always in the hypotheses of biaxiality with $\varepsilon_x = 0$, the (4.4) will yield

$$\varepsilon_{y,p} = \left(1 - \nu^2\right) . \frac{\left(\sigma_y - \sigma_s\right)}{E_p} \tag{4.9}$$

Equation (4.9) is analogous to Eq. (4.6) valid for the elastic case. However, in the plastic regime, as has already stated in Sect. 1.4 by Eq. (1.5), strains will occur without volume change, ΔV_p. since $\Delta V_p/V$ can be expressed through the three plastic strain components

$$\frac{\Delta V_p}{V} = \varepsilon_{x,p} + \varepsilon_{y,p} + \varepsilon_{z,p} = 0 \tag{4.10}$$

it can be deduced that, since strains ε_x and ε_z are related to ε_y by Eq. (4.2), in the plastic field the Poisson's coefficient ν must be worth exactly 0.5 at variance with 0.3 of the elastic domain. Numerical comparison between Eqs. (4.9) and (4.8) yields

$$\varepsilon_{y,p,biax} = 0.75 \cdot \varepsilon_{y,p,biax} \tag{4.11}$$

Therefore, the simple appearance of σ_x that prevented the contraction $\varepsilon_x = -\nu\varepsilon_y$, bringing the generic element of material in a biaxial state, reduced the strain ε_y by 10% in the elastic field and by 25% in the plastic domain. Assume, now, to continue to increase the stress σ_x up to the value of the σ_y, which will produce a strain $\varepsilon_x > 0$. Let's always start with the elastic case. With this new condition Eq. (4.4) provides

$$\begin{aligned}
\varepsilon_x &= \varepsilon_y \\
\varepsilon_y &= (1-\nu) \cdot \frac{\sigma_y}{E} \\
\varepsilon_z &= -2\nu \cdot \frac{\sigma_y}{E}
\end{aligned} \tag{4.12}$$

and in the plastic domain ($\nu = 0{,}5$):

$$\varepsilon_{y,p,biax} = 0.5 \cdot \varepsilon_{y,p,uniax} \tag{4.13}$$

The growth of σ_x has further reduced the deformation ε_y bringing it from 75%, Eq. (4.11), to 4.50% of the uniaxial one. Let's push the analysis further by imagining moving from the biaxial stress state just described to a triaxial one, $\sigma_z \neq 0$ that prevents the contraction in the z “””’direction, $\varepsilon_z = 0$, leaving the system in a plain strain condition. On the element of material will now act three stresses σ_x, σ_y and σ_z different from zero. Equation (4.4) yield

$$\begin{aligned}
\varepsilon_x &= \frac{1}{E} \cdot \left[\sigma_x - \nu\left(\sigma_y + \sigma_z\right)\right] \neq 0 \\
\varepsilon_y &= \frac{1}{E} \cdot \left[\sigma_y - \nu(\sigma_x + \sigma_z)\right] \neq 0. \\
\varepsilon_z &= \frac{\nu}{E} \cdot \left[\sigma_z - \nu\left(\sigma_x + \sigma_y\right)\right] = 0
\end{aligned} \tag{4.14}$$

As first case, we consider the one in which the σ_x is only 50% of the σ_y: $\sigma_x = 0.5$ σ_y. From the third of (4.14) we obtain

$$\sigma_z = 1.5 \cdot \nu \cdot \sigma_y \tag{4.15}$$

And for $\nu = 0{,}3$:

$$\sigma_z = 0.45 \cdot \sigma_y \tag{4.16}$$

and substituting in the second of Eq. (4.14) yields

$$\varepsilon_y \sim 0.7 \cdot \varepsilon_y / E = 0.7 \cdot \varepsilon_{y,uniax} \tag{4.17}$$

and in the plastic domain, being $v = 0{,}5$

$$\varepsilon_{y,p,triax} = 0.375 \cdot \varepsilon_{y,p,uniax} \tag{4.18}$$

Therefore, the onset of a triaxial stress state that prevents the contraction ε_z, reduces to about a third the deformation $\varepsilon_{y,p}$. It will be noted that the transition from the simple uniaxial case to a biaxial one and then to the final triaxial stress state, leads to the progressive reduction of the deformations of the individual elements of the material and therefore, ultimately, of the ductility of the material. Let's continue the analysis by further increasing the stress σ_x until it reaches the value of σ_y. From the third of Eq. (4.14), with $\sigma_x = \sigma_y$, it yields

$$\sigma_z = 2 \cdot \nu \cdot \sigma_y \tag{4.19}$$

and substituting in the second of Eq. (4.14)

$$\varepsilon_y = \frac{\sigma_y}{E} \cdot \left(1 - \nu - 2\nu^2\right) \tag{4.20}$$

which yields

$$\varepsilon_{y,triax} = 0.52 \cdot \varepsilon_{y,uniax} \tag{4.21}$$

while in the plastic domain with $\mathrm{v} = 0.5$

$$\varepsilon_{y,p,triax} = \frac{\left(\sigma_y - \sigma_s\right)}{E_p} \cdot \left(1 - \nu - 2\nu^2\right) \tag{4.22}$$

therefore

$$\varepsilon_{y,p,triax} = 0 \tag{4.23}$$

The fundamental result expressed by Eq. (4.23) is that a generic element brought beyond its elastic limit in a state of complete triaxial stress, will no longer deform even if the material is extremely ductile and capable of undergoing ultimate strains of the order of 50–60%, in a uniaxial tensile test. It has become brittle as a consequence of the triaxial stress state, also called of plane deformation or strain because the deformations occur only on one direction, $\varepsilon_x \neq 0$, $\varepsilon_y \neq 0$ and $\varepsilon_z = 0$. The area under the stress–strain curve, Eq. (1.1), which provides a measure of the ductility of the material, will be restricted to the elastic component only and, therefore, will be worth practically zero.

$$A = \int_o^{\varepsilon_e} \sigma \mathrm{d}\varepsilon = \frac{1}{2}\sigma_S \varepsilon_e \approx 0 \tag{4.24}$$

4.3 Triaxiality Factor

It is convenient to represent the results obtained in the previous paragraph in a stress–strain diagram, σ-ε. In the first case examined, the generic element is subjected to simple traction. The stress–strain diagram is therefore that of Fig. 4.5, indicated by line OSA. Moving on to the second case, the biaxial one with $\sigma_x < \sigma_y$, and using the subscript *1,2* and *3* instead of *x, y* and *z* to indicate primary stresses, the maximum plastic deformation $\varepsilon_{y,2,}$, Eq. (4.11), will be reduced by 25%; the corresponding path on the σ-ε diagram is the OSB line of Fig. 4.5. Continuing to grow the σ_1 until it reaches the value of σ_2, third case, the corresponding maximum deformation $\varepsilon_{y.3}$ will reach just 50% of the uniaxial one, Eq. (4.13). The new characteristic σ-ε is therefore the OSC line. It should be borne in mind that, as indicated by lines OSB and OSC of Fig. 4.5, in both cases the value of the yield strength σ_s at which the material begins to deform plastically did not change. To verify this, we can use the Tresca criterion of the maximum shear tension τ_{max}. It assumes that plastic slip takes place when the maximum shear stress reaches the yield strength τ_s measured in a uniaxial tensile, which is evidently equal to σ_s / 2. This theory, first published by Tresca in 1865 [5], has proved valid in predicting the beginning of slip in a multiaxial stress state in ductile materials. We chose the Tresca criterion simply because it's easier to use than von Mises' theory. The shear stresses σ_{ij} are obtained from the Mohr circles relative to the stress state under examination, biaxial or triaxial, as schematized in Fig. 4.6. With σ_z being the lesser of the three main stresses, we have that the maximum shear stress, $\tau_{y,z}$, is

$$\tau_{y,z} = \tau_{\max} = \frac{\sigma_y - \sigma_z}{2} \tag{4.25}$$

According to the Tresca theory, slip in the material starts when the τ_{max} given by Eq. (4.25) reaches the uniaxial shear strength, $\tau_s = \sigma_s/2$, being σ_s the uniaxial yield stress. Therefore

$$\tau_{\max} = \frac{\sigma_y - \sigma_z}{2} = \frac{\sigma_s}{2} \tag{4.26}$$

or

$$\sigma_y - \sigma_z = \sigma_s \tag{4.27}$$

It is evident that in a biaxial state of stress, in which a component is necessarily zero, namely the σ_Z, the maximum shear stress must coincide with $\sigma_y/2$, as shown in Fig. 4.6 and slip will take place when the main stress σ_y has reached the yield stress σ_s, exactly as in the case of uniaxial stress state. At variance, in the triaxial or plain strain condition with $\varepsilon_z = 0$ and $\sigma_x = 0.5\text{y}\cdot\sigma$, the third case treated, where $\sigma_z = 0.45\cdot\sigma_y$, as indicated by (4.16), the maximum shear stress is

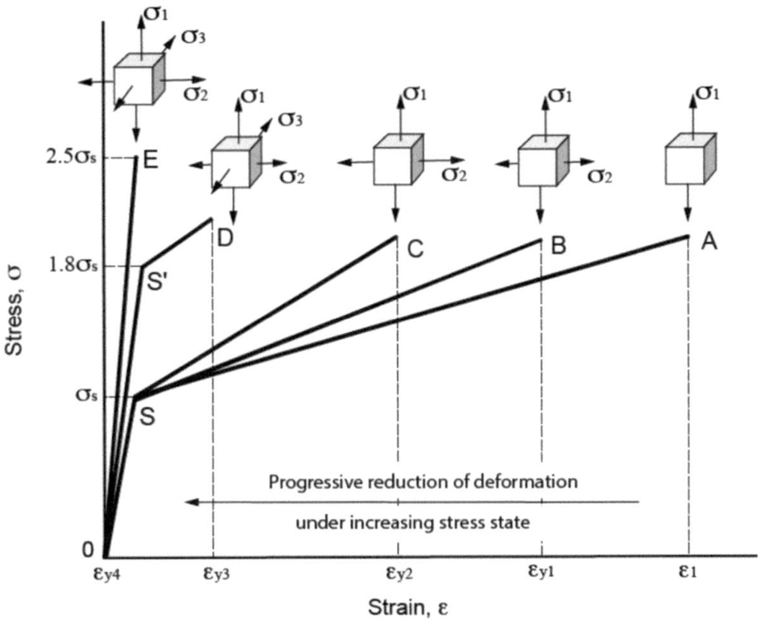

Fig. 4.5 Progressive reduction of the strain to failure and increase of the yield stress of a ductile material in relation to stress state. Stresses σ_x, σ_y and σ_z are shown as σ_1, σ_2 and σ_3 since they are principal stresses

$$\tau_{\max} = \frac{(\sigma_y - \sigma_z)}{2} = \frac{0.4}{2}\sigma_y \tag{4.28}$$

This implies that yield will occur when Eq. (4.28) equals $\sigma_s/2$ or

$$\sigma_y = 1.8 \cdot \sigma_s. \tag{4.29}$$

Equation (4.29) indicates that the stress in the y direction σ_y can reach 1.8 times the uniaxial yield strength σ_s before the material begins to yield. Under this triaxial condition the maximum deformation $\varepsilon_{y,p}$ is, Eq. (4.18), only 0.375 times the corresponding uniaxial yield strength, so that the stress–strain characteristic will be given by line OS'D of Fig. 4.5, in which $\varepsilon_{y,p}$ is indicated with $\varepsilon_{y,4}$ (IV case). Finally, the transition to the triaxial case in which the σ_x equals σ_y was characterixed by the total impossibility to enter the plastic deformations domain, Eq. (4.23). As soon as the elastic limit is reached, the material will immediately fracture. Under these conditions, however, as Eq. (4.19) indicates, the σ_z is $0.6\sigma_y$, therefore τ_{max} is

$$\tau = \frac{(\sigma_y - \sigma_z)}{2} = \frac{0.4}{2}\sigma_y \tag{4.30}$$

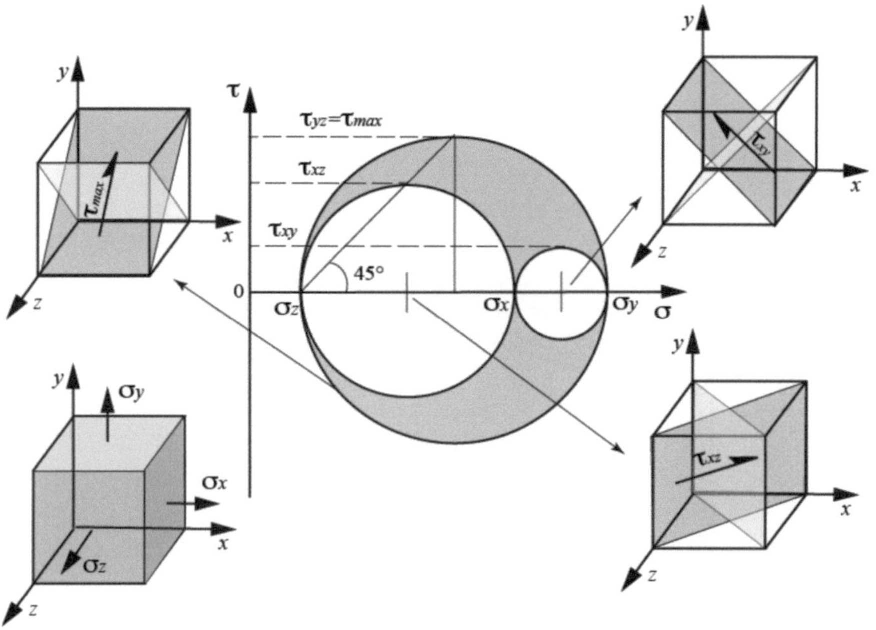

Fig. 4.6 Representation of Mohr circles for a triaxial state of stress σ_x, σ_y and σ_z. Planes of action of the shear stress components $\tau_{xy} = \tau_{max}$, $\tau_{xz} = \tau_{max}$ and $\tau_{yz} = \tau_{max}$ are marked in gray

which states that the new elastic limit is

$$\sigma_y = 2.5 \cdot \sigma_s \tag{4.31}$$

Equation (4.31) tells us that the axial stress σ_y can further grow to become 2.5 times larger than the uniaxial yield stress σ_s. beyond that point the material can no longer go due to its inability to flow plastically. The corresponding ε_y is worth only 50% of the monoaxial elastic one, Eq. (4.21). Therefore, the corresponding stress–strain characteristic is line OE of Fig. 4.5. It should be noted that, if one assumes for the Poisson coefficient, v, the value 0.33 that many authors consider for steels, instead of the usual 0.3, Eq. (4.19) would yield $\sigma_z = 0{,}66 \cdot \sigma$ and Eq. (4.31) would give $\sigma_y = 3 \cdot \sigma_s$. The curves of Fig. 4.5 were obtained through a simplified schematization of the behavior of the material. However, they are conceptually valid and visualize how the progressive transition from a state of uniaxial tension to a triaxial one of plain strain, on the one hand greatly increases the tensile strength of the material and, on the other hand, drastically reduces its ductility and therefore also its toughness, that is, the ability to withstand brittle fracture. Very indicative are, in this respect, the researchers conducted by Lüdwick already in 1926 [5] on smooth and notched specimens of mild steel having the same resisting cross section, subjected to simple traction, Fig. 4.7. It is very interesting to note the analogy between these curves and those of Fig. 4.5, obtained with simplified calculations. The analogy is not just a coincidence. Notches, in fact, introduce a complex state of stress, more or less severe, depending on the notch severity. The strong reduction in ductility combined with an increase in the failure stress that is observed in specimens 4, 5 and 6 of Fig. 4.7 is precisely due to the onset of an increasingly high state of triaxiality. Specimen 6 is very close to the stress state described by line OE of Fig. 4.5. Its yield stress has jumped from about 300 MPa to about 700 MPa. Remarkable is the loss of ductility of a mild steel whose maximum strain reduces from a significative 30%, relative to the uniaxial stress state, to practically zero for the high triaxial stress state. It can be noted that in specimen 2 of Fig. 4.7 the yield strength has not changed compared to case 1, but the deformation at failure reduced significantly. It has a stress–strain characteristic similar to lines OSB or OSC of Fig. 4.5. In this specimen the triaxial stress state that begins to be seen in specimen 4 has not occurred yet. The increment of the yield stress due to a triaxial stress state, $\sigma_{s'}$, with respect to the uniaxial yield stress, σ_s, is defined as *constraint factor*, c_m

$$c_m = \frac{\sigma_{s'}}{\sigma_s} \tag{4.32}$$

For the maximum degree of triaxiality shown in Figs. 4.5 and 4.7 the constraint factor is 2.5 that can also reach 3 with a Poisson factor equal to 3.3. The degree of triaxiality can be quantified using the so-called *Triaxiality factor* or TF defined as the ratio of the hydrostatic component σ_m acting at a material point, average of the three main stresses σ_x, σ_y and σ_z or σ_1, σ_2 and σ_3, and the equivalent stress $\sigma_{eq.}$.

The equivalent stress, σ_{eq}, also called effective stress, is the von Mises stress, defined as the uniaxial stress that produces the same octahedral shear stress generated by the triaxial state σ_1, σ_2 and σ_3, and is given by [6]

$$\sigma_{eq} = \frac{1}{\sqrt{2}} \cdot \sqrt{(\sigma_1 - \sigma_2)^2 + (\sigma_2 - \sigma_3)^2 + (\sigma_1 - \sigma_3)^2} \tag{4.33}$$

The triaxiality factor TF is, therefore, defined as

$$\begin{aligned} TF &= \frac{\sigma_m}{\sigma_{eq}} \\ &= \frac{\sqrt{2}}{3} \frac{\sigma_1 + \sigma_2 + \sigma_3}{\sqrt{(\sigma_1 - \sigma_2)^2 + (\sigma_2 - \sigma_3)^2 + (\sigma_1 - \sigma_3)^2}} \end{aligned} \tag{4.34}$$

Figure 4.8 shows, labeled as Theory (Milella), the relative ductility expressed as the ratio of the strain in the multiaxial condition to the uniaxial strain, $\varepsilon_{y,mult}/\varepsilon_{y,uniax}$, obtained for the 5 cases previously analyzed and shown in Fig. 4.5, as a function of the triaxiality factor TF. As TF increases, the ductility of the material, drastically reduces and tends to zero when TF exceeds the value of two and the strain cannot exceeds the elastic strain. Figure 4.8 also shows the experimental results obtained by Manjoine[7] on five different materials: carbon steel 0.35 C and 0.25 C, annealed brass and aluminum alloys type 2024-T3 and 7075-T651 and those obtained by

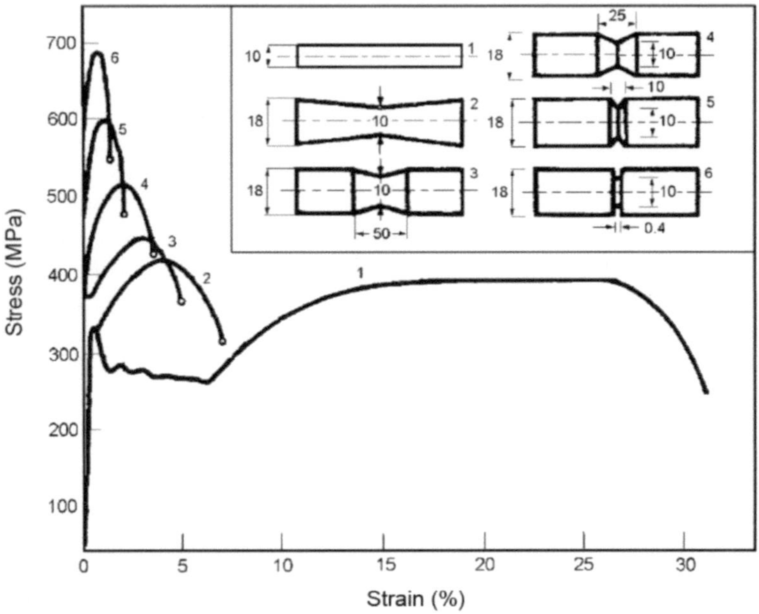

Fig. 4.7 Experimental results obtained by Lüdwick on smooth and notched specimens of mild steel showing the stress–strain characteristic, σ-ε, in relation to the applied local stress state achieved through the variation of the type of notch [5] (specimen dimensions expressed in mm)

Hancock and Mackenzie [8] on notched specimens of high-strength Q-1 steel taken in the LT and ST direction, see Fig. 3.28, and HY 130 steel, in the LT direction.

The TF = 0 case is obtained for a multiaxial stress state in which one or two primary stresses are negative. A pure torque, for example, yields TF = 0, that is, the pure shear increases the ductility of the material making it double compared to the case of single-axial traction. The behavior of the HY 130 steel seems to overestimate the loss of ductility of the material. However, the equivalent plastic strain considered in the experiments was that at the initiation of failure, when holes start to form in the area of maximum triaxiality and not the plastic strain achieved at failure. The data of Fig. 4.8 seem to follow a trend described by the equation

$$RD = (3 \cdot \mathrm{TF})^{-1}. \tag{4.35}$$

where *RD* is the relative ductility ratio, $RD = \varepsilon_{y,mult} / \varepsilon_{y,uniax}$, The maximum theoretical value of the TF is ∞ corresponding to a hydrostatic stress condition where $\sigma_1 = \sigma_2 = \sigma_3$ that cannot produce deformation by slip. However, already with $\sigma_3 = 0.62 \cdot \sigma$ and $\sigma_2 = \sigma_1$, OE line of Fig. 4.5, the ductility reduces practically to zero while the yield stress goes to 2.5 σ_y or 3.0 σ_y depending on the value of ν. For this situation the TF factor is equal to 2.16.

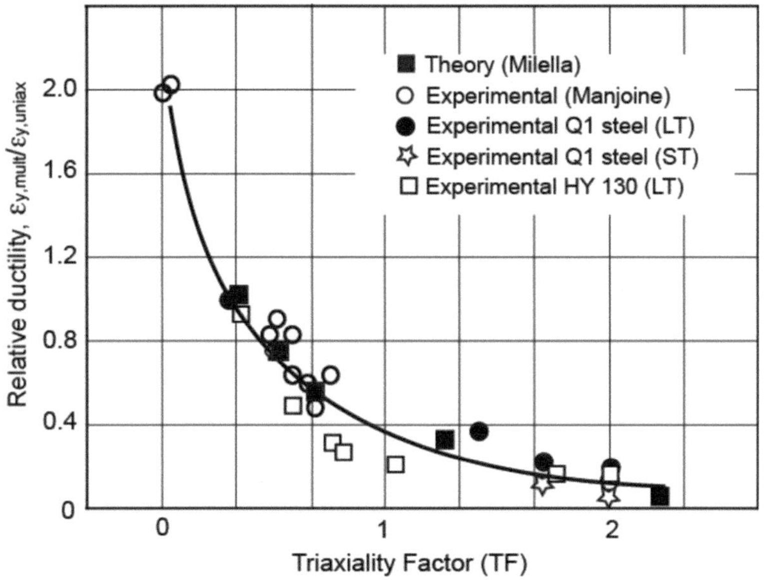

Fig. 4.8 Relative ductility versus Triaxiality Factor (TF) (exp. data obtained by Manjoine are taken from [7] those on Q1 and HY 130 steels from Hancock and Mackenzie [8])

Similar results were obtained by the author (Milella et. al.) [9] on cylindrical specimens containing circumferential notches with radii of curvature equal to 1.25 mm, 2.0 mm, 4.0 mm and 10.0 mm, respectively. The results, shown in Fig. 4.9, are bounded by an upper bound curve of Eq. (4.34) and a lower bound curve

$$RD = 2^{(1-3\cdot\mathrm{TF})} \tag{4.36}$$

4.4 Size Effect on Stress State

We have discussed, in the previous paragraph, the effect of a multiaxial stress state on fracture and how it can change the ductility of a material. In this paragraph we will show how biaxial and triaxial states of stress are generated ahead of a discontinuity and, in particular, of a sharp notch and a crack as a simple consequence of size.

First, it will be shown in a simple fashion, but conceptually valid, why a traction in the x direction must arise ahead of a notch or crack in simple traction along the y direction. Consider a body containing a notch subjected to a nominal external tension σ_n, as shown in Fig. 4.10.

In the absence of any discontinuity all material points are equally stressed and therefore equally deformed: same elongation ε_y and same contraction ε_x with the

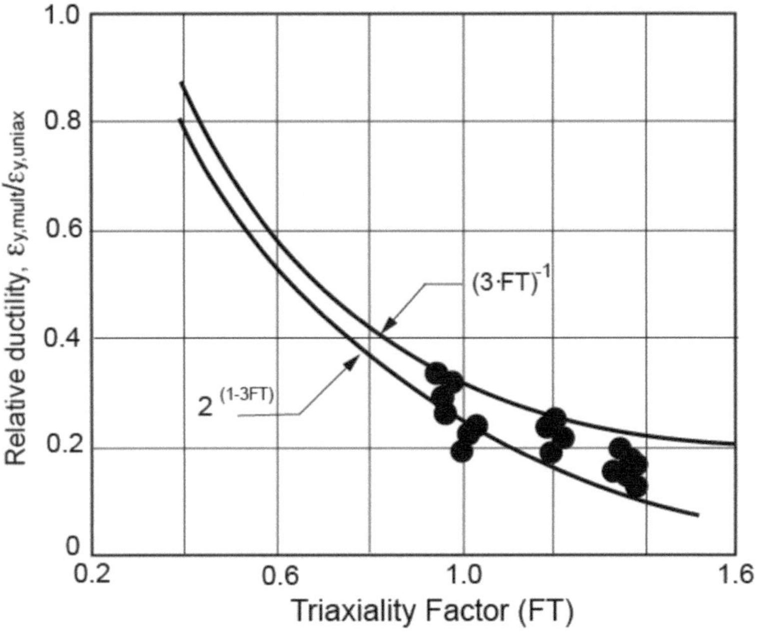

Fig. 4.9 Relative ductility versus Triaxiality Factor (TF). Experimental results obtained by the author *et.al* on specimens of A 533 steel containing notches of 1.25 mm, 2.0 mm, 4.0 mm and 10.0 mm radius, respectively [9]

resulting effect of having a longer body with a reduced section, within the elastic limits. The presence of the notch modifies the local stress field, σ_y, which is no longer uniform, but presents a stress gradient, as shown in Fig. 4.10. The stress field gradient will result in a similar strain field gradient that will be larger the sharper the notch. The result is that every element of material ahead of the notch, as, for instance, element *a*, *b*, *c* and *d*, of Fig. 4.10, will be constrained in its lateral contraction in the *x* direction by the next contiguous element that wants to contract less. This causes the birth of interface stresses in the *x* direction, as shown in Fig. 4.10 between any element as *a*, *b*, *c* and *d*. Therefore, a stress gradient in the *y* direction will also produce a stress gradient in the orthogonal direction *x*. The strength of this second stress field, σ_x, depends on three factors, namely the intensity of the notch stress field σ_y, which is the forcing function, the opening of the notch (blunting) and the extension of the two volumes, A and B of Fig. 4.10, behind and ahead of the notch, respectively, where the stress is practically zero, volume A, or the stress intensification vanishes, volume B. It must be considered that the volume between A and B is practically negligible since, as will be shown in Chap. 8, the crack tip stress intensification vanishes very soon. These two volumes are necessary to exert the constraint action that opposes the free contraction in the *x* direction of individual elements *a*, *b*, *c* and *d* of Fig. 4.10. They act as fixed *walls* within which the high deformation field must be confined to generate a strong compatibility reaction in the *x* direction.

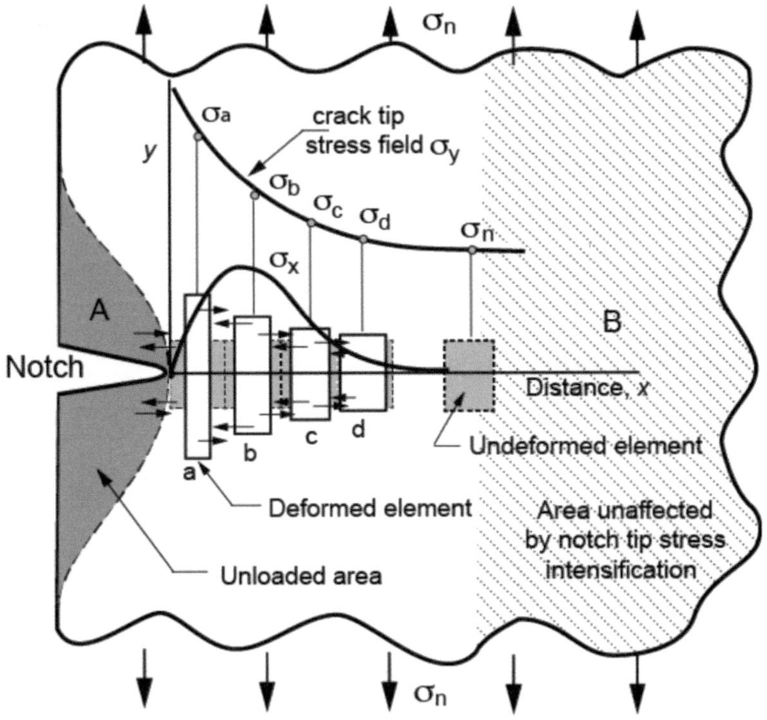

Fig. 4.10 Schematic of notch tip stress field σ_y that produces different elongations in the y direction and, therefore, different contractions along the x direction that must be accommodated through mutual actions which generate the second component of stress, σ_x

Therefore, the maximum biaxial effect will be reached with the most proper combination of deep notch or crack with volumes A and B, as schematically shown in Fig. 4.11a.

At variance, a long crack in a shallow body will not generate a high gradient stress field because the volume beyond the crack, volume B, will yield plastically, Fig. 4.11c. In the same fashion, a very short notch or crack will not generate a high reaction because the volume behind, volume A, cannot exert any restrain action, Fig. 4.11b. The role of notch sharpness is also clear. In a blunt notch the stress field σ_x must start from zero because on the free surface there is no external stress acting, it reaches a maximum and then drops to zero. However, if the notch is very sharp (crack) the R reaction can be so high as to elevate the σ_x to a level close to the σ_y stress. In this case the lowest stress will be σ_z. Based on the theory of elasticity [10], a circular hole in an infinite plate subject to remote traction σ_n generates a biaxial stress field given by

$$\sigma_y = \sigma_n\left(1 + \frac{1}{2}\frac{a^2}{r^2} + \frac{3}{2}\frac{a^4}{r^4}\right)$$

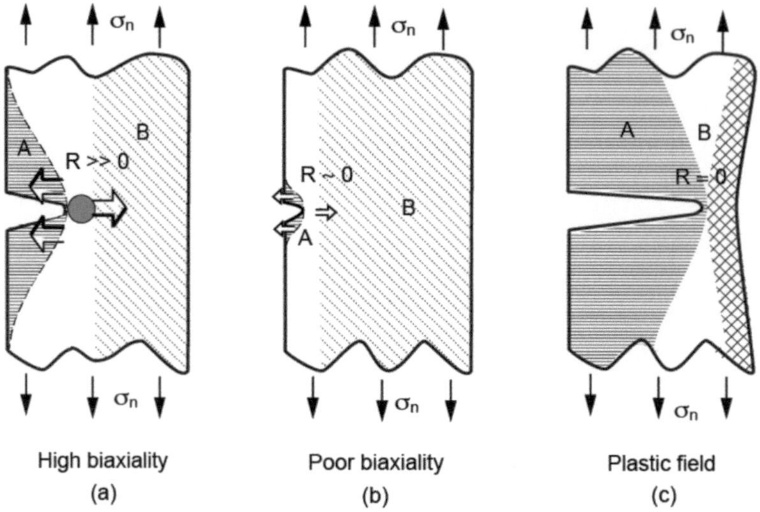

Fig. 4.11 Schematic of notch-specimen combination effect on the degree of biaxiality generated ahead of the notch. Case **a** the reaction R in the x direction reaches its maximum. Case **b** there is not enough unloaded area, A, behind the notch tip to exert a large R reaction and case **c** though the area A is very large a large reaction R cannot occur because the area ahead of the notch tip fully yielded

$$\sigma_x = \sigma_n \frac{3}{2}\left(\frac{a^2}{r^2} - \frac{a^4}{r^4}\right)$$
$$\tau_{xy} = 0 \tag{4.37}$$

where a is the radius of the hole and r the distance from the center, as shown in Fig. 4.12 together with the distribution of the stress fields σ_x and σ_y. As indicated by Eq. (4.37) and shown in Fig. 4.12, the normal stress σ_y has a gradient that from the maximum value $3{\cdot}\sigma_n$ drops to the nominal value σ_n within a distance equal to about four to five times the radius a of the hole while the stress σ_x starts from zero, reaches a maximum at a distance $r \approx 1.4/a$ and decreases rapidly. Note how the increase in the elastic field extinguishes within a distance equal to 3–4 times the characteristic size of the disturbance that has produced it which, in this case, is the size a of the hole, but already at twice that size the effect of the perturbation is practically over, in accordance with the Principle of Saint Venant. As it can be seen, a circular hole is not really capable of generating a high biaxial stress state. If we consider an elliptical one with b and a as semi-major and minor axis, respectively, then the stress fields would be much higher, but will always have the same trend. Inglis [11] found that at the front of an elliptical hole the stress σ_y was

$$\sigma_y = 1 + 2\frac{b}{a} \tag{4.38}$$

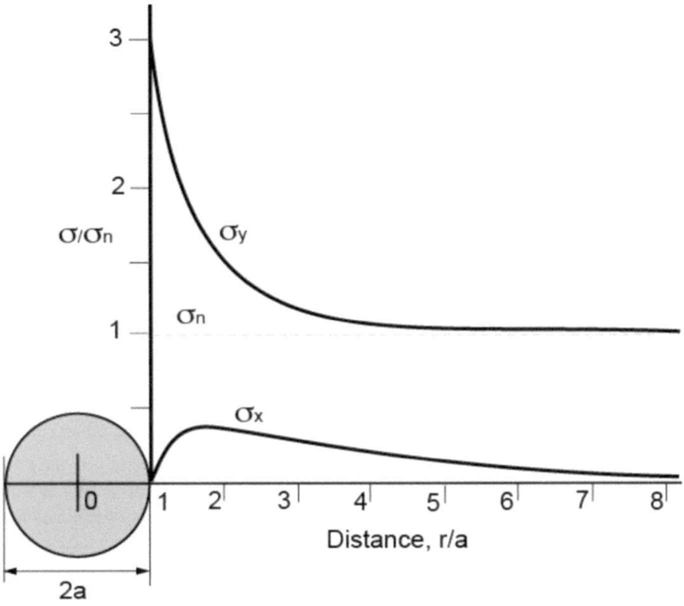

Fig. 4.12 Stresses σ_x and σ_y ahead of a circular hole in an infinite plate subjected to a remote traction stress σ_n

which is equal to three when $a = b$.

4.5 Thickness Triaxiality

It has been illustrated in the previous paragraph, albeit briefly and only from a phenomenological point of view, how the uniaxial stress state always turns into a biaxial stress state, more or less marked, ahead of a notch or crack. We will now see, always phenomenologically, how the thickness of the member containing the discontinuity that originates biaxiality plays a fundamental role transforming the biaxial stress state into a triaxial stress state in which the value of the third component of stress, σ_z, depends on both the degree of biaxiality and the thickness. Let's imagine, therefore, having the usual body containing a surface notch, subjected to a traction stress in the *y* direction, as schematized in Fig. 4.13. The *x, y and z* axes are those shown in the figure. A small plastic enclave forms at the notch tip because of high stress concentration that will necessarily tend to contract in the *z* direction. To understand what happens in front of the notch we will refer to the two cases of modest and high biaxiality represented by $\sigma_x = \nu\sigma_y$ and $\sigma_x = \sigma_y$ already described by Eqs. (4.5) and (4.12) that can be considered typical of sharp cracks (see Chap. 8).

In the two cases taken as a reference, we will have that the ε_z contractions are, respectively

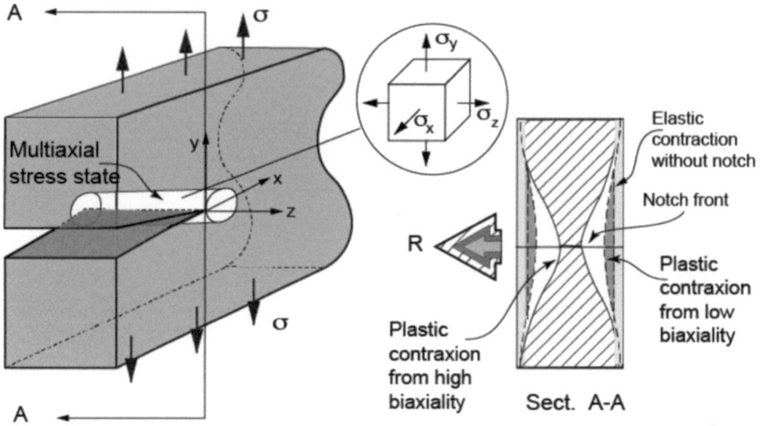

Fig. 4.13 As the degree of biaxiality generated by the notch increases, that is, as the notch tip radius decreases, the value of the lateral contraction ε_z in the plastic zone increases, which will recall increasingly large masses of material, which in turn will develop increasingly large R reactions

$$\begin{aligned} \varepsilon_z &= -\frac{\nu\varepsilon_y}{(1-\nu)} \quad per \quad \sigma_x = \nu \cdot \sigma_y \\ \varepsilon_z &= -\frac{2\nu\varepsilon_y}{(1-\nu)} \quad per \quad \sigma_x = \sigma_y \end{aligned} \tag{4.39}$$

In the plastic field the Poisson coefficient, ν, is worth exactly 0.5, to satisfy the Eq. (4.10) and then just on the notch tip, where the local plastic yield occurred, is

$$\begin{aligned} \varepsilon_z &= -\varepsilon_y \quad per \quad \sigma_x = \nu\sigma_y \\ \varepsilon_z &= -2\varepsilon_y \quad per \quad \sigma_x = \sigma_y \end{aligned} \tag{4.40}$$

Equations (4.40) clearly show the effect of degree of biaxiality on the transverse plastic contraction ε_z that becomes twice and four times larger than in the monoaxial case in which is $\varepsilon_z = 0.5\varepsilon_y$, and 3.33 times or 6.66 times larger than the elastic uniaxial one far from the discontinuity in which ε_z will stop at just $0.3\varepsilon_y$. This particular situation produced by the edge notch, is schematized in Fig. 4.13 on section AA in front of the notch. The figure shows the hourglass shape that section A-A should take if the plastic contraction ε_z, expected by the second of Eq. (4.40), were to be fully implemented, which, as mentioned, is 6.66 times greater than the elastic one. However, such plastic deformations will never be possible in large thicknesses since all the elastic volume opposes. The elastic volume and its stiffness are enormously larger than that tiny plastic enclave that is forcing it and will prevent it from contracting. The resistance, R, to the contraction will therefore be greater the greater the thickness of the body containing the notch. If we assimilate, as a first approximation, the plastic volume ahead of the notch to a cylinder of length equal to the thickness of the body, t, Fig. 4.13, it is, approximately

$$\Delta z = \Delta t = \varepsilon_z \cdot t \tag{4.41}$$

Therefore, a σ_z component will be born, located just in the notch tip area and with it the triaxial stress state will occur.

On the contrary, when the body is slender there will be just few material to be affected by the Δz contraction of the plastic zone at the root of the notch, even when the biaxiality is very strong. The R reaction will, therefore, be irrelevant and the body under the action of the nominal tension σ will tend to contract along the z direction presenting the typical neck formation, as schematized in Fig. 4.14a. There will still be a state of plain stress, beingσ_x and σ_z practically zero. It is clear that starting from a thin member and progressively increasing the thickness until it completely inhibits lateral contractions, Fig. 4.14b, the entire scenario of possible conditions will be obtained between the state of plain stress, ductile, and that of plain strain, brittle, including the cases described by the broken OS'D line of Fig. 4.5 or by curves 2, 3, 4 and 5 of Fig. 4.7.

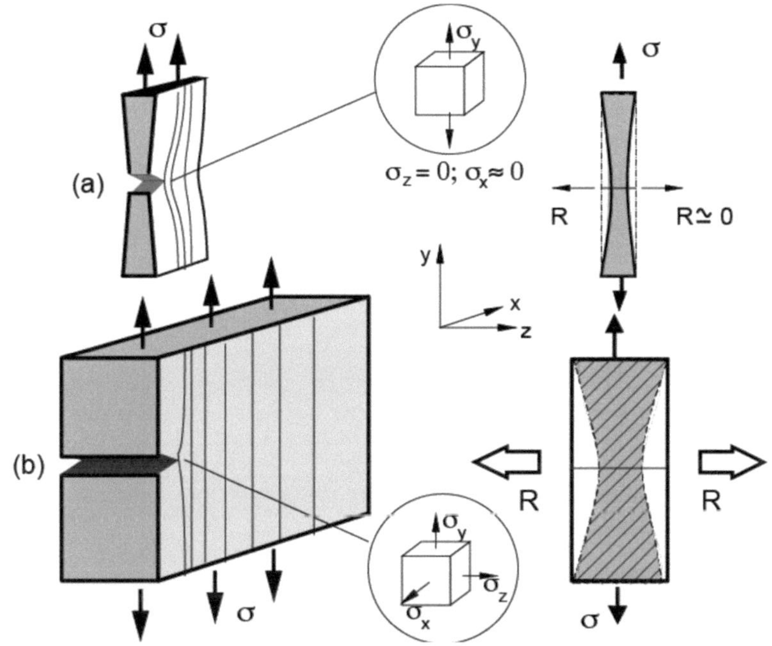

Fig. 4.14 Schematic of the thickness effect in the generation of a plane strain effect or triaxial tension state: **a** thin body, state of plane stress and absence of reaction along both x and z with necking formation, **b** thick body, state of plain strain with strong reaction R along the z direction and without necking formation

4.6 Stress State and Fracture

Treating the subject of fracture, it is of paramount importance to fully understand the role of the plastic zone at the crack tip, because there is always a crack, or a sharp notch associated with fracture. This small plastic enclave is the only barrier the material can exert against fracture. To let it grow also means to let plastic stress fields develop, open the crack tip and change a crack into a notch: the difference is total. To restrain this small plastic pimple from growing and confine it to a negligible size is to keep the crack sharp and lethal. It is worth recalling that a crack is a discontinuity with a tip of about 0.005 mm radius, i.e., just 5 micron. Therefore, this plastic zone must remain in this range of dimensions; it cannot be larger than 50 to 100 micron to allow brittle fracture to occur. This are, in practice, the conditions that we call Small-Scale Yielding (SSY). We will see, in Chap. 6, that these are precisely the dimensions of the plastic zone ahead of the crack tip in all events of brittle fracture, i.e., practically inessential and not capable of altering the crack tip geometry and the elastic stress field.

The loss of such conditions will be referred to as Large Geometry Change (LGC) indicating the *blunting* of the crack tip. To this purpose, let's consider a body in the form, for example, of an infinite plate containing an elliptical hole. Let σ_n be the value of the nominal stress to which the plate is subjected, as shown in Fig. 4.15.

The theory of elasticity states that at the edges of the hole the nominal stress, σ_n, tends to increase, because of stress concentration, reaching its maximum, σ_{max}, at the tip where it will be worth

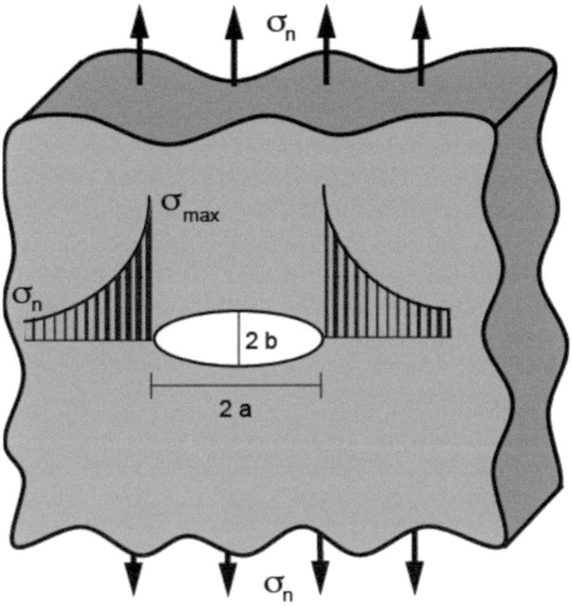

Fig. 4.15 Raise of the stress state at the edges of an elliptical hole in a sheet subjected to external traction σ_n

$$\sigma_{\max} = \sigma_n \left(1 + 2\frac{a}{b}\right) \tag{4.42}$$

a and b being the values of the major and minor semi-axes, respectively. By introducing the radius of curvature ρ of the ellipse, which is

$$\rho = b^2/a \tag{4.43}$$

Equation (4.42) becomes

$$\sigma_{\max} = \sigma_n \left(1 + 2\sqrt{\frac{a}{b}}\right). \tag{4.44}$$

On the edges of the ellipse the nominal stress σ_n is, therefore, amplified by the factor $(1 + 2\,\sqrt{a/\rho})$. If the elliptical hole were very flat *($b \ll a$)*, its radius of curvature ρ will be extremely small and will become almost zero in a very sharp flaw that we call crack, for which b and ρ can be assumed to be almost zero. In this case, the σ_{max}, according to Eq. (4.44), will tend to an infinite value, Fig. 4.16. In practice, the threshold between a notch and a crack is placed around some tens of microns. Since, as shown in Sect. 4.4, the amplitude of the second component of stress, σ_x, depends on the steep gradient of the first component,σ_y, it is logical to expect that from a σ_y going to infinity also σ_x will tend to infinity.

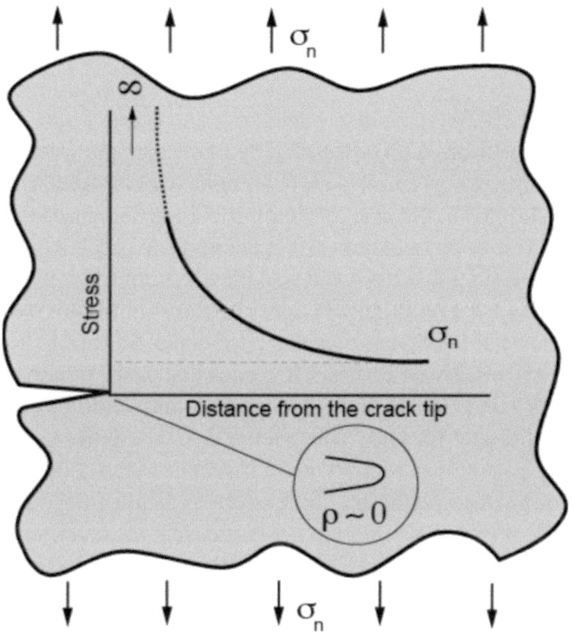

Fig. 4.16 At the tip of a very sharp notch (radius of curvature ρ ~ 0), in an ideally elastic material subjected to remote traction σ_n the stress tends to ∞

This is precisely what the theory of elasticity and Linear Elastic Fracture Mechanics predict for the first two component of stress σ_x and σ_y in an infinite plate containing a sharp crack loaded in traction that are exactly equal to each other and their trend is asymptotic, according to $1/\sqrt{r}$, with r being the distance from the crack tip, as it will be shown in Chap. 8.

Therefore, as the distance r tends to zero, that is, at the tip of the crack, the stress will tend to infinity, whatever the value of the nominal stress, σ_n, as shown in Fig. 4.16. No real material is able to withstand an infinite stress. Therefore, the presence of a crack in a perfectly elastic material subjected to traction, even very modest, should inevitably lead to fracture. Practice, fortunately, belies the drama of such a linear elasticity outcome. In real cases, in fact, once the stress reaches the yield strength, σ_s, the material starts to slip and strain harden, limiting the effect of the stress concentration. Again, the yield strength will be indicated by the symbol σ_s (the subscript s stands for *slip*) to differentiate from the stress in the y direction that in LEFM is traditionally assumed to be the main component of stress. The maximum stress is, then, truncated at the yield stress value, σ_s, or, in strain hardening materials, it continues to grow but with a lower slope, depending on the strain hardening exponent, up to a maximum but finite value, whereas the strain may grow further, but it always remains finite. The result is the creation of a crack tip plastic zone. Question, then, is: can a crack continue to be sharp if there is a plastic enclave at its tip? Or, recalling what we have said about the size of the plastic zone, can the plastic zone be confined within a millimeter or less without swelling? LEFM states that the radius, r_p, of the crack tip plastic zone, approximated with a circle, is given by, see Eq. (8.1)

$$r_p = \frac{a}{2}\left(\frac{\sigma_n}{\sigma_s}\right)^2 \tag{4.45}$$

Equation (4.45) states that by increasing the yield strength of a material by a factor 3 using, for example a high strength steel of 1000 MPa instead of a low carbon steel of 333 MPa, the size of the plastic zone decreases by a factor of 9. Moreover, a full triaxiality by thickness effect allows σ_s to reach 2.5 times the uniaxial yield strength, as per Eq. (4.31) or whatever the constraint factor C, Eq. (4.32) may be. This yield strength elevation by triaxiality further contributes to the reduction in the plastic zone size by about 56 times so that a small plastic zone of 1 or 2 mm, which is already large to cause a brittle fracture, becomes just 0.017 mm to 0.034 mm, precisely the size for SSY condition to apply. The same result would be achieved with a BCC ferritic steel well below the NDT, see Sect. 5.2.1, where the yield stress becomes as high as the ultimate stress. Mohr's circle is that of Fig. 4.17a with the plastic zone that remains confined to negligible dimensions typical of SSY condition, and the stress state is that of Fig. 4.17b. Figure 4.18a presents the crack tip plastic zone observed in a steel plate, 63 mm wide, 10.66 mm thick, of Fe-3%Si, containing a central crack of 11 mm, loaded at 70% of the fracture toughness K_{Ic} of the material to avoid brittle fracture [12]. The plastic zone was revealed using the Morris reagent etching technique. This steel is particularly brittle, having a toughness, K_{Ic}, of only 33 MPa$\sqrt{\text{m}}$ at room temperature. Therefore, the system can be considered under plain strain conditions. The plastic zone is not precisely a circle but shows two symmetrical lobes extending at 60° over and below the crack plane, as expected by the theory, see Chap. 8. Its furthest extent directly in front of the crack tip is only 0.1 mm, as expected using Eq. (4.45).

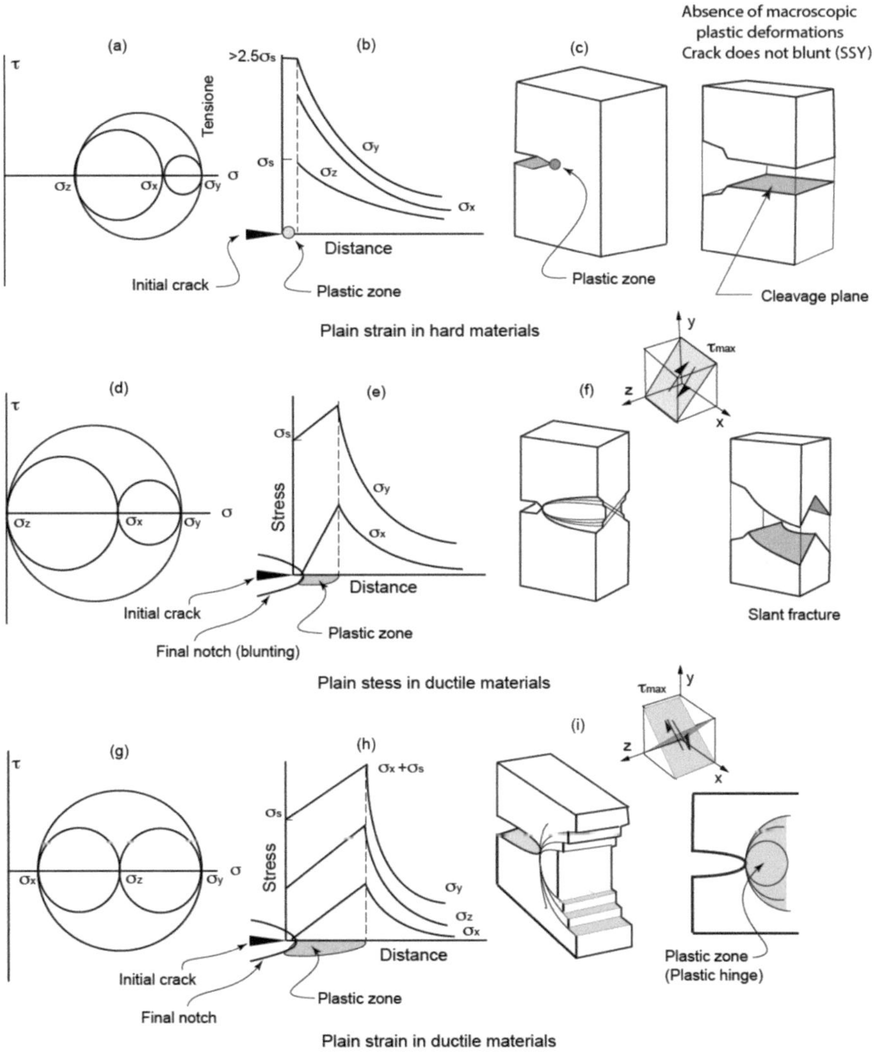

Fig. 4.17 **a** Triaxial stress state or plain strain condition in thick, high strength materials that allow the crack to remain sharp (SSY) and causes brittle fracture, **d** biaxial stress state or plain stress conditions typical of thin, low strength materials that causes blunting (LGC) and a slant or ductile fracture, and **g** triaxial stress state in low strength highly ductile materials that allows LGC and fracture by general yield

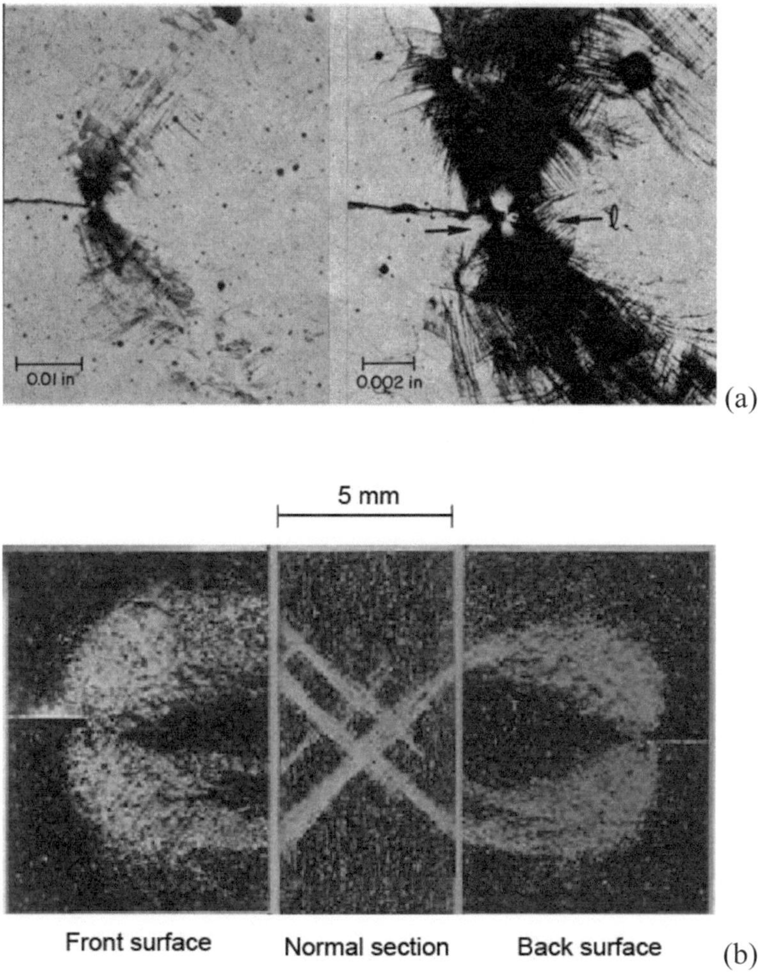

Fig. 4.18 Crack tip plastic zones revealed on the surface and on the section of Fe-3Si steel plates by etch-pitting electrolytically in Morris reagent [12, 13]. **a** Specimen 63 mm wide, 10.66 mm thick loaded at about 70% of the fracture toughness of the steel under plain strain conditions [12]. The plastic zone consists of two lobes expanding at 60° over and below the crack plane. It is very small with the furthest extension in front of the crack tip of only 0.1 mm. **b** Specimen 0.5 mm thick loaded at 0.9 the yield stress under plain stress conditions. The plastic zone has swelled to the edge [13]

Increasing the applied load, dealing with FCC metals or alloys that do not have a transition temperature and are fundamentally ductile or with BCC materials well above the NDT, there is no way to keep the size of the plastic zone at negligible levels, even considering the triaxiality increase of the yield strength. The crack blunts offering an external surface so that the second component of stress, σ_x, must necessarily start from zero, at the notch tip, as shown in Fig. 4.17e and will never reach the value of σ_s but will remain below, depending on how much the crack has opened.

As a result, the maximum hydrostatic elevation will no longer occur at the tip of the crack but at a certain, albeit small, distance from it, right at the limit of the plastic zone, while the maximum deformations remain on the tip. The light growth of the σ_y stress over the yield is due the possible strain hardening of the material. The low biaxiality does not allow a full triaxial condition to develop in particular when the specimen is not sufficiently thick. In not thick and ductile materials, it may happen that the plastic zone extends to the outer edge, Fig. 4.17f, and this spread will lead to a simultaneous increase in the apical blunting with further reduction of the amplitude of the stress. Now a brittle fracture cannot happen at all. The reduced thickness does not give rise to the third stress component, $\sigma_z = 0$, so that the maximum shear stress, that is given by the semi-difference between σ_y and σ_z as per the Tresca criterion, assures that the yield strength of the material will remain exactly that of the uniaxial case, σ_s. The Mohr's circles are those of Fig. 4.17d and the plastic slip will take place on planes perpendicular to the *yz* plane, then passing through the *x*-axis. This new situation is depicted in Fig. 4.17f. An example of such a plastic zone is presented in Fig. 4.18b [13] for a plate of Fe-3Si steel, 5 mm thick. The normal section, etched by the Morris reagent, shows precisely the slip planes crossing at 90° to each other, while on the surface, front and back, the plastic zone appears with two lobes, specular with respect to the crack plane, projecting to the border. If the thickness increases as to prevent transverse contraction to a certain degree, then the stress component σ_z will appear but this time it will be the σ_x stress to have the lowest value, Fig. 4.17h. The Mohr circles become those of Fig. 4.17g and slip will take place in the planes perpendicular to the *yx* plane, that is, passing through the *z*-axis, Fig. 4.17i. Now the largest of the stress semi differences is given by $(\sigma_y–\sigma_x)/2$. This condition contributes to the opening of the crack, but now there will also be a certain increase in the yield strength of the material from triaxiality, equal to $\sigma_x + \sigma_s$. Therefore, triaxiality will allow the growth of the σ_y component. However, with ductile materials this effect may not prevent the plastic zone from spreading and form a plastic hinge, Fig. 4.17g, i, with a LGC result as shown in Fig. 4.21.

4.7 Experimental Evidence

We have seen, in the previous section, how the stress state ahead of a crack can vary depending on the ductility and size and of the apical plastic zone. It is precisely the plastic zone the true and only defense any material has got against brittle fracture. A crack cannot trigger a brittle fracture if this plastic zone is too large. To understand what the term *large* actually means, it must be recalled that the radius of a crack is no larger than 0.005 mm. Therefore, a plastic zone of just one millimeter radius can be considered already large because it blunts the crack that becomes a notch. If the ductility of the material is not high, as in high strength steels or ferritic steels at very low temperatures, the crack tip plastic zone remains confined within less than a millimeter. The crack tip does not blunt, remaining sharp. This conditions is known as SSY (Small Scale Yielding). The effects of these two opposite conditions can be observed in Fig. 4.19 which refers to some tests [14] run on cylindrical specimens of medium–high strength steel forgings, of the type adopted in the construction of rotor shafts. The tests were conducted at room temperature. The yield strength of the material, σ_s, was equal to 565 MPa and the ultimate, σ_u, was 765 MPa, rather high. The specimens were provided with a 45-deg circumferential V-groove with a root radius of 0.076 mm, very sharp though not exactly a crack interesting 30% of the diameter. The Charpy-V-50 per cent ductility, see Fig. 3.2b, was reached at 150 °C, therefore the general behavior at room temperature was rather brittle. It can be observed that with a diameter of only half a centimeter the failure stress is 1.8 times higher than the uniaxial yield stress, as observed in Fig. 4.5 for the 0S'D case and in Fig. 4.7 for case 5 where the specimen diameter was 10 mm. This actually means that the sharp notch has generated a triaxial stress state, but not too high similar to that of case (g) of Fig. 4.17 where a plastic zone can develop. The notch had a beneficial effect on the strength of the specimen. As the diameter of the test specimen increases the ratio σ/σ_s between the failure stress and the uniaxial yield strength σ_s decreases continuing to be higher than one as long as the specimen diameter remains below 6 cm. The situation changes as the diameter of the specimen containing the circumferential crack increases. Above 6 cm the ratio σ/σ_s falls below unity. With 10 cm diameter, the resistance is now only 80% of the uniaxial yield strength. The behavior of the material has totally changed because the larger size introduced a very high triaxiality at the bottom of the notch that did not allow the development of a large plastic zone. Now the initial crack remains sharp since the plastic zone has a negligible size, SSY condition, and the fracture is brittle, case of Fig. 4.17a.

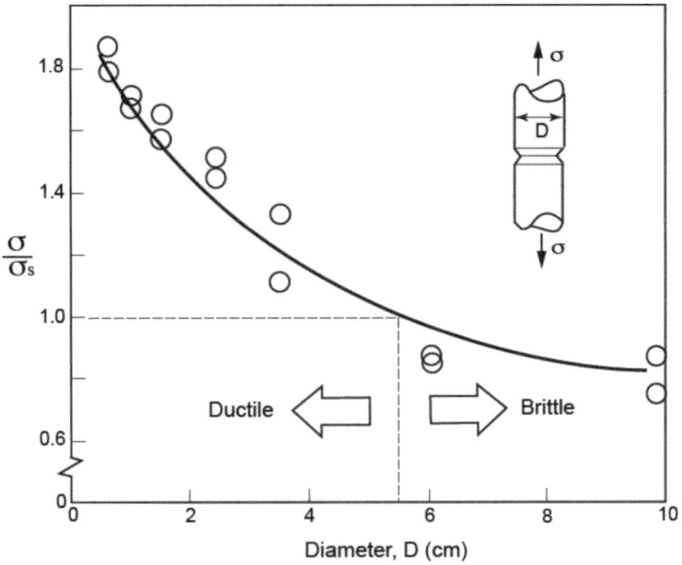

Fig. 4.19 Failure stress to yield strength ratio, σ/σ_S, versus the diameter of a cylindrical specimen of steel forging having a circumferential sharp notch (adapt. from [14])

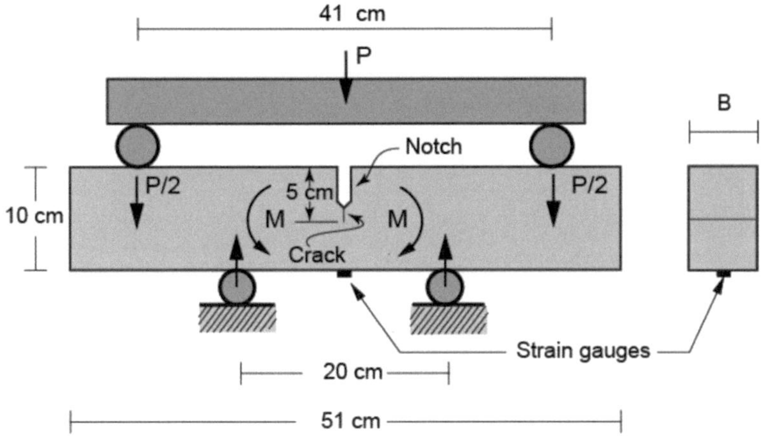

Fig. 4.20 Pre-cracked specimen of carbon steel type A 533, of variable thickness B, subjected to pure bending, used by Wilson and Begley [15] for the study of the thickness effect

Equally instructive are the results obtained by Wilson and Begley [15] on specimens of steel type A 533 B whose thickness ranged from 12.7 mm to 203 mm. The specimens were subjected to four-point bending, as shown in Fig. 4.20 for the 51 cm long specimen. The samples contained a sharp surface crack 50% depth, obtained by fatiguing a machined notch.

The experimental results are shown in Fig. 4.21a, b as applied load *P* normalized to the thickness *B*, *P/B*, versus longitudinal strain ε_x and through thickness strain ε_z, (contractions), respectively [15]. It can be seen the gradual transition from a condition of plain stress, relative to the thickness of only 17.2 mm to that of plain strain or stress triaxiality, relative to the 203 mm thick specimen. Under plain stress condition the departure from linearity starts at about 5250 N/m which corresponds to a yield stress of about 450 MPa. In plain strain the contraction σ_z for $B = 203$ mm is zero and the yield stress jumps to about 1.8 the uniaxial yield strength, revealing a full state of triaxiality, as expected. But the fracture was not brittle at all. This can be clearly seen in Fig. 4.22 [15]. The external appearance of several specimens that did not break in two reveals large plastic deformations. The initial sharp crack has turned into a blunt notch. Triaxiality has not prevailed over the metallurgical ductility of the steel. The crack blunted lowering the second component of stress, σ_x, case (g) of Fig. 4.17. Tests were run at room temperature, i.e. at 40 °C above the NDT of the material that was −15 °C.

Concluding Remarks
A plain strain state or triaxial stress state acting at the tip of a crack is a necessary condition to trigger a brittle fracture but it is not sufficient. It is also necessary that the plastic zone ahead of the crack remains negligible in size (Small Scale Yielding

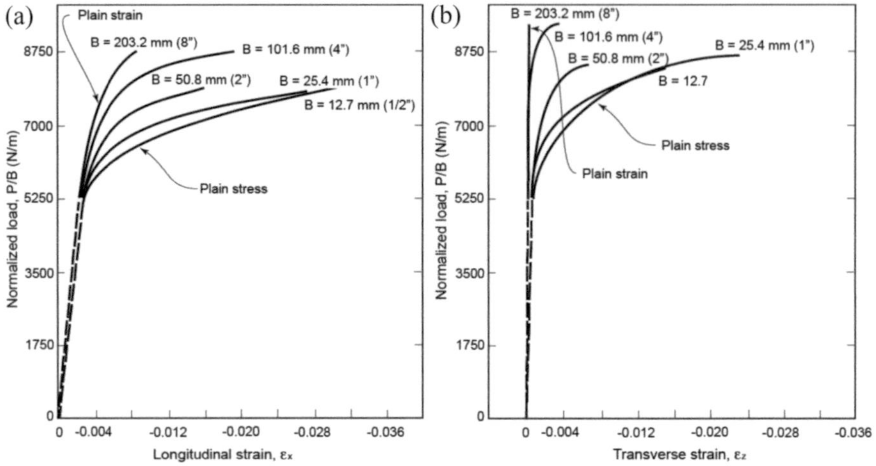

Fig. 4.21 Normalized load P/B versus: **a** longitudinal strain and **b** transverse strain (from [15])

Fig. 4.22 Specimens of A 533 GrB Cl1 steel of variable thickness from 1/2" (12.7 mm) to 8" (203.2 mm) with a central notch carrying a fatigue crack, subjected to pure bending at room temperature [15]. There was no brittle fracture even in the 203 mm thick specimen, since the swelling of the plastic zone opened the crack (blunting) leading to a LGC condition. Note the presence of large longitudinal deformations but also the lack of apparent transverse plastic deformations (triaxiality)

condition). This last condition is rather easier to achieve with steels and alloys of high-very high strength than in metals and alloys with high metallurgical ductility such as austenitic stainless steels. In these latter steels, the growth of the plastic zone, even in large thicknesses, produces the blunting of the crack tip that becomes a large notch (Large Geometry Change). So far, it has never been observed a single case of brittle fracture in austenitic stainless steels. In ferritic steels and alloys having a BCC lattice and, therefore, a brittle to ductile transition temperature NDT, blunting always occurs at temperatures over the NDT or RT_{NDT} preventing brittle fracture. At 30 °C above the RT_{NDT} a brittle fracture will never occur whatever the thickness of the workpiece may be. Even in the case where the plain strain condition is reached, the spread of the plastic zone (LGC) does not allow brittle fracture to occur.

References

1. Wessel, E.T., Wilson, W.K., Clark, W.G., Jr., Development and application of LEFM in Westinghouse, Scient. Papaer 74-1E7-MSLRA-P1 (1974)
2. Manjoine, M. J., Biaxial brittle fracture tests. J. Basic Eng. Trans. ASME, 293–298 (1965)
3. Whitman, G.D.: Historical summary of the Heavy-Section Steel Thecnology Program and some related activities in light water reactor pressure vessel safety research, NUREG/CR-4489, ORNL-6259 (1986)
4. Roos, E.: Determination of fracture mechanics parameters of small and large scale specimens. Proceedings of the Second CSNI Workshop on Ductile Fracture Test Methods, NUREG/CP-0064, CSNI Report No. 105 MEA-2313, April 17–19, 1985
5. Lüdwick, P.: Z. Ver. Dent. Ing. (1926)
6. Nadai, A.: Theory of Flow and Fracture of Solids, 2nd ed., Vol. 1, McGrow-Hill Book Co, New York (1950)
7. Manjoine, M.J.: Multiaxial Stress and Fracture, Fracture, vol. 3. New York, Academic Press, Eng. Fund. and Environmental Effects (1970)
8. Hancock, J.W., Mackenzie, A.C.: On the mechanisms of ductile failure in high strength steels subjected to multi-axial stress-states. J. Mech. Phys. Solids **24**, 147–169 (1976)
9. Bonora, N., Iacoviello, F., Milella, P.P., Modellazione e Verifica del Comportamento Elasto-Visco-Plastico di Acciai Per Contenitori Metallici e Recipienti in Pressione Sottoposti ad Elevati Strain Rates, Rapporto Tecnico Università di Cassino-APAT, 15 novembre, 1999
10. Timoscenko, S.P., Goodier, J.N.: Theory of Elasticity, McGraw-Hill, p. 90 (1970)
11. Inglis, C.E.: Transactions Institute of Naval Architecture, LV 1, p. 219 (1913)
12. Hann, G.T., Mincer, P.N., Rosenfield, A.R.: The Fe-3Si steel etching technique for local strain measurements, Experimental Mechanics, pp. 248–253 (1971)
13. Rosenfield, A.R., Dai, P.K.: Hann, G.T.: Crack Extension and Propagation Under Plain Stress, Pro. First Internatl. Conf. Shenday, Japan, **1**, p. 223 (1966)
14. Wundt, B.M.: A Unified Interpretation of Room Temperature Strength of Notched Specimens as Influenced by Their Size, paper No. 59 MET, ASME Met. Eng. Conference, Albany (1959)
15. Wilson, W.K., Begley, J.A.: Variable Thickness Study of Edge Cracked Bending Specimens, ORNL WCAP-8237 (1974)
16. Tresca, H.: Mémoires Scientifique savants Academie Sci., 18, p.733, Paris (1868)

Chapter 5
Temperature and Strain Rate Effects

5.1 Effect of Temperature

It has been shown in the previous Chapter how the stress state has an impact on the ductility of materials, though the crystalline structure still remains the microscopic base that determines the intrinsic attitude to slip of materials. Among all different crystal lattices, BCC, FCC, HCP and BCT, in general FCC metals are more ductile, exhibit higher toughness and have not got a brittle to ductile transition temperature or NDT, opposite to BCC metals (see Chap. 3). Besides the stress state, among the many non-intrinsic factors that affect the toughness of materials, the most important are temperature and strain rate. In this chapter we will analyze the temperature effect and its impact on material properties and brittle fracture. At variance with the thickness, temperature is known to have a beneficial effect on ductility and toughness, in particular for BCC materials. If at a certain temperature a test piece fails by general yielding, upon increasing the thickness also the temperature must be raised to compensate and get the same effect as before, otherwise general yielding would not be reached. This is shown in Fig. 5.1 for notched mild steel bars of various thickness loaded in bending at different temperatures [1].

5.1.1 Temperature Effect on BCC Crystal Lattice Materials

The typical temperature dependence of the yield strength σ_s and ultimate strength σ_u of a BCC crystal lattice material is presented in Fig. 5.2. Also indicated in the figure is the trend of the true stress σ_t with temperature (see Sect. 1.6) and the ductility expressed either as reduction of area, RA, or as deformation at maximum load ε_{max}. The characteristic feature is the exponential growth of both σ_s and σ_u with decreasing temperature starting from about 0.2 T_m, where T_m indicates the melting temperature of the metal or alloy. Above this temperature the strength seems to remain almost

P. P. Milella, *Fundamentals of Fracture Mechanics*,
https://doi.org/10.1007/978-3-032-23081-2_5

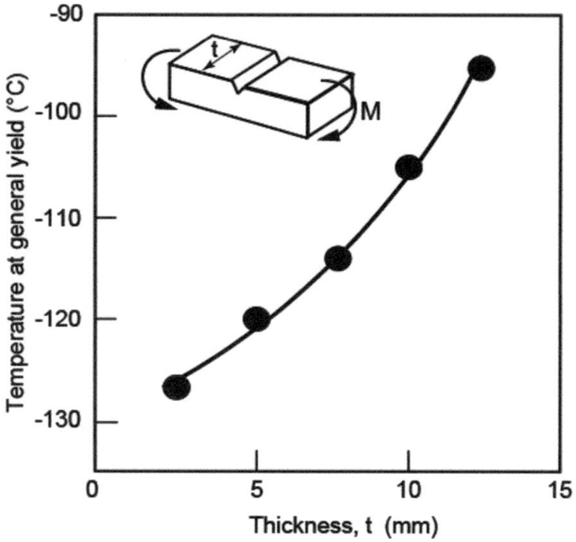

Fig. 5.1 Effect of specimen thickness on the temperature at which failure occurs by general yield in notched bars of mild steel [1]

constant up to a point where it starts to fall down because of diffusion processes that activate a creep mechanism of deformation that begins to take a dominant role.

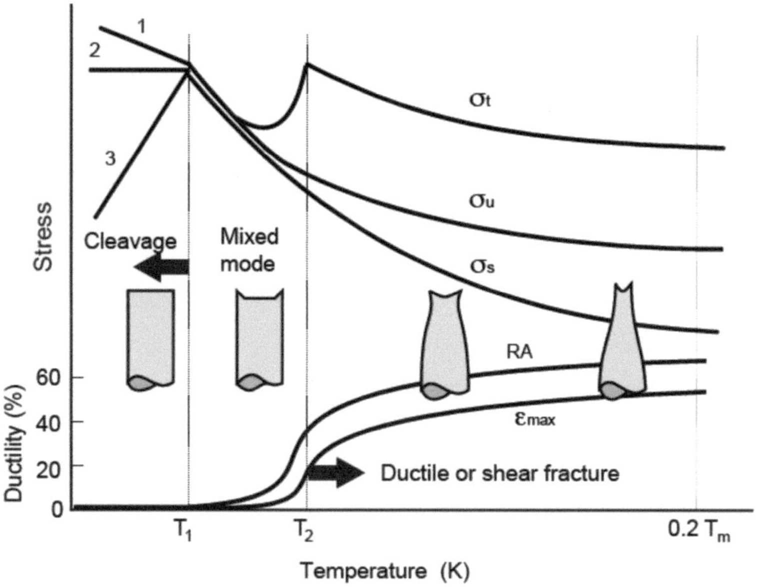

Fig. 5.2 Schematic of dependence of true stress, σ_t, ultimate strength, σ_u, and yield strength, σ_s, on temperature in BCC materials. Also shown is the material ductility

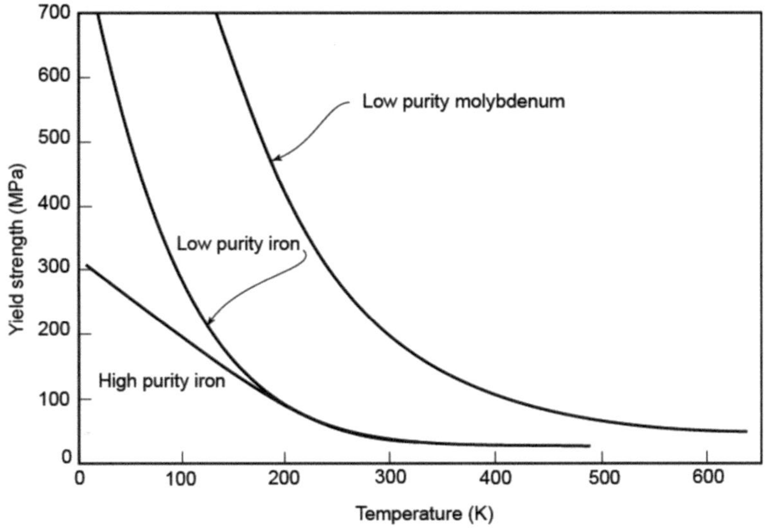

Fig. 5.3 Yield strength variation with temperature for iron and molybdenum (adapted from [2])

This characteristic feature is shown in Fig. 5.3 for polycrystalline iron and molybdenum [2]. It can be seen how the average trend of the yield strength in iron remains practically constant from room temperature to about 500 K while for molybdenum this athermal trend starts at about 550 K. Consider that the melting temperature of molybdenum is about 2,625 °C, much higher than that of iron. Really impressive is the yield strength increase with decreasing temperature in both iron and Molybdenum, while in high purity iron is not so large, indicating the importance of interstitial element and foreign phases.

Interstitial atoms can raise the yield strength by blocking dislocation motion and this effect increases with decreasing temperature. The high sensitivity of the yield strength of BCC materials to temperature is due to their narrow dislocations, as explained in Sect. 2.10. The Peierls-Nabarro barrier that exerts a strong barrier to dislocation motion are temperature dependent and contribute to the buildup of a thermal component of the yield strength. Starting from the high temperature range, three characteristic zones can be distinguished in Fig. 5.2. The first is roughly extending between T_2 and 0.2 T_m. Here the material has gained full ductility, as indicated by the deformation at maximum load ε_{max}. Its behavior is completely ductile, and deformations are activated by shear stresses. This condition is schematized in Fig. 5.2 by the extensive necking presented by specimens at failure. True stress σ_t is much higher than ultimate strength σ_u because of ductility (see Eq. 1.18) and run almost parallel to the latter. The yield strength σ_s grows at a higher rate than the ultimate strength with decreasing temperature. Therefore, the two strengths get closer and closer and eventually become equal at T_1 in the very low temperature range. In steels, the T_2 temperature can be placed between $-100°$ C and $-150°$ C and a typical T_1 is about $-180°$ C. In the second zone, between T_2 and T_1, the specimen fails by mixed mode. In the proximity of T_2 the ductile component is dominant while towards T_1 cleavage is the main component of failure. As such, the fracture zone presents either the typical shear lips along the external surface that witness the ductile component of failure, more or less pronounced depending on the temperature, or a complete flat surface that is indicative of a total loss of ductility, when the temperature is closer to T_1. In this temperature range, the true stress presents a trend opposite to that of the ultimate strength, decreasing and approaching the value of σ_u till they become equal at T_1 because ductility is completely disappeared, at least macroscopically. Below this last temperature there is no apparent ductility. Three different behaviors are now possible, shown in Fig. 5.2 by the three lines marked 1, 2 and 3. Line 1 is indicative of a further growth of the strength, but at a lower rate, while line 2 indicates a constant growth with decreasing temperature. In both cases responsible for the slowdown or complete arrest in the strength growth is twinning, see Sect. 2.14. As the temperature decreases, more and more grains develop twins. Generally, in the competition between twin and dislocation motion as primary mechanism of deformation, it is this latter to prevail.

However, as temperature T_1 is approached, twinning becomes more and more dominant reducing the growth rate of the yield strength, curve 1 of Fig. 5.2. This behavior is typical of medium–low strength steels [3–6] and is shown in Fig. 5.4 for a SAE 1020 steel [7]. This growing percentage of twinning with decreasing temperature can be seen in Fig. 5.5 [3] for a mild steel and a manganese steel. It can be seen how the yield strength σ_y, the ultimate strength σ_u and the true stress σ_t of both steels reach a common value of about 700 MPa and at $-200°$ C that can be considered the T_1 temperature for these steels. The temperature T_2 is about $-145°$ C. The onset of twinning in some grains starts at about this last temperature. At $-200°$ C twinning has already interested 25–26% of grains. Still slip by dislocation motion is the dominant deformation mechanism, but below $-200°$ C twinning plays an important role in the deformation mechanism starting to slow down the growth of the yield strength. Note that the difference between the two steels that have practically the same yield and ultimate strength, is ductility; the manganese steel is more ductile than mild steel. This difference can be understood through light microscopy. The grain size of the manganese steel is equal to that of mild steel, but the low manganese steel has grain boundary carbide films up to about 8 μm in thickness the other, instead, contains much thinner carbide films. In the competition between slip and twinning it may happen that twinning totally prevails and cuts off completely the growth process of yielding by slip because twinning develops at a stress $\sigma_{y,twin}$ lower than the corresponding $\sigma_{y,slip}$. When this is the case, the T_1 temperature is precisely T_{qs} of Fig. 2.44 for the quasi-static loading condition typical of a traction test.

From this point down, temperature is no longer the controlling parameter because twinning is temperature independent, see Sect. 2.14, and the yield stress becomes constant, line 2 of Fig. 5.2. This behavior is typical of molybdenum [8] and tungsten [9]. Line 3, at last, is typical of medium–high strength steels. This behavior can be seen in a NiMoV forging, as shown in Fig. 5.6 [10], which is a martensitic steel having a BCT crystal structure. High strength materials are characterized by small grains. Grain size is important for the onset of twinning. Twins generally occur in

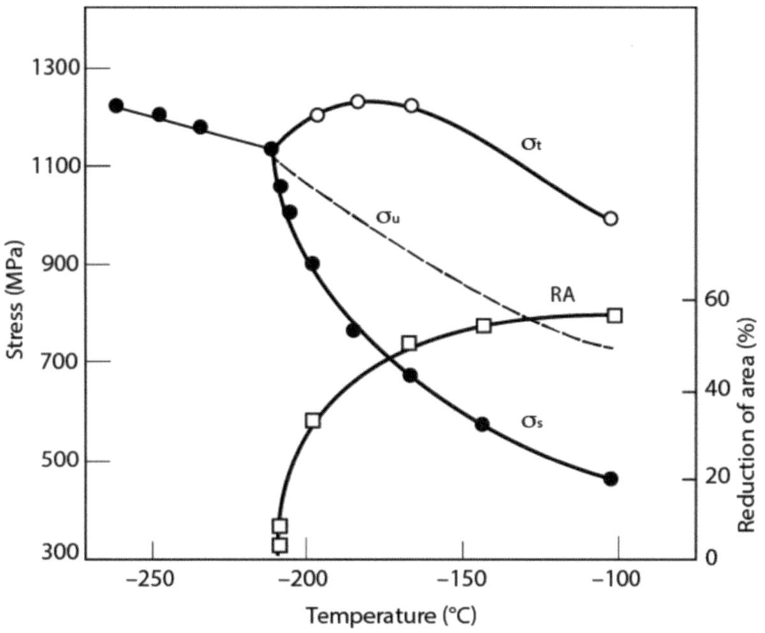

Fig. 5.4 The effect of testing temperature on the yield strength σ_s, true stress σ_t at failure and ductility, as RA, in a SAE 1020 steel (adapted from [7])

large grains, larger than 60 μm, and, in practice, are not observed at all in small size grains. However, a decrease of temperature shifts the grain size favorable to the onset of twinning toward lowerer values, as documented in Fig. 2.44. For this reason, in high strength martensitic steels twinning can prevail over slip. For these types of steel, the strength reduction below the T_l temperature is due to microcracks that once formed in a grain lead to instantaneous fracture because of the extremely low toughness of the material. As said in Sect. 2.17, micro twinning can easily reach 1 μm, which actually means that, if blocked, it is equivalent to a pileup of ~10^4 dislocations.

This explains why the relative stress fields are so high and why twinning is so an important cause of crack nucleation in coarse grain BCC and BCT metals at low temperatures. Looking at the stress–strain curves obtained by a standard traction test, mechanical strength response of BCC crystal lattice metals at different temperatures takes the form presented in Fig. 5.7 Note how a temperature decrease reduces the ductility and, therefore, the toughness of the metal and plays the same role as the carbon content shown in Fig. 1.16. This is the case of 99.95% purity molybdenum wires, containing 140 ppm C, 6 ppm O and 8 ppm N, whose stress–strain characteristic as function of temperature is shown in Fig. 5.8 [11].

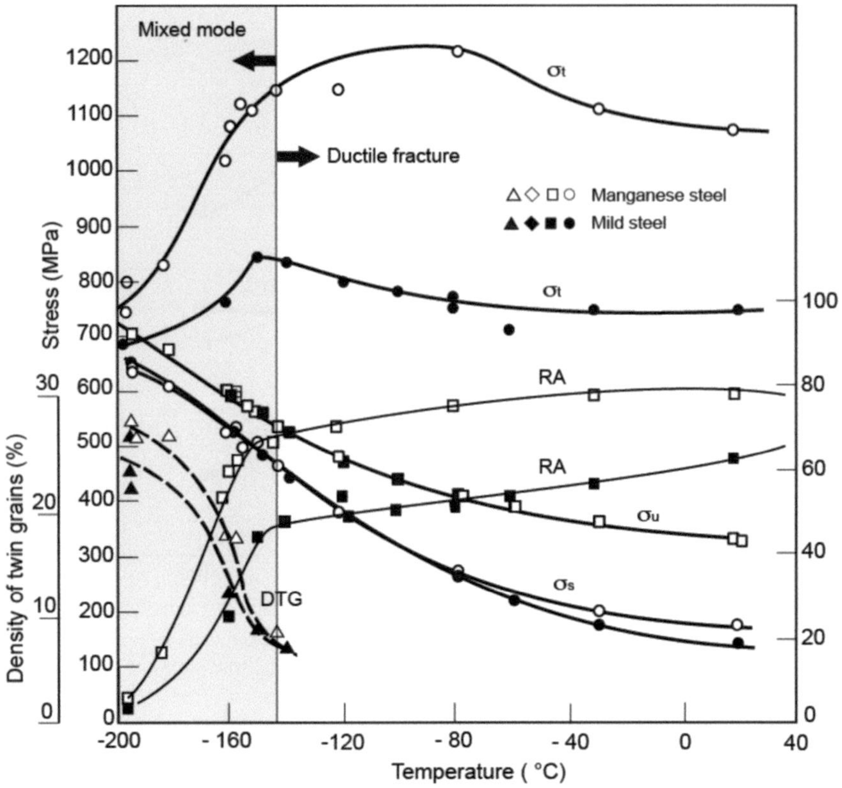

Fig. 5.5 Trend of yield strength σ_s, ultimate strength σ_u, true stress σ_t and ductility RA with temperature in mild steel and manganese steel. Also shown is the percentage of twinned grains (DTG) with temperature (adapted from [3])

5.1.2 Temperature Effect on FCC Crystal Lattice Materials

The behavior of materials having an FCC crystal structure appears to be completely different from that of BCC metals, just presented in previous section.

The main difference stems from dislocations that in FCC materials are wide (see Fig. 2.33). A wide dislocation is characterized by an almost negligible Peierls-Nabarro barrier and Cottrell atmosphere as well, which is to say that they have almost no thermal component of the yield stress.

A typical stress–strain temperature dependence of FCC high purity metals is shown in Fig. 5.9. The trend is opposite to that of Fig. 5.7. Now the yield stress is practically constant, independent of temperature while ductility, expressed as area under the stress–strain curve, is increasing with decreasing temperature. This characteristic behavior can be seen in Fig. 5.10 for high purity aluminum [12]. About the same behavior can be seen in Fig. 5.11 for another FCC high purity metal: silver [13]. Again, the yield strength is almost constant while the true fracture stress is varying

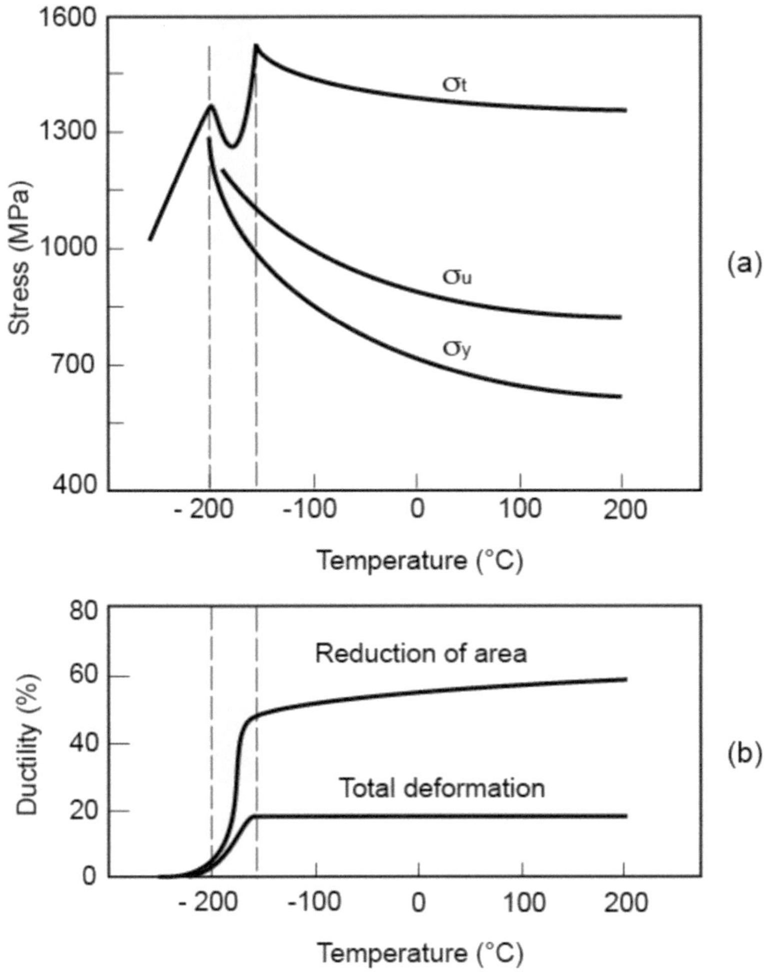

Fig. 5.6 Variation of true stress, ultimate strength, yield strength and ductility in a NiMoV forging (adapted from [10])

inversely with temperature. Parallel to that, the ductility at failure is increasing with decreasing temperature. At -253 °C the deformation at fracture is more than 50% while at room temperature is only 30%. At variance with BCC crystal structure materials whose ductility almost fall to zero at very low temperatures, the ductility of this materials never falls to zero. These metals remain ductile also at very low temperatures; they may rather increase their ductility. This is the reason why FCC crystal lattice materials are used for cryogenic applications.

Another way to present the data is that of Fig. 5.12 that refers to high purity nickel and Fig. 5.13 that pertains to OFHC copper [14]. Again, we notice the trend opposite to that of Figs. 5.4 and 5.5 relative to two BCC materials. The yield strength and the

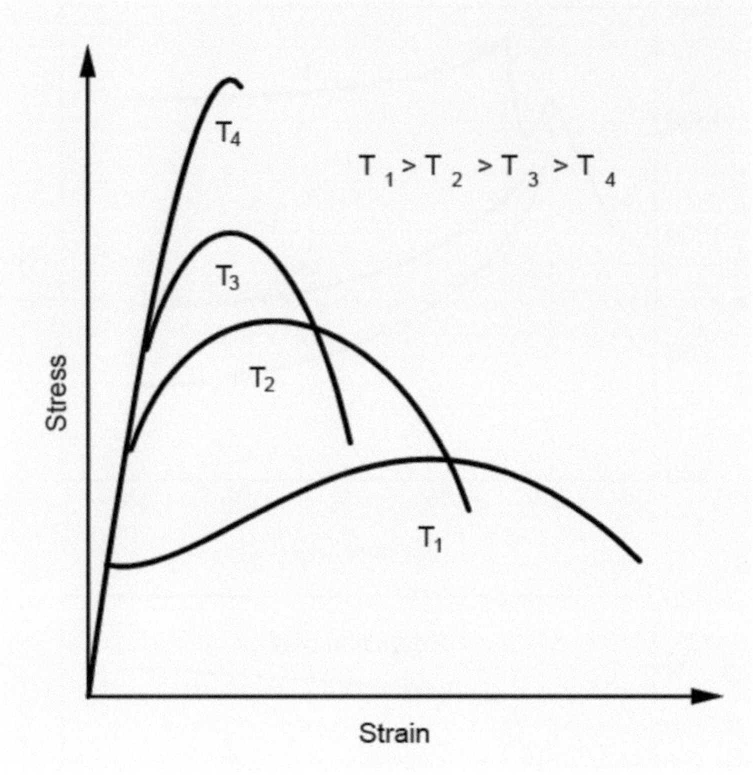

Fig. 5.7 Effect of temperature on the stress–strain characteristic of a BCC crystal lattice material

ultimate strength do merge to a common value, but this time it happens at very high temperatures while they are far apart at very low temperatures. The ultimate strength is sharply varying with temperature while the yield strength is practically stable from −200° C to 300° C, denoting the absence of a thermal activated component.

The true stress that in BCC metals disappears at very low temperatures because of the absence of ductility now has a very high value right at very low temperature. Things may change moving from high purity FCC metals to their alloys. The presence of fine dispersed particles in the FCC matrix that elevate the strength of pure metals by precipitation hardening changes their mechanical behavior. For instance, if zinc, magnesium and copper, in particular, are added to high purity aluminum shown in Fig. 5.10, after tempering to get the 7075-T6 alloy, the behavior changes from that of pure aluminum, as shown in Fig. 5.14 [15].

As can be seen, impurities introduce in the alloy a certain, though minimal, thermal component that is almost absent in the pure metal. This should not come unexpected since impurities that act as barriers to dislocation motion obey to thermal diffusion laws. The yield strength of 7075-T6 increases from 480 MPa at room temperature to about 650 MPa at the very low temperature of 296 °C, with a 26% increase.

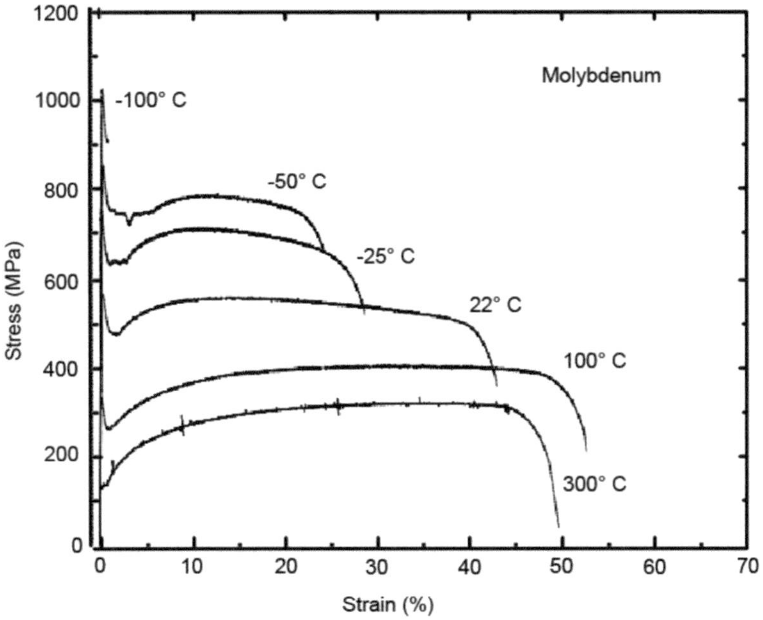

Fig. 5.8 Effect of temperature on the stress–strain characteristic of low carbon arc cast molybdenum (adapted from [11])

Ductility decreases but still remains high and practically constant even at very low temperatures. This behavior is shown also by other aluminum alloys, as it can be seen in Fig. 5.15 [15]. This flat trend of the ductility, practically independent of temperature, that remains quite large also at very low temperatures reveals the absence of NDT and toughness independent of temperature. The same behavior can be seen in austenitic stainless steels. Austenite has an FCC crystal structure. Figure 5.16 presents the mechanical behavior of AISI 304 SS [16–18]. It is interesting to note how the yield strength remains practically constant from −200° C to about 70 °C while the ultimate strength goes from about 400 MPa at 200 °C to 1675 MPa at −250 °C; a four-fold jump. Impressive is the elongation that even below zero remains constantly at 40%. This alloy cannot be fractured by cleavage. A crack under tension will soon become a notch by tip blunting. As to HCP materials, their behavior can be placed between that of BCC and that of FCC crystal lattice materials, anyhow closer to BCC than to FCC metals and alloys. Figure 5.17 presents the strength and ductility response of α-β 6Al-4V titanium alloy [14]. Aluminum stabilizes the α-phase (HCP structure) while vanadium stabilizes the β-phase (BCC structure). The sharp elevation in strength can be seen with decreasing temperature and the reduction of ductility that seems to go down to zero at very low temperatures. Stress–strain curves of annealed α-titanium at different temperatures are shown in Fig. 5.18 [19].

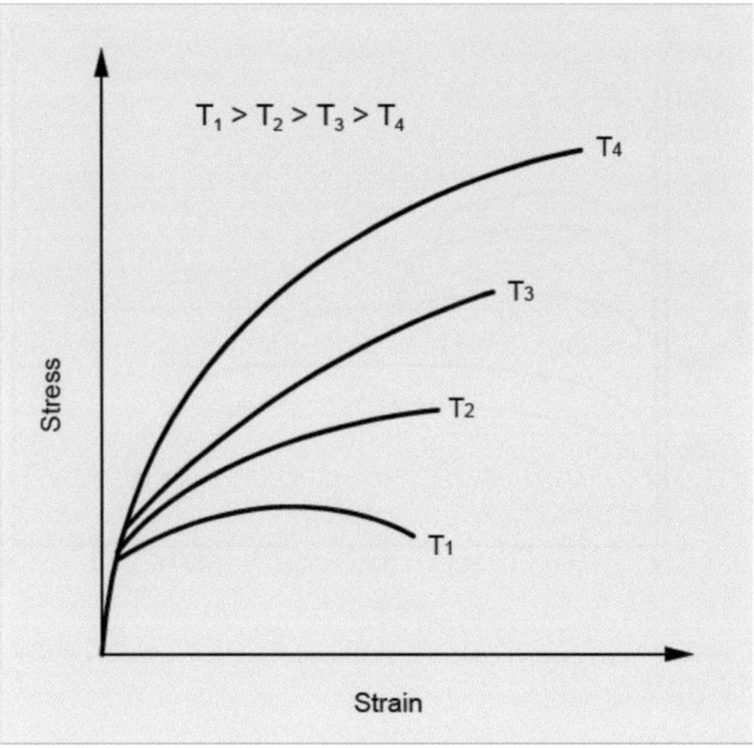

Fig. 5.9 Effect of temperature on the stress–strain characteristic of an FCC crystal lattice material

5.2 Thermal and Athermal Component of Flow Stress

We have seen in Sect. 2.15 Eq. (2.28) that the yield strength, σ_s, of a polycrystalline metal can be split in two terms one of which, $kd^{-1/2}$, depends on the grain size d and is independent of temperature while the other, σ_o, depends on temperature and represents the thermal component, σ_T, of the yield strength. Let's try to have a better look at these two components and develop a model-based phenomenology that is a rather fruitful approach when dealing with very complicated phenomena. During a quasi-static traction test in which the strain rate is relatively low (10^{-4}-10^{-3} s^{-1}, see Sect. 1.13), the linear elastic phase terminates with the first massive movement of dislocations. At the beginning, the density of dislocations in the virgin material in its annealed condition may be rather low, of the order of 10^7–10^8 cm^{-2}, as seen in Sect. 2.1 The first barrier that these relatively few dislocations must overcome in order to start moving is crystal friction. We have seen that the energy barrier due to the lattice resistance of a perfect crystal lattice without dislocations is exceptionally high, about as large as the shear modulus (see Sect. 2.6). With such large a shear strength, metal cannot be worked at all. We owe to intercrystalline defects the ability to work metals and the progress achieved. Due to these defects, dislocations are sitting on a

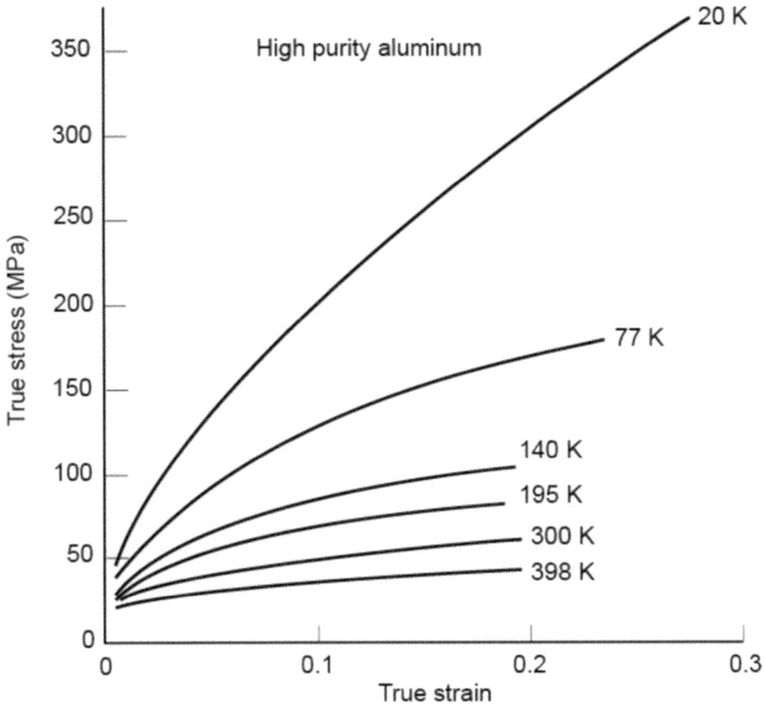

Fig. 5.10 Effect of temperature on the true stress-true strain characteristic of high purity aluminum (adapted from [12])

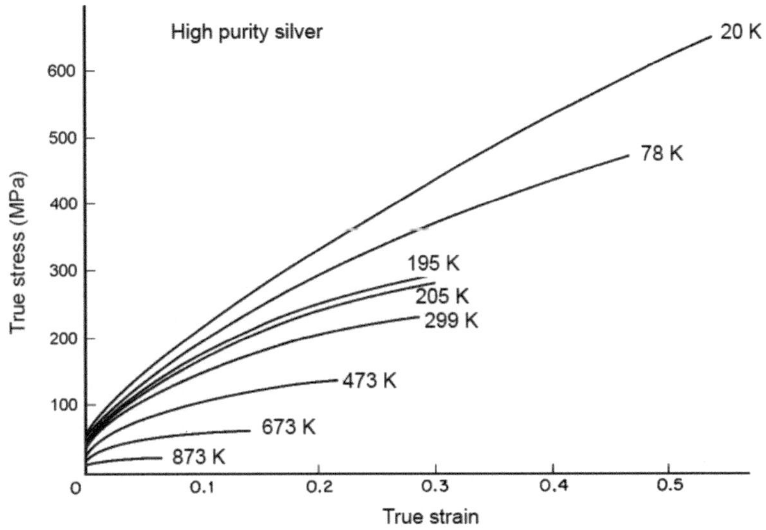

Fig. 5.11 Effect of temperature on the true stress-true strain characteristic of high purity silver (adapted from [13])

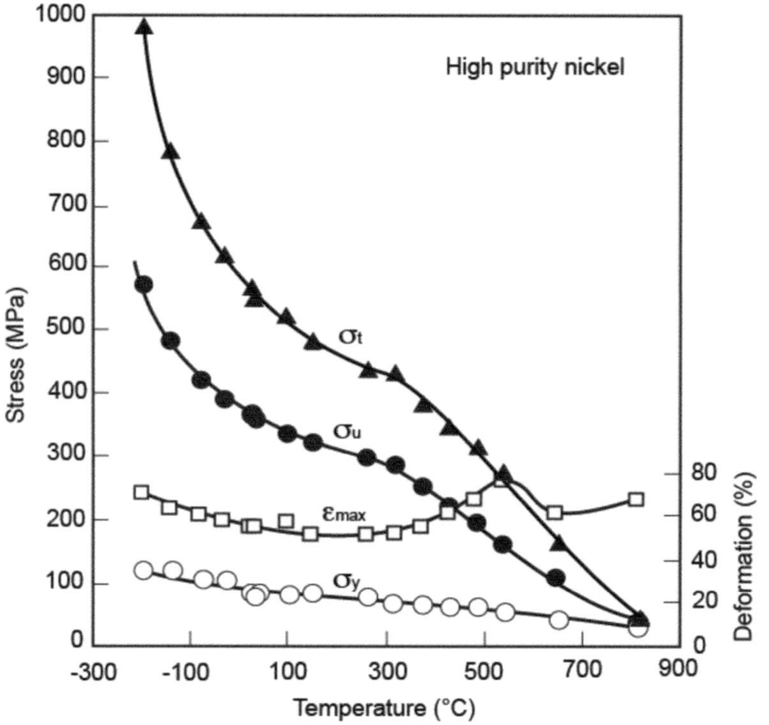

Fig. 5.12 Variation of true stress, ultimate strength, yield strength and deformation at maximum load in high purity nickel (constructed with data from [14])

metastable equilibrium in the Peierls well whose eight is the Peierls-Nabarro stress, τ_{PN} (see Sect. 2.10). In BCC crystal lattice materials that have narrow dislocations (see Sect. 2.10), this barrier is rather high, though much smaller than the friction stress of a perfect crystal, but also thermally activated which means that its eight can be reduced or even eliminated by a proper temperature increase. Temperature provides part if not all the energy needed to overcome the obstacle, as described in Sect. 2.11. This is shown in Fig. 5.3 for iron and molybdenum whose yield strength sharply reduces till almost vanishing with increasing temperature. At variance, FCC metals that have wide dislocations have almost negligible Peierls-Nabarro stress, τ_{PN} so that slip can proceed much more freely on the many slip systems available, almost independently of temperature or with a negligible effect that still remains, as shown in Figs. 5.9, 5.10 and 5.15. If we look at the barrier in terms of Cottrell's atmosphere, to move dislocations requires the concurrent displacement of the associated atmosphere (see Sect. 2.11). In BCC crystal materials this component of the yield strength can be large and thermally activated up to the point where it vanishes completely as temperature becomes so high to dissolve the Cottrell's atmosphere, while it is almost negligible in FCC alloys. In polycrystalline materials, it is likely that the first plastic slips are activated on some crystals on the surface of the specimen where triaxiality

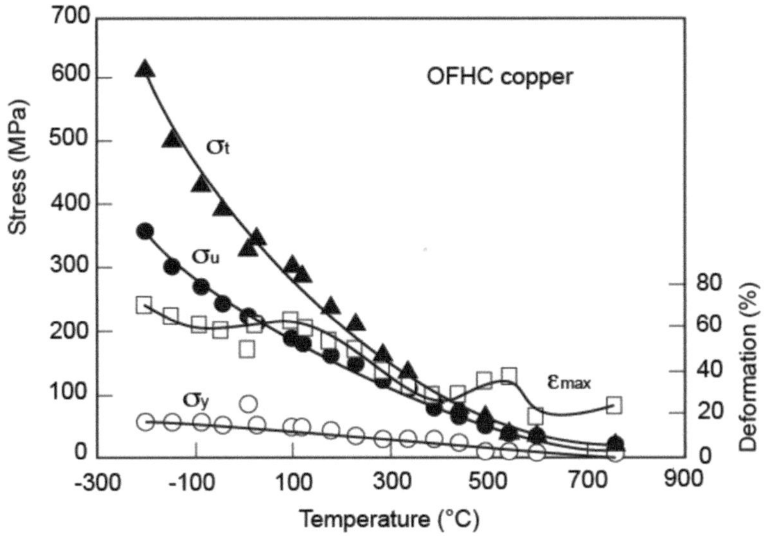

Fig. 5.13 Variation of true stress, ultimate strength, yield strength and deformation at maximum load in OFHC copper (constructed with data from [14])

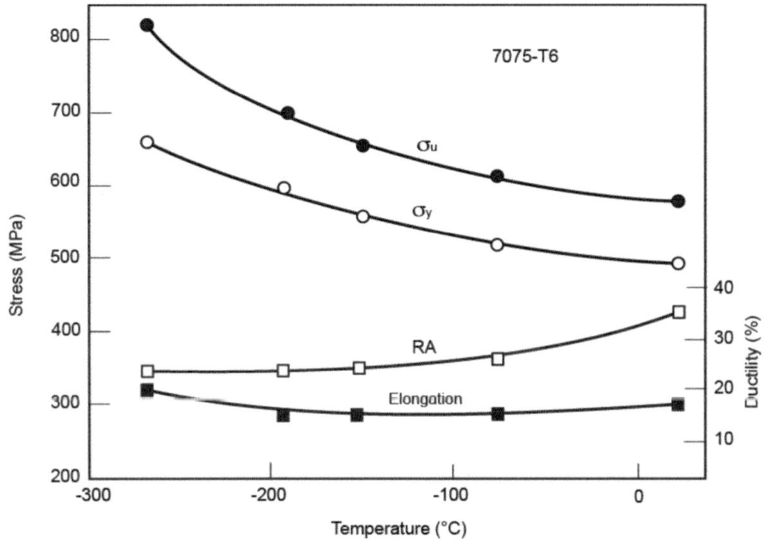

Fig. 5.14 Variation of ultimate strength, yield strength and ductility in 7075-T6 aluminum alloy (constructed with data from [15])

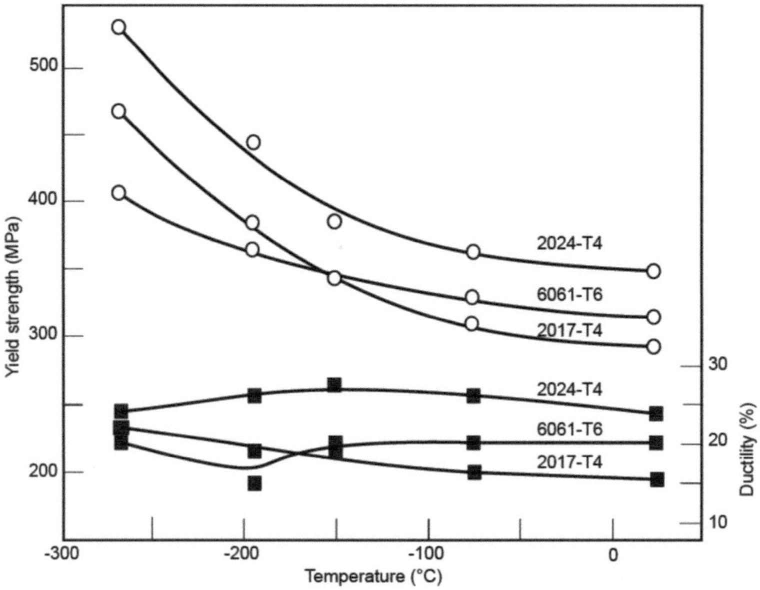

Fig. 5.15 Variation of the yield strength and ductility with temperature in several aluminum alloys (constructed with data from [15])

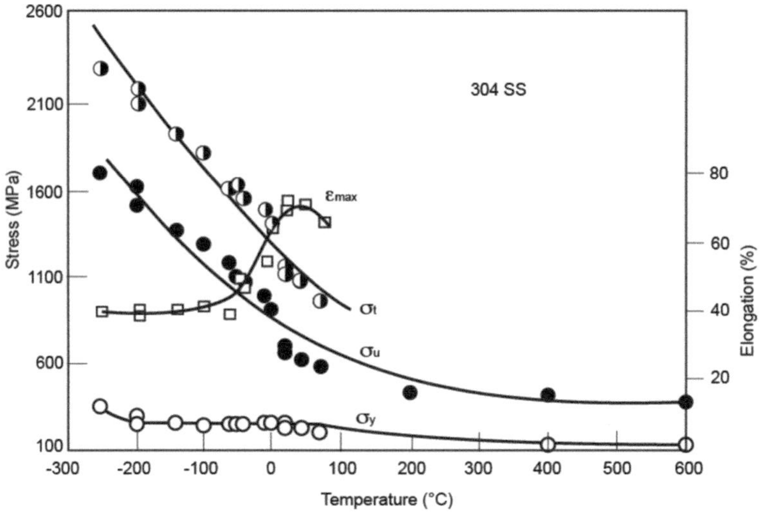

Fig. 5.16 Variation of the yield strength and elongation with temperature in 304 stainless steel (constructed with data from [16–18])

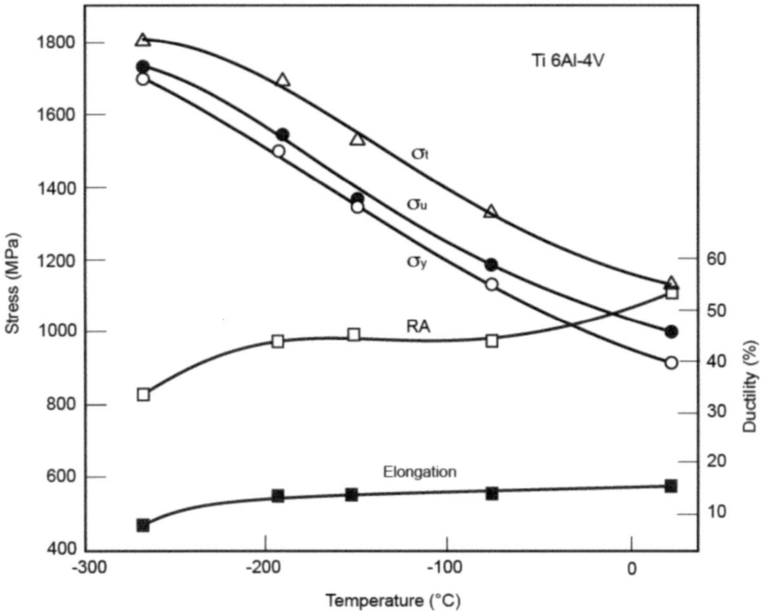

Fig. 5.17 Variation of true stress, ultimate strength, yield strength and ductility in Ti 6Al-4V alloy (adapted from [15])

cannot develop and slip is not restricted by neighboring crystals as in the center of the specimen section. These *free to slip* crystals are those better oriented with respect to the external load, meaning that they have the largest Schmidt factor and require the lowest critical resolved shear stress to slip, τ_{crss} (see Sect. 2.4). These first plastic slips happen when the whole section is still well in the elastic regime and may be almost inessential; it just increases the density of dislocations. Continuing to increase the applied load, more and more grains yield. Once a massive slip process initiates it must propagate to other grains to produce a macroscopic plastic strain and in doing so dislocations must move toward grain boundary. The motion of dislocations to the grain boundary accounts for the commonly observed pre-yield microstrain. First the upper yield stress appears with the formation of the initial Lüders lines and band (see Sect. 2.2). The lower yield stress is then obtained through the propagation of the Lüders lines and band. Here is when the athermal component of the yield strength, $kd^{-1/2}$, enters into action. Grain boundaries, in fact, are another barrier that blocks dislocation. However, even before reaching the grain borders, the motion of dislocations is obstacle limited. Dislocations moving through the matrix encounter barriers that are within the grain itself. These obstacles are voids, vacancies, foreign phases, inclusions or precipitates and in general, any foreign particle. Dislocations can either cut through the particles or bend around and bypass them. It is questionable whether or not their action on this initial dislocation motion can be thermally activated. It

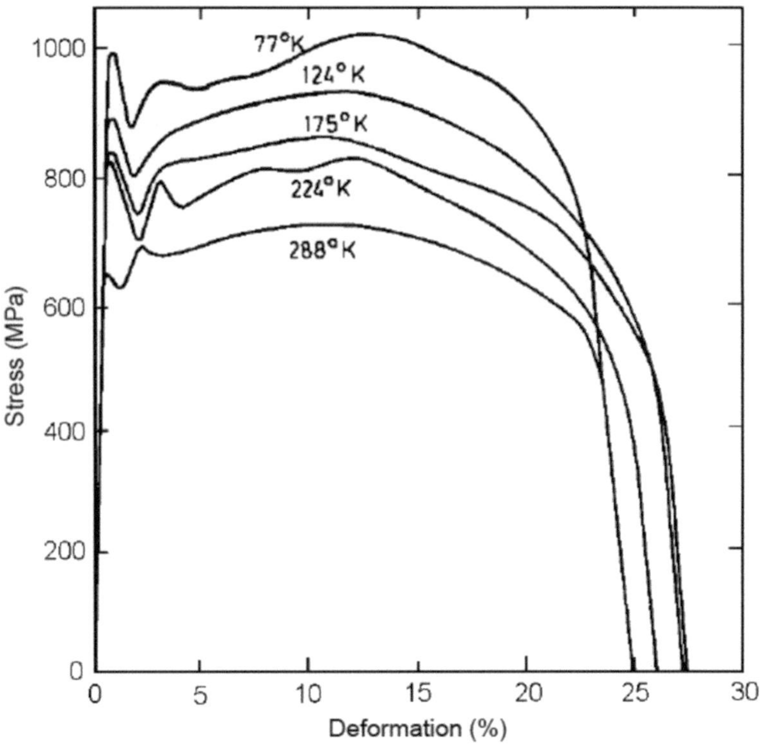

Fig. 5.18 Stress–strain characteristic of annealed α-Titanium at various temperatures (adapted from [19])

may be argued that their mobility within the crystal lattice is governed by a diffusion process and, as such, also their possible blocking action on dislocations should be thermally activated. From a theoretical point of view above 0 K any problem becomes a problem of atomic or molecular kinetic. As matter of fact, testing high purity FCC crystal lattice metals such as aluminum, silver, OFHC copper and nickel, either single crystal or polycrystalline, it can be noted that there is not any significant variation in the yield strength in a temperature range from −253 °C to 100 °C, as previously shown. Consider that, just as an example of temperature dependence, the yield strength of molybdenum, which has a BCC crystal lattice, decreases from 1500 MPa at −200 °C to about 140 MPa at 300 °C, a tenfold decrease! In high purity metals foreign particles are nearly absent, bur vacancies still exists. Voids and vacancies are thought to lock dislocations by forming atmospheres around them in the same way as solute atoms do [20] or, even more intriguing, can facilitate dislocation movements in aluminum through what is referred to as the lubrication effect [21]. If we look at aluminum alloys whose strength is precisely due to dispersed coherent particles, their yield strength may be varying by no more than 20 to 30% from −270 °C to 100 °C. This is the case of 2024-T4 or 6061-T6 or 7075-T6 aluminum alloys, already seen in Figs. 5.14 and 4.15. It must be considered that the shear modulus normally increases with decreasing temperature. Therefore, if we consider that in the same temperature range the shear modulus of those alloys vary by some 10%, then the possible effect of temperature on the blocking action of vacancies and particles on moving dislocation and, therefore, on the yield strength will reduce to 10 or 20% at most. Furthermore, if we subtract the athermal component of the yield strength, what is left may be varying with temperature by less than 10%. Small interstitial particles are more effective in raising the yield strength of a material than large substitutional elements. Interstitial particles feed the thermal component of the yield strength while large particles, whether single atoms or cluster or phases, block dislocations as almost immobile obstacles. This different effect can be seen in Figs. 5.19 and 5.20.

Figure 5.19 [22] shows the dependency of the critical resolved shear stress on temperature for several grades of zirconium differing in the amount of interstitial oxygen. As it can be seen, the effect of oxygen concentration is to increase the temperature-dependence of the critical resolved shear stress and, therefore, of the yield strength.

At variance, larger substitutional solutes (8 percent Al) tend to add a temperature-independent component to the yield strength increasing the level of the yield stress by roughly the same amount at all temperatures, as it can be appreciated in Fig. 5.20 [23]. This diagram also shows a third curve corresponding to the commercially pure state (CP) and alloy in the precipitation-hardened condition. Note that also in this case the yield stress is raised by roughly the same amount al all temperatures. We

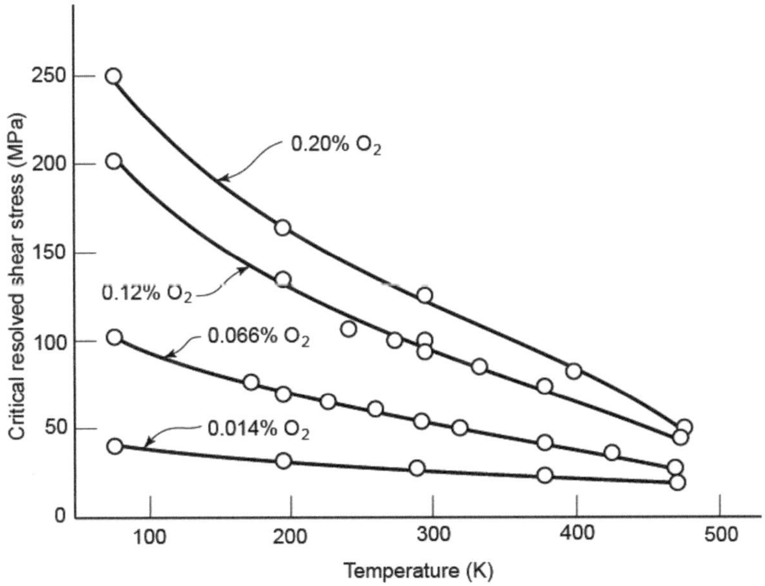

Fig. 5.19 Effect of oxygen on the temperature-dependence of the critical resolved shear stress of zirconium crystals (adapted from [22])

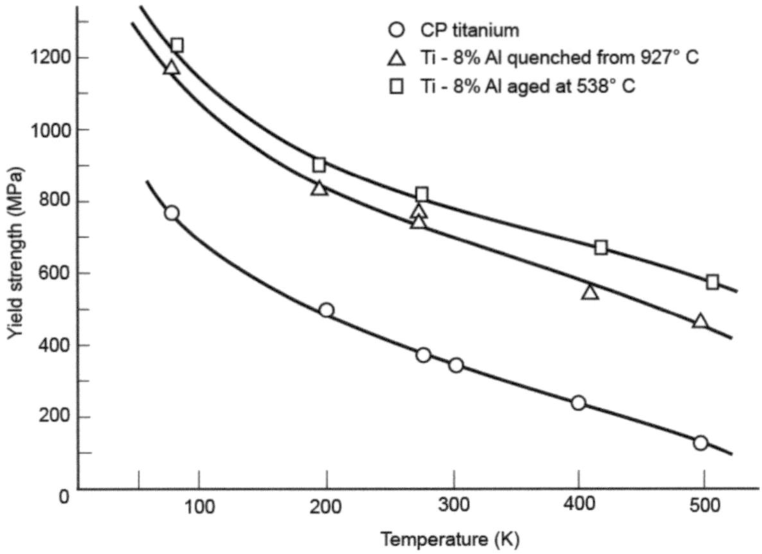

Fig. 5.20 Temperature-independent increase of the athermal component of the yield strength in titanium due to Al alloying (adapted from [23])

shall conclude that the addition of Al has no influence on the thermal component of the yield strength. Therefore, in BCC and HCP metals the action of small impurities is strong and strongly dependent on temperature, as shown in Fig. 5.3, because they can migrate towards dislocations and interact, while larger barriers are almost motionless and not sensitive to temperature. Conversely, FCC crystal lattice materials do not have a temperature activated component of the yield strength till the annealing-recrystallization temperature is reached, which changes the material properties.

For this reason, the transition between elastic and plastic behavior appears smooth and not-well defined, practically undetectable on the stress–strain curve (see Sect. 1.4, Fig. 1.16). At variance with that, temperature can strongly affect the ultimate strength. We can now make explicit the two components, σ_T thermally activated and σ_{at} athermal, of the yield strength

$$\sigma_s = \sigma_T + \sigma_{at} = (\sigma_{Ctrl})_T + \sigma_{obs} + kd^{-1/2} \tag{5.1}$$

where σ_{Ctrl} is the drag stress due to the Cottrell's atmosphere and σ_{obs} is the resistance to dislocations motion exerted by all other relatively large obstacles as precipitates, inclusions and foreign phases that is not quite affected by temperature, at least till temperature becomes so high that diffusive process become important. Introducing the deformation by twining in Sect. 2.14, we have seen that there exist two limits to the athermal component of the yield strength, an upper component and a lower one. The latter is reached when the applied strain rate $\dot{\varepsilon}$ slows down to an extremely low value, of the order of 10^{-9}-10^{-8} s^{-1}, the former when the strain rate is attained by shock waves generated by detonations, with $\dot{\varepsilon}$ higher than 10^9 s^{-1}. This upper limit is due to twinning that is almost independent of temperature and not so much sensible to strain rate as slip, while the lower limit is due to a slip process so slow that there is no obstacle really capable of blocking dislocation motion; it is so slow that it is almost, forgive me the paradox, a motion apparently motionless. The material does not even feel that something is moving in it. Obstacles to dislocation motion have different periodicity, spacing and length so that they will contribute in different manner to the creation of drag resistance to slip. Grains, for example, have a characteristic length d of the order of 100–1000 Å, much longer than that of σ_{Ctrl} whose range is about 10 Å. Therefore, it is good custom to group the components of the yield strength that appear in Eq. (5.1) into a short range component σ_{SR}, which is mostly thermally activated, and a long range component σ_{LR}, typically the grain size, which is athermal

$$\sigma_s = \sigma_{SR} + \sigma_{LR} = (\sigma_{Ctrl})_T + \sigma_{obs} + kd^{-1/2}. \tag{5.2}$$

Continuing to load and strain the material beyond the Lüders strains, things become more complicated. New obstacles are continuously created along their course. Also these obstacles have different periodicity, spacing and length so that they will contribute in different manner to the creation of a continuous drag resistance to slip. Plastic deformation increases the vacancy density and the dislocation density as well, as explained in Sect. 2.12. The enormous increase in the dislocation density results in a continuous garbling that creates tangles and forests. Tangles generate and fill grains of sub-grains or cells. Their resistance to dislocation motion is similar to that of grains, but with a different strength since the $d^{-1/2}$ term is replaced by the correspondent $c^{-1/2}$, where c is the cell size ad it is $c < d$. However, misorientations between the different cells are very small in comparison to those of regular grains and misorientation is an obstacle itself. The result is that dislocations move easier across cell boundaries than across grain boundaries. Moving dislocations now have to cut through forests of dislocation that oppose or block their motion and overcome tangles and other long range barriers. In this chaotic entangling, sessile dislocations are also created that will not move anymore and block other dislocations. Hardening is the overall result which is also the cause of further obstacles to the dislocations motion. But things do not end here. In addition to these physical obstacles there are some others and more complex types of barriers due to the crystal lattice itself. A crystal lattice is made of atoms in equilibrium positions that are not still but vibrate. These vibrations generate an elastic wave known as phonon wave that propagates through the crystal lattice. Dislocation motion produces a sort of continuous compression and dilatation of the surrounding zones that generates an elastic wave. These elastic waves can interfere with phonon waves creating what is known as phonon drag force to dislocation motion. In a well-known paper, Kocks et al. [24] indicated that a dislocation of length l had a frequency of vibration given by

$$\nu_o = \frac{\nu b}{4l} \tag{5.3}$$

where v is the frequency of vibration of an atom. The expression is similar to that for the frequency of a string, where frequency depends on string length. Atoms vibrate at about 10^{13} cycles per second. Since a dislocation is longer, it will vibrate at a much lower frequency. For $l = 250b$, a reasonable value for deformed metals, the frequency of vibration will be $v = 10^{10}\ \mathrm{s}^{-1}$. However, the dislocation spacing l can be related to the density ρ

$$l \cong \rho^{-\frac{1}{2}}. \tag{5.4}$$

A metal in its annealed condition may have a dislocation density of 10^7–10^8 cm^{-2}. Cold work is able to increase the number of dislocations by a factor as large as 10,000 to 1,000,000. This actually means that dislocations in heavily deformed conditions can possibly approach the atomic vibration frequency. Also, free electrons can interfere with moving dislocations in the same manner as the phonon waves. However, their effect becomes noticeable only below 100 K. In 1954 Seeger made the hypothesis that the entire flow stress of crystals consists of two components, an athermal component σ_{at} which is proportional to the Young's modulus and a thermal component σ_T that depends on temperature T and strain rate $\dot{\varepsilon}$. The effect of temperature over those two components is schematized in Fig. 5.21 [25]. Up to about 0.2 T_m, where T_m is the absolute melting temperature of the material, there is a strong temperature effect, afterward an almost flat region appears that terminates at about 0.7 T_m when the flow stress decreases dramatically, as we have already seen in the previous sections, because the material enters the annealing temperature field where dynamic recovery or dynamic recrystallization starts to act producing softening. Note how the flow curves merge in A at very low temperature where the upper yield strength is reached, and this limit is due to twinning (see Sect. 2.14 and Fig. 2.44). Therefore, we may describe the entire flow stress σ of any material by separating the different components of strength in the more general fashion as already done for the yield strength

$$\sigma = \sigma_{at} + \sigma_T. \tag{5.5}$$

The thermal component, σ_T, depends on the crystal structure, temperature and strain rate $\dot{\varepsilon}$ and can be split, in turn, in two further components so that Eq. (5.5) becomes

$$\sigma = \sigma_{Ctrl}(T, \dot{\varepsilon}, structure) + \sigma_{LR}(\dot{\varepsilon}, structure) \tag{5.6}$$

where, again, σ_{LR} is the long range component of the strength to differentiate from σ_{Ctrl} whose range, as said, is about 10 Å and represents, therefore, the short range barrier. Equation (5.6) can be written to evidence the short-range component as

$$\sigma = \sigma_{SR}(T, \dot{\varepsilon}, structure) + \sigma_{LR}(\dot{\varepsilon}, structure). \tag{5.7}$$

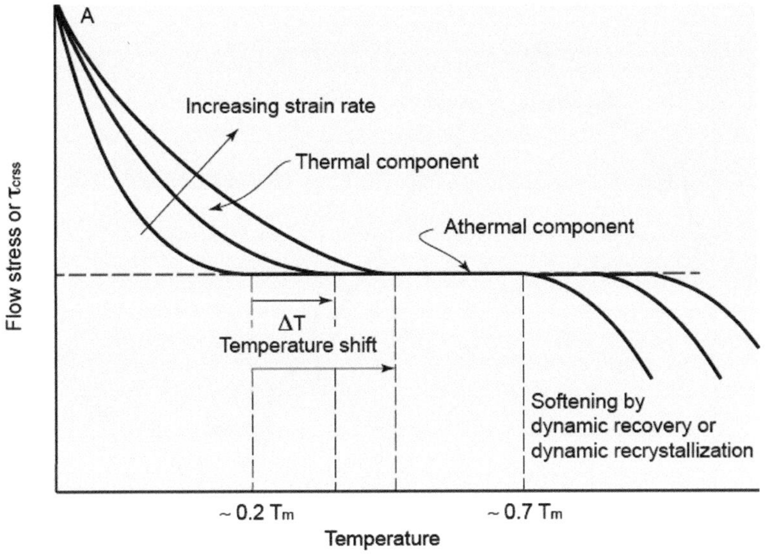

Fig. 5.21 Schematic of the flow stress dependence on temperature (modified from [25])

This representation of the flow stress in terms of shear stress can be schematized as shown in Fig. 5.22 [26] as a big and wide hill with short-range obstacles, shown as little *pimples*. The athermal or long-range part of the flow stress represents obstacles whose free energy is so high that thermal energy does not provide any significant contribution to the slip process. As to this term, the energy needed to overcome obstacles is entirely mechanical and is provided by the external work done on dislocations.

5.3 Strain Rate Effect

The strain rate effect on the strength of materials has received rather little attention in the past. Besides the scientific interest, on the practical ground there was not such a large demand of knowledge on dynamic response of materials. Earthquake loading and marine collisions would result in strain rates ranging from 0.1 s^{-1} to 10 s^{-1} and land transport or aircraft undercarriage would go from 10 s^{-1} to 10^3 s^{-1}. It has been increasing interest in military applications, explosion and ballistic producing strain rates of the order of 10^4 to over 10^6 s^{-1}, the use of impact resisting materials and, in particular, the metalworking technology that has broaden the study of the metal response to dynamic loading conditions. Typical deformation speed and strain rates applied in the metalworking processes are indicated, as an example, in Table 5.1.

As can be seen, most metalworking processes enter the field of high to very high strain rates. This is also why computer code producers have lately addressed the

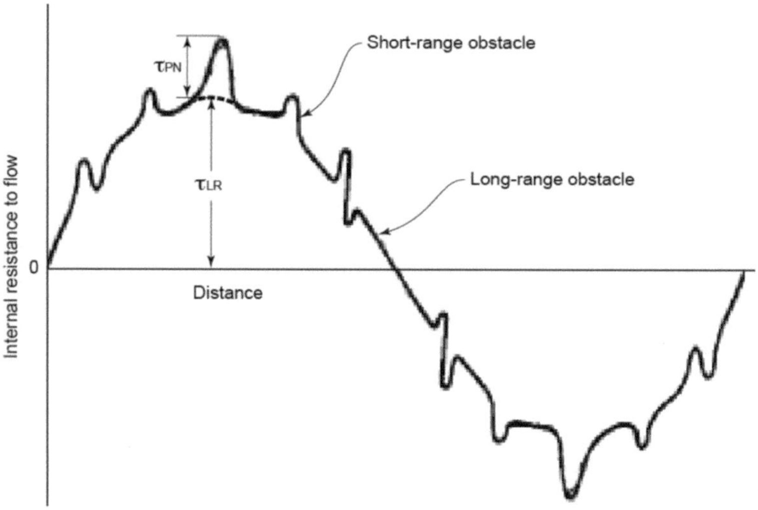

Fig. 5.22 Schematic representation of different barriers encountered by dislocation as it moves (adapted from [26])

Table 5.1 Typical ranges of strain, deformation speed and strain rates in metalworking

Process	True strain	Deformation speed (m/s)	Strain rates (s^{-1})
Cold working			
Forging, rolling	0.1–0.5	0.1–100	$1–10^3$
Wire & tube drawing	0.05–0.5	0.1–100	$1–10^4$
Explosive forming	0.05–0.2	10–100	$10–10^5$
Hot & warm working			
Forging, rolling	0.1–0.5	0.1–30	$1–10^3$
Extrusion	2–5	0.1–1	$10^{-1} – 10^2$
Machining	1–10	0.1–100	$10^3–10^6$
Sheet-metal forming	0.1–0.5	0.05–2	$1–10^2$
Superplastic forming	0.2–3	$10^{-4}–10^{-2}$	$10–4–10^{-2}$

issue of strain rate effects willing to offer a new numerical tool to FE users and this is why we shall address with particular attention the subject. We have seen in Sect. 2.11 how in BCC crystal lattice material, in particular, the motion of dislocations under the action of a shear stress is restrained by the Cottrell atmosphere. When dislocations are at rest there will be no interaction at all between the dislocation and its atmosphere. As the dislocation moves, its atmosphere of interstitial atoms starts to exert a drag force opposing the motion. The basic mechanism through which this interaction operates is basically related to the different activation mode: dislocations are moved by shear stresses while interstitials atoms of their atmospheres respond to the atomic diffusion law controlled by temperature, as has been explained in the previous section. Accordingly, this resistance to motion reduces more and more as the dislocation velocity decreases. Eventually, at extremely low velocity, dislocations can move without any apparent resistance because the mobility of interstitials by atomic diffusion becomes comparable to that of dislocations. When this happens, atmospheres can form and dissolve around moving dislocations without any lag. Likewise, the anchor action on dislocations also reduces when their velocity becomes extremely high. At very high velocity, in fact, the atoms of the atmosphere were left behind because too slow, have no time to catch the running dislocation. Dislocations, then, can move through the crystal structure freely without any impediment, as if they would not have any atmosphere at all. In both cases, the concept itself of atmosphere loses any meaning. There are only atoms in solid solution. Apart from those two extreme cases, the problem arises with the first, statically formed atmosphere that is left behind by moving dislocations and has to dissolve in the metal matrix, because no other atmosphere has time to form during the high velocity motion of the dislocation. Between the two extremes, rest or very low velocity and extremely high dislocation velocity, the maximum drag stress will be achieved.

This is schematically shown in Fig. 5.23a [27]. It is clear that since dislocation velocity also depends on the applied strain rate, $\dot{\varepsilon}= d\varepsilon/dt$, the same dependence of the drag resistance on applied strain rate will be observed, as shown in Fig. 5.23b. By varying the temperature, the general trend shown in Fig. 5.23a will still be observed with the maximum drag stress always the same, because it corresponds to the maximum blockage action, but shifted toward higher velocities, as shown in Fig. 5.23c. Finally, the direct effect of temperature at constant strain rate is shown in Fig. 5.23d. It has precisely the same trend seen in Fig. 5.23a, except that temperature decreases on the T axis. At very low temperature atomic mobility is so poor that the atmosphere makes almost a solid block with the dislocation. Since the former doesn't move, the latter will not move. This is why ferritic steels, even mild steels, regain strength by lowering the temperature till they reach have the same strength of very hard-high-strength steels. These last steels, in fact, having a martensitic structure which is not precisely a BCC crystal structure, see Sect. 2.1, do not have a strong component of the yield strength thermally activated. Their strength is almost independent or very few dependent on temperature because it is already in the uppermost range of values and is there is not that much room left to temperature to further increase it. Continuing with BCC crystal lattice materials, at very high temperatures atomic mobility is so high that the Cottrell atmosphere vanishes and does not exist anymore. This is because its interstitial atoms escape everywhere and the atmosphere, formed at lower temperatures, simply dissolves. This relatively simple phenomenological description of the drag resistance to dislocation motion yet contains all salient features associated with more sophisticated modelling and can be used to derive an expression for the drag stress. It can be assumed [28] that the elemental variation $d\sigma_s$ of the yield strength, σ_y, due to a temperature variation dT is proportional to the variation $d\lambda$ of the ratio $\lambda = n_C/n_{ss}$ of the concentration of interstitial atoms of the Cottrell atmosphere, n_C, to that, n_{ss}, of interstitials in solid solution

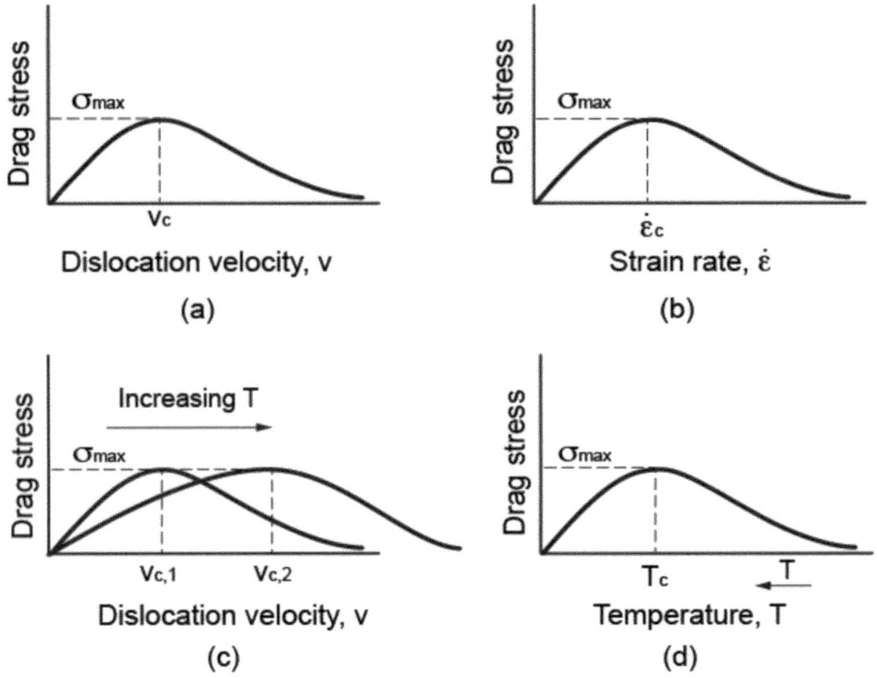

Fig. 5.23 Variation of the drag stress with, **a** the dislocation velocity, **b** the strain rate, **c** the dislocation velocity at two different temperatures and d) the temperature at constant strain rate (modified from [27])

$$\frac{d\sigma_s}{\sigma_s} = \frac{d\lambda}{\lambda}. \tag{5.8}$$

Integrating it yields

$$\int_{T_o}^{T} \frac{d\sigma_s}{\sigma_s} = \int_{T_o}^{T} \frac{d\lambda}{\lambda} \tag{5.9}$$

where T_o is a reference temperature. Solving Eq. (5.9) it is

$$\ln[\sigma_s(T_o)] - \ln[\sigma_s(T)] = \ln[\sigma_s(\lambda_o)] - \ln[\sigma_s(\lambda)] \tag{5.10}$$

in which $\sigma_s(T_o)$ and $\lambda\ (T_o)$ are a reference stress and a reference concentration, respectively. The concentration of interstitials can be expressed through an equation of the type of Eq. (2.17)

$$n_C = a \cdot e^{-\frac{b}{kT}}\ ;\ n_{ss} = c \cdot e^{-\frac{d}{kT}}\ \lambda = \frac{a}{c} \cdot e^{\frac{g}{kT}} \tag{5.11}$$

in which a, b, c and d are constants characteristic of the material and k is the Boltzmann constant equal to $N{\cdot}R$, where R is the gas constant equal to 2 cal/mole-K and N the Avogadro number. The exponent $g = d\text{-}b$ of the λ parameter is positive since the concentration n_C of interstitials in the Cottrell atmosphere is larger than that of the same interstitials n_{ss} in solid solution: $d\text{-}b > 0$. Substituting λ given by Eq. (5.11) into Eq. (5.10) it yields

$$\ln \sigma_s(T) = \frac{B}{T} + A \tag{5.12}$$

with

$$A = \left[\ln \sigma_s(T_o) - \tfrac{g}{kT_o}\right]; \; B = \tfrac{g}{k} \tag{5.13}$$

or

$$\sigma_s(T) = C \cdot e^{\frac{B}{T}} \tag{5.14}$$

were C is a constant characteristic of the material

$$C = e^A = \sigma_s(T_o) \cdot e^{-\frac{g}{kT_o}} \tag{5.15}$$

Equations (5.12) or (2.14) provides the dependence on temperature of the component of the yield stress, σ_s, due to the dislocation atmosphere. Note that Eq. (5.12) is a line in a semi-logarithmic scale $ln(\sigma_s)$-$1/T$, whose angular coefficient is B and A is the known term. We may say, then, that materials that have a thermal component of the yield strength, such as BCC crystal structure metals and alloys, follow a linear trend in a $ln(\sigma_s)$-$1/T$ scale.

This is shown in Fig. 5.24 for 14 steels ranging from a low strength mild steel to a very high strength 18% Ni cold worked. Extrapolating to zero the trend lines of Fig. 5.24 yields the value of constant A in Eq. (5.12). However, $1/T = 0$ means an infinite temperature, which is physically impossible since the upper value of T is the melting temperature T_m of the material.

Actually, the real upper value of T is ~ $0.2T_m$, as indicated in Fig. 5.21, since beyond that value the temperature dependence of the yield strength ceases, at least at a quasi-static strain rate (10^{-4} to 10^{-3} s^{-1}). For steels, $0.2T_m$ ~ 300° C so that trend lines should terminate at about $1/T$ = 0.0015–0.0017 1/K, as indicated in Fig. 5.24. For high strength steels, values of constant A vary from 898 to 1640 MPa. For medium–low strength steels, such as A 533B, SAE 1020 and A 302 B, A varies from 300 to 400 MPa. It is interesting to note how the slopes, i.e., the B factor of Eq. (5.12) of the various trend lines, are roughly following three main courses, depending on the strength and, therefore, structure of the steel. The higher slope, of the order of B ~ 125 K^{-1}, competes to low strength alloys, essentially low carbon ferritic steels, a medium slope of the order of B ~ 70–100 K^{-1} is typical of medium

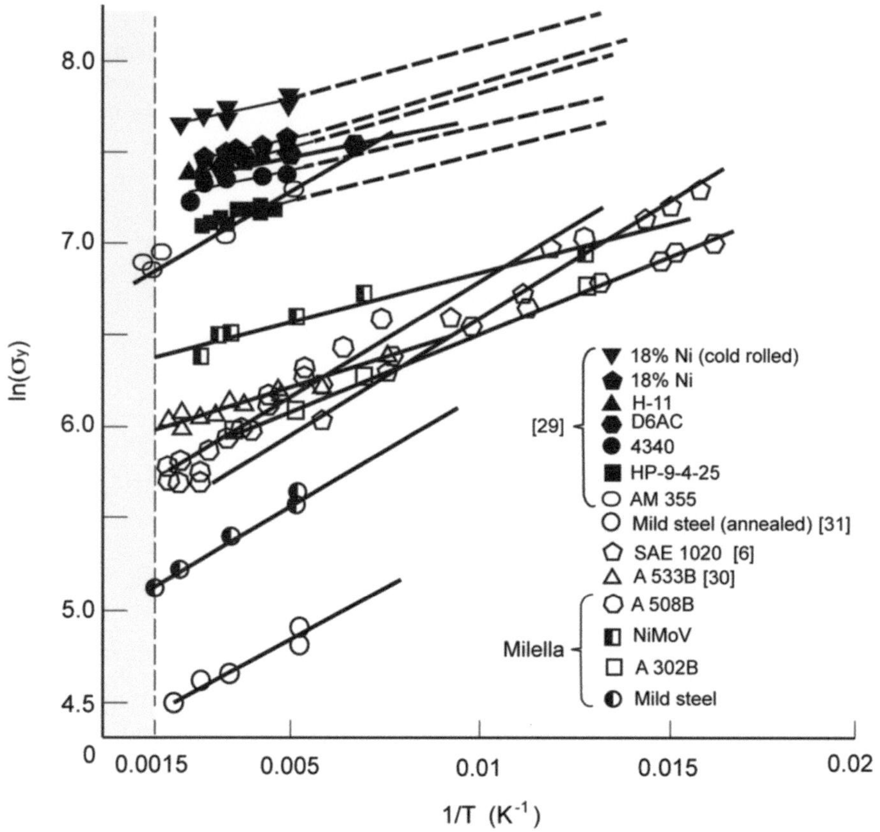

Fig. 5.24 1/T linear dependence of the natural logarithm of the yield strength for selected steels (data from Milella and [6, 28–31])

strength steels, ferritic-pearlitic steels, and the lower slope, of the order of $B \sim 30–50$ K^{-1}, characterizes high strength martensitic steels.

Low strength BCC metals regain strength with a much higher rate than high strength martensitic steels. It goes without saying that the 1/T linear dependence of $ln(\sigma_y)$, typical of BCC crystal structure materials that have a temperature dependent component of the yield strength, cannot be seen in FCC materials. These last metals and alloys, as we have discussed, do not have a thermally activated component that follows an Arrhenius type relation and, therefore, the $ln(\sigma_y)$ cannot be linear with 1/T. Figure 5.25 presents the dependence on 1/T of the natural logarithm of the yield strength for a series of FCC materials. Note the complete difference with those materials shown in Fig. 5.24. It's all another thing.

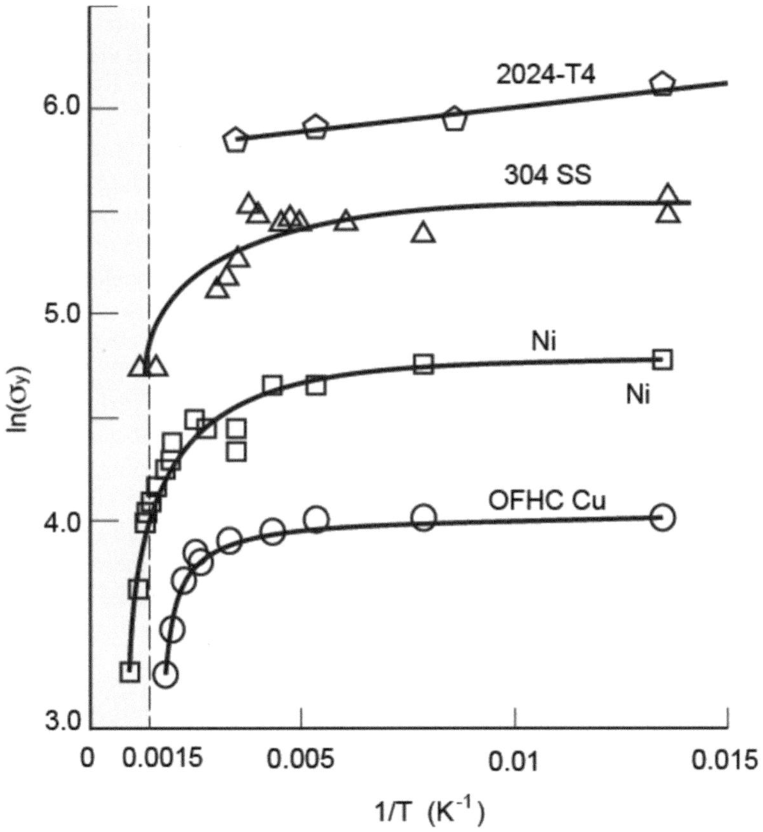

Fig. 5.25 Non-linear dependence of the natural logarithm of the yield strength on 1/T for selected FCC metals and alloys

5.4 Dislocation Dynamics

In the previous section it has been shown how the drag stress exerted by the Cottrell atmosphere depends on the dislocation velocity and, therefore, on the strain rate. Fundamentally, the motion of interstitials forming the Cottrell atmosphere and that of dislocations obey different lows and can be rather in conflict with each other.

This conflict becomes more and more severe as the mechanical factor (applied force) that moves dislocations is applied at a strain rate much too high with respect to temperature that moves interstitials. Let's return to the concept of dislocation vibration already introduced in Sect. 5.2, Eq. (5.3). It must be said, however, that it doesn't exist, at the moment, a unified theory that can describe the phenomenon. Let's try to start from the famous composite picture of Campbell and Ferguson of Fig. 5.26 [31] that presents the shear strength of an annealed mild steel as a function of both temperature T and shear strain rate $\dot{\gamma}$. Note the field of variability of these two parameters that is particularly wide, from – 80 °C to 440 °C the first and from 10^{-3} s^{-1} to 10^5 s^{-1}, the second. Three regions can be identified where the steel shows a different and characteristic behavior. Region I bounds a domain in which temperature is so high, in excess of 493 K = 220 °C, that both strain rate and temperature have no effect at all. Fundamentally, in this region the Cottrell atmosphere dissolves or the Peierls-Nabarro barrier is completely filled up by thermal energy. We are over the $0.2{\cdot}T_m$ temperature of Fig. 5.21. In this region, athermal deformation processes are dominant, in which the dislocation motion is influenced by long range barriers such as grain boundaries, precipitations and second phases. The yield stress follows the Young's modulus variation with temperature. Short range barriers, which are thermally activated, heavily intervene in Region II that roughly covers the interval $10^{-3} < \dot{\gamma} < 10^3$ s^{-1} s. Here, strain rate effects are also significant. It can be seen, in fact, that at a strain rate of 10^3 s^{-1} the shear strength triplicates from 100 MPa at 713 K (440 °C) to even 340 MPa at 193 K (−78 °C). Beyond the strain rate of regions I and II ($\dot{\gamma} > 10^3$ s^{-1}) Region III opens. Here the stress required to overcome barriers is so high that no thermal assistance is needed. A characteristic feature must be caught. Both at a very low and very high strain rate the trend curves seem to merge into two different common points. These are the lower athermal component and the upper athermal component of the yield strength, respectively. Their meaning has been discussed in Sect. 2.14 and shown in Fig. 2.44. Let's go back to Region II of Fig. 5.26. As a first approximation we may think that the motion of a dislocation develops in a viscous fashion that behaves like a Newtonian fluid. As such, the force F per unit length that opposes the motion can be represented as

$$F = C \cdot \upsilon \tag{5.16}$$

where v is the velocity of the dislocation and C a coefficient of viscous damping that does not depends on v since we have supposed the fluid to have a Newtonian behavior. The force F per unit length can be expressed as the product of the shear stress, τ, times the burgers vector b, therefore

$$\tau \cdot b = C \cdot \upsilon. \tag{5.17}$$

The shear strain γ can be expressed in terms of dislocation density ρ (dislocations per unit area l^2) as

$$\gamma = \rho \cdot b \cdot l \tag{5.18}$$

which, deriving with respect to time, yields

$$\frac{d\gamma}{dt} = \rho \cdot b \cdot \frac{dl}{dt} \dot{\gamma} = \rho \cdot b \cdot \upsilon \tag{5.19}$$

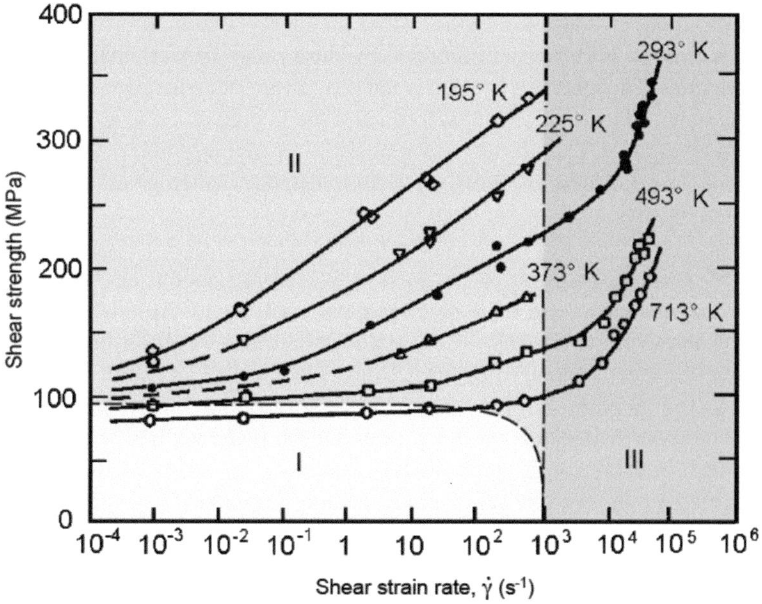

Fig. 5.26 Shear strength dependence on the shear strain rate and temperature for mild steel (adapted from [31])

where v is the dislocation velocity. The shear strain γ can be converted into a tension strain, ε,, through an orientation factor M that is usually taken equal to 3.1 for BCC crystals and 2.75 for FCC crystal

$$\dot{\varepsilon} = \frac{d\varepsilon}{dt} = \frac{1}{M}\rho \cdot b \cdot \upsilon. \tag{5.20}$$

Solving for v, substituting in Eq. (5.17) and recalling that the shear stress is half the normal stress, $\tau = 1/2\ \sigma$, it yields

$$\sigma = \frac{2CM}{\rho b^2}\dot{\varepsilon}. \tag{5.21}$$

What Eq. (5.21) is actually saying is that if the motion of dislocations occur through a viscous type damping in a Newtonian fluid (constant C independent of velocity v), the yield strength must depend on the strain rate $\dot{\varepsilon}$. This seems not be the case, as experimental data like those of Fig. 5.26 indicate. In Region II of the diagram the relationship between stress and strain rate is linear, but on a logarithmic scale which actually suggests a relationship pf the type

$$\sigma = K\dot{\varepsilon}^m. \tag{5.22}$$

Let's try, then, to address the issue from an energy standpoint. In Region II the dislocation motion is strongly influenced by short range barriers that have periodic lattice potential. This periodicity makes the movement of dislocations a kind of stop and go sequence. The dislocation will run and then come to a barrier where it stops for a waiting time t_w. Then it will move again for a running time t_r and stop again. The overall time required to perform a certain distance is then given by

$$\Delta t = t_w + t_r. \tag{5.23}$$

Mark Meyers made use of a nice, ironical analogy to visualize the point by comparing the event to La Jolla (CA) traffic where the waiting time at the traffic lights greatly exceeds the driving time. Since I use to live in Rome, Italy, I understand perfectly the point. For the others, Fig. 5.27 [25] may well visualize the point. The time interval Δt depends on the applied stress. If the applied stress σ_2 is large enough, such barriers can be immediately overcome $t_w \sim 0$. At lower stresses the waiting time t_w may become relatively large and dominate: $t_w > > t_r$. Effectively, Wood and Clark [32], using mild steel specimens of yield strength equal to 252 MPa, found that by applying a sudden stress of 259 MPa to the material it was necessary to wait 6 s before yielding occurred, while with a stress of 357 MPa the time shrank to 5 ms.

As a first approximation, we can then neglect t_r with respect to t_w, which is the inverse of a frequency

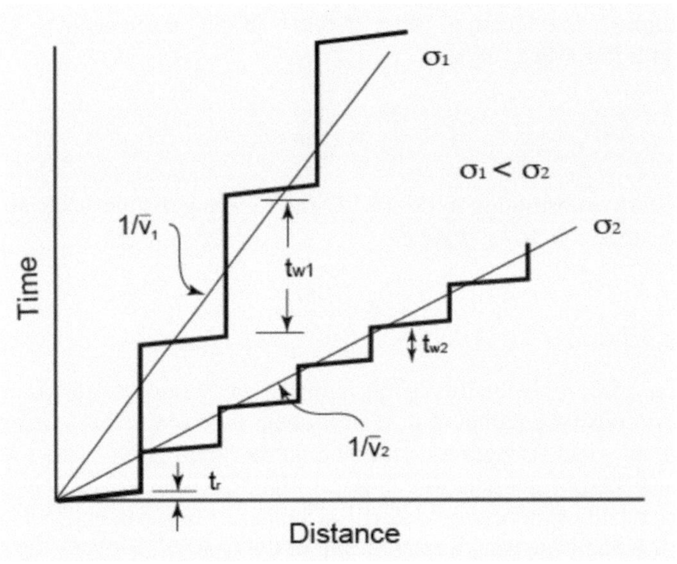

Fig. 5.27 Schematic representation of dislocation motion by a stop and go sequence at two stress levels (adapted from [25])

$$t_w \approx \Delta t = \frac{1}{\nu} \tag{5.24}$$

As reference frequency we can take that of successful jumps of dislocations encountering a barrier, given by Eq. (2.17) and write

$$\Delta t = \nu_o \exp\left(-\frac{E}{kT}\right) \tag{5.25}$$

Expressing the velocity v in Eq. (5.20) as $\Delta l/\Delta t$, substituting Δt with its expression (5.25) and solving for $\dot{\varepsilon}$ we have

$$\dot{\varepsilon} = \frac{\nu_o \rho b \Delta l}{M} \exp\left(-\frac{E}{kT}\right) = \dot{\varepsilon}_o \exp\left(-\frac{E}{kT}\right) \tag{5.26}$$

where Δl is the distance between two consecutive barriers. Solving for the energy E it is

$$E = \ln \frac{\dot{\varepsilon}_o}{\dot{\varepsilon}} kT \tag{5.27}$$

Relation (5.27) is particularly important. It says that the energy available to jump a barrier depends on temperature, as it was already known (see Eq. 5.17), but also on the inverse of the strain rate $\dot{\varepsilon}$, which actually means that by increasing the strain rate, the thermal energy effectively available to overcome the barrier reduces. This complies with the experimental evidence that at high strain rates, over 10^3 s^{-1} for the steel in Fig. 5.26, temperature seems to lose any effect. Gilman e Johnston [33, 34] showed for the first time that as the shear stress is increased from the yield stress, the dislocation velocity also increase, following a diagram of the type shown in Fig. 5.28 that refers to nickel, but can be considered of general type. Three different slopes can be recognized that can be described by an equation of the type

$$\upsilon = C_1 \tau^m. \tag{5.28}$$

The three regions are characterized by decreasing values of the exponent m with $m_1 > m_2 > m_3$ and $m_2 = 1$. They refer to three different mechanisms governing dislocation motion and plastic deformation that we have been discussing in the previous sections. In the first region, $m_1 > 1$, the resistance to dislocation motion is exerted by the Peierls-Nabarro barrier and associated Cottrell atmosphere and is thermally activated (see Sect. 2.11 and 4.1). Drag resistance due to tangles, grain boundaries, cells formation and dislocation cutting dislocation forests is responsible for the second region where $m = 1$. This is the region of long-distance barriers, temperature is not effective in helping dislocation motion. Finally, the third region is the most intriguing and difficult to describe. We enter the mysterious world of phonon and free electrons resistance (see Sect. 5.3). As the dislocation velocity approaches its upper limit value, which is

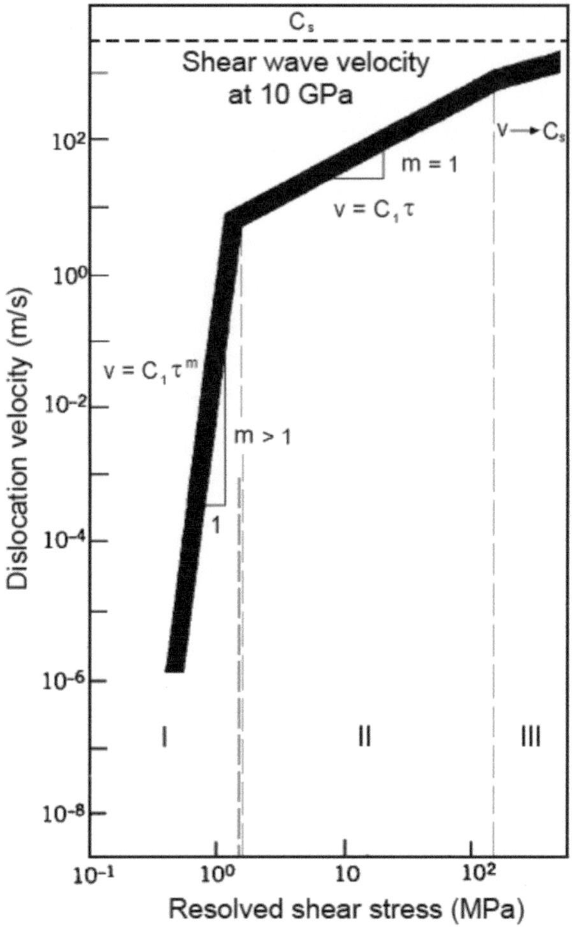

Fig. 5.28 Schematic representation of the relationship between dislocation velocity and resolved shear stress showing the three characteristic regions

considered to be the sound velocity in solid materials v_s (see Fig. 5.27) also relativistic effects may enter into action. This is because if we assume that a dislocation moves and accelerates, we must also assume that it has a mass that allows writing an equation of Newtonian type

$$F = m \cdot a = \tau \cdot b. \tag{5.29}$$

The limiting value given to dislocation velocity, both screw and edge dislocations, derives from the fact that they cannot propagate faster than the waves they generate. The relativistic effects were originally proposed by Frank [35] who considered the speed of sound like that of light and posed

$$m = \frac{m_o}{\sqrt{1 - \upsilon^2/\upsilon_s^2}}. \tag{5.30}$$

As the shear wave velocity is approached, the mass and, therefore, the energy goes to infinity. Already at 0.8–0.9 v_s an almost infinite energy is required to move dislocations. Interesting enough, if the diagram of Fig. 5.28 is inverted reporting on the abscissa axis *log(v)* and on the ordinate axis the resolved shear stress the result would be a diagram of the type of that of Fig. 5.26.

5.5 Strain Rate Sensitivity

It has become clear, from the preceding sections, that the higher the strain rate, $\dot{\varepsilon}=d\varepsilon/dt$, the higher the corresponding inability of the material to deform plastically and, therefore, the higher the yield strength. Actually, it is the entire flow stress that is affected. Figure 5.29 is an example of stress–strain response of a carbon steel type A 533 GrB subjected at − 60 °C to several strain rates varying from 0.001 s^{-1}, which is representative of a quasi-static traction test, to 531 s^{-1} [36]. As can be seen, the effect of strain rate is to make the alloy harder and harder, leaving almost unaffected the deformation at maximum stress. However, the yield strength grows at a higher rate than the ultimate. Eventually, the two stresses must merge at a certain strain rate. This occurrence is shown in Fig. 5.30 for a mild steel [37]. The yield strength and the ultimate strength converge toward a common value for $\dot{\varepsilon}>10^2$ s^{-1}. Note that the same instance occurs in BCC metals and alloys as the temperature decreases below a certain very low value, as shown in Fig. 5.2.

The entire stress–strain curve dependence on the strain rate is offered in Fig. 5.30 for a mild steel with strain rates up to 1750s^{-1} [38].

Data such as those of Fig. 5.32 suggest a relationship of the type already considered, see Eq. (5.22), analyzing Region II of Fig. 5.26

$$\sigma_s = A \cdot (\dot{\varepsilon})^m \tag{5.31}$$

where A is a material constant and *m* is the so called strain rate sensitivity exponent or index. Note the similitude of Eq. (5.31) with the expression given by Hollomon for the flow stress, Eq. (1.19). As matter of fact, it was Hollomon [38] to propose an equation like Eq. (5.31) that becomes a linear relationship between the yield stress and the logarithm of strain rate. This occurrence is shown in Fig. 5.31 for a series of high strength steels [39] including a modified HP9-4–20 austenitic stainless steel.

These values are somehow smaller than those characterizing low-to-medium strength steels. It is worth noting what has been already said about high strength steels; their yield strength is already so high at slow strain rate that there is not that much room left to any possible action to further increase it. In a 10^6 fold strain rate increase, the yield strength is varying by a mere 150 MPa (~19%) with an even smaller increase in the ultimate strength over the same range.

This particular feature can be better seen in Fig. 5.33 [40]. Here, the strain rate exponent has been plotted as a function of the quasi-static yield strength. Note the enormous difference in the strain rate exponent between mild steels and high strength

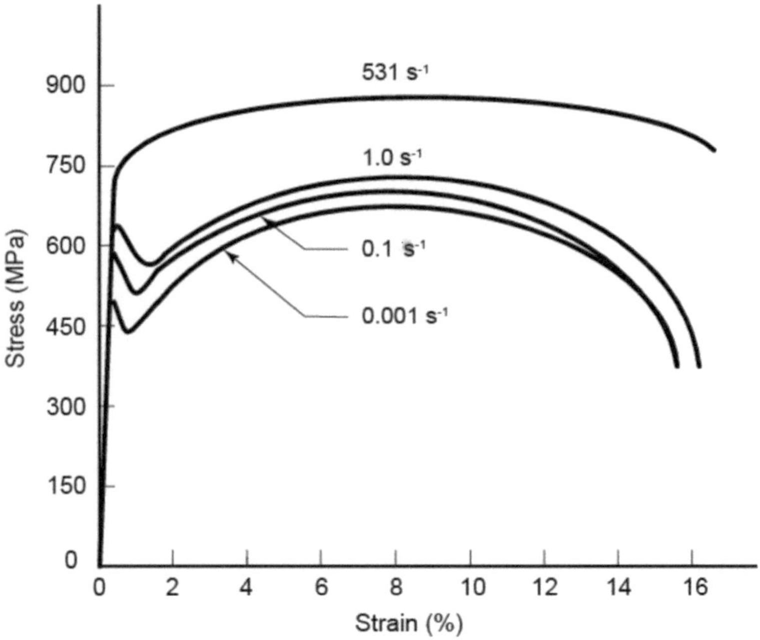

Fig. 5.29 Stress–strain responce for low-alloyed carbon steel, type A 533 GrB, showing effect of strain rates $\dot{\varepsilon}$ at a temperature of $-$ 60° C [36]

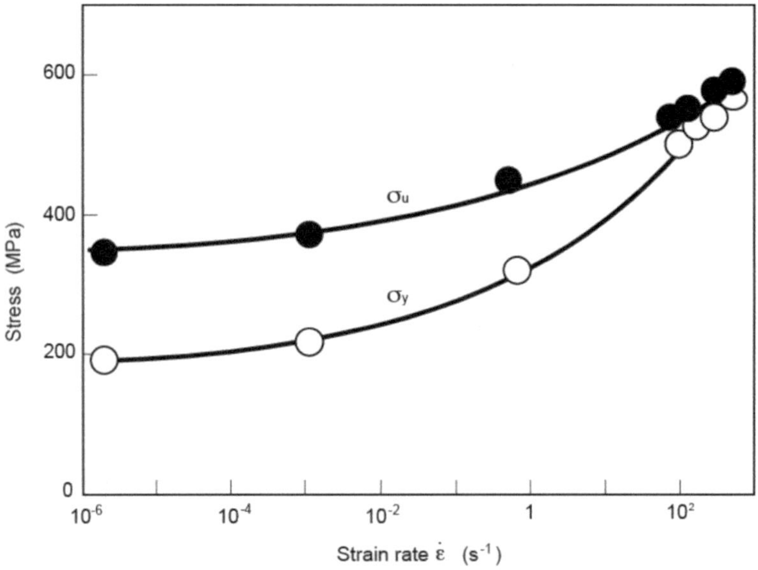

Fig. 5.30 Variation of yield and ultimate strength with strain rate for a mild steel (adapted from [37])

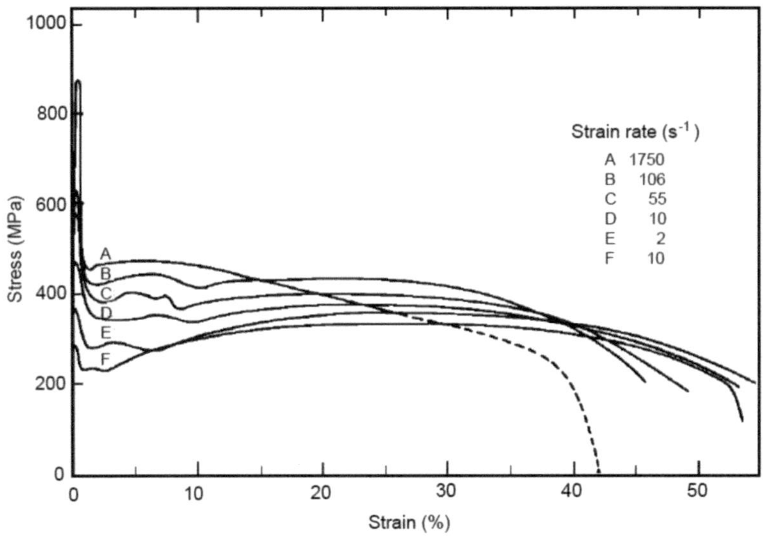

Fig. 5.31 Stress–strain characteristic for a mild steel for strain rates up to 1750s^{-1} (adapted from[30])

steels. The dependence on the yield strength is of exponential type. For a mild steel having 200 MPa yield strength it is $m \sim 0.5$–0.6 while for high strength steel of 900 MPa yield, m is almost zero. The strain rate exponent can be experimentally inferred through a couple of tests run at two different strain rates $\dot{\varepsilon}_1$ ed $\dot{\varepsilon}_2$, that yield two values of the yield stress, $\sigma_{s,1}$ and $\sigma_{s,2}$

$$\frac{\sigma_{s,1}}{\sigma_{s,2}} = \left(\frac{\dot{\varepsilon}_2}{\dot{\varepsilon}_1}\right)^m \tag{5.32}$$

which yields the value of m

$$m = ln\frac{\sigma_{s,1}}{\sigma_{s,2}} / \ln\frac{\dot{\varepsilon}_2}{\dot{\varepsilon}_1} \tag{5.33}$$

For body centered cubic (BCC) metals, the strain rate sensitivity is in the yield stress dependence as compared with the face centered cubic (FCC) in which the same sensitivity is in the strain hardening characteristic, in a very similar fashion as the sensitivity to temperature, see Fig. 5.7 as being opposite to Fig. 5.9.

An important consequence is that for BCC crystal structure materials an opposite influence on ductility occurs, as measured by the strain at maximum stress that decreases with increasing strain rate while it increases with increasing strain rate for FCC metals. This seems to be in contradiction with what just said and shown in Fig. 5.29. Hexagonal close packed (HCP) metals are shown to follow either the

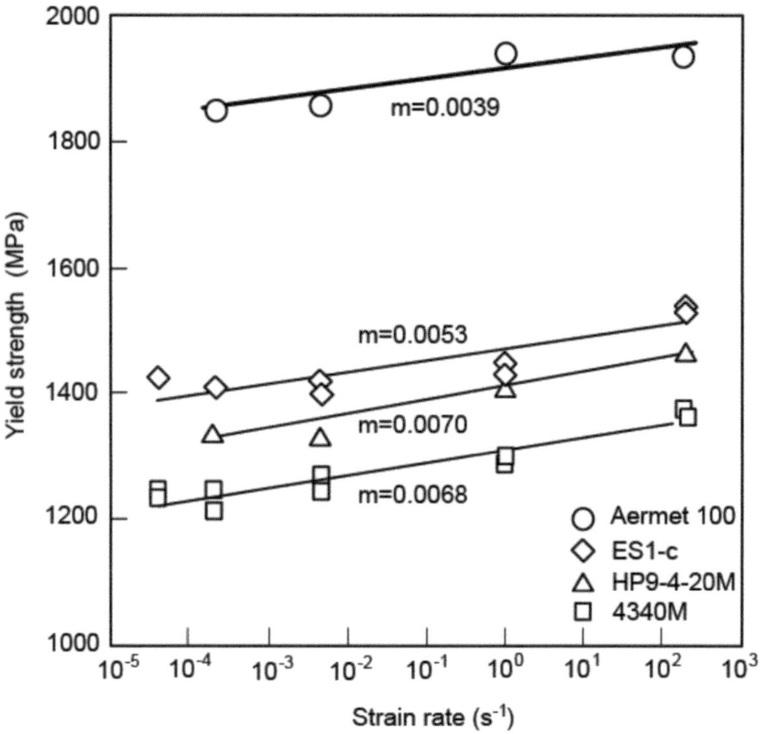

Fig. 5.32 Linear trend of the yield stress with the logarithm of the strain rate for selected high strength steels whose slope is defined by the strain rate sensitivity index (adapted from [39])

FCC or the BCC case. An example of the ductility reduction obtained by Hoge and Mukherjee [41] for tantalum is shown in Fig. 5.34. The experimental data were obtained at constant strain rate of 0.0001 s^{-1} at various temperatures from an extremely low 22 K (−251° C) up to 298 K (25° C). Also shown is the locus of maximum load stress. Figure 34b reports, for comparison, the corresponding reduction of ductility at different strain rates, from a quasi-static condition of 0.0001 s^{-1} to a real dynamic condition of 4900 s^{-1}, while temperature is kept constant at 298 K. The role of strain rate as opposite to that of temperature is clearly visible. They both work as to reduce the ductility of the material, though with opposite effect: increasing the strain rate is equivalent to decreasing the temperature. However, this effect is not necessarily seen in all materials. Low-alloyed steels, such as A 533 Gr B does not happen to show such ductility reduction, at least up to 531 s^{-1}, as shown in Fig. 5.29. FCC crystal structure materials also show a strain rate sensitivity. The behavior of an FCC material such as high purity aluminum or austenitic stainless steel, type 304, is shown in Figs. 5.35 [42] and 5.36 [43], respectively. In the aluminum case, strain rate plays an inverse role as temperature; it increases strength and ductility. In the 304 steel case, the increase in the strain rate produces a decrease of ductility, though

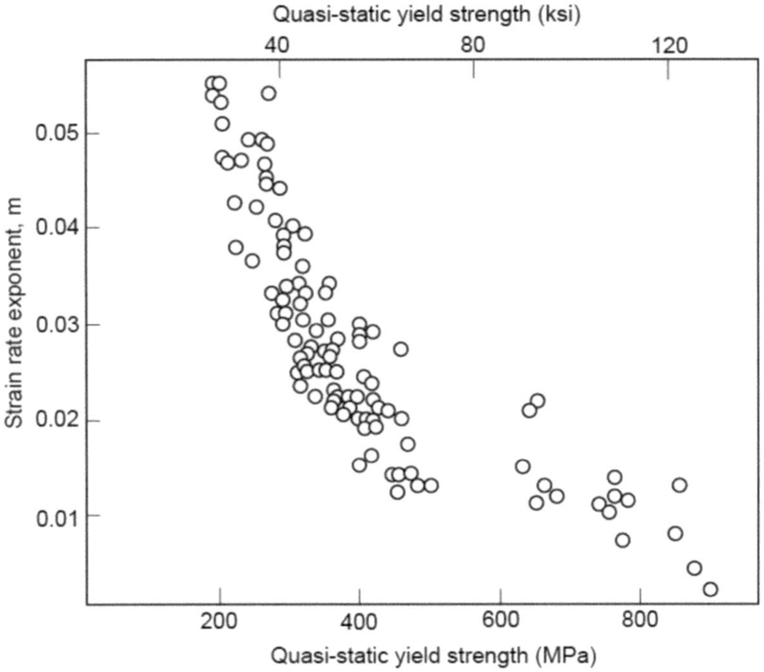

Fig. 5.33 Correlation between the strain rate exponent, *m*, and the quasi-static yield stress of steels (adapted from [40])

the maximum strain rate is not that high. This is a behavior typical of BCC type materials. The apparent discrepancy can be partially resolved just remembering that austenite at room temperature exists in a metastable condition. Dynamic loadings can partially change austenite into martensite, which modifies the crystal structure from FCC into a body-centered tetragonal closer to a BCC crystal structure. However, this question of phase transformation needs a further discussion because it is a rather intriguing matter. Actually, that's not the only intriguing matter in the overall story. We shall, probably, ask ourselves whether non-intriguing arguments still exist. Easy things seem to have gone, if they ever existed. The more we know, the more we need to know. Martensite transformation does happen under high strain rate loadings, but in smaller percentage than under quasi-static conditions. This effect can be seen in Fig. 5.37 [44] for high strength austenitic stainless steel EN 1.4318 at room temperature. At 10^3 s^{-1} and over, the volume fraction of austenite transformed into martensite has halved with respect to quasi-static loading conditions. This is because high strain rates introduce adiabatic conditions that increase the specimen temperature, in particular in the shear bands. This deformation-induced heating inhibits phase transformation. Zener and Hollomon [45] were the first to recognize that an increase of strain rate tends to change the condition from isothermal to adiabatic.

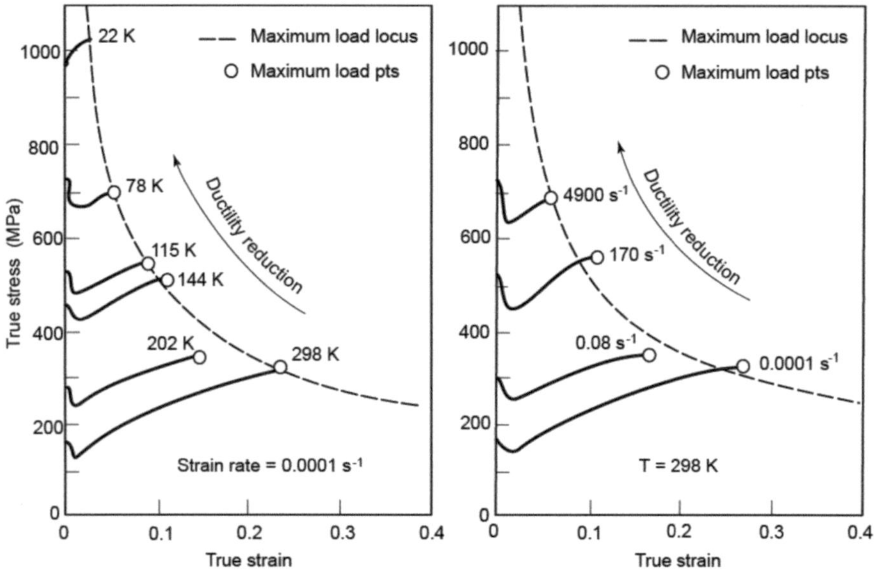

Fig. 5.34 Tensile stress–strain curves for tantalum: **a** at fixed strain rate for different temperatures and **b** at fixed temperature for different strain rates (adapted from [41])

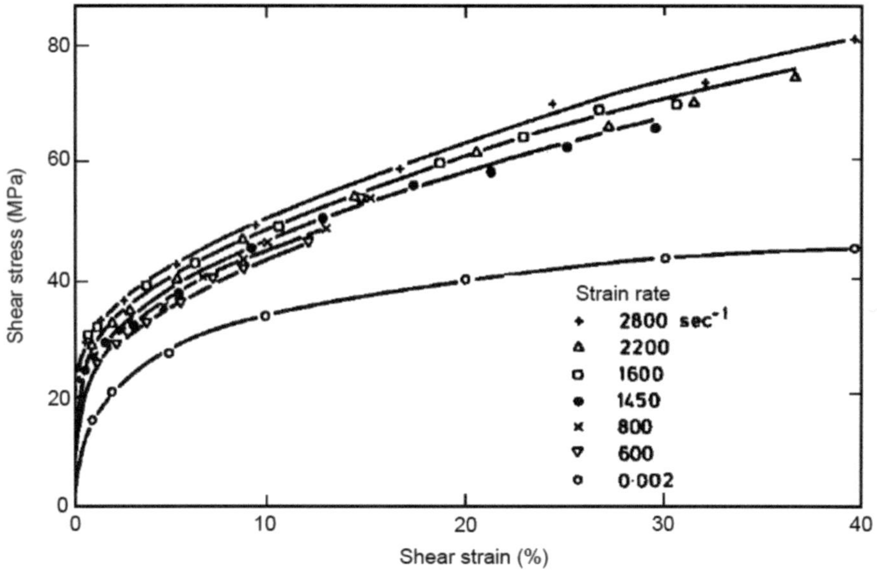

Fig. 5.35 Stress–strain curves for high purity aluminum tested in torsion at different strain rate and RT [42]

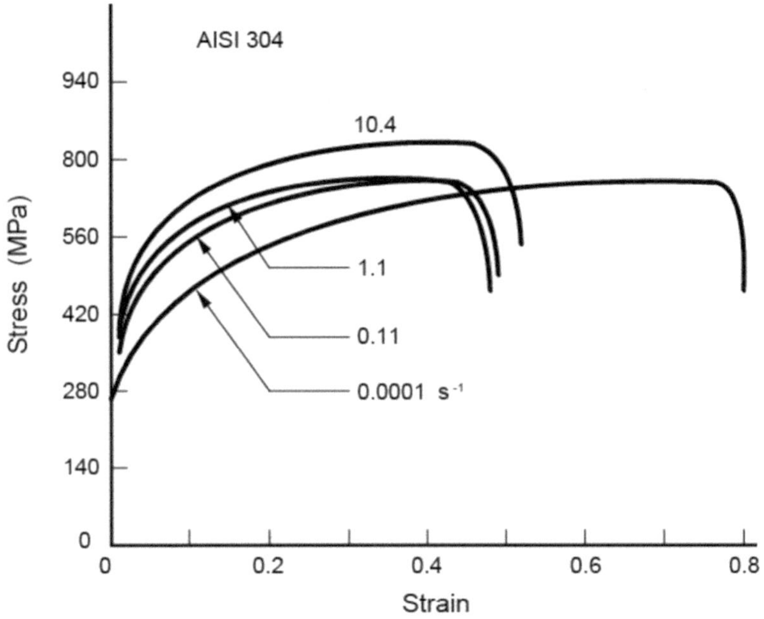

Fig. 5.36 Stress–strain curves at different strain rate for AISI 304 SS [43]

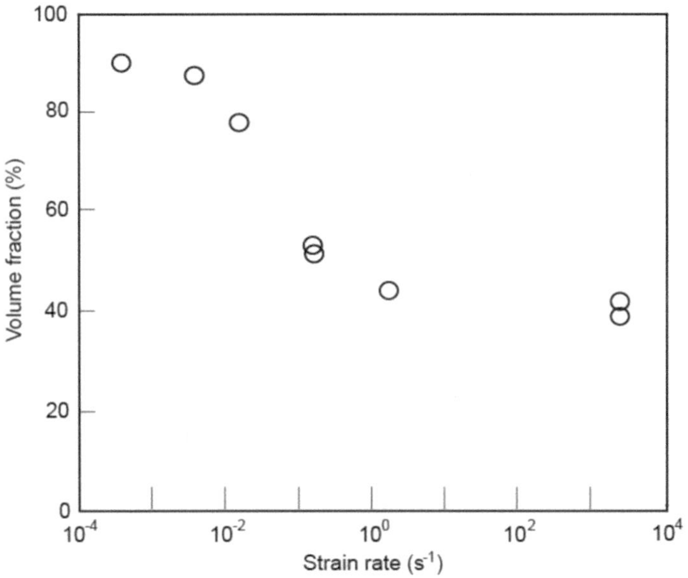

Fig. 5.37 Total volume fraction of transformed martensite in EN 1.4318 at different strain rates (adapted from [44])

They formulated the hypothesis that when the material undergoes an adiabatic strain increment the stress increases by strain hardening and drops because of the associated temperature increase. The temperature increase may prevail over the strain hardening effect. However, if during the test the strain rate changes from high to low value, the martensite volume fraction returns to increase. A remarkable increase of ductility with increasing strain rate can be seen in Fig. 5.38 for an Al–Mg alloy, type 5086 [46]. Note the different behavior between high purity aluminum of Fig. 5.34 and this aluminum alloy. The serrated appearance of the flow curves at 2500 s^{-1} is indicative of a dynamic strain aging that is rather common in aluminum alloys. Dynamic strain aging describes a phenomenon of yield stress and strain hardening increase that occurs in certain alloys during the test itself, from which the term dynamic. This is also the cause of a negative strain rate sensitivity when compared to the flow curves at 25 °C and strain rates of 0.001 s^{-1} and 0.1 s^{-1}.

It is clear that the previously described lattice friction based on Peierls-Nabarro barrier or the Cottrell atmosphere does not apply to these materials since their thermal component of stress is negligible.

In notched specimens the effective strain rate at the notch tip is higher than for unnotched test pieces because of the local stress and strain intensification. The maximum strain rate ahead of the notch is equal to the notch concentration factor multiplied by the strain rate in the unnotched section. There is an enormous difference between the strain rate in a smooth specimen subjected to a standard tensile test and that of a notched bar impact test. For example, the strain rate at the notch tip in a

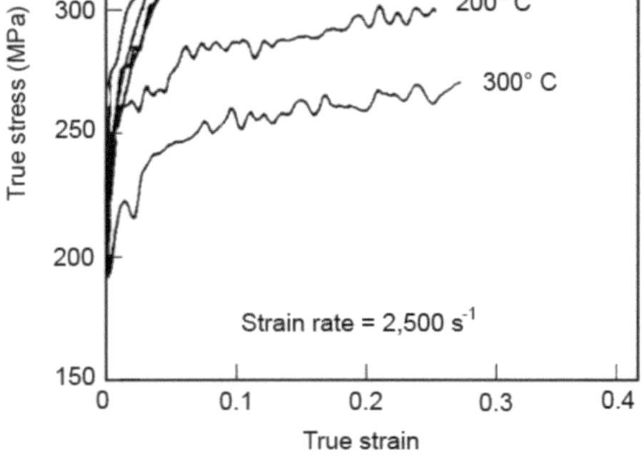

Fig. 5.38 Stress–strain curves for a Al–Mg alloy (type 5086) tested in tension at different temperatures and strain rates (adapted from [46])

Charpy-V specimen (see Chapter 2) can be judged to be about $10^3\ s^{-1}$, which means about ten million times the strain rate of a standard tensile test.

5.6 Constitutive Equations

We have largely discussed, in the previous sections, how the mechanical response of a material depends on the combined effect of temperature T and strain rate $\dot{\varepsilon}$. It is a fact, however, that to catch the complexity of such a combined effect, enclosed in a diagram such as, for example, that of Fig. 5.26, by a single equation valid for all metals and alloys is extremely difficult and, in many respects, almost impossible. Many attempts have been made in pursuit of such a single equation, known in the technical literature as the *constitutive equation* or mechanical equation of the solid state since it provides the link between mechanical quantities as stress σ and strain ε and physical quantities as temperature T and strain rate $\dot{\varepsilon}$. The definition of equation of the solid state is made in analogy to the equation of ideal gas law that correlates the three fundamental variables that define the state of a gas: pressure, volume and temperature. Fundamentally, the constitutive equation should bear the form

$$\sigma = f(\varepsilon, \dot{\varepsilon}, T). \tag{5.34}$$

Since plastic deformations are irreversible and depend on the particular path followed to reach them, i.e., on the strain history, Eq. (5.34) should be better written as

$$\sigma = f(\varepsilon, \dot{\varepsilon}, T, \textit{deformation history}) \tag{5.35}$$

Probably, the first attempt at developing such an equation was done by Lüdwik [47] at the beginning of last century, who postulated the following logarithmic equation for the flow stress

$$\sigma = \sigma_1 + \sigma_o ln(\dot{\varepsilon}_p / \dot{\varepsilon}_{o,p}) \tag{5.36}$$

where σ_1, σ_o and $\dot{\varepsilon}_{o,p}$ are material constants. In particular, σ_1 is the yield stress corresponding to a strain rate $\dot{\varepsilon}_{o,p}$ and σ_o is the strain rate sensitivity m (see Sect. 4.3). Zener and Hollomon in 1944 [45], Manjoine in 1944 [48], Cottrell in 1948, '49,'53,'57 and '58 [49–54], Hollomon back in 1947 [55], Yokobori in 1953 [56] and later Orowan [57], Seeger [58, 59], Campbell [60, 61], Conrad [62–64], Petch [65, 67] and many others have tried to infer that equation through an empirical way, simply based on the interpolation of experimental data, or a physical one based on the concept of activation energy or dislocation models. Often, the proposed equations were not even congruent. Early attempts concentrated on the equivalence between temperature and strain rate and the concept of activation energy. This last aspect is well evidenced in the famous equation by Zener and Hollomon [45] at a fixed strain

$$\sigma = f\left(\dot{\varepsilon} \cdot e^{Q/RT}\right) \tag{5.37}$$

where R is the gas constant equal to 8,309 J/mole·K or 1,987 cal/mole·K, Q is the activation energy. Note that the dependence of the yield strength on the strain rate $\dot{\varepsilon}$ is linear, as predicted by Eq. (5.21) assuming a viscous damping. We have been observing that this behavior is not necessarily true, certainly it is not true in FCC crystals and an exponential relationship of the type proposed by Hollomon [55], see Eq. (5.22), is more likely to occur. The difference between the two expressions, linear and exponential, for the entire flow curve is shown in the schematic of Fig. 5.39. The exponential trend of the flow stress with strain rate, Eq. (5.22), predicts a difference $\Delta\sigma$ in the flow stress at two different strain rates $\dot{\varepsilon}_1 < \dot{\varepsilon}_2$ proportional to the flow stress, while with the linear relationship this difference is independent of the flow stress.

This difference can be observed in Fig. 5.29 that refers to a carbon steel type A 533 GrB and Fig. 5.35 that refers to aluminum. However, Eq. (5.37) predicts that the stress required to produce yielding under dynamic conditions depends upon the parameter P also known as the Zener-Hollomon parameter

$$P = \dot{\varepsilon} \cdot e^{Q/RT} \tag{5.38}$$

which is a combination of temperature and strain rate, independent of each other. A given stress may thus correspond to a high strain rate at high temperature or to a low strain rate at low temperature. A verification of Eq. (5.37) and the equivalence between temperature and strain rate against experimental data is shown in Fig. 5.40 for high purity aluminum and its alloys. On the ordinate axis it is shown the parameter

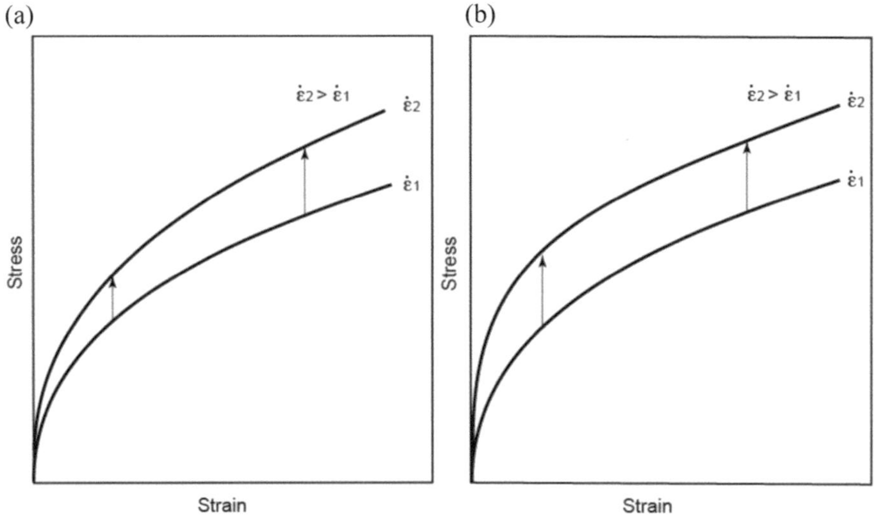

Fig. 5.39 Prediction in the flow stress at two different strain rates by **a** a power low relationship between stress and strain rate, **b** a linear relationship

P, while the yield strength is normalized to the Young's modulus. The activation energy Q for aluminum is equal to 142 kJ/mole. The agreement is excellent also considering the large number of experimental data. However, it must be considered that all data refer to low strain rates and temperatures over 100° C.

This actually means that the real $\dot{\varepsilon}$—T field investigated turns out to be somewhat narrow and cannot really represent a base for the demonstration of validity of Eq. (5.37) for any operating combination $\dot{\varepsilon}$—T. Particularly poor is the field of low temperatures and high strain rates. The same parameter P is plotted in Fig. 5.41 [45] versus both the yield and the ultimate strength for a cast steel. The agreement is excellent, but again the $\dot{\varepsilon}$—T field scanned by experimental results is rather poor.

More recently, Johnson and Cook [68], Johnson et al. [69], Follansbee [70], Zerilly-Armstrong [71–73], Follansbee and Kocks [74], Nemat-Nasser et al. [75–77] and Milella [28, 78, 79] have developed other equations on either empirical or physical basis. In particular, the model developed by Johnson and Cook [69] or *J–C* model, also known as *working horse* because of its preferred and heavy use, has obtained a great success for its simplicity. It is an empirical model for the entire flow curve and takes the form

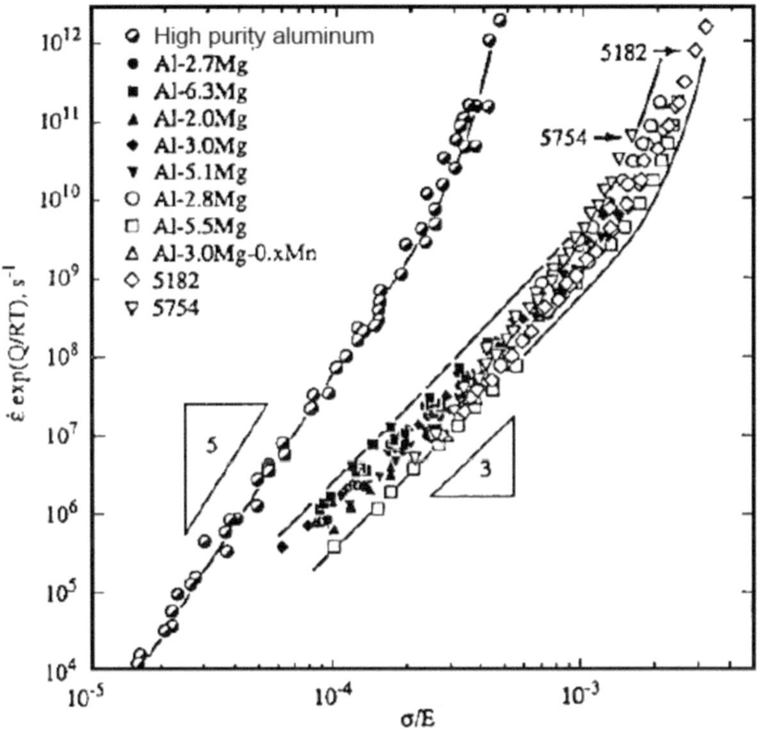

Fig. 5.40 Demonstration of equivalence of temperature and strain rate for high purity aluminum and some alloys based on the Zener parameter P

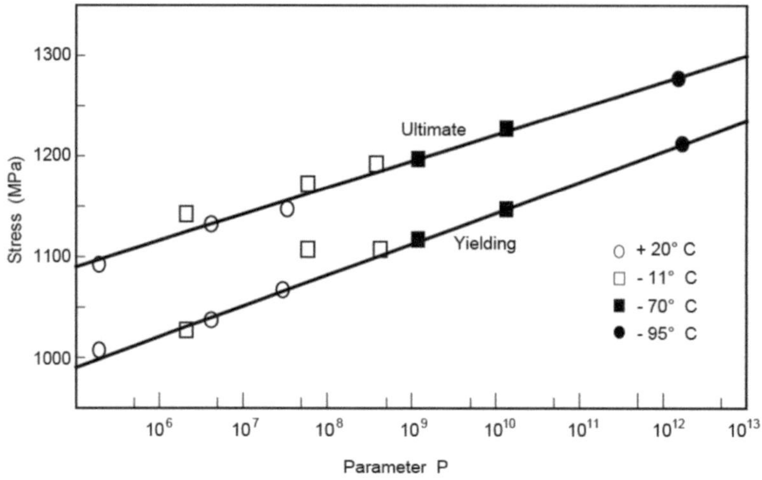

Fig. 5.41 Demonstration of equivalence of temperature and strain rate for a cast steel based on the Zener parameter *P* (adapted from [45])

$$\sigma_e = \left(A + B\varepsilon^n\right) \cdot \left(1 + C \cdot ln\dot{\varepsilon}^*\right) \cdot \left(1 - T^{*m}\right) \tag{5.39}$$

where σ_e is equivalent von Mises stress, ε the equivalent strain, $\dot{\varepsilon}^* = \dot{\varepsilon}/\dot{\varepsilon}_o$ is a non-dimensional ratio with $\dot{\varepsilon}_o$ a reference strain rate (normally that corresponding to a quasi-static loading condition 10^{-4}-10^{-3} s^{-1}) and $T^*$the homologous temperature

$$T^* = \frac{T - T_o}{T_m - T_o} \tag{5.40}$$

where T_o is a reference temperature (the temperature at $\dot{\varepsilon}_o$) and T_m the melting temperature, in degree Kelvin. *A*, *B*, *C* and *m* are material characteristics to be determined. The first term in bracket of Eq. (5.39) takes care of the dependence of the flow stress on strain ε (equivalent) in the hardening domain, the second term considers the effect of strain rate and the third and last term that of temperature. Their value is reported in Table 5.2.

However, the $J - C$ model has some drawbacks.

- it assumes that the shape of the flow curve is independent of the temperature and strain rate and the ultimate or maximum stress always occurs at the same deformation. We have direct experimental evidences that this is not always true (see, for example, Fig. 5.35).
- it assumes a linear increase in the flow stress with the logarithm of strain rate. Also in this case we have direct experimental evidences that this is not always correct; at strain rates of 10^3 s^{-1} and above, especially for BCC materials, substantial deviation occur,

Table 5.2 Constants of the Johnson and Cook equation for several materials

Material	Properties				Costants of Eq. (4.39)				
	Hardness	Density Kg/m^3	Sp. He. J/kgK	T_m (K)	A MPa	B MPa	n	C	m
OFHC copper	F-30	8960	383	1356	90	292	0.31	0.025	1.09
Brass (cartriges)	F-67	8520	385	1189	112	505	0.42	0.009	1.68
Nickel 200	F-79	8900	446	1726	163	648	0.33	0.006	1.44
Armco iron	F-72	7890	452	1811	175	380	0.32	0.060	0.55
Pure iron	F-83	7890	452	1811	290	339	0.40	0.055	0.55
Steel 1006	F-94	7890	452	1811	350	275	0.36	0.022	1.00
Alluminum 2024	B-75	2770	875	775	265	426	0.34	0.015	1.00
Alluminum 7039	B-76	2770	875	877	337	343	0.41	0.010	1.00
Steel 4340	C-30	7830	477	1793	792	510	0.26	0.014	1.03
S-7 tool steel	C-50	7750	477	1763	1539	477	0.18	0.012	1.00
Tungsten alloy	C-47	17,000	134	1723	1506	177	0.12	0.016	1.00
DU-0.75Ti	C-45	18,600	117	1473	1079	1120	0.25	0.007	1.00

- the temperature and strain rate effects are independently considered and multiplied by each other. Also this may not be necessarily true.

Moreover, there is a possible effect that makes the first prerequisite fail also for materials that seem to follow it. The effect we are referring to is caused by the small dimensional deviations that are found to grow during deformation. The local stress in the smallest cross-section is slightly higher than in other regions. Therefore, the local strain is higher there and the smallest cross-section further decreases amplifying the effect. This trend is fostered by the adiabatic character of the high strain deformation process since shear bands are even more softened by the deformation heat (see Sect. 2.13). The overall effect is a continuously increasing deformation at which the maximum load is reached, as can be seen in Fig. 5.42 [80] for a 1% maximum deviation of the cross-sectional area of the specimen.

At variance with the *J–C* model, the constitutive equations proposed by Zerilli-Armstrong [71–73] are based on the thermally activated dislocation motion described in Sect. 2.11. The equations show an excellent match with experimental results.

The different response of BCC and FCC crystal materials, discussed in previous sections, is also taken into consideration

$$BCC \rightarrow \sigma = C_1 \exp(-C_3 T + C_4 T ln\dot{\varepsilon}) \; FCC \rightarrow \sigma = C_2 \varepsilon^{1/2} \exp(-C_3 T + C_4 T ln\dot{\varepsilon}) \tag{5.41}$$

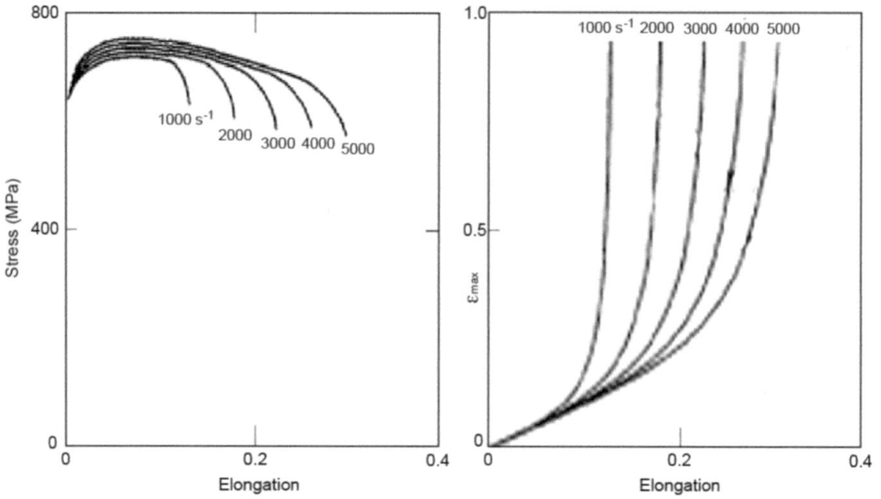

Fig. 5.42 Influence of geometrical imperfection on the flow stress response at different strain rates. The maximum strain is reached at higher and higher deformations with increasing strain rate (adapted from [80])

where the four constant C_1, C_2, C_3 and C_4, three for each equation, are material characteristics. In particular, constants C_1 is the stress at $T = 0$. It is worth noting the particular structure of Eq. (5.41); temperature intervenes twice: first as an independent term, $- C_3T$, separated from the strain rate effect, secondly as a term multiplying the logarithm of the strain rate. This actually means that temperature has a direct effect on the strain rate. Figure 5.43 shows Zerilli-Armstrong predictions for copper and iron, prototypical FCC and BCC crystal structures, respectively [25]. Two characteristic features shall be noted. In the first place, for both materials the trend curves at different temperatures seem to be merging into a common point at very high strain rate, $2.02 \cdot 10^7$ s^{-1} for iron and $3.75 \cdot 10^{10}$ s^{-1} for copper, respectively. However, while curves for iron are very steep, those relative to copper are almost flat. In the second place, the same merging seems to occur for iron at very low strain rate, $\dot{\varepsilon} << 10^{-5}$ s^{-1}. This seems to comply with what has been said in Sect. 2.15 about the existence of a lower athermal component for both very high and very low strain rates, see Fig. 2.45. Therefore, 1250 MPa must be regarded as the upper athermal component and ~ 200–250 MPa as the lower athermal component of the yield strength for iron. As to copper, while there is a definite common upper limit of the yield strength for all temperatures at about 350 MPa, it seems not exist a lower athermal component common to all strain rates for the very simple reason that FCC crystal materials do not have it.

The dependence of the yield strength, σ_s, on temperature inferred from an Arrhenius type approach leads to an equation of the type (14) which is indicative of a linear relationship between $ln(\sigma_s)$ and $1/T$. This is shown in Fig. 5.24 [28] for fourteen steels tested under quasi-static conditions. It is now appropriate to verify whether

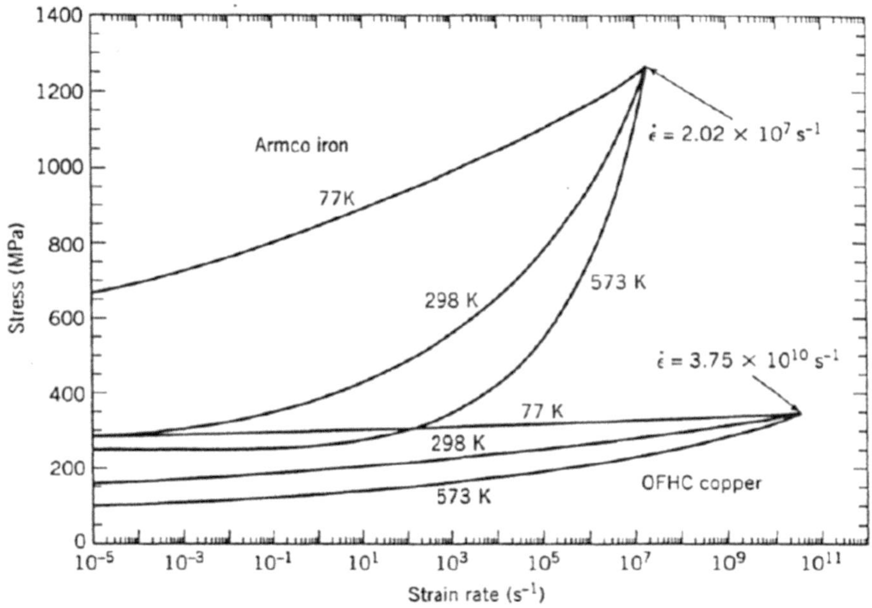

Fig. 5.43 Zerilli-Armstrong prediction of temperature and strain-rate dependence of typical BCC (Armco iron) and FCC (copper) metals [25]

the same linear relation is maintained by varying the applied strain rate, $\dot{\varepsilon}$. In 1997 the author made this check for A533 grade B steel, mild steel, niobium, titanium and other metals [28]. The results are shown in Fig. 5.43 for A533 Gr B steel. The data analyzed were obtained by Kanninen, et al.. Effectively, the linear trend is kept for the four strain rates considered that were 10^{-3}, 10^{-1}, 1 and 540 s^{-1}. The strain rate $\dot{\varepsilon}$ is affecting the slope *m* of the various straight lines that increases with increasing $\dot{\varepsilon}$. It is interesting to observe that all straight lines seem to converge for $1/T \sim 0$. This common point provides the value of constant *A* in Eq. (5.12) and *C* in Eq. (5.14). If this point belongs to all straight lines, it will be shared also by the horizontal line having zero slope which, being independent of temperature, must be referring to the athermal component of the yield strength, Fig. 5.24. Therefore, the stress corresponding to such a common point is the athermal component of the yield strength. For the case of A 533 Gr B steel, $A = 5.8$ and the athermal component of the yield strength is

$$\sigma_a = e^{5.8} = 330MPa \tag{5.42}$$

Even more interesting is the fact that in a semi-logarithm diagram, *m*-*ln*($\dot{\varepsilon}$), all slopes *m* are also aligned along a straight line, as shown in Fig. 5.44, whose equation is

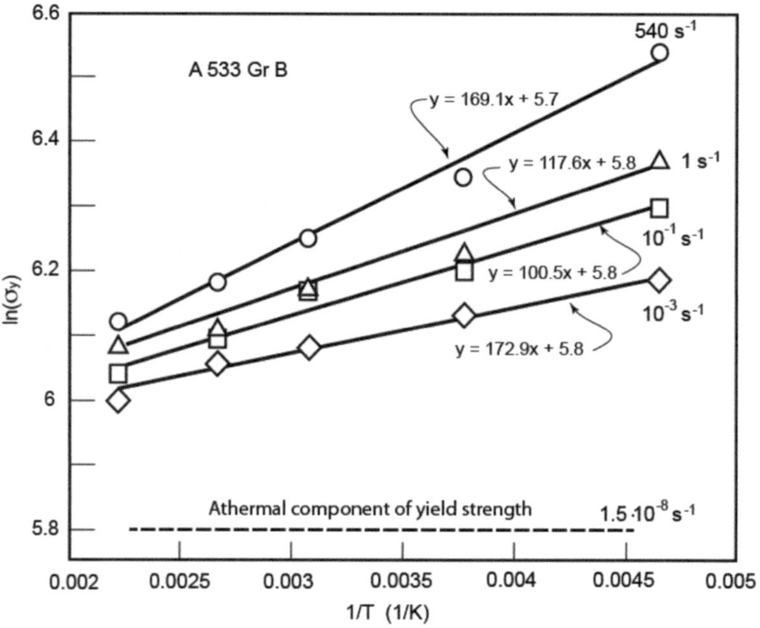

Fig. 5.44 Experimental data obtained by Milella for A 533 Gr B steel show that the linear trend $\log(\sigma_y)$ vs 1/T is maintained independently of the applied strain rate [28]. All lines converge toward a common point that represents the athermal component of the yield stress

$$m = 6.9 \cdot ln(\dot{\varepsilon}) + 124 \tag{5.43}$$

Therefore, the angular coefficient B of Eq. (5.12) or Eq. (5.14) is not constant but varies with the strain rate according to Eq. (5.43) by posing $B = m$. The result is that Eq. (5.14) becomes

$$\sigma_y = 330\left(e^{\frac{124}{T}}(\dot{\varepsilon})^{\frac{6.9}{T}}\right) \tag{5.44}$$

This is the equation of the solid state for A 533 Gr B steel valid in Region II of Fig. 5.24 where there is a component of the yield strength thermally activated. Note that the straight line of Fig. 5.45 goes to $m = 0$ for $ln(\dot{\varepsilon}) = -18$. This means that the strain rate corresponding to the horizontal line (the athermal component of yield strength) is $\dot{\varepsilon}$=exp(−18) = 1.5·10^{-8} s^{-1}. At this strain rate to reach the 0.002 yield strength in a tensile test takes about one and a half day.

By applying the same procedure to the annealed mild steel of Fig. 5.26 with strain rates from 10^{-3} s^{-1} to 200 s^{-1} we get the diagram of Fig. 5.46. Again, data points are aligned along straight lines of increasing slope with the applied strain rate and again all lines are pointing back toward a common point for $1/T$ ~ 0. The value of this point that provides the value of the constant A in Eq. (5.12) and C in Eq. (5.14) is 4.2 and the athermal component of the yield strength is

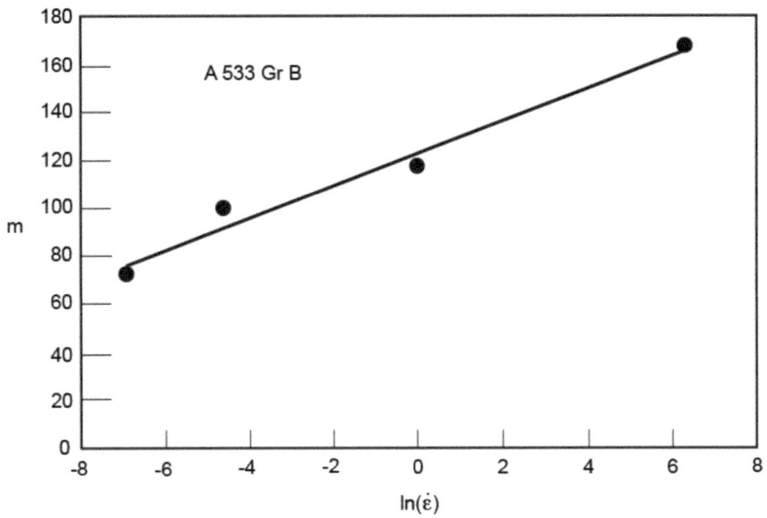

Fig. 5.45 The values of the slopes m of the lines of Fig. 5.44 are aligned along a straight line in a m-$ln(\dot{\varepsilon})$ diagram

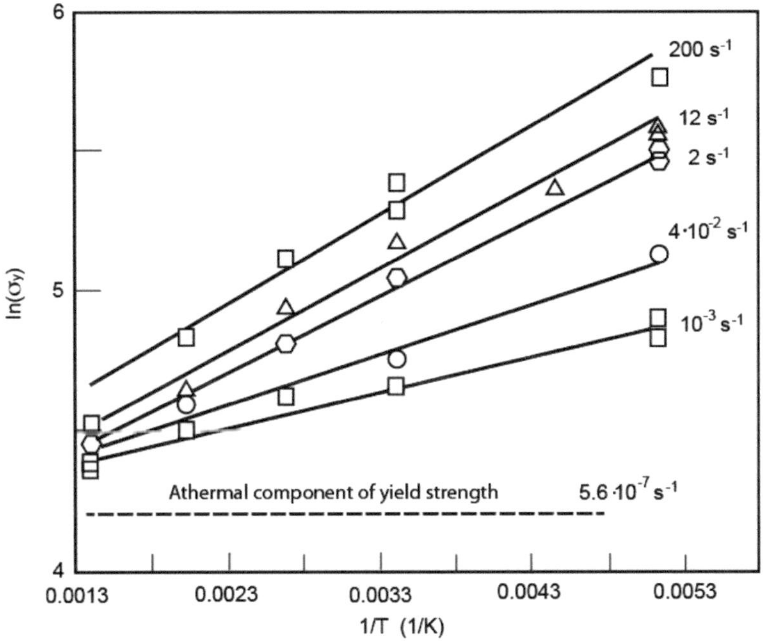

Fig. 5.46 Verification of the linear trend log(σ_y) vs 1/T at different strain rates for annealed mild steel

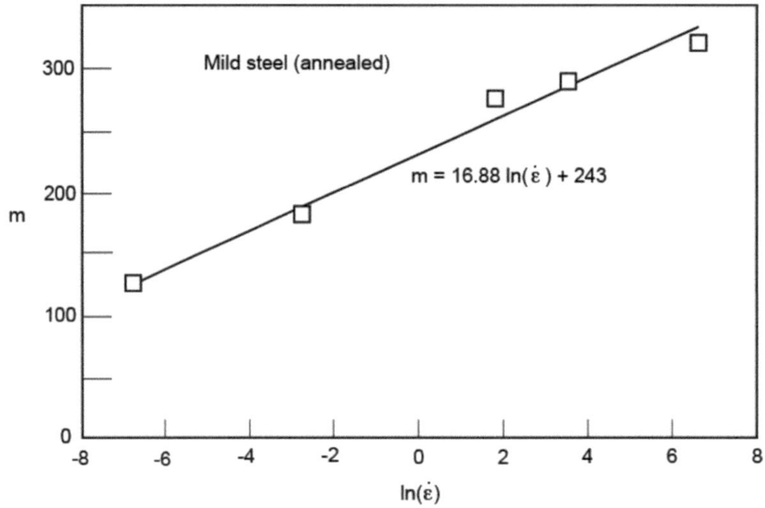

Fig. 5.47 The values of the slopes m of the lines of Fig. 5.46 are aligned along a straight line in a m-$ln(\dot{\varepsilon})$ diagram

$$\sigma_a = e^{4.2} = 67MPa \tag{5.45}$$

Again, plotting the slopes of the lines of Fig. 5.46 in a m versus $ln(\dot{\varepsilon})$ diagram we get another straight line, as shown in Fig. 5.47, whose equation is

$$m = 16.88 \cdot ln(\dot{\varepsilon}) + 243 \tag{5.46}$$

Therefore, the solid state equation for the mild steel considered for strain rates up to 200 s^{-1} is

$$\sigma_y = 67\left(e^{\frac{243}{T}} (\dot{\varepsilon})^{\frac{16.88}{T}}\right) \tag{5.47}$$

Figures 5.44 and 5.46 show that as temperature increases, the response of the material at different strain rates comes closer and closer as if temperature were *washing-out* the strain rate effect. The opposite happens at high strain rates, where the material is more sensitive at high temperatures, and the mechanical response gets closer and closer as temperature decreases. This can be seen for A 533 B steel in Figs. 5.48 and 5.49 for the mild steel considered where higher strain rates have been included [82].

Note that for A 533 Gr B steel all lines at low strain rates merge at $1/T = 0.002$ that is about 770 °C or 0.5 the melting temperature T_m which seems to be in contrast

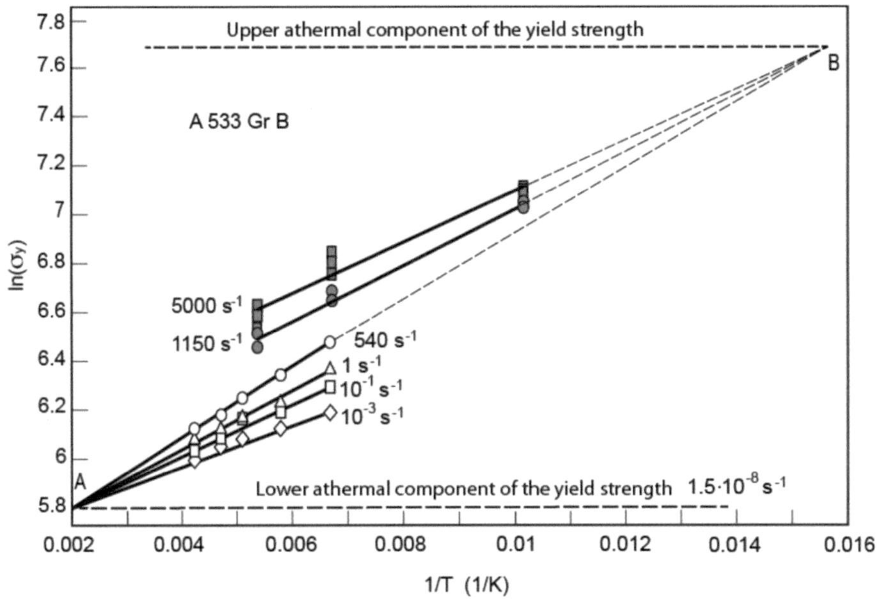

Fig. 5.48 High and low strain rate dependence of $\ln\sigma_y$ versus 1/T for A 533 B steel [82]

with what has been said in Sect. 5.3 (see Fig. 5.20) about the temperature at which the yield strength becomes temperature independent.

However, the contrast is just apparent since the temperature at which the different strain rates curves flatten down is not actually varying with strain rate between 0.2 and 0.5 T_m but is the same, only reached asymptotically at lower strain rates. For a generic material having BCC crystal structure Eq. (5.44) takes the form

$$\sigma_y = C\left(e^{\frac{M}{T}} (\dot{\varepsilon})^{\frac{N}{T}}\right) \tag{5.48}$$

with C, M and N material characteristics. The new element that Eq. (4.48) bears is that temperature affects directly also the strain rate. However, Figs. 5.48 and 5.49 open the question of whether or not the plot should be left open. In other words, the straight lines representing the material behavior at different temperatures and strain rates appear to diverge and run without any common end in the directions opposite to the merging points A and B that represent the lower and upper athermal component of the yield strength. This seems not be physically acceptable, while it is rather convincing the opposite, i.e., that the diagram of material behavior must have a closed form, which actually means that all lines shall joint at both extremities, as shown schematically in Fig. 5.50. There will be a transition strain rate at which the corresponding straight line will join both points A and B. The strain rate that seems to mark this watershed in the material behaviour is about 1000 s^{-1}. If this is the case, then a new formulation of the solid state equation can be written that takes the form:

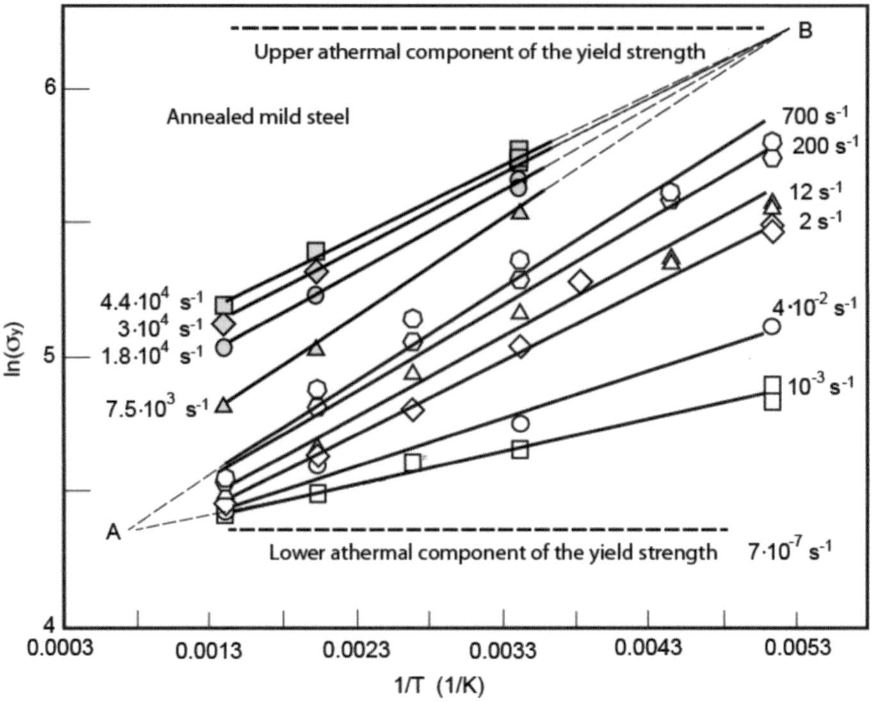

Fig. 5.49 High and low strain rate dependence of the $\ln\sigma_y$ versus 1/T for annealed mild steel

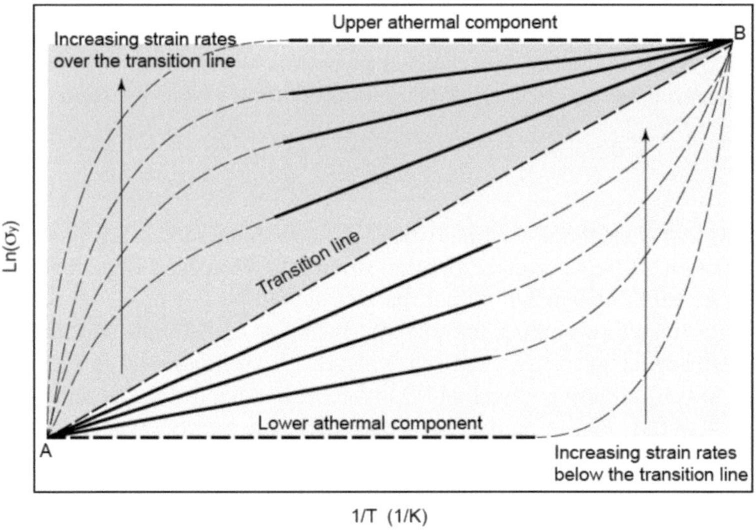

Fig. 5.50 Overall trend of $ln(\sigma_y)$ vs $1/T$ at various strain rates forming a closed diagram

$$\frac{\sigma_y - \sigma_o}{o_{max} - \sigma_o} = 1 - \left[1 - \frac{ln(t/t_{min})}{ln(t_{max}/t_{min})}\right]^m \tag{5.49}$$

where m is a strain rate exponent, σ_{max} the maximum value of the yield strength achievable at any strain rate at the minimum temperate T_{min}, i.e., the upper athermal component, σ_o the lower athermal component of the yield strength, $t = 1/T$, $t_{max} = 1/T_{min}$ and $t_{min} = 1/T_{max}$. T_{min} is the temperature T_1 at which the ductility of the material goes to zero and the yield strength becomes equal to the ultimate strength, as shown in Fig. 5.2.

For carbon steels T_{min} is about − 200 °C. T_{max} is the temperature at which the yield strength of the material starts to flatten afterward it drops abruptly, as schematically shown in Fig. 5.21. Equation (5.49) can be re-written as:

$$\begin{aligned} 1 - \frac{\sigma_y - \sigma_o}{\sigma_{\max} - \sigma_o} &= \left[1 - \frac{\ln(t/t_{\min})}{\ln(t_{\max}/t_{\min})}\right]^m \\ \frac{\sigma_{\max} - \sigma_y}{\sigma_{\max} - \sigma_o} &= \left[1 - \frac{\ln(t/t_{\min})}{\ln(t_{\max}/t_{\min})}\right]^m \quad ; \end{aligned} \tag{5.50}$$

or

$$\begin{aligned} \ln\left(\frac{\sigma_{\max} - \sigma_y}{\sigma_{\max} - \sigma_o}\right) &= m\left\{\ln\left[1 - \frac{\ln(t/t_{\min})}{\ln(t_{\max} - t_{\min})}\right]\right\}. \\ &= m \cdot \ln\left[\ln\left(\frac{t_{\max}}{t_{\min}}\frac{t_{\min}}{t}\right)\frac{1}{\ln(t_{\max}/t_{\min})}\right] \\ &= m \cdot \ln\ln\left(\frac{t_{\max}}{t}\right) - m \cdot \ln\ln\left(\frac{t_{\max}}{t_{\min}}\right) \end{aligned} \tag{5.51}$$

Equation (5.51) describes a cluster of straight lines whose center C has co-ordinates $C \equiv (a;0)$. Therefore, all straight lines described by Eq. (5.51) must pass through $y = 0$. Figure 5.50 is a compendium of results obtained by the author for several steels, niobium and α-titanium [82] at different temperatures and strain rates, using Eq. (5.50) procedure. There are several clues arising from Fig. 5.51.

First, the strain rates sensitivity is not the same for all metals, as already discussed. α-Titanium, for example, is particularly sensitive even at high temperatures while steels, in general, are less prone to react to high strain rates, especially at high temperatures. Niobium is somehow in between the two; less sensitive than α-titanium but certainly more than steel. The diagram also includes data coming from OFHC copper, which is a FCC crystal lattice material. For what it has been said in Sect. 5.3 copper should not have a thermally activated component of the yield strength. However copper is highly sensitive to strain rate. This can be seen in Fig. 5.50 where the yield strength of copper jumps up even at very high temperature, much more than α-titanium.

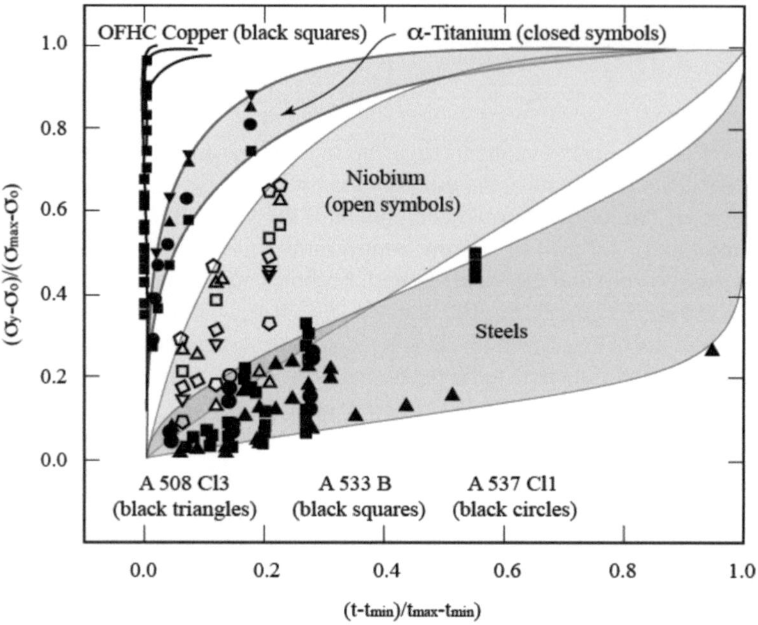

Fig. 5.51 Dependence of the yield strength σ_y on $t = 1/T$ for several metals at different strain rates

Concluding Remarks

Temperature and strain rate have a strong and opposing effect on BCC metal behavior. The temperature effect is due to the fact that these materials have a thermally activated component of the yield stress associated to the dislocation motion. At very low temperatures, dislocations are locked down so tightly that their motion becomes practically impossible elevating the yield stress. In medium–low strength steels yield stress at very low temperatures reaches impressive values typical of high strength steels that do not show the same thermal effect almost at all. Strain rate has the opposite effect sharply increasing the yield stress with increasing rate of load application. Dislocations are always responsible for that elevation since the cannot move at the same rate of high strain rate applied to the material. At variance with BCC behavior, FCC materials are almost not sensitive to temperature variations and the yield stress remains almost constant within a large temperature range. This effect has been already seen for the Charpy V energy, see Sect. 3.4, and will be seen with the fracture toughness, K_C.

References

1. Smith, E.: Cleavage fracture in Mild Steel, Intern. J. of Fracture Mechanics (1966)
2. Conrad, H., Hayes, W.: Correlation of the thermal component of the yield stress of body centered cubic metals, Rep. N. TDR-169(3240–11)TN-3,Comm. Space Syst. Div. USAF (1963)
3. Oates, G.: The effect of hydrostatic stress on cleavage fracture in a mild steel and a low-carbon manganese steel, Lab. Rep. No RD/L/R 1481, Central Electric Research Laboratory, (1967)
4. Owen, W.S., Cohen, M., Averbach, B.L.: Brittle fracture of mild steel in Tension at –196° C. Transactions ASM **50**, 634 (1958)
5. Hahn, G.T., Averbach, B.L., Owen, W.S., Cohen, M.: Initiation of cleavage microcracks in polycrystalline iron and steel. Fracture, Technology Press, MIT and John Wiley and Sons, N.Y., p. 91 (1959)
6. Eldin, A.S., Collins, S.C.: Fracture and Yield stress of 1020 steel at low temperature. J. Appl. Phys. **22**, 1296–1297 (1951)
7. Eldin, S.a., Collins, S.C.: Fracture and yield stress of 1020 steel at low temperature, J. Appl. Phys., **22**, 1296–1297 (1951)
8. Bechtold, J.H., Shewmon, P.G.: The flow and fracture characteristics of annealed tungsten. Transactions ASM **46**, 397 (1954)
9. Wessel, E.T.: The Ductile-to-brittle Transition in Molybdenum, The Metal Molybdenum, ASM, Cleveland, p. 241 (1958)
10. Wessel, E.T.: The influence of pre-existing sharp cracks on brittle fracture of a NiMoV forging steel. Transactions ASM **52**, 277 (1960)
11. Meimei, Li, Byun, T.S., Hashimoto, N., Snead, L.L., Zinkle, S.J.: The temperature dependence of yield stress for neutron-irradiated molybdenum. J. Nucl. Mater., **371**, 53–60 (2007)
12. Carreker, R.P., Jr., Hibbard, W.R., Jr., Tensile deformation of aluminium as a function of temperature, strain rate and grain size, G.E. Research Lab., WADC Tech. Rep. 55–113, (1955)
13. Carreker, R.P., Jr.: Tensile deformation of silver as a function of temperature, strain rate and grain size, G.E. Research Lab., WADC Tech. Rep. no. RL-1225 (January 1955)
14. Jenkins, W.D., Digges, T.G., Johnson, C.R.: Tensile properties of nickel, copper and 70-percent-copper-30-percent-nickel alloys at high temperatures. J. Res. Natl. Bur. Stand., **58**(4) (1957)
15. Hickey, C.F., Jr.: Mechanical properties of titanium and aluminium alloys at cryogenic temperatures, Watertown Arsenal Lab., Technical Report WAL TR 540.2/1, (1962)
16. Antoun, B.R.: Temperature effects on the mechanical properties of annealed and HERF 304L Stainless steel, Sandia National Laboratories, Report SAND2004–3090, November (2004)
17. North American Stainless, Flat Product Stainless Steel Grade Sheet (2014)
18. Austenitic Steels: Mechanical Properties at Cryogenic Temperatures, Penflex Engineering Bulletin N. 106, 17 February (2009)
19. Harding, J.J.: Arc. Mechaniki Stosowanej **27**, 715 (1975)
20. Nabarro, F.R.N.: Theory of crystal dislocations, Oxford University Press, 1st Ed. (1967)
21. Benoit, W., Gremaud, G., Quenet, B.: Mater. Sci. Eng., A **164**, 42 (1993)
22. Soo, P., Higgins, G.T.: The deformation of zirconium-oxygen single crystals. Acta Met. **16**, 177–186 (1968)
23. Evans, K.R.: The embrittlement of Ti-8 pct Al alloys. TMA-AIME **242**, 648 (1968)
24. Kocks, U.F., Ashby, M.F., Argon, A.A.: Thermodynamics and kinetics of slip, Prog. Mat. Sci., **19** (1975)
25. Meyers, M.A.: Dynamic behavior of materials, John Wiley and Sons, Inc., Chapter 14 (1994)
26. Conrad, H.: The athermal component of the flow stress in crystalline solids. Mater. Sci. Eng. **6**(4), 265–273 (1970)
27. Reed-Hill, R.E.: Physical Metallurgy Principles, D. Van Nostran Co., **341** (1973)
28. Milella, P.P.: Temperature and strain rate dependence of mechanical behavior of body-centered cubic structure materials, TMS-Annual Meeting, Chicago (1998); On the dependence of the yield strength of metals on temperature and strain rate; The mechanical equation of solid state, CP620, Shock Compression of Condensed Matter, 2002 American Institute of Physics, 642–645 (2001)

29. Wessel, E.T., Clark, W.G., Wilson, W.K.: Engineering methods for the design and selection of materials against fracture, Westinghouse Research Laboratories, Technical report (1966)
30. Pugh, C.E., Heavy section steel technology program, section 2.4, Kanninen, M.F. et. al., NUREG/CR-4219, **3**(1), (1986)
31. Campbell, J.D., Ferguson, W.G.: The temperature and strain-rate dependence of the shear strength in mild steel, Philosophical Magazine, **21**, 63–82 (1970), Appl. Mech. Rev., **45**, p. 519 (1992)
32. Wood, D.S., Clark, D.S.: The time delay for the initiation of plastic deformation at rapidly applied constant stress. Proc. Am. Soc. Testing Mat. **49**, 717–737 (1949)
33. Gilman, J.J., Johnston, W.G.: Dislocations and mechanical properties of crystals, vol. 116. Wiley, New York (1957)
34. Johnston, W.G., Gilman, J.J.: Dislocation velocity, dislocation density and plastic flow in litium fluoride crystals. J. Appl. Phys. **30**(2), 129–144 (1959)
35. Frank, F.C.: On the equations of motion of crystal dislocations. Proc. Phys. Soc. **62A**, 131–136 (1949)
36. Manjoine, M.J.: Influence of rate of strain and temperature on the yield stress of mild steels. J. Appl. Mech. **11**(4), A211–A218 (1944)
37. Dowling, A.R., Harding, J.: Tensile properties of mild steel high strain rates. In: Proc. 1st HERF Conference, University of Denver, **2**, 7.3.1. (1967)
38. Hollomon, J.H., Jaffe, L.D.: Ferrous Metallurgical Design, vol. 112. John Wiley & Sons, New York (1947)
39. Boyce, B.L., Crenshow, T.B., Dilmore, M.F.: The strain rate sensitivity of High-strength high-Toughness steels, SANDIA Report, SAND2007–0036, Unlimited Release (2007)
40. Saxena, A., Chatfield, D.A.: High strain rate behaviour of some hot and cold rolled low carbon steels, SAE paper 760209, Detroit (1976)
41. Hoge, K.G., Mukherjee, A.K.: The temperature and strain rate dependence of the flow stress of tantalum. J. Mater. Sci. **12**, 1666–1672 (1977)
42. Tsao, M.C.C., Campbell, J.D.: Plastic shear properties of metals and alloys at high strain rates, Oxford University, Dep. Eng. Sci., Report N. 1055/73 (1973)
43. Wilkowski, G.M., et al.: Fracture of circumferentially cracked type 304 stainless steel pipes under dynamic loading, 2nd ASTM Inter, pp. 6–9. Symposium on Elastic-Plastic Fracture Mechanics, Philadelphia, PE (October (1981)
44. Isakov, M.: Strain rate history effects in metastable austenitic stainless steel, Thesis for the Degree of Science in Technology, Tampere University of Technology (2012)
45. Zener, C., Hollomon, J.H.: Effect of Strain Rate upon Plastic Flow of Steel. J. Appl. Phys. **15**, 22–32 (1944)
46. Vagenhofer, M., Ericson-Natishan, M., Armstrong, R.W.: Influence of strain rate and grain size on yield and serrated flow in commercial Al-Mg alloy 5086. Scripta Mater. **41**, 1177–1184 (1999)
47. Lüdwik, P.: Über den Einfluss der Deformations-geschwindigkeit bei bleibenden Deformationen mit besonderer Berücksichtigung der Nachwirkungserschein nungen. Physik. Z. **10**, 411–417 (1909)
48. Manjoine, M.J.: Influence of rate of strain and temperature on yield stress of mild steels. Jour. Appl. Mech. **11**(4), A211-218 (1944)
49. Cottrell A.H.: Effects of solute atoms on the behaviour of dislocations. Ph. Soc., p. 30, (1948)
50. Cottrell, A.H., Bilby, B.A.: Dislocation theory of yielding and strain aging of iron. Ph. Soc. **62**, 49 (1949)
51. Cottrell A.H.: Dislocations and plastic flow in cristals. Oxford, (1953)
52. Cottrell A.H.: Deformation of solids at high rates of strain. In: Conf. On Prop. Of Mat. At High Rates of Strain, pp. 1–12, (1957)
53. Cottell, H.: The properties of materials at high rate of strain. Inst. Mech. Eng., London (1957)
54. Cottrell, A.H.: Theory of brittle fracture in steels and similar metals. Trans. AIME **212**, 192–203 (1958)
55. Hollomon, J.H.: The mechanical equation of state. Trans. AIME **171**, 535–545 (1947)

56. Yokobori, T.: The cottrell bilby theory of yielding of Iron. Ph. Rev. **88**, 1423 (1953)
57. Orowan E.: Dislocations and mechanical properties. Disl. In Metals, AIME, (1954)
58. Seeger, A.: The generation of lattice defects by moving dislocations and its applications to temperature dependence of the flow stress in FCC crystals. Phil. Magaz. **46**, 1194 (1955)
59. Seeger, A.: The temperature dependence of the critical shear stress and of work hardening of metal crystals. Phil. Magaz. **45**, 771 (1954)
60. Campbell, J.D.: Yield delay time of mild steel. Trans. ASM **51**, 659 (1959)
61. Campbell, J.D.: The dynamic yielding of mild steel. Acta Metall., 1, p. 706, (1953)
62. Conrad, H.: On the mechanism of yielding and flow in Iron. Jour. Iron Steel Inst. **198**, 364 (1961)
63. Conrad, H., Hayes W.: Correlation of the Thermal Component of the Yield Stress of the BCC Metals. Aerosp. Corp. (1962)
64. Conrad, H., Wiedersich, H.: Activation energy for deformation of metals at low temperatures. Acta Metall. **8**, 128 (1960)
65. Petch, N.J.: The cleavage strength of policrystals. Jour. Iron Steel Inst. **174**, 25 (1953)
66. Petch, N.J.: The fracture of metals. Progr. In Metal Phys. **5**, 1 (1964)
67. Petch, N.J., Helsop, J.: The stress to move a free dislocation in alpha iron. Phil. Magaz. **47**, 886 (1956)
68. Johnson, G.R., Cook W.H.: In: Proc. 7th Inter. Symp. Ballistic, Am. Def. Prep. Org. (ADPA), Netherlands (1983)
69. Johnson, G.R., Hoegfeldt, J.M., Lindolm, U.S., Nagy, A.: ASME Jour. Eng. Mater. Tech. **105**, 42 (1983)
70. Klopp, R.W., Clifton, R.J., Shawki, T.G.: Mech. Mater. **4**, 633 (1985)
71. Follansbee, P.S.: In Metallurgical Applications of Shock-Waves and High Strain Rate Phenomena, eds. Murr, L.E., Staudhammer, K.P., Meyers, M.A., Dekker, New York, p. 451 (1986)
72. Zerilli, F.J., Armstrong, R.W.: Jour. Appl. Phys. **61**, 1816 (1987)
73. Zerilli, F.J., Armstrong, R.W.: Jour. Appl. Phys. **68**, 1580 (1990)
74. Zerilli, F.J., Armstrong, R.W.: Acta Met. Mat. **40**, 1803 (1992)
75. Follansbee, P.S., Kocks, U.F.: Acta Metall. **36**, 81 (1988)
76. Nemat-Nasser, S., Obata, M.: Proc. Royal Society of London **A407**, 343 (1986)
77. Nemat-Nasser, S., Okinaka T., Ni L., J. Mech. Solids, **46**(6), p. 1038 (1998)
78. Nemat-Nasser, S., Li, Y.: Acta Met. Mater. **46**(2), 565 (1998)
79. Milella, P.P.: Temperature and strain rate dependence of mechanical behavior of body-centered cubic structure materials. TMS Fall Meeting '98, Chicago, Illinois, 11–15 October (1998)
80. Bonora, N., Milella, P.P.: Advanced modelling for ductile behavior of material under high strain rates and temperature using continuum damage mechanics and solid state equation. US AIR FORCE, AFORS Research Contract F61775–00-WE029, Final Report (2001)
81. El-Magd, E.: Mechanical properties at high strain rates. J. Phys. IV **04**(C8), 149–170 (1994)
82. Pugh, C.E.: Heavy-section steel technology program semiannual progress report for October 1985-March 1986, NUREG/CR-4219, 3, No. 1, ORNL/TM-9593/V3&N1 (1986)
83. Milella, P.P., Bonora, N.: Strain rate and temperature effect in ductile failure process: characterization and modelling using continuum damage mechanics, US-AIR FORCE Research Contract, Final Report, E.O.A.R.D., 223/231 Marylebone Rd., London NW1 5TH—England (2001)

Chapter 6
Morphological Aspects of Fracture

6.1 Introduction

We have analyzed in Chap. 4 the stress state ahead of a crack or a notch and its influence on the fracture process. We will now move our attention to the morphological aspects that the different stress state have on the fracture appearance, whether produced by a notch or a crack. This will complete the analysis of stress state impact on material behavior.

It has long been recognized that some correspondence exists between macroscopic fracture appearance and loading conditions responsible for failure and between macroscopic and microscopic features as well. Therefore, attempts have been made to relate fractographic features to macroscopic fracture appearance not only in terms of properties such as toughness but also stress conditions and derive quantitative relationships. It may be worth to initiate this study by saying that macroscopically any failure can be described through three terms that define the mechanism of failure, the behavior of the material and its external appearance, both macroscopically and microscopically as well. The mechanisms of failure can be basically grouped into 5 fundamental types:

1. *Cleavage*, or net severance on crystallographic planes,
2. *Void growth by plastic slip* and coalescence produced by the shear stresses,
3. *Fatigue*,
4. *Corrosion*,
5. *Creep*.

The behavior of the material will be *brittle*, *ductile* or *mixed mode* depending on the degree of deformation associated with the failure. In this respect, brittle failure and cleavage do not define the same thing, although they are often, but improperly, used as synonyms. The cleavage, in fact, identifies a breaking mechanism and not a behavior of the material. For example, in the rupture that occurs in BCC metals below the NDT the behavior is certainly brittle, since no plastic deformation is present, at

P. P. Milella, *Fundamentals of Fracture Mechanics*,
https://doi.org/10.1007/978-3-032-23081-2_6

least on a macroscopic scale, and the failure mechanism is cleavage because it occurs in each grain by net severance along crystallographic planes. For BCC materials in the T_2 temperature region, Fig. 5.2, the failure mechanism may still be cleavage, but activated by plastic slips, which are no longer negligible so that we have a brittle behavior with more or less ductile islets. This leads some authors to speak of *semi-brittle* behavior caused by an improperly called *quasi-cleavage* mechanism. In fact, it is more appropriate to speak of cleavage that is not totally brittle. Tempered martensite, which has a tetragonal structure Fig. 2.6, is typically subjected to this type of failure [1]. So far, there is no evidence that anyone has observed true cleavage in FCC metals. The one important class of metals in which cleavage is the most frequently observed type of failure is the BCC. In the same way there will be a ductile or a brittle fatigue failure. It is often preferred to indicate the behavior of the material as fibrous, rather than ductile and radial instead of brittle. Also in this case, the use is improper since the terms *fibrous* and *radial* belong to the topography of the fracture, or to its external appearance as it can be seen at naked eye or at most with an optical microscope at low magnifications. This all is schematically shown in Fig. 6.1a that schematizes the failure of a smooth cylindrical steel specimen without any surface defect put in traction at low temperature or at room temperature but having a very large section. On the surface, three characteristic zones can be distinguished at naked eye. The first is the origin of rupture at the center where a plain strain condition generated by thickness triaxiality reaches its maximum, see Sect. 4.4. Conversely, grains on the surface are necessarily under a uniaxial stress state and can deform freely. This central area appears flat but dull, fibrous, with small crests and voids, revealing a certain amount of ductile rupture.

From the center, it propagates toward the rim of the specimen along radial directions where it arrives flat and terminates with more or less pronounced shear lip whose size depends on the temperature or on the thickness of the specimen. These features can be seen in Fig. 6.1(b) for a very thick steel specimen [1]. The relative amount of fibrous, radial and shear lips has been found to depend on the strength of material and the test temperature, as shown in Fig. 6.2 [1]. Fracture may also initiate from the rim or external surface.

However, when this is the case, it is always because somewhere on the rim there was a preferential initiation site such as a large surface inclusion or an initial surface defect, as in the case of Fig. 6.3 that shows the fracture origin on the surface of a hard steel specimen from a very small indentation.

6.2 Fracture in Notched Specimens

When a notch is introduced in a specimen or structure the process of fracture initiation and propagation is somewhat different. As discussed in Sect. 4.4, a notch always introduces a biaxial stress state in which the first component of stress normal to the notch plane, σ_y, reaches a maximum at the notch tip whereas the second component, σ_x, must necessarily start from zero, reaches a maximum and then drops to zero

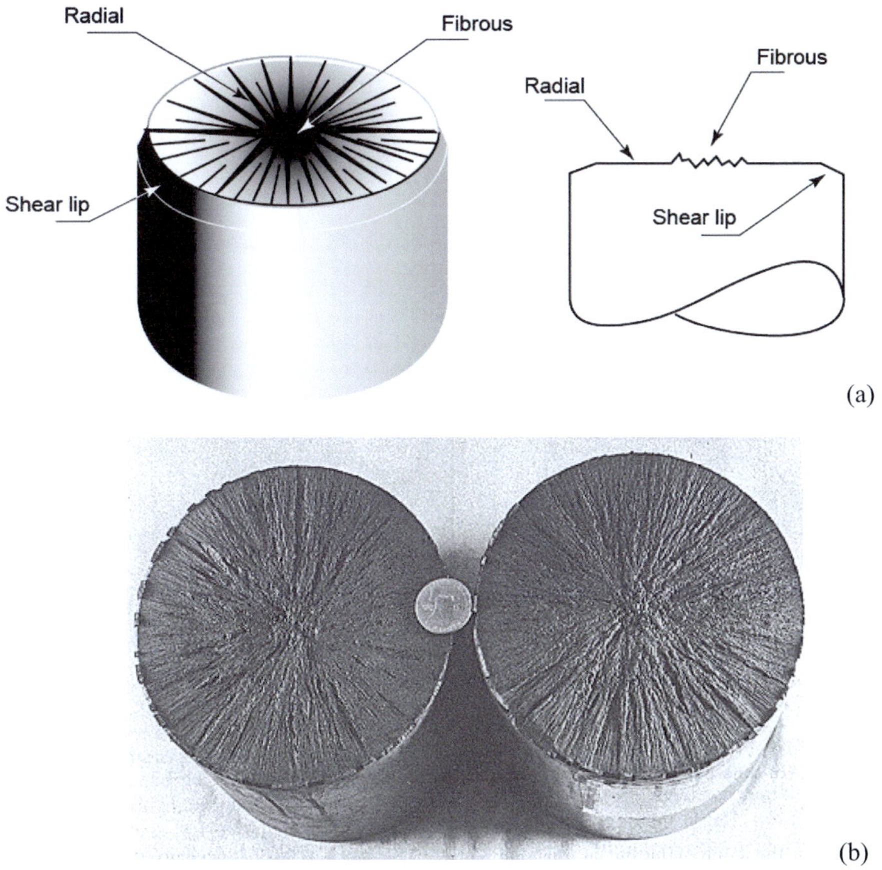

Fig. 6.1 **b** Typical flat fracture appearance of smooth, large tensile bar [1]. Shear lip is almost inexistent. **a** Schematic of the fibrous, radial and shear lip zones

again. This is a completely different scenario with respect to the crack where the two components are almost identical. If there is enough thickness, a third component, σ_z, will appear. The maximum triaxiality is likely to occur at a distance from the notch tip that depends on the geometry and the sharpness of the notch, while the point of maximum strain always lies at the notch tip [2–5]. This is shown schematically in Fig. 4.18g. It is now well recognized that the origin of the fracture, either ductile or brittle, is likely to occur right at the point of maximum triaxiality.

Consider, now, a set of three cylindrical specimens of carbon steel, type 22NiMoCr37, of 20 mm diameter, 525 MPa yield strength, having a circumferential V-shaped notch of 2 mm radius, 4 mm and 10 mm, respectively and a reduced section of 10 mm common to all specimens, as shown in Fig. 6.4. Figure 5(b) [6] shows the FE calculated longitudinal stress component, σ_y, through the section of the specimen with a 2 mm radius. A notch with a 2 mm radius is far from being

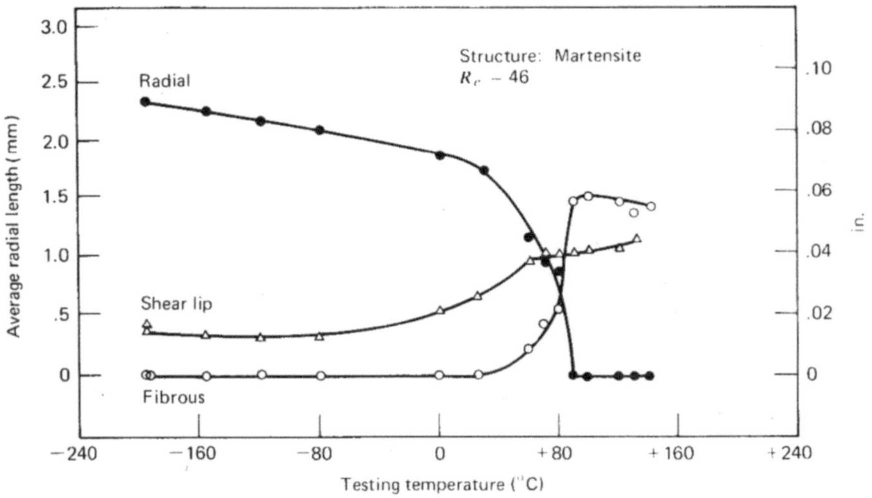

Fig. 6.2 Effect of temperature on relative size of fracture zones for 4340 steel treated to Rc 46 [1]

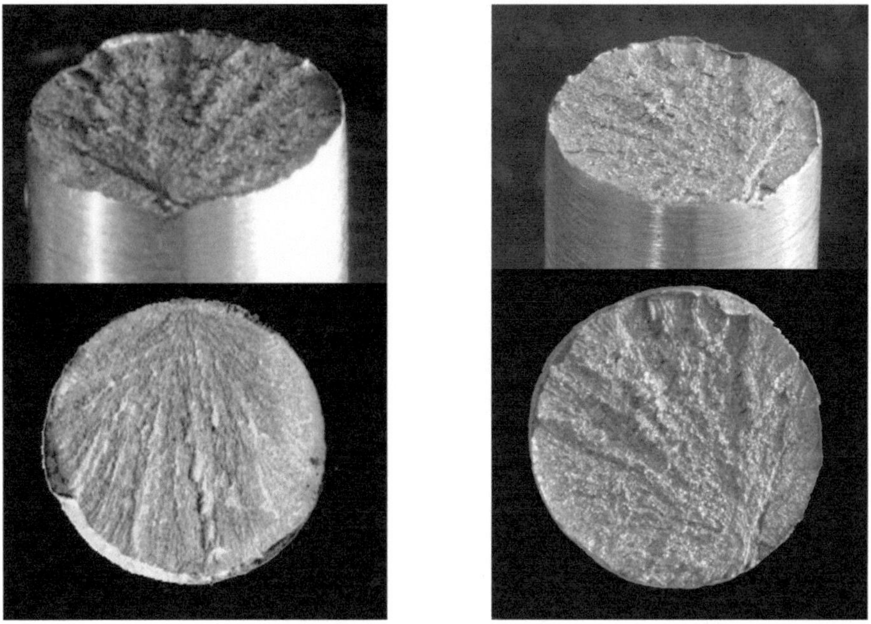

Fig. 6.3 Typical radial flat fracture with some residual shear lip in two high strength steel initiated from a small surface defect. All radii run back to the initiation point. The fibrous zone is practically absent (Milella)

a sharp notch or a crack. The maximum stress is reached at the notch root in the elastic regime, curve *a* of Fig. 6.5b, but moves toward the center of the specimen as the yield strength is exceeded, reaching the center at maximum load, Fig. 6.5g. The triaxiality factor (FT) follows the same trend, reaching its maximum at the center of the specimen for all three specimens [5–8]. Specimens with the circumferential notch of 4 mm and 10 mm radius show the same trend though with a much lower TF. At variance with the TF, the maximum strain is always reached at the notch root.

Therefore, at very low temperature where brittle fracture is more likely to occur, we may expect fracture to start at the center of the specimen where the TF is maximum. Effectively, tests run at –90 °C have shown that brittle fracture always started at the center of the specimens and propagated towards the free surface, though with same interesting differences depending on the root radius. This is shown in Fig. 6.6 [8] for the three specimens. Figure 6.6a refers to the specimen with 2 mm notch radius

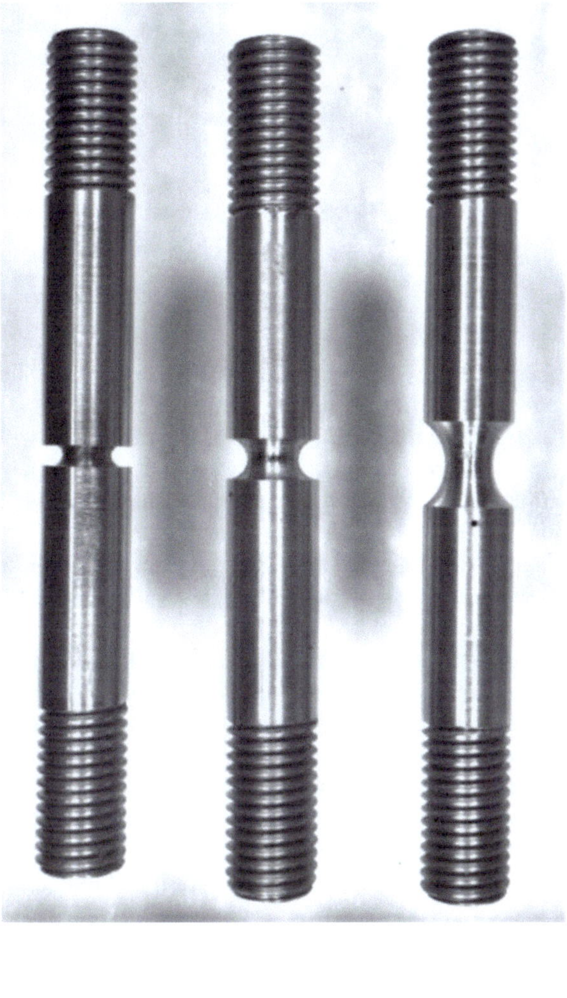

Fig. 6.4 Set of three cylindrical specimens of carbon steel, type 22NiMoCr37, of 20 mm diameter, having a circumferential V-shaped notch of 2 mm radius, 4 mm and 10 mm, respectively

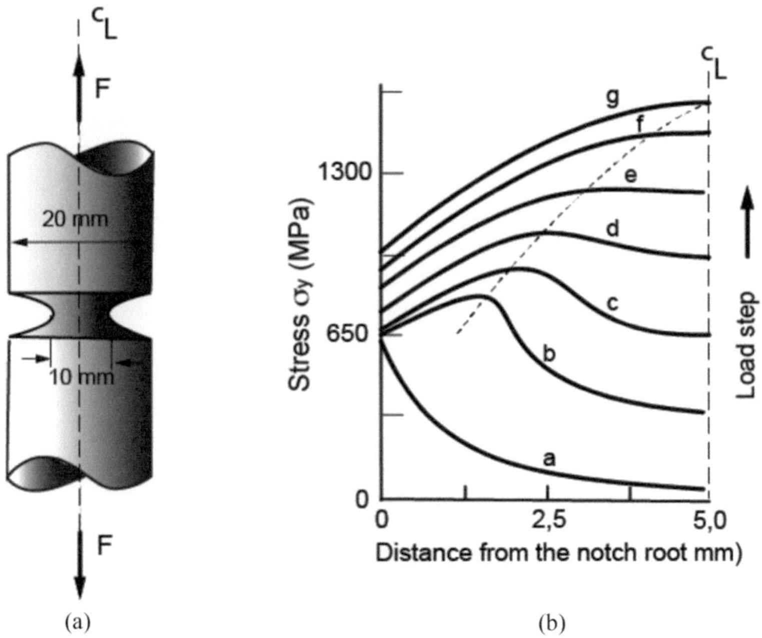

Fig. 6.5 **a** circumferentially notched traction specimen (2 mm notch radius) and **b** Finite Element results showing the stress distribution across the thickness at increasing load steps. Triaxiality also follows the same trend [6]

and shows a very irregular, ruth and dull surface characterized by radial crests and a central area, where fracture initiated, coarse and slightly raised. This scenario that reflects the turmoil of the material, is indicative of a rather steep triaxiality gradient. At larger magnification, Fig. 6.6b, each flat facet, within areas that resemble lettuce leaves, indicates cleavage in individual grains. They are separated by little white stripes that reveal tearing between different cleavage planes. On each facet some marks are visible. They are known as *river patterns* since they converge towards a single line resembling tributaries to a river and, much like a river, indicate the direction of crack propagation. However, cleavage is not completely flat but seems to be largely undulated, with hills and valleys, with secondary cracks and tear ridges showing evidence of dimples fracture (ductile) on the crests. These characteristic features are shown in Fig. 6.6g. Moving to the larger notches, the ruth and dull surface becomes more and more regular with lower hills and valleys. The lower triaxiality makes the fracture appear smoother, Figs. 6.6c and e. at larger magnification, the flat brittle facets appear as islets surrounded by waves of ductile tearing that appear as a sponge tissue. Some secondary cracks may also appear between grains. This is the sign that fracture is not totally brittle which many authors refer to as quasi-cleavage failures. It may also appear some localized intergranular separations, Fig. 6.8. The initiation point was an inclusion, namely a carbide or a particle clump. G. Irwin has called these *isolated cleavage regions* of fracture initiation ICRs [7]. We will use this acronym. The fracture started at the center and propagated towards the rim where some radii appear. The presence of a ductile component in a brittle fracture shall not surprise. It is a fact that brittle fracture is always preceded by some plastic flow. Plastic flow is always needed to start a brittle fracture. For example, it has been found that in iron at very low temperature, − 196 °C, where brittle cleavage dominates, compression yielding occurs under the same stress that produces brittle fracture in traction [9]. Since the yield strengths in compression and in traction coincide, it must be concluded that brittle fracture cannot occur until a plastic slip, even minimum, takes place. We have already seen a Charpy V-Notch specimen broken at—90 °C showing signs of ductile fracture preceding the brittle separation by cleavage, Fig. 3.12. It is the inhomogeneous nature of slip responsible for the fracture initiation. Slip bands, Fig. 2.12, are activated by shear strains somewhere in the crystal, where the stress conditions are most favorable.

When they start moving on crystallographic planes, they initiate a process of dislocations multiplication so that trains of dislocations will end up stacking, for example, at grain junctions, which constitute an area of discontinuity, a real barrier, especially if rich in precipitates, such as carbides (cementite), sulfides (MnS, FeS), silicates, nitrogen, hydrides, various oxides such as aluminates etc.. This leads to the formation of a very high local stress state that can overcome the strength of a possible particle precipitated on the grain boundary. The mechanism of crack initiation is schematically shown in Fig. 6.8 for a grain boundary precipitate, but the foreign phase may also be internal. At this point, the external stress, that was necessary to activate dislocations motion and their multiplication, becomes almost inessential in this local stress concentration, as it is negligible compared to the pile-up stress, σ_{pl} (Fig. 6.7).

$$\sigma_{p_l} \approx n \cdot \tau \tag{6.1}$$

In a large-grained metal, a single grain has dimensions of the order of one tenth of millimeter. Dislocations are, on the other hand, line defects of interstitial dimensions, therefore of the order of the Å. Even in small grains, therefore, hundreds of dislocations can be activated and pile-up. The weight of the applied stress σ disappears, then, against the much more consistent multiplier effect offered by *n*. It took Stroh [11] to present the first hypothesis that under the action of these enormous, localized stresses microcracks could enucleate. McMaon and Coen [12] working with thin sheets of metal in which the lack of triaxiality prevented the yield strength elevation, found that almost all local and isolated episodes of brittle fracture, observed at the scanning electron microscope SEM, were associated with carbide failure. On the other hand, iron carbides are nearly ideally elastic hard materials, having a strength of about 4100 MPa and the elastic modulus close to that of steel [13]. Therefore, they fracture at about 2% strain, which is just beyond the beginning of the strain-hardening range for most structural steels.

In a carbon steel of medium resistance, even at very low temperature the yield strength does not exceed 1000–1200 MPa, which in plain strain conditions generated

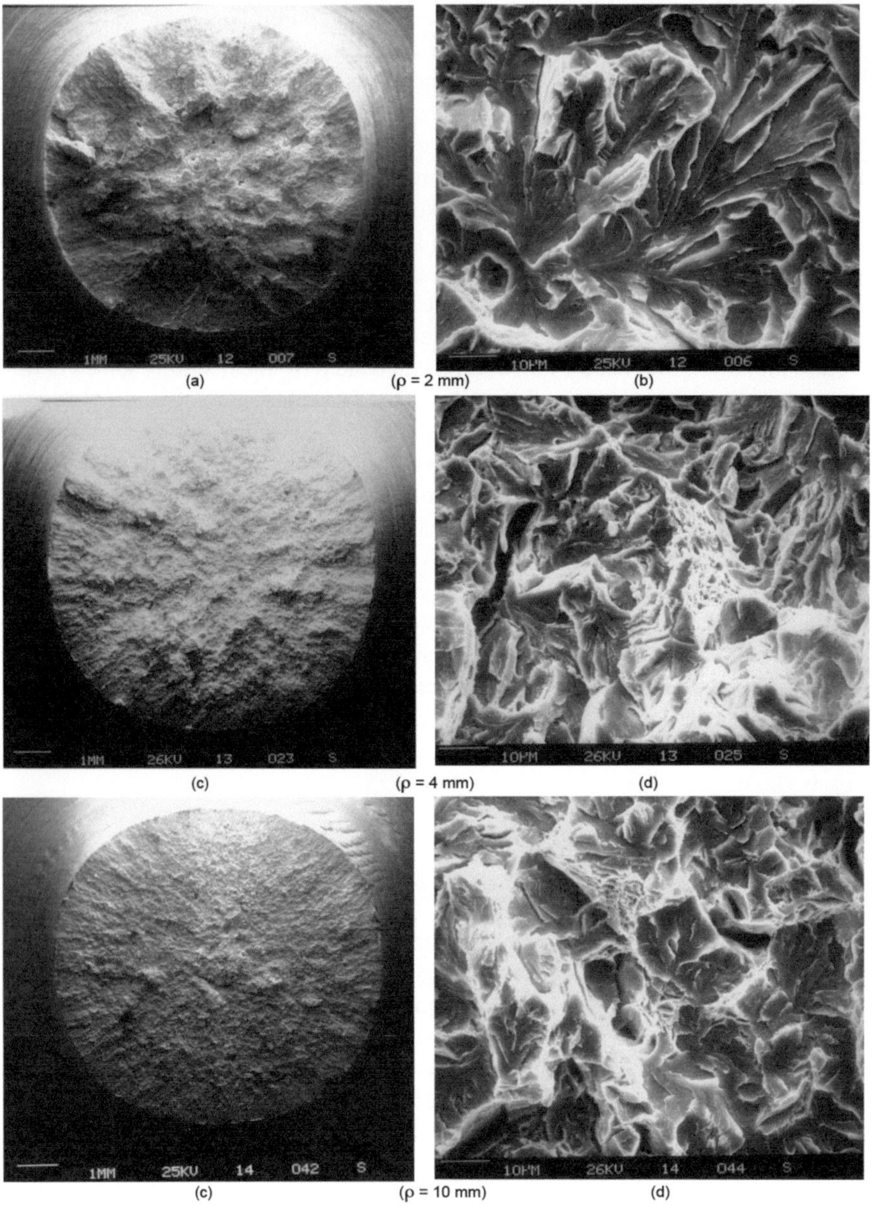

Fig. 6.6 SEM images of round specimens of 22NiMoCr37 steel having a circumferential notch of 2 mm, 4 mm and 10 mm radius, respectively, broken at –90 °C [7]

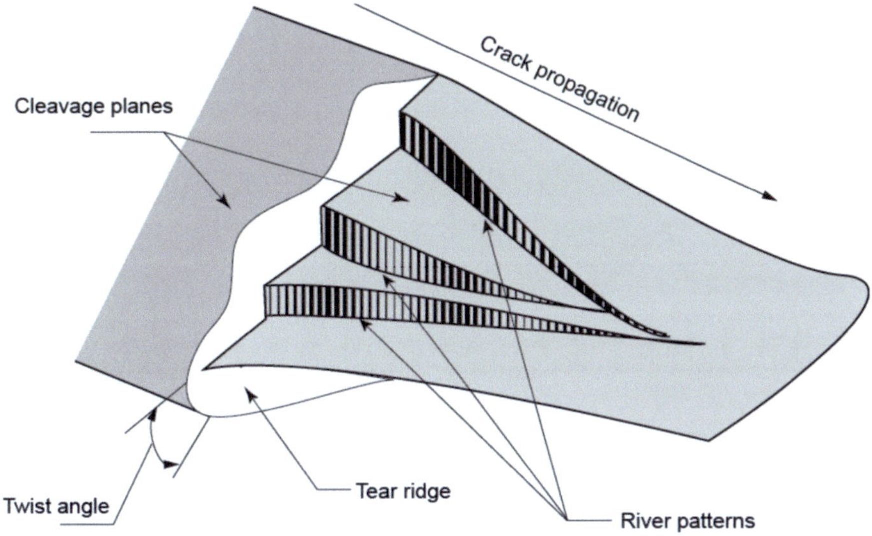

Fig. 6.7 Schematic of formation of river patterns as a result of a cleavage crack crossing a twist boundary between grains and possible tear ridge formation revealing a ductile component of fracture

Fig. 6.8 Evidence of intergranular separation on specimen of Fig. 6.4a [7]

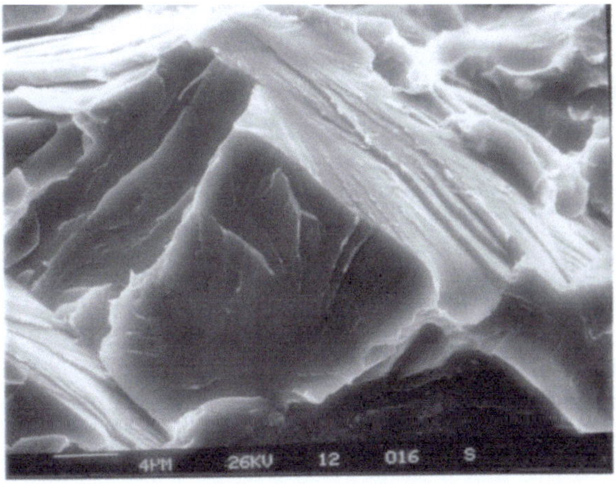

by triaxiality, see Fig. 4.8, can rise up to 3000–3600 MPa, but never as high as 4100 MPa to break carbides. Therefore, the only way to explain the carbide fracture is precisely through the mechanism of dislocations piling up, just described. However, triaxiality is necessary to assure the brittle behavior of the matrix, otherwise carbides would debond and create a void, typical of ductile fracture. For the same reason high triaxiality is not needed in high strength steels or high strength aluminum alloys for brittle fracture to occur and for the very same reason brittle fracture will never occur in soft austenitic steels. As the size of the grains increases, also the overall

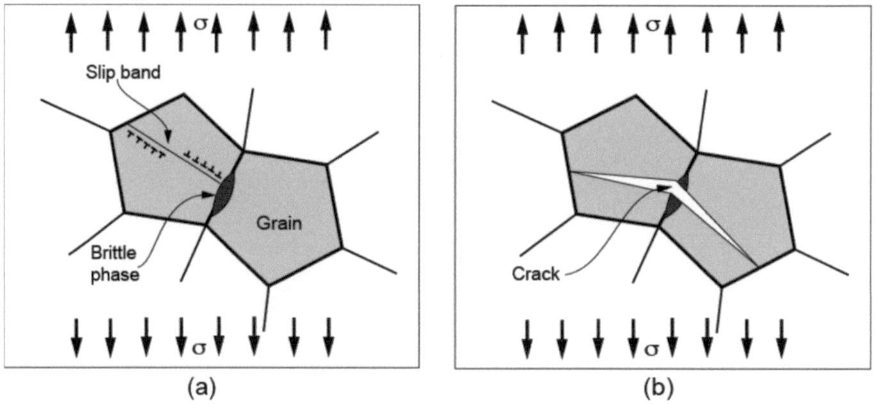

Fig. 6.9 Schematic of the mechanism of crack initiation by piling-up of dislocations against a grain boundary precipitate

length of grain boundary increases and so does the concentration of the second, brittle phases present in the metal or isolated cleavage regions, ICRs, of fracture initiation. These two factors, i.e., the number *n* of dislocations piling up and brittle phases concentration, lead to the conclusion that coarse-grained alloys are, in general, more prone to brittle fracture from ICRs than fine-grained do. An example of fracture initiated at an ICR is shown in Fig. 6.9 for a A 508 carbon steel, 620 MPa yield strength, at a temperature close to the RT_{NDT} [31]. Note that the particles were too small, 0.1 to 0.2 μm, and did not break but debonded from the matrix. Also note the sponge-tissue formation all around indicating a ductile fracture component due to the temperature not well below the transition temperature.

It should be emphasized that today this mechanism of initiation is considered by far the most important in polycrystalline metals. Figure 6.9b is a very interesting example of cracks initiated at second-phase, oxide particles in ductile high-strength steel D6aC, 1400 MPa yield strength, obtained by Zackay et al. [14]. Complete fracture was prevented only because the test temperature of the material was high enough to assure ductile behavior of the surrounding matrix.

The failure mechanism envisaged by Stroh was, in fact, a variant of the one that was first proposed by Zener [15] who did not consider the forcing action on foreign particles, but a simple coalescence of dislocations, that is, of intercrystalline voids, stacked along an obstacle, as schematized in Fig. 6.11a. Similarly, Cottrell [16] suggested that the initial crack could be generated by the coalescence of two dislocation trains activated on different crystalline planes that merged at a grain boundary, as shown in Fig. 6.11b. Twinning is a very important cause of crack nucleation in coarse-grained BCC metals at low temperature, in particular, where it becomes the preferred mode of plastic slip, as we have been discussing in Sect. 2.14. The blockage of the twins at grain boundaries or other twins can result in cleavage cracks, as shown in Fig. 6.12 [17].

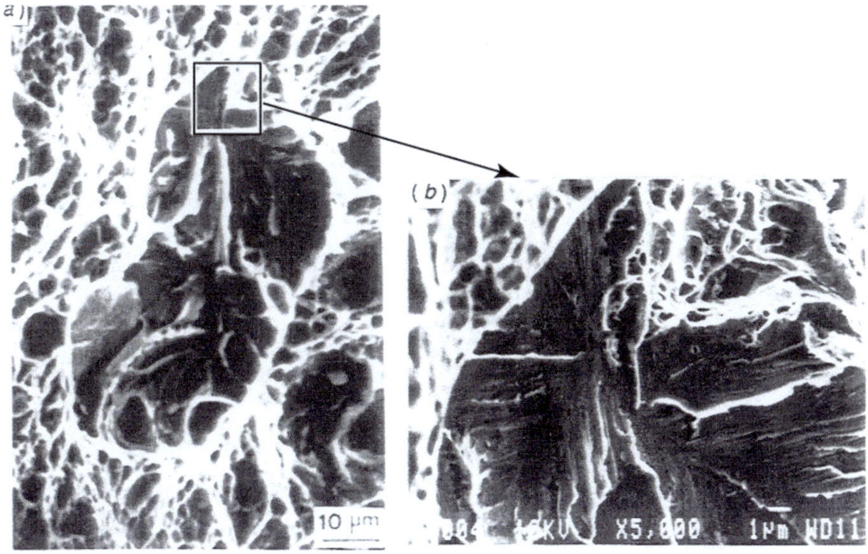

Fig. 6.10 **a** Typical ICR in A 508 steel that triggered failure at a temperature not suitable for complete cleavage and **b** enlargement of box in **a** [31] **b** Crack nucleation at second-phase particles in ultra-high strength steel, type D6aC. Figure 6.7b shows a particle debounding from the soft matrix followed by cracking [13]

6.3 The Origin of Ductile Fracture

The morphology already seen in Fig. 6.3 completely changes when the three specimens of Fig. 6.4 are tested at 100 °C. At this temperature we are well above the NDT of the material (about—30 °C). At this temperature the fracture is completely ductile. Void growth and coalescence is the normal mechanism of ductile fracture in FCC metals or in BCC metals at temperature over the NDT. The initiation mechanism is very much the same as for brittle fracture, i.e., the piling-up of dislocations against a brittle phase or ICRs, except that this time, because of the large ductility of the matrix, the foreign particles may not fracture, as shown in Figs. 6.9 and 6.10a, but detach from the matrix creating voids. These voids grow in size during the loading phase and eventually coalesce according to the general scheme shown in Fig. 6.13.

Note that this process of void formation, coalescence and ligament rupture is fed also by all individual particles of finer size dispersed in the matrix ($\approx$ 1μm) that would have not contributed to the brittle fracture at low temperatures because too small.

Under ductile conditions, these small particles, that are more resistant to cracking, simply debond from the matrix and favor the bridging process between larger inclusions. This process of void formation, growth and coalescence, reduces progressively the area of the resisting section that, eventually, fails by ductile overload. It is now well established that this process of foreign phases decohesion and voids growth in

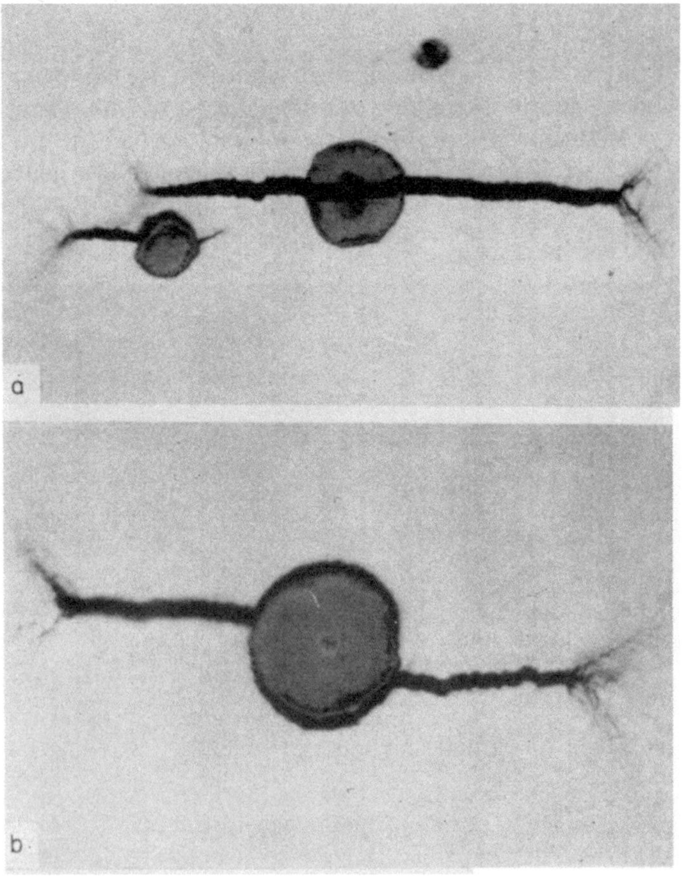

Fig. 6.10 (continued)

the crystalline matrix are responsible for the ductile failure of materials under stress [18, 19]. Cottrell [20] found that as the content of inclusions in metals decreased, the percentage of area reduction at failure also increased and this was a clear sign of the negative impact of inclusions. Without them, a pure metal could deform and even reach a 100% elongation. At failure, all specimens of Fig. 6.4 show the characteristic sign of ductile fracture: necking with cup-and-cone formation: Fig. 6.14a shows a cone, Figure 6.14c and e the cup [8]. The cup presents a central area, coarse and rough, typical of ductile tearing, where the fracture initiated and a circumferential crown at 45° indicating fast fracture by shear (shear lips). At SEM, Fig. 6.14b, d and f, the central region appears like a sponge tissue in which each individual dimple is the seat of a foreign particle that detached and produced a void that grew under stress [8]. Some of them are still visible as little balls in the cavity they produced. A better view of these foreign particles can be seen in Fig. 6.15 [8]. In Figs. 6.14b, d and e

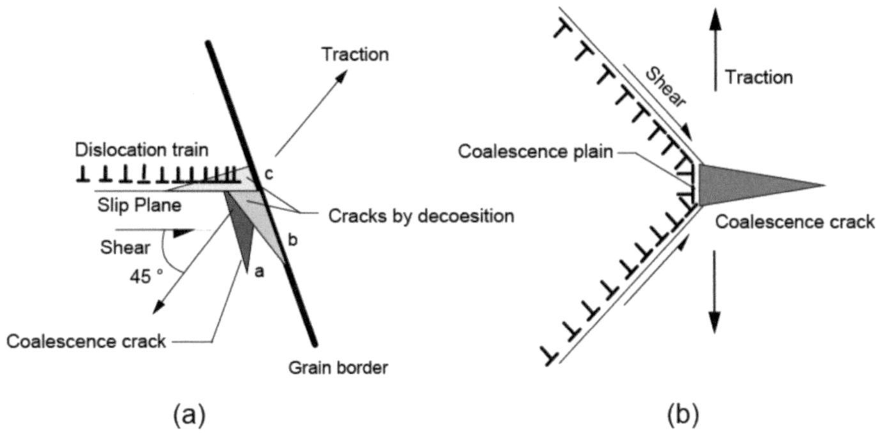

Fig. 6.11 Possible mechanisms of microcracks generation: **a** coalescence of a train of dislocations with grain edge stacking, case **a**, decohesion of crystals along the edge, case **b** or along a sliding plane (crack c) as in HCP lattice; **b** coalescence of two converging trains of dislocation with the formation of one minute crack

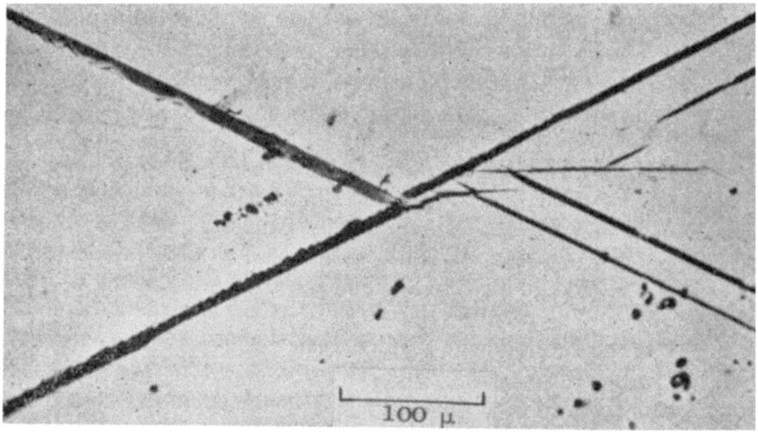

Fig. 6.12 Cleavage cracks at the intersection of twins in silicon iron [17]

it is possible to see some little areas that appear flat like fish scales. They represent islets of brittle fracture or intergranular decohesion.

In a smooth round specimen, the progression of necking, voids formation and coalescence would appear as shown in Fig. 6.16. It is during the phase of necking, when deformations stat to concentrate in a limited region and the initially smooth specimen start to change its shape resembling a notched specimen that the maximum triaxiality occurs at the center, as already said. Therefore, it is there that voids start to form, Fig. 6.16a, grow and coalesce, Fig. 6.16b, till fracture occurs, Fig. 6.16c. Ductile fracture will always start from the center of the specimen. Evidence of this

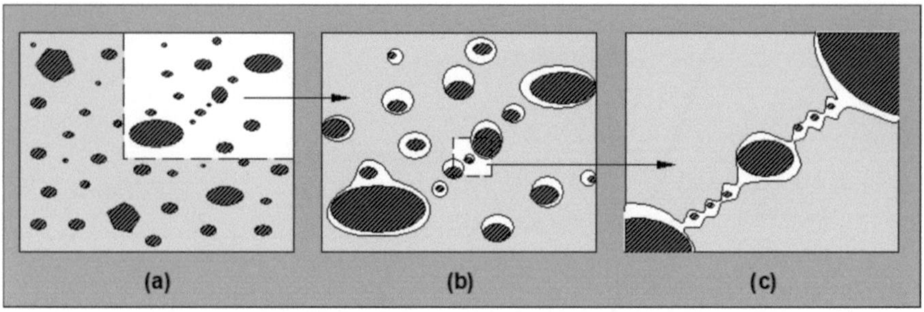

Fig. 6.13 Schematic of void formation and coalescence: **a** distribution of segregated inclusions or second phases in the crystalline matrix at rest, **b** debonding of the inclusions from the matrix and voids formation under traction and **c** void coalescence by legaments failure as the applied load increases (courtesy of Bonora)

appears in Fig. 6.17 which refers to a smooth electrolytic copper cylindrical specimen [23]. As the applied load increases, the necking starts, Fig. 6.16a, which favors the generation of pores in the center of the section, Fig. 6.17b. These voids increase in number and size till they coalesce, Fig. 6.17c, leading to fracture by overload, giving rise to the so-called cup-and-cone. In the shear lip area fracture occurs by sudden shear stress overload. A typical cup-and-cone ductile fracture can be seen in Fig. 6.18 for two smooth specimens of zircalloy at room temperature.

In a very ductile material, with low yield strength and strong hardening, such as austenitic stainless steel, remarkable large plastic slips occur that can be seen at naked eye on the external surface.

This is shown in Fig. 6.19 that refers to an initially smooth cylindrical specimen of austenitic stainless steel broken at high temperature. The signs of plastic non-uniform elongation are visible on the surface, initially perfectly smooth and now very corrugated and irregular.

It should also be noted that the section, initially circular, has become elliptical because of loss of symmetry due to the wide flow that occurs in an uncontrolled manner. The mathematical theory of growth and subsequent coalescence of holes formed by inclusion decohesion was developed by McClintock [21, 22]. McClintock postulated that in a material containing an elliptical hole that grows under increasing stress state, fracture occurs when the semiaxis of the hole approaches half of the mean spacing between two adjacent cavities. Rice and Tracy [23] further developed the McClintock model of cavity growth in a ductile medium arriving to the well-known equation for the growth rate $\dot{R}/R_0$

$$\dot{R}/R_0 = 0.283 \cdot \dot{\varepsilon}_{eq} exp\left(\frac{3\sigma_m}{2\sigma_{eq}}\right) \tag{6.2}$$

where R_o is the initial void radius, $\dot{\varepsilon}_{eq}$ is the equivalent plastic strain rate, σ_m the hydrostatic stress and σ_{eq} the equivalent stress. Note that in Eq. (6.2) the term in parenthesis σ_m/σ_{eq} is the triaxiality factor, FT, introduced in Sect. 4.3, Eq. (4.34).

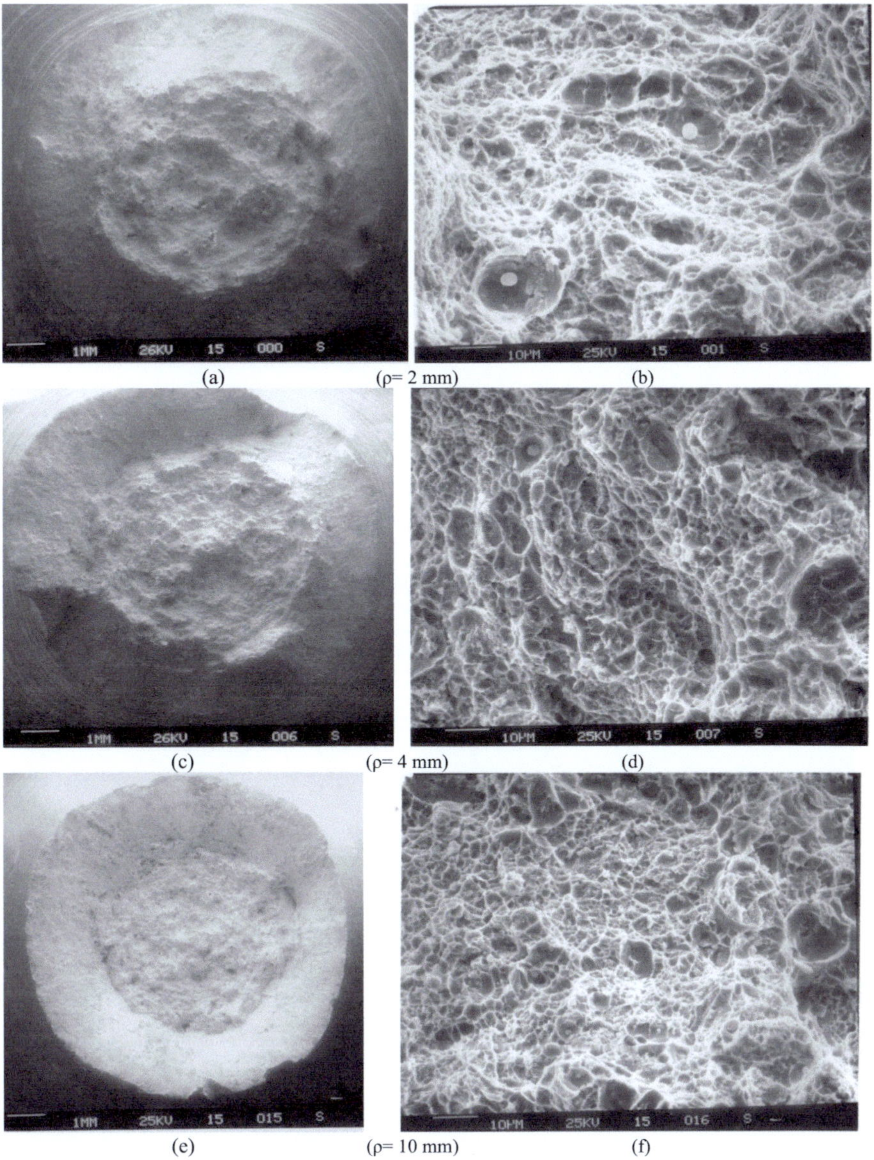

Fig. 6.14 SEM images of round specimens of type 22NiMoCr37 steel, carrying a circumferential notch of 2 mm, 4 mm and 10 mm radius, respectively, broken at 100 °C. Particles clinging to cavity sides is visible [7]

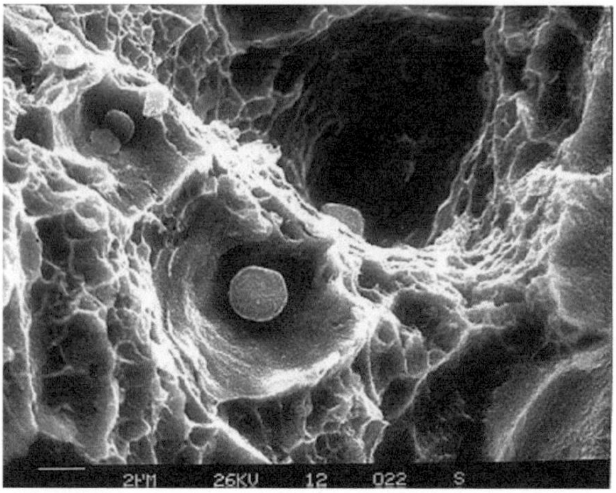

Fig. 6.15 SEM appearance, at larger magnification, of ductile fracture in type SA 533 GrB Cl1 or 22NiMoCr37 carbon steel of Fig. 6.14, tested at 100 °C. Some dimples, become caverns, still contain half of the foreign particle that broke in two (Milella)

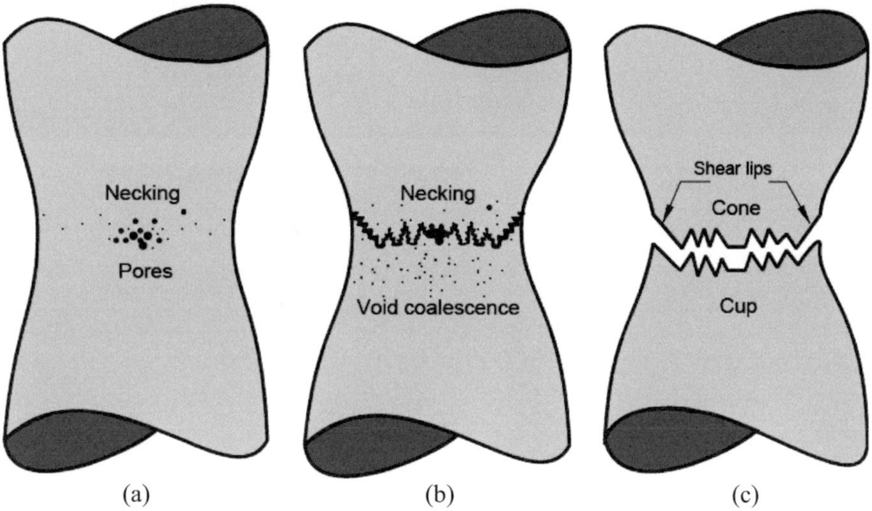

Fig. 6.16 Schematic of the formation of pores at the center of the necking in a cylindrical, smooth specimen (**a**) and their coalescence with macrocrack formation (**b**) and final rupture in the form of cup and cone (**c**)

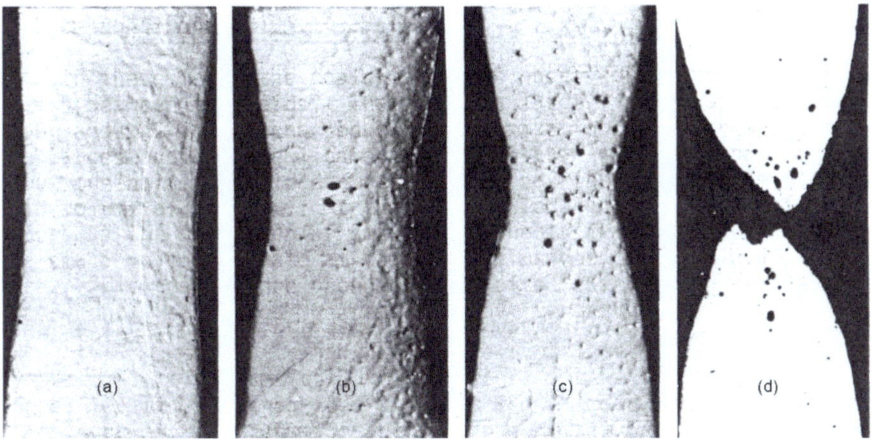

Fig. 6.17 **a** smooth cylindrical specimen of pure copper in traction, **b** the appearance of the necking generates a triaxial state in the center of the specimen that favors the formation of pores, **c** the porosity increases with increasing load up to **d** the coalescence of the cavities that leads to the collapse of the resistant section [22]

6.4 The Origin of Fracture in Members Containing a Crack

We have discussed, in the previous sections, the issue of brittle and ductile fracture initiation in members with and without notches. We have seen a crescendo in the evolution of damage: submicroscopic line defects (dislocations) that produce microscopic defects (microcracks or voids at grain level), which grow and cause macroscopic fracture. It can be thought, then, that the first two phases are skipped when a crack is already present in the material. It is a fact that the presence of a crack does not change the general picture previously described except for two main reasons. First, at variance with a notch, now the maximum triaxiality is reached right at the crack tip and not at the center of the specimen. Secondly, ahead of a real crack the first component of stress, σ_x, equals the value of the second component, σ_y, see Chap. 7, Eq. (7.114), that results in the maximum biaxiality possible. If the thickness is large enough, the third component or transverse stress, σ_z, will rise to produce the maximum triaxiality possible. Therefore, the material along the crack front is potentially brittle. Nevertheless, also in this case, a plastic slip is still possible, though in limited amount, because there will always be a plastic slip, even in the case of completely brittle fracture. As pointed out by Orowan [24], the first crack growth event is precisely a localized ductile tearing. Following this initial, very limited plastic tearing the fracture convert to cleavage, causing catastrophic failure. To this end, the works of Low [9], Knott [25] and others [26–30] have highlighted this now recognized and confirmed circumstance. Figure 6.20 shows a detail obtained at the scanning electron microscope [28] of the fracture surface of a cylindrical specimen of carbon steel, type A 508 Cl3, containing a sharp circumferential notch of 0.13 mm radius, made even sharper by a precompression that generated a real crack. The

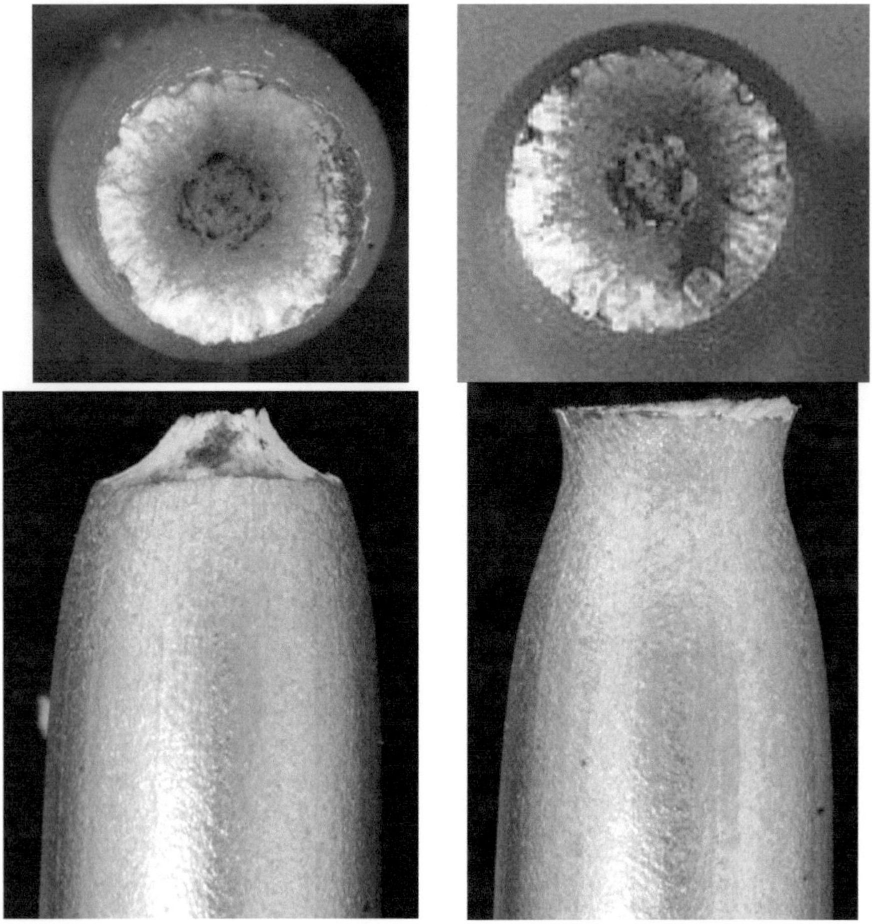

Fig. 6.18 Typical ductile fracture, at RT, with cup-and-cone formation at necking in a smooth, cylindrical specimen of zircalloy. Shear lips of fast fracture by shear stresses can be seen (Milella)

specimen had been tested under dynamic traction by the fall of a weight. The crack front appears in the upper area of the picture. Ahead of the crack front, it is visible the very small zone of ductile separation that appears as a spongy and porous tissue like that of Fig. 6.15. The extension of this plastic zone that we know always exists at the crack front, is extremely small of the order of tens of microns. Beyond this area, bordered in Fig. 6.20 by a dotted line, the cleavage fracture appears with its typical feature of lettuce leaves, Fig. 6.6. The box marks the point where the brittle fracture initiated, a ICRs. This point, as it can be seen, is just outside the plastic zone at its border where the elastic triaxiality reaches its highest value. This sequence of phases is described in Fig. 6.21 in which it is schematized the plastic zone, of limited extension Δa immediately ahead of the crack and the area of brittle, unstable fracture. It is also shown the position, at a distance X from the border of the plastic zone, of

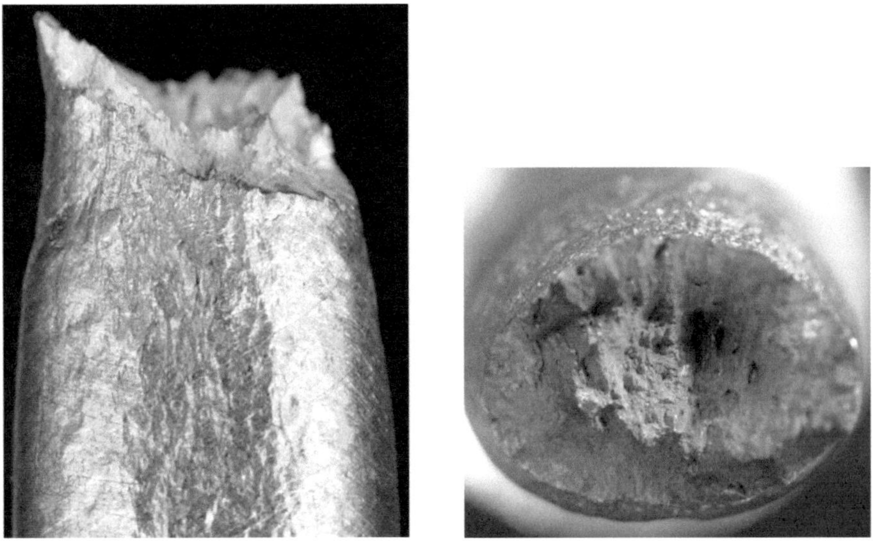

Fig. 6.19 External appearance of an austenitic stainless steel specimen, broken in traction at a temperature higher than RT (Milella)

the primary site where fracture started, usually a carbide, sulfide or other small brittle phases, and the possible so-called secondary sites that act as sites of further origin of brittle fractures. This secondary sites are shown in Fig. 6.22 for a 7075-T651 aluminum alloy [29]. The observation of well-defined initiation sites in all of the specimens containing a crack supports the concept of "*weak points*" along the crack front and the consequential scatter and size-dependence of the fracture toughness in the transition region, as it will be discussed in Chap. 15. As long as this blunting is small, almost negligible, the crack tip remains sharp and SSY conditions prevail favoring brittle fracture. Hahn and Rosenfield [30] found that the large, typically 1 to 10 μm Fe-, Si-, and Cu-bearing inclusions in aluminum alloys cracked easily as the matrix ahead of the crack deformed plastically. Virtually all of these particles in the path of the extending crack rupture.

They also found that the toughness of the aluminum alloys was approximately varying as $(f_c)^{-1/6}$ where f_c was the volume fraction of these inclusions [29]. The ductile propagation, Δa, that precedes cleavage has the effect of opening and blunting the crack, Fig. 6.23, therefore it must be kept substantially small, almost neglecting its presence, if we want to maintain the crack sharp and have brittle fracture. When this plastic zone swells too much fracture turns into a ductile tearing mode. The figure also shows the Crack Tip Opening Displacement (CTOD) that represents the crack root radius and its relationship with the blunting. The larger Δa, the larger the blunting of the specimen at fracture. Figure 6.24a shows the SEM fractography of a cruciform beam specimen, 610 × 91x102 mm, uniaxially loaded, of A533B Cl.1 carbon steel, containing a central surface crack 112 mm long, 10 mm deep, tested at $RT_{NDT}= -25$ °C, with the measurements of Δa and X, respectively [31]. The low

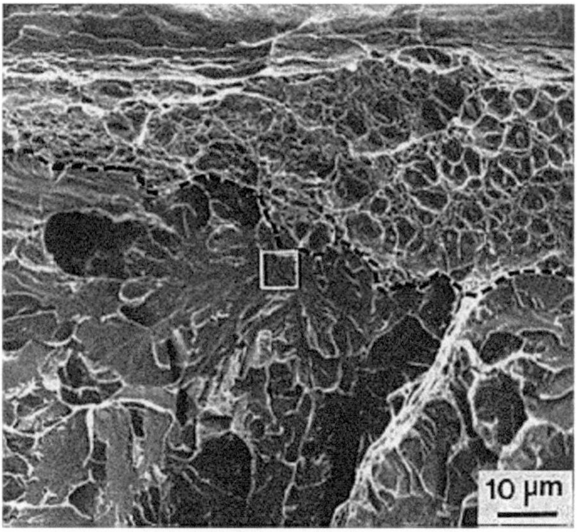

Fig. 6.20 Detail, obtained at the scanning electron microscope, of the fracture surface of a cylindrical specimen with a circumferential crack, which shows the apical plastic zone of spongy appearance (upper part) immediately below the crack front, followed by the cleavage fracture. The boundary between the two areas is highlighted by a dotted line. The small square marks the point of initiation of the brittle fracture [28]

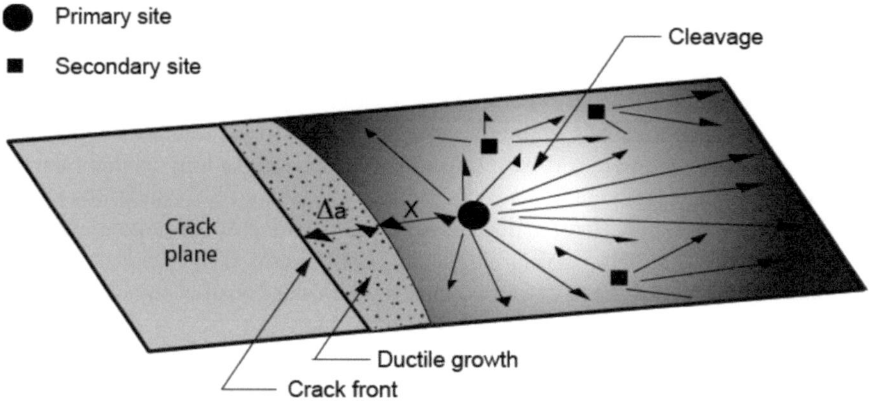

Fig. 6.21 Schematic of the fracture surface of a body containing a crack showing the plastic zone of extension Δa at the front of the crack and the primary site, at distance X, that triggered cleavage. Also sown are possible secondary sites of fracture initiation. The plastic zone must be small enough to be almost neglected in order to leave the crack front sharp

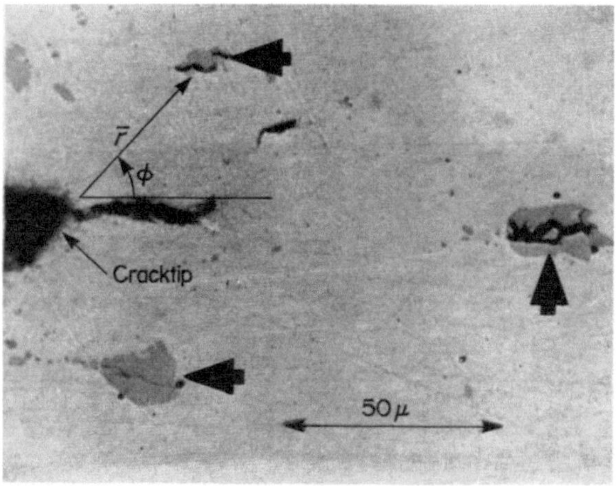

Fig. 6.22 Cracked particles ahead of the crack tip in a 7075-T651 aluminum alloy [29]

test temperature, T = −47 °C, resulted in a very low precleavage ductile tearing Δa of only 11 μm and relatively low toughness (K_{Ic} = 109 MPa$\sqrt{}$m). The plastic zone size ahead of the crack tip can be considered irrelevant, almost inexistent and typical of SSY conditions for true cleavage and full brittle fracture. The initiation site or cleavage origin, between 75 and 120 μm from the precleavage tearing, was an inclusion, as it can be seen in Fig. 6.24b. For comparison, it is worth noting the SEM fractographies for a 1T-C(T) specimen of the same steel but tested at 8 °C above the RT_{NDT},T = −17 °C, Fig. 6.25. The precleavage ductile tearing swell from a mere 11 μm of the previous case to 342 mm and so did the toughness that jumped to 253 MPa$\sqrt{}$m, more than double. A precleavage ductile tearing, Δa, equal to 342 μm means that the crack blunted too much (CTOD ≈ 2 Δa) becoming a sharp notch and the measured toughness of 253 MPa$\sqrt{}$m cannot be considered valid. Nevertheless, also in this case the cleavage origin was an inclusion located in the box area of Fig. 6.25a, enlarged in Fig. 6.25b. This kind of fracture that shows also some ductile tearing on the brittle fracture surface is often called quasi-brittle fracture.

Table 6.1 is a compilation of fractographic data Δa and X from experiments run by M. C. Rao and R. K. Nanstad [31] at ORNL on cruciform beam and 1T-CT specimens of SA-533 GrB Cl 1 carbon steel at different temperatures.

In all tests run at a temperature below the RT_{NDT} of the material, the ductile growth was below 30 μm that resulted in a negligible crack tip blunting (necessary for SSY conditions to apply) and the measured toughness remained always below 110 MPa$\sqrt{}$m. Cases 4, 5, 6 and 8, where the measured toughness was over 160 MPa$\sqrt{}$m with a Δa in the range of 70 μm, the double than in the previous cases, are exceptions that reveal a blunting, though still confined and a non-completely brittle fracture. This toughness elevation over the expected value was due to the crack length, too short to allow a full brittle fracture, see Sect. 9.17. In all other cases where the test temperature was over the RT_{NDT} of the material, the toughness was very high, over 250 MPa$\sqrt{}$m, that comply with Δa growths over 300 μm.

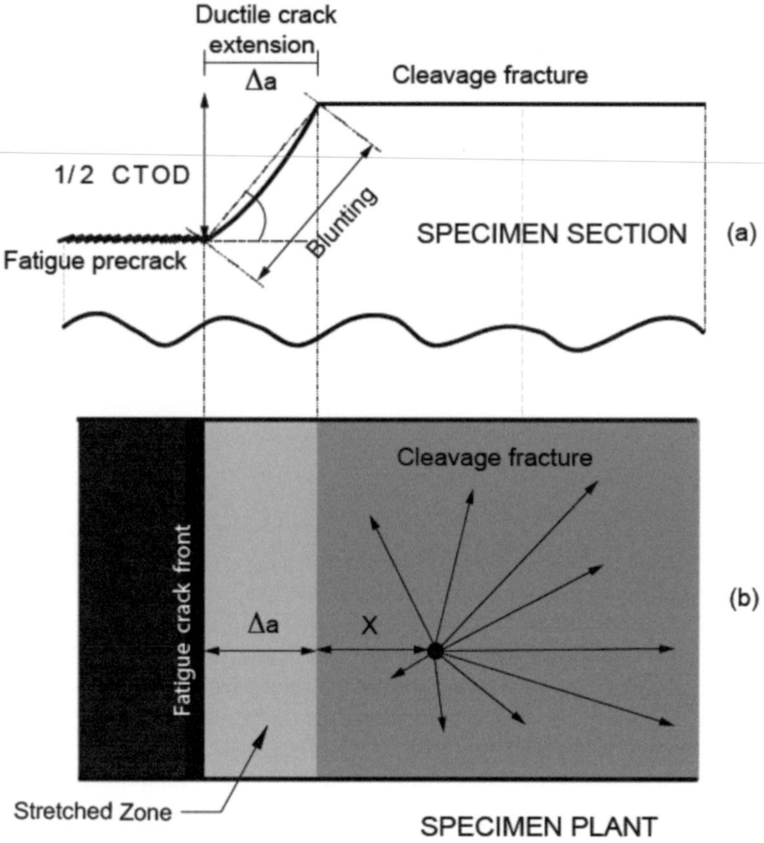

Fig. 6.23 Schematic drawing showing how the ductile crack extension Δa affects the shape of an initially sharp crack

Data obtained at ORNL on SA 533 steel and listed in Table 6.1 are plotted in relation to the apparent toughnessK_c of the material in Fig. 6.26. The dependency of the apparent toughness on Δa, which actually means on the crack root radius, is evident.

6.5 Crack Tip Radius

G. Irwin was one of the first to recognize and study experimentally the issue of crack tip radius, ρ, on the apparent toughness of a material, K_c. He tested a high strength steel, 0.4C, 5.0Cr, 1.3Mo, 0.5V, of 1700 MPa yield strength, obtaining the results shown in Fig. 6.27 [32].

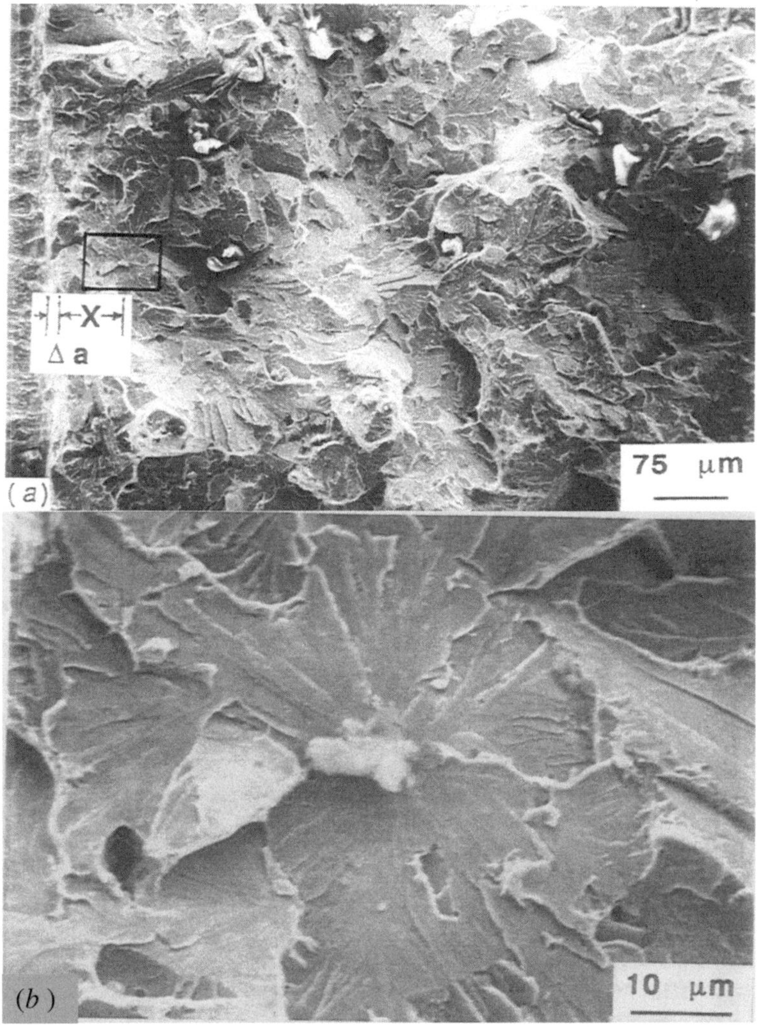

Fig. 6.24 SEM fractographies of the fracture surface of a beam specimen of SA-533 GrB Cl 1 carbon steel containing a through-crack, broken at—25 °C the RT_{NDT} of the material. The plastic zone, Δa, and the initiation site at distance X (in the small box) are indicated. **b** is an enlargement of the box area where showing the small inclusion that triggered cleavage. Note the radii emanating from the cleavage origin [31]

Irwin used flat specimens 1.5 mm thick, containing a central crack. Note that with such a high strength steel that almost doesn't show an RT_{NDT} and has a toughness as low as 30 to 50 MPa$\sqrt{}$m, 1.5 mm were thick enough to maintain a plain strain condition. He found a lower bound toughness of only 30 MPa$\sqrt{}$m when a sharp crack with a root radius up to 0.01 mm was used. For root radii over 0.01 mm the apparent toughness started to grow to an even fivefold higher 150 MPa$\sqrt{}$m with 0.09 mm

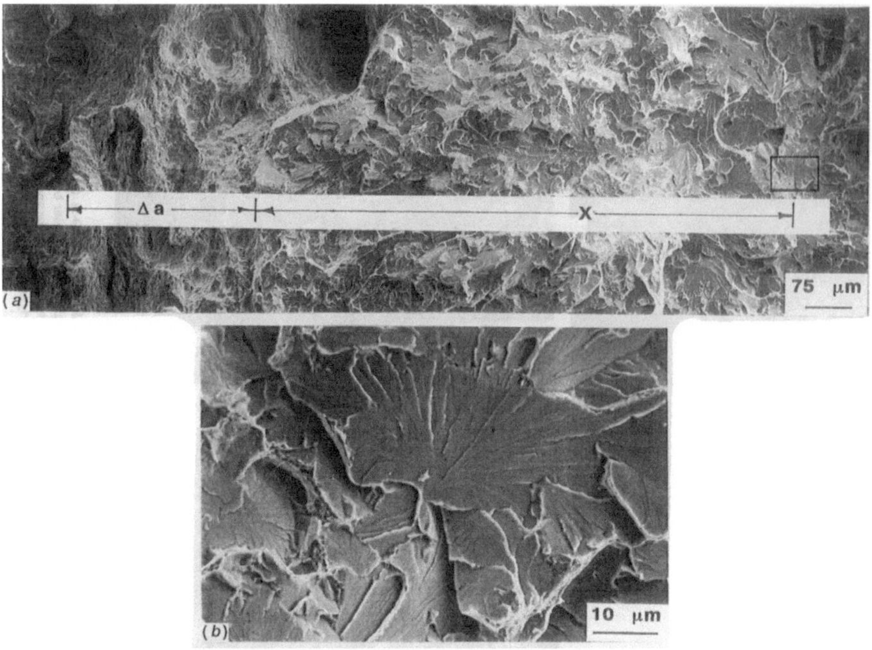

Fig. 6.25 **a** SEM fractograph of the fracture surface of a 1 T-C(T) specimen of SA-533 GrB Cl 1 carbon steel, tested at 8 °C over the RT_{NDT} of the material [31] showing the large ductile growth, Δa, that precedes brittle fracture triggered by the rupture of an inclusion located at a distance X from the limit of the plastic zone (area in the box); **b** enlargement of the area in the box, site of initiation of the quasi-brittle fracture. A large cavern is visible in the upper area at the border of the ductile growth and large areas of dimple separation are within the X distance

radius that still seems to be very sharp but is not sharp enough, also for that steel. These results confirm those previously seen, Fig. 6.26, for the low alloy steel SA 533 where the crack tip blunting was due to the plastic tearing that always precedes the unstable cleavage propagation. Analogous results were obtained by Kraft and Sullivan [33]. They were using CC(T) panels (Central Crack panels in Traction, see Fig. 9.8), 1,5 mm thick, of 0.4C 5Cr VMo steel of 1700 MPa yield strength, Fig. 6.28. The minimum toughness, K_{Ic}, was measured with cracks having a tip radius, ρ, no larger than 0.005 mm and ranged from about 20 MPa$\sqrt{m}$ to 45 MPa$\sqrt{m}$. These values can be considered the true toughness, K_{Ic}, of the steel.

As the crack tip radius increases, also the apparent toughness increases arriving to a fourfold or even eightfold value of 165 MPa$\sqrt{m}$ with a notch radius of 0.1 mm that cannot be considered a sharp crack. This happens in high strength materials when the thickness is too small to constraint the plastic zone or in BCC materials when the transition temperature RT_{NDT} is eexceeded and the metal starts to show a ductile behavior (see Sects. 6.7 and 5.2.1). However, it also happens when the crack is not deep enough to generate a full reaction in the x direction and produce biaxiality, Fig. 4.12b. a loss of biaxiality results in a parallel loss of triaxiality, though

Table 6.1 Summary of fractographic data and toughness measurements obtained at ORNL [31]

K_c # (MPa)		TRT_{NDT} (°C)	Δa (µm)	X (µm)	$\Delta a + X$ (µm)
1	109	22	24	75-120	99-144
2	109	−25	11	70	81
3	71,5		20	100	120
4	160*	22	63	75-150	138-213
5	162*	−24	74	200	274
6	162	10	70	170	240
7	152	10	150	150	300
8	261*	−7	270	220	490
9	276	10	305	880	1185
10	253	8	342	550-950	892-1292
11	223	10	510	160	670
12	333*	8	535	350	885
13	382*	9	750	750	1500
14	253		1000	1333	2333
15	352	10	1550	1150	2700
16	358	10	2100	1300	3400

*shallow crack

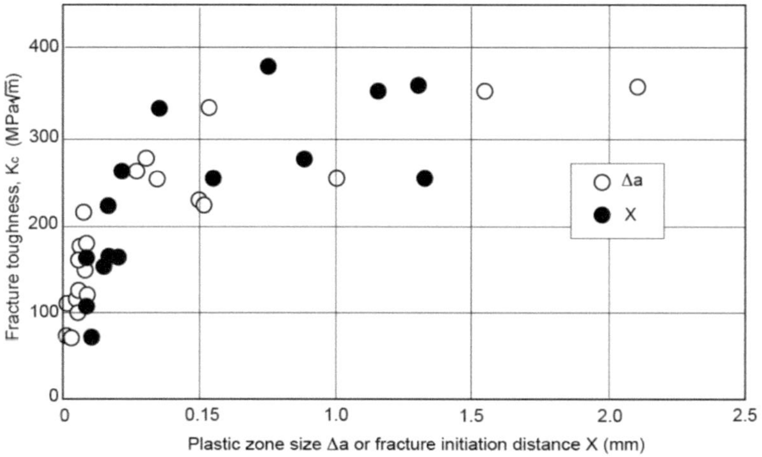

Fig. 6.26 Plot of the apparent fracture toughness K_c versus the plastic zone extension Δa or the fraction initiation distance X for specimens of A533B Cl.1 low alloy steel tested at ORNL

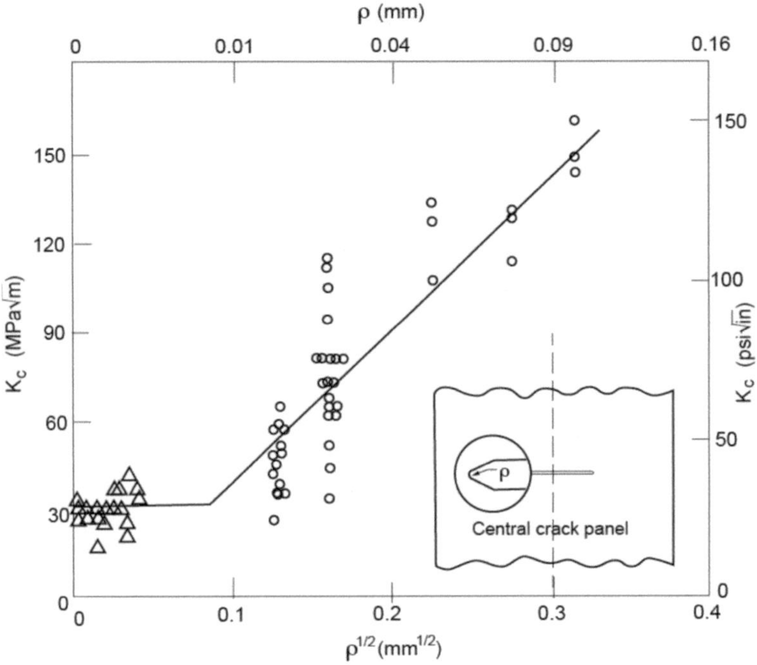

Fig. 6.27 Effect of notch root radius on the apparent fracture toughness K_c of a high strength steel [32]

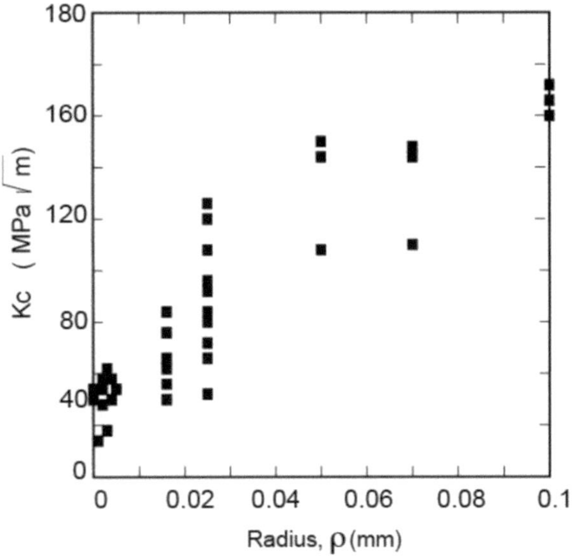

Fig. 6.28 Trend of the apparent *Kc* toughness with crack tip radius, *ρ*, in a 0.4 C 5Cr V Mo steel (adapted from [33])

in thick members. This appears to be the case of Fig. 6.29 for a SEC specimen, Fig. 9.13, having a shallow crack that failed by ductile fracture. The external appearance reveals the characteristic dimple fracture by void growth and coalescence, which is a relatively slow process compared to cleavage, followed by a sudden failure by shear overload of the remaining resistant cross section with the so-called shear lip formation, as and schematized in Fig. 6.30.

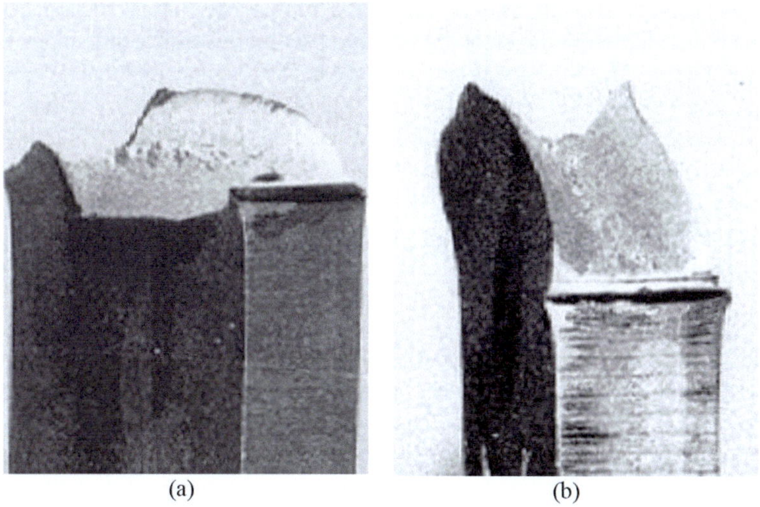

Fig. 6.29 Ductile fracture in a bending specimen of carbon steel containing an edge shallow crack tested at a temperature over the RT_{NDT}. In **a** it can be seen a central region of void coalescence followed by fast oblique shear fracture (shear lip). In **b** the same scenario occurs with double shear lip formation

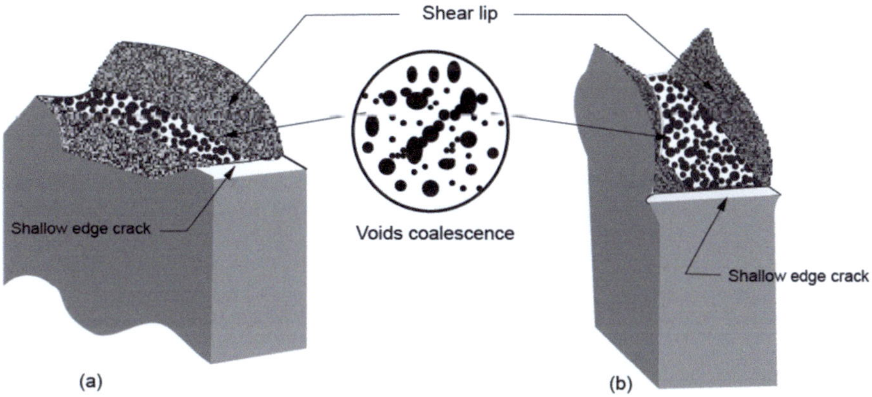

Fig. 6.30 Schematic of the ductile fracture of Fig. 6.28 showing the central area of dimple growth and coalescence and the shear lip formation revealing sudden failure by overload

6.6 Maximum Principal Stress Theory

In an early study, Ritchie, Knott and Rice [4] developed a relationship between tensile stress and cleavage fracture toughness in a mild steel under plain-strain conditions. In their study, unstable cleavage fracture is postulated to occur when the maximum principal stress exceeds a critical value over a characteristic distance ahead of the crack tip, determined as twice the grain size of the material. The use of a sharp-crack model necessitated that the critical stress be achieved over some microstructurally significant distance from the crack tip to avoid predictions of fracture under vanishingly small loads due to the stress singularity. Merkle [34] asserted that a description of stress and strain fields required to model the onset of cleavage fracture can be attained only by considering the blunting of the crack tip. Crack-tip blunting leads to finite crack-tip root radius, with zero stress normal to the blunted free surface. Therefore, the maximum hydrostatic stress occurs at a small distance ahead of the notch tip, while the maximum strain occurs at the notch tip. Finite deformation analyses performed by McMeeking [35] and McMeeking and Parks [36] indicate that the point of maximum stress is approximately $3\delta_t$ ahead of the crack-tip, where δ_t is the crack-tip opening displacement (CTOD). This observation is perfectly consistent with what has been presented in Sect. 6.4. Anderson and Dodds [37] constructed a correlation procedure to remove the geometry dependence of cleavage fracture-toughness values for single-edge-notched bend specimens (SEN(B)), see Sect. 9.5, of A36 steel for a range of crack depth. This procedure utilizes a local stress-based criterion for cleavage fracture and detail plain-strain finite-element analysis. From dimensional analysis of SSY implies that the maximum principal stress, σ_1, ahead of the crack type written as

$$\frac{\sigma_1}{\sigma_0} = f\left(\frac{J^2}{\sigma_0^2 A}\right) \tag{6.3}$$

where σ_o is the yield strength, J the J-integral, see Sect. 10.6, and A is the area enclosed by the contour on which σ_1 is a constant. The local fracture criterion is based upon achieving a critical volume $V(\sigma_1)$ within which the principal stress is greater than σ_1. For a specimen subjected to plain-strain conditions, the volume is equal to the specimen thickness B times the area within the σ_1 contour on the midplane (V = BA). Thus Eq. (6.3) is the appropriate normalization for small-scale yielding solutions when using the latter fracture criterion based on process volume or area, see Sect. 9.8.

6.7 Phenomenology of Brittle Fracture

We can now reassume the entire phenomenology of brittle fracture. In high strength alloy of low toughness and in BCC alloys at a temperature close or below the RT_{NDT}, brittle fracture occurs by cleavage. it is a stress-controlled phenomenon triggered by the high intensity elastic stress field existing ahead of the crack root that reaches its maximum at the interface with the very small plastic zone of extension Δa, the so-called stretch zone or SZ, that is always present, Figs. 6.18 and 6.19. It is there that a foreign phase first breaks and the entire grain with the inclusion breaks, along with the small plastic zone behind. Once this small initial enclave breaks, the stress field quickly moves forward to affect new virgin material as schematized in Fig. 6.31. Thermodynamic reasons lead us to believe that this process proceeds with constant absorption of energy per unit of time, which is equal to the elastic plus plastic energy stored in the elements that brake. This is confirmed by DT tests that in case of cleavage fracture yield a completely flat curve, Fig. 3.50. As a result, the growth rate of a brittle fracture must necessarily be uniform. It is important to consider that the load transfer from one element to the next happens dynamically, as if a spring were suddenly broken. It is estimated that the strain rate in the ferrite close to a foreign particle that breaks and initiates the fracture is possibly as high as 10^6 s^{-1}[38]. Considering that the extension $\Delta a + X$ is equal to about 300 μm see Table 6.1, and that the fracture proceeds at the sound velocity in the material, which is of about 5.000 m/s-6.000 m/s in most metals, the time needed to break the initial strained ligament is of the order of 10^{-8} s.

Considering that brittle failure occurs at just some fraction of percent deformation, the load transfer to the next element is, therefore, dynamic with a strain rate $\dot{\varepsilon}$ extremely high, of the order of 10^6 to 10^7 s^{-1}. This strain rate is by far higher than the initial strain rate imposed by a possible dynamic loading. This makes crack propagation easier than initiation. We have, today, ample experimental evidence of this [39, 40]. Once initiated, cleavage proceeds as a dynamic phenomenon that propagates over the entire material fracturing foreign phases that no longer bear the importance they had before, as initiators.

Concluding Remarks

> From all studies and metallurgical evidence, it can be concluded that plastic deformation, though small, always precedes brittle fracture and that cleavage is directly associated with the inhomogeneous nature of plastic flow. Inhomogeneous plastic deformation, caused by either slip or twin band, nucleates a cleavage crack by fracturing a hard particle at grain boundary or within a grain boundary because of the general inability of hard particles to deform plastically. The stress required to break inclusions and second phases is provided by the piling-up of dislocations against them. In most FCC lattice materials, the intrinsic high ductility of the matrix does not allow the fracture

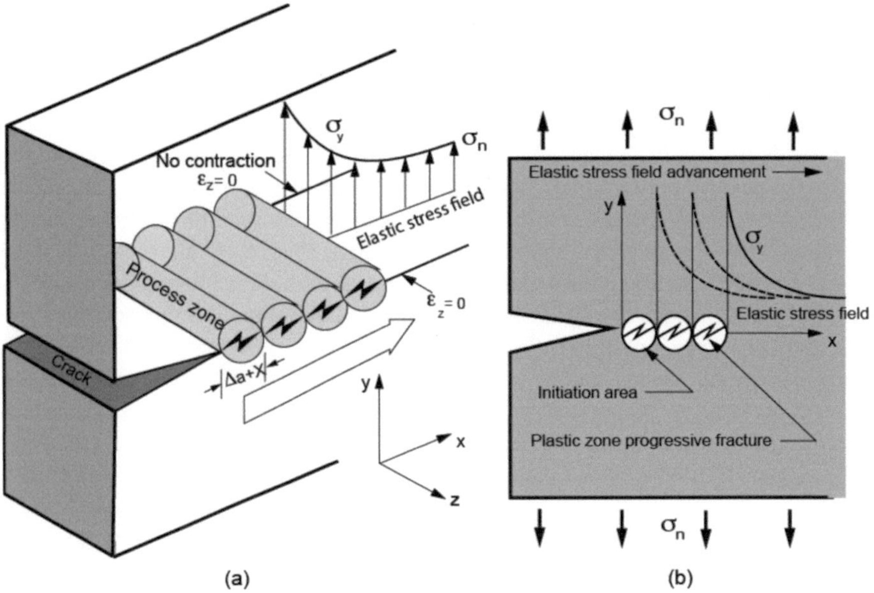

Fig. 6.31 Schematic of the progressive rupture of the process zone ahead of the crack and the consequent advancement of the elastic field. The size of the very small apical plastic zone has been increased for convenience

of foreign particles that under the piling-up pressure debond from the matrix without fracturing creating voids. Under such a condition the initial crack blunts. Voids growth and coalescence result in a progressive reduction of the resisting area that leads to the final ductile fracture by sudden overload. Void formation and growth occur at the point of maximum triaxiality, i.e., at the center of a notched member or at the tip of a crack. It is a fact that brittle fracture by cleavage has never been seen in FCC materials that fracture by void formation and coalescence. This type of failure also occurs in BCC materials, but only at temperatures over the RT_{NDT} where the matrix is soft. Brittle fracture by cleavage is sustained by an entire elastic stress field, plastic strains are negligible. Energy is released at a constant rate during propagation and it is almost entirely elastic energy. True cleavage is associated with low toughness values, never in excess of 150-180 MPa$\sqrt{m}$ in BCC steels. Typical fracture toughness associated with cleavage are in the range of 45 to 100 MPa$\sqrt{m}$.

References

1. Larson, F.R., Carr, F.L.: Trans. ASM **55**, 599 (1962)
2. Beachem, C.D., Brown, B.F., Edwards, A.J., NRL Memorandum Ref. 1432, June (1963)

3. Manjoine, M.J., Multiaxial stress and fracture, Fracture an Advanced Treatise, H. Liebowitz Ed., **III**, pp. 265–309 (1971)
4. Ritchie, R.O., Knott, J.F., Rice, J.R., On the relationship between critical tensile stress and fracture toughness in mild steel, J. Mech. Phys. Solids, **21**, pp. 395–410 (1973)
5. Bonora, N., Marchetti, M., Maricchiolo, C., Milella, P.P., Pini, A., Cleavage fracture prediction of K_{IC} in three point bending specimens of a nuclear steel using local approach criteria, 17th MPA Conference, Stuttgart, Germany, October (1991) (Published in Nuclear Engineering and Design)
6. Bonora, N., Marchetti, M., Maricchiolo, C., Milella, P.P., Pini, A., Local approach criteria: an application of cavity growth theory to ductile fracture in low alloy steel, ASME, Vol. 137, Giovanola, J. H. and Rosakis, A. J. Ed., Book No. H00741, pp. 217–231 (1992)
7. Irwin, G.R., Zhang, X., Cleavage Behavior in nuclear vessel steels, NUREG/CR-4219, Vol. 10, No. 1, pp.12–21 (1993)
8. Bonora, N., Iacoviello, F., Milella, P.P., Modellazione e Verifica del Comportamento Elasto-Visco-Plastico di Acciai Per Contenitori Metallici e Recipienti in Pressione Sottoposti ad Elevati Strain Rates, Rapporto Tecnico Università di Cassino-APAT, 15 Novembre (1999)
9. Low, J.R., I.U.T.A.M. Madrid Colloquium, Deformation and Flow of Solids, p. 60, Springer Verlag, Berlin, (1956)
10. Koehler, J.S., Phys. Rev., **85**, p. 408 (1944)
11. Stroh, A.N., The Formation of Cracks as a Result of Plastic Flow, Proceedings of Royal Society, London, **A 223**, p. 404 (1954)
12. McMahon,C.J., Cohen M., ACTA Metallurgica, Vol. 23, pp. 593–604, June (1965)
13. Argon, A.S., Formation of Cavities from Nondeformable Second Phase Particles in Low temperature Ductile Fracture", ASME, Journal of Engineering Materials Technology, **98**, pp. 60–68, (1976)
14. Zackay, V.F., Gerberich, W.W., Parker, E.R., Structural mode of fracture, fracture an advanced treatise, Liebovitz, H. Ed. Vol. 1, Chapt. 6, pp.395–440 (1968)
15. Zener, C.: The Micro-Mechanism of Fracture ", Fracturing of Metals, pp. 3–31, AMS, (1948)
16. Cottrell, A.H.: Theory of brittle fracture in steel and similar metals. Trans. Met. Soc., AIME, **212**, p. 192, (1958)
17. Hull, D., Acta Metallurgica, **9**, p. 191 (1960)
18. Tipper, C.F., Metallurgia, **39**, p.133 (1949)
19. Crussard, C., Plateau, J., Tamhancar, R., Henry, G., Lajeunesse, D.: A comparison of ductile and fatigue fracture. Fracture, p. 524, Averbach, Felbeck, Hahn and Thomas Editors, John Wiley and Sons, N.Y. (1959)
20. Cottrell, A.H.: Theoretical aspects of fracture, fracture, p. 20, Averbach, Felbeck, Hahn and Thomas Editors, John Wiley and Sons, N.Y. (1959)
21. McClintoc, F.A.: A criterion for ductile fracture by growth of holes, J. Appl. Mechanics (1968)
22. McClintoc, F.A.: Plasticity aspects of fracture, Fracture an Advanced Treatise, H. Liebowitz Ed., **III**, pp. 47–225 (1971)
23. Argon, A.S., Im, J., Safoglu, R.: Cavity formation from inclusion in ductile fracture, metallurgical transactions, **6A**, pp. 825–837 (1975)
24. Orowan, E.: Energy Criteria for Fracture, Weld. J. research Supp., (1955)
25. Knott, J.F., Cottrell, A.H.: Notch Brittleness in Mild Steel, Journal of Iron and Steel Institute, pp. 249–260, (1963)
26. Worthington, P.J., Smith, E.: Acta Metallurgica, pp. 14–35 (1966)
27. Petch, N.J.: The Fracture of Metals, In Prog. Metal Phys., **5**, p. 1 (1954)
28. Fourney, W.L., Irwin, G.R., Barker, D.B., Dally, J.W., Schwartz, C.W., Zhang, X.J.: Investigation of Damping and Cleavage-Fibrous Transition in Reactor-Grade Steel at the University of Mariland, W. R. Corwin, US NRC NUREG /CR-4219, **5**(1), pp. 32–61 (1988)
29. Van De Kasteele, J.C.W., Broek, D.: The failure of large second Phase Particles in a cracking aluminum alloy. Eng. Fract. Mech. **9**, 625–635 (1977)
30. Hahn, G.T., Rosenfield, A.R.: Metallurgical factors affecting fracture toughness of aluminum alloys, metallurgical transactions, **6a**, pp. 653–668, (1975)

31. Rao, M.C., Nanstad, R.K., Fracure Surface Evaluation, W.E., Pennel Ed., Heavy Section Steel Technology Program, Semiannual Progress Report for October 1992-March 1993, NUREG/CR-4219, **10**(1), pp. 29–40, (1994)
32. Irwin, G., Srawley, J.E.: Progress in the development of crack toughness fracture tests, paper presented at the Fracture Phenomena and Testing of Materials at Würzburg, Germany, March 16–17, (1961)
33. Kraft, J.M., Sullivan, A.M.: Effects of speed and temperature on crack toughness and yield strength in mild steel, NRL Report 5776 (1962)
34. Merkle, J.G.: An Examination of Size Effects and Data Scatter Observed in Small-Specimen Cleavage Fracture Toughness Testing, US NRC NUREG/CR-3672 (1984)
35. McMeeking, R.M.: Finite deformation analysis of crack-tip opening in elastic-plastic materials and implications for fracture, J. Mech. Phys. Solids, **26**, pp. 357–381 (1977)
36. McMeeking, R.M., Parks, D.M.: On the criteria for J-Dominance of crack-tip field s in large-scale yielding. ASTM STP **668**, 175–197 (1979)
37. Anderson, T.L., Dodds, R.H: Specimen-size requirements for fracture toughness in the transition region, report MM-6586–90–5 texsas A&M university, college station, texas, mechanics and material center, (1990)
38. Shoemaker, A.K., Rolfe, S.T.: J. Eng. Fract. Mech., **2**, pp. 319–339 (1971)
39. Mager, T.R.: Nucl. Eng. Des., **17**(1), pp. 76–90 (1971)
40. Grigory, S.C.: Test of 6-Inch thick tensile specimens, HSSTP-TR-23, (1973)

Chapter 7
Linear Elastic Fracture Mechanics

7.1 Introduction

Fracture is a term that identifies the rupture of materials where a crack is present. Fracture mechanics is the science that studies these kind of ruptures. Basically, we can distinguish among three different types of fracture, shown schematically in Fig. 7.1. In case (a) the plastic zone that we know always precedes brittle fracture, see Sect 6.4, is very small, almost negligible in size, of the order of hundreds of microns. A full plain strain condition is attained. This is the condition that we have defined as Small-Scale Yielding or SSY. The system containing the crack does not feel at all the presence of this plastic zone, the crack remains a crack, i.e., it does not blunt and open and the system behavior will continue to be fundamentally elastic. The fracture morphology appears flat, crystalline and bright without any plastic deformations, at least on a macroscopic scale. This is precisely the domain of Linear Elastic Fracture Mechanics or LEFM, covered in this chapter. Its field of application is limited to extremely hard and brittle materials or ductile materials used at temperatures below *NDT*. Its fundamental result is that it is always possible to identify an elastic stress field around the crack tip that is self-similar, i.e., independent of the crack size, position and shape of the system. The intensity of this stress field can be described by a single parameter that G. Irwin called *Stress Intensity Factor* or K. The K formulation accomplishes the long-lasting search for the mysterious link among the stress, σ, the crack length, a, and the toughness of the material, depicted in Fig. 1.5. As the size of the crack front plastic zone increases and plain strain condition starts to vanish, also this elastic stress field vanishes which will no longer be self-similar and described by a single parameter, the stress intensity factor K. The system begins to show some elasto-plastic behavior. The impact of the plastic zone size can no longer be neglected, and its existence will deeply influence the process of fracture. Now, we have to face a ductile or semi-brittle fracture. The shear component of fracture is, in fact, equal and often greater than the brittle or cleavage component, as shown schematically in Fig. 7.1b where shear lips appear. These type of fracture is studied

P. P. Milella, *Fundamentals of Fracture Mechanics*,
https://doi.org/10.1007/978-3-032-23081-2_7

by the Elastic–Plastic Fracture Mechanics or EPFM which we will discussing in Chap. 10. Its fundamental results is that within the plastic zone, just in a limited area at the crack tip, it is always possible to identify an elastic–plastic self-similar stress field that can be described by a single parameter, the Rice *J-Integral* parameter, that plays the same role as the *K-Factor* in LEFM. By further increasing the ductility of the material through, for example, a temperature increase or a reduction of thickness, a new situation arises where the plastic zone swells over the entire section ahead of the crack, as shown schematically in Fig. 7.1c.

Now the fracture is by general yield and the plastic slips cover the entire section of the specimen. This is the field of General Yielding Fracture Mechanics, GYFM. It is difficult to say whether in such a condition the presence of a crack affects the

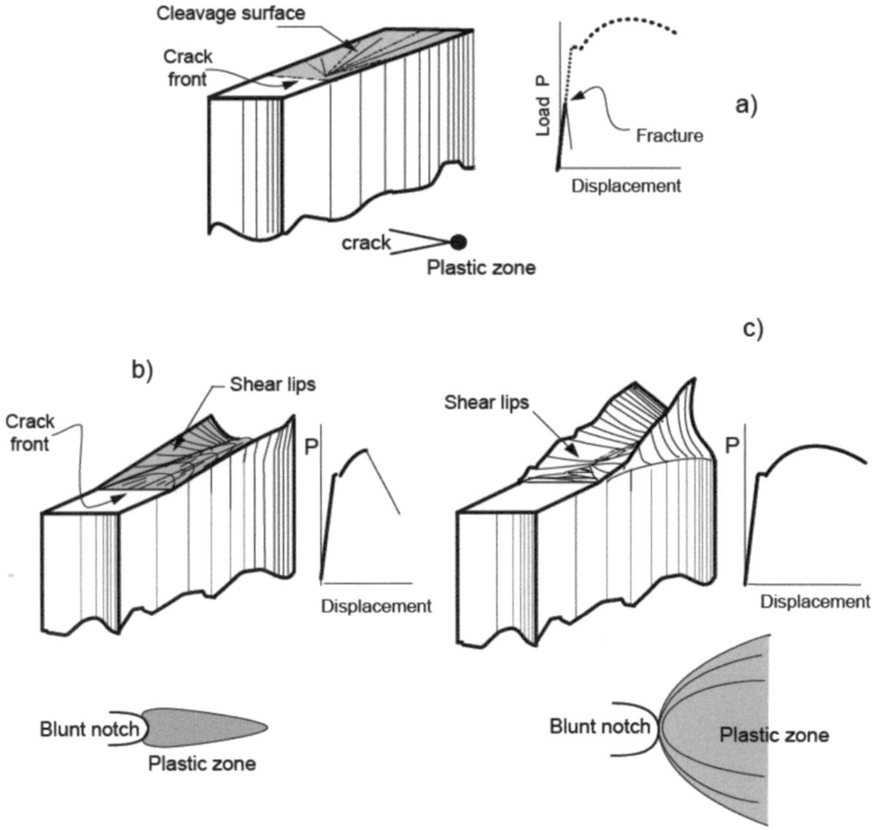

Fig. 7.1 **a** Brittle fracture by cleavage. Very small, almost inexistent, plastic area at the crack tip. The fracture surface appears flat and crystalline, and the load–displacement diagram is linear, yield strength is not reached; **b** ductile fracture. The plastic zone enlarges, and the crack blunts and becomes a notch. Visible plastic deformations appear as shear lips. Fracture occurs beyond the elastic limit; **c** ductile fracture by general yield. The plastic zone swells covering the entire section of the body, failure is by general yield

strength of the body in a manner different by the simple reduction of the net section area. Experiments so far conducted on slabs or pipes of small thickness or ductile material, indicate that, in fact, failure occurs by plastic collapse of the net section containing the initial crack that becomes a large notch. It is assumed that failure occurs at a stress level equal to the flow stress, σ_f, defined as half the sum of the yield strength, σ_s, and the ultimate strength, σ_u, of the material

$$\sigma_f = \frac{\sigma_s + \sigma_u}{2} \tag{7.1}$$

7.2 Ideal Tensile Strength

Cracks cause fracture at a stress level lower than that measured in a traction test using smooth specimens. However, also smooth specimens fail at a stress level lower than the theoretical or ideal tensile stress. The reason is precisely the same: cracks, but in the latter case on a smaller scale, i.e., microscopic crack-like defects always present in real materials. The mechanism is always the same: localized stress around microscopic crack-like defects that reduce the ideal strength of materials precisely as macroscopic cracks do with the normal strength of materials that we measure with crack-fee specimens. Cracks, either macroscopic or microscopic, are, therefore, responsible for the loss of resistance of materials and their mechanism of action is precisely the same. To this purpose, let's introduce, albeit briefly, the concept of theoretical tensile strength of a homogeneous and isotropic material with a crystalline structure. Suppose, then, to consider a crystalline solid and focus our attention on a generic crystallographic plane imagining that it consists of an arrangement of atoms such as that of Fig. 7.2a. The atoms of our hypothetical material are at an equilibrium distance d_o from each other. Consider a generic pair of atoms, such as the couple 3–3' of Fig. 7.2a, that face the plane of trace A–A and express their bond energy U as a function of the mutual distance d. In general, the relation of this energy to the distance d is shown in Fig. 7.2b. The bond energy U assumes a minimum, U_o, at the equilibrium distance between the atoms in the crystal lattice, d_o. In this position the interaction force, F, between atoms will be null. This force grows if we try to move atoms apart, reaching a maximum $F_{\max}$ at the energy U inflection point.

From here on, the force F will begin to decay very quickly, until it becomes zero when the distance between atoms tends to infinity, Fig. 7.2b, in practice this occurs at several times the equilibrium distance. Even the energy U drops rapidly when atoms are moved apart, so that it can be said that already at distances of the order of three or four times d_o there is no longer any significant interaction. The area under the curve of force F, dashed in Fig. 7.2b, represents the work L necessary to be spent on moving apart two atoms, such as 3 and 3', from their equilibrium position to a distance were, in practice, any interaction between the two ceases. The sum of all works L_i necessary to separate all the couples, i, of atoms that face on plane A–A

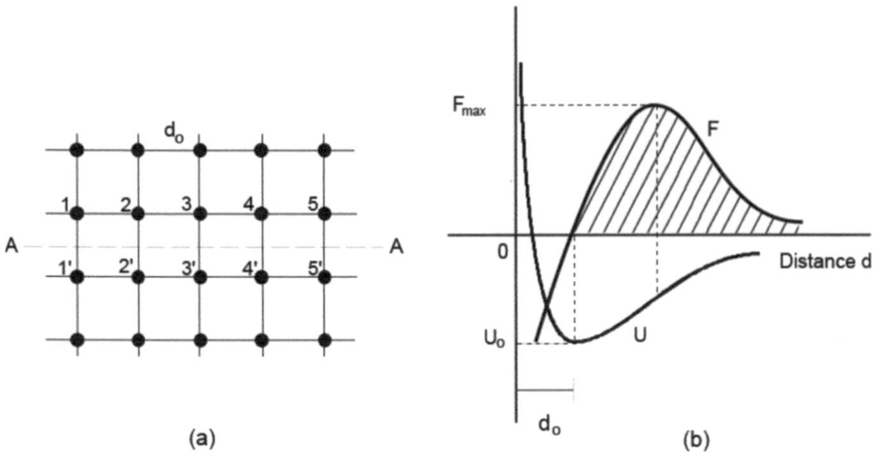

Fig. 7.2 **a** Schematic arrangement of atoms in a crystal lattice; **b** trend of the bond energy U and the interaction force F with distance between adjacent atoms

of Fig. 7.2a represent the energy required to split the body in two halves. When this occurs, two new surfaces are produced and since each free surface, even in solid bodies, contains energy in the form of surface energy, it can be asserted that the work spent to break the body in two will be equal to the energy accumulated on the two new surfaces originated with the fracture, where it is assumed that the two parts separate without any kinetic energy, that is, in an almost static way. Let's look more closely at the work L required to separate a pair of atoms such as the 3–3', i.e., at the area under the curve F of Fig. 7.2b. We give it a more convenient representation. Translate the origin of the y axis by a distance equal to the equilibrium lattice spacing, d_o, so that the strain may be written as x/d_o, and express the ordinate as F/d_o^2, as shown in Fig. 7.3, d_o^2 being the unit surface. The area under this curve represents the work that will be required to break the atom bonds per unit area. As a first approximation it can be identified with the dashed one A in Fig. 7.3, neglecting the contribution of the tail that tends to zero very rapidly. The new stress–strain curve is usually approximated by a half sine wave of wavelength λ with the initial slope representing the Young's modulus of the material, $E = \sigma \cdot d_o/x$, as proposed by Orowan [1]. The area under the curve is therefore

$$A = \frac{\lambda}{\pi} \cdot \sigma_{\max} \tag{7.2}$$

$\sigma_{\max}$ being the amplitude of the sinusoid, equal to $F_{\max}/d_o^2$. For what has been said, at failure it must equal the value of the surface energy of the two surfaces created

$$\frac{\lambda}{\pi}\sigma_{max} = 2\Gamma \tag{7.3}$$

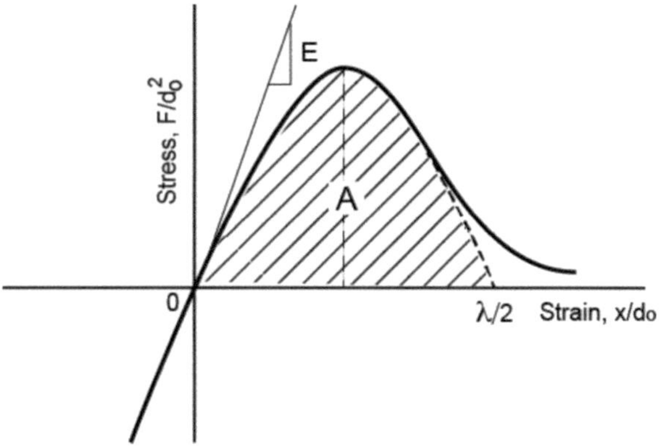

Fig. 7.3 Schematic representation of stress–strain curve at atomic level

Γ being the surface energy of the material per unit area. Referring to Fig. 7.3 we can write

$$\sigma = \sigma_{\max} \sin\left(2\pi \frac{x}{\lambda}\right) \tag{7.4}$$

which for very small values of x can be approximated with

$$\sigma = \sigma\left(2\pi \frac{x}{\lambda}\right)_{max}. \tag{7.5}$$

For small values of x it is also (Orowan hypothesis)

$$E = d_o \cdot \frac{\sigma}{x} \tag{7.6}$$

E being the Young's modulus of the material. Substituting Eqs. (7.5) in 7.4) we have

$$E = 2\pi \frac{d_o}{\lambda} \cdot \sigma_{\max} \tag{7.7}$$

from which we obtain

$$\lambda = \frac{2\pi\, \mathrm{d_0}}{E} \sigma_{max} \tag{7.8}$$

Which, replaced in Eq. (7.2) leads to

$$\sigma_{max}^2 \frac{d_0}{E} = \Gamma \tag{7.9}$$

or

$$\sigma_{\max} = \sqrt{\frac{E \cdot \Gamma}{d_o}} \tag{7.10}$$

Equation (7.9) is the Orowan expression for the cleavage fracture stress. Since for most materials Γ is worth about 0,01 $E\ d_o$ [2], Eq. (7.9) yields

$$\sigma_{\max} = \sqrt{0.01 \cdot E^2} = 0.1 \cdot E. \tag{7.11}$$

Tough obtained through an approximation of a really complex interaction between atoms, in particular the assumption of a sinusoidal force low, Eq. (7.9) remains a valid measure of the order of magnitude of what must be the maximum theoretical tensile strength of a material [4]. Kelly has reviewed [4] these kinds of calculations concluding that the Orowan estimate may be higher by a factor 2. The theory predicts a maximum strength equal to about one tenth of the Young's modulus, Eq. (7.10). Practice, on the other hand, teaches that the strength of materials is much lower, normally less than 10–100 times what is indicated by Eq. (7.10). Table 7.1 shows the values of the Young's modulus, E, of the surface energy, Γ, the theorical or ideal fracture strength, σ_{th}, given by Eq. (7.9) and the real fracture strength, σ_u, of some materials. It can be seen the enormous difference between the real strength, and the ideal strength, the former being some two orders of magnitude lower than the latter.

Table 7.1 Surface energy and theoretical fracture stress

	E	Γ	d_o	σ_{th}	$E\sigma_{th}$	σ_u
	(GPa)	(J/m^2)	(nm)	(MPa)		(MPa)
Aluminum	70	1.2	0.4	14,491	4.8	110
Iron	190	2	0.28	36,839	5.1	360
Magnesium	44	1.2	0.32	12,845	3.4	170
Cobalt	211	1.6	0.25	36,748	5.7	
Silicon	165	1.2	0.54	19,149	8.6	
Glass	65	1.2	0.2	19,748	3.3	50*
Nickel	170	2.2	0.35	32,689	5.2	
Copper	122	1.7	0.36	24,002	5.1	200
Graphite	200	1.0	0.2	31,623	6.3	

* Compression

7.3 The Griffith's Theory

It was Griffith [5] who first gave a possible explanation to the apparent discrepancy between the theoretical and the real strength of materials suggesting that in all materials, even in those so-called homogeneous and isotropic, there were always defects, albeit very small, responsible for the loss of the ideal strength.

He began his studies by investigating the fracture behavior of glass, an extremely brittle material that he choused as close to an ideally elastic material. His theory, first expounded in 1920, is based on energy balance and can certainly be considered as the historical basis of Fracture Mechanics. The results obtained by the English scholar are based on the evaluation of the variation of the elastic deformation energy of a body into which a crack is introduced. His analysis can be considered complete and exact, only if applied to ideally elastic materials. The studies he conducted on glass provided valid experimental confirmation. Consider an infinite flat plate of elastic material subject to uniform biaxial stress, σ, at infinity into which a through-crack of length, $2a$, is subsequently introduced. Let t be the thickness of the plate, as shown in Fig. 7.4.

Let U be the potential energy of the system, where U_o is the initial potential energy prior to introducing the crack. Upon introducing the central crack, the energy of the system, U, can be written as

$$U = U_o - U_D + U_A \tag{7.12}$$

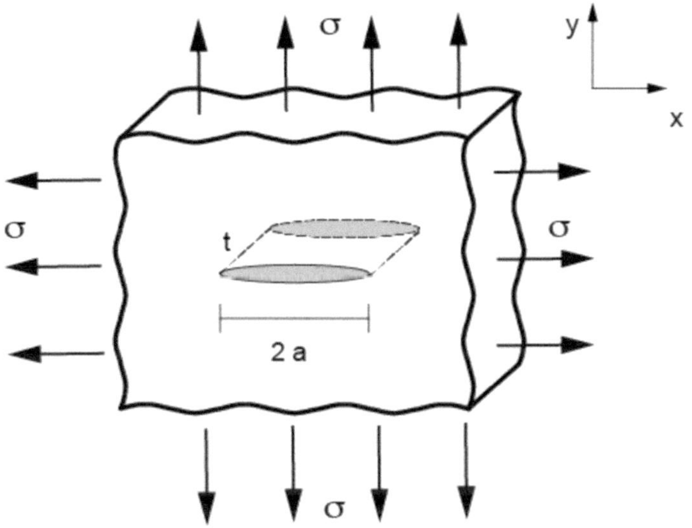

Fig. 7.4 Infinite plate of thickness t under remote biaxial tension stress, in which a central crack of length $2a$ is introduced

where U_D represents the decrease in potential energy of deformation (strain energy and boundary force work) and U_A the increase in surface energy due to the new surface introduction of area $4 \cdot a \cdot t$. Therefore, the surface energy term U_A can be put in the explicit form as

$$U_A = 4 \cdot a \cdot t \cdot \Gamma \tag{7.13}$$

with Γ being the surface tension energy (energy per unit surface) of the material. Griffith employed the stress analysis of Inglis [6] in computing the potential energy of deformation

$$U_D = \frac{\pi \sigma^2 a^2 t}{E} \tag{7.14}$$

where E is Young's modulus of the material. Equation (7.11) thus becomes

$$U = U_0 - \frac{\pi \sigma^2 a^2 t}{E} + 4at\Gamma \tag{7.15}$$

Let's look at the function $f(U)$ given by Eq. (7.14) and at its derivative with respect to the flaw length a. Where $\partial U/\partial a = 0$, the function $f(U)$ assumes a maximum or a minimum depending on whether the second derivative, $\partial^2 U/\partial a^2$, is positive or negative. The point of maximum represents a condition of instability while the minimum defines a condition of stability. Deriving with respect to a and equating to zero we can identify the condition for which the energy U assumes a maximum or a minimum:

$$\frac{\partial U}{\partial a} = 4t\Gamma - \frac{2\pi a \sigma^2}{E} \cdot t = 0 \tag{7.16}$$

or

$$\frac{\pi a \sigma^2}{E} = 2\Gamma \tag{7.17}$$

In plain strain, see Eq. (4.6)

$$\frac{\pi a \sigma^2 (1 - \nu^2)}{E} = 2\Gamma \tag{7.18}$$

The second derivative is worth

$$\frac{\partial^2 U}{\partial a^2} = -\frac{2\pi \sigma^2}{E} \cdot t \tag{7.19}$$

Therefore, the second derivative is always negative which means that Eq. (7.16) is a point of maximum that defines the instability condition

$$\sigma\sqrt{a} = \sqrt{\frac{2 \cdot E \cdot \Gamma}{\pi}} \tag{7.20}$$

or

$$\sigma\sqrt{\pi a} = \sqrt{2 \cdot E \cdot \Gamma} \tag{7.21}$$

When the length, $2a$, of the crack introduced in the infinite plate subject to the external biaxial stress σ satisfies Eq. (7.20) or makes the first term be larger than the second, the crack is unstable and propagates spontaneously through the plate.

7.4 The Energy Release Rate

Equation (7.16) defines the stability of a central through crack in a plate remotely loaded using mathematical treatment. However, a different approach can be followed that bears a more engineering appearance. Return to the energy balance of Eq. (7.11). If we imagine an incremental increase of the length $2a$ of the crack that creates a new surface dA, under equilibrium condition, we can write

$$\frac{\mathrm{d}U}{\mathrm{d}A} = \frac{d}{dA}(U_0 - U_D) + \frac{\mathrm{d}U}{\mathrm{d}A} = 0 \tag{7.22}$$

or

$$\frac{dU_D}{\mathrm{d}A} = \frac{\mathrm{d}U_A}{\mathrm{d}A} \tag{7.23}$$

The energy term U_D has been assessed through the Inglis analysis, Eq. (7.13), while U_A is given by Eq. (7.12) so that we can write

$$\frac{\pi a \sigma^2}{E} = 2\Gamma \tag{7.24}$$

The left-hand member can be interpreted to be the elastic energy per unit crack surface area produced made available by an infinitesimal crack extension da that creates new energy dA, and is defined as the energy release rate G

$$G = \frac{\partial E}{\partial A} = \frac{\pi a \sigma^2}{E} \tag{7.25}$$

The dimensions of the G term are those of a work per unit area: $\mathrm{J/m^2}$. The right-hand side of Eq. (7.23) is the material resistance to crack extension, *R*, *i,e.*, the energy absorbed during crack extension per unit area of new surface created

$$R = \frac{\mathrm{d}U_A}{\mathrm{d}A} \tag{7.26}$$

$$R = 2\Gamma \tag{7.27}$$

therefore

$$G = R \tag{7.28}$$

is the condition for the instability of a crack, *i.e.*, if a virtual crack extension *da* of the crack of length $2a$ in the plate biaxially loaded at remote results in a whole redistribution of the internal stress and deformation fields, such that the volume integral of the deformation energy density over the entire system decreases and the energy release rate G following such a decrease equals the energy absorption rate *R*, the crack becomes unstable.

Equation (7.23) can be put in the form

$$K^2 = 2E\Gamma \tag{7.29}$$

with

$$K = \sigma\sqrt{\pi a} \tag{7.30}$$

G. Irwin will call the term $K = \sigma\sqrt{\pi a}$,stress intensity factor. The corresponding fracture stress σ in Eq. (7.23) is the critical stress, σ_c. It is rather interesting to observe how Eq. (7.23) provides the analytical key we were looking for in Sect. 1.1, Fig. 1.5, since it sets the interdependence of the three factors: *crack size a-stress level* σ *-property of the material (toughness).* In Eq. (7.23) the toughness appears for the first time as a material property, not just a word or a concept, having physical dimensions, MPa $\sqrt{m}$ since it is given by the term

$$K_c = \sigma_c\sqrt{\pi a} \tag{7.31}$$

For the first time, toughness became a measurable quantity. Equation (7.30) is the basis of Linear Elastic Fracture Mechanics. When, instead of the infinite plate with a central crack, we have a workpiece of any finite dimension and shape with a crack of any length in any position, Eq. (7.30) continues to be valid, losing its closed-form, assuming a more general form

$$K_c = \sigma_c\sqrt{\pi a} \cdot f(a/W) \tag{7.32}$$

where $f(a/W)$ is a geometrical non dimensional factor, greater than 1, that depends on the length a of the crack and some width, W, of the system. To verify Eq. (7.23), put in the form

$$\sigma\sqrt{a} = \sqrt{\frac{2E\Gamma}{\pi}} \tag{7.33}$$

Griffith used glass since it appeared very closed to an ideally elastic material. He evaluated the surface tension energy of the glass, Γ, measuring its value in the liquid phase and then, reducing the temperature, in its viscous phase at temperatures lower and lower to reach the solid state where, finally, he estimated Γ by extrapolation. He obtained, thus, for the second member of Eq. (7.32) the value:

$$\sqrt{\frac{2E\Gamma}{\pi}} = 0.15MPa\sqrt{m}\left(137lb/in^{3/2}\right) \tag{7.34}$$

Then, he run tests on tubes and spheres of glass with prepared cracks, which were visible, since glass is transparent, and, therefore, measurable in length, and from ten different tests he obtained for the left term of Eq. (7.32)

$$\sigma\sqrt{a} = 0.25 \div 0.26\,MPa\sqrt{m}(229 \div 251lb/in^{3/2}) \tag{7.35}$$

The two values expressed by Eqs. (7.33) and (7.34) can be considered in good agreement with each other, tough non exactly equal, considering the difficulty to evaluate the surface tension energy of glass and neglecting any ductile deformation energy involved which, though glass is brittle, was not negligible at all. The results obtained gave Griffith proof that his theory was substantially correct. Perhaps the most convincing aspect of the goodness of the theory lies precisely in the fact that the experimental value given by Eq. (7.34) was almost twice as high as the theoretical and critical value foreseen by Eq. (7.34). Glass, in fact, is considered to be close to an ideally brittle material, but there is clear evidence that it breaks with some plastic deformation, though extremely small [7], in good agreement with what has already been said for brittle metals (Sect. 6.4) which always show some plastic deformation that precedes cleavage fracture. This means that the value of Eq. (7.33) assessed using only the surface energy Γ underestimates the real value of the energy U_A absorbed by the system, Eq. (7.12), missing the plastic deformation energy term E_p, which is even larger than Γ. The approach followed by Griffith holds the granitic solidity of the energetic approach. It bypasses the difficulty to calculate the actual stress and strain distribution near the crack tip and arrives to set a condition for instability using the nominal stress, σ, far from the crack tip.

The value of the elastic deformation energy U_D given by Eq. (7.13) can be deduced through relatively simple considerations, although not strictly exact. Consider, to this purpose, a plate containing a crack of length $2a$ subjected to a remote tension stress. We may visualize the stress field around the crack through the stress lines, Fig. 7.5, using the hydraulic methodology of the flow lines.

The stress field lines get closer and closer at the crack tip indicating a stress concentration but leave e central area above and below the crack practically empty, i.e., without stress. This central area can be thought of as having an elliptical shape. Its minor axis is equal to the length of the crack, $2a$, the major axis length can be

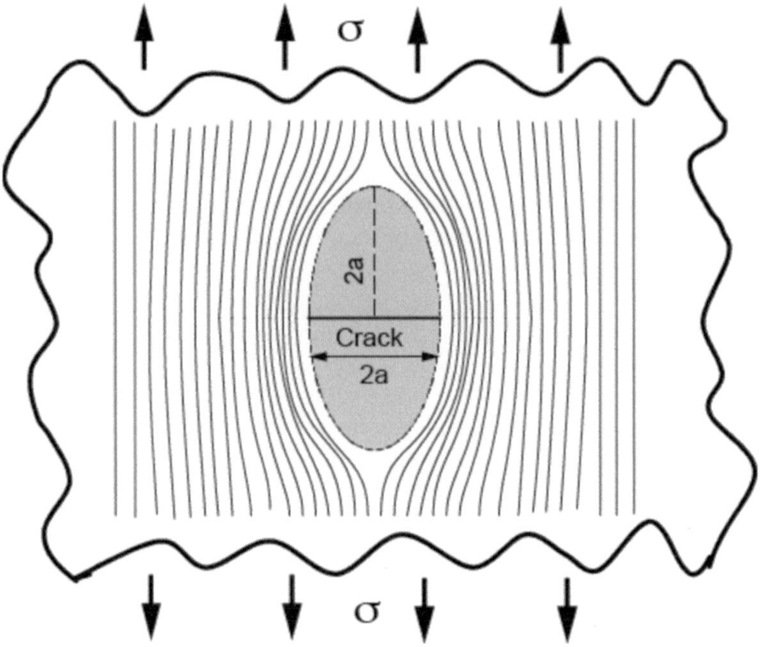

Fig. 7.5 Flow lines representing the stress field around a crack of length $2a$ in a fixed end plate loaded in tension. The central area (in gray) is practically unloaded

deduced from the Saint–Venant principle [8] according to which a local disturbance (the crack) is no longer felt at *distances which are large enough in comparison with the linear dimensions of the disturbance itself. In our case we can say that* at a distance equal to two or three times the crack size, at most, its effect is no longer felt. Therefore, we may consider the length of the major semi-axis equal to $2a$. Volume V of this central portion is given by the area of the ellipse times the thickness s of the plate

$$V = 2\pi a^2 s \tag{7.36}$$

Therefore, the elastic deformation energy, U_D, contained in this central volume before the introduction of the flaw is

$$\begin{aligned} U_D &= \frac{1}{2}\sigma \cdot \varepsilon \cdot V \\ &= \frac{\pi a^2 \sigma^2}{E} \cdot s \end{aligned} \tag{7.37}$$

Which is precisely that found by Inglis, by analysis, Eq. (7.16).

7.5 The True Strength of Materials

We can now return to the original issue of theoretical strength of materials σ_{th} that cannot be reached because of crack-like defects, as second phases or inclusions, always existing in real materials. Consider the Griffith crack in the fixed end plate of Fig. 7.4 that can be assumed to have an elliptical shape, with major axis $2a$ and the minor axis $2b$. Due to stress concentration, when the plate is subjected to the nominal stress σ_n, at the tip of the Griffith crack the local stress,σ_l, will reach the value given by [6]

$$\sigma_l = \sigma_n\left(1 + \frac{2a}{b}\right) \tag{7.38}$$

and recalling that the radius of curvature, ρ, of the tip of the ellipse is

$$\rho = \frac{b^2}{a} \tag{7.39}$$

Equation (7.37) becomes

$$\sigma_l = \sigma_n\left(1 + 2\sqrt{\frac{a}{\rho}}\right) \tag{7.40}$$

The local stress, σ_l, increases as the radius of curvature, ρ, decreases and, theoretically, it will reach its maximum value, $\sigma_{\mathrm{l}max}$, as the radius ρ approaches zero

$$\begin{aligned}\sigma_{l,max} &= \lim_{\rho\to 0}\left[\sigma_n\left(1 + 2\sqrt{\tfrac{a}{\rho}}\right)\right] \\ &\approx \lim_{\rho\to 0}\left[\sigma_n\left(2\sqrt{\tfrac{a}{\rho}}\right)\right]\end{aligned} \tag{7.41}$$

From a practical point of view, the minimum value that the crack tip radius, ρ, can reach is d_o, so that Eq. (7.40) becomes

$$\sigma_{l,max} = \left[\sigma_n\left(2\sqrt{\tfrac{a}{\mathrm{d}_o}}\right)\right] \tag{7.42}$$

Recalling Eq. (7.23) that provides the fracture stress in the Griffith crack end substituting in Eq. (7.41) yields

$$\sigma_{l,max} = 2\sqrt{\frac{2E\Gamma}{\pi\,\mathrm{d}_o}} \tag{7.43}$$

very close to the ideal tensile stress obtained by Orowan using an approach so far from that of Griffith.

7.6 The Strength Condition $G = R$ for Real Materials

Consider the quantities G and R given by Eqs. (7.24) and (7.26)

$$G = \frac{\pi a \sigma^2}{E} \tag{7.44}$$

$$R = 2 \cdot \Gamma \tag{7.45}$$

They represent the rate of release and absorption of energy per unit thickness, respectively, during an elementary act of crack propagation. The critical condition expressed by equality (7.27) can be written as

$$G_c = R \tag{7.46}$$

to underline that Gc it is the critical value of G that corresponds to the critical stress σ_c that causes instability of the crack. It is also called *Driving Force,* though it is not a force but transfers the idea of something that drives the crack open. We can represent graphically the terms that appear in Eq. (7.47), as done in Fig. 7.6. The rates of released or absorbed energy, G and R, appears as straight lines. The driving force G is a line that passes through the origin of the axes and has the angular coefficient given by σ^2. The resistance R is a constant that does not depend on a, Fig. 7.6. It will therefore be represented by a horizontal half-line that starts with the value a_o relative to the initial semi-length of the flaw present in the system. The "state" of the system is identified by the relative position of the lines G and R. As the applied stress σ increases, the slope of the G line also increases. When G becomes equal to R the crack becomes unstable. This is the critical condition, $G_c = R$, because further growth of the crack, however small, would release energy in excess of that absorbed. The stresses σ_1 and σ_2 that give rise to intersections 7.1 and 7.2 are not capable of giving fracture. They would cause instability and fracture with cracks of semi-length a_1 and a_2, respectively. For several years the original Griffith concept based on the Inglis solution was dormant because of its fundamental hypothesis of ideally brittle material. However, in 1948 Irwin [9] proposed that the Griffith theory could be modified and applied to real materials that are not as brittle as glass by considering the energy balance between the strain energy release rate and the plastic strain work rate required for crack extension. About a year later, in 1949, Orowan [10, 11] suggested a modification of Griffith's theory very similar to that proposed by Irwin, even though he still considered the theory suitable for very brittle materials. In 1952, Irwin and Kies [12] showed that the modified form of the Griffith concept in

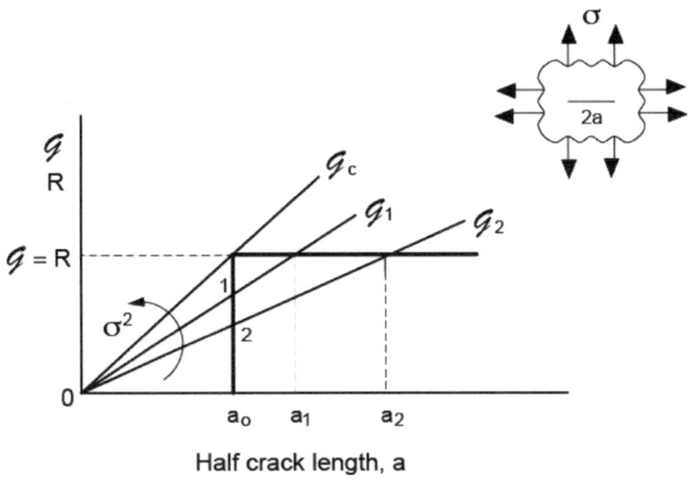

Fig. 7.6 Graphic representation of the stability or instability condition for an ideally brittle material

which the plastic work was replacing the surface tension could be widely employed in fracture strength assessment as long as fracture occurred prior to general yielding. This led to the introduction of the term E_p, representing the plastic energy absorbed, in Eq. (7.12). On the fracture surface of a carbon steel specimen broken by cleavage in a totally brittle manner at very low temperature there is, in fact, enough plastic deformation to neglect the surface energy term Γ.

It has been calculated that the plastic deformation energy accumulated on the broken surfaces is equal to about $2{\cdot}10^3$ J/m^2 against just few joules/m^2 of surface tension energy [11].

Therefore, Griffith's theory can still apply to real materials if the term E_p substitutes Γ, by which the condition of instability, Eq. (7.19), becomes

$$\sigma\sqrt{a} = \sqrt{\frac{2EE_p}{\pi}} \tag{7.47}$$

The resistance curve, R, will still continue to be independent of a, as it was for ideally brittle materials but at a level 1000 times higher. Introducing in Eq. (7.48) the numerical values of E and E_p typical of a carbon steel for brittle conditions $\left(E = 2.1 \cdot 10^5 MPa, E_p = 2 \cdot 10^3 joules/m^2\right)$ a toughness of about 20 MPa $\sqrt{m}$ is obtained. Effectively, at very, very low temperatures, well below the NDT, a toughness of the order of 30–35 MPa $\sqrt{m}$ is measured on carbon steels, in agreement with Eq. (7.48). However, it must be recalled that the plastic energy is not only distributed on the two fracture surfaces produced by crack extension, but it also accumulates in the plastic area on the crack front. As will be shown in Chap. 8, the radius of this plastic zone, r_p, depends on the square of the length of the crack, a^2, therefore, the rate $\partial E_p/\partial a$, i.e., the resistance R will no longer be constant but will grow according to some nonlinear law as a increases, as shown in Fig. 7.7. As long as the material exhibits a totally brittle behavior with a crack tip plastic zone of negligible size, this dependence on a will be modest, almost negligible and linear, as it is controlled mainly by the energy E_p accumulated on the fracture surfaces. But as the temperature and ductility of the metal increase, the plastic deformation energy at the crack tip will increase its weight until the spread of the plastic enclave becomes uncontrolled and its effect preponderant over all other effects and this will make the R curve assume a nonlinear trend. This is indicative of a continuous regain of the metals fracture resistance capability with the progressive growth of the crack, analogous to the phenomenon of mechanical hardening. This is schematically shown in Fig. 7.7, in which various R curves related to different metal toughness are qualitatively reported, identified by the operating temperature T with respect to the *NDT*, at least for BCC lattice materials.

The driving force also will no longer be represented by a straight line for the very same reasons. The energy released during crack extension becomes predominantly plastic, coming mainly from the crack tip plastic zone, which introduces the upward nonlinear trend of G, as shown in Fig. 7.7. In comparison, the release of elastic deformation energy $\partial U_D/\partial a$ that would be available from the unloaded area, Fig. 7.5, is extremely small. It is evident, then, that for high toughness materials in which the contribution to the driving force G and to the resistance R comes mainly from the plastic deformation energy, the condition of instability will no longer be represented by Eq. (7.47), becoming

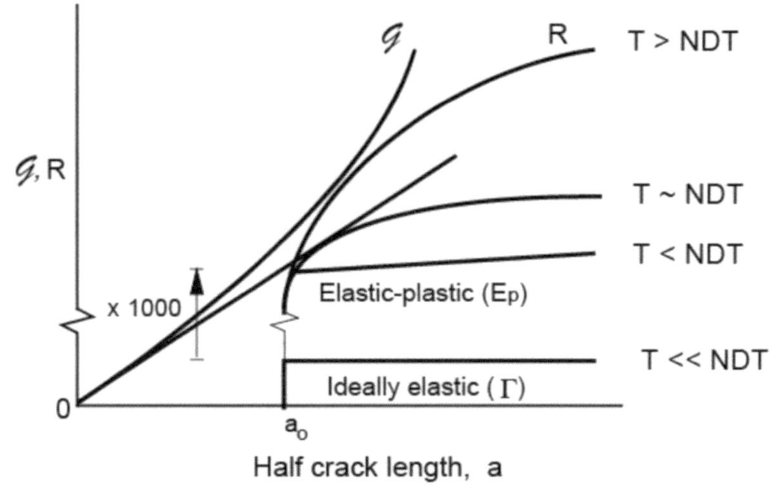

Fig. 7.7 Schematic of driving force G and resistance R as function of crack length a. For an ideally elastic material in which the only resistance to crack extension is provided by surface tension energy Γ, R is represented by a straight line, Fig. 7.6. As the toughness increases or the temperature increases exceeding NDT, the plastic energy contribution, E_p, increases becoming dominant. For high toughness materials or temperatures over the NDT both R and G assume a non-linear trend

$$\frac{\partial G}{\partial a} = \frac{\partial R}{\partial a} \tag{7.48}$$

i.e., by the tangency of G to R, as shown in Fig. 7.7.

7.7 Reversibility of a Crack

Griffith's theory is based on the assessment of potential energy changes in a system containing a flaw. For an ideally elastic solid, such as Griffith's solid, potential energy changes following the introduction of a crack of semi-length a can be expressed, per unit thickness, by the relation, see Eq. (7.14)

$$U = U_0 + 4 \cdot a \cdot \Gamma - \frac{\pi a^2 \sigma^2}{E} \tag{7.49}$$

In a plot U, a, Eq. (7.50) is represented by a curve of the type shown in Fig. 7.8. In plain strain, the energy U assumes a maximum at those values of a, which we have indicated as critical, a_c a_c, points A and B of Fig. 7.8, for which the first derivative $\partial U/\partial a$, Eq. (7.17), is

$$\frac{\pi a_c \sigma^2 \left(1 - \nu^2\right)}{E} = 2\Gamma \tag{7.50}$$

or when

$$\sigma = \sqrt{\frac{2E'\Gamma}{\pi a_c}} \tag{7.51}$$

As the stress, σ, increases the maximum of Eq. (7.50) is reached for ever smaller values of both U and a, as shown in Fig. 7.8. Since the stability condition of a system is given by the minimum potential energy, it is clear that if the crack semi-length is smaller than a_{1c}, $a_s < a_{1c}$, ora_{2c}, $a_s < a_{2c}$, the crack will never grow spontaneously since its growth is opposed by the potential barrier ΔU_1 or ΔU_2 but it will go back to zero and close, bringing the system into a condition of stability to which the minimum potential energy corresponds $U_o = 0$. This means that, according to Griffith's theory, any crack whose length is shorter than the critical one must close spontaneously. In other words, in an ideally elastic material, in which the two forms of energy at stake, the elastic and the surface tension, are reversible there can be no crack or defect smaller than the critical one. This, however, clashes, apparently, with the most common evidence of the existence of flaws of any size smaller than the critical one. Griffith himself, presenting his theory, pointed out that any flaw of length smaller than the critical one should remain indefinitely in the system, at rest. In other words, the satisfaction of the condition of minimum potential energy,

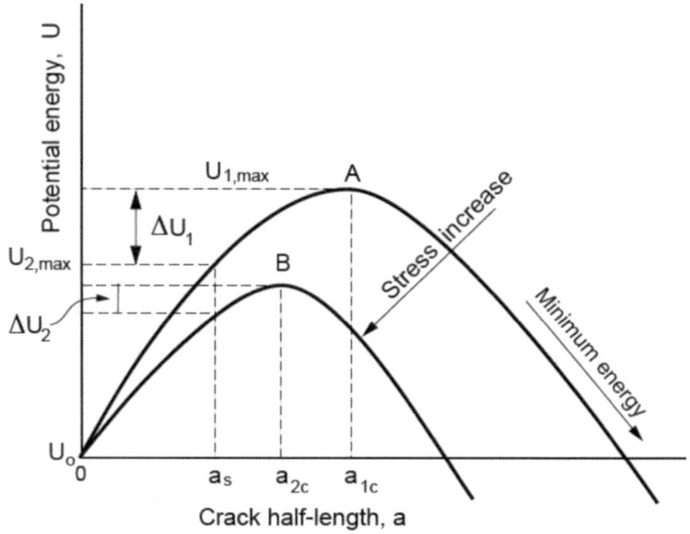

Fig. 7.8 Trend of the potential energy U in an elastic solid containing a flaw of length $2a$ subjected to an external stress σ. As the stress, σ, increases, the energy curve flattens down and contracts to the left

therefore of stable equilibrium, would work only in one direction: from the point of maximum energy to the right, i.e. towards instability. Griffith attributed this to the fact that his flaw was of elliptical type with a radius of curvature of the tip, ρ, that, though small, was always $\rho > d_o$, being d_{or} the interatomic distance, Fig. 7.9.

Under these conditions, the atomic bonds broken by the introduction of the crack into the system are no longer able to heal and close the crack. Russian Frenkel [13] attempted to explain the irreversibility of a flaw existing in an ideally elastic body. He formulated the hypothesis, however correct, that the flaw was not of elliptical

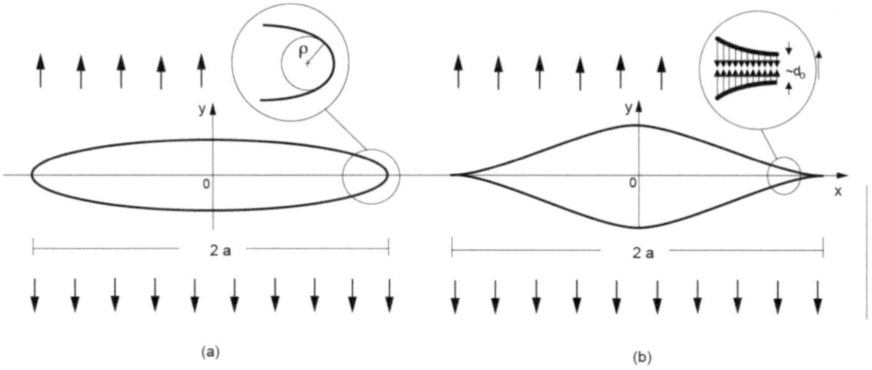

Fig. 7.9 **a** Griffith's flaw, **b** Frenkel crack

shape, but gradually closed on the edge, as already hypothesized by Dugdale and Barenblat, see Sect. 8.3 Fig. 8.4, until it reached precisely the interatomic distances d_o, as shown in Fig. 7.9. With this hypothesis, the potential energy of the system, $U(a)$, derives from the contribution of elastic deformation energy, surface tension energy and also from the accumulation of potential energy on the flaps, stretched but still in contact, of the crack that is worth:

$$E_p = \int_a^\infty \int_0^y f \cdot dxdy \tag{7.52}$$

being a the semi-length of the crack and f the force that molecular bonds in tension, but not yet broken, exert per unit area. For a system wedge loaded as such that of Fig. 7.10, Frenkel calculated the expression of the total energy $U(a)$ as

$$U = \frac{F^2 \mathrm{d}_0}{Ea} + \Gamma a - \frac{F^2 a^3}{6EI} \tag{7.53}$$

where $F^2 d_o / Ea$ represents the value of the integral (7.53) and the last term in the second member represents the elastic potential energy due to the application of the force F, therefore:

$$\frac{F^2 a^3}{6EI} = \frac{1}{2} \cdot F \cdot y_0 \tag{7.54}$$

with y_o being the displacement y at the center crack (origin of the axes) and I the inertia moment of the section, $\mathrm{I} = Bh^3/12$.

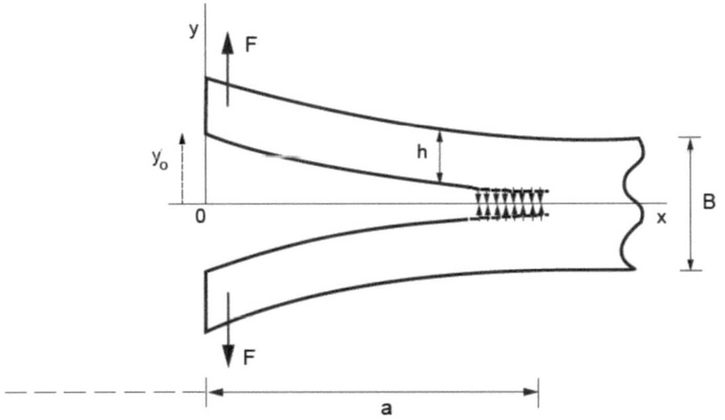

Fig. 7.10 Frenkel model for the evaluation of the potential energy associated with the presence of a crack

The expression (7.55) is shown in Fig. 7.11 for increasing values of the applied force F. As can be clearly seen, the potential energy of the system now has a relative stable minimum, $a_{1,s}, a_{2,s\ldots}$, that no longer falls into the origin, that is, for $a = 0$. This time a flaw of length $2a$ less than the critical one a_c will no longer slip into the origin closing up completely but will move into a relative energy well corresponding to a state of stability of the system. As external force F increases, the critical value, a_c, reduces and the points a_c and a_s are getting closer and closer until they coincide. However, the most plausible reason to explain the spontaneous non-closure of a flaw shorter than the critical one appears to be another, besides the instantaneous reaction of metal with air and, in particular, with oxygen that will cover the two halves of the crack with an oxide film that prevents the crack from healing and closing. In Griffith's theory, two energies have been considered, the elastic energy and the surface tension energy, which are reversible and, therefore, interchangeable: one can turn into the other and vice versa. Under these hypotheses there is no doubt that if a crack is not critical it will necessarily have to close since there is an excess of surface tension energy gained compared to the lost elastic energy.

In other words, if we hypothesize a virtual decrease $-da$ of the crack from a_0'' to a_0', the crack will regain the elastic deformation energy dE_e but not all plastic energy dE_p, keeping the entire permanent plastic deformation. This means that the elementary act of crack reclosing brings the system into an energy situation higher than that competing to the crack length a_0'. Physically it is as if the tip of a crack is in a well of potential from which it can only come out by increasing the stress state. The situation described is shown schematically in Fig. 7.12.

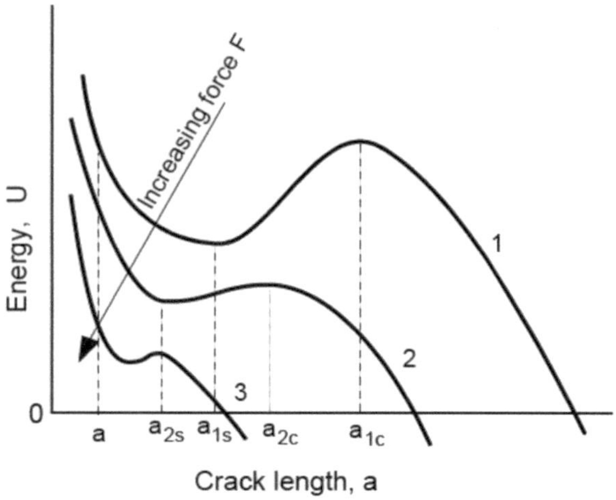

Fig. 7.11 Trend of the potential energy U of the system of Fig. 7.10 as a function of the length of crack, a, for increasing applied force F

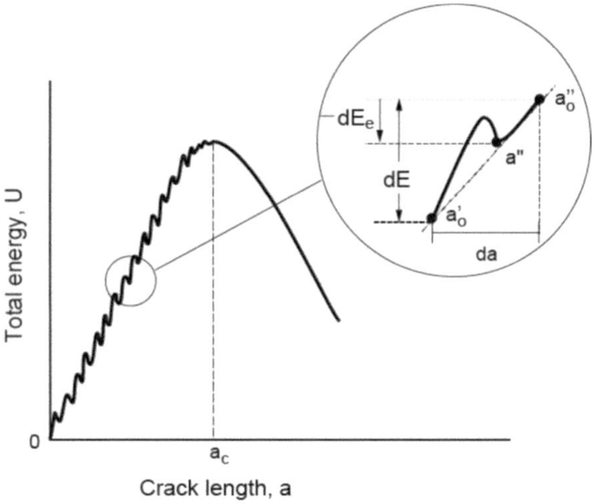

Fig. 7.12 In a real material the elementary reclosure – *da* from a generic crack length a_0'' would bring the system into a local energy well

7.8 Stress Analysis of Cracks

Another approach to assess the stability of a crack in a body subjected to external forces is to perform a study of the stress field at the crack tip aiming at deriving a closed-form expression. In LEFM the stress analysis is conducted on the basis of the theory of elasticity, therefore assuming isotropic perfectly elastic behavior. There are three fundamental mode of opening a crack, schematized in Fig. 7.13 commonly referred to as MODE I loading, where the principal load is applied normal to the crack plane and tends to open the crack, MODE II that refers to in-plane shear loading that tends to slide one crack face with respect to the other and MODE III, that refers to out-of-plane shear. Any real situation can be reconducted to one of these three modes or a combination of two or of all three of them. Historically, most of the researchers' interest has been devoted to the study of MODE I because it represents the most common form of crack opening and certainly the most critical as it requires a lower nominal stress to cause fracture, ceteris paribus. In other words, using the usual energy concepts, this mode remains associated with the higher rate of release of deformation energy by a virtual crack extension. Let's refer, therefore, to MODE I. Consider an elementary parallelepiped of material of dimensions *dx, dy* and *dz* subjected to a non-uniform stress field whose components are shown in Fig. 7.14. We express with $\overline{X}, \overline{Y}, \overline{Z}$ the average values of the components of the mass forces. The equilibrium condition of the parallelepiped of Fig. 7.14 will be expressed as

$$\sum F_x = \sum F_y = \sum F_z = 0 \tag{7.55}$$

F_x, F_y ed F_z being the components of all forces acting on the element with respect to the three directions *x*, *y* and *z*.

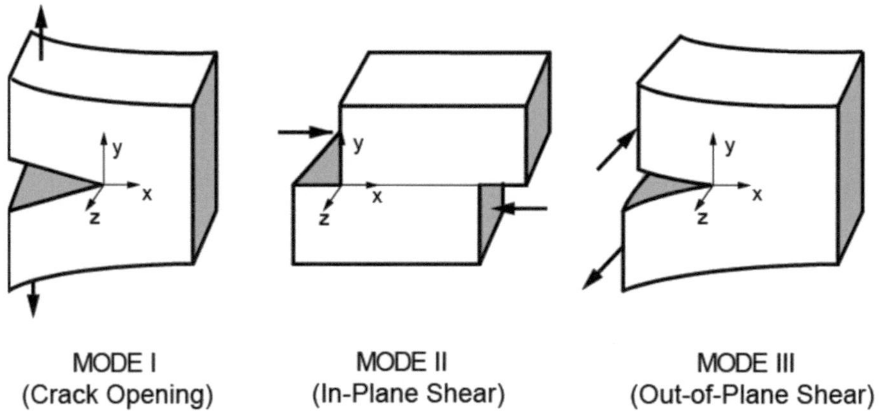

Fig. 7.13 Schematic representation of the three fundamental modes of crack opening

Fig. 7.14 Diagram of the stress components in a reference unit element in a coordinate system *x*, *y* and *z*. The small numbers identify the 6 faces of the element

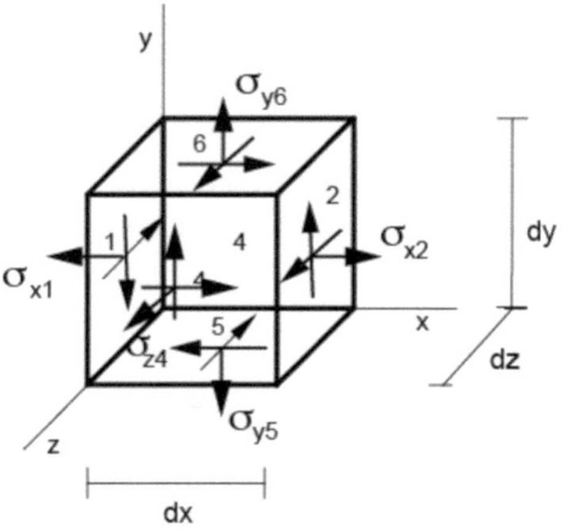

We explain the first of the three sums (7.55) assigning to each stress σ and τ a number that indicates the face of the element on which it acts, as shown in Fig. 7.14

$$\begin{aligned}\sum F_x = 0 = (\sigma_{x2} - \sigma_{x1})dydz + (\tau_{xy6} - \tau_{xy5})dxdz+\\ +(\tau_{xz4} - \tau_{xz3})dxdy + \overline{X}dxdydz\end{aligned} \tag{7.56}$$

Dividing by $dxdydz$

$$\frac{\sigma_{x2} - \sigma_{x1}}{dx} + \frac{\tau_{xy6} - \tau_{xy5}}{dy} + \frac{\tau_{xz4} - \tau_{xz3}}{dz} + \overline{X} = 0 \tag{7.57}$$

As dimensions dx, dy e dz tend to zero Eq. (7.58) becomes

$$\frac{\partial \sigma_x}{\partial x} + \frac{\partial \tau_{xy}}{\partial y} + \frac{\partial \tau_{xz}}{\partial z} + \overline{X} = 0 \tag{7.58}$$

Similarly explaining the other two summations in Eq. (7.56) we obtain the system of equilibrium equations

$$\begin{aligned}
&\frac{\partial \sigma_x}{\partial x} + \frac{\partial \tau_{xy}}{\partial y} + \frac{\partial \tau_{xz}}{\partial z} + \overline{X} = 0 \\
&\frac{\partial \sigma_y}{\partial y} + \frac{\partial \tau_{xy}}{\partial x} + \frac{\partial \tau_{yz}}{\partial z} + \overline{Y} = 0 \\
&\frac{\partial \sigma_z}{\partial z} + \frac{\partial \tau_{xz}}{\partial x} + \frac{\partial \tau_{yz}}{\partial y} + \overline{Z} = 0
\end{aligned} \tag{7.59}$$

in which $\overline{X}$, $\overline{Y}$, $\overline{Z}$ are infinitesimals of a higher order with respect to stresses σ_x, σ_y and σ_z and can be overlooked in the first approximation. The system (7.60) provides three equations in 6 unknowns and is, therefore, completely insufficient to solve the elastic problem. Its solution is possible through the compatibility condition that expresses the congruence of the deformations ε_x, ε_y, ε_z, γ_{xy}, γ_{yz}, γ_{xz} without discontinuity. The continuous distribution of deformations means, in mathematical terms, that they are expressed by differentiable functions with continuous derivatives or that the order of the partial derivations will be irrelevant. Let us then examine the elastic problem in two dimensions, for simplicity, distinguishing between the two fundamental states of stress: *plain strain* and *plain stress* for the so important consequences that they have on the fracture process. Let's start with the two-dimensional problem in a plain state of deformation. In this case the strains, stresses and mass forces are independent of the third z-coordinate and it will be possible to assume:

$$\varepsilon_z = \gamma_{xz} = \gamma_{yz} = \tau_{xz} = \tau_{yz} = 0 \tag{7.60}$$

The equilibrium equations then reduce to

$$\begin{aligned}
&\frac{\partial \sigma_x}{\partial x} + \frac{\partial \tau_{xy}}{\partial y} + \overline{X} = 0 \\
&\frac{\partial \sigma_y}{\partial y} + \frac{\partial \tau_{xy}}{\partial x} + \overline{Y} = 0
\end{aligned} \tag{7.61}$$

which is a system of two equations in three unknowns. Now let consider the three deformations

$$\varepsilon_x = \frac{\partial u}{\partial x};\ \varepsilon_y = \frac{\partial v}{\partial y};\ \gamma_{xy} = \frac{\partial u}{\partial y} + \frac{\partial v}{\partial x} \tag{7.62}$$

Since the displacements u and v are, as we have said, continuous with continuous derivatives, the condition of congruence of the deformations will allow to write

$$\begin{aligned}
\frac{\partial^2 \varepsilon_x}{\partial y^2} &= \frac{\partial^3 u}{\partial x \partial y^2} \\
\frac{\partial^2 \varepsilon_y}{\partial x^2} &= \frac{\partial^3 v}{\partial x^2 \partial y} \\
\frac{\partial^2 \gamma_{xy}}{\partial x \partial y} &= \frac{\partial^3 v}{\partial x^2 \partial y} + \frac{\partial^3 u}{\partial x \partial y^2}
\end{aligned} \tag{7.63}$$

or

$$\frac{\partial^2 \varepsilon_x}{\partial y^2} + \frac{\partial^2 \varepsilon_y}{\partial x^2} = \frac{\partial^2 \gamma_{xy}}{\partial x \partial y} \tag{7.64}$$

Equations (7.64) and (7.65) apply and have been written without regard to Hooke's law and independently of it. Through Hooke's law, Eq. (7.65) provides the third missing equation in Eq. (7.62). In case of plain strain, the stress–strain relations given by Hooke's law are

$$\begin{aligned}
\varepsilon_x &= \frac{1}{E}[\sigma_x - \nu(\sigma_y + \sigma_z)] \\
\varepsilon_y &= \frac{1}{E}[\sigma_y - \nu(\sigma_x + \sigma_z)] \\
\varepsilon_z &= \frac{1}{E}[\sigma_z - \nu(\sigma_x + \sigma_y)] = 0 \\
\gamma_{xy} &= \frac{2(1+\nu)}{E}\tau_{xy} \\
\sigma_z &= \nu(\sigma_x + \sigma_y)
\end{aligned} \tag{7.65}$$

Substituting Eq. (7.66) in the compatibility or congruence Eq. (7.65) yields

$$2\frac{\partial^2 \tau_{xy}}{\partial x \partial y} = (1-\nu)\left(\frac{\partial^2 \sigma_x}{\partial y^2} + \frac{\partial^2 \sigma_y}{\partial x^2}\right) - \nu\left(\frac{\partial^2 \sigma_x}{\partial x^2} + \frac{\partial^2 \sigma_y}{\partial y^2}\right) \tag{7.66}$$

Equation (7.67) combined with the equilibrium Eq. (7.62) finally provides the sought system of three equations in three unknowns. Around 1860 an English astronomer and mathematician, G.B. Airy [14] showed that the elastic problem could be solved by introducing a function ϕ, which he called *Airy function*, that satisfies the equilibrium equations identically. He posed

$$\tau_{xy} = -\frac{\partial^2 \phi}{\partial x \partial y} \tag{7.67}$$

with ϕ continuous function of x and y. Assuming that mass forces are conservative

$$\frac{\partial \overline{X}}{\partial y} = \frac{\partial \overline{Y}}{\partial x} \tag{7.68}$$

and therefore there is a potential Ω such that

$$\overline{X} = -\frac{\partial \Omega}{\partial x} \overline{Y} = -\frac{\partial \Omega}{\partial y} \tag{7.69}$$

then the equilibrium Eq. (7.62) become

$$\begin{aligned} \frac{\partial}{\partial x}\left(\sigma_x - \frac{\partial^2 \varphi}{\partial y^2} - \Omega\right) &= 0 \\ \frac{\partial}{\partial y}\left(\sigma_y - \frac{\partial^2 \varphi}{\partial x^2} - \Omega\right) &= 0. \end{aligned} \tag{7.70}$$

The simplest solution is, of course

$$\begin{aligned} \sigma_x &= \frac{\partial^2 \varphi}{\partial y^2} + \Omega \\ \sigma_y &= \frac{\partial^2 \varphi}{\partial x^2} - \Omega \\ \tau_{xy} &= -\frac{\partial^2 \varphi}{\partial x \partial y}. \end{aligned} \tag{7.71}$$

Substituting this solution in the congruence Eq. (7.67) yields

$$\frac{\partial^4 \varphi}{\partial x^4} + 2\frac{\partial^2 \varphi}{\partial x^2 \partial y^2} + \frac{\partial^4 \varphi}{\partial y^4} = -\frac{1-2\nu}{1-\nu}\left(\frac{\partial^2}{\partial x^2} + \frac{\partial^2}{\partial y^2}\right)\Omega. \tag{7.72}$$

or

$$\nabla^4 \varphi = -\frac{1-2\nu}{1-\nu}\nabla^2 \Omega \tag{7.73}$$

where ∇^4 is the biharmonic operator

$$\begin{aligned} \nabla^4 = \nabla^2(\nabla^2) &= \left(\frac{\partial^2}{\partial x^2} + \frac{\partial^2}{\partial y^2}\right)\left(\frac{\partial^2}{\partial x^2} + \frac{\partial^2}{\partial y^2}\right) \\ &= \frac{\partial^4}{\partial x^4} + \frac{2\partial^4}{\partial x^2 \partial y^2} + \frac{\partial^4}{\partial y^4} \end{aligned} \tag{7.74}$$

Note that if the potential Ω is a harmonic function, that is a solution of the Laplace equation, whose operator is called Laplacian, then being $\nabla^2 \Omega = 0$ we have

$$\nabla^4 \phi = 0 \tag{7.75}$$

and then the Airy function will also be a harmonic function as a solution of the Laplace equation. Equation (7.76) thus expresses, in synthetic form, the condition of plain strain in the two-dimensional elastic problem. In plain stress conditions it is

$$\sigma_z = \tau_{xz} = \tau_{yz} = 0 \tag{7.76}$$

The equilibrium Eq. (7.72) still hold while the compatibility condition provides a larger series of relations

$$\begin{aligned}
&\frac{\partial^2 \varepsilon_x}{\partial y^2} + \frac{\partial^2 \varepsilon_y}{\partial x^2} = \frac{\partial^2 \gamma_{xy}}{\partial x \partial y}; 2\frac{\partial^2 \varepsilon_y}{\partial x \partial z} = \frac{\partial^2 \gamma_{xy}}{\partial y \partial z} \\
&\frac{\partial^2 \varepsilon_y}{\partial z^2} + \frac{\partial^2 \varepsilon_z}{\partial y^2} = 0; 2\frac{\partial^2 \varepsilon_x}{\partial y \partial z} = \frac{\partial^2 \gamma_{xy}}{\partial x \partial z} \\
&\frac{\partial^2 \varepsilon_z}{\partial x^2} + \frac{\partial^2 \varepsilon_x}{\partial z^2} = 0; 2\frac{\partial^2 \varepsilon_z}{\partial x \partial y} = -\frac{\partial^2 \gamma_{xy}}{\partial z^2}
\end{aligned} \tag{7.77}$$

If we introduce the further condition that σ_x, σ_y and τ_{xy} are independent of z then the compatibility Eq. (7.78) reduce to the first one as in the case of plane strain. In practice this means that the stresses σ_x, σ_y and τ_{xy} must be constant along the thickness z. An exact solution to the problem [14] indicates, however, that tensions are distributed parabolically along the thickness of the body. If, however, the thickness is very small, which is also necessary for a condition of plain stress, then the stresses will assume modest variations that will disappear more and more as the thickness itself decreases, making the assumption legitimate. The equations of equilibrium and congruence are therefore the same as those of the problem of plain strain. However, the relationships given by Hooke's law change, which are now

$$\begin{aligned}
\varepsilon_x &= \frac{1}{E}(\sigma_x - \nu\sigma_y) \\
\varepsilon_y &= \frac{1}{E}(\sigma_y - \nu\sigma_x) \\
\varepsilon_z &= -\frac{\nu}{E}(\sigma_x + \sigma_y)
\end{aligned} \tag{7.78}$$

which, introduced into the congruence Eq. (7.65) yield

$$\left(\frac{\partial^2 \sigma_x}{\partial y^2} + \frac{\partial^2 \sigma_y}{\partial x^2}\right) - \nu\left(\frac{\partial^2 \sigma_x}{\partial x^2} + \frac{\partial^2 \sigma_y}{\partial y^2}\right) = 2(1+\nu)\frac{\partial^2 \tau_{xy}}{\partial x \partial y} \tag{7.79}$$

Instead of Eq. (7.67) valid for the plain strain case. Introducing the Airy stress function that will have the same formulation (7.68) since the equilibrium equations are the same, gives

$$\frac{\partial^4 \varphi}{\partial x^4} + 2\frac{\partial^2 \varphi}{\partial x^2 \partial y^2} + \frac{\partial^4 \varphi}{\partial y^4} = -(1-\nu)\left(\frac{\partial^2 \Omega}{\partial x^2} + \frac{\partial^2 \Omega}{\partial y^2}\right) \tag{7.80}$$

or

$$\nabla^4 \varphi = -(1-\nu)\nabla^2 \Omega \tag{7.81}$$

Therefore, if $\nabla^2 \Omega = 0$,:

$$\nabla^4 \phi = 0 \tag{7.82}$$

exactly as in the two-dimensional elastic problem in plain strain.

7.9 Holomorphic Functions of Complex Variables

The choice of the Airy function that will be biarmonic, that is, will satisfy Eqs. (7.76) or (7.83) and the boundary conditions of the particular problem, is one of the most difficult problems of engineering. It is appropriate, at this point, to introduce a particular type of functions of great utility in the study of stress analysis problems in cracked bodies: the *holomorphic functions*. Holomorphic are the functions of complex variable $z = x + iy$ that are continuous, derivable and with continuous derivative. Following the traditional notation of complex numbers, x will be the real part and y the complex one. If we denote with $f(z)$ such a function, we will have:

$$\begin{aligned} \frac{\partial f(z)}{\partial z} &= \lim_{\Delta z \to 0} \frac{f(z+\Delta z) - f(z)}{\Delta z} \\ &= \lim_{\Delta z \to 0} \frac{f(x+\Delta x, y+\Delta y) - f(x,y)}{\Delta x + i\Delta y} = f'(z) \end{aligned} \tag{7.83}$$

The limit operation (7.84) can be performed by alternately leaving y and x constants and giving an increment Δx and Δy.

$\varDelta y$ to the other variable so that

$$\begin{aligned} f'(z) &= \lim_{\Delta x \to 0} \frac{f(x+\Delta x, y) - f(x,y)}{\Delta x} = f_x'(z) \\ f'(z) &= \lim_{\Delta y \to 0} \frac{f(x, y+\Delta y) - f(x,y)}{i\Delta y} = \frac{1}{i} f_y'(z) \end{aligned} \tag{7.84}$$

from which

$$f_x'(z) = \frac{1}{i} f_y'(z) \tag{7.85}$$

Equation (7.86) is of considerable importance. In fact, called $f(z)$ a holomorphic function of the type $f(z) = \phi(z) + i\psi(z)$ in which $\phi(z)$ and $\psi(z)$, respectively real and complex part of the function $f(z)$, are both holomorphic functions of the complex variable z, we have

$$\begin{aligned} \frac{\partial}{\partial x} f(z) &= \frac{\partial \phi}{\partial x} + i\frac{\partial \psi}{\partial x} = f'(z) \\ \frac{\partial}{\partial y} f(z) &= \frac{\partial \phi}{\partial y} + i\frac{\partial \psi}{\partial y} = i \cdot f'(z) \end{aligned} \tag{7.86}$$

and

$$i\left(\frac{\partial \phi}{\partial x} + i\frac{\partial \psi}{\partial x}\right) = \frac{\partial \phi}{\partial y} + i\frac{\partial \psi}{\partial y} \tag{7.87}$$

Since it is $i\, i = -1$, we have

$$\begin{aligned} \frac{\partial \phi}{\partial x} &= \frac{\partial \psi}{\partial y} \\ \frac{\partial \phi}{\partial y} &= -\frac{\partial \psi}{\partial x} \end{aligned} \tag{7.88}$$

Equations (7.89) represent the Cauchy-Riemann system of equations from which differentiating the first with respect to x and the second equation with respect to y, we obtain:

$$\begin{aligned} \frac{\partial^2 \phi}{\partial x^2} &= \frac{\partial^2 \psi}{\partial x \partial y} \\ \frac{\partial^2 \phi}{\partial y^2} &= -\frac{\partial \psi}{\partial x \partial x} \end{aligned} \tag{7.89}$$

and eliminating ψ

$$\begin{aligned} \frac{\partial^2 \phi}{\partial x^2} &= -\frac{\partial^2 \phi}{\partial y^2}; \\ \frac{\partial^2 \phi}{\partial x^2} &+ \frac{\partial^2 \phi}{\partial y^2} = 0 \end{aligned} \tag{7.90}$$

Equation (7.91) ensures that any holomorphic function is harmonic in that it satisfies the Laplace equation and can therefore be used as a possible solution to the elastic problem. Thus, we can say that any holomorphic function of complex variable $f(z) = \phi(z) + i\psi(z)$ satisfies Eq. (7.76) and for this reason it will also be an Airy function that can be used in Eqs. (7.72) to determine stresses σ_x, σ_y and τ_{xy}. So will any combination: $xf(z)$, $yf(z)$, $x\,\partial f/\partial x$, $y\,\partial f/\partial y$, $xf(z) + yf(z)$ etc. Among these, a particular function is that expressed by

$$\Phi = \varphi_1 + x\varphi_2 + y\varphi_3 \tag{7.91}$$

with ϕ_1, ϕ_2 and ϕ_3 always holomorphic functions. The peculiarity of Eq. (7.92) lies in the fact that, given

$$\psi(z) = \phi_2 + i\phi_3 \tag{7.92}$$

and introduced another holomorphic function $\chi(z)$ whose real part is ϕ_1, one can rewrite Eq. (7.92) as

$$\Phi = \mathrm{Re}[(x - iy)\psi(z) + \chi(z)] \tag{7.93}$$

where *Re* indicates the real part of the function in square brackets. Equation (7.94) is known as the Goursat function [15]. It allows to express the Airy function in the most general form possible by means of two holomorphic functions. Goursat's method was used by Musckhelishvili [16] for the solution of the problem of cracked bodies.

7.10 Stress Analysis of Cracks

As a first case we examine precisely the Griffith crack, or the sharp crack in an infinite plate subjected to a state of biaxial traction $\sigma_x = \sigma_y = \sigma$, already considered in Sect. 7.3. The direct solution of the elastic problem is difficult. A little less difficult is, however, the choice of the Airy function that satisfies the particular boundary conditions that the problem requires, even if this, as mentioned in the previous paragraph, is still configured as one of the most difficult problems of engineering. Referring to Fig. 7.15, the boundary conditions in the case considered are expressed by the fact that the stress σ_y must be zero for $y = 0$ and $-a < x < +a$ (free surface) and equal to the remote stress $\sigma_2 = \sigma$ for x tending to infinity.

Moreover, the stress σ_y intensifies when x approaches $\pm a$. The latter condition suggests a form of the Airy function of the type:

$$\sigma \frac{x^2}{x^2 - a^2} \tag{7.94}$$

This function, in fact, becomes equal to σ for $x \to \infty$, as required, and grows as $|x|$ to $|a|$. The other condition, that is, $\sigma_y = 0$ for $y = 0$ and $-at < x < +a$, is obtained by placing the term multiplying σ under square root

$$\frac{\sigma x}{\sqrt{x^2 - a^2}} \tag{7.95}$$

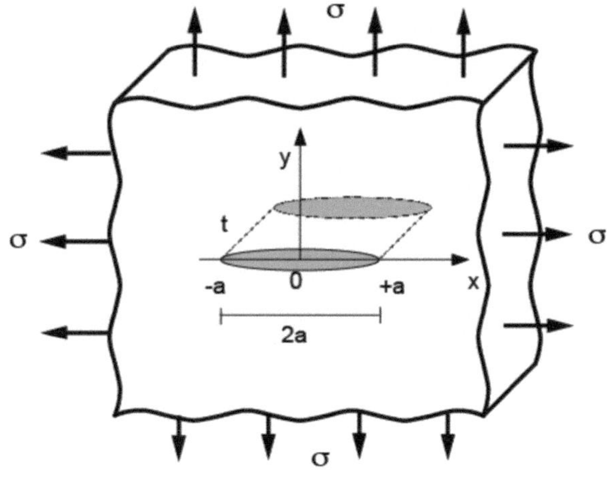

Fig. 7.16 Reference system for the elastic problem

By doing that, the proposed function becomes imaginary for $-a < x < +a$ becoming a function of complex variable whose real part will be zero for $y = 0$. Dealing with a function of complex variable is of extreme importance not only because it allows us to easily respond to a precise condition around the problem, but also because we know (see Sect. 7.9) that these functions are certainly harmonic and therefore any of them or their combination, will be an Airy function. Westergaard [17], who was among the first to address the problem of the exact solution of the Griffith crack, suggested an Airy function structured as follows

$$\Phi = \overline{Z} + yi\overline{Z} \tag{7.96}$$

where

$$Z = \frac{\partial \overline{Z}}{\partial z}, \overline{Z} = \frac{\partial \overline{\overline{Z}}}{\partial z}, Z' = \frac{\partial Z}{\partial z} \tag{7.97}$$

with $Z(z)$ given by Eq. (7.96)

$$Z(z) = \frac{\sigma z}{\sqrt{z^2 - a^2}} \tag{7.98}$$

It can be noted that the Westergaard function is a special case of that of Goursat (7.94) if we pose

$$\psi = \frac{1}{2}\overline{Z}, \chi = \overline{\overline{Z}} - \frac{1}{2}z\overline{Z} \tag{7.99}$$

Recalling Eq. (7.72) and the properties of the holomorphic functions of complex variable (7.87) and (7.88) we obtain:

$$\begin{aligned} \sigma_x &= Z - yiZ' \\ \sigma_y &= Z + yiZ' \\ \tau_{xy} &= -y \cdot Z' \end{aligned} \tag{7.100}$$

The second Eq. (7.101) reflects the boundary condition sought: $\sigma_y = 0$ for $-a < x < +a$ and $y = 0$. In fact, when $y = 0$ the σ_y is reduced to the only real part of $Z(z)$. Developing Eq. (7.99) for $y = 0$ we have

$$\sigma_y \frac{x\sqrt{x^2 - a^2}}{x^2 - a^2} \tag{7.101}$$

whose real part, for $-a < x < +a$, is worth just zero. When, on the other hand, is $|x| > |a|$ then Eq. (7.96) or Eq. (7.99), represents a real function that for $y = 0$, always based on the second of Eqs. (7.101), gives precisely the value of the σ_y sought

$$\sigma_y = \sigma \frac{x}{\sqrt{x^2 - a^2}} \tag{7.102}$$

We now move the origin of the coordinates x, y on the front of the crack by introducing the new variable $r = x - a$, Eq. (7.103) become:

$$\sigma_y = \sigma \cdot \frac{(a + r)}{\sqrt{2ar}} \cdot \left(1 + \frac{r}{2a}\right)^{-\frac{1}{2}}$$

and expanding in series the square root

$$\sigma_y = \sigma \cdot \frac{(a + r)}{\sqrt{2ar}} \cdot \left(1 - \frac{r}{4a} + \frac{3}{32}\frac{r^2}{a^2} + \cdots\right) \tag{7.103}$$

and truncating the Mac Laurin series expansion to the first term, which is possible if we assume $r/a << 1$, we obtain that in the close surroundings ($r/a << 1$) of the crack tip the stress σ_y is worth

$$\sigma_y = \sigma\sqrt{\frac{a}{2r}} \tag{7.104}$$

G. Irwin [18] suggested a more convenient form for the $Z(z)$ function by introducing a significant change of variables, passing from the cartesian coordinates x, y to polar ones r, θ, and moving the origin of the reference system to the front line of the crack, as shown in Fig. 7.16

$$z = a + r \cdot e^{i\theta} \tag{7.105}$$

with

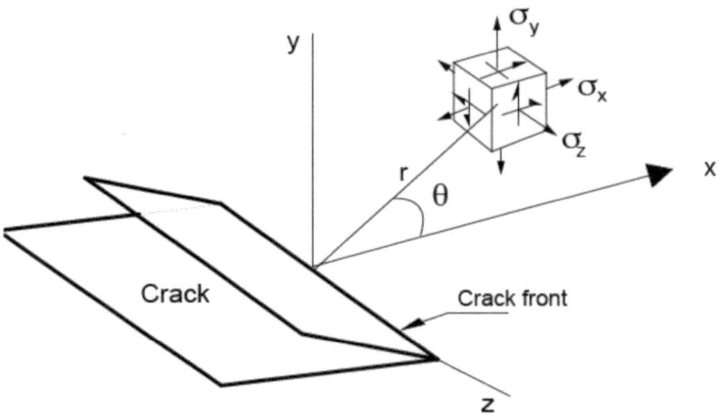

Fig. 7.17 Polar coordinates used by *G*. Irwin

$$r^2 = (x-a)^2 + y^2, \tan\theta = \frac{y}{x-a}, y = r \cdot \sin\theta \tag{7.106}$$

Assuming, once again, that r/a is negligible with respect to the unit: $r/a << 1$, the real part of the first two Eqs. (7.101) becomes

$$Z = \sigma\sqrt{\frac{a}{2r}} \cdot \cos\frac{\theta}{2} \tag{7.107}$$

with the imaginary part given by

$$yiZ = \sigma\sqrt{\frac{a}{2r}} \cdot \sin\frac{\theta}{2}\cos\frac{\theta}{2}\sin\frac{3\theta}{2} \tag{7.108}$$

therefore

$$\sigma_x = \sigma\sqrt{\frac{a}{2r}} \cdot \cos\frac{\theta}{2}\left(1 - \sin\frac{\theta}{2}\sin\frac{3\theta}{2}\right) \tag{7.109}$$

Similarly, we will have for the other two stresses σ_y and τ_{xy}:

$$\begin{aligned} \sigma_y &= \sigma\sqrt{\frac{a}{2r}} \cdot \cos\frac{\theta}{2}\left(1 + \sin\frac{\theta}{2}\sin\frac{3\theta}{2}\right) \\ \tau_{xy} &= \sigma\sqrt{\frac{a}{2r}} \cdot \sin\frac{\theta}{2}\cos\frac{\theta}{2}\cos\frac{3\theta}{2} \end{aligned} \tag{7.110}$$

The previous hypothesis $r/a << 1$ is extremely important because Eqs. (7.110) and (7.111) provide the complete stress field in a very narrow area around the crack tip for any value of r and θ. Outside that very limited region those equation are no longer valid. Ahead of the crack along the x-axis, i.e. for $\theta = 0$, the two stresses σ_x and σ_y are equal to each other

$$\sigma_x = \sigma_y = \sigma\sqrt{\frac{a}{2r}} \tag{7.111}$$

This particular solution ($\theta = 0$), already expressed by Eq. (7.105), is analogous to that found by Inglis already in 1913 [19]

$$\sigma_x = \sigma_y = \frac{\left(1 + \frac{r}{a}\right)}{\sqrt{2\frac{r}{a} + \left(\frac{r}{a}\right)^2}}\sigma \tag{7.112}$$

when we limit the stress field to a small circle around the crack front, that is, for $r/\mathrm{a} << 1$. G. Irwin grouped the product $\sigma\sqrt{\pi a}$ into a single term K_{I} which he called "*stress intensity factor*" in which the index I underlines, as already mentioned, MODE I of crack opening Fig. 7.13

$$K_{\mathrm{I}} = \sigma\sqrt{\pi a} \tag{7.113}$$

then Eqs. (7.110) and (7.111) can be written as

$$\begin{aligned}
\sigma_x &= \frac{K_I}{\sqrt{2\pi r}} \cdot \cos\frac{\theta}{2}\left(1 - \sin\frac{\theta}{2}\sin\frac{3\theta}{2}\right) \\
\sigma_y &= \frac{K_I}{\sqrt{2\pi r}} \cdot \cos\frac{\theta}{2}\left(1 + \sin\frac{\theta}{2}\sin\frac{3\theta}{2}\right) \\
\tau_{xy} &= \frac{K_I}{\sqrt{2\pi r}} \cdot \sin\frac{\theta}{2}\cos\frac{\theta}{2}\cos\frac{3\theta}{2}
\end{aligned} \tag{7.114}$$

or just

$$\sigma_{ij} = \frac{K_I}{\sqrt{2\pi r}} \cdot f_{ij}(\theta) \tag{7.115}$$

to which must be added, for the biaxial case in plain strain:

$$\begin{aligned}
\sigma_z &= \nu(\sigma_x + \sigma_y) \\
\tau_{xz} &= \tau_{yz} = 0
\end{aligned} \tag{7.116}$$

From Eqs. (7.63), (7.66) and (7.100) we can get the expressions for the displacements u and v:

$$u = \frac{(1+\nu)(1-2\nu)}{E}(\overline{Z} - yiZ)$$
$$v = \frac{2(1-\nu^2)}{E}(i\overline{Z} - yZ) \tag{7.117}$$

from which we derive that for $y = 0$, i.e., on the x-axis, it is

$$v = \frac{2(1-\nu^2)}{E} \cdot \sigma \cdot \sqrt{a^2 - x^2} \tag{7.118}$$

By changing coordinates, Eqs. (7.106) and (7.107), and retaining the usual $r/a <<$ 1 approximation, we get

$$u = \frac{K_I}{G}\sqrt{\frac{r}{2\pi}} \cos\frac{\theta}{2}\left(1 - 2\nu + \sin^2\frac{\theta}{2}\right)$$
$$v = \frac{K_I}{G}\sqrt{\frac{r}{2\pi}} \sin\frac{\theta}{2}\left(2 - 2\nu - \cos^2\frac{\theta}{2}\right) \tag{7.119}$$

G being the shearing modulus of elasticity equal to $E/2(1+\nu)$. Some very important comments should be made about Eq. (7.114).

1. Equation (7.114) are valid only near the crack tip where the $1/\sqrt{r}$ singularity dominates the stress field. In this region the stress field around the crack is self-similar.
2. The stress intensity factor K_I completely defines the amplitude of the crack tip singularity. Therefore, Linear Elastic Fracture Mechanics is based on a single parameter description of the entire crack tip conditions which makes the K_I parameter be one of the most important concepts in fracture mechanics.
3. Equation (7.114) derive from the assumption $r/a << 1$, having truncated the Mac Laurin series expansion of the term $(1 + r/2a)^{-1/2}$ to the first constant term, see Eq. (7.104). This means that Eq. (7.114) are exact only asymptotically, in a very small surrounding of the crack tip where r is so small that the terms r/a, $(r/a)^2$ and those of higher order can be neglected. The region where Eq. (7.114) describe the crack-tip stress field is defined as *singularity-dominated zone*, Fig. 7.17.
4. Since the stress field shows a singularity of type $1/\sqrt{r}$ its amplitude will reach an infinite value at the tip of the crack ($r = 0$). In real materials that yield, this leads to the formation of a small plastic zone ahead of the tip, as described in Chap. 8. Therefore, Eq. (7.114) will still be valid for real materials only if this plastic zone is so small as to be negligible compared to the maximum value of r for which condition 2 is valid. In practice, this means that the size of the plastic zone must be negligible.
5. Equation (7.114) were obtained for plain strain conditions for which $\sigma_z = \nu(\sigma_x + \sigma_y)$. They can also be used in plain stress conditions simply by placing $\sigma_z = 0$. However, since the dimension of the plastic zone increases with plane

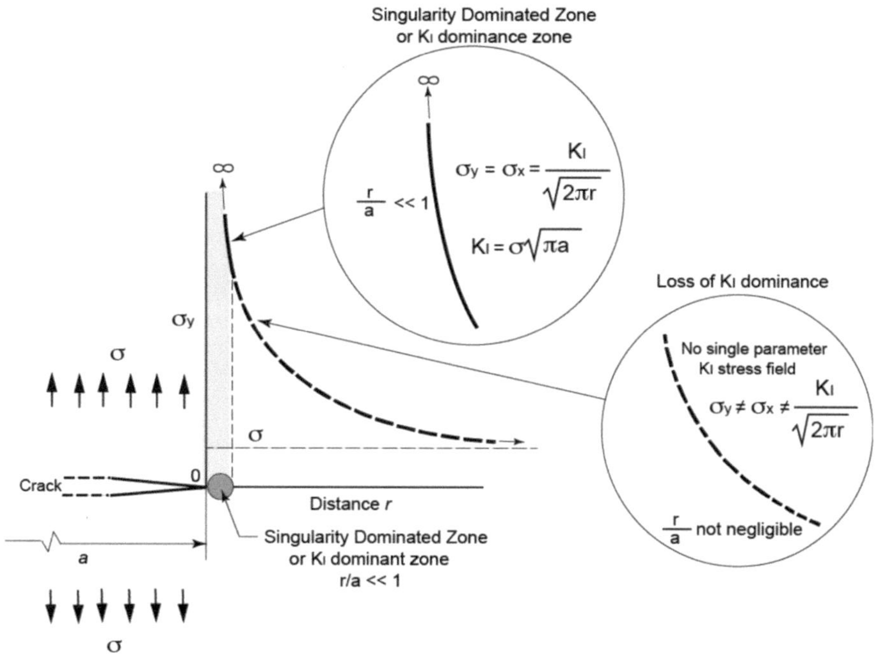

Fig. 7.17 The K_I dominance stress field exists only at a very short distance from the crack tip (condition $r/a << 1$). For *r*/*a* ratios no longer negligible, though small, the K_I dominance is lost and the two stress components σ_y and σ_x are no longer equal to each other

stress conditions, see Sect. 8.5, special care must be taken not to fall within restriction 4. which in plain stress can hardly be satisfied. One must also recall the hypothesis made in Sect. 7.8 of independence of σ_x, σ_y and τ_{xy} from z in the case of plane stress to restrict the compatibility relations (7.78) to the first one and obtain Eq. (7.83) to which solutions (7.114) refer. This means, that, at variance with the plain strain condition, the thickness of the test piece must still be small, so as to keep the stress constant throughout the thickness.

6. Equation (7.114) have been obtained for a crack in an infinite plate, biaxially loaded. If the external load were uniaxial and if the crack were not central but edge crack, Eq. (7.114) may no longer be valid and σ_x would no longer be equal to σ_y.

7.11 Cracks in Finite Plates

We still recall that Equation (7.114) have been obtained for a crack in an infinite plate. If the geometry were finite, as always in practice, border effects must be considered that change the simple, close-form expression of the stress intensity factor $K_{\text{I}} = \sigma\sqrt{\pi a}$.This is schematically shown in Fig. 7.18.

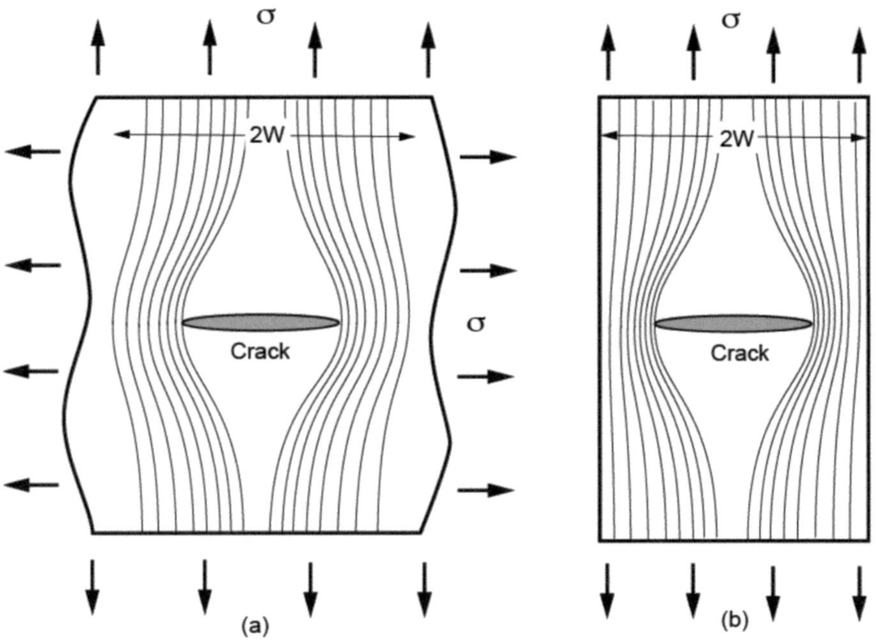

Fig. 7.18 Border effects in a finite geometry. **a** Lines of force around a trough-wall crack in an infinite geometry, **b** in a finite width geometry. Finite geometries exert a constraint on the lines of force increasing the amplitude of the crack tip stress field K_I

In an infinite geometry, Fig. 7.18a, the lines of force are free to divert around the crack whose effect is felt beyond a distance $2W$ while in the finite geometry of width $2W$, Fig. 7.18b, borders constrain the lines of force that cannot expand and become thicker, increasing the stress field at the crack tip. The stress intensity factor K_I cannot be the same for the two cases. In a finite geometry the amplitude of the crack tip stress field will no longer be given by the closed-form solution found by Irwin and the stress intensity factor $K_I = \sigma\sqrt{\pi a}$ needs to be adjourned as

$$K_I = \sigma f(a, W) \tag{7.120}$$

where $f(a, W)$ is a non-dimensional function of the system geometry, usually obtained by finite-element analysis or boundary conditions, that has a polynomial expression whose value is generally greater than 1. Most configurations for which there is a closed-form solution consist of a crack of simple shape in an infinite geometry. The term "infinite" is rather relative and not absolute. A relatively small crack in a large plate behaves as if the structure were infinite. Also, in cases where two length dimensions are needed to characterize the crack as, for instance, for an elliptical or semielliptical crack of $2c$ and $2a$ size, as shown in Fig. 7.19. In these cases, besides the function $f(a, W)$ also a flaw shape parameter Q is needed obtained from an elliptical integral [16]

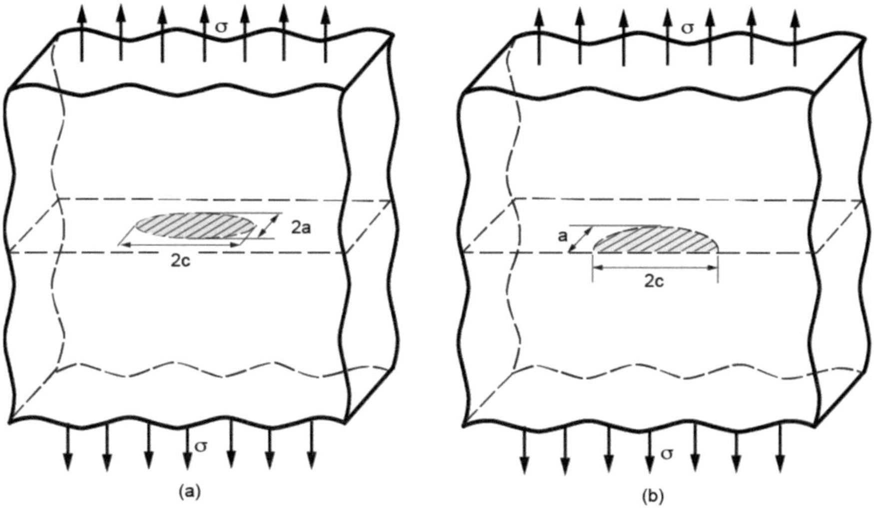

Fig. 7.19 **a** embedded elliptical crack in a finite plate under remote traction, **b** surface semielliptical crack in a finite plate under remote traction

$$\Psi = \int_0^{\pi/2} \sqrt{1 - \frac{c^2 - a^2}{c^2} \sin^2 \phi d\phi} \tag{7.121}$$

A series expansion of Eq. (7.121) yields.

$$\Psi = \frac{\pi}{2}\left[1 - \frac{1}{4}\frac{c^2 - a^2}{c^2} - \frac{3}{64}\left(\frac{c^2 - a^2}{c^2}\right)^2 - \dots\right] \tag{7.122}$$

which can be approximated by

$$Q = \Psi^2 = 1 + 1.464\left(\frac{a}{c}\right)^{1.65} \tag{7.123}$$

The final solution is

$$K_I = \sigma\sqrt{\frac{\pi a}{Q}} f(\phi) \tag{7.124}$$

for the embedded elliptical crack, Fig. 7.19a and

$$K_I = \lambda\sigma\sqrt{\frac{\pi a}{Q}} f(\phi) \tag{7.125}$$

The value of the two functions $f(\varphi)$ and Q for the two cases shown in Fig. 7.19 are given in Appendix A to Chap. 14.

7.12 Material Toughness

If the Irwin stress intensity factor K_I completely defines the stress field ahead of a crack in an infinite plate under remote traction σ or in a finite plate of thickness W multiplying K_I by the nondimensional function $f(a, W)$, Eq. (7.120), then we shall expect that increasing the external load till fracture occurs at σ_c that we call critical, the corresponding value K_{Ic} will measure the toughness of the material. The toughness of a material can be measured using any specimen containing a crack for which the corresponding function is known, see Chap. 9 and Sect. 7.13. Interesting enough, until the introduction of K_I toughness had been just a word indicating the generic property of a material to withstand brittle fracture. With the introduction of the Irwin stress intensity factor, K_I, and its critical value K_{Ic}, toughness became for the first time a physical property having a precise dimension of MPa $\sqrt{m}$ or $\mathrm{MNm}^{-3/2}$, suitable of being measured. This represented an epochal turning point in the history of Fracture Mechanics.

Since Eq. (7.114) were obtained for a crack in an infinite plate subjected to a remote state of biaxial traction $\sigma_x = \sigma_y = \sigma$ they must provide precisely the value of the nominal stress σ for r tending to ∞. It can be seen, however, that Eq. (7.114) are worth 0 and not σ for $\theta = 0$ and r tending to ∞. Therefore, to describe the field outside the very narrow domain delimited by the condition $r/\mathrm{a} << 1$ it will be necessary to use the other terms of the series, and the simple K_I will no longer be sufficient. However, as has already been described extensively in Sect. 6.4, in brittle cases, which are precisely those treated by LEFM, fracture originates in the first grains at the vertex of the crack, and it is wright there that Eq. (7.114) are valid. What happens far from that tiny apical area is absolutely inessential for the stability of the defect itself. The Irwin stress intensity factor, K_I, that appears in Eq. (7.114) is just a far relative of the well-known stress concentration factor $k = \sigma_{max}/\sigma$ that describes the maximum stress in the zones of geometric discontinuity in which the nominal stress σ is amplified and should not be confused with the latter.

7.13 Similitude Principle

The existence of a self-similar stress field ahead of the crack tip characterized by a single parameter, i.e., the Irwin stress intensity factor, K_I, that completely defines the stress, strain and displacement conditions in the singularity-dominated zone and the assumption that by reaching a critical value, K_{Ic}, that stress field causes the fracture of the material, allows to postulate that fracture will always occur in any specific material precisely at that critical value, K_{Ic}, independently of the crack size, geometry and loading conditions of the cracked body. The K_{Ic} is a geometry-independent material constant that we call toughness. Since K_I is the only parameter that in Eq. (7.114) contains information on the load conditions, the geometry of the system and the size of the crack, it is clear that crack instability can only depend on its critical value that varies from material to material and is specific for any material. In traditional engineering applied to non-cracked bodies it is the local stress or stress at a point that defines the condition for failure: when the *local stress*, σ, equals the strength of the material, σ_u, i.e., $\sigma = \sigma_u$, the material fails or, in cases of multiaxial stresses, when the equivalent stress σ_e at a point equals the yield strength σ_s, i.e., $\sigma_e = \sigma_s$, the material yields or fails when $\sigma_e = \sigma_u$.

Dealing with a crack, it is not just the stress at a point but an entire stress field whose amplitude is defined by K_I that determines the instability conditions: when the stress amplitude of this entire *local stress field*, K_I, equals the critical stress field, K_{Ic}, i.e., when $K_I = K_{Ic}$, then the material fractures. This is the great difference between the conventional engineering approach to failure and fracture mechanics approach to cracked bodies fracture. Perhaps, the difference is more apparent than real. The K_I dominated stress field has, in fact, such limited an extension, i.e., it is so localized (condition $a/r << 1$, Fig. 7.17) that the entire stress field can almost be considered a stress at a point. However, beside this subtle arguing, the fundamental conclusion is that we can also apply the similitude principle saying that what happens in a small specimen at the tip of a crack will also happen at the tip of a crack in a large specimen or in a real structure, provided that the corresponding K_I be the same. Equal K_I results in equal crack tip conditions which in real materials that yield, also means equal crack tip plastic zone size, Fig. 7.20. Therefore, we can measure the toughness, K_{Ic}, of a material using a convenient small specimen and argue that any structure of the same material will fail at the same $K_I = K_{Ic}$.

7.14 Mode II and Mode III

For Mode II crack opening, Fig. 7.13, in which the in-plain shear stress, τ, operates, the same general procedure as for Mode I can be followed. The most suitable Airy function in this case is of the type

$$\phi_{II} = -y\overline{Z}_{II} \tag{7.126}$$

in which the subscript II recalls the Mode II of crack opening. With this Eq. (7.72) provide the values of the three stress components:

$$\begin{aligned} \sigma_x &= 2i\phi_{II}(z) + y\phi'_{II}(z) \\ \sigma_y &= -y\phi'_{II}(z) \\ \tau_{xy} &= \tau_{yx} = \phi_{II}(z) - yi\phi'_{II}(z) \end{aligned} \tag{7.127}$$

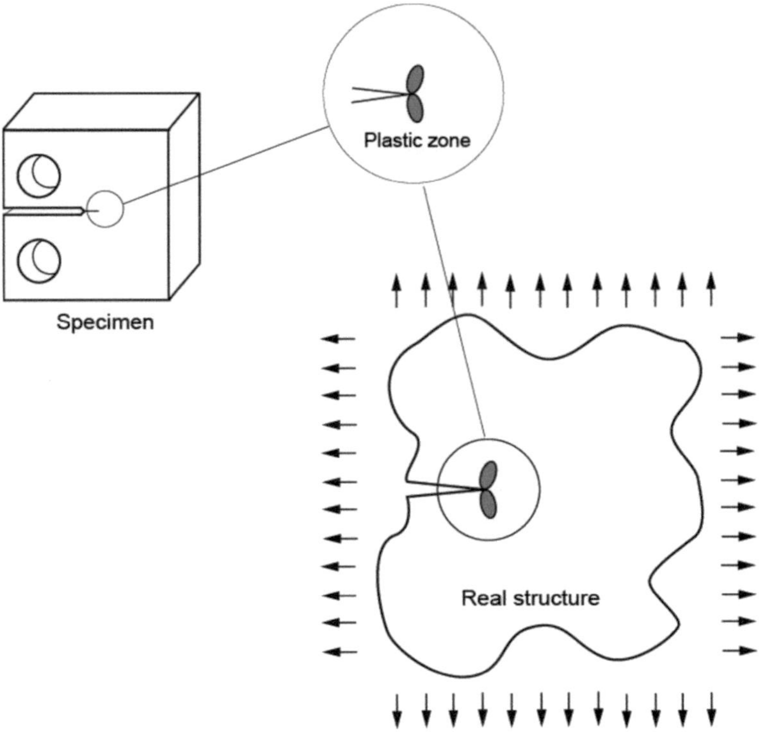

Fig. 7.20 Similitude principle: conditions existing at the crack tip of a small specimen at a given K_I, i.e., stress, strain, displacements and plastic zone size and shape, exist precisely at the crack tip of a real structure loaded at the same K_I level which implies that when the small specimen fractures at a certain K_{Ic} value also the real structure will fail at the same K_{Ic} level

Choosing for the holomorphic function of complex variable $Z(z)$ a form similar to that adopted for case I:

$$Z(z) = \frac{\tau}{\sqrt{1 - (a/z)^2}} \tag{7.128}$$

finally, we have that for $r/a << 1$ the approximate solution is

$$\begin{aligned}
\sigma_x &= \frac{K_{II}}{\sqrt{2\pi r}} \cdot \sin\frac{\theta}{2}\left(2 + \cos\frac{\theta}{2}\cos\frac{3\theta}{2}\right) \\
\sigma_y &= \frac{K_{II}}{\sqrt{2\pi r}} \cdot \sin\frac{\theta}{2}\cos\frac{\theta}{2}\cos\frac{3\theta}{2} \\
\tau_{xy} &= \frac{K_{II}}{\sqrt{2\pi r}} \cdot \cos\frac{\theta}{2}\left(1 - \sin\frac{\theta}{2}\sin\frac{3\theta}{2}\right)
\end{aligned} \tag{7.129}$$

where $K_{II} = \tau\sqrt{\pi a}$ and with displacements *u, v* given by:

$$
\begin{aligned}
u &= \frac{K_{II}}{G}\sqrt{\frac{r}{2\pi}}\sin\frac{\theta}{2}\left(2-2\nu+\cos^2\frac{\theta}{2}\right)\\
v &= \frac{K_{II}}{G}\sqrt{\frac{r}{2\pi}}\cos\frac{\theta}{2}\left(-1+2\nu+\sin^2\frac{\theta}{2}\right)
\end{aligned}
\tag{7.130}
$$

Finally, for Mode III crack opening, Fig. 7.12, with out-of-plane shear τ in which displacements take place only in the z direction and therefore, $u = v = 0$:

$$
\begin{aligned}
\tau_{xy} &= \tfrac{1}{G}\tfrac{\partial z}{\partial x}\\
\tau_{yz} &= \tfrac{1}{G}\tfrac{\partial z}{\partial y}
\end{aligned}
\tag{7.131}
$$

with equilibrium equations that boil down to

$$
\frac{\partial\tau_{xz}}{\partial x}+\frac{\partial\tau_{yz}}{\partial y}=0 \tag{7.132}
$$

which combined with Eq. (7.97) leads to

$$
\frac{\partial^2 z}{\partial x^2}+\frac{\partial^2 z}{\partial y^2}=\nabla^2 z=0 \tag{7.133}
$$

Placing

$$
z=\frac{1}{G}iZ_{III} \tag{7.134}
$$

and choosing the function *Z(z)* as in case II, Eq. (7.122) we obtain

$$
\begin{aligned}
\tau_{xz} &= -\frac{K_{III}}{\sqrt{2\pi r}}\sin\frac{\theta}{2}\\
\tau_{yz} &= \frac{K_{III}}{\sqrt{2\pi r}}\cos\frac{\theta}{2}\\
\sigma_x &= \sigma_y = \sigma_z = \tau_{xy} = 0\\
z &= \frac{K_{III}}{G}\left(\sqrt{\frac{2r}{\pi}}\sin\frac{\theta}{2}\right)
\end{aligned}
\tag{7.135}
$$

Note that, in cases of mixed-mode of aperture, Mode I + Mode II + Mode III, the stress at any point in the region of validity of Eq. (7.114) can be obtained by simply adding the respective individual contributions

$$
\sigma_{ij}^{(total)}=\sigma_{ij}^{(I)}+\sigma_{ij}^{(II)}+\sigma_{ij}^{(III)} \tag{7.136}
$$

it is not the same with the stress intensity factors, K_I, K_{II} and K_{III} that cannot be added.

7.15 Cracks and Notches

We have already discussed the difference between a sharp crack and a notch and shown how the apparent toughness of a material depends on the notch tip radius, Figs. 6.26 and 6.27. Equations (7.114) can further clarify this difference since they predict the equality between σ_y and σ_x ahead of a crack. This is a new surprising situation that results from the boundary conditions posed to the Airy stress function by saying that the stress σ_y must be zero for $y = 0$ and $-a < x < +a$ (free surface) and equal to the remote stress for x tending to infinity, see Sect. 7.10. This condition, $\sigma_y = \sigma_x$, results in the maximum biaxiality possible. Dealing with a notch, the stress field situation is much different in that the stress σ_x must necessarily start from zero because a notch has a free surface, therefore the notch tip is under a plain stress condition. Paris et al. [20] have presented a stress field solution for notches given by

$$
\begin{aligned}
\sigma_x &= \frac{K_I}{\sqrt{2\pi r'}} \cos\frac{\theta'}{2}\left(1 - \sin\frac{\theta'}{2}\sin\frac{3\theta'}{2}\right) - \frac{K_I}{\sqrt{2\pi r'}}\frac{R}{2r'} cos\frac{3\theta'}{2} \\
\sigma_y &= \frac{K_I}{\sqrt{2\pi r'}} \cos\frac{\theta'}{2}\left(1 + \sin\frac{\theta'}{2}\sin\frac{3\theta'}{2}\right) + \frac{K_I}{\sqrt{2\pi r'}}\frac{R}{2r'} cos\frac{3\theta'}{2} \\
\tau_{xy} &= \frac{K_I}{\sqrt{2\pi r'}}\left(\sin\frac{\theta'}{2}\cos\frac{\theta'}{2} cos\frac{3\theta'}{2}\right) - \frac{K_I}{\sqrt{2\pi r'}}\frac{R}{2r'} sin\frac{3\theta'}{2}
\end{aligned}
\tag{7.137}
$$

where K_I is given by

$$
K_I = \frac{\sqrt{\pi R}}{2}\sigma_{max} \tag{7.138}
$$

with σ_{max} being the maximum stress at the notch base (stress concentration). Symbols are shown in Fig. 7.21. Equations (7.121) clearly show that the two components of stress, σ_y and σ_x, are not equal for $r' = R/2$, at the notch tip. The σ_x component starts from zero, as expected for a blunt notch, reaches a maximum and then drops to zero. It is worth noting that Eq. (7.121) still have a single parameter, K_I, to represent the notch tip stress field. However, this stress field depends on the notch radius R and so does its critical value, K_{IQ}, or apparent toughness, as said in Sect. 6.5. Therefore, the use of a fracture mechanics approach in case of a notch of a given tip radius requires the knowledge of the apparent toughness of the material, K_{IQ}, corresponding to that radius.

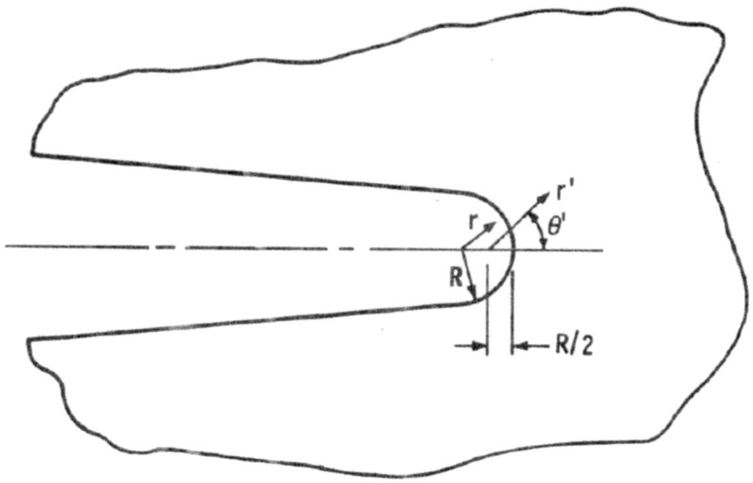

Fig. 7.21 Blunt notch. Symbols used in Eq. (7.137)

7.16 The Energy Release Rate Related to Elastic Compliance

We have introduced, in Sect. 7.4, the energy release rate G. In 1956 G. Irwin [21] provided a new expression using an energy approach to fracture that was basically equivalent to the Griffith concept but with some more convenient application in fracture mechanics, see Chap. 11. His energy release rate G defines the strain energy and boundary force work, $U_o - U_D$, Eq. (7.11), made available for an increment of crack extension, *da*, that creates a new free surface *dA*

$$G = -\frac{dU}{dA} \tag{7.139}$$

The Irwin approach can be examined for its wide applications using the concept of elastic compliance, *C*, of a body (inverse of spring constant or plate stiffness)

$$C = \Delta/P \tag{7.140}$$

were Δ is the displacement caused by the applied load *P* in its own direction, Fig. 7.22. Since G is the derivative of a potential it was also called *crack driving force*, though dimensionally G is not a force. The potential energy *U* is defined as

$$U = E - L \tag{7.141}$$

were *E* is the elastic energy stored in the body (the area under the stress–strain curve) and *L* the work done by the external force, *P*. This work is

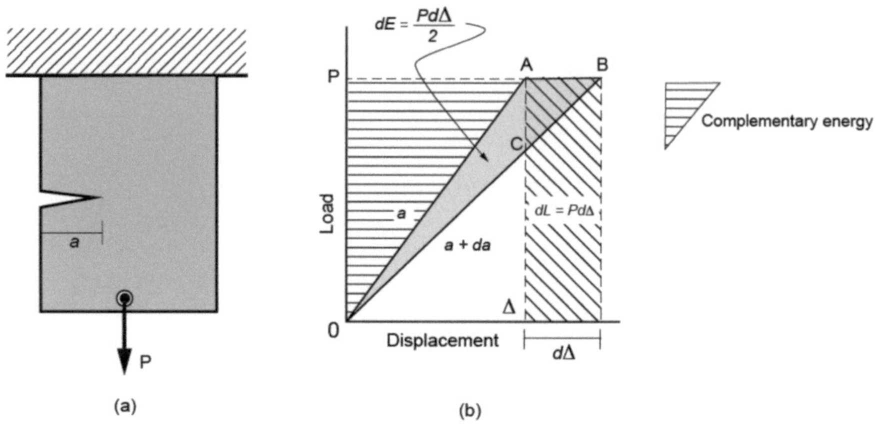

Fig. 7.22 Cracked panel under a fixed load P condition

$$L = P\Delta \tag{7.142}$$

The stored energy, *E*, is always positive and contributes to the energy release, –*dE*, during a crack extension, *da*

$$E = \int_0^{\Delta} P\mathrm{d}\Delta = \frac{P\Delta}{2} \tag{7.143}$$

Therefore, Eq. (7.141) yields

$$U = -\frac{1}{2} \cdot P\Delta = -\frac{1}{2}P^2C \tag{7.144}$$

being $\Delta = CP$, Eq. (7.140). Note that this energy is the complimentary energy of the system, as shown in Figs. 7.22 and 7.23 by the horizontal dashed area OPA. Using the compliance *C*, it is

$$G = \frac{P^2}{2} \cdot \frac{\partial C}{\partial A} \tag{7.145}$$

It should be noted that Eq. (7.145) does not contain any term involving *dP*. This implies that the method of load application, such as constant force, fixed displacement or springiness of the testing machine, does not affect the available energy rate, G, for an infinitesimal crack extension, *da*. This independence of G from loading conditions, whether fixed load *P* or fixed displacement *Δ*, can be seen comparing Figs. 7.22 with 7.23.

Under a fixed load condition, Fig. 7.22a, the energy released, *dE*, during a virtual crack extension, *da*, is the area of the triangle 0AB (shaded area in Fig. 7.22b) which is equal to half the work of the external force, *dL* = *PdA* and exactly equal to

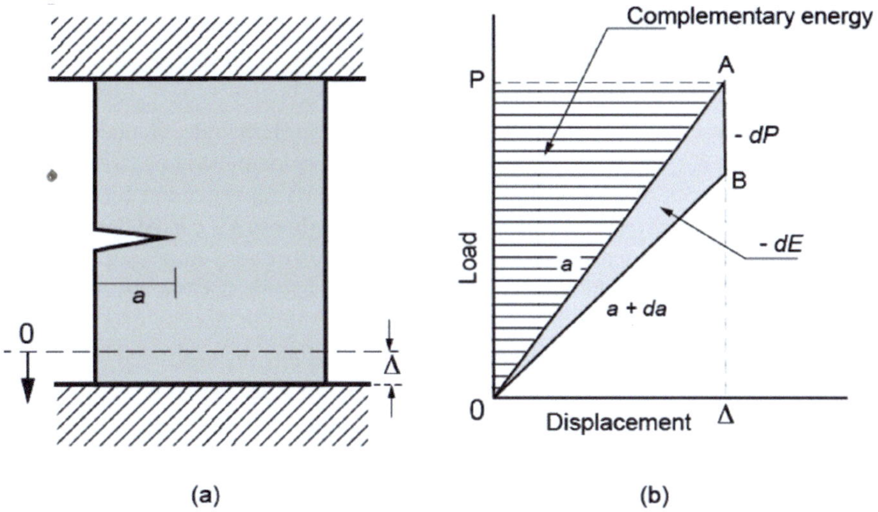

Fig. 7.23 Cracked panel under fixed displacement Δ condition

the difference of the complementary energies of two panels having slightly different cracks length equal to a and $a+da$, respectively, dashed area 0PB minus the area 0PA of Fig. 7.23b). Under a fixed end condition, Fig. 7.23a, the same energy release rate is $-dE$, area 0AB of Fig. 7.23b, again equal to the difference of the complimentary energies of two panels having slightly different cracks length equal to a and $a + da$, respectively, dashed area 0PA minus the area 0PAB0 of Fig. 7.23b, which is equal to the former, besides the sign, since the triangle ABC of Fig. 7.22b is a differential of the second order, $dP \cdot d\Delta$, and can be neglected with respect to dE. Then, the two energies are equal but have different signs since in one case U is an energy taken from a work done by the external load while in the second case it represents an energy released from the deformation energy stored in the system. While in the first case the system is similar to a tank in which the water (elastic deformation energy) is fed at a rate double that at which it is released, in the second case the water is continuously released without being refilled so that eventually the tank completely empties, i.e., the crack will necessarily stop growing. Equation (7.145) may be used as the basis of experimentally determining G and, therefore, the toughness K_{Ic} of a material, see Eq. (7.151). As it will be shown in Sect. 9.4, this technique involves measuring the compliance, C, of the specimen used, i.e., the slope of the elastic load–displacement curve, for various crack length and computing its derivative,dC/dA.

7.17 Relationship Between Driving Force G and Stress Intensity Factor K_I

In earlier sections, the equivalence between the energy release rate, G, and the stress intensity factor, K_I, was noted. This similarity is not accidental and bears very interesting consequences on practical grounds since G in its most general form is not a linear elastic parameter as K_I. This means that the toughness K_{Ic} can be derived from an elastic–plastic parameter that is not subjected to restrictions imposing triaxiality and, therefore, adequate thickness. The relationship can be derived by noting that G and K_I are both measures of the elastic field effect in the vicinity of a crack tip, tending to cause crack extension. Attention is, therefore, focused on the crack tip region and the tendency for the body to supply elastic energy to the crack extension process.

Since both G and K_I are not affected by method of application of external forces, see previous section, we may now regard the external load as applied by fixed grips and compute the work required to close an increment, Δa, of the crack adjacent to the tip, as schematized in Fig. 7.24. Taking a unit thickness of the body, the work per unit area in closing the crack increment Δa is

$$\frac{\partial U_c}{\partial a} = G = \lim_{\Delta a \to 0} \frac{2}{\Delta a} \int \frac{\sigma_y v}{2} dx \tag{7.146}$$

where, as shown in Fig. 7.22, v is the displacement in the y direction of each individual point of the crack surface during the opening and growth Δa, and σ_y is the surface stress which must be supplied to the increment in order to close it. This stress is precisely that at the opening of the crack given by the second of Eq. (7.114) for $\theta = 0$

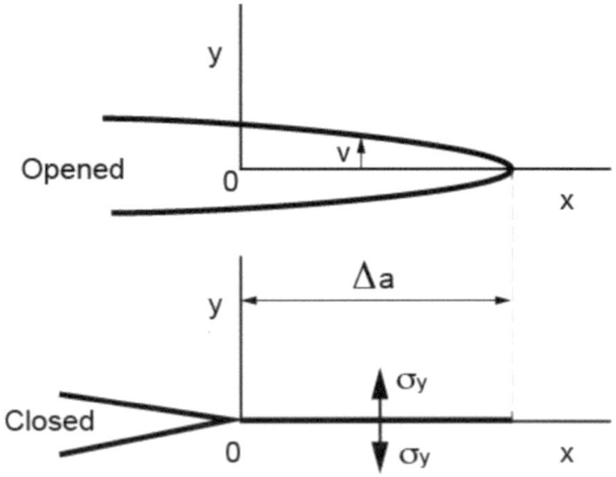

Fig. 7.24 Coordinates for the work of crack closure

$$\sigma_y = \frac{K_I}{(2\pi r)^{1/2}} \tag{7.147}$$

Under plain strain conditions, the displacements v of the crack surface were found by Paris and Sih [22]

$$v = \frac{2K_I}{E}\sqrt{\frac{2(\Delta a - x)}{\pi}}\left(1 - v^2\right)(plain\,strain) \tag{7.148}$$

v being Poisson coefficient. Substituting Eq. (7.148) in Eq. (7.146) and integrating we get

$$G = \frac{K_I^2}{E}\left(1 - v^2\right)(plain\,stress) \tag{7.149}$$

Again, the term $\left(1 - v^2\right)$ disappears under plain stress conditions

$$v = \tfrac{2K_I}{E}\sqrt{\tfrac{2(\Delta a - x)}{\pi}} \tag{7.150}$$

which leads to

$$G = \frac{K_I^2}{E}(plain\,stress) \tag{7.151}$$

7.18 General Solution with Conformal Mapping Functions

The expression of the stress field ahead of a crack found by Westergaard and Irwin, Eq. (7.114), represents a very special solution of the elastic problem of a solid affected by a crack. This is not only because, as has been said, Eq. (7.114) were obtained by truncating the Mac-Laurin series expansion of the binomial $(1 + r/2a)^{-1/2}$ to the first constant term equal to one, but also and above all because the reference geometry was that of an infinite plate loaded biaxially. Irwin himself [18] prudently pointed out that if the plate had a finite width, then an edge effect would be felt which, because of its distance from the end of the crack, would alter the stress σ_x by a quantity $-\sigma_{xo}$ difficult to evaluate theoretically

$$\sigma_x = \frac{K_I}{\sqrt{2\pi r}} - \sigma_{xo} \tag{7.152}$$

It can be thought that the particular monoaxial case in which the remote external component σ_{xo} is missing is obtainable from the biaxial one found by Irwin simply by exploiting the principle of superposition of the effects, always valid in the linear field, applying a remote stress $-\sigma_{xo}$ equal, less than the sign, to the external one of the biaxial case, with which the solution (7.144) will be found. Indeed, Bonora et al. [23, 24] arrived at this solution using the method of conformal transformations or conformal mapping. Conformal transformations allow to transfer the elasticity problem from the geometric plane *x,y* into an auxiliary domain, called precisely conformal, where it is easier to solve the system of associated differential equations. This new conformal domain belongs to a complex field ζ in which all the points of the geometric domain *x,y* that make up the geometry under consideration in the specific problem are transported. Muskhelishvili [16] studied the problem of the stress field intensification in front of a hole by determining the transport function, also called the mapping function, with which the profile of the hole is transformed into a unitary circle of the complex domain ζ and the material plane ends within this circle. The mapping function that provides this transport inside the unit circle of each point of the domain outside an elliptical hole, whose semi-major axis lies on the x-axis, is given by:

$$z = R \cdot \left(\frac{1}{\zeta} + m\zeta \right) \tag{7.153}$$

having assumed

$$R = \frac{a+b}{2}, m = \frac{a-b}{a+b} \tag{7.154}$$

where a and b are the semi-major and minor axes, respectively. The case of an sharp crack of length $2a$ derives directly from that of the elliptic hole simply by placing $b = 0$ with which m becomes equal to one and R to $a/2$. With this, solving Eq. (7.153) with respect to ζ yields

$$\zeta = \frac{z - \sqrt{z^2 - a^2}}{a} \tag{7.155}$$

It has been already said, that every linear combination of holomorphic functions is an Airy function satisfying the biharmonic equation and, in particular, the most general function, the Goursat function, which was used by Muskhelishvili

$$\Phi(z) = \text{Re}[\bar{z}\varphi(z) + \chi(z)] \tag{7.156}$$

where $\bar{z} = x - iy$ is the conjugate of z, $\psi = \partial\chi/\partial z$ and *Re*, according to the other usual convention of complex numbers, represents the real part of the function in parentheses. Recalling Eq. (7.72), neglecting, Ω, and Eq. (7.89) we get

$$\sigma_x + \sigma_y = 2 \cdot \left[\overline{\phi'}(z) + \phi'(z) \right] \tag{7.157}$$

from which the expressions for the stresses σ_x, σ_y and τ_{xy} are derived

$$\begin{aligned}
\sigma_x &= \mathrm{Re}\left[2\cdot\phi'(z) - \bar{z}\cdot\phi''(z) - \psi'(z)\right] \\
\sigma_y &= \mathrm{Re}\left[2\cdot\phi'(z) + \bar{z}\cdot\phi''(z) + \psi'(z)\right] \\
\tau_{xy} &= \mathrm{Im}\left[\bar{z}\cdot\phi''(z) + \psi'(z)\right]
\end{aligned} \tag{7.158}$$

Moving on to the complex functions $\phi(\zeta)$ and $\psi(\zeta)$ for the problem under consideration we obtain the expressions [25]

$$\begin{aligned}
\phi(\zeta) &= \frac{\sigma_{yo}R}{4}\cdot\left[\zeta + \frac{\left(2e^{2i\alpha} - m\right)}{\zeta}\right] \\
\psi(\zeta) &= -\frac{\sigma_{yo}R}{2}\cdot\left[\frac{\zeta}{e^{2i\alpha}} + \frac{e^{2i\alpha}}{m\zeta} - \frac{\zeta\left(e^{2i\alpha} - m\right)\left(1+m^2\right)}{m\left(\zeta^2 - m\right)}\right]
\end{aligned} \tag{7.159}$$

where σ_{yo} represents the remote stress applied in the y-direction inclined at a generic angle α. Recalling Eqs. (7.155) and (7.158) we obtain that for $\alpha = \pi/2$, i.e. for an external stress σ_{yo} acting on the y-axis

$$\begin{aligned}
\sigma_x &= \mathrm{Re}\left[\frac{z}{\sqrt{z^2 - a^2}} - \frac{iya^2}{\left(\sqrt{z^2-a^2}\right)^3} - 1\right]\cdot\sigma_{yo} \\
\sigma_y &= \mathrm{Re}\left[\frac{z}{\sqrt{z^2 - a^2}} + \frac{iya^2}{\left(\sqrt{z^2-a^2}\right)^3}\right]\cdot\sigma_{yo} \\
\tau_{xy} &= \mathrm{Im}\left[1 + \frac{2iya^2}{\left(\sqrt{z^2-a^2}\right)^3}\right]\cdot\frac{\sigma_{yo}}{2}
\end{aligned} \tag{7.160}$$

Changing the reference system introduced by Irwin, Eqs. (7.106) and (7.107), and (7.160) for $\theta = 0$ finally become

$$\begin{aligned}
\sigma_x &= \left[\frac{r+a}{\sqrt{r^2 + 2ra}} - 1\right]\cdot\sigma_{yo} \\
\sigma_y &= \frac{r+a}{\sqrt{r^2+2ra}}\cdot\sigma_{yo} \\
\tau_{xy} &= 0
\end{aligned} \tag{7.161}$$

Note that the second of Eq. (7.161) is identical to Eq. (7.104) obtained by Westergaard before series expansion. The first equation, on the other hand, contains a term $-\sigma_{yo}$ that represents the missing remote stress. Introducing the condition $r/a << 1$ we get Eq. (7.114) already found by Irwin less than the term σ_{yo} on the σx which becomes equal to Eq. (7.152). The term $-\sigma_{yo} = -\sigma_{xo}$ clearly cannot alter the value of the field at the crack tip where stresses intensify to become even infinite in ideally elastic materials, so that one could think that it should have no effect on the instability of the crack itself which depends solely on what happens immediately below the notch.

7.19 The Elastic *T*-stress

Equations (7.114) describing the crack tip elastic stress field have been obtained by truncating to the first constant term the series expansion of the function

$$\left(1+\frac{r}{2a}\right)^{-1/2} \tag{7.162}$$

This, it has been said, confines the field of validity of Eq. (7.114) to a very narrow area ahead of the crack tip in which $r/a << 1$. Moreover, the geometry considered was that of an infinite plate with a central crack, remotely loaded in both x and y directions by a remote stress σ_o. This makes the use of those equations even more restricted and, consequently, also restricts the concept of LEFM as a single parameter theory: the K_I. Also, the general solution of the elastic problem, obtained through the use of the mapping functions, always applied to an infinite plane geometry, but loaded in uniaxial traction along the y direction, Eq. (7.161), clearly shows that the problem of cracked bodies is not definable with a single parameter. The case of a finite geometry introduces further variations due to border effects that will be felt on the stress σ_x through the insurgence of a stress σ_{xo}, Eq. (7.160), difficult to quantify theoretically, which will be different from the σ_o missing in the x direction. This had already been pointed out in 1957 by G. Irwin [18] who even went so far as to conclude that the LEFM could not be, in these cases of finite geometry, a theory with a single parameter. Of course, the presence of a singularity of the type $1/\sqrt{r}$ that brings the stress field to asymptotic unlimited values may, at first sight, induce to believe that any other additive term, $\pm\sigma_{xo}$, undetermined but of finite value anyway, be less and less important getting closer and closer to the crack tip. However, it must be kept in mind that the case of limitless singularity exists only in an ideally elastic domain, but not in real elastic–plastic materials where plastic strains develop at the crack tip creating a plastic zone that cuts the peak stress to the yield stress, Fig. 8.1, or at most to a somewhat higher value, in case of strain hardening materials, but still finite. However, despite all these limitations, LEFM can still be considered a valid single parameter theory under SGC or SSY conditions, where the crack tip plastic zone may be neglected and the stress field expressed by K_I, if Eq. (7.114) are rewritten in expanded form of infinite series

$$\sigma_{ij} = \frac{K_I}{\sqrt{2\pi r}} f_{ij}(\theta) + T\delta_{1j}\delta_{1i} + \ldots \tag{7.163}$$

δ_{ij} being the Kronecker delta. Equation (7.163) is normally known as the Irwin-Williams series [26, 27]. The stress T, or T-stress, is a nonsingular parameter that represents a uniform stress field, independent of both the distance r from the crack tip and the anomaly θ and is parallel to the plane of the crack itself. The terms of the series following the T-stress that contain the powers of r of order higher than $1/2$ can also be conveniently grouped into a single parameter $T_{ij}(r)$ which is nonzero near the outer surface of the body containing the crack and tends to zero when r tends to zero. In the initial formulation by Larsson and Carlsson [28] followed by that of Betegón and Hancock [29] and that of Al-Ani and Hancock [30], the series (7.163) is truncated to the second term, the T-stress, which operates only on the stress σ_x making it vary from σ_y. Although the Irwin Williams series (7.163) is strictly valid in the linear elastic field, we can say that in a finite geometry and in SGC conditions it is only the terms K_I and T that control the state of stress near the crack tip. At higher distances it is only the T-stress that dominates and therefore it must be $|T| << \sigma_s$ if we want to constrain plasticity without letting it spread. If we consider that, according to Eq. (7.163), far from the crack the stress σ_x can take over the σ_y so as to neglect the latter, in a state of complete constriction by thickness, $\sigma_z = 2\nu\sigma_x$, Eq. (4.19), we have that, adopting the von Mises criterion, the stress σ_x can grow up to $1.125{\cdot}\sigma_s$ before activating plastic slips and $|T|$ has to remain lower than $1.125{\cdot}\sigma_s$ [31]. If σ_y is equal to $\sim$ $0.5\sigma_s$ then the T-stress can rise further as long as it remains less than $1.8{\cdot}\sigma_s$, see Eq. (4.29). In the whole family of values that T can assume for the various geometric configurations, the solution $T = 0$ clearly refers to the SGC or SSY condition contained in an infinite geometry with a central crack. The further T moves away from 0 in the negative field, the further the structure or specimen moves away from the SSY conditions. The T-stress, therefore, must represent a measure of the degree of triaxiality existing at the crack tip on which the confinement of the plastic zone depends. In fact, the higher T, the more σ_x deviates from σ_y. If T assumes large negative values, we will lose the equality between σ_x and σ_y so decisive for the onset of a complete triaxiality and for this reason we will get closer and closer to the conditions of low triaxiality that exist, for example, in front of shallow cracks (see Sect. 9.13). Since, as already mentioned in Sect. 8.5, the shape and extent of the crack tip plastic zone depends on the ratio between σ_x and σ_y, i.e., on the degree of biaxiality, we should expect in equal measure an influence of the parameter T on this zone. Since T varies linearly with the applied external stress, it must be proportional to $K_I/\sqrt{\pi a}$. For what has been said, the degree of biaxiality in the x, y plane will then be related to the dimensionless parameter β

$$\beta = \frac{T\sqrt{\pi a}}{K_I} \tag{7.164}$$

High positive values of β define states of high plastic constriction and, therefore, of high triaxiality that occur in large thicknesses or, in any case, at the center of a specimen or a structure where high triaxiality conditions always exist if the initial degree of biaxiality is high. This can also be the case of a finite geometry, but loaded remotely in a biaxial manner, especially with $\sigma_x > \sigma_y$. A$\beta = 0$ will mean $T = 0$

and, therefore, will also indicate a situation of SSY in which K_I is the only significant parameter defining the crack tip stress field. Finally, $\beta < 0$ will be indicative of low biaxiality with consequent loss of triaxiality. Larsson and Carlsson [28], through FE calculations, were the first to find that, indeed, the parameter β was related to the shape and size of the crack tip plastic zone. They examined four different geometries typical of fracture mechanics, shown in Fig. 7.25, used as specimens for the $K_{Ic,}$ measurement: the panel with a central crack CC(T) and the panel with double lateral crack DEC(T) both in traction (see Sect. 9.2), the single edge crack specimen in three point bending or SE(B) (see Sect. 9.7), also known as TP(B), *three point bending*, and the compact specimen C(T) (see Sect. 9.3). The β values found for the four geometries with the crack length to specimen depth ratio $a/w = 0.5$ are shown in Table 7.2. Sham [32] and Leevers and Radon [33] have extended the study to a wider field of *a/w* ratios, ranging from 0.05 to 0.9, and with two new geometries, the panel with a single edge crack SEC(T) in traction and the four point bending bar SEC(B). Table 7.3 summarizes the results for the five geometries CCP(T), DEC(T), SEC(T), SEC(B) and TP(B). The result was that those geometries loaded in traction (T) had negative values of the T-stress while the others loaded in bending (B) gave positive results, with the exception of shallow cracks $a/w \leq 0.3$ and the three points bending that are very close to the condition $T = 0$ of an SSY situation. It can therefore be said that, in general, specimens or structures with the ligament ahead of the crack in bending have higher degrees of triaxiality. As a further evidence, Fig. 7.26 presents a comparison between the triaxiality factors FT, see Sect. 4.3, obtained by Milella and Bonora for a C(T) specimen, Fig. 7.2d, and a tensile panel with single edge crack SEC(*T*), half panel of Fig. 7.25b, this last loaded in uniaxial and biaxial traction. The graph shows that the *TF* drops at the boundary of the plastic zone, reaching its maximum well within the high-deformation plastic with the C(T) curve higher than that of the SEC(T) panel in uniaxial traction. The *TF* of SEC(T) gains triaxiality when loaded biaxially, reaching that of C(T) for $\sigma_x = \sigma_y$. This confirms what has been said about a geometry subjected to biaxial traction that goes from a negative T-stress to $T = 0$, SSY condition, that even becomes positive if the applied stress σ_x exceeds σ_y. What, however, may seem surprising is the very high triaxiality that is achieved with the C(T) specimen, which, as shown in Sect. 9.3 and Fig. 9.8, is, in fact, a biaxially loaded specimen. The T-stress has, obviously, an effect on the crack tip plastic zone that shows differences in shape and extension.

Those found by Larsson and Carlsson [28] for the four geometries of Fig. 7.25 and Table 7.2 are shown in Fig. 7.27, in coordinates x, y normalized to $(K_I/\sigma_s)^2$, together with that competing to SSY conditions, for a value of $K_I - 0, 6\sigma_s\sqrt{a}$ very close to, but below, the ASTM limit for SSY condition. The enormous variation of the plastic zone with the degree of triaxiality can be observed for the various specimens linked to the T-stress. The two geometries with T < 0 have plastic zones greater than that which competes to the condition of SSY, $T = 0$, while those with $T > 0$ are smaller and located inside the first. Since the coordinates x and y have been indexed, when the plastic zone depends only on K_I its shape and size must not change whatever the geometry, the load and, therefore, the K_I applied, but it is evident that in our case the conditions of SSY do not exist. The three-point bending bar specimen

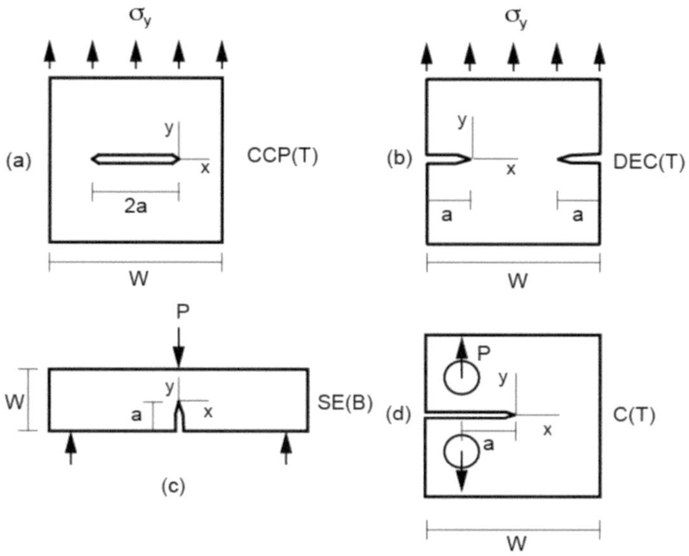

Fig. 7.25 Geometries considered by Larsson and Carlsson in the study of T-stress and β parameter

Table 7.2 *Values of the parameter* $\beta = T\sqrt{\pi a}/K_I$ *for various geometries and* $a/w = 0.5$

Specimen	$T\sqrt{\pi a}/K_I$
CC(T) panel in traction	−1.040
DEC(T) panel in traction	−0.255
TP(B) Three point bending	0.058
C(T) Compact specimen	0.516

Table 7.3 Values of the parameter $\beta = T\sqrt{\pi a}/K_I$ for various geometries and a/w ratios

a/w	CCP(T)	DEC(T)	SEC(T)	SE(B)	SEC(B)
0.05	−1.000	−0.469		−0.430	
0.10	−1.000	−0.469	−0.464	−0.361	−0.363
0.20	−1.004	−0.473	−0.434	−0.233	−0.228
0.30	−1.012	−0.479	−0.371	−0.090	−0.073
0.40	−1.029	−0.484	−0.277	0.061	0.092
0.50	−1.059	−0.480	−0.153	0.217	0.262
0.60	−1.113	−0.463	0.007	0.379	0.433
0.70	−1.214	−0.432	0.210	0.553	0.610
0.80	−1.409	−0.376	0.501	0.786	0.839
0.90	−1.894	−0.283	1.030	1.227	1.267

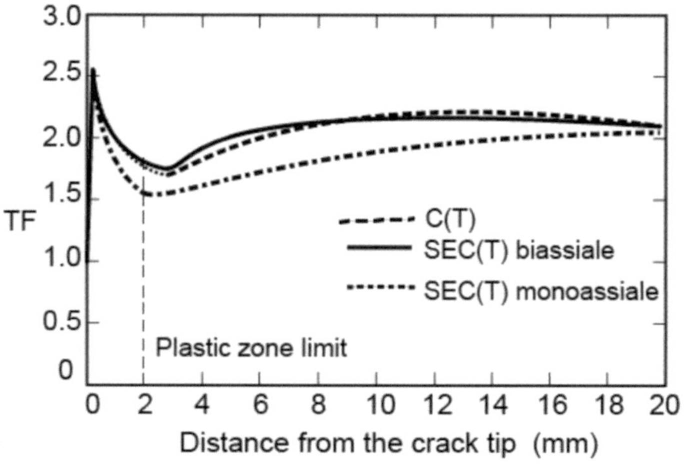

Fig. 7.26 Triaxiality factor, TF, ahead of a crack in a C(T) and in SEC(T) specimen, this last loaded in simple traction and biaxial traction

Fig. 7.27 Plastic zone sizes in the four geometries considered in Table 7.2 plus the SSY condition, indexed to the nondimensional distances $y/(K_I/\sigma_s)^2$ and $x/(K_I/\sigma_s)^2$ (taken from [28])

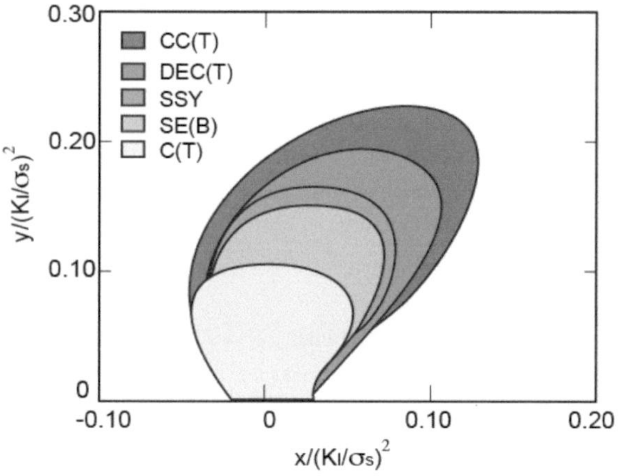

SE(B) is among all the geometries considered the one that comes closest to the SSY condition while the compact C(T), has the greatest triaxiality. Note how its axis is oriented precisely at 90°, as provided for the infinite panel subjected to biaxial remote traction in conditions of flat deformation and shown in Fig. 8.8.

References

1. Orowan, E.: Fracture and strength of solids. Report on Progress of Physics, Phys. Soc., London **12**, 48 (1949)
2. Cottrell, A.H.: Tewksbury Symposium on Fracture, 1. University of Melbourne (1963)
3. Knott, J.F.: Fundamentals of fracture mechanics. Butterworth 94–98 (1973)
4. Kelly, A.: Proc. Roy. Soc.,Ser. A **282**, 1388 (1964)
5. Griffith, A.A.: The Phenomena of Rupture and Flow in Solids, Transactions, Royal Society of London, vol. 221 (1920)
6. Inglis, C.E.: Stress in a plate due to the presence of cracks and sharp corners. Trans. Inst. Nav. Arch. **55**, 219–230 (1913)
7. Shand, E.B.: J. Am. Ceram. Soc. **44**, 21 (1961)
8. De Saint Venant, B.: Mémoires des Savants Etrangers **14** (1855)
9. Irwin, G.R.: Fracture Dynamics, pp. 147–166. Fracturing of Metals, American Society Metals, Cleveland (1948)
10. Orowan, E.: Fracture and Strength of Solids, Report on Progres in Physics, Phys. Soc., London, vol. 12, p. 185 (1949)
11. Orowan, E.: Energy criteria for fracture. Repr. Weld. J. Res. Suppl. 1–4, March (1955)
12. Irwin, G.R., Kies, J.A.: Fracture and fracture dynamic. Weld. Res. Suppl. **31**(2), 95–100 (1952)
13. Frenkel, Y.I.: Theory of reversible and non-reversible cracks in solids. NACA, Technical Memorandum 1387 (1955)
14. Airy, G.B.: British Assoc. Advance, Scientific Report (1862)
15. Goursat, É: Cours d'Analyse Mathématique, 3-me éd., Paris (1927)
16. Muskhelishvili, N.I.: Some Basic Problems in the Theory of Elasticity, Nordhoff P. Ltd, Groningen, Netherlands (1953)
17. Westergaard, H.M.: Bearing pressure and cracks. J. Appl. Mech. **61**, A49–A53 (1939)
18. Irwin, G.R.: Analysis of stresses and strains near the end of a crack traversing a plate. J. Appl. Mech. **24**, 361–364 (1957)
19. Inglis, C.E.: Stress in a plate due to the presence of cracks and sharp corners. Trans. Inst. Nav. Arch. **55**, 219–230 (1913)
20. Creager, M., Paris, P.C.: Elastic field equations for blunt cracks with reference stress corrosion cracking. Inter. J. Fract. Mech. **3**, 247–252 (1967)
21. Irwin, G.R.: Onset of fast crack propagation in high strength steels and aluminum alloys. Proc. Sagamore Res. Conf. **2**, 289–305 (1956)
22. Paris, P.C., Sih, G.C.M.: Stress Analysis of Cracks, ASTM Special Technical Publication No. 381, pp. 30–83 (1965)
23. Bonora, N., Costanzi, M. Marchetti, M.: Stress and Stain Field Near and Far-off a Crack Tip in Isotropic, Orthotropic and Composite Materials, American Society for Composites, ASC-6, Cleveland, Ohio, 1993
24. Bonora, N., Costanzi, M. Marchetti, M.: On closed form solution for the elastic stress field around holes in orthotropic composite plates under in-plane stress conditions. Publ. Int. J. Compos. Struct. (1994)
25. Barenblat, G.I.: The mathematical theory of equilibrium of cracks in brittle fracture. Adv. Appl. Mech. **7**, 55 (1962)
26. Timoshenko, S.P., Goodier, J.N.: Theory of Elasticity, McGraw-Hill Book C., Third Ed., p. 165 (1970)
27. Williams, M.L.: On the stress distribution at the base of a stationary crack. J. Appl. Mech. **24**, 109–114 (1957)
28. Larsson, S.G., Carlsson, A.J.: Influence of non-singular stress terms and specimen geometry on small-scale yielding at crack tips in elastic-plastic materials. J. Mech. Phys. Solids **21**, 263–277 (1973)
29. Begetón, C., Hancock, J.W.: Two parameter characterization of elastic-plastic crack-tip fields. J. Appl. Mech. **58**, 104–110 (1991)

30. Al-Ani, A.M., Hancock, J.W.: J-dominance in short cracks in tension and bending. J. Mech. Phys. Solids **39**, 23–43 (1991)
31. Parks, D.M.: Advances in Characterization of Elastic-Plastic Crack-Tip Fields, pp. 59–98. Topics in Fracture and Fatigue, Springer Verlag (1992)
32. Sham, T.L.: The determination of the elastic T-Term using higher order weight functions. Int. J. Fract. **48**, 81–102 (1991)
33. Leevers, P.S., Radon, J.C.: Inherent stress Biaxiality in various fracture specimen geometries. Int. J. Fract. **19**, 311–325 (1982)

Chapter 8
Crack Tip Plastic Zone

8.1 Introduction

Equations (7.114) have been derived using the theory of elasticity that applies to ideally perfect elastic materials and does not recognize the occurrence of plastic deformations. Real materials are far enough from behaving ideally perfectly elastic. However, Linear Elastic Fracture Mechanics (LEFM) is still applicable to real materials that yield if the amount of local plasticity that precede fracture is negligible so that the crack does not blunt and does not turn into a notch, remaining a sharp crack. This very particular situation has been called Small-Scale Yielding (see Sect. 4.1) or SSY. In these conditions linear elastic stress field concept and equations may still be applied and the crack tip stress field can be fully described in terms of the Irwin stress intensity factor, K, a unique single valued parameter which combines the variables of stress, σ, defect size, a, and geometry of the system, $f(a/W)$, Eq. (7.120). Since plastic deformations occur at the crack tip, it is convenient to focus the attention to the size and shape of the crack tip plastic zone. The size and shape of this zone can be estimated by two fundamental methods: the Irwin approach [1] that uses the elastic stress field equations to assess the elastic–plastic boundary and the strip-yield model developed independently by Dugdale [2] and Burdekin [3]. With the Irwin approach, Eq. (7.114) can be used to assess the precise shape of the plastic zone ahead of the crack tip using the Tresca or the von Mises yield criterion under plain strain and plain stress conditions. The following paragraphs will discuss in detail the crack tip plastic zone issue.

P. P. Milella, *Fundamentals of Fracture Mechanics*,
https://doi.org/10.1007/978-3-032-23081-2_8

8.2 The Irwin Hypothesis

The $1/\sqrt{r}$ singularity of the crack tip stress field leads, in real materials, to a local plastic yielding that will redistribute stress and strains around the crack tip and cut to a finite value the infinite peak stress that the elastic theory predicts. It is, therefore, appropriate to ask what shape and size will have the area that yields plastically. Irwin accounted for crack tip plasticity by defining a plastic zone of circular shape. An estimation of the extension, r_{p}, of this plastic zone can be simply obtained by using the second of Eq. (7.114) putting $\theta = 0$ and σ_y equal to the yield stress σ_s and solving for $r = r_{\mathrm{p}}$

$$r_P = \frac{1}{2\pi}\left(\frac{K_I}{\sigma_s}\right)^2 \tag{8.1}$$

This is shown in Fig. 8.1 relative to a body containing a crack subjected to a state of remote traction, σ. It is easy to realize, however, that the distance r_{p} given by Eq. (8.1) is only a first-order estimate of the real plastic zone size. In fact, by truncating the stress field σ_y to the yield strength, σ_s, we lose the area A, dashed in Fig. 8.1, that represents an internal reaction force, F_{in}, that in an ideally elastic material contributes to balance the external applied force F. To restore the balance between internal and external forces this amount of missed reaction, F_{in}, has to be recovered by the system

$$F_{in} = F\sigma_s \cdot r_p \tag{8.2}$$

where F is the force per unit thickness corresponding to the entire area under the curve $K_{\mathrm{I}}/\sqrt{2\pi r}$ between $r = 0$ and $r = r_p$

$$\begin{aligned} F &= \int_0^{r_p} \sigma_y dr = \int_0^{r_p} \frac{K_I}{\sqrt{2\pi r}} dr \\ &= \sqrt{\frac{2}{\pi}} K_I \cdot \left(r_p\right)^{\frac{1}{2}} \\ &= 2\sigma_s r_p \end{aligned} \tag{8.3}$$

Equation (8.2) then yields

$$F_{in} = \sigma_s \cdot r_p \tag{8.4}$$

which is twice the area of the base rectangle, A', of Fig. 8.1. Therefore, to recover the missing force it will be just enough to translate the elastic stress field by an amount r_p gaining the area B of Fig. 8.1 that has just been seen to be equal to the missing area A, as shown in Fig. 8.1. This would increase the size of the plastic zone, ρ, whose

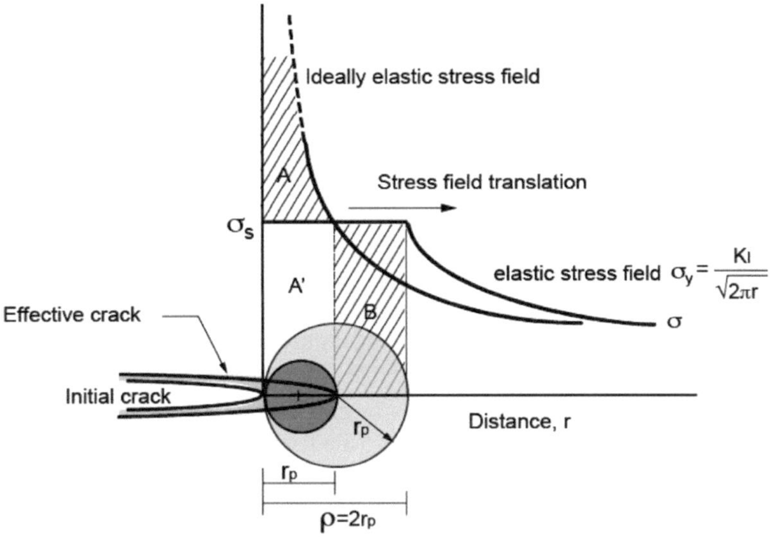

Fig. 8.1 First and second order estimate of the plastic zone according to Irwin. The missing area A is recoveredby translating the elastic stress field by an amount equal to r_p, A = B

radius, $\rho/2$, will precisely be equal to r_p

$$r_p = \frac{\rho}{2} = \frac{1}{2\pi}\left(\frac{K_I}{\sigma_s}\right)^2 \tag{8.5}$$

i.e., its extension ρ will be twice the first-order estimate r_{p} given by Eq. (8.1).

$$\rho = \frac{1}{\pi}\left(\frac{K_I}{\sigma_s}\right)^2 \tag{8.6}$$

It must be borne in mind, however, that the value of the stress at which a material starts to yield depends on the stress state applied. Under conditions of complete triaxiality or plane strain, such as those characterizing brittle fracture, the yield strength can rise up to 2.5 or 3.0 times the monoaxial yield strength σ_s, depending on whether the Poisson coefficient v is equal to 0.3 or 0.33 (see Sect. 4.3). By using the factor of 3 to elevate the plain strain yield stress, Eq. (8.5) will be written as

$$\begin{aligned} r_p &= \frac{1}{2\pi}\left(\frac{K_I}{3\sigma_s}\right)^2 \\ &= \frac{1}{18\pi}\left(\frac{K_I}{\sigma_s}\right)^2 \end{aligned} \tag{8.7}$$

In real cases, the condition of complete triaxiality is obtained only at the center of a specimen or work piece containing a crack, but not on its surface, Fig. 8.2, where the third stress component, σ_z, cannot exist (see also Fig. 3.36). Therefore, Eq. (8.7) must be written in the most general form

$$r_p = \frac{1}{\alpha\pi}\left(\frac{K_I}{\sigma_s}\right)^2 \tag{8.8}$$

in which α is a parameter that takes into account the actual degree of triaxiality mediated through the entire thickness between the limit value of 18, which corresponds to an $FT > 2$ (see Sect. 4.3), and 2, which competes to a plain stress situation with $FT <$ 0,6. For plain strain condition, Irwin [1] suggested an average increase in the uniaxial yield stress σ_s equal to $\sqrt{2\sqrt{2}} = 1.68$ by which α becomes ~6. To give a proper value to the radius of the plastic zone according to the Irwin expression (8.7) with $\alpha = 6$ it may be interesting to consider that for a very brittle hard steel having a toughness of

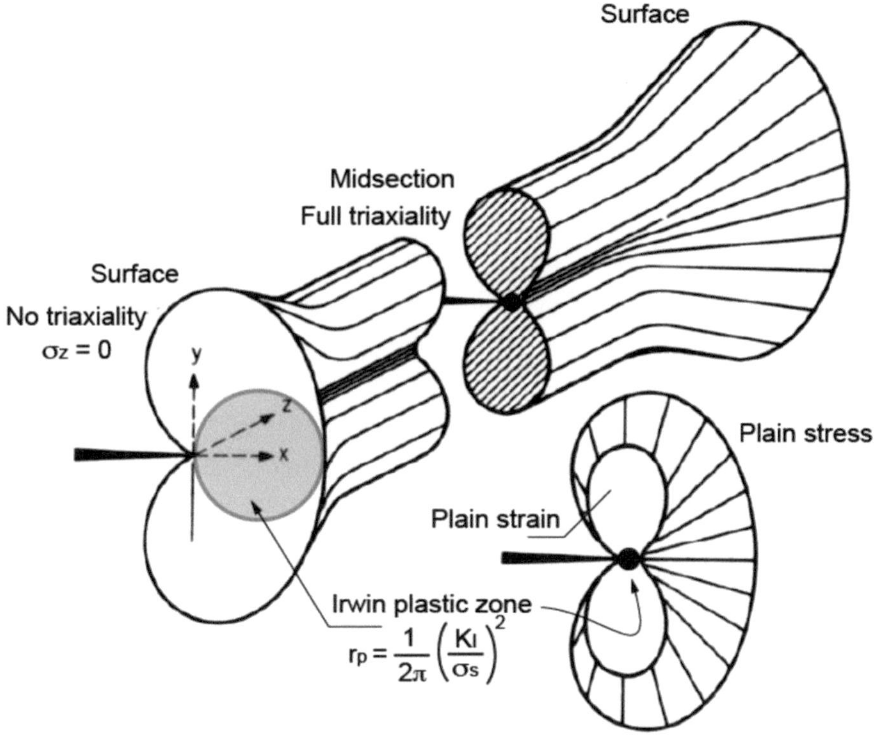

Fig. 8.2 Schematic of the plastic zone at the front of a through-thickness crack in a plate loaded in traction. The plastic zone swells on the surface where triaxiality is lost and contracts at midsection. Also shown is the difference between actual plastic zone in plain stress and plain strain conditions, respectively, and the Irwin plastic zone of circular shape (integrated from [4])

only 50 MPa$\sqrt{}$m and yield stress equal to 1,200 MPa, Eq. (8.8) gives $r_p = 0.09$ mm. For a ductile steel like SA 533 B at −22 °C below the RT_{NDT} where the toughness can be equal to 110 MPa $\sqrt{m}$ and the yield strength is about 600 MPa, the radius of the plastic zone is: $r_p = 1.7$ mm which is much larger than the experimental values ranging from about 11–24 μm, see Table 6.1 row 1 and 2, indicating that a value $\alpha =$ 18 is more acceptable. It is quite evident that the translation of the elastic stress field by a quantity r_p, Fig. 8.1, results in rather longer effective crack. The plasticity at the crack tip causes the elastic portion of the specimen to respond as if the crack were slightly longer than it is. Thus, the apparent crack tip is actually somewhere within the plastic zone. Irwin [1] gave a reasonable estimate of the apparent increment of the of crack length under plain strain conditions using Eq. (8.8) with $\alpha = 6$, i.e. equal to the plastic zone radius. The effective crack length then becomes

$$a_{eff} = a + r_p \tag{8.9}$$

as shown in Fig. 8.18. Work by Hult and McClintok [5] indicates that the elastic stress field surrounding the small plastic zone in antiplane stress conditions is exactly the same as the stress field of a perfectly elastic case provided that the crack tip advanced to the center of the plastic zone. Therefore, Eq. (7.129) continue to be valid also for real materials and we shall suppose that also Eq. (7.114) and (123) are still valid with the same provision. The effective stress intensity factor $K_{I,eff}$ is obtained by substituting a_{eff} to a in the K_I expression (7.114a)

$$K_{I,eff} = \sigma\sqrt{\pi a_{eff}} \tag{8.10}$$

The issue of parameter α has been long discussed. Figure 8.3 shows the values of the plastic constraint factor, *pcf*, and α as function of the relative toughness K_c/σ_s of the material obtained by Hahn and Rosenfield [6] processing the experimental data obtained by Kraft on cracked mild steel specimens broken at cryogenic temperatures between −196° and −84 °C. Also shown are the values of parameter α obtained by Milella [7] analyzing crack growth data in compact C(T) specimens, Sect. 9.3, of 11 different alloys at room temperature where toughness is much higher. Interesting enough, the experimental data obtained by Milella fit the trend reported by Hahn and Rosenfield extending in the higher toughness domain. Experimental data are reported also as function of the ratio σ_m/σ_s between the maximum notch tip stress, σ_m, averaged through the thickness, and that of the uniaxial yield strength, σ_s, called plastic constraint factor, *pcf*. The relation between the *pcf* and α is $\alpha = 2{\cdot}(pcf)^2$. It is significant to note that the experimental data of Fig. 8.3 indicate that for σ_m/σ_s equal 1, that is to say in conditions of plain stress with no triaxiality effect at all, the parameter α takes precisely the value 2, while it approaches the theoretical maximum of 18 when the triaxiality ratio becomes 3. It may appear difficult to understand why at very low K_c/σ_s ratios, the plastic constraint factor is low increasing with increasing toughness. The actual reason is that at very low toughness and high strength the plastic zone is so small that the high triaxiality state competes only to a very small region ahead of the crack tip, and just at the center of the specimen, but that high triaxiality is enough to trigger brittle, cleavage fracture. This also indicate that high triaxiality may not be necessary over large portion of material when the toughness is very poor. In materials with low toughness and high strength, such as 4340, 12Ni-5Cr-3Mo or 7079-T6 which are represented by the lower black squares in Fig. 8.3 or in carbon steel at extremely low temperatures, represented by the open triangles, the ε_z contraction may remain confined within a few hundredths of a millimeter if not even within a few microns.

and this will not allow the achievement of very high values of average triaxiality through the entire thickness. These are really SSY conditions. These materials do not really need a very high triaxiality or a complete plain strain condition to undergo a brittle fracture since their toughness is at its minimum possible, around 40 MPa$\sqrt{m}$. At variance, the high *pcf* or triaxiality of the high range K_c/σ_s ratios shown in Fig. 8.3 are the consequence of the reaction of the material plastically deformed over a large region and can be possible only with large specimens. Now, greater ductility means larger lateral contraction, ε_z, over larger portion of material, see Sect. 4.5, and, therefore, greater attempt to force elastic masses to contract that react more vigorously. However, the fracture cannot be brittle but ductile because the SSY

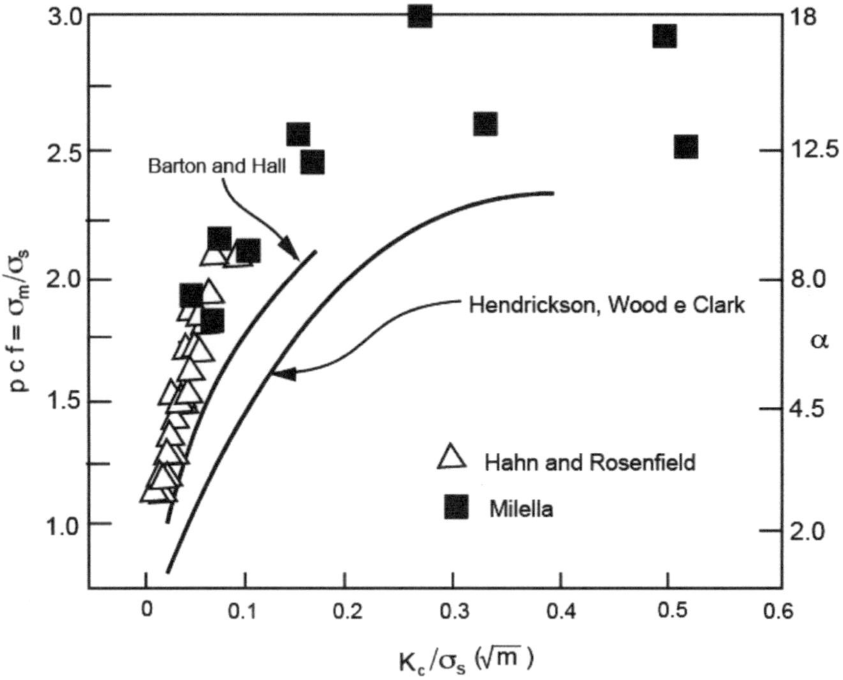

Fig. 8.3 Trend of the parameter α as function of the ratio K_c/σ_s obtained by Hahn and Rosenfield [5] for a carbon steel between −196° C and −84 °C, and by Milella [6] for 11 different alloys at room temperature. Also shown is the trend line obtained by Barton and Hall [7] for sharp hyperbolic notches and by Hendrickson, Wood and Clark [8] for less sharp hyperbolic notches

conditions are completely lost. The data obtained by Barton and Hall [8] run parallel to those of Hahn e Rosenfield, Fig. 8.3, but slightly on the right. They are coming from specimens containing two deep and sharp lateral hyperbolic notches that are not exactly cracks, having a radius of curvature of the tip equal to 1/50 of the length. Also shown, in Fig. 8.3, is the trend obtained by Hendrickson et al. [9] with specimens with a double hyperbolic notch, but with a notch tip radius, ρ, even wider and equal to 1/15 of the notch depth. In both cases the reported K_c was non actually valid since notches, though sharp, are not sharp enough to be considered a crack, see Sect. 7.13. The Irwin plastic zone adjustment does not consider the strain hardening effect. As shown in Fig. 8.1, the stress ahead of the crack tip has been truncated at the yield stress value σ_s. It is clear that the strength recovery due to strain hardening may reduce the actual crack tip plastic zone size. Also the circular shape of the plastic zone at the crack tip proposed by Irwin is only a convenient first approximation of reality, as shown in Fig. 8.2. It is valid as long as the applied stress does not exceed the value of about $0.3{\cdot}\sigma_s$, in other words as long as LEFM is valid. Under this condition, also the strain hardening effect may be disregarded. Beyond this value, Irwin's hypothesis of the plastic zone tends to underestimate the real extent of the plastic enclave and other models, such as that of Dugdale, Sect. 8.3, or Rice, Table 10.1, are necessary to describe the plastic stress–strain fields ahead of the crack.

8.3 The Strip-Yield Model

The strip-yield model proposed by Dugdale [2] and Barenblatt [3], provides an interesting interpretation of the crack tip plastic zone. The geometry considered in the early analysis was that of an infinite plate containing a through crack of length $2a$. The material is assumed to behave as elastic-perfectly plastic (non-hardening material) under plain stress conditions.

In this model the two narrow crack front plastic zones represent a sort of physical continuation of the crack that vanishes at a certain distance ρ from the tip of the physical crack, Fig. 8.4a. To determine the length ρ of the plastic zone it is assumed that the crack of length $2a$ actually spreads elastically to a distance 2ρ, but the crack has been closed up in the plastic zone by an internal tensile stress which acts across the crack faces in the region $|a| \leq |x| \leq |a+\rho|$, Fig. 8.4b. Therefore, this internal tensile stress that was equal to the yield stress, σ_s, must now be replaced by a compressive stress state of the same magnitude, Fig. 8.4b. Superimposing this internal stress field on the elastic stress field of the crack in the presence of an externally applied tensile stress σ and imposing the restriction that the stress at the end of the plastic zone be finite, brings to the Dugdale solution

$$\frac{a}{a+\rho} = 2\sin^2\left(\frac{\pi\sigma}{4\sigma_s}\right) \tag{8.11}$$

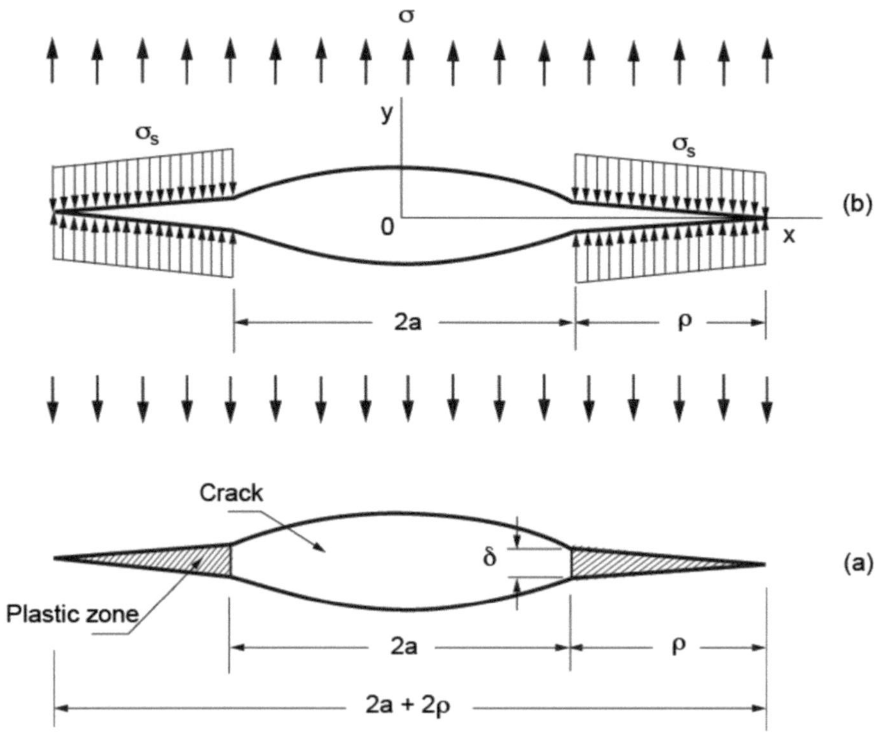

Fig. 8.4 **a** Dugdale strip-yield model of elastic–plastic deformation ahead of a crack of length $2a$ under plain stress loading. Two crack tip plastic zones of length ρ are formed, which are the physical continuation of the crack. **b** A new crack of effective length $2a + 2\rho$ is assumed to exist where the crack tip plastic zones have been opened substituted by an external compression stress state equal to the yield stress of the material, σ_S. Also shown is the Crack Tip Opening Displacement, δ, or CTOD

Note that the model does not foresee any stress singularity at the crack tip so that the closure stresses must cancel the crack tip singularity. Stone and Burdekin [10] obtained an analogous expression

$$\frac{a}{a+\rho} = \cos\left(\frac{\pi\sigma}{2\sigma_s}\right) \tag{8.12}$$

A similar expression was also obtained by Smith [11] and Bilby et al. [12]. Cosine in Eq. (8.12) can be expanded in a Taylor series and for small values of σ/σ_s ratio the higher-order terms can be neglected obtaining

$$\rho = \frac{\pi^2 a \sigma^2}{8\sigma_s} \tag{8.13}$$

For small σ/σ_s ratios we can suppose that LEFM still applies and introduce the Irwin stress intensity factor K_I, Eq. (7.114a), so that Eq. (8.13) becomes

$$\rho = \frac{\pi}{8}\left(\frac{K_I}{\sigma_s}\right)^2 \tag{8.14}$$

very close to the Irwin's expression for r_p, under plain stress conditions, Eq. (8.6), since $\pi/8 = 0.39$ and $1/\pi = 0.32$.

A comparison of the plastic zone size ratios $\rho/(\rho + a)$ obtained with the Stone-Burdekin expression, Eq. (8.12) , and the experimental results reported by Dugdale [2], relating to foils 1.27 mm thick and 12.7 cm wide of mild steel with C = 0.05, N = 0.013 and Mn = 0.40 percent, having a yield strength of 193 MPa is shown in Fig. 8.5. The plates contained both internal and edge slits. The plastic zone had been evidenced using the Fry reagent. The agreement is rather good, considering that Eq. (8.11) refers to an infinite plate of non-strain hardening material. Also shown in Fig. 8.5 is the Irwin solution, Eq. (8.9) in plain strain, $\alpha = 6$, recalling that $\rho = 2r_p$. According to the Dugdale plastic zone, the effective length of the crack would become $a_{eff} = a + r$ so that the effective stress intensity factor, Eq. (8.10), becomes (Fig. 8.6).

$$K_{eff} = \sigma\sqrt{\pi a \cdot sec\left(\frac{\pi\sigma}{2\sigma_s}\right)} \tag{8.15}$$

Fig. 8.5 Trend of the ρ/(a + ρ) ratio as a function of σ/σ_s according to Dugdale and Irwin. Also marked are the experimental results obtained by Dugdale on panels with external and internal cracks

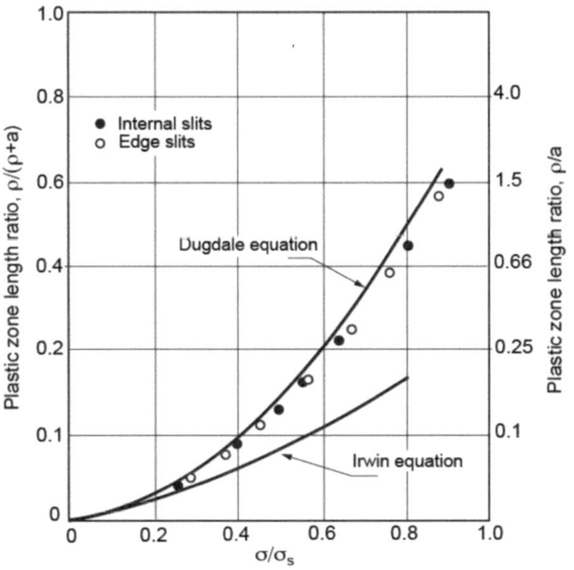

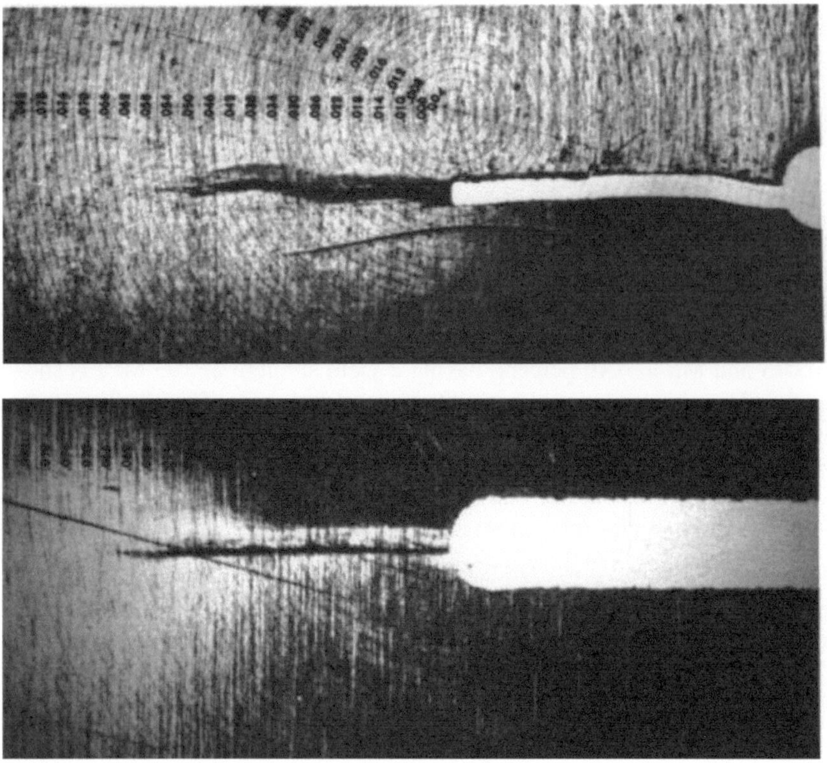

(a)

(b)

Fig. 8.6 Evidence of the plastic zone **a** in a 0.1 mm thick copper foil containing a crack 7 mm long with a tip radius of 0.05 mm and **b** in a sheet of mild steel 0.15 mm thick with crack of 0.23 mm of tip radius [13]

However, Eq. (8.15) overestimate the actual K_{eff}. Stone and Burdekin [10] obtained a more accurate estimate of the K_{eff} given by

$$K_{eff} = \sigma\sqrt{\pi a}\left[\frac{8}{\pi^2} in\,sec\left(\frac{\pi\sigma}{2\sigma_s}\right)\right]^{\frac{1}{2}} \tag{8.16}$$

According to the strip-yield model, K_{eff} is infinite at yield. The plastic zone extends across the entire plate to the border. Thus, the strip-yield model has a built-in limit load criterion for perfectly plastic nonhardening materials. This model has been used to derive the original R-6 FAD described in Sect. 14. An example of plastic zone in foils is shown in Figure 8.6 [13]. However, the strip-yield model that seems to well describe the plastic zone pattern ahead of a crack tip in foils bears little resemblance to actual plastic zone in metals of larger thickness.

8.4 Stress Field Around a Crack

Knowing the equations of the stress field ahead of the crack, Eq. (7.114), it is possible to make some first important considerations. Let's start by looking at the stress along the x-axis for which is $\theta = 0$. In the very narrow field of validity of Eq. (7.114) the stresses σ_x and σ_y are exactly equal to each other. This is a completely new situation that does not occur in any blunted notch for which the stress component,σ_x, always starts from zero for $x = 0$ and always remains below the σ_y normal to the crack plane, Figs. 4.16a, d. On the other hand, in developing Eq. (7.114), the only boundary conditions had been imposed to the σ_y which had to be equal to zero for $-a < x < +a$. This is equivalent to recognizing a free surface of the crack on the xz plane, without external forces, but not on the yz plane normal to the direction of action of σ_x. In practice, this means that the size of the crack tip is so small, in the limit equal to only 2–3 times the atomic spacing d_o compared to the 10^3–10^4 times larger scale of the continuous through which stresses are transmitted, that it has not got any free surface and therefore does not allow condition $\sigma_x = 0$ to apply. In other words, the crack is supposed to be extremely flat and sharp, with a radius of curvature of the tip close to zero, apparently without any free surface capable of exerting a tip reaction in the x direction. This effect bears a great importance and consequence on the practical ground of fracture strength as it precisely realizes the condition of highest biaxiality possible that generates, in turn, the maximum degree of triaxiality in bodies of large thickness, Fig. 4.16 g. However, the condition of equality of the first two stress components, σ_x and σ_y, is lost for $\theta \neq 0$. Figure 8.7 shows the trend, as a function of the θ angle, of the stresses σ_x and σ_y indexed to $K_I / \sqrt{2\pi r}$, which is the trend of the dimensionless factor $f_{ij}(\theta)$ of Eq. (7.115) containing the terms in θ.

Also shown in Fig. 8.7 are the third stress component σ_z, which in the case of plain strain is non-zero and equal to $\nu(\sigma_x + \sigma_y)$, the shear stress τ_{xy}, and the principal stresses σ_1 and, σ_2

$$\begin{aligned}
\sigma_1 &= \frac{\sigma_x + \sigma_y}{2} - \frac{1}{2}\sqrt{(\sigma_x - \sigma_y)^2 + 4\tau_{xy}^2} \\
\sigma_2 &= \frac{\sigma_x + \sigma_y}{2} + \frac{1}{2}\sqrt{(\sigma_x - \sigma_y)^2 + 4\tau_{xy}^2} \\
\sigma_3 &= \nu \cdot (\sigma_1 + \sigma_2)
\end{aligned} \tag{8.17}$$

It is evident the sudden drop of the stress σ_x with the angle θ. In particular, for $\theta = 60°$ the shear stress τ_{xy} if zero and the σ_y reaches its maximum value coinciding with the principal stress σ_2 and it is along this direction that in real materials plastic deformations are likely to occur, see Fig. 8.12. As yield criterion to assess where the plastic slip will take place in a real material, we can use the Tresca criterion or of the maximum shearing stress τ_{max}

$$\tau_{\max} = \frac{\sigma_2 - \sigma_{\min}}{2} \tag{8.18}$$

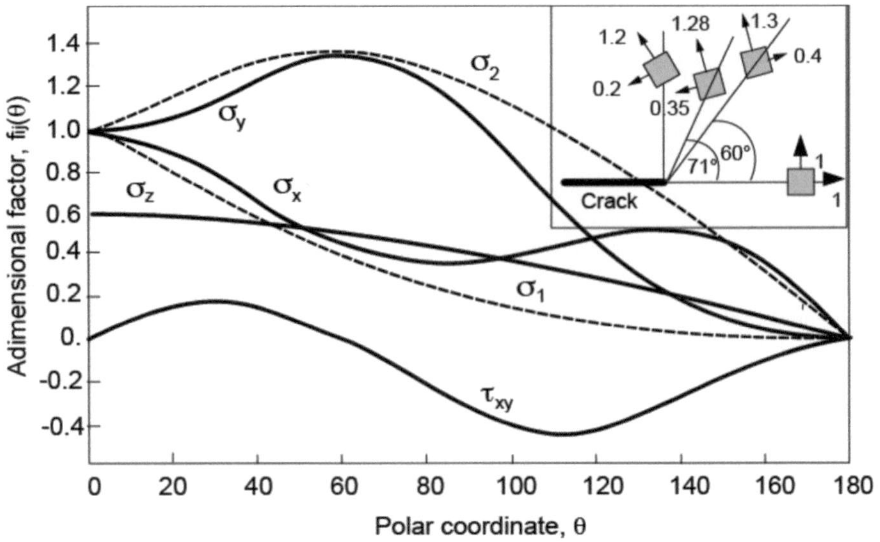

Fig. 8.7 Variation of stresses σ_x, σ_y, σ_z,τ_{xy} and principal stresses σ_1 and σ_2, indexed to $K_I/\sqrt{2\pi r}$, factor $f_{ij}(\theta)$, as function of the polar angle θ. In the box the relative values and direction of the principal stresses σ_1 and σ_2 are shown for three significant directions with respect to the reference plane $\theta = 0$

where σ_{min} is the smallest stress between σ_2 and σ_3, and that of von Mises of the octahedral stress, related to the equivalent stress σ_e, $\tau_{oct} = \sqrt{2}/3\, \sigma_e$

$$\sigma_e = \frac{\sqrt{2}}{2}\left[(\sigma_1 - \sigma_2)^2 + (\sigma_1 - \sigma_3)^2 + (\sigma_2 - \sigma_3)^2\right]^{1/2} \tag{8.19}$$

The two shear stresses are shown in Fig. 8.8 for the two conditions of plane strain plane stress in which the third stress component σ_z is null.

It can be seen that in plain strain conditions the maximum value of the equivalent shear stress is reached at 90° with respect to the *x*-axis or to the crack plane, either the Tresca or the von Mises criterium is applied. It is along this direction, on both sides of the crack plain, that the plastic zone is likely to develop, Fig. 8.9. Under conditions of plane stress the σ_{min} is given by the σ_3 which is null. In this case plastic slip will take place between 60° and 71° depending on whether we consider the Tresca criterion or the von Mises one.

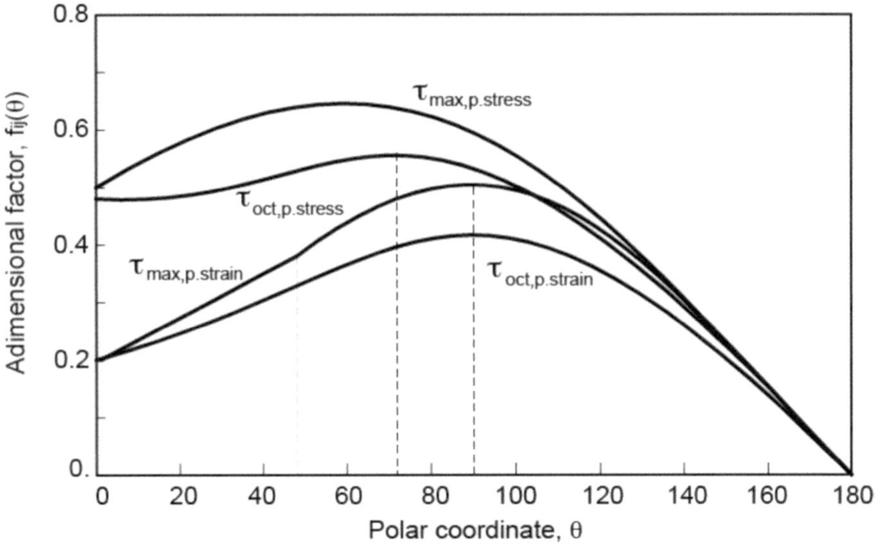

Fig. 8.8 Trend of the dimensionless Tresca maximum shear stress, τ_{max}, and of the Von Mises octahedral stress, $\tau_{,oct}$, under conditions of plain stress, subscript *p, stress*, and plain strain, subscript *p, strain*

8.5 The Shape of the Plastic Zone

The shape of the crack tip plastic zone is actually quite different from the simple circular one hypothesized by Irwin and illustrated in the paragraph 8.2. A first attempt to evaluate its form and real dimensions can start, once again, from the knowledge of the equations of the stress field ahead of the crack, Eq. (7.114), from which it is possible build the locus of points delimiting the plastic area where the material has undergone a plastic flow. To this end, we adopt, as usual, the yield strength criteria of von Mises, Eq. (8.16), and Tresca, Eq. (8.15) and write the principal stresses given by Eq. (7.114) and the distance r normalized to the applied stress σ and the crack length, a, respectively

$$S_1 = \frac{\sigma_1}{\sigma}, \quad S_2 = \frac{\sigma_2}{\sigma}, S_3 = \frac{\sigma_3}{\sigma} \qquad (8.20)$$
$$R = \frac{r}{a}$$

Therefore

$$S_1 = \frac{1}{2\sqrt{2R}}\left[A + B + \sqrt{(A - B)^2 + 4C^2}\right]$$

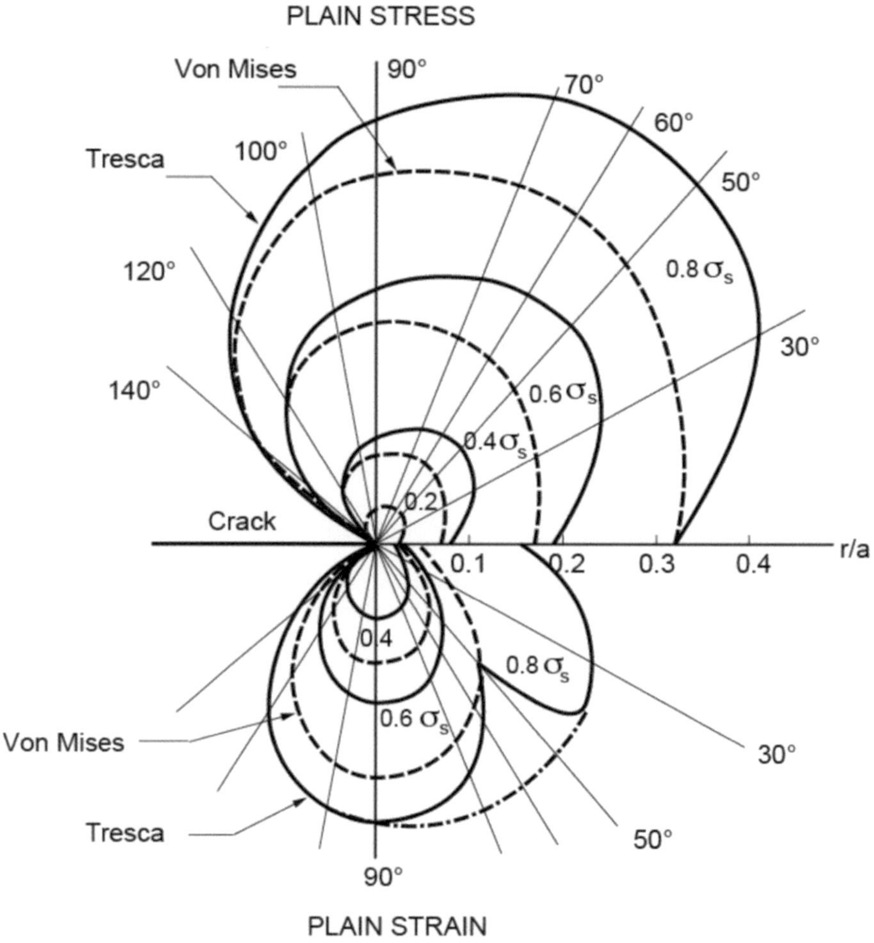

Fig. 8.9 Crack tip plastic zone shape and size in an infinite geometry with a central flaw under remote loading in plain stress conditions (upper part) and plain strain conditions (lower part), according to the yield strength criteria of von Mises (dotted curve) and Tresca (full curve), for various values of the remote stress expressed as a fraction of the yield strength σ_s

$$S_2 = \frac{1}{2\sqrt{2R}}\left[A + B - \sqrt{(A - B)^2 + 4C^2}\right] \tag{8.21}$$

$$S_3 = \frac{\nu}{\sqrt{2R}}(A + B) or 0$$

where *A*, *B* and *C* are the terms expressing the angular functions $f_{\mathrm{ij}}(\theta)$, Eq. (7.115)

$$A = \cos\frac{\theta}{2} \cdot \left(1 - \sin\frac{\theta}{2}\sin\frac{3\theta}{2}\right)$$
$$B = \cos\frac{\theta}{2} \cdot \left(1 + \sin\frac{\theta}{2}\sin\frac{3\theta}{2}\right) \tag{8.22}$$
$$C = \sin\frac{\theta}{2}\cos\frac{\theta}{2}\cos\frac{3\theta}{2}$$

The dimensional term S_3 that appears in Eq. (8.21) is zero in plain stress and equal to $\nu \cdot (A + B)/\sqrt{2R}$ in plain strain. We first adopt the Tresca criterion, in plain stress conditions. The maximum tear stress, τ_{max}, will be equal to $\sigma_2/2$ and the plastic slip will start when τ_{max} will be equal to half of the yield strength, σ_s. Remembering the second of Eq. (8.21) and solving with respect to R gives

$$R = \frac{1}{8S_s^2} \cdot \left(A + B - \sqrt{(A - B)^2 + 4C^2}\right)^2 \tag{8.23}$$

where S_s is the yield stress σ_s indexed to the remote stress σ. Equation (8.23) provides the locus of points delimiting the plastic zone at the crack tip under plain stress conditions, adopting the Tresca yield strength criterion. This contour is shown in Fig. 8.9 for various values of external stress expressed as a fraction of the yield strength. The growth of the plastic zone with the applied stress is evident. The maximum growth takes place along a 60° to 70° direction for which the shear stress is maximum, see Fig. 8.8. Under plain strain conditions the maximum shear stress, τ_{max}, is

$$\tau_{max} = \frac{\sigma_2 - \sigma_1}{2} \tag{8.24}$$

and the distance R to the plastic border is

$$R = \frac{1}{2S_s^2} \cdot \left[(A - B)^2 + 4C^2\right] \tag{8.25}$$

Equation (8.25) is represented in the lower half of Fig. 8.9 for several values of the applied stress with respect to the yield strength. Note that at each stress level, under plain strain conditions the crack tip plastic zone is much smaller than that in plane state stress, as already said. The maximum extension will be, this time, at 90° with the crack plane where the shear stress reaches its maximum, τ_{max}, Fig. 8.8. The characteristic lobe at 50° that has been traced in Fig. 8.9 only for $\sigma = 0{,}8 \cdot \sigma_s$ is the result of the fact that up to an angle of 47° the minimum stress is the σ_z (see Fig. 8.7) after which σ_1 becomes the minimum stress. Let us now adopt the von Mises criterion of the maximum octahedral stress, to establish the conditions of incipient plastic slip, first in plain stress conditions Eq. (8.21) in which σ_3 is equal to zero. Normalizing to the yield stress σ_s, equating to $\sqrt{2/3}\, S_s$ to impose the yield strength condition and solving with respect to R we obtain

$$R = \frac{A^2 + B^2 - AB + 3C^2}{2S_s^2} \tag{8.26}$$

Equation (8.27) it is also shown in Fig. 8.9 for the same stress ratios adopted for the Tresca criterion. Dimensions of the plastic zone according to von Mises are slightly smaller than the corresponding Tresca profiles. The maximum extension is along a polar angle of 71°, as already pointed out in Sect. 8.3 (see Fig. 8.8). As last case let's analyze the plain strain case, using the von Mises criterion, Eq. (8.16). Operating the usual normalization to the remote stress, equating to $\sqrt{2/3}\ S_s$ and solving with respect to R we have

$$R = \frac{A^2 + B^2 - AB + 3C^2 + (A+B)^2(\nu^2 - \nu)}{2S_s^2} \tag{8.27}$$

Equation (8.27) is also shown in Fig. 8.9 for comparison. As in the case of plane strain according to the Tresca criterion, the maximum extension of the plastic lobe is achieved along a 90° direction but remains a little bit smaller.

Interestingly enough, the plastic zone has been significantly reduced compared to the plain stress case. At 0.8 times the yield strength σ_s the extent of the plastic zone in front of the crack, $\theta = 0$, is equal to only 5% of the crack size, a, and six times smaller than in the case of plane stress. The plastic zone shapes just found and shown in Fig. 8.9 are relative to an infinite plate under biaxial remote stress, $\sigma_x = \sigma_y = \sigma$. If load conditions change to a simple end traction (uniaxial load) the plastic zone shape will also change.

Based on an analysis by Sneddon [14] of an infinite panel with a central crack of length $2a$ under an internal pressure $p = \sigma$, Rooke [15] assessed the solution for the panel with a central crack remotely loaded in traction without internal pressure using the principle of superposition, as shown in Fig. 8.10. The principle of superposition enables us to derive the solution of a given case from known cases. In the case of a central crack in an infinite panel under remote tension σ, case (a), we can write

$$K_I^a = K_I^b + K_I^c = K_I^b + 0 \tag{8.28}$$

therefore

$$\begin{aligned}
\sigma_x &= \sigma\left[\frac{r}{\sqrt{r_1 r_2}}\cos\left(\theta - \frac{1}{2}\theta_1 + \theta_2\right) - \frac{r\cdot\sin\theta}{a}\left(\frac{a^2}{r_1 r_2}\right)^{3/2}\sin\frac{3}{2}(\theta_1+\theta_2) - 1\right] \\
\sigma_y &= \sigma\left[\frac{r}{\sqrt{r_1 r_2}}\cos\left(\theta - \frac{1}{2}\theta_1 + \theta_2\right) + \frac{r\cdot\sin\theta}{a}\left(\frac{a^2}{r_1 r_2}\right)^{3/2}\sin\frac{3}{2}(\theta_1+\theta_2)\right] \\
\tau_{xy} &= \sigma\frac{r\cdot\sin\theta}{a}\left(\frac{a^2}{r_1 r_2}\right)^{3/2}\cos\frac{3}{2}(\theta_1+\theta_2)
\end{aligned} \tag{8.29}$$

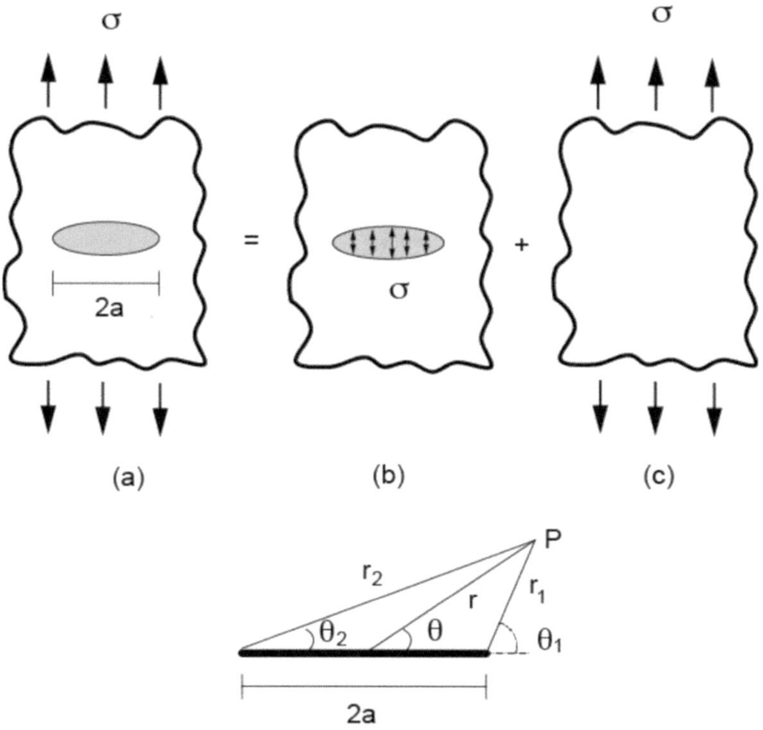

Fig. 8.10 Determination of K_I for a central crack panel under remote tensionσ by means of the principle of superposition and coordinates used

with σ_x, σ_y and τ_{xy} defining the components of the stress field in the x and y directions for a generic point P. The polar coordinates r, θ, θ_1 and θ_2 are shown in Fig. 8.10. Using notations (8.19), but with three distinct values R, R_1 and R_2 relative to the three rays r, r_1 and r_2 of Fig. 8.10, and the angular parameters A, B and C, Eq. (8.21), that are worth

$$\begin{aligned} A &= \frac{R}{(R_1 R_2)^{1/2}} \cdot \cos\left(\theta - \frac{1}{2}\theta_1 + \theta_2\right) \\ B &= \frac{R \cdot \sin\theta}{(R_1 R_2)^{3/2}} \cdot \sin\frac{3}{2}(\theta_1 + \theta_2) \\ C &= B \cdot \cot\frac{3}{2}(\theta_1 + \theta_2) \end{aligned} \tag{8.30}$$

Equations (8.29) can be written as.

$$\begin{aligned} S_x &= A - B - 1 \\ S_y &= A + B \\ S_{xy} &= C \end{aligned} \tag{8.31}$$

that now are

$$\begin{aligned} S_1 &= \frac{1}{2}(2A - 1) + \sqrt{\frac{1}{4}(2B + 1)^2 + C^2} \\ S_2 &= \frac{1}{2}(2A - 1) - \sqrt{\frac{1}{4}(2B + 1)^2 + C^2} \\ S_3 &= 0 \ , \ S_3 = \nu(S_1 + S_2) \end{aligned} \tag{8.32}$$

where the third equation reflects the two possible conditions of plane strain and plain stress, respectively. Apply the Tresca criterion. The identification of the plastic boundaries this time is not as immediate as for the biaxial case. Stresses, in fact, are non-separable functions of R and θ. An iterative approach is therefore required. The results are shown in Fig. 8.11. With respect to Fig. 8.9, two immediate considerations can be done: first: the plastic zone in the uniaxial case in plain stress conditions extends in the x direction ahead of the crack much more than in the biaxial one, in practice the distance over which the plastic zone extends is worth twice as much. Second: under plain strain conditions the extension of the plastic zone in the plain of the crack is practically the same, but the plastic lobe is no longer extended at 90°, but at about 60°.

This occurrence is confirmed by experimental evidence shown in Fig. 8.12a, b, which presents the development of the plastic lobes in front of a crack in the central section of a thick carbon steel plate loaded in traction and in a room-temperature-cured epoxy material [16].

Liebowitz et al. [17] analyzed the effect of load biaxiality on the shape and extension of the loci of equal maximum equivalent shear near the ends of a central crack in an infinite plate using the theory of elasticity. They found that increasing the second component of stress, σ_x, parallel to the crack plane, from $-2\sigma_y$ (compression) to $3\sigma_y$ the shape and inclination of the stress field changed, moving from forward to backward as shown in Fig. 8.13. Note that for $\sigma_x = \sigma_y$ (k = 1) the lobes extend along a 90° direction. Their finding was confirmed by photoelastic isochromatics for load biaxiality ranging from $\sigma_x = 0$ to $\sigma_x = 1.3\ \sigma_y$, as shown in Fig. 8.14. [17]. Rice and Rosengren [18] obtained that, for a strain hardening material subjected to biaxial traction, the plastic lobes were inclined at 100°. Changing the second component of stress from traction to compression, the lobe went back, down to 45° [19].

These findings confirm that the plastic lobes developed ahead of the crack tip reduce their size and move from low angle inclination toward high angle values as the degree of biaxiality-triaxiality increases, as schematically shown in Fig. 8.15. It is worth noting that this reduction and moving backwards of the plastic zone reduces the protection exerted on the material ahead of the crack tip increasing the probability of brittle fracture. If we move from Mode I of opening to Mode II in which the crack

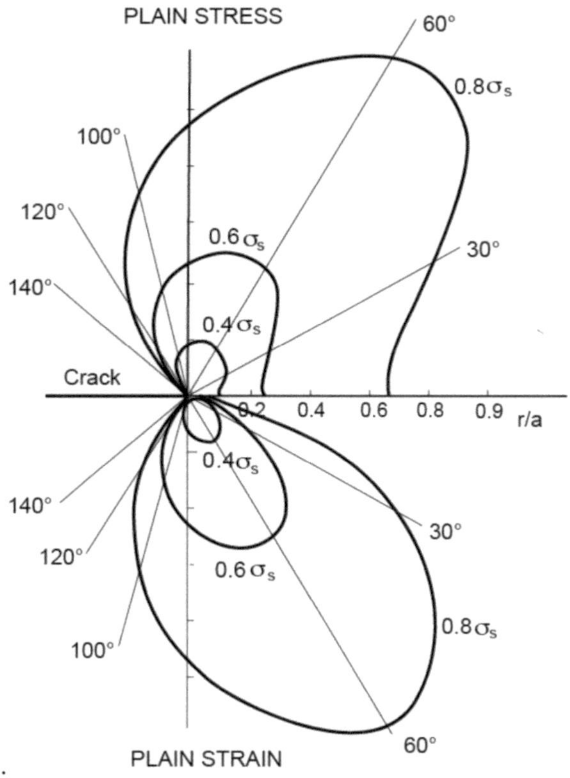

Fig. 8.11 Trend of the plastic zone in a panel with a central crack loaded under remote traction, σ, for various values of the σ/σ_s ratio according to the Tresca criterion, in conditions of both plain strain and plain stress conditions

is opened by the shear forces acting in its own plane, Fig. 7.15b, the shape of the plastic zone changes radically. Using Eq. (7.123), we obtain the graph of Fig. 8.16a for the two conditions of plane strain and plain stress, according to the von Mises criterion. Finally, in Mode III, the plastic zone becomes perfectly circular, as shown in Fig. 8.16b.

In deriving the plastic zone boundaries shown in the previous figures some sort of errors have been introduced. The first stems from the warning already expressed about Eq. (7.114) regarding their origin: the validity of Eq. (8.114) must be limited to a very close distance from the crack tip so that the ratio r/a is negligible, i.e., $r/a < < 1$. Beyond that point, or for a crack whose size is not negligible with respect to the body size, the Irwin stress intensity factor K_I cannot be applied. Secondly, by limiting the stress to the yield stress, Eq. (8.1), and truncating the stress field to that stress level some extra load has to be carried by the material outside the plastic zone, as done with the Irwin plastic radius shown in Fig. 8.1. Thirdly, as well as materials are not perfectly elastic, they are also not elastic-perfectly plastic, some hardening

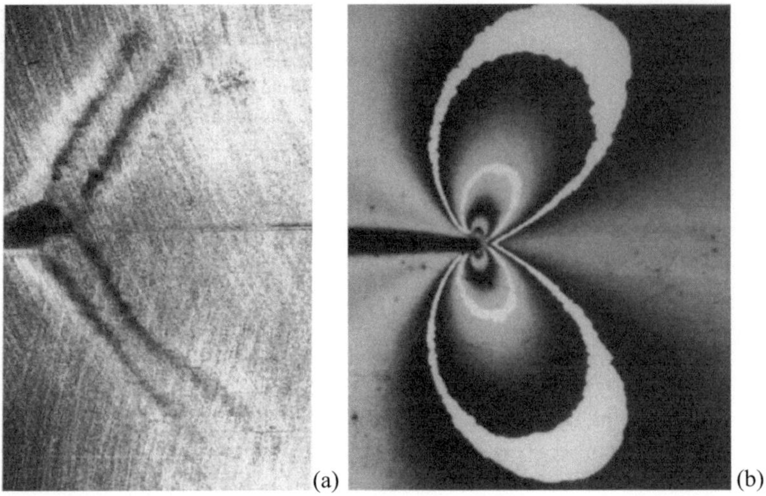

Fig. 8.12 Development of plastic lobes at 60° with the crack plane in **a** a carbon steel plate and **b** photoelastic fringe patters digital image in a room-temperature-cured epoxy material under remote traction [16]

Fig. 8.13 Contours of constant shear stress as a function of biaxiality: **a** k = σ_x/σ_y = 3, **b** k = 1, **c** k = 0 and **d** k = -2 [17]

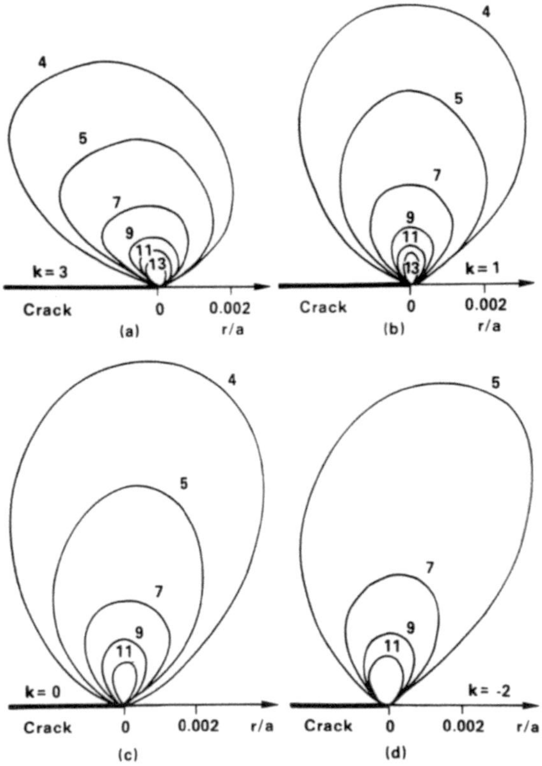

always occurs redistributing the stress state. This hardening mast be very low, as in very hard materials, or may disappear under very high plain strain conditions, Fig. 4. 8.

This is why isochromatics treatment of the subject, such those shown in Figs. 8.12 and 8.14, seems to comply fairy well with forecasts: simply because they are based on linear elastic stress distribution and do not consider any hardening or plastic deformation. It is argued that as the elastic stress field develops ahead of the crack tip, so does the plastic one. Correcting all these errors is not an easy job. Rice and Rosengren [18] treated the subject of strain hardening.

The results they obtained are shown in Fig. 8.17. There is a clear effect of strain hardening on the plastic zone size and shape with the farthest boundary decreasing more and more with increasing strain hardening, always resembling a lobe but moving backwards from about 90° to 100°: strain hardening seems to be protecting the crack tip.

8.6 Plastic Zone According to the Slip Line Field Theory

Slip line field theory applies to elastic-ideally plastic materials under plain strain conditions [19, 20]. Consider a central crack in an infinite medium subjected to remote traction. According to this theory, in an undefined extension of material around the crack develops the so-called Prandtl field, Fig. 8.18, represented by planes of maximum shear stress that are oriented at 45 degrees to the axes of principal stress and, therefore, at 90° between each other, lines α and β of Fig. 8.18. It is

$$\tau_{\max} = \tau_s - \frac{\sigma_s}{2} = \gamma_s G \tag{8.33}$$

where σ_s is the slip stress or yield stress, γ_s the shear strain and G is the shear modulus equal to $E/2(1 + \nu)$, with E being the Young's modulus and ν the Poisson's coefficient. Three regions are identifiable in the Prandtl field. In the first, indicated by the letter A, the principal stresses are obviously constant as the shear stress is constant and equal to τ_s. The τ_{max} lines run, as said, at 45° with respect to the x and y axes. This means that the stresses in the x and y axes are principal, $\sigma_x = \sigma_1$ and $\sigma_y = \sigma_2$. Since zone A faces the free surface of the crack, the σ_y must be zero.

The σ_x, on the other hand, will be equal to $2\tau_s$ while the σ_z, which arises due to the thickness effect, will be equal to$\nu\sigma_x$ and since in the plastic domain is $\nu = 0.5$ we will have, for the generalized Hooke's law

$$\begin{aligned} \sigma_x &= \sigma_1 = 2 \cdot \tau_s = \sigma_s \\ \sigma_y &= \sigma_2 = 0 \\ \sigma_z &= \sigma_3 = \tau_s = \tfrac{\sigma_s}{2} \end{aligned} \tag{8.34}$$

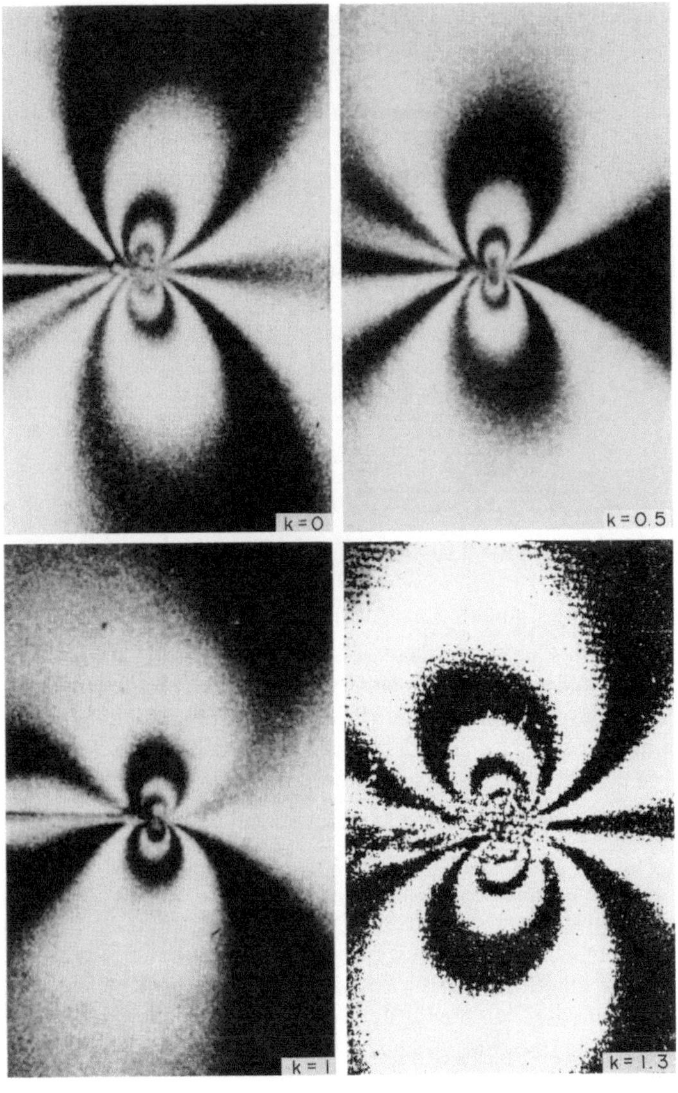

Fig. 8.14 Variation of the shear stress isochromatics with applied load biaxiality [17]

In Region B, stresses are also constant and principal. Since the flow lines that start from the free surface of the crack in Region A arrive in front of it after a π/2 rotation, the stresses σ_x and σ_y will switch undergoing a hydrostatic rise equal to $2\cdot\tau_s\cdot(\pi/2)$ so that

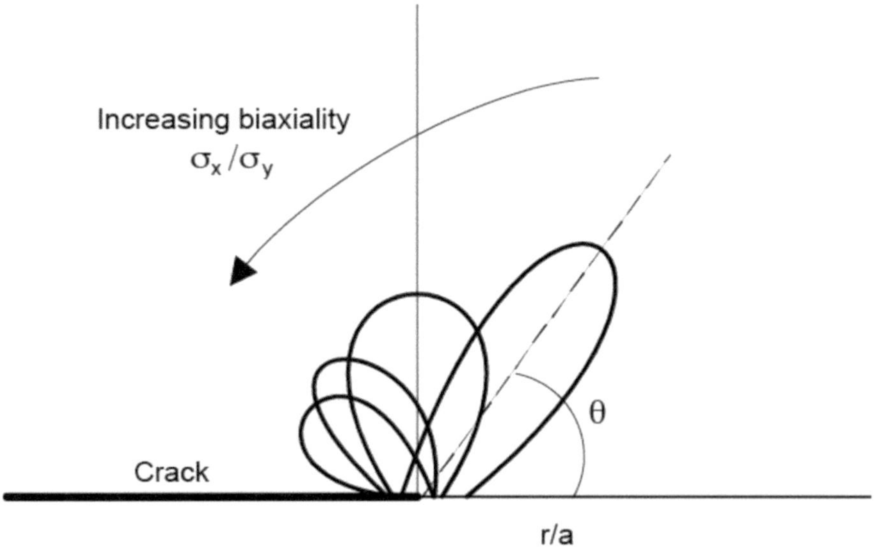

Fig. 8.15 The increase of the biaxiality ratio σ_x/σ_y leads to a progressive reduction of size and shift of the orientation of the plastic lobes towards increasing values of θ

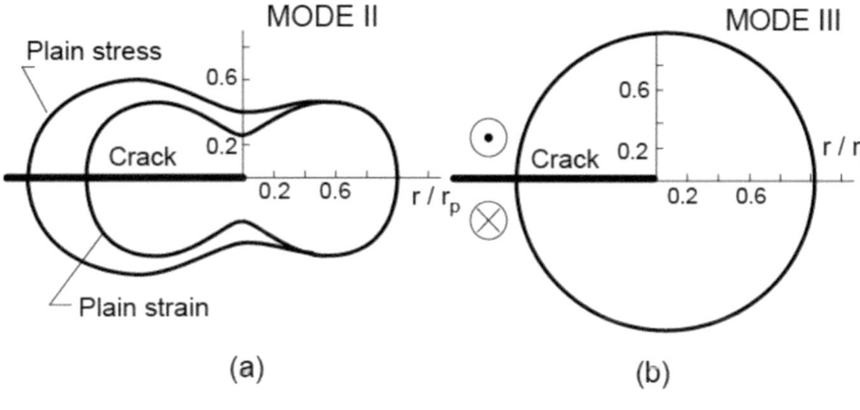

Fig. 8.16 Shape of the plastic zone in Mode II of crack opening **a** and III **b**, respectively, according to the von Mises criterion

$$\begin{aligned}\sigma_x &= \pi \cdot \tau_s \\ \sigma_y &= (2+\pi)\tau_s \\ \sigma_z &= (1+\pi)\tau_s\end{aligned} \tag{8.35}$$

In these two regions the lack of a focus point of the slip lines indicates that there can be no concentration of deformation which, consequently, will remain very low. For a strain hardening material, the stress rise at the crack front may continue well

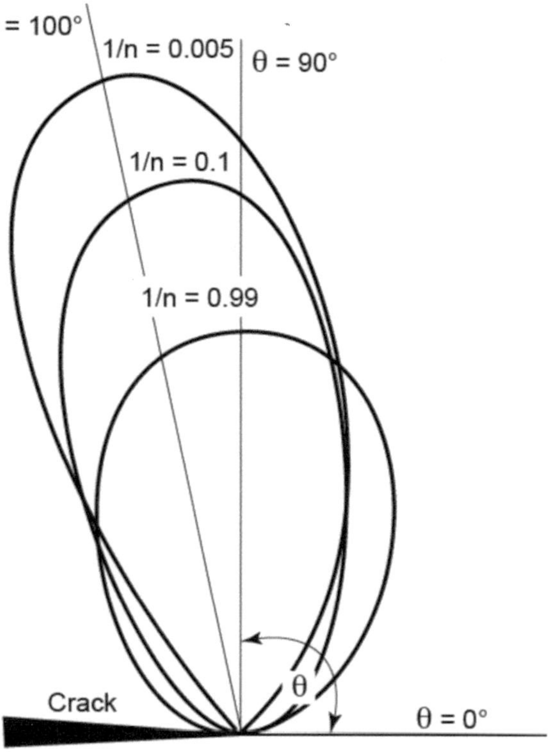

Fig. 8.17 Approximate location and shape of elastic–plastic boundary ahead of the crack tip (taken from [18])

beyond the values given by Eq. (8.35). The situation is different in region C where the merging of the slip lines towards the tip of the crack results in an unlimited growth of the deformation $\gamma_{\theta r}$ according to $1/r$ [20, 21]. Using an angular reference system r, θ and z, it will be represented as [18]:

$$\gamma_{r,\theta} = \frac{\gamma_s R(\theta)}{r} \tag{8.36}$$

where γ_s is the yield strain, Eq. (8.30), and $R(\theta)$ an indeterminate function which can be regarded as the distance along which the strain remains equal to the yield value or as the elastic–plastic boundary function analogous to r_p. In the central region C there will be a situation of hydrostatic stress state in which the three angular components r, θ and z are

$$\sigma_r = \sigma_\theta = \sigma_z = \tau_s\left(1 + \frac{3\pi}{2} - 2\theta\right) \tag{8.37}$$

The singularity of the strain $\gamma_{r,\theta}$ can lead to assume [22] that it also exists in the product

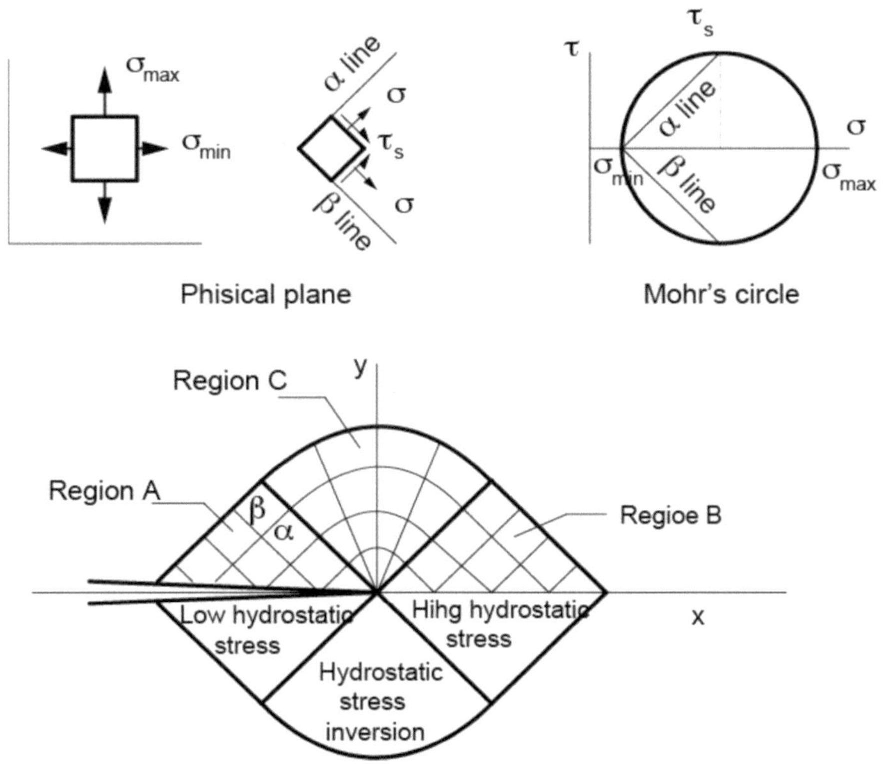

Fig. 8.18 Prandtl stress field ahead of a crack according to the slip line field theory

$$\sigma_{ij}\varepsilon_{ij} = \frac{f(\theta)}{r} \tag{8.38}$$

There are no rigorous demonstrations of Eq. (8.38). However, it is considered acceptable [22]. On the other hand, the crack tip stress state derived with the theory of elasticity, shows a singularity according to $1/\sqrt{r}$, Eq. (7.114), as the deformation field $\varepsilon_{\mathrm{ii}} = \partial u_{\mathrm{ii}}/\partial x_{ii}$ that is proportional to $1/\sqrt{r}$ leading to the dependence of the product $\sigma{\cdot}\varepsilon$ on $1/r$. In materials with work hardening n, using Ramberg Osgood notation (1.42), Eq. (8.38) implies that notch tip stresses and deformations are proportional to [21, 22]

$$\sigma_{ij} \propto J^{\frac{1}{1+n}} \cdot r^{-\frac{1}{1+n}} f_{ij}(\theta, n)$$

$$\varepsilon_{ij} \propto J^{\frac{1}{1+n}} \cdot r^{-\frac{1}{1+n}} g_{ij}(\theta, n) \tag{8.39}$$

where J is the Rice integral, see Sect. 10.6, equal to the energy release rate G, Eq. (7.45), which under plain strain conditions and within the MFLE limits is equal to

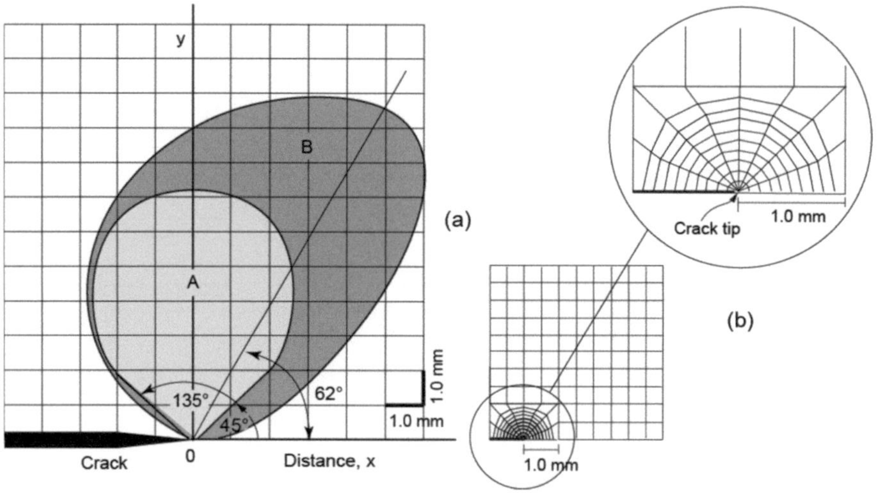

Fig. 8.19 **a** Plastic zones resulting from FE calculations for a C(T) specimen of SA 533 steel under plain strain conditions. The A domain, confined within 135° and 45°, results from first yielding and corresponds to Region C of the Prandtl stress field of Fig. 8.18, the B domain is the plastic zone at 100 MPa$\sqrt{}$m for the strain hardening condition. **b** Crack tip mesh used (Milella and Bonora)

$K^2 \cdot (1 - \nu^2)/E$. Equations (8.39), known as the HRR or Hutchinson-Rice-Rosengren field, will be treated in more details in Sect. 10.7.

FE calculations were made by Milella and Bonora [23] on a C(T) specimen of SA 533 B steel of 420 MPa yield strength, using a crack tip mesh of 0.1 mm, Fig. 8.19b, for two different material conditions, first elastic-perfectly plastic with the yield strength, σ_s, equal to 420 MPa and then with a strain hardening $n = 7$ under plain strain condition to assess the crack front plastic zone shape.

The specimen was loaded at three different levels of stress intensity factors K_I equal to 50 MPa$\sqrt{}$m, 75 MPa$\sqrt{}$m and 100 MPa$\sqrt{}$m, respectively. Two different plastic zone were obtained at each K_I value, one for the elastic-ideally plastic case and one considering the strain hardening effect. Figure 8.19a shows the shape of the plastic zone for the elastic-ideally plastic situation and that of the steel with strain hardening, at $K_I = 100$ MPa. The first zone, between 45° and 135° of aperture, corresponds to region C of the Prandtl field where the stresses σ_x and σ_y, according to Eq. (8.35), should be equal to 660 and 1080 MPa, respectively, and remain constant. The constancy of the value of σ_{ij}, however, is also predicted by Eq. (8.39) for $n = 10$. FE calculations yielded $\sigma_x = 700$ MPa and $\sigma_y = 1100$ MPa, respectively, in excellent agreement. Figure 8.19 is also an indication of the strain hardening effect that is not considered by the slip line field theory. A strain hardening $n = 10$ is typical of that carbon steel, Fig. 1.31, lower than that competing to a high strength steel for which $n =$ 16–20 but higher than that competing to ductile materials like, for instance, austenitic stainless steels. High strength alloys develop a much smaller plastic zone

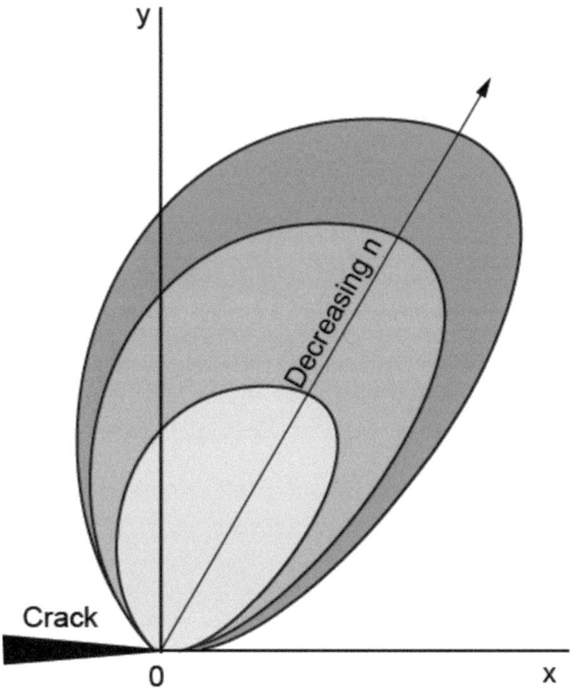

Fig. 8.20 Effect of strain hardening on the size and shape of the crack tip plastic zone in a C(T) geometry

than low alloy steels or ductile alloys such as austenitic stainless steels. Figure 8.20 schematizes this dependence for a C(T) geometry.

A very important consequence, apparently a paradox, of the existence of the Region C in the Prandtl stress field, Fig. 8.18, where the distortion γ grows without any limit according to $1/r$, Eq. (8.36), is that the tip of the crack will stay close, as belonging to Region B, but will open up in Region C, slipping along radial directions. The displacement along the y direction is [18]

$$\frac{\partial v}{\partial \theta} = \gamma_s \cdot R(\theta) \cdot \sin\theta \tag{8.40}$$

from which it can be derived that the opening of the crack tip, δ_t, also called *CTOD,* or Crack Tip Opening Displacement, Fig. 8.21, will be worth

$$\delta_t = 2v = 2 \cdot \gamma_s \cdot \int_{\pi/4}^{3\pi/4} R(\theta)\, \sin\theta\; d\theta \tag{8.41}$$

The integration limits express precisely the condition that it is only in the central Region C of the Prandtl field, contained between $\pi/4 \leq \theta \leq 3\pi/4$, Fig. 8.18, that $R(\theta)$ is nonzero. The value of the *CTOD* will therefore depend on the form of that

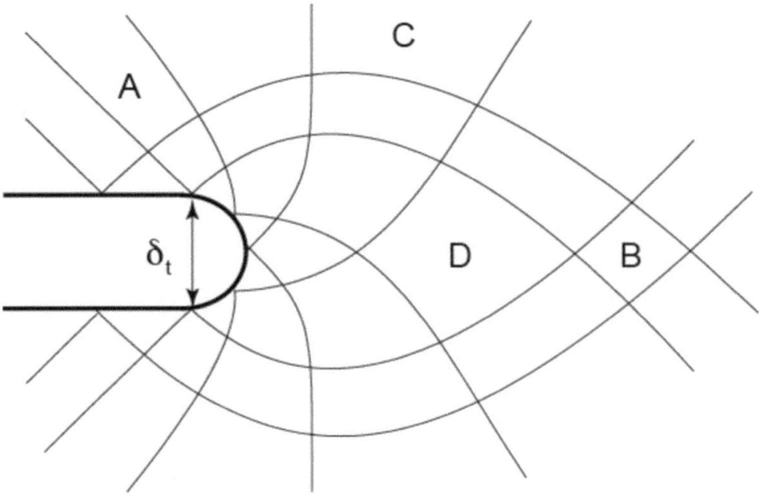

Fig. 8.21 Slip-line field in a blunted notch corresponding to a LSY

function. Guided by the studies of Hahn and Rosenfield [24] on the shape of the crack front plastic zone, Rice [25] assumed for $R(\theta)$ the following expression

$$R(\theta) - R_{\max} \cos\left(\theta - \frac{\pi}{2}\right) \tag{8.42}$$

The use of the Rice integral around the Prandtl field [18] supported by numerical results obtained by Rice and Johnson [26] gave for $R_{\max}$ the following approximation

$$R_{\max} = \frac{3(1 - \nu)}{4\sqrt{2}(2 + \pi)} \left(\frac{K_I}{\tau_s}\right)^2 \tag{8.43}$$

Recalling that the shear stress is linked to the uniaxial tensile stress by Eq. (8.33), following the Tresca criterion, we obtain

$$\begin{aligned} R_{\max} &= \frac{12(1 - \nu)}{4\sqrt{2}(2 + \pi)} \left(\frac{K_I}{\sigma_s}\right)^2 \\ &= 0.206 \cdot \left(\frac{K_I}{\sigma_s}\right)^2 \end{aligned} \tag{8.44}$$

Equation (8.44) can be expressed in terms of R/a as a function of the ratio $(\sigma/\sigma_s)^2$ by using the K_I expression $\sigma\sqrt{a}$ and σ_s is, as usual, the yield strength or slip stress of the material. For $\sigma/\sigma_s = 0.8$, Eq. (8.44) thus gives the value $R/a = 0.13$ which is precisely the theoretical value found by Rooke [15] for the case of plane deformation according to the Tresca criterion and shown in Fig. 8.11. We assumed, as usual in cases of plastic slip, $\nu = 0.5$. Adopting, instead, the von Mises criterion for which is $\sigma_s = \sqrt{3}\,\tau_s$, we get

$$\begin{aligned} R_{\max} &= \frac{2(1-\nu)}{4\sqrt{2}(2+\nu)}\left(\frac{K_I}{\sigma_s}\right)^2 \\ &= 0.155 \cdot \left(\frac{K_I}{\sigma_s}\right)^2 \end{aligned} \tag{8.45}$$

The 0.155 coefficient that appears in Eq. (8.45) is precisely the value found by Levy et al. [27], through a numerical analysis. For $K_I = 100\ \text{MPa}\sqrt{\text{m}}$ and $\sigma_s = 420$ MPa, Eq. (8.45) gives for $R_{\max}$ the value of 8.87 mm against ~10 mm of extension of the plastic lobe obtained by Milella and Bonora and shown in Fig. 8.19, a value exactly halfway between what is foreseen by Eqs. (8.44) and (8.45) suggesting a multiplicative coefficient of $(K_I/\sigma_s)^2$ equal to 0.180. Expressing the R_{max} through Eqs. (8.45) and (8.41) provides the following value of CTOD, δ_t

$$\begin{aligned} \delta_t &= \frac{2\left(1-\nu^2\right)}{2+\pi}\frac{K_I^2}{E\tau_s} = 0.674 \cdot \frac{K_I^2}{E'\sigma_s} \\ &= 4.35 \cdot \frac{\sigma_s}{E'} \cdot R_{\max} \end{aligned} \tag{8.46}$$

where, as usual, $E' = E/(1-\nu^2)$ is Young's modulus in plane strain. Levy et al. [27] found a numerical value of the multiplicative constant of $K^2{}_I/E'\sigma_s$ equal to 0.425. Adopting the Tresca criterion, Eq. (8.41) yields the value

$$\delta_t = 0.777 \cdot \frac{1}{E} \cdot \left(\frac{K_I^2}{\sigma_s}\right) \tag{8.47}$$

against the value of 0.735 ÷ 0.64 of the multiplicative constant found in Sect. 8.7, Eqs. (8.60) and (8.61) for values of the factor α equal to 6 and 10, respectively. Therefore, Eqs. (8.46) and (8.47) provide the value of the CTOD in plane strain or SSY condition where the blunting of the crack tip is negligible. This blunting is not at all sensitive to the work hardening of the material, at variance with the crack tip stresses and strains. It is worth remembering that the simplest expression that can be adopted for the function *R(θ)* is obviously the one that takes the form of a constant *R*. With this the boundary function of the elastic–plastic zone will become an arc of circle of radius *R* bounded, above and below the crack, within sector $\pi/4 \leq \theta \leq 3/\pi 4$. The value of this constant will be given by [18]:

$$\begin{aligned} R &= \frac{2\pi(1-\nu)}{\sqrt{2}(2+\pi)}\left(\frac{\sigma}{\sigma_s}\right)^2 a \\ &= \frac{2\pi(1-\nu)}{\sqrt{2}(2+\pi)}\left(\frac{K_I}{\sigma_s}\right)^2 \end{aligned} \tag{8.48}$$

with a being the half crack length, or

$$R = \frac{1}{2.31\pi}\left(\frac{K_I}{\sigma_s}\right)^2 \tag{8.49}$$

Equations (8.48) and (8.49) provide the minimum value that $R(\theta)$ can take. It is very close to the size of the Irwin plastic radius, r_p, given by Eq. (8.6) with the only difference that in the latter is a complete circumference that develops all around the crack tip while the former refers to a pair of lobes above and below the crack tip. Increasing the external load will eventually open the tip of the crack, *CTOD*, just enough to lose the SSY conditions and enter the so-called *Large Scale Yielding*, LSY, or *Large Geometry Change*, LGC, condition where the crack tip blunts so much that linear elastic fracture mechanics cannot apply any longer and the Irwin K_I factor loses its prerogative to be the only parameter to describe the stress field ahead of the crack become a notch. The transition from SSY to LSY condition is decisive for the fracture strength of a material as it completely changes the reference frame which is no longer that of Fig. 8.18, but the other shown in Fig. 8.21. This time, as we will describe in detail in Sect. 5.2, the slip-line field will develop from the rounded tip of the notch in the form of logarithmic spirals defining a new region, region D of Fig. 8.21, of very high deformation. The possible fracture will no longer be totally brittle, as it is controlled by deformations and no longer by stress. An attempt to evaluate the value of the applied K_{I} at which this transition from *SCG* to *LCG* begin can be made through Eqs. (8.46) and (8.47). We know that the crack tip radius of the Charpy-V notch specimen, equal to 0.25 mm, is not enough sharp to be used in a fracture mechanics specimen, see Sect. 3.1 and Fig. 3.3, even though the load is dynamically applied. Experience has shown (see Sect. 6.4) that in practice under quasi-static loading conditions the radius, ρ, of the notch tip must be limited to 0.1 mm at most to represent a real crack and obtain a valid toughness measurement, K_{Ic}, representative of a totally brittle fracture or cleavage. This means that considering the *CTOD* just as the diameter, 2ρ, of the crack tip, we can use Eq. (8.47) and put $\delta_t/2$ equal to the maximum allowable radius $\rho = 0.1$ mm to get the maximum allowable or valid $K_{I,max}$

$$\delta_t = 2\rho = 0.1 \cdot 10^{-3}\mathrm{m}$$

$$K_{I,max} = \sqrt{\frac{\delta_t E' \sigma_s}{0.777}} \tag{8.50}$$

For a medium strength carbon steel of 500 MPa yield strength, σ_s, and a Young's modulus $E' = E/(1\text{–}\nu^2)$ Eq. (8.50) gives

$$K_{I,max} \approx 120\ MPa\sqrt{m} \tag{8.51}$$

Therefore, it is around 120 MPa$\sqrt{}$m that linear elastic fracture mechanics should start to lose its validity as a one single parameter theory, namely the K_I. In this medium strength steel it is worthless to increase the thickness of a specimen to comply with size requirements, Eq. (9.49), to measure a toughness in excess of 120–130 MPa$\sqrt{}$m. On the other hand, carbon steels have a transition temperature, see Chap. 3, and to measure 120 to 130 MPa$\sqrt{}$m of toughness it is necessary to go over the RT_{NDT} of the material where the steel is no longer brittle. Any crack would turn into a notch and SSY conditions would be lost for LSY.

8.7 The Shape of a Crack Under Stress

A first, immediate assessment of the shape of a crack under a remote stress state can be carried out using the results Irwin elastic stress field for a center crack in an infinite plate under remotely loaded in traction. Equation (7.118) provides the value of the displacement v in the y direction of each point of the crack from the center, x = 0, to the tip, $x = a$

$$v = \frac{2(1-\nu^2)}{E} \cdot \sigma \cdot \sqrt{a^2 - x^2} \tag{8.52}$$

Equation (8.52) may be written as

$$\frac{v^2}{C} + \frac{x^2}{1} = a^2 \tag{8.53}$$

where is C equal to $(2\sigma/E')2$ and $E' = E/(1-\nu^2)$ is the Young's modulus in plain strain. Equation (8.53) is the equation of an ellipse indicating that a central trough crack in an infinite plate under remote stress assumes an elliptical shape. Figure 8.22 shows the profile of half crack under an external traction equal to 130 MPa.

The maximum displacement, $v(0)$, reached at the center of the crack, called *COD* or Crack Opening Displacement, Fig. 8.23, is worth 11.8 mm. If the crack were an edge crack with a depth, a, of 10 cm in a slab of very large width, the opening at the crack mouth, also called Crack Mouth Opening Displacement or CMOD, would have been larger because of the lack of symmetry and constraint that would introduce a bending effect. This time the shape of the crack would be parabolic. This last case is shown in Fig. 8.22b for several traction loads. Returning to Eq. (8.52), we can say that the maximum aperture, $\delta = COD$, at the center of the crack for $x = 0$, is

$$\begin{aligned} COD = \delta = 2 \cdot v|_{x=0} \\ = 4 \cdot a \cdot \left(1-\nu^2\right) \cdot \frac{\sigma}{E} \end{aligned} \tag{8.54}$$

or

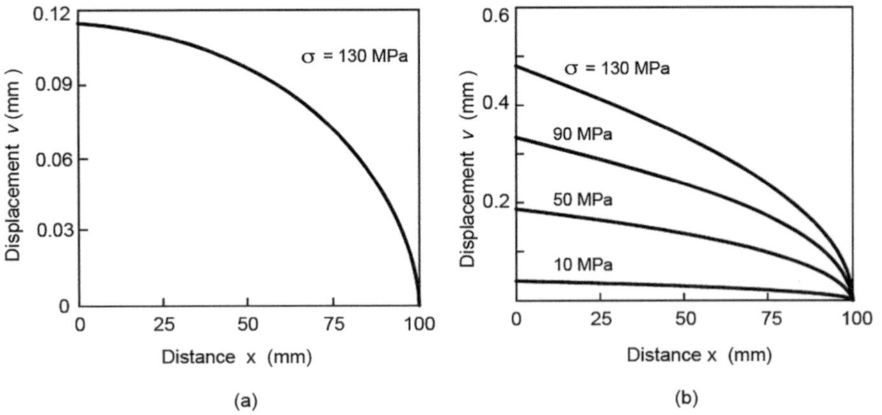

Fig. 8.22 **a** Profile of a central crack in an infinite plate remotely loaded at 130 MPa and **b** parabolic shape of a surface crack in a very wide panel loaded in traction at four different values

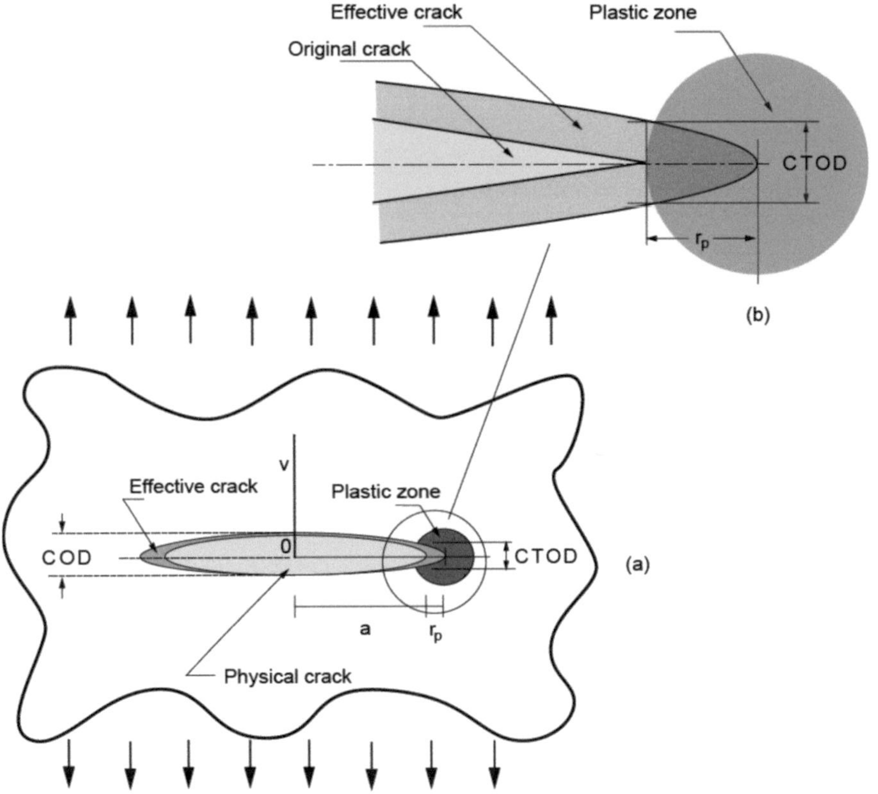

Fig. 8.23 **a** Schematic view of both the physical or original crack and the effective crack that includes half of the plastic zone. Also shown is the COD and the CTOD. **b** Detail of the crack tip with its plastic zone

$$COD = \delta = 4 \cdot a \cdot \frac{\sigma}{E'} \tag{8.55}$$

where E′ is the Young's modulus in plane strain. Considering the stress intensity factor K_I equal to $\sigma\sqrt{\pi a}$, Eq. (8.55) becomes

$$\delta = \frac{4}{\pi}\frac{K_I^2}{E'}\frac{1}{\sigma} \tag{8.56}$$

Equation (8.52) states that for $r = a$, i.e., at the crack tip, the opening is zero. With linear elastic materials the tip of a crack under remote loading does not open at all. In real materials that yield the plastic zone formed at the crack front allows the crack tip to open. Following the Irwin hypothesis of an effective crack a_{eff} longer than the physical one by a quantity equal to the radius of the plastic zone, r_{p_s}, $a_{eff} = a + r_p$, as shown schematically in Fig. 8.23, we have that in real materials it is

$$\begin{aligned} \delta &= 4\,\tfrac{\sigma}{E'}\,(a + r_p) \\ &= \tfrac{4}{\pi}\,\tfrac{K_I^2}{E'}\left(\tfrac{1}{\sigma} + \tfrac{\sigma}{\alpha\sigma_s^2}\right) \end{aligned} \tag{8.57}$$

With these assumptions Eq. (8.59) yields

$$\begin{aligned} v|_{x=a} &= 2\,\tfrac{\sigma}{E'}\,\sqrt{(a + r_p)^2 - a^2} \\ &\approx 2\,\tfrac{\sigma}{E'}\,\sqrt{2ar_p} \end{aligned} \tag{8.58}$$

Recolling Eq. (8.6) for the radius we have

$$\delta_t = 2 \cdot v|_{x=a} = \tfrac{4}{\pi}\sqrt{\tfrac{2}{\alpha}}\,\tfrac{K_I^2}{E'\sigma_s} \tag{8.59}$$

in which δ_t represents the Crack Tip Opening Displacement (*CTOD*). Since under plain strain conditions α is worth ~6, as suggested by Irwin, it is

$$CTOD = \delta_t = 0.735\,\tfrac{K_I^2}{E'\sigma_s} \tag{8.60}$$

and for higher values of the constraint factor $\alpha = 8 \div 10$ (Fig. 8.2):

$$CTOD = \delta_t = 0.57 \div 0.64\,\frac{K_I^2}{E'\sigma_s} \tag{8.61}$$

compared with the value 0.425 numerically found by Levy et al. [26]. In its more general form, Eq. (8.60) or Eq. (8.61) can be written as

$$CTOD = \delta_t = \frac{K_I^2}{m\sigma_s E'} \tag{8.62}$$

where m represents the plastic constraint factor which elevates the uniaxial yield stress, σ_s, under a plain strain condition of triaxiality, Fig. 8.3 and Sect. 4.3. the ISO 12135 Standard [28] provides an explicit form for m

$$m = 4.9 - 3.5 \cdot \left(\frac{\sigma_s}{\sigma_u}\right) \tag{8.63}$$

which depends on the yield stress to the ultimate stress ratio, σ_s/σ_u. An analogous relationship is given by the ASTM 1820 Standard [29]

$$m = 3.62 - 4.21 \cdot \left(\frac{\sigma_s}{\sigma_u}\right) + 4.33\left(\frac{\sigma_s}{\sigma_u}\right)^2 + 2.0\left(\frac{\sigma_s}{\sigma_u}\right)^3 \tag{8.64}$$

This last Standard also suggests replacing the yield stress, σ_s, in Eq. (8.62) with the flow stress of the material, σ_f given by Eq. (10.21).

The most striking consequence of the crack tip plastic zone formation is not the adjourned crack length from a to $a + r_p$ but, rather, the opening of the crack tip which blunts. If this blunting were too large, the crack loses its sharpness turning into a notch. Therefore, as already said, the plastic zone size cannot grow beyond a reasonably small value. Recalling Figs. 6.26 and 6.27, we can assume that already at 0.02 mm of radius the crack sharpness is loss yielding a toughness 3 to 4 time larger than that measured with a real crack having 0.005 mm of radius.

Concluding Remarks

> Real materials that undergo plastic deformations and do not behave in an ideally elastic fashion develop a plastic zone at the crack tip that, however small, cut the apical peak stress by yielding. This crack tip plastic zone, whose size depends on the yield stress and working conditions, being larger under plain stress state than in plain strain condition, makes the crack be longer than the physical one and, which is extremely relevant, allows the crack tip to open. The amount of the crack tip opening, called CTOD, depends on the yield stress of the material and on the plastic constraint exerted by the apical stress state, being larger with decreasing triaxiality.

References

1. Irwin, G.R.: Plastic zone near a crack and fracture toughness. Sagamore Res. Conf. Proc. **4**, 63 (1961)
2. Dugdale, D.S.: Yielding in steel sheets containing slits. J. Mech. Phys. Solids **8**, 100–104 (1960)
3. Barenblatt, G.I.: The mathematical theory of equilibrium of cracks in brittle fracture. Adv. Appl. Mech. **VII**, 55–129 (1962)

4. Srawley, J.E.: Linear elastic fracture mechanics-a review of principles and methods. In: Proceedings of the Symposium on Fracture Touhgness Concepts for Weldable Structural Steels, p. A7, Risley (1969)
5. Hult, J.A.H., McClintoc, F.A.: Elastic-Plastic stress and strain distribution around sharp notches under repeated shear. In: 9th Int. Congress of Applied Mechanics, Vol. 8, pp. 51–58, Brussels (1957)
6. Hahn, G.T., Rosenfield, A.R.: Experimental determination of plastic constraint ahead of a sharp crack under plain-strain conditions. Trans. ASM **59**, 909–919 (1966)
7. Milella, P.P.: A Fatigue Crack Growth Theory Based Upon Energy Considerations", IAEA Specialists' Meeting on Subcritical Crack Growth, pp. 484–508, Freiburg, 13–15 May (1981)
8. Barton, F.W., Hall, W.J.: A Study of Brittle Fracture Initiation in Mild Steel, Ship Structure Committee Report SSC-147 (1963)
9. Hendrickson, J.A., Wood, D.S., Klark, D.S.: The Initiation of Brittle Fracture in Mild Steel. In: Transactions of the ASM, 39th Conferencess, vol. 50, pp. 656–676 (1958)
10. Stone, D.E.W., Burdekin, F.M.: 1st Progress Report on Fracture Mechanics Work Being Carried out at BWRA under a DSIR Earmarked Grant (1964)
11. Smith,E.: The spread of plasticity from a crack: an approach based on the solution of a pair of dual integral equations. CEGB, Res. Lab., Lab. Note N. RD/L/M31/62 (1962)
12. Bilby, B.A., Cottrell, A.H., Swinden, K.H.: The spread of plastic yield from a notch. Proc. Roy. Soc. A., **273**, 304–314 (1963)
13. Liu, H.W.: Fracture Criterion of Cracked Metallic Plate, California Institute of Technology, Pasadena, GALCIT SM 63–29 (1963)
14. Sneddon, I.N.: The distribution of stress in the neighborhood of a crack in an elastic solid. Proc. Royal Soc. **187**, 229 (1946)
15. Rooke, D.P.: Elastic Yield Zones Round a Crack Tip, Royal Aircraft Establishment, Technical Note CPM. 29 (1963)
16. Wang, W.: A digital imaging algorithm for extracting stress intensity factor from the photoelastic fringe pattern. Eng. Fract. Mechanics **36**(4), 690 (1990)
17. Eftis, J., Jones, D.L., Liebowitz, H.: Load biaxiality and fracture: synthesis and summary. Eng. Fract. Mechanics **36**(4), 551–552 (1990)
18. Rice, J.R., Rosengren, G.F.: Plain strain deformation near a crack tip in a power low hardening material. J. Mech. Phys. Solids **16**, 1–12 (1968)
19. Hill, R., The Mathematical Theory of Plasticity, Oxford University Press (1950)
20. Prager, W., Hodge, P.G. Jr.: Theory of Perfectly Plastic Solids. John Wiley and S. (1951)
21. Rice, J.R.: A path Independent Integral and the Approximate Analysis of Strain Concentration by Notches and Cracks, Brown University ARPA SD-86, Report E 39, Providence, R.I., May (1967)
22. Hatchinson, J.W.: Singular behavior of the end of a tensile crack in a hardening material. J. Mech. & Phys. Solids **16**, 13 (1968)
23. Milella, P.P., Bonora, N.: nonpublished FE calculations
24. Hahn, G.T., Rosenfield, A.R.: Experimental determination of plastic constraint ahead of a sharp crack under plain strain conditions. Trans. ASM **59**, 909–919 (1966)
25. Rice, J.R.: J. Appl. Mech. **35**, 379 (1968)
26. Rice, J.R., Johnson, M.A.: The Role of Large Crack Tip Geometry Changes in Plain Strain Fracture, Inelastic Behavior of Solids, pp. 641–672, McGrow-Hill, New York (1977)
27. Levy, N., Marcal, P.V., Ostergren, W.I., J.R.: Small scale yielding near a crack in plain strain: a finite element analysis. Int. J. Fract. Mech. **7**(2), 143–156 (1971)
28. International Standard ISO 12135: Metallic materials. Unified method of tests for the determination of quasistatic fracture toughness, Third Edition (2021)
29. ASTM E 1820 08a: Standard Test Method for Measurements of Fracture Toughness

Chapter 9
Experimental Linear Elastic Fracture Mechanics

9.1 Introduction

The introduction of the G. Irwin stress intensity factor, K_I, provided the analytical tool to assess the impact of a crack on the integrity of a structure and, for the first time, made it possible the measurement of that property of materials known as fracture toughness, K_{Ic}. The puzzle of the interconnection of the three elements of Fig. 1.5, i.e., stress, crack size and toughness, seemed to be solved thanks to K_I, a unique single valued parameter at least for plain strain conditions where the plastic zone ahead of the crack tip is negligible, the so called SSY conditions. However, stress intensity factor expressions initially available were only those relative to an infinite plate with a central crack under remote loading, Eqs. (7.114), (7.123) and (7.124). Real cases with geometries different from the infinite plate, containing cracks of any type and size required new, specific K_I solutions, though a small crack in a very large test piece could approximate the infinite geometry. Early efforts in developing K_I solutions for other geometries were supported by the rocket industry, because of the high incidence of brittle fracture of high-strength steels used, and by electrical industries, having problems of structural integrity of heavy-section equipment, in particular large rotating machinery. Also, the new nuclear power generation industry with its demanding need for safety, using large and thick reactor pressure vessels and piping required new, accurate K_I solutions. Much of the early experimental work that started around 1955 was focused on the fracture behavior of large shafts for turbine and generators which were made of medium–high strength forging steels of about 700 MPa yield strength.

Historically, the electric industry employed spin burst test of large and thick discs with sharp machined notches emanating from the central bore [1–3], Fig. 9.1 [2]. These specimens were run at increasing rotational speed till the disc either went into pieces in a brittle manner, Fig. 9.3, or failed by general yielding, depending on the test temperature and, therefore, on material behavior. With the introduction of LEFM the test was interpreted in terms of the new stress intensity factor, K_I. It

P. P. Milella, *Fundamentals of Fracture Mechanics*,
https://doi.org/10.1007/978-3-032-23081-2_9

was also recognized the need for a very sharp notch, since toughness measurements, K_{Ic}, made with specimens of the same type having different notch tip radii were yielding different K_{Ic} values, increasing as the notch root radius increased, as already discussed in Sect. 6.4 and shown in Figs. 6.26 and 6.27. This finding changed the technique for sharpening the machined notch crack tip from the earlier procedure of jeweler's saw-cutting of the notch tip that could introduce a thin slot of radius $\rho = 0{,}15$ mm, to the more sophisticated electro-discharge machining (EDM) with a wire diameter as small as 15 μm followed by fatigue pre-cracking that introduced an almost natural sharp crack. To visualize and fully understand the difference between a thin slot and a natural fatigue crack it is convenient to see the effects of the loading sequence on a specimen containing a thin slot, Fig. 9.2a, compared to a similar specimen with a flaw fatigue pre-cracked, Fig. 9.2b [4].

In the first case, Fig. 9.2a, the slot opens leading to a COD (Crack Opening Displacement) and a CTOD (crack Tip Opening Displacement), which is by far larger than that associated to a natural crack, Fig. 9.2b. Since CTOD is a measure of material toughness, see Sect. 10.2, the toughness measured with the specimen containing the slot will be much larger than that measured with the sharp fatigue crack, which yields the real toughness, K_{Ic}, of the material. Returning to the spinning disc with a central crack of total length $2a$, Fig. 9.1, we have that the stress state at any point has a radial component, σ_r, and a tangential one, σ_t, given by [5]:

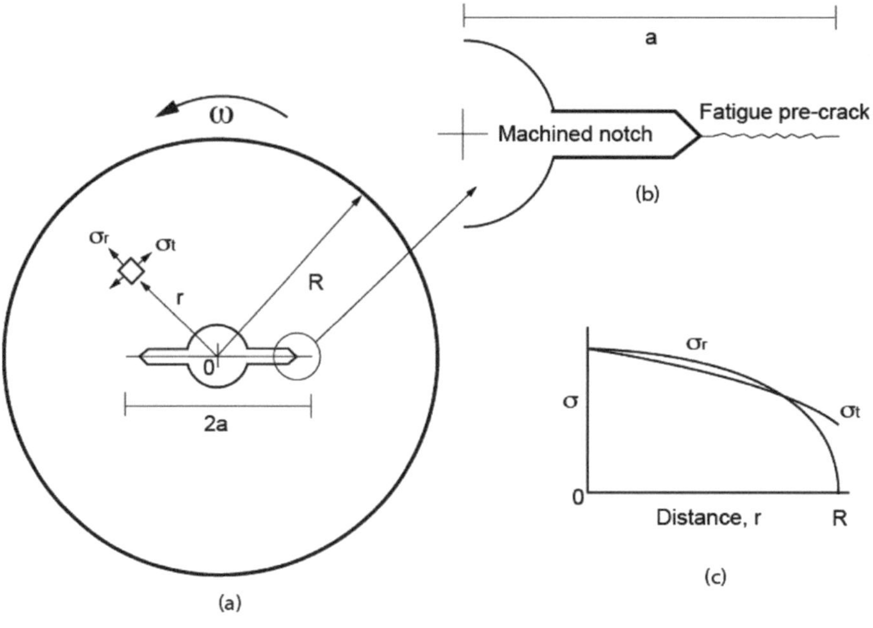

Fig. 9.1 **a** Spin test specimen, **b** detail of the notch tip with machined notch and fatigue pre-crack, **c** plot of the stress state, σ_r and σ_t, with distance r

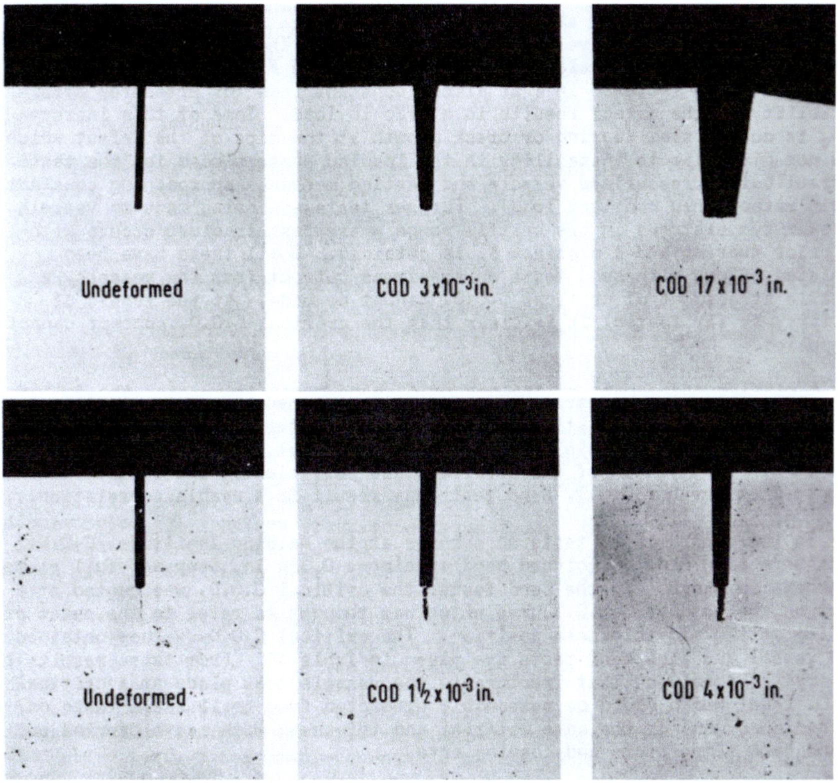

(a)

(b)

Fig. 9.2 Loading sequence on a specimen containing: **a** a thin slot compared to that **b** on an equal specimen whose slot has been fatigue pre-cracked to produce a natural crack. The slot blunts and opens while the crack does not [3]

$$\sigma_r = \frac{3+\nu}{8}\rho\omega^2\left(R^2 - r^2\right)$$
$$\sigma_t = \frac{3+\nu}{8}\rho\omega^2 R^2 - \frac{1+3\nu}{8}\rho\omega^2 r^2 \tag{9.1}$$

The amplitude of the stresses σ_r and σ_t, depends on the density of the material, ρ, the square of the angular velocity of rotation, ω, and the distance r.

An example of rotating spin test that resulted in a final fracture is shown in Fig. 9.3. With the developments of LEFM and the techniques for a very sharp crack, the analytical expression of K_I for a cracked disc was found using the *boundary collocation* method [6]

$$K_I = \sigma_o \cdot \sqrt{\pi a}\, f(a/R) \tag{9.2}$$

with

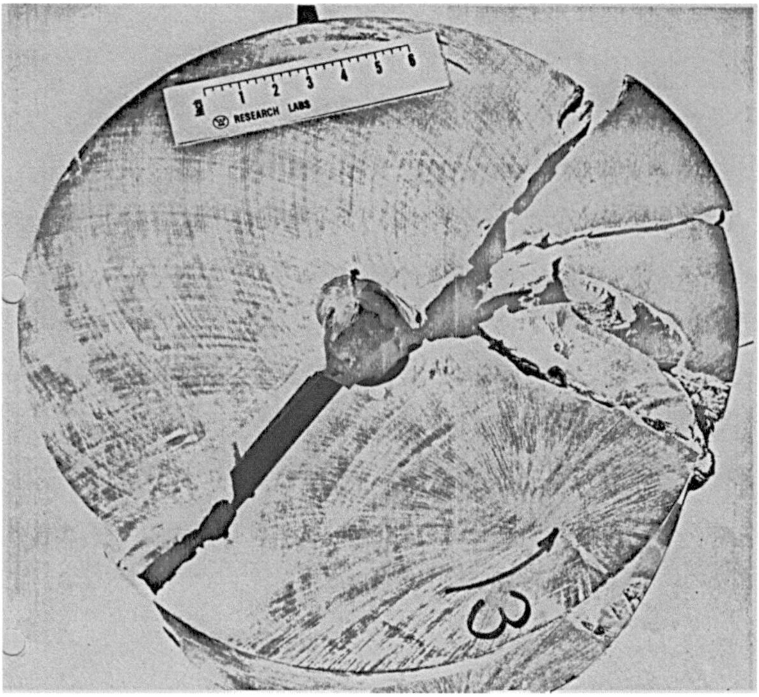

Fig. 9.3 Rotating spin test. Scale model of a flywheel made of A 533 grade B class 1 carbon steel, 21" diameter, containing a 5" crack at fracture at 13,000 rpm (from Westinghouse Nuclear)

$$f(a/R) = 1 + 0.0569\frac{a}{R} - 0.3877\left(\frac{a}{R}\right)^2 + 3.8544\left(\frac{a}{R}\right)^3 - 7.6061\left(\frac{a}{R}\right)^4 + 5.4447\left(\frac{a}{R}\right)^5 \quad (9.3)$$

$2a$ being the length of the crack, Fig. 9.1, and σ_o the maximum tangential stress, σ_t, at the center of the disk, given by Eq. (9.1) for $r = 0$. The use of the rotating disc soon declined but, interesting enough, today we are witnessing a return of the spin disc in the form of a rotating hollow cylinder [7]. The rotating cylinder can, in fact, provide information on both the toughness and accuracy of the FE calculations applied to large structures with full thickness, even greater than 15 cm, biaxially loaded, as are the large pressure vessels in use in the chemical or nuclear industry, replacing the pressure with stresses deriving from centrifugal forces. At the time when the spin test was used as a fracture mechanics K_{Ic} specimen, other geometries were also exploited in which a sharp fatigue precrack was introduced. Among these, some early tests used plate type, double edge-notched specimens. The basic idea that a large and thick geometry was necessary to have plain strain conditions, see Chap. 4, was inspiring the use of this new specimen whose dimensions were quite large, as it can be seen in Fig. 9.4.

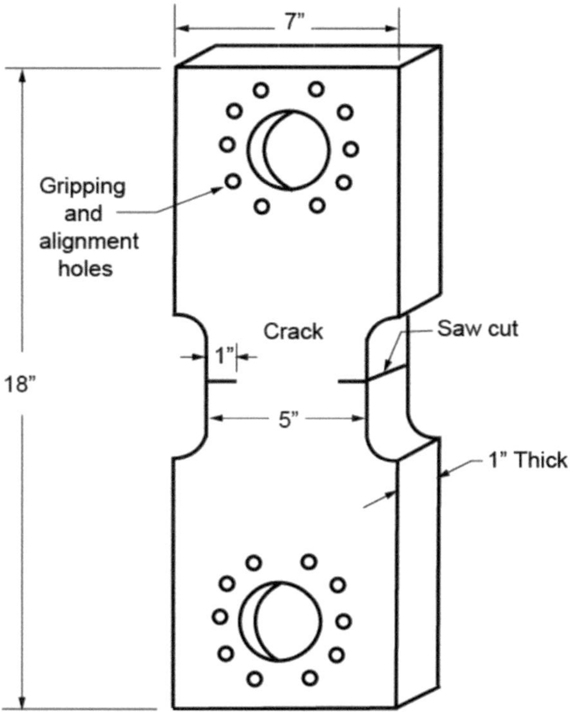

Fig. 9.4 General configuration of large double-edge notch tensile specimen

However, such a large specimen required very large testing rigs not commonly available in all research centers. Therefore, new specimens of smaller dimensions were introduced.

9.2 Notched Panels. CC(T), DEC(T) and SEC(T) Specimens

Parallel to the experiments on rotating discs, new fracture mechanics tests for measuring the material toughness, K_{Ic}, were developed using small, cracked panels. These flat specimens had edge or central notches made sharp using a knife or a jeweler's saw or EDM followed by a fatigue precracking procedure to obtain a natural sharp crack, as described in the previous section. Precracking has to be performed at relatively low ΔK with respect to the K_{Ic} expected to be measured in order to avoid the formation of an initial crack front plastic zone wider than that set up during the test that could affect the result. Three types of panels were originally considered that are shown in Fig. 9.5. The first panel, Fig. 9.5a, called CC(T), *Central Crack in Traction* or traction panel, had a central through wall crack, loaded in remote traction. The second panel, DEC(T) or *Double Edge Crack in Traction*, Fig. 9.5b, had two

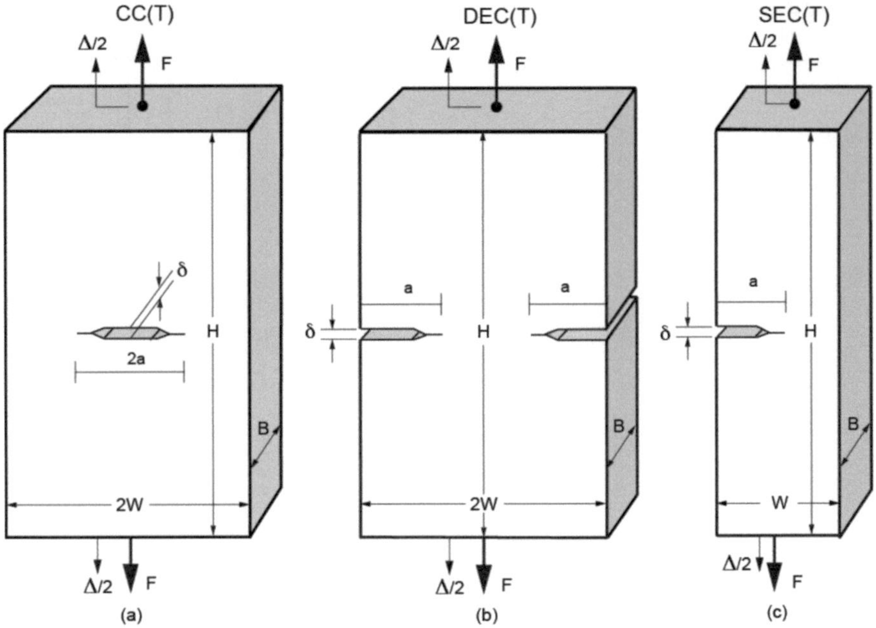

Fig. 9.5 Early types of flat or panel specimens in traction used for toughness measurement: **a** CC(T), **b** DEC(T) and **c** SEC(T)

lateral edge cracks and the third panel, SEC(T) or *Single Edge Crack in Traction*, Fig. 9.5c, had a single lateral crack.

In the acronym of the specimen, the letter in brackets indicates the type of load that can be traction (T) or bending (B). The stress intensity factor solution, K_{I}, COD, displacement load line displacement Δ and compliance C for the three panels remotely loaded are given in the following.

CC(T) Central Crack Panel in Traction [8]

K_I solution:

$$K_I = \frac{F\sqrt{\pi a}}{2BW} \cdot f(a/W) \tag{9.4a}$$

or

$$K_I = \sigma\sqrt{\pi a} \cdot f(a/W) \tag{9.4b}$$

with

$$f(a/W) = \left[1 - 0.025(a/W)^2 + 0.06(a/W)^4\right]\sqrt{\sec\frac{\pi a}{2W}} \tag{9.5}$$

Crack opening COD or δ (at the center):

$$\delta = \frac{2Fa}{BWE\prime} V_1(a/W) \tag{9.6}$$

with $E' = E/\left(1 - \nu^2\right)$ and:

$$V_1(a/W) = -0.071 - 0.535(a/W) + 0.169(a/W)^2 + 0.02(a/W)^3 - 1.071 \frac{1}{(a/W)} \ln(1 - a/W) \tag{9.7}$$

Load line displacement, Δ**,** at remote points (H/W $\geq$ 6).

The load line displacement, Δ, is the sum of the displacement of the panel due to the presence of the crack, Δ_{crack}, plus the elastic displacement of the panel without crack, $\Delta_{no-crack}$:

$$\Delta = \Delta_{crack} + \Delta_{no-crack} \tag{9.8}$$

$$\Delta_{\mathrm{crack}} = \frac{2Fa}{BWE'} V_2(a/W) \tag{9.9}$$

with $\alpha = (a/W)$

$$V_2 = -1.071 + 0.250\alpha - 0.357\alpha^2 + 0.121(\alpha)^3 - 0.047(\alpha)^4 + 0.008(\alpha)^5 - 1.071 \frac{1}{\alpha} \ln(1 - \alpha) \tag{9.10}$$

and

$$\Delta_{no-crack} = \frac{\sigma}{E} H \tag{9.11}$$

Load Line Compliance, C:

$$C = \frac{\Delta}{P} = \frac{\Delta_{crack} + \Delta_{no-crack}}{P} \tag{9.12}$$

DEC(T) Double Edge Crack Panel in Traction [8]

K_I **solution**:

$$K_I = \frac{F\sqrt{\pi a}}{2BW} \cdot f(a/W) \tag{9.13}$$

with

$$f(a/W) = \frac{1.122 - 0.561(a/W) - 0.205(a/W)^2 + 0.471(a/W)^3 - 0.190(a/W)^4}{\sqrt{1-(a/W)}} \tag{9.14}$$

Crack opening COD or δ (at the edges)

$$\delta = \frac{2Fa}{BWE'} V_1(a/W) \tag{9.15}$$

$$V_1\left(\frac{a}{W}\right) = \frac{1}{\left(\frac{\pi a}{2W}\right)} \left[0.459\left(\sin\frac{\pi a}{2W}\right) - 0.065\left(\sin\frac{\pi a}{2W}\right)^3 - 0.007\left(\sin\frac{\pi a}{2W}\right)^5 + \cos h^{-1}\left(\sec\frac{\pi a}{2W}\right)\right] \tag{9.16}$$

Load line displacement, *Δ*, at remote points (H/W ≥ 6), $\Delta = \Delta_{crack} + \Delta_{no-crack}$, Eq. (9.8):

$$\Delta_{crack} = \frac{2Fa}{BWE'} V_2(a/W) \tag{9.17}$$

$$V_2(a/W) = \frac{1}{\left(\frac{\pi a}{2W}\right)} \left[0.0629 - 0.061\left(\cos\frac{\pi a}{2W}\right)^4 - 0.0019\left(\cos\frac{\pi a}{2W}\right)^8 + ln\left(\sec\frac{\pi a}{2W}\right)\right] \tag{9.18}$$

Load line compliance, C:

it is always given by Eq. (9.12) using the proper value for Δ,being $\Delta = \Delta_{crack} + \Delta_{no-crack}$, with Δ_{crack} given by Eq. (9.17) and $\Delta_{no-crack}$ by Eq. (9.11)

SEC(T) Single Edge Crack Panel in Traction [8]

***K_I* solution**:

$$K_I = \frac{F\sqrt{\pi a}}{BW} \cdot f(a/W) \tag{9.19}$$

with

$$f(a/W) = 1.12 - 0.231\left(\frac{a}{W}\right) + 10.55\left(\frac{a}{W}\right)^2 - 21.72\left(\frac{a}{W}\right)^3 + 30.39\left(\frac{a}{W}\right)^4 \tag{9.20}$$

Crack opening COD or δ (at the edge)

$$\delta = \frac{4Fa}{BWE\prime} V_1(a/W) \tag{9.21}$$

$$V_1(a/W) = \frac{1.46 + 3.42\left(1 - \cos\frac{\pi a}{2W}\right)}{\left(cos\frac{\pi a}{2W}\right)^2} \tag{9.22}$$

Load line displacement, *Δ*, at remote points (H/W $\geq$ 6), $\Delta = \Delta_{crack} + \Delta_{no-crack}$, Eq. (9.8):

$$\Delta_{crack} = \frac{4Fa}{BWE\prime} V_2(a/W) \tag{9.23}$$

$$V_2(^a/_W) = \frac{\frac{a}{W}}{\left(1 - \frac{a}{W}\right)^2}\left[0.99 - \frac{a}{W}\left(1 - \frac{a}{W}\right)\left(1.3 - 1.2\frac{a}{W} + 0.7\frac{a^2}{W^2}\right)\right] \tag{9.24}$$

Load line compliance, C: it is always given by Eq. (9.12) using the proper value for Δ, being $\Delta = \Delta_{crack} + \Delta_{no-crack}$, with Δ_{crack} given by Eq. (9.23) and $\Delta_{no-crack}$ by Eq. (9.11).

As already said in Sect. 9.8, Eq. (8.9), G. Irwin suggested to increase the initial length, a, of the crack by an amount Δa equal to the radius r_p of the plastic zone

$$\Delta a = r_p = \frac{1}{\alpha\pi} \cdot \left(\frac{K_I}{\sigma_s}\right)^2 \tag{9.25}$$

with $\alpha = 6$, as discussed in Sect. 8.2. Considering this increase Δa of the initial length a of the crack, for the CC(T) panel under remote traction stress σ the stress intensity factor K_I becomes

$$K_I = \sigma\sqrt{w}\sqrt{\tan(u)}$$
$$u = \frac{\pi}{w} \cdot \left(a + \frac{K_I^2}{\alpha\pi\sigma_s^2}\right) \tag{9.26}$$

With this correction the assessment of K_{I} becomes the result of an iterative process in which an expected initial value K_{Io} is given and then it proceeds by subsequent approximations to the final evaluation of the actual K_{I}. A comparison of the toughness, K_{Ic}, measured on a 3 mm thick 7075 T6 aluminum sheet with and without the correction (9.26), with $a/W = 2$, for different values of the depth to thickness ratio, *W/B*, is given in Fig. 9.6.

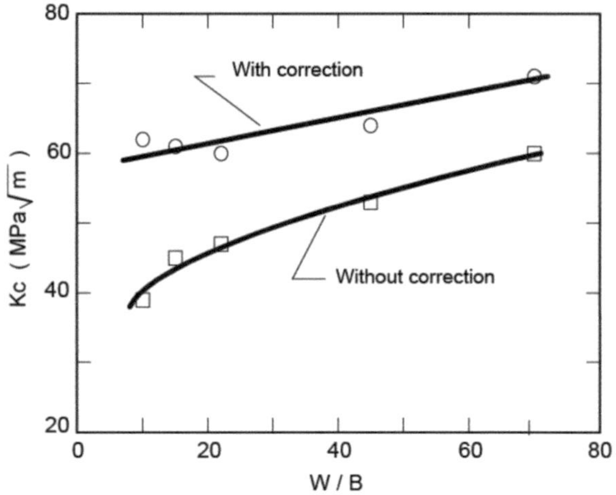

Fig. 9.6 Trend of the apparent toughness K_c measured with a CC(T) panel of Al 7075-T6, 3 mm thick, with and without the Irwin correction for different *W/B* ratios

9.3 Biaxial Fracture Specimens

It was soon recognized that the edge-notched plate specimens of Fig. 9.5 had a low degree of restraint (triaxiality) to provide a full plain strain condition, unless used in very thick sections. However, thick specimens took too large a room in the limited space available in a nuclear power reactor pressure vessel for surveillance programs. To control the radiation embrittlement of the pressure vessel steel exposed to the high energy neutron flux, E $\geq$ 1 MEV (EV = Electron-Volt), hundreds of surveillance specimens had to be introduced in the narrow gap between the vessel steel wall and the central core containing the nuclear fuel bars. These specimens were withdrawn at each scheduled reactor shut down and tested in hot cells facilities to measure the change in fracture toughness of the steel with operation time. Moreover, the stress state in the pressure vessel wall was biaxial, while in panel-like specimens the stress was uniaxial. This biaxial stress state results in a larger triaxiality factor, see Sect. 4.3, and a consequent loss of ductility. Therefore, it was thought that new specimens were needed.

In 1964 Manjoine, at Westinghouse, developed a new specimen to simulate the biaxial stress state existing in the belt line of a pressure vessel, Fig. 9.7 [9]. Along section 1-1 of the Manjoine specimen of area S_1, Fig. 9.7, the stress, σ_1, generated at any point by the applied force F is due to bending, $M_1 = F \cdot b$, and the traction F

$$\sigma_1 = \frac{Fb_1}{W_1} + \frac{F}{S_1} \tag{9.27}$$

while on section 2-2 of area S_2 the stress at the crack tip, σ_2, due to the moment is

$$\sigma_2 = \frac{Fb_2}{W_2} \tag{9.28}$$

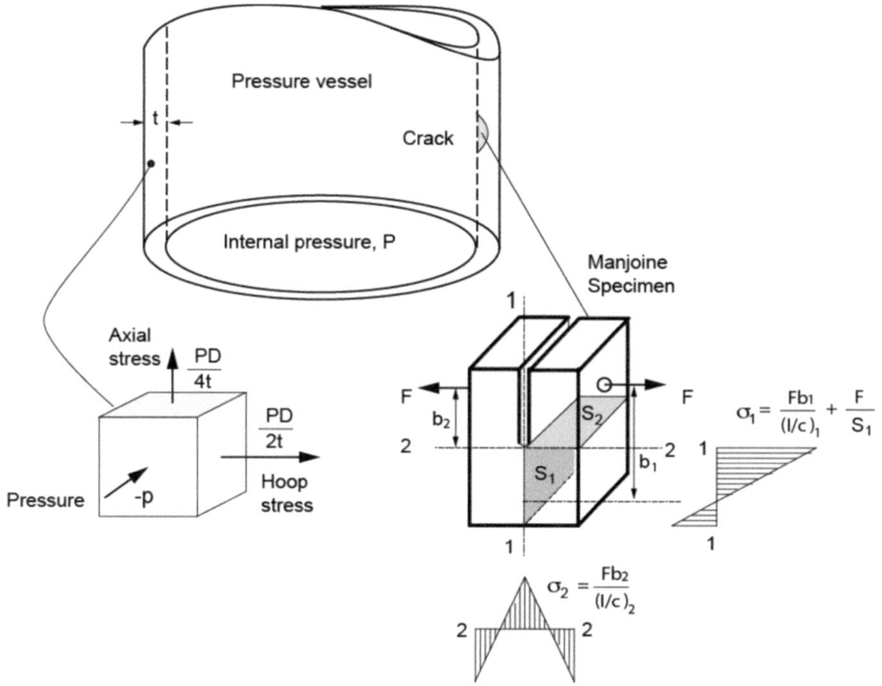

Fig. 9.7 Stress at a point in the pressure vessel wall and in the Manjoine biaxial test specimen (adapted from [9])

where W_1 and W_2 are the resistance modules of sections S_1 and S_2, respectively. This new specimen is shown in Fig. 9.8a. The specimen is known as the X type WOL or Wedge Opening Loaded or simply WOL-X Type. In 1964 a more refined determination of the stress intensity factor, K_I, for the WOL-X Type specimen was obtained experimentally [10] through the compliance method described in the next paragraph, Eq. (9.37).

Later in 1965 a major analytical study was performed that yielded excellent agreement [11]. The K_I expression for the X-Type WOL specimen is

$$K_I = \frac{F}{B\sqrt{a}} f\left(\frac{a}{W}\right) \tag{9.29}$$

or

$$K_I = \frac{F}{B\sqrt{W}} h\left(\frac{a}{W}\right) \tag{9.30}$$

The two functions *f(a/W)* and *h(a/W)* are plotted in Fig. 9.9. These specimens, as said, were put into nuclear reactors for irradiation damage surveillance study. However, when very tough materials were used, excessive plastic bending of the

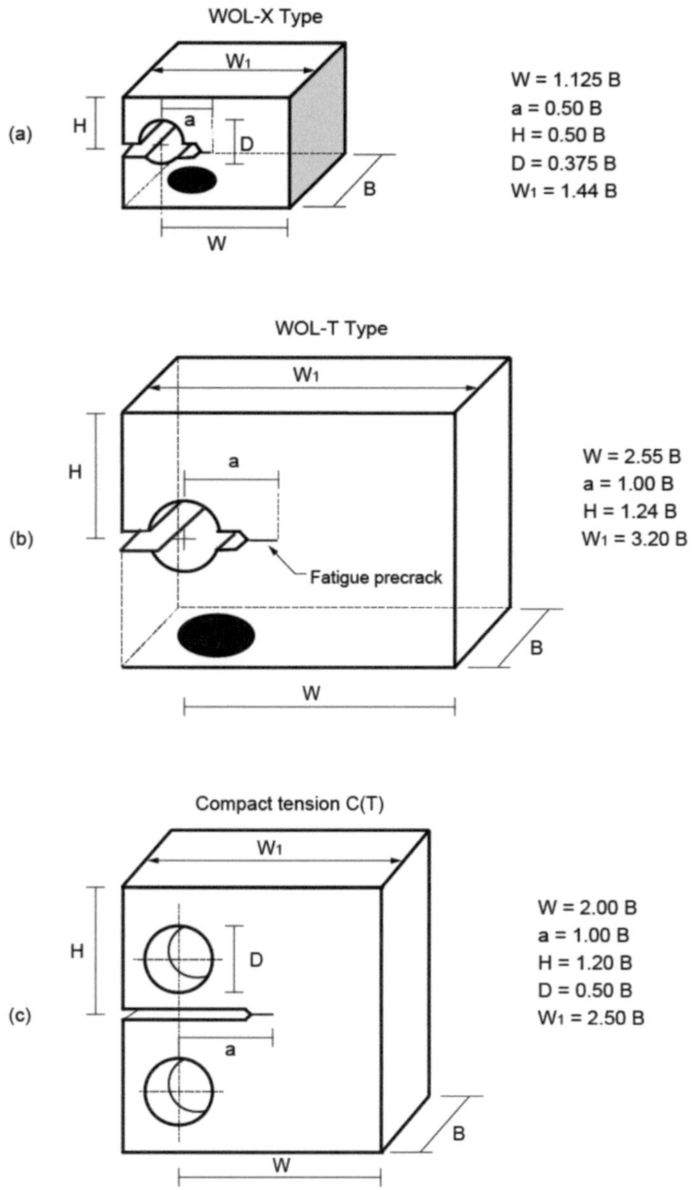

Fig. 9.8 **a** X-Type WOL specimen, **b** T-Type WOL specimen and **c** C(T) or compact specimen developed by Manjoine et al. at Westinghouse

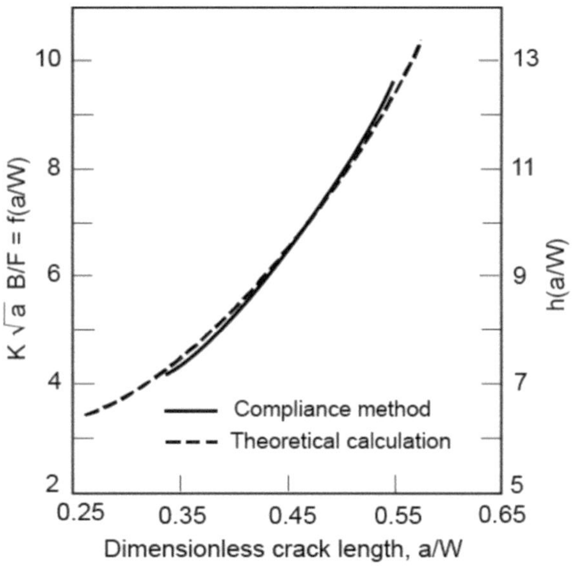

Fig. 9.9 Comparison of non-dimensional stress intensity factors obtained experimentally by the compliance method with those obtained analytically by the boundary collocation method for WOL-X Type specimen

specimen arms was observed. To reduce the occurrence of this plastic bending in 1966 the specimen design was modified to what was referred to as the WOL-T Type specimen, Fig. 9.8b [12].

For this new specimen, Novak and Rolfe [13] measured the compliance C and determined the nondimensional function $g(a/W) = BE'C$ while Saxena and Hudak [14] used the analytical method of the boundary elements to evaluate the K_I expression. Equations (9.29) and (9.30) also apply to the WOL-T Type specimens. The relative $h(a/W)$ multiplied by $\sqrt{(a/W)}$ is plotted in Fig. 9.10 together with the function $g(a/W)$.

9.4 Compact C(T) Specimen

Finally, after long accurate testing and analysis of WOL-X Type and WOL-T Type specimens, Westinghouse researchers arrived to the C(T) or Compact Specimen in Traction, Fig. 9.8c.

The C(T) specimen represents one of the most preferred and used fracture mechanics specimens ever. It became an ASTM standard specimen and appeared in the first ASTM E 399 Standard Test Method [15].

C(T) specimen K_I solution [15]:

$$K_I = \frac{F}{\sqrt{BB_N W}} \cdot h(a/W) \tag{9.31}$$

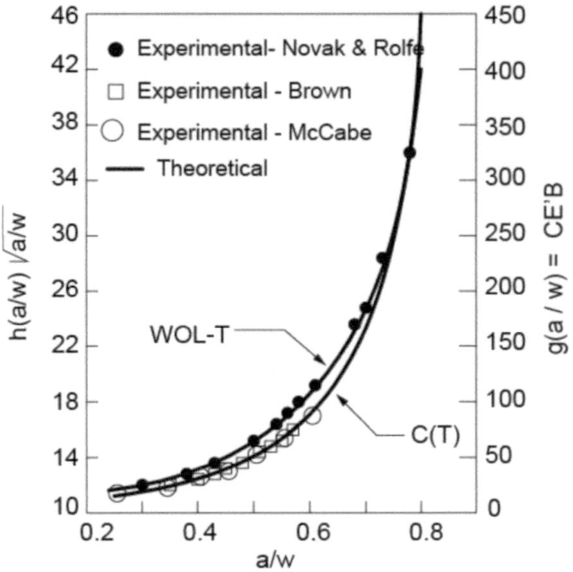

Fig. 9.10 Plot of the functions $h(a/W)\sqrt{a/W}$ and $g(a/W)$ for WOL-T Type and C(T) specimens, respectively

where B_N is the reduced thickness between side grooves, see Sect. 9.6. Without side grooving $B_N = B$.

$$h(a/W) = \frac{2 + \frac{a}{W}}{\left(1 - \frac{a}{W}\right)^{3/2}}\left[0.886 + 4.64\left(\frac{a}{W}\right) - 13.32\left(\frac{a}{W}\right)^2 + 14.72\left(\frac{a}{W}\right)^3 - 5.60\left(\frac{a}{W}\right)^4\right] \quad (9.32)$$

The values of the function $h(a/W)$ are listed in Table 9.1.

C(T) Load Line Opening Displacement Δ, Fig. 9.11b or c:

$$\Delta = \frac{F}{EB'} \cdot g\left(\frac{a}{W}\right) \quad (9.33)$$

with $E' = E/\left(1 - \nu^2\right)$, B' is a reduced thickness given by

$$B' = B - \frac{(B - B_N)^2}{B} \quad (9.34)$$

and the function $g(a/W)$ is [16]

$$g\left(\frac{a}{W}\right) = \frac{15.8}{1 - \left(\frac{a}{W}\right)^2}\left[0.121 + 1.21\left(\frac{a}{W}\right) - 0.159\left(\frac{a}{W}\right)^2 - 1.47\left(\frac{a}{W}\right)^3 + 1.30\left(\frac{a}{W}\right)^4\right] \quad (9.35)$$

or [17]

$$g\left(\frac{a}{W}\right) = \left(\frac{W + a}{W - a}\right)^2$$

Table 9.1 Non dimensional factors for the calculation of K_I and load line displacement Δ for a Compact C(T) specimen

a/w	h(a/w)	g (a/w)	a/w	h(a/w)	g (a/w)
0.450	8.34	10,95	0.555	11.56	14,06
0.455	8.46	11,08	0.560	11.77	14,24
0.460	8.58	11,20	0.565	11.98	14,42
0.465	8.70	11,33	0.570	12.20	14,60
0.470	8.83	11,47	0.575	12.42	14,80
0.475	8.96	11,60	0.580	12.65	14,99
0.480	9.09	11,74	0.585	12.89	15,19
0.485	9.23	11,87	0.590	13.14	15,39
0.490	9.37	12,01	0.595	13.39	15,60
0.495	9.51	12,15	0.600	13.65	15,82
0.500	9.66	12,30	0.605	13.93	16,04
0.505	9.81	12,44	0.610	14.21	16,26
0.510	9.96	12,59	0.615	14.50	16,49
0.515	10.12	12,74	0.620	14.80	16,73
0.520	10.29	12,90	0.625	15.11	16,97
0.525	10.45	13,05	0.630	15.44	17,22
0.530	10.63	13,21	0.635	15.77	17,48
0,535	10.80	13,38	0.640	16.12	17,74
0.540	10.98	13,54	0.645	16.48	18,02
0.545	11.17	13,71	0.650	16.86	18,29
0.550	11.36	13,88	0.655	17.25	18,58

$$\left[2.163 + 12.219\left(\frac{a}{W}\right) - 20.065\left(\frac{a}{W}\right)^2 - 0.9925\left(\frac{a}{W}\right)^3 + 20.609\left(\frac{a}{W}\right)^4 - 9.9314\left(\frac{a}{W}\right)^4\right] \tag{9.36}$$

It is worth noting that Eq. (9.36) provides values that are also function of the crack length, *a*. The values of the function *g(a/W)* given by Eq. (9.35) are listed in Table 9.1.

C(T) Crack Mouth Opening Displacement CMOD, δ (at the front face) [16], Fig. 9.11a or Fig. 9.34

$$\delta = \frac{F}{E'B'} \cdot g\left(\frac{a}{W}\right) \tag{9.37}$$

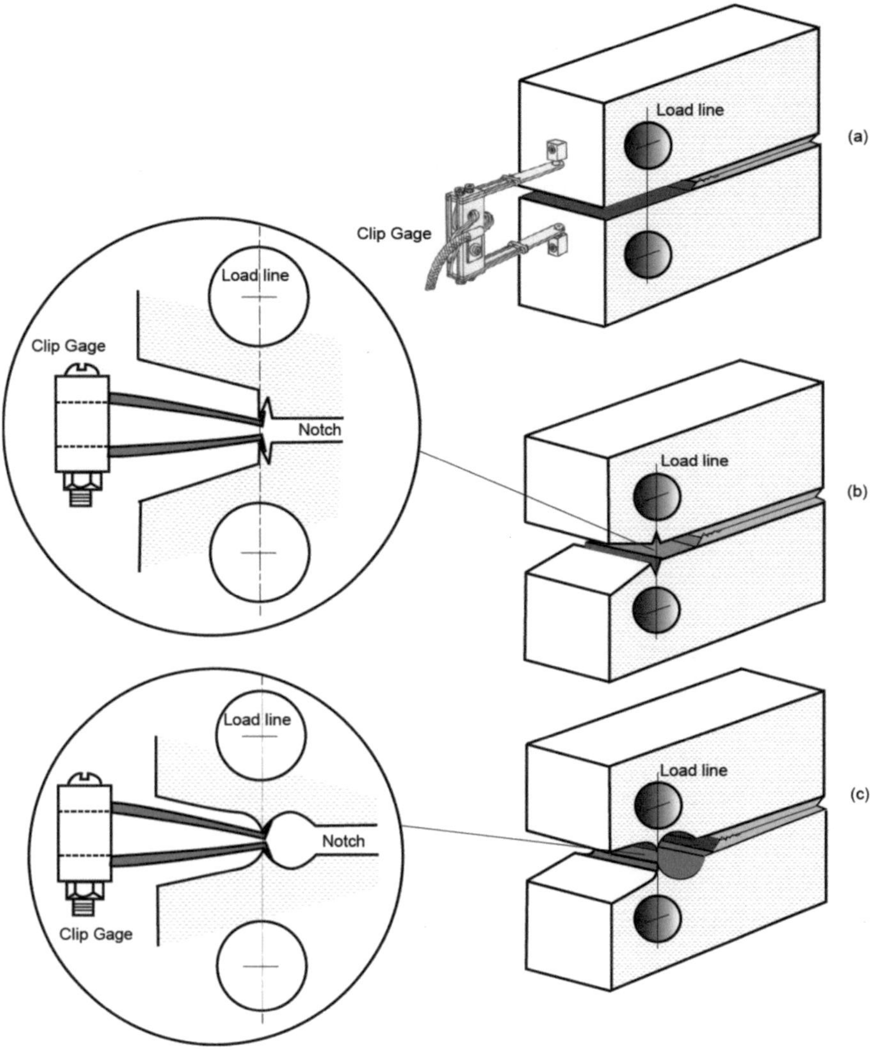

Fig. 9.11 **a** Clip gage on the external surface of C(T) compact specimen to measure COD or δ, also called crack mouth opening displacement (CMOD), **b** and **c** clip gage mounted on the load line to measure load line displacement Δ

Under linear elastic behavior Δ and δ are the same and the $g(a/W)$ function in Eq. (9.36) is that of Eq. (9.35). This equivalence is lost after excessive rotation of the arms.

9.5 Single Edge Notch in Bending, SEN(B)

At the same time when panel and compact specimens were developed, a prismatic specimen with a single edge crack subjected to three points bending load, called SEC(B) Single Edge Crack in Bending or Single Edge Notch SEN(B) or simply SE(B), was introduced. Its standard proportions [15] are shown in Fig. 9.12 and 9.13a. However, the thickness B can be W/2 ≤ B ≤ W.

The specimen takes great advantage from the amplifying action exerted by the bending arms, reducing the amplitude of the applied force *F*. However, the degree of triaxiality at the crack front is not as high as in the C(T) specimen.

SEN(B) specimen K_I solution [15, 16]:

$$\frac{K_I}{\sqrt{2\pi r}} \cos\frac{\theta}{2} f(\theta) = \frac{N\lambda}{2Ct} \tag{9.38}$$

where B_N is the reduced thickness between side grooves, see Sect. 9.9. Without side grooving $B_N = B$.

The function $h(a/W)$ is given by [14, 15, 17, 18]

$$h\left(\frac{a}{W}\right) = \frac{3}{2}\frac{S}{W}\sqrt{\frac{a}{W}} \cdot \frac{1.99 - \alpha(1-\alpha)\left(2.15 - 3.93\alpha + 2.7\alpha^2\right)}{(1+2\alpha)(1-\alpha)^{\frac{3}{2}}} \tag{9.39}$$

with $\alpha = a/W$. Values given by Eq. (9.39) are listed in Table 9.2 for various a/W ratios and shown in Fig. 9.14 in graphical form. Similar values, within a maximum 5–6% deviation for 0.2 < *a/w* < 0.6, were found for the *h(a/w)* function by Gross and Srawley in 1965 [19] using the Boundary Collocation method.

SEN(B) Load Line Opening Displacement *Δ*, Fig. 9.13

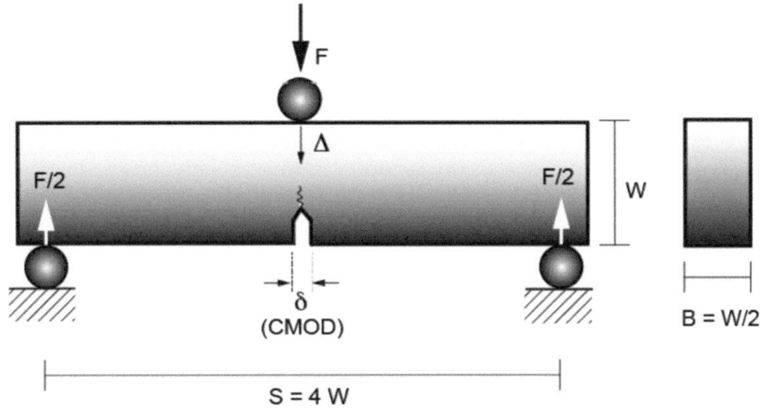

Fig. 9.12 SEN(B) specimen loaded in three-point bending and its standard proportions

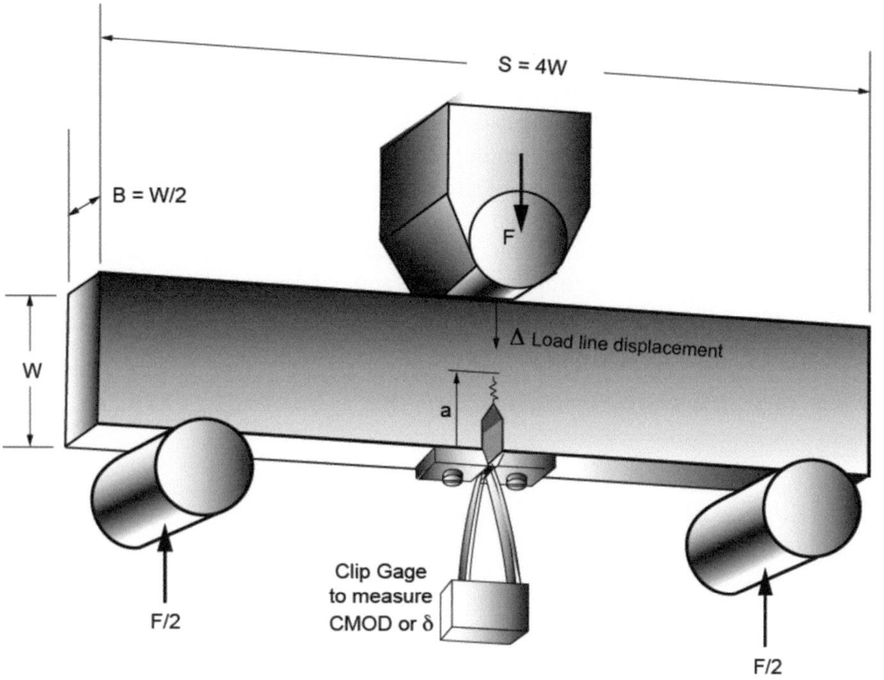

Fig. 9.13 SEN(B) specimen loaded in three-point bending with clip gage to measure CMOD or δ

$$\Delta = \frac{F}{EB'} \cdot g\left(\frac{a}{W}\right) \tag{9.40}$$

with *B'* given by Eq. (9.34) and *g(a/W)*

$$\frac{S^3(1-\nu^2)}{W^3}\left[0.25 + 0.6\left(\frac{W}{S}\right)^2(1+\nu)\right] + 1.5\left(\frac{S}{W}\right)^2\left(\frac{\frac{a}{W}}{1-\frac{a}{W}}\right)^2$$
$$\left[5.58 - 19.57\left(\frac{a}{W}\right) + 36.82\left(\frac{a}{W}\right)^2 - 34.94\left(\frac{a}{W}\right)^3 + 12.77\left(\frac{a}{W}\right)^4\right] \tag{9.41}$$

or [17]

$$g\left(\frac{a}{W}\right) = \left(\frac{S}{W-a}\right)^2\left[1.193 - 1.98\left(\frac{a}{W}\right) + 4.478\left(\frac{a}{W}\right)^2 - 4.443\left(\frac{a}{W}\right)^3 + 1.739\left(\frac{a}{W}\right)^4\right] \tag{9.42}$$

SEN(B) Crack Mouth Opening Displacement CMOD or δ [20], Fig. 9.13

$$\delta = \frac{F}{BE'} \cdot V\left(\frac{a}{W}\right) \tag{9.43}$$

Table 9.2 Non dimensional factors for the calculation of K_I and load line displacement Δ for a SEN(B) specimen

a/w	h(a/w)	V (a/w)	a/w	h(a/w)	V (a/w)
0.450	2.29	26,93	0.555	3.20	48,83
0.455	2.32	27,65	0.560	3.25	50,36
0.460	2.35	28,40	0.565	3.31	51,94
0.465	2.39	29,16	0.570	3.37	53,58
0.470	2.43	29,96	0.575	3.43	55,29
0.475	2.46	30,78	0.580	3.50	57,06
0.480	2.50	31,63	0.585	3.56	58,91
0.485	2.54	32,51	0.590	3.63	60,82
0.490	2.58	33,43	0.595	3.70	62,81
0.495	2.62	34,37	0.600	3.77	64,88
0.500	2.66	35,35	0.605	3.85	67,03
0.505	2.70	36,36	0.610	3.92	69,26
0.510	2.75	37,41	0.615	4.00	71,58
0.515	2.79	38,50	0.620	4.08	73,99
0.520	2.84	39,63	0.625	4.16	76,50
0.525	2.89	40,80	0.630	4.25	79,10
0.530	2.94	42,02	0.635	4.34	81,80
0,535	2.99	43,28	0.640	4.43	84,61
0.540	3.04	44,59	0.645	4.53	87,53
0.545	3.09	45,95	0.650	4.63	90,56
0.550	3.14	47,37	0.655	4.73	93,71

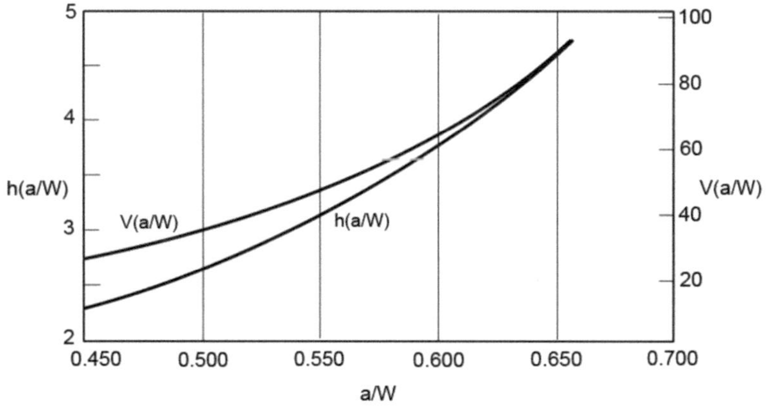

Fig. 9.14 Non dimensional functions *h(a/W)* and *V(a/W)* for a SEN(B) specimen

$$V\left(\frac{a}{W}\right) = 24\frac{a}{W}$$
$$\left[1.45 - 2.18\left(\frac{a}{W}\right) + 13.71\left(\frac{a}{W}\right)^2 - 5.96\left(\frac{a}{W}\right)^3 - 36.9\left(\frac{a}{W}\right)^4 + 70.7\left(\frac{a}{W}\right)^5\right]$$

The function *V(a/W)* is listed in Table 9.2.

9.6 Round Notch Bar in Traction RNB(T)

The Round Notch Bar in Traction, RNB(T), Fig. 9.15, is indeed the specimen that provides the highest degree of triaxiality, nevertheless it is also the least used in fracture toughness measurements.

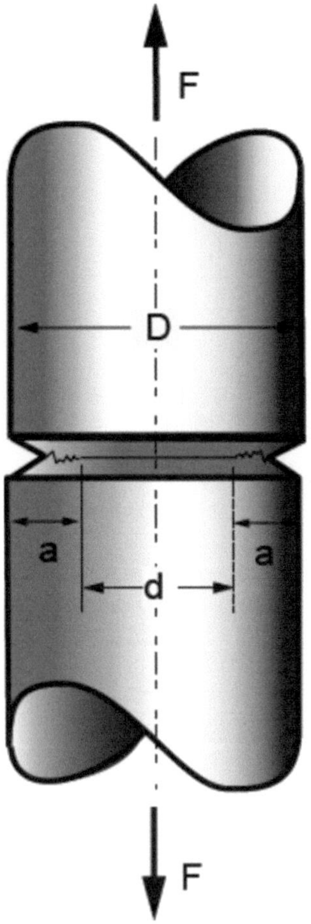

Fig. 9.15 Round Notch Bar in traction or RNB(T) specimen

Probably this is due to the initial difficulty in getting the proper K_I non-dimensional factor and to the amount of material required that needs more loading capacity than any of the other types of specimens. This specimen was first studied by G. Irwin et al. already in 1958 [21]. Wundt [22] and Lubahn [23] added an experimental verification to the theoretical solution [24, 25] which is particularly difficult for the mathematical implications.

RNB(T) specimen K_I solution [25]:

$$K_I = \frac{4F\sqrt{D}}{\pi d^2} h\left(\frac{d}{D}\right) \tag{9.44}$$

with D being the nominal diameter of the rod and d the diameter of the reduced section, Fig. 9.15. The non-dimensional function *h(d/D) is* given by [25]

$$h^2\left(\frac{d}{D}\right) = 0.172 - 0.8\left(\frac{d}{D} - 0.65\right)^2 \tag{9.45}$$

The non-dimensional function, $[h(d/D)]^2$, is plotted in Fig. 9.16 versus *d/D*. Also shown are the experimental points obtained by Irwin [21] and Lubahn [23].

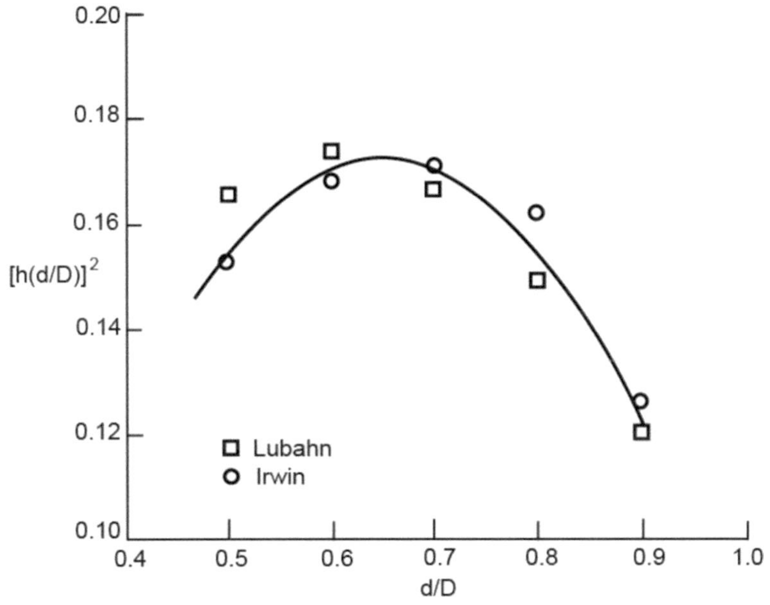

Fig. 9.16 Trend of the dimensionless function $h^2(d/D)$ for a round notched bar in traction, RNB(T), for varying d/*D* ratios. Also shown are the experimental results obtained by Irwin [21] and Lubahn [23]

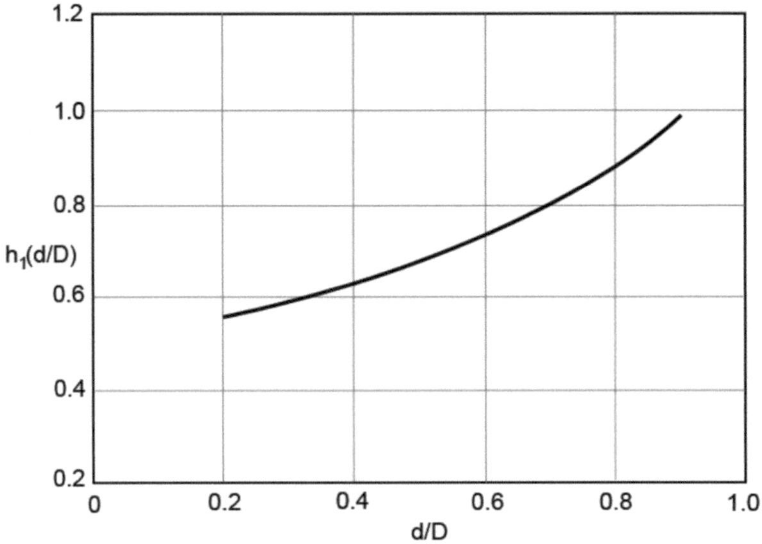

Fig. 9.17 Trend of the dimensionless function $h_I(d/D)$ for a round notched bar in traction, RNB(T), for varying *d/D* ratios

More recent studies by Benthem and Koiter [24] proposed the following expression for K_I.

RNB(T) specimen K_I solution [24]:

$$K_I = \frac{4F}{\pi d^2}\sqrt{\frac{ad}{D}}h_1\left(\frac{d}{D}\right) \tag{9.46}$$

with:

$$h_1\left(\frac{d}{D}\right) = \left[1 + \frac{1}{2}\left(\frac{d}{D}\right) + \frac{3}{8}\left(\frac{d}{D}\right)^2 - 0.363\left(\frac{d}{D}\right)^3 + 0.731\left(\frac{d}{D}\right)^4\right] \tag{9.47}$$

The non-dimensional function $h_I(d/D)$ given by Eq. (9.44) is plotted in Fig. 9.17.

9.7 Fracture Toughness Assessment Using Specimen Compliance

To measure the toughness of a material it may be convenient to recall Eq. (7.145)

$$\mathcal{G} = \frac{F^2 \partial C}{2B \partial a} \tag{9.48}$$

that relates the energy release rate, $\mathcal{G}$, to the compliance C of the specimen of thickness B that can be, for instance, the C(T) specimen of Fig. 9.8c or the SEN(B) specimen of Fig. 9.13 or any other. In a specimen having side grooves, see Sect. 9.6, that reduce the thickness on the crack plane to B_N, Eq. (9.48) becomes

$$\mathcal{G}=\frac{F^2}{2B_N}\frac{\partial C}{\partial a} \tag{9.49}$$

The elastic compliance C can be expressed through a nondimensional function *g(a/W)* of the ratio *a/W* of crack length, *a*, to the specimen size, *W*

$$g(a/W) = CEB \tag{9.50}$$

where E is the Young's modulus. Equation (9.49) then becomes:

$$\mathcal{G}=\frac{F^2}{2\,EWBB_n}\frac{1}{}\frac{\partial g(x)}{\partial a} \tag{9.51}$$

with $x = a/W$. Recalling the equivalence between $\mathcal{G}$ and K_I under plain strain conditions, Eq. (7.141)

$$\mathcal{G}=\frac{K_I^2}{E}\left(1-\nu^2\right) \tag{9.52}$$

or simply

$$\mathcal{G}=\frac{K_I^2}{E'} \tag{9.53}$$

with

$$E' = E/\left(1-\nu^2\right) \tag{9.54}$$

E' being the Young's modulus for plain strain conditions. Combining Eq. (9.51) with Eq. (9.51) we get:

$$K_I = F\sqrt{\frac{1}{2BB_N W\left(1-\nu^2\right)}\frac{\partial g(x)}{\partial x}} \tag{9.55}$$

Recalling Eq. (9.50), since $F = \Delta/C$, where Δ is the displacement on the load line, Eq. (9.55) can be written as

$$K_I = \frac{\Delta}{g(x)}E\sqrt{\frac{B}{2B_N W\left(1-\nu^2\right)}\frac{\partial g(x)}{\partial x}} \tag{9.56}$$

or

$$K_I = \Delta E \sqrt{\frac{B}{B_N W}} f(x) \tag{9.57}$$

with

$$f(x) = \frac{1}{g(x)} \sqrt{\frac{1}{2(1-\nu^2)} \frac{\partial g(x)}{\partial x}} \tag{9.58}$$

or, alternatively, Eq. (9.28) can be written as a function of the applied force F as

$$K_I = \frac{F}{\sqrt{BB_N W}} h(x) \tag{9.59}$$

where

$$h(x) = f(x) \cdot g(x) \tag{9.60}$$

In practice, given a specimen of any geometry containing a crack whose compliance is C, which is a function of the crack length a, by measuring the displacement Δ on the load line or the force, F, it is possible, through Eq. (9.57) or Eq. (9.59), to determine the value of the stress intensity factor K_I of that body or specimen. The value of K_I at which, under an applied critical force F_c, fracture occurs yields the K_{Ic}, i.e., the toughness of the material is:

$$K_{Ic} = \frac{F_c}{\sqrt{BB_n W}} h(x) \tag{9.61}$$

9.8 The CCWL or CCA Specimen

The procedures described in the previous paragraph based on the elastic compliance, C, can be used for the calibration of a new specimen used for the experimental measurement of crack arrest toughness, K_{Ia}, see Chap. 12.

The specimen is the so-called CCWL Compact Crack-Line Wedge-Opening Load or simply Compact Crack Arrest (CCA). The CCA specimen, of the compact type, is shown in Fig. 9.18. The CCWL specimen is loaded by the insertion of a wedge between two split pins that opens the crack in Mode I until the critical displacement Δ is reached on the load line or a critical CMOD, δ, between points A and B of Fig. 9.18, and brittle fracture occurs. The specimen is clearly loaded at fixed displacement so that during the unstable propagation of the crack, as mentioned in Sect. 7.11, the deformation energy stored in the system will decrease progressively and eventually the release rate becomes equal to the energy absorption rate, arresting the crack.

This is why this kind of specimen is used in crack arrest touhgness measurement, as described in Chap. 12. Equations (9.59) and (9.60) provide a practical tool for measuring the applied K_I and, therefore, the toughness of the material at arrest. For the CCWL specimen of Fig. 9.18, the function *g(a/w)*, Eq. (9.48), has been assessed through the experimental measurement of its compliance *C*. To this end, a series of CCWL specimens containing cracks of different lengths, a_i, have been used. Each specimen was loaded at a force F_i and the corresponding displacement δ_t between points A and B, Fig. 9.18, measured. The value of the corresponding displacement on the load line, *Δ*, can be assessed multiplying the CMOD δ_t by a corrective factor. This factor, *Δ*/δ, for the CCWL specimen is ~0.77 (see Table 9.3), practically constant for every *a/w*. Using this experimental procedure, Milella and Pini [26] determined, point by point, i.e., for each crack length, a_i, the compliance $C(a/W) = \Delta(a/W)/F$ of the CCWL specimen and then the analytic function of interpolation, *g(a/w)*, expressed by a polynomial of the 4th degree in $x = a/w$:

$$g(x) = \frac{5.073 + 16.568x + 19.192x^2 - 79.053x^3 + 63.687x^4}{(1 - x)^2} \tag{9.62}$$

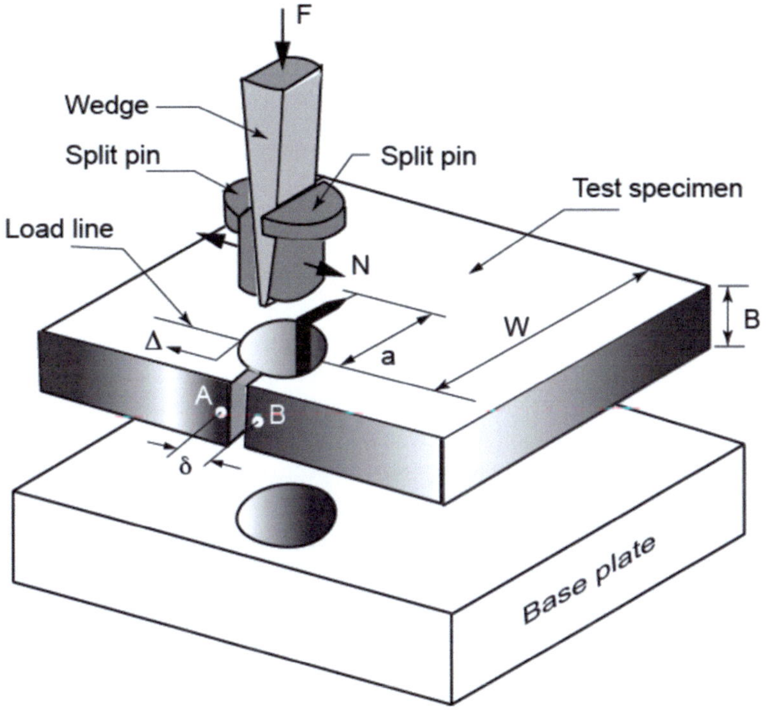

Fig. 9.18 CCWL type specimen loaded by wedge insertion

Table 9.3 Values of the experimental function, $g_{ex}(a/w)$, of the interpolation function g(a/w) and h(a/w) and f(a/w) obtained by Milella and Pini and of the theoretical $g_{N\delta}$ (a/w) e g_{Ny}(a/w) for several values of the ratio a/W

a/W	g_{ex}(a/w)	g(a/w)	h(a/w)	f(a/w)	$g_{N\delta}$(a/w)	g_{Ny}(a/w)	Δ/δ
0.25	16.88	16.76	5.26	0.40	15.64	12.30	0.786
0.30	20.12	20.72	5.79	0.36	19.90	15.51	0.779
0.35	26.36	25.53	6.42	0.32	25.12	19.37	0.771
0.45	39.58	39.08	8.15	0.27	39.73	30.34	0.763
0.50	50.19	49.01	9.53	0.25	50.33	38.42	0.763
0.55	64.17	62.55	11.1	0.23	64.59	49.36	0.764
0.60	82.76	81.88	13.5	0.21	84.63	64.85	0.766
0.65	112.7	111.09	16.9	0.19	114.3	87.91	0.769
0.70	159.7	158.34	21.8	0.18	161.0	124.0	0.770
0.75	240.0	241.52	29.4	0.16	240.8	186.5	0.774
0.80	390.0	405.53	42.6	0.14	392.4	305.7	0.779

Ripling and Crosley [27] following a similar procedure, but using a specimen with side grooves 20% deep, found

$$g(x) = \frac{9.85 - 0.17x + 11.0x^2}{(1 - x)^2} \tag{9.63}$$

The two calibration functions are shown, for comparison, in Fig. 9.19. The figure also shows the experimental points obtained by Milella and Pini [26] interpolated by the relative curve *g(a/w)*.

The agreement can be considered very good. Newman [28] obtained the compliance of the specimen using a FE calculation. His results are shown in Fig. 9.20 together with the experimental results of Milella and Ripling, normalized to $(1 - x)^2$. Again, the agreement is quite excellent for cracks longer than 0.4W. The actual force, *N*, acting on the load line and opening the specimen can be easily inferred from the force *F* applied to the wedge using the procedure described in Sect. 12.6, Eq. (12.27). Using the function *g(x)* given by Eq. (9.62), it is possible to obtain both *f(x)* and *h(x)* through Eq. (9.56) and (9.60), respectively.

$$f(x) = \sqrt{\frac{\left(26,714 + 54,952x - 237,15x^2 + 333,801x^3 - 127,374x^4\right)(1 - x)}{2\left(1 - \nu^2\right)\left(5,073 + 16,568x + 19,192x^2 - 79,053x^3 + 63,687x^4\right)^2}} \tag{9.64}$$

$$h(x) = \sqrt{\frac{26,714 + 54,952x - 237,15x^2 + 333,801x^3 - 127,374x^{4^4}}{2\left(1 - \nu^2\right)(1 - x)^3}} \tag{9.65}$$

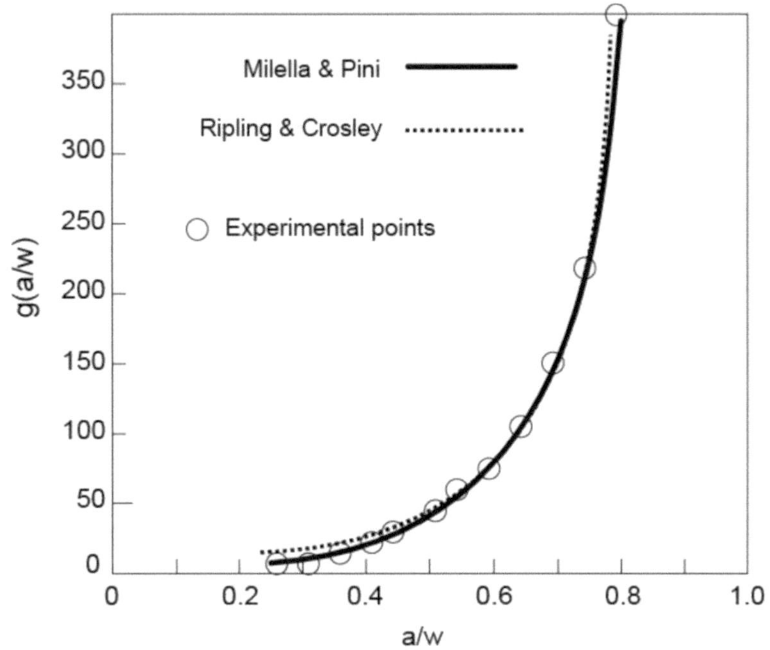

Fig. 9.19 Comparison of experimental points obtained by Milella and Pini with the calibration curve obtained by Ripling and Crowsley for the CCWL specimen

Table 9.3 shows the experimental data obtained by Milella and Pini, their interpolation values, *g(a/w)*, expressed by Eq. (9.62), the calibration functions *f(a/w)* and *h(a/w)* derived from *g(a/w)*, Eqs. (9.64) and (9.65), and both the theoretical function $g_{N\delta}(a/w)$ and $g_{N\Delta}(a/w)$ obtained by Newman with FE analysis for $\delta = CMOD$ and Δ measured on the load line, obtained using the ratio Δ/δ shown in the last column. The three calibration functions are presented in Fig. 9.21. Once the functions *f(x)* and *h(x)* are known, it is possible to assess the applied K_I using Eqs. (9.59) and (9.60). It is worth noting how the function *f(a/W)*, Fig. 9.21, decreases with increasing *a/W* values, i.e., with increasing crack length. This means that under fixed displacement conditions provided by the wedge opening, the stress intensity factor K_I decreases with crack length, indicating a continuous decrease of the deformation energy stored in the specimen.

9.9 Side Grooving

There are several reasons why side grooving is introduced in fracture mechanics specimens, Fig. 9.22. First, it provides a preferential path of propagation to the fatigue precrack avoiding out-of-plain deviations that would activate MODE II crack

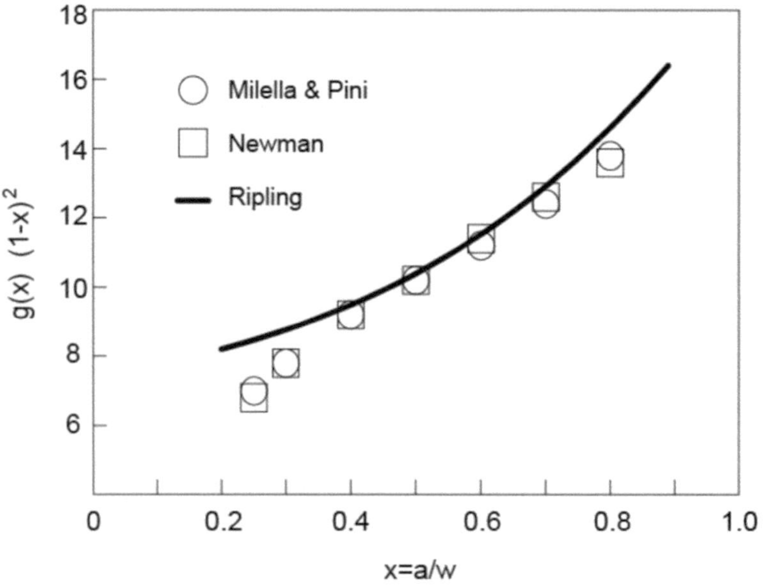

Fig. 9.20 Comparison of the experimental calibration functions obtained by Milella and Pini and by Ripling and Crosley with the FE analysis by Newman, normalized to ($1\!-\!x^2$), for a CCWL specimen

opening, Fig. 7.14, and even MODE III in not thick enough specimens due to possible warping. Secondly, it avoids crack tunneling, Fig. 11.6, which is a non-uniform growth of the crack along the entire propagation front. Finally, side grooves introduce a local stress concentration effect that generates the triaxiality lost on the free surface and, therefore, it increases the apparent thickness, *B*, of the specimen.

This would give the specimen an effective thickness, B_N, higher than the nominal one, *B*. This last effect can be seen in the tests run by Freed on 7075-T6 aluminum using C(T) specimens 6.3 mm and 25,4 mm thick, respectively, Fig. 9.23 [29]. The 6.3 mm thick specimens were a little less thick than the minimum required for a valid test, see Sect. 9.12, and yielded a toughness much higher than that obtained with the 25.4 mm specimen: 55 MPa$\sqrt{m}$ against 30–35 MPa$\sqrt{m}$.

However, by side grooving the smaller specimens the toughness measured decreased arriving to yield the same value obtained with the larger and valid specimens. This occurred already with side grooves 20% of the gross thickness *B*. From that depth on, side grooving effect saturated. Note that valid specimens yielded the same values independently of whether or not they were side grooved. Indeed, 20% side grooving is considered to be the most proper depth.

When side grooves are machined, the effective thickness becomes

$$B_{eff} = \sqrt{BB_N} \tag{9.66}$$

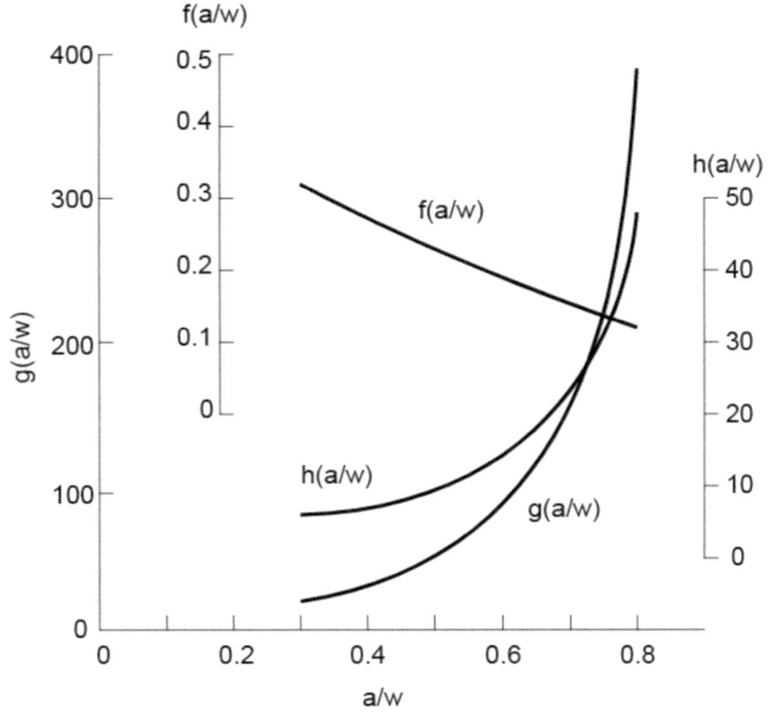

Fig. 9.21 Plot of the functions *f(a/W)*, *h(a/W)* and *g(a/W)* obtained by Milella and Pini for the CCA specimen [26]

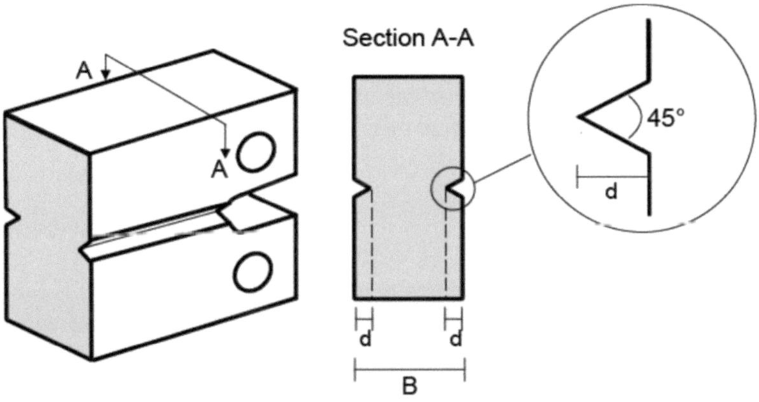

Fig. 9.22 C(T) specimen with side grooves

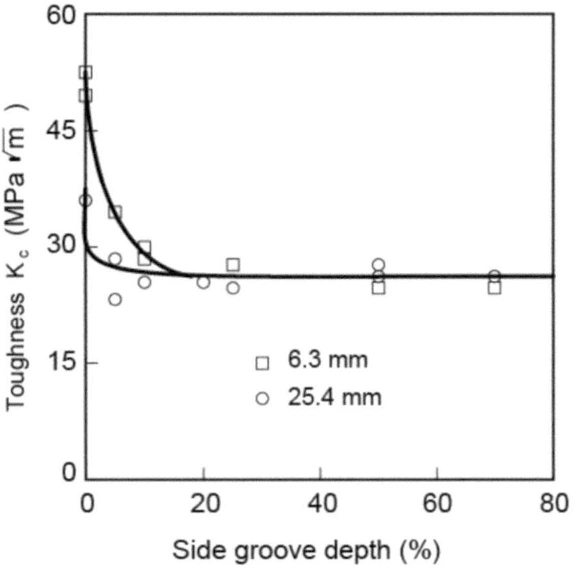

Fig. 9.23 Trend of the apparent toughness measured with specimens 6.3 and 25.4 mm (1in.) thick of 7075-T6 aluminum, as function of the depth of side grooves [29]

9.10 CTOD Testing

The first CTOD test standard was published in the UK in 1979 by the British Standard as BS 5762 [30]. Several years later, in 1981 ASTM published an American version of the CTOD standard as ASTM E813 [31]. The British standard used conversion of crack mouth opening displacement (CMOD) to CTOD using rotation factor, which is also known as plastic hinge model, see Sect. 10.2 and Fig. 10.10. However, the standard ceased in 1991; the geometric CTOD measurement does not exist anymore. Presently, elastic CTOD is calculated indirectly from a K_I measurement using Eq. (8.62), as per ASTM E1820 or E1290 [32]:

$$CTOD = \delta = \frac{K_I^2 (1 - \nu^2)}{mE\sigma_s} \tag{9.67}$$

where $E/(1 - \nu^2) = E'$ is the Young's modulus in plain strain and m is the plastic constraint factor given by Eq. 8.63. This, actually, raises the question of why CTOD should still be used when we already have the value of the stress intensity factor K_I?

9.11 K_I Stress Field Evaluation Using Photoelastic Analysis

Photoelasticity is an experimental technique of stress and strain analysis which uses the temporary birefringence exhibited by most transparent media when they are subjected to strain. Birefringence is a phenomenon in which a ray of light passing through a given material experiences two refractive indices. This can be thought of as dividing the light into two beams with orthogonal vectors that are orientated to the directions of the principal strains in the plane normal to the light path. The magnitude of the refractive indices at each point in the material is directly related to the stress state at that point.The difference in the refractive indices leads to a relative phase retardation between the two components. On exit from the medium the two light beams will be elliptically polarized, i.e., they will be out of phase with one another except where the principal strains are equal. When the beams are combined by a polarizer, an interferogram is produced and fringes appear. The fringes in the interferogram, known as isochromatics, are contours of the relative retardation of the two beams which is directly proportional to the integral along the light path, t, of the difference in principal strains. The instrument employed is known as a polariscope and typically consists of a light source illuminating the component of interest on which a transparent photoelastically effective layer has been deposited, as shown in Fig. 9.24. Fringe orders, N, can be assigned based on the stress-optic law

$$N = (\sigma_1 - \sigma_2)\frac{Ct}{\lambda} \tag{9.68}$$

where C (m^2/N) is the optical stress constant and t (m) is the thickness of the layer. The position of these fringes is a function of the wavelength of the light, λ(m), as well as the principal stresses $\sigma_{1,2}$. Isochromatic fringes are lines of constant principal stress difference, $(\sigma_1 - \sigma_2)$. If the source light is monochromatic these appear as dark and light fringes, whereas with white light illumination-colored fringes are observed.

The fringes give the entire stress distribution and allow the measurement of stress direction and magnitude at any point. Figures 9.25 and 9.23b and c provide three examples of fringe patterns obtained by Fourney [33] and Sanford et al. [34, 35] on a SEC(B) specimen, a center-crack panel, CC(T) and a Compact C(T) specimen, respectively. From a mechanical point of view Eq. (9.68) states the direct proportionality between the birefringence in the plane orthogonal to the *kth* principal direction of stress and the shear stress in the same plain

$$\tau_{\max} = (\sigma_1 - \sigma_2)/2 \tag{9.69}$$

Recalling Eqs. (7.114), Eqs. (9.68) and (9.69) yield (Figs. 9.26 and 9.27)

$$\frac{K_I}{\sqrt{2\pi r}} \cos\frac{\theta}{2} f(\theta) = \frac{N\lambda}{2Ct} \tag{9.70}$$

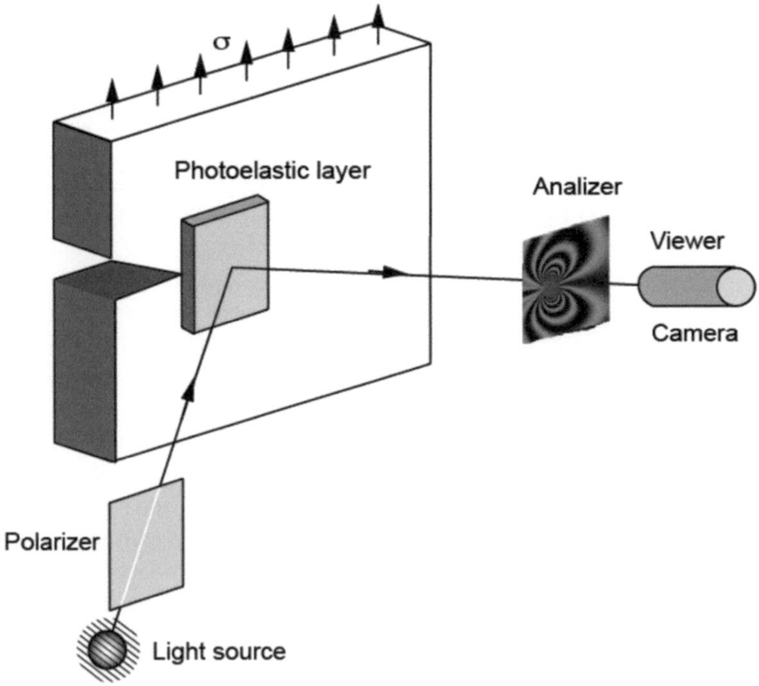

Fig. 9.24 Schematic of test arrangement for photoelastic analysis

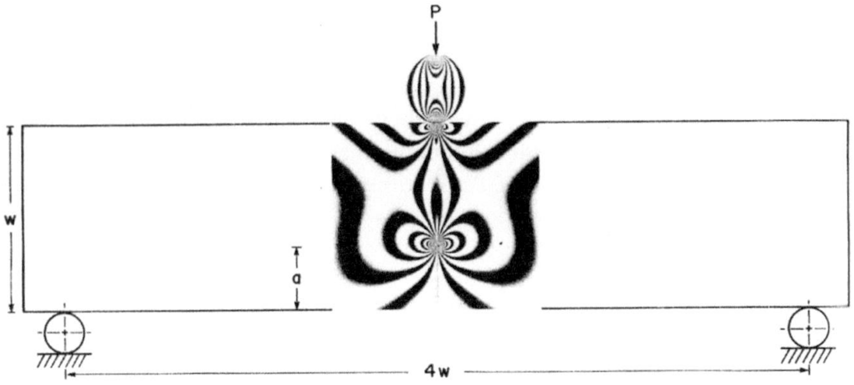

Fig. 9.25 Isochromatic fringes created around a crack in the central part of a SEC(B) specimen [33]

Fig. 9.26 Isochromatic fringes at the crack tip of a CC(T) specimen [34]

Therefore, by measuring the second term in Eq. (9.70) it is possible to infer K_I. The experimental usefulness of such a relation thus becomes clear: the maximum birefringence is proportional to the maximum shear and accordingly can be related to stress intensity factor K_I. To-day, the difficulty in observing and count each individual fringe is overcome in digital photoelasticity where fringe patterns, that contain the whole field stress information in terms of the difference of principal stresses (isochromatics), and their orientation (isoclinics) are captured as a digital image which is processed for further quantitative evaluation. This is well visible in Fig. 9.28 that shows the difference between a fringe pattern observed with the normal procedure and the same pattern after being processed using mean values of computed coefficients. Fringe orders are easily recognized.

Fig. 9.27 Isochromatic fringes at the crack tip of a C(T) specimen [35]

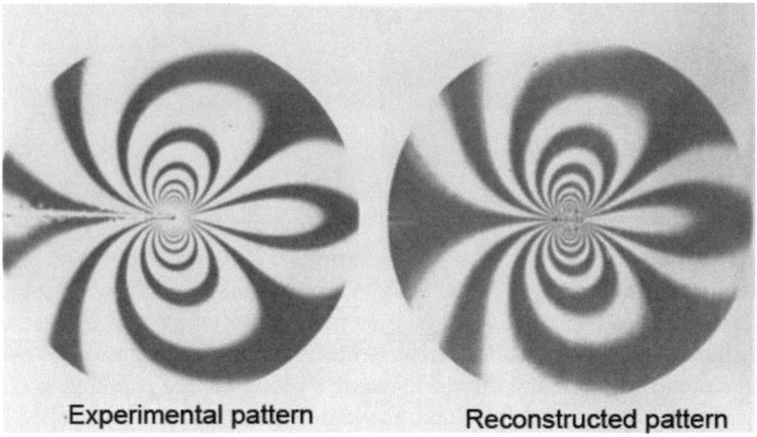

Fig. 9.28 Fringe pattern around a crack in the SEC(B) specimen of Fig. 9.20b and reconstructed pattern (modified from [33])

9.12 The Size Requirement

We have already treated the subject of thickness and its effect on the triaxiality and brittle fracture of metals from a qualitative point of view, see Sect. 4.4. With the introduction of LEFM and the consequent possibility to measure fracture toughness with small specimens, the question of adequate thickness became important also from the quantitative point of view. How thick should a fracture mechanics specimen be to be considered sufficiently thick to yield a valid toughness measurement? This question became soon very important. This section will provide a brief presentation of the long and slow way followed by researchers toward the definition of a minimum thickness of a specimen to yield a valid fracture toughness measurement. Imagine a fracture that propagates in a body initially thin but of increasing thickness, as shown in Fig. 9.29 it will be noticed a progressive change in the external appearance of the fracture. Started as a tearing induced by shear stress and therefore slant, Fig. 9.29a, the fracture, as it progresses through larger and larger thicknesses, changes into an increasingly flat brittle fracture, reducing the ductile areas or shear lips that eventually disappear.

This is shown schematically in Fig. 9.29c which also shows the typical trend of the load-deformation curve for the various thicknesses and the so-called ductile or shear fracture factor *SFF* Fig. 9.29b, defined as:

$$SFF = \frac{B - f}{f} \tag{9.71}$$

where B is the thickness of the body and f the extension of the brittle or cleavage area.

In thin bodies fracture is always slant or oblique, on planes at 45° with the applied force and the corresponding SFF = 100%, Fig. 9.29b. The stress–strain diagram is that of Fig. 9.29c-A. The concept of thin body is relative as it depends, as it is easy to guess, on the toughness of the material. Thus, a very brittle material, such as high strength steels or some aluminum alloys, can break through a net cleavage mechanism already with thicknesses of 0.50 cm that will therefore be "*large*" and not "*thin*" as it would be, instead, for a ductile one that will require 15–20 cm to develop the same type of fracture. In the following it will be clearly defined what is meant by small or large thickness which will necessarily be associated with the specific toughness of the material. G. Irwin [36] related the toughness of a material, K_{Ic}, to the amount of lateral contraction in the transverse direction, δ_z, through the relation

$$K_{Ic} = \sqrt{\sigma_s \delta_z E} \tag{9.72}$$

where σ_s is the yield stress and E is the Young's modulus of the material. Since the minimum value of toughness K_{Ic} actually measurable in steels is equal to about 30–40 MPa$\sqrt{m}$ and the highest yield stress rarely exceeds 2000 MPa, it follows that

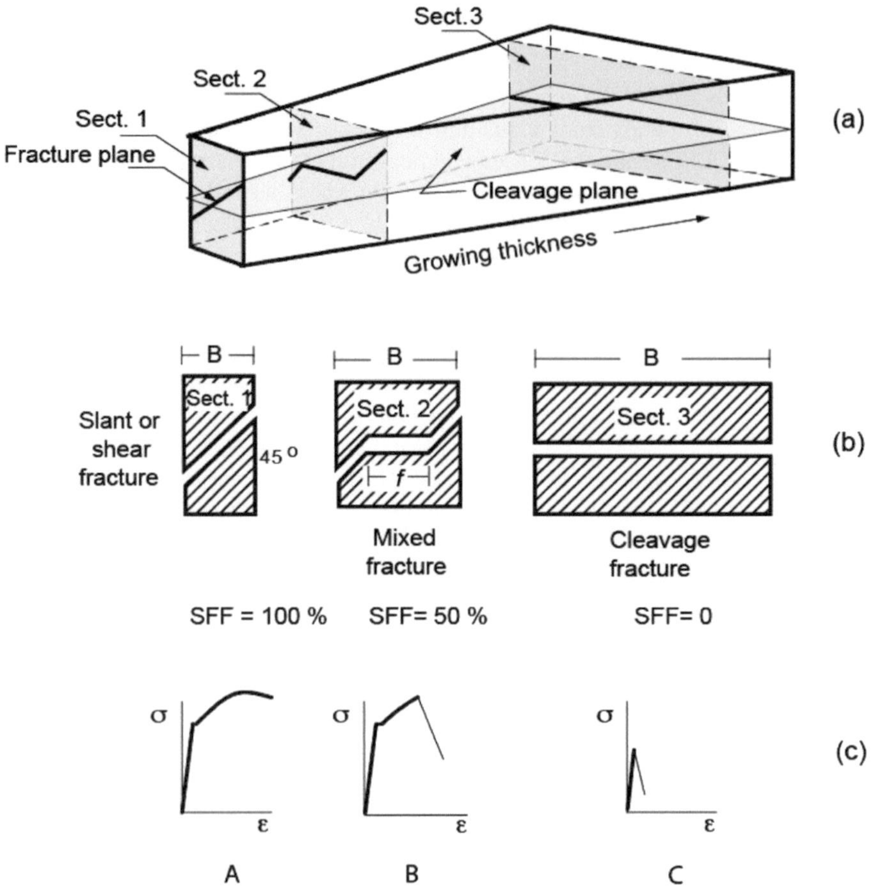

Fig. 9.29 Increasing the thickness of a body in which fracture propagates, the mode and appearance of crack propagation changes from a slant, 45° degrees shear stress induced fracture to a completely flat one of brittle nature, Fig. 9.20a and b. The corresponding stress–strain diagram, Fig. 9.22c, changes from a full ductile behavior to a brittle one

δ_z can never be zero but, at minimum, will be worth about $3\cdot 10^{-3}$–$4\cdot 10^{-3}$ mm, very close to zero. Equation (9.72) implies, as expected, that the contraction, δ_z, is higher in tough and lo strength materials. Let us introduce the factor β

$$\beta = \frac{2\pi r_P}{B} \tag{9.73}$$

which defines the relationship between the radius of the crack tip plastic zone, r_p, and the thickness B. When fracture occurs, the factor β reaches its critical value which will be indicated by β_c or β_{Ic} for a true brittle fracture with a valid toughness K_{Ic}. Recalling Eq. (8.6) it yields

$$\beta = \frac{1}{B} \cdot \left(\frac{K_c}{\sigma_s} \right)^2 \tag{9.74}$$

When the plastic zone size is equal to the thickness, $r_p = B$, it is $\beta = 2\pi$. This value marks the transition from brittle to ductile type fracture. This is clearly visible in Fig. 9.30 which refers to a series of burst tests carried out on high-strength steel cylinders of about 30 cm diameter and 2 mm thick containing small surface cracks [37]. The figure shows the relationships between the circumferential burst stress to the yield strength ratio, σ_{brst}/σ_s, as a function of the parameter β_c. The value of β_c was calculated for each test through Eq. (9.73). The data refer to five steels with ten different values of yield stress that varied from a minimum of 1,250 MPa to a maximum of 1,650 MPa. Results presented in Fig. 9.30 indicate that below $\beta = 2\pi$ the fracture of pressure vessels always occurs at a stress lower than the yield stress of the materials used. At variance, above $\beta = 2\pi$ fracture occurs at a stress higher than the yield stress, indicating the ductile nature of the event. Therefore, the brittle to ductile transition can be identified by $\beta_c = 2$. Further validation of this finding is given by a series of fracture experiments conducted on plates of steel, titanium and aluminum of variable thickness containing a central through crack subjected to a traction stress normal to the plane of the crack. Results are shown in Fig. 9.31.

In this case the factor β_c is indexed to the percentage of shear stress or SFF, Eq. (9.71). The thickness of the plates was ranging from 1.8 to 5.6 mm at room temperature [36]. It can be seen that for $\beta_c < 2$ the fracture is always brittle. As β_c

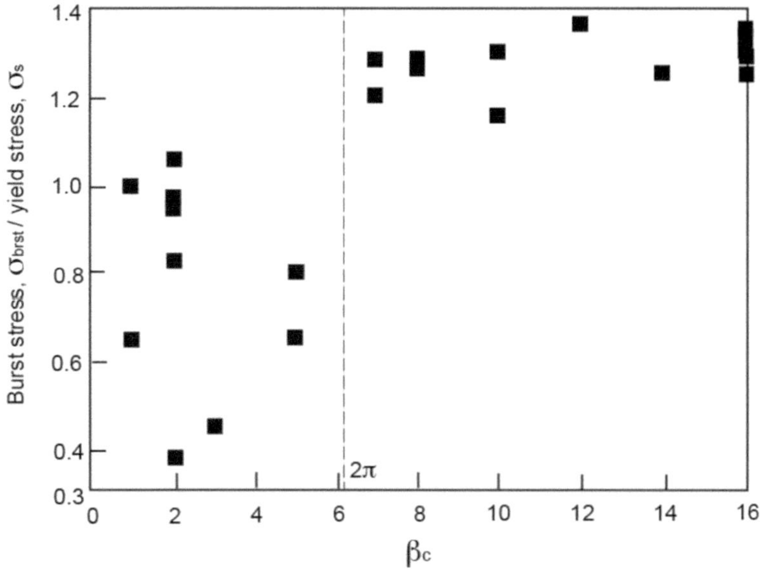

Fig. 9.30 Trend of the circumferential burst stress to the yield strength ratio, σ_{brst}/σ_s, as a function of the parameter β_c for a series of high strength steel vessels. Above $\beta = 2\pi$ the fracture becomes totally ductile (data from [37])

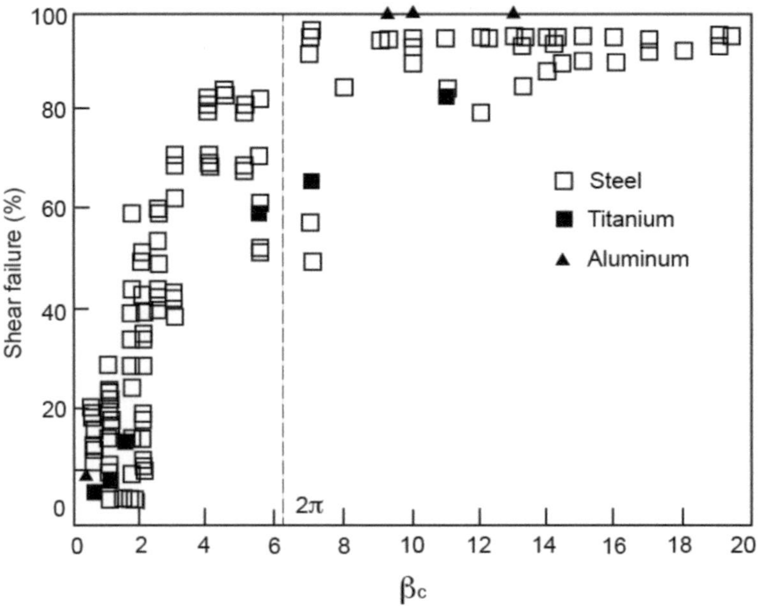

Fig. 9.31 Trend of the percentage of ductile or shear fracture, as a function of the β_c factor for a series of steels, titanium and aluminum alloys. For β_c lower than 2 the fracture is brittle. Over 2π the fracture is ductile (adapted from [38])

increases, the percentage of ductile fracture also increases until, having reached 2π the fracture is totally ductile.

The same trend will be observed on the toughness K_{Ic}. An example is shown in Fig. 9.32 for an 18% Ni maraging steel with high tensile strength, $\sigma_s = 1700\text{MPa}$ [37]. It can be seen how the apparent toughness, K_c, decreases as the thickness of the specimen increases. The term *apparent toughness*, K_c, indicates a toughness that may not be the real one and needs to be verified because of insufficient thickness. While for thicknesses below 10 mm, open circle, the apparent toughness is very high, $K_c > 140$ MPa$\sqrt{}$m, as the thickness increases over 15 mm plain strain conditions are approached and the toughness drops to 80 MPa$\sqrt{}$m that can be considered the true toughness of the material K_{Ic}.The figure also shows three points (solid circles) where a brittle fracture event occurred at local level, single grains, but did not propagate throughout the entire specimen because the local fracture was confined by the surrounding ductile matrix. This phenomenon is called pop-in, Fig. 9.38. Figure 9.32 also shows the locus of points having $\beta = 0.4$, Eq. (9.74).

This curve separates the zone where the toughness K_c seems to be depending on the thickness from that where it becomes independent and reaches a lower limit that represents the true toughness of the material, K_{Ic}. Another example is given in Fig. 9.33 which shows the trend of the toughness K_c of an aluminum alloy, type 7075-T6 with yield strength equal to 497 MPa, as function of the specimen thickness [38]. It can be seen that for thicknesses greater than 12 mm the toughness K_c reaches

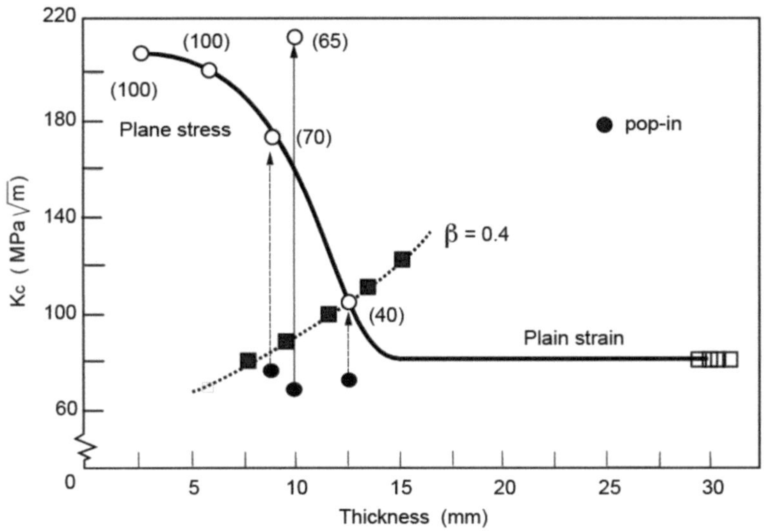

Fig. 9.32 Trend of the apparent toughness K_c in maraging steel 18% Ni specimen as function of the thickness. The values in parentheses indicate the percentage of ductile fracture. The solid circles represent events of fracture at local level (pop-in) confined by the ductility of the matrix. Also marked by solid squares is the locus of points with $\beta = 0.4$ constant crossing the toughness curve at the foot of the ductile to brittle transition. The true toughness is $K_{Ic} = 80$ MPa$\sqrt{}$m [34]

a minimum value. This lower limit indicates that a definitive condition of plain strain state occurred. Characteristic is also the drop of the toughness for very small thicknesses before the apparent maximum of ~100 MPa$\sqrt{}$m. This is because for such a small thickness, less than 1 mm, specimens cannot be geometrically stable. Figure 9.33 also shows the trend of β equal to 2 and 0.4, respectively. It defines an area where the material changes its behavior from ductile to brittle. One more example of the dependence of toughness on thickness is presented in Fig. 9.34 for a titanium alloy type 6 A1-6 V-2Sn aged for 4 h at 600 °C, having a yield strength of 1218 MPa [39].

Once again it can be observed that the apparent toughness K_c reaches its minimum value of about 30 MPa$\sqrt{}$m when the thickness grows over 10 mm. From all the examples presented it turns out that the line $\beta = 0.4$ is a suitable index to mark the transition from a ductile to a full brittle behavior of materials. Recalling Eq. (9.74) we obtain that the corresponding thickness B is:

$$B = 2.5\left(\frac{K_I}{\sigma_s}\right)^2 \tag{9.75}$$

An extensive experimental study on this issue [40–44] has confirmed the adequacy of the thickness requirement expressed by Eq. (9.75) which has led to the ASTM (American Society for Testing and Materials) Standard E-399-74 [45] requirement

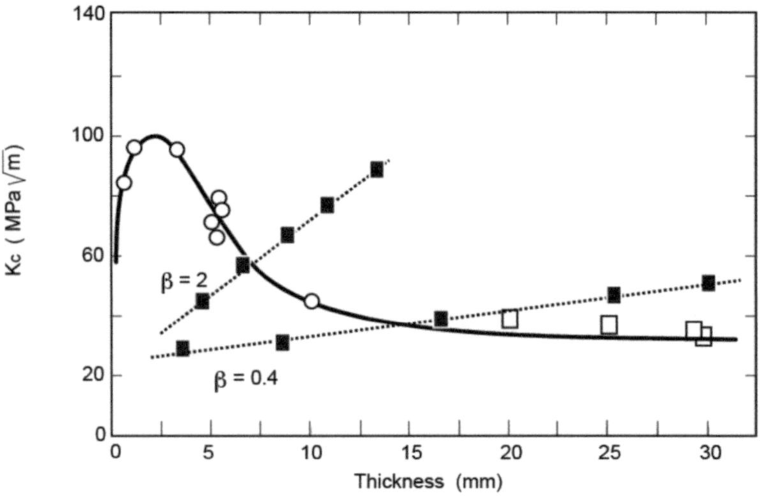

Fig. 9.33 Trend of the apparent toughness K_c of an aluminum alloy, type 7075 T6, with thickness variation. The lines of constant β equal to 2 and 0.4 mark the transition from ductile to brittle behavior [35]

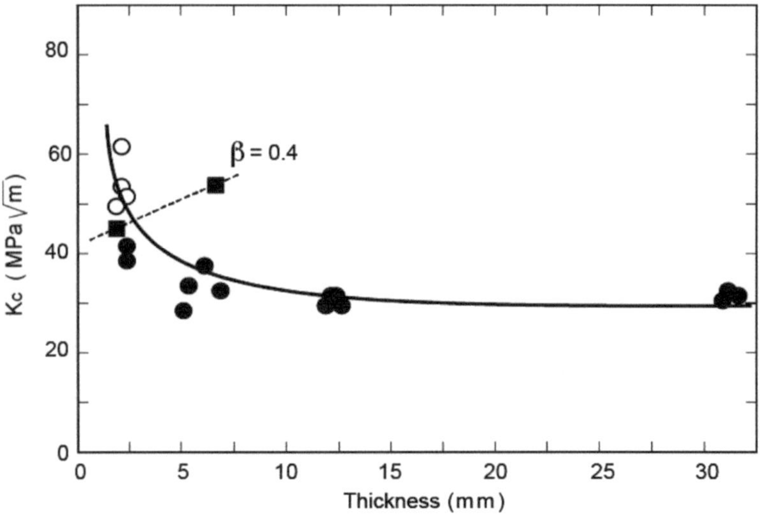

Fig. 9.34 Trend of apparent toughness K_c for a titanium alloy 6A1 6 V 2Sn with thickness. Open circles indicate unacceptable toughness values for insufficient thickness. Beyond line $\beta = 0.4$ the toughness drops to the lower limit [36]

Table 9.4 Measurement capability of C(T) fracture mechanics specimens and their dimensions

Type C(T) Size (inch)	K_{Ic}/σ_s $(\sqrt{m})$	$(K_{Ic}/\sigma_s)^2$ (m)	Thickness (cm)	Hight H (cm)	W (cm)
0.5 T-CT	0.07	0.005	1.27	3.04	3.17
1 T-CT	0.1	0.01	2.54	6.09	6.35
2 T-CT	0.14	0.02	5.08	12.2	12.7
3 T-CT	0.17	0.03	7.62	18.3	19.0
4 T-CT	0.20	0.04	10.16	24.4	25.4
6 T-CT	0.247	0.06	15.24	36.6	38.1
8 T-CT	0.285	0.08	20.32	48.8	50.8
10 T-CT	0.318	0.10	25.4	61.0	63.5
12 T-CT	0.349	0.12	30.48	72.2	76.2

for all specimens dimensions

$$a, B, (W - a) \geq 2.5\left(\frac{K_I}{\sigma_s}\right)^2 \tag{9.76}$$

to yield a valid K_{Ic}. Table 9.4 lists the testing capability of C(T) specimens expressed in terms of specific toughness, K_{Ic}/σ_s, as a function of their dimensions according to the ASTM thickness requirements, Eq. (9.75). Dimensions H and W are specified in Fig. 9.7c. Table 9.4 indicates that to measure the toughness of a brittle alloy such as the titanium alloy of Fig. 9.34, having a yield strength of 1218 MPa and a possible toughness of about $K_c = 30$–40 MPa$\sqrt{\text{m}}$, which yields a ratio K_c/σ_s between 0.025 $\sqrt{\text{m}}$ and 0.033 $\sqrt{\text{m}}$, we need using a small 0.5 T-C(T) specimen 0.5 inch thick.

Conversely, if we want to measure the toughness of a ductile carbon steel of 500 MPa yield strength having a provisional toughness at room temperature of about 150 MPa$\sqrt{\text{m}}$ that yields a ratio K_c/σ_s of 0.3 $\sqrt{\text{m}}$ we need a huge specimen 8 T-C(T) 8 inch thick (20.32 cm). However, with such a large toughness the corresponding CTOD is about 0.13 mm, see Eq. (9.67), which raises the question of whether the crack is still a crack or a notch which invalidates the use of linear elastic fracture mechanics and the K_I.

9.13 Fracture Mechanics Test

The first standardized test method for K_{Ic} measurements was published in 1970 as ASTM standard E 399 [45]. It marked a turning point in the linear elastic fracture mechanics application. The first step in running a fracture mechanic test is the choice of the specimen and its dimensions. Table 9.4 can be used to establish a first-choice dimension of the specimen based on a first-guess toughness of the material used.

The first test results will confirm the choice otherwise it will be labeled as K_Q. The measurement of a valid K_{Ic} fracture toughness implies a brittle behavior of the material. This means that, whatever the specimen chosen may be, the load displacement diagram must show a linear relationship between load, P, and displacement, V, without any significant nonlinearity at the maximum load.

The initial machined V-notch with a root radius generally not smaller than 0.08 mm, is not enough sharp to provide the required natural crack with a tip radius of the order of 0.005 mm, (see Figs. 6.26 and 6.27). To this purpose the initial machined notch must be fatigue cycled to generate a transgranular fatigue crack that resembles a natural crack, Fig. 9.2b. It is important to understand that the fatigue cycling must be imposed at a ΔK_I so low as to avoid the formation of an initial too large plastic zone incompatible with linear fracture mechanics requirements. It is recommended to fatigue cycling at a ΔK_I not exceeding 60% at most of the provisional K_Q value of the material. This limitation also ensures that very brittle materials do not fracture during the fatigue cycling when the first crack appears. Since the toughness of a material, as the Charpy V-notch energy (resilience) see Figs. 3.26, 3.27 and 3.28, depends on the orientation and direction of crack propagation with respect to a reference direction, care must be done to the picking direction of the specimen from the plate either rolled or forged. Figure 9.35 identifies those directions by two letters. The first letter indicates the direction of the load that opens the specimen. The second letter corresponds to the direction of crack propagation. Figure 9.36 shows a compact type specimen with loading grips and strain gages applied on the front face measuring a CMOD (Crack Mouth Opening Displacement) at a distance z from the front face that must be taken into consideration, Fig. 9.36b. Displacements can be measured directly on the load line, as shown in Fig. 9.11b and c.

Caution must be taken to ensure that the fatigue pre-cracking has a linear front. To this purpose, the initial notch can be machined with a V-chevron shape, as shown in Figs. 9.36b and 9.42, or the specimen can be side grooved, see Sect. 9.9. Actually, ASTM E 399 standard does not permit side grooving, but the new ASTM E 1820

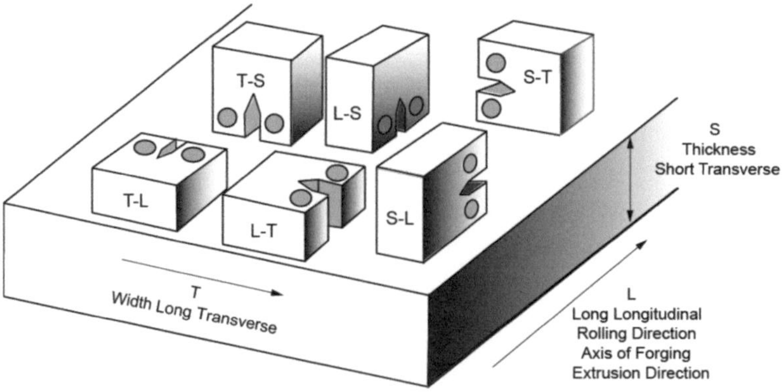

Fig. 9.35 Crack plane orientation identification code for specimens taken from a plate

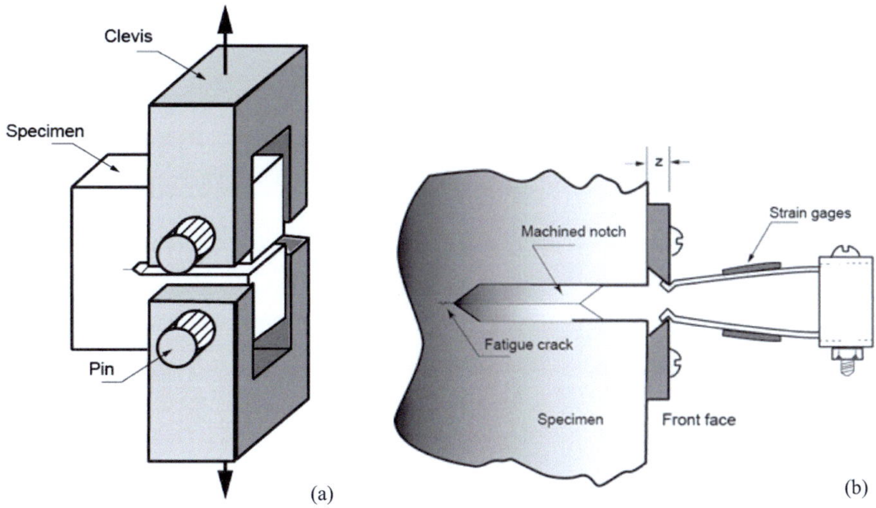

Fig. 9.36 **a** Compact specimen on the testing rig; **b** double cantilever clip gage detail to measure the crack mouth opening displacement (CMOD)

does allow side grooves in K_{Ic} tests. The specimen is pulled open to failure while the load, *P*, versus displacement, *V*, is recorded. The test record may be one of the three types of load–displacement acceptable diagrams shown in Fig. 9.37.

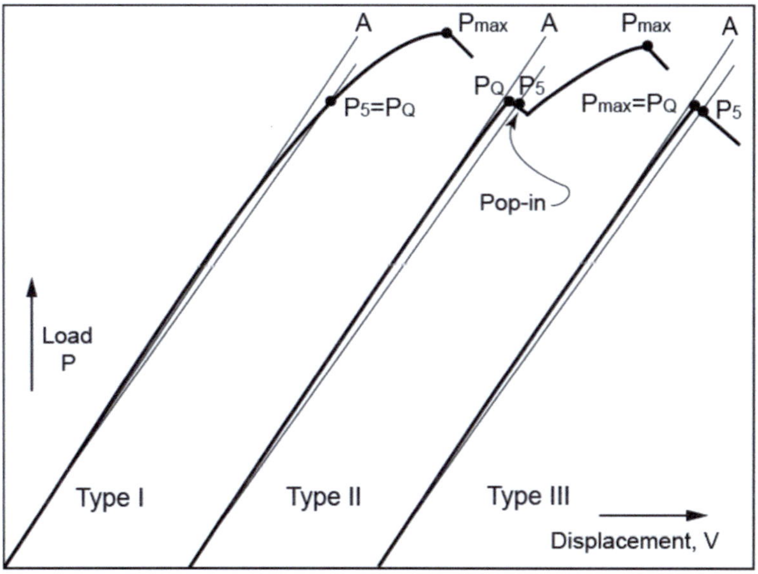

Fig. 9.37 Principal types of load–displacement record

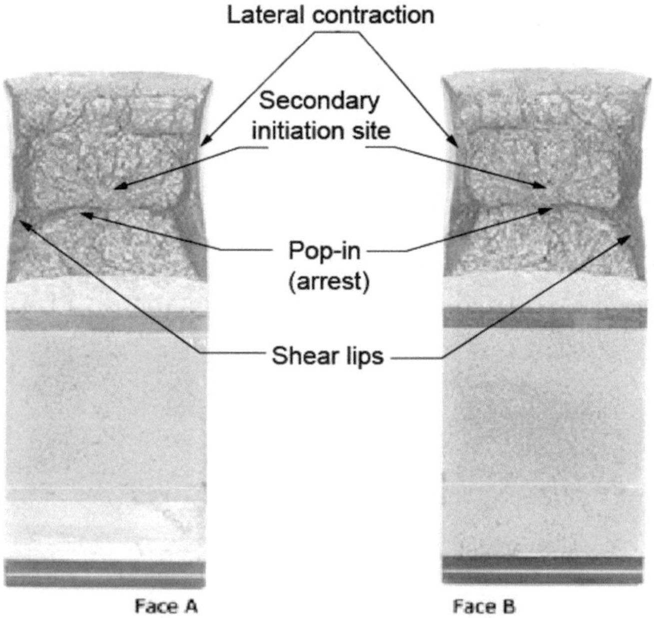

Fig. 9.38 Fracture surface of 1"CT specimen of weld metal on P 690 QL1 steel showing a pop-in event that arrested and restarted from a secondary site resulting in the final ductile fracture (Courtesy of ATB RIVA CALZONI SPA, Brescia, Italy)

The procedure to calculate the toughness K_Q or K_{Ic} of the material is the following (see Fig. 9.37):

1. draw a slope OA following the linear part of the curve,
2. draw a secant line OP having a slope 5% less than OA (5% secant) to find load P_5, Fig. 9.37 Type I,
3. if the load at every point on the test record which precedes P_5 is lower than P_5, then P_5 is P_Q,
4. if the load seems to drop after the intersection, point P_5 (Type II or III diagram of Fig. 9.35) is P_Q that shall be used in the K_Q calculation,
5. if the test record shows a maximum, P_{max}, not exceeding 1.10·P_Q, i.e., $P_{\max}/P_Q \leq 1.10$, then P_Q is confirmed otherwise the diagram cannot be considered linear, and the test is not valid.

Once the load P_Q is obtained, calculate K_I for a C(T) specimen as:

$$K_{Ic} = \frac{P_Q}{B\sqrt{W}} \cdot h(a/W) \tag{9.77}$$

where the function *h(a/W)* is given by Eq. (9.32). For a SEN(B) specimen K_I is calculated using Eq. (9.38) with the load $F = P_Q$ and the $h(a/W)$ function given by

Eq. (9.39). If any one of the validity requirements is not met, including the thickness requirement Eq. (9.75) or (9.76), test is not valid, and the toughness measured is indicated as K_Q. The little drop observed in Fig. 9.37 Type II diagram is due to a cleavage fracture event that occurs at local level (grain or grains) but doesn't result in the unstable fracture of the whole specimen because this is prevented by the surrounding matrix that continue to withstand the applied load. This occurrence is called pop-in. An example of pop-in as it appears on the surface of a 1"CT specimen of weld metal on P 690 QL1 steel is shown in Fig. 9.38.

9.14 Effect of Temperature on Toughness

Temperature, it has been said, plays a very important, if not the most important role in the brittle fracture processes, particularly in ferritic steels and alloys that have a BCC structure and show a Transition Temperature, NDT (see Sect. 3.2). The temperature affects the resilience and ductility of materials (see Sects. 3.4 and 5.2) and, therefore, also their toughness. The way temperature operates is through two distinct and complementary actions, linked to the same basic mechanism: the dilution of the Cottrell atmosphere, which makes plastic slip under stress easier (see Sects. 2.12 and 2.13). The first action is purely metallurgical, having to do with the intrinsic ductility of the material related to the width of dislocations, Fig. 2.33. In materials having narrow dislocations the intrinsic ductility, in fact, grows with increasing temperature. Extremely modest at low temperatures where atomic mobility is restrained, the ductility of crystalline grains becomes so high above the NDT as to prevent any cleavage event. Under these conditions any fracture specimen, though thick enough to comply with size requirement, Eq. (9.55), will yield a fracture toughness value K_Q or K_c much too high to sustain an elastic stress field described by a single parameter, namely the K_I, if it still exist. CFC materials and, in particular, austenitic stainless steels have high intrinsic ductility, because of their large dislocations. They have not got a NDT and, therefore, are almost insensitive to temperature, Fig. 5.16, see also the concluding remarks of Chap. 6. The second action, instead, is of a pure geometry change. High toughness, over 150–200 MPa$\sqrt{m}$, results in high CTOD, as predicted by Eq. 9.67. At 200 MPa$\sqrt{m}$ a 500 MPa yield strength material develops an elastic CTOD of about 0,23 mm. This actually means that the crack tip original sharpness is completely lost, and the crack turned into a notch. SSY condition is lost and LSY prevails. This combined temperature effects on the toughness, K_{Ic}, is visible in Fig. 9.39 for a low carbon steel type SA 533 Grade B, Class 1, in use for the construction of nuclear pressure vessels [46, 47], along with the trend of the yield stress, σ_s.

Specimens used for the toughness measurement were of C(T) type or WOL-X type, Fig. 9.7, and had a thickness ranging from one inch (2.54 cm) to 12 inches (30.5 cm). At –18 °C and –4 °C where the ratio $K_{Ic}/\sigma_s < 0.17$, specimens of 4" and 6" were used, well complying with size requirements listed in Table 9.4. They yielded valid results of 95 MPa$\sqrt{m}$ and 101 MPa$\sqrt{m}$, respectively. At 10 °C, about 30 °C

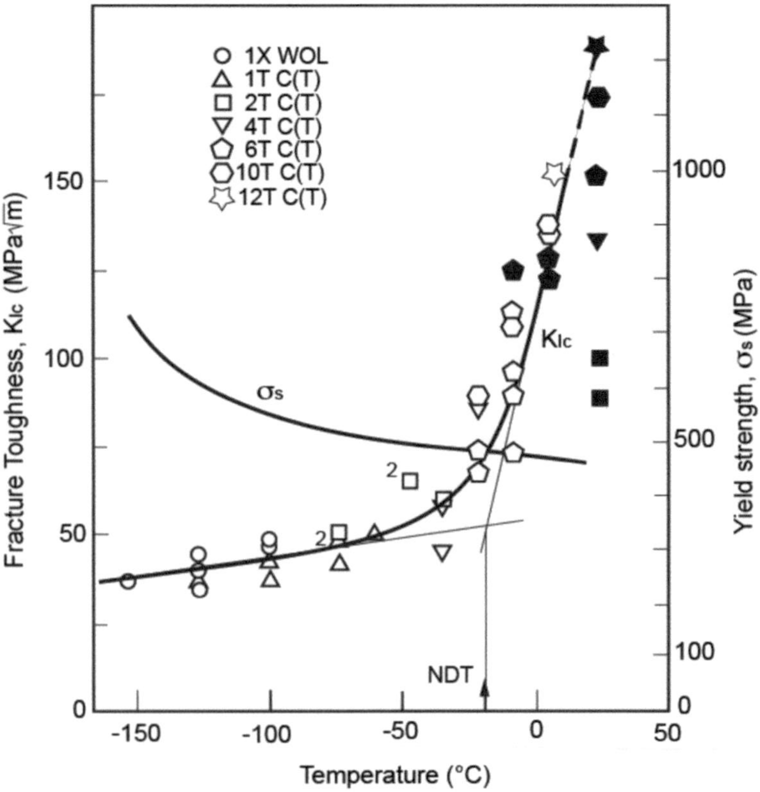

Fig. 9.39 Temperature dependence of the yield strength, σ_S, and fracture toughness, K_{Ic}, of low carbon steel SA 533 GrB Cl1 in use in the nuclear industry, obtained with specimens of increasing thickness. Open points are valid tests, closed points do not satisfy size requirements (combined from [46, 47])

over the transition temperature of the material, NDT, it was necessary to use a 10" C(T) specimen that yielded a valid toughness of about 152 MPa$\sqrt{}$m, valid according to the ASTM size requirement, but not necessarily valid because of the large CTOD. The fracture appearance is shown in Fig. 9.40 [46]. The three specimens present the same feature typical of a brittle fracture: flat and brilliant without any sign of ductile deformation or tearing besides some very modest shear lips. A completely different picture appears at room temperature, Fig. 9.41, where a 12" C(T) specimen was used [46].

The specimen yielded a toughness of 209 MPa$\sqrt{}$m, very high value that for the 450 MPa yield stress results in a K_c/σ_s ratio of 0.46 requiring a much higher thickness of at least 54 cm (21"), see Table 9.5 and Eq. (9.75). Question arises as to whether a K_I stress field still exists and LEFM applies at 50 °C over the NTD of the material. For this type of material having a BCC structure we know that already at 30 °C over the NDT brittle fracture is no longer possible. Therefore, increasing the thickness,

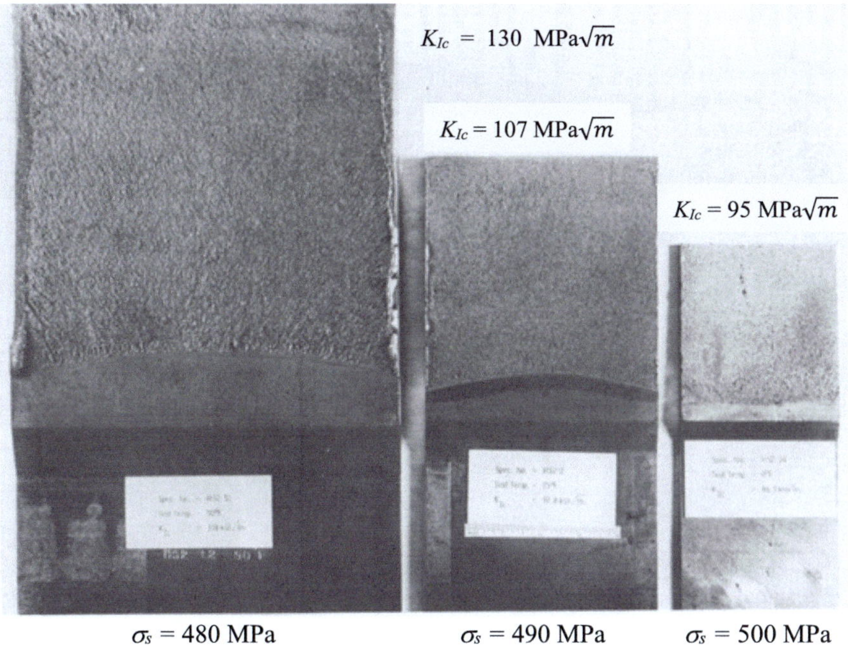

Fig. 9.40 Fracture surface appearance of 10", 6" and 4" thick compact specimens tested at 10, –4 and –18 °C, respectively. Temperature dependence of the steel toughness requires increasing thickness with temperature. Fractures appear flat, predominantly brittle with marginal shear lip [47]

B, of the specimen to comply with size requirements would have no meaning at all. It should be noted that the greatest dispersion of experimental data is obtained near and over the transition temperature, NDT, where metallurgical variability is responsible of the data scatter [48]. The larger metallurgical variability is due to the larger process volume associated with the larger size. For this reason, in this area the effect of specimen volume becomes very important not only from the point of view of triaxiality, but also from the metallurgical and statistical point of view. The brittle to ductile transition temperature, NDT, that clearly appears in the K_{Ic} versus Temperature curve at about –25 °C, through the-sudden rise of toughness, can be derived from the intersection of the two straight lines tangent to the lower and upper portion of the toughness curve, respectively, as shown in Fig. 9.42.

Other examples of temperature dependence of the fracture toughness, K_{Ic}, is shown in Figs. 9.43 and 9.44 for steels type NiCrMoV and CrMoV for rotors, respectively [49].

All these steels have a transition temperature, NDT, clearly visible and determined with the tangents intersection technique previously described and shown in the respective figures. It is interesting to note that all curves terminate at 100 °C at most or, which is a better indication, at about 60 °C over each respective transition temperature, NDT. This is because, as discussed in Chap. 3, at 60 °C over the NDT

Fig. 9.41 Fracture appearance of a 12" C(T) specimen broken at room temperature yielding a non-valid toughness of 209 MPa$\sqrt{m}$. There are clear signs of ductile deformation and necking [47]

of the material no cleavage fracture event is possible. However, already at 20 °C or 30 °C over the NDT, LEFM is no longer a single parameter theory and the stress intensity factor, K_{Ic}, loses its validity unless used with another parameter or term, as it will be said in Sect. 10.10. Therefore, measuring a fracture toughness at those temperatures is not meaningful at all. Figure 9.45 shows a series of 1 T C(T) specimens of 38NiCrMoV, type ASTM 4340, broken at –70, –50, –20, –10 and 25 °C, respectively (Fig. 9.46).

It is interesting to note how while at – 70 °C the fracture appearance is completely flat and without any visible plastic deformation, indicating a brittle fracture occurrence by cleavage, already at –20 °C shear lip appears on both the free surfaces that grows more and more at higher temperatures and eventually, at 25 °C, turns the cleavage into a ductile type failure even though a central area still remains brittle showing radii emanating from a central little nail at the interface with the fatigue pre-cracking.

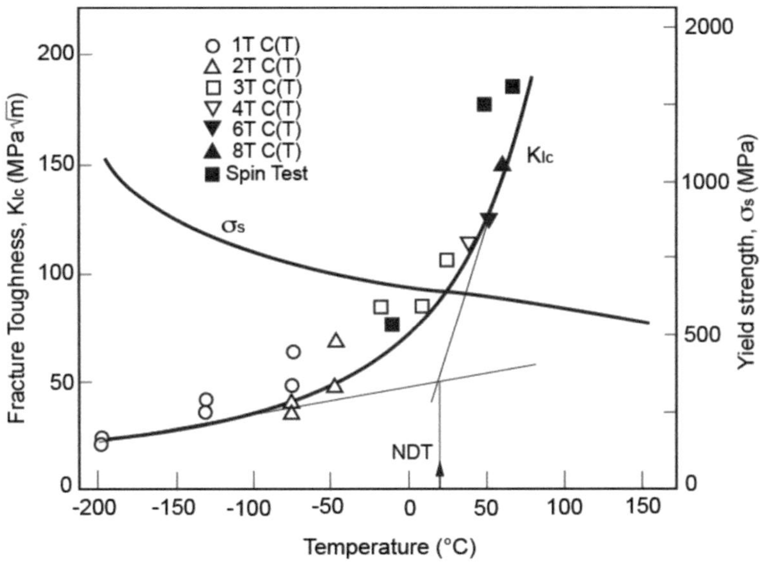

Fig. 9.42 Temperature dependence of the yield strength, σ_s, and plain strain toughness, K_{Ic}, of a NiCrV alloy forging, obtained with specimens of increasing thickness (taken from [47])

Table 9.5 Available stress intensity factor solutions for semielliptical surface flaws in a cylinder with $R/t = 10$

a/c a/t	0.1	0.2	0–3	0.4	0.5	1.0
0.1	■		◊		◊	
0.2	■	□	◊■	□	◊	□
0.25			■●◆			
0.4	■		◊		◊	
0.5	■	□	●◆	□		□
0.6	■		◊		◊	
0.8	■	□	●◆	□		□

■ Angeloni, Peano, □ Newman, Raju, ◊ Milella, Sampietri, ● Mc Gowan, Raimond, ◆ Heliot, Labbens

The 1 T C(T) specimen cannot be used at those last two temperatures. As said, at variance with the BCC structure materials, the FCC metals and alloys do not show the transition temperature, NDT, from brittle to ductile fracture. This is shown in Figs. 9.49 for 7079-T6 aluminum. However, the same trend is also seen in Fig. 9.47 [50] for a high toughness HP-9-4-25 low-alloy high strength steel with a high concentration of cobalt and nickel, used for structural components such as rocket motor casings, armor plate, pressure vessels and gears [50]. The KIc trend is practically

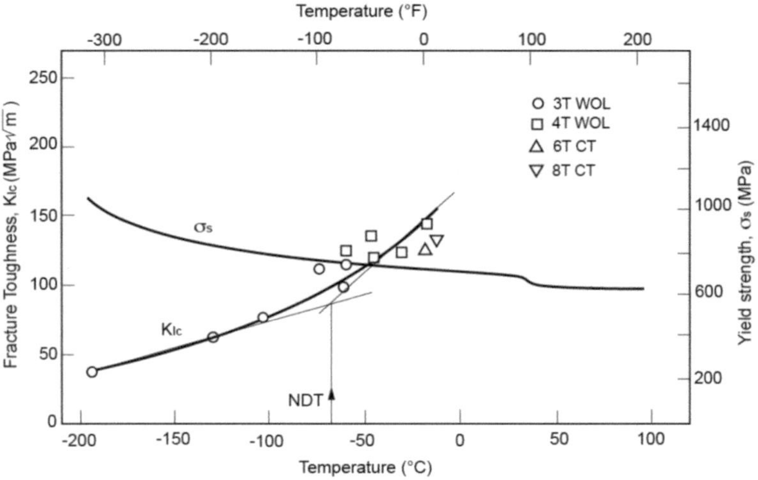

Fig. 9.43 Temperature dependence of the yield strength, σ_s, and plain strain toughness, K_{Ic}, of a NiCrMoV alloy forging, obtained with specimens of increasing thickness (taken from [49])

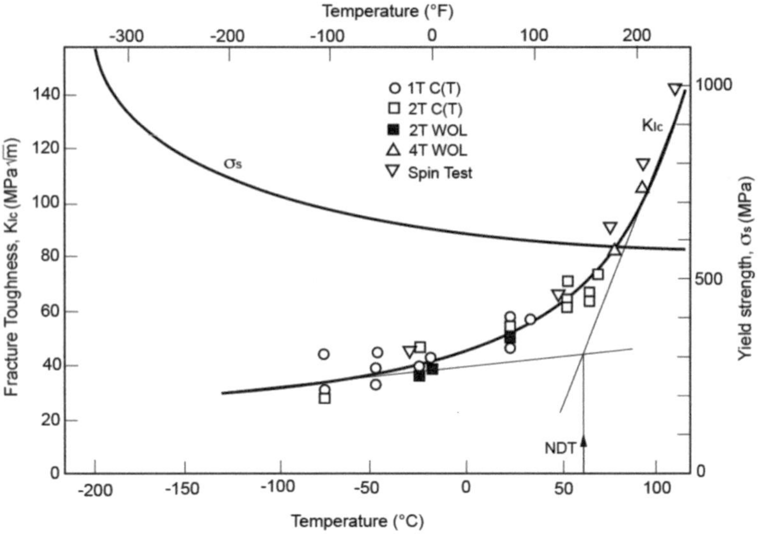

Fig. 9.44 Temperature dependence of the yield strength, σ_s, and plain strain toughness, K_{Ic}, of a CrMoV alloy forging, obtained with specimens of increasing thickness (taken from [49])

flat. Finally, some temperature variation, though small, was observed in the alpha–beta titanium high-strength alloys as shown in Fig. 9.48 [50]. The alpha phase has a BCC structure while the beta phase has FCC structure.

Fig. 9.45 Fracture appearance of 1 T C(T) specimens of 38NiCrMoV broken at, –70, –50, –20, –10 and 25 °C, respectively. Note the V shaped notch that allowed the introduction of a straight front fatigue pre-cracking. Already at –50 °C it can be seen the presence of some shear lip on both free surfaces that grows much larger at –10 and 25 °C where fracture is no longer completely brittle. Also note the lateral contraction at 25 °C

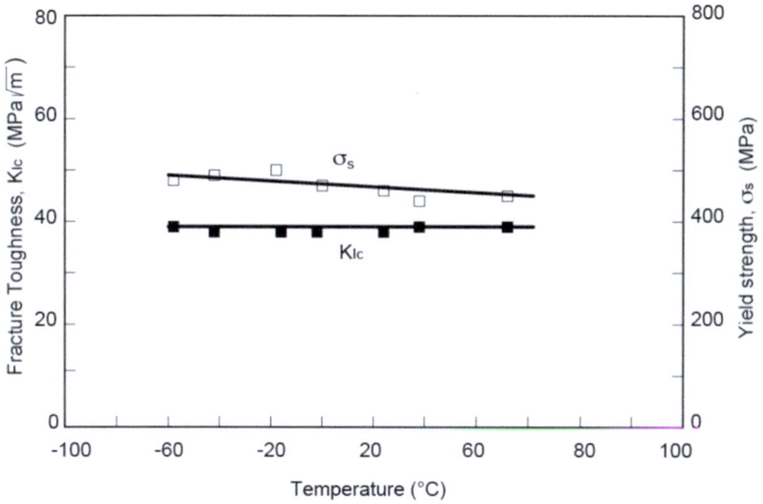

Fig. 9.46 Temperature dependence of the yield strength, σ_s, and plain strain toughness, K_{Ic}, in 7079-T6 aluminum alloy. There is no evidence of any brittle to ductile transition temperature NDT (data from [50])

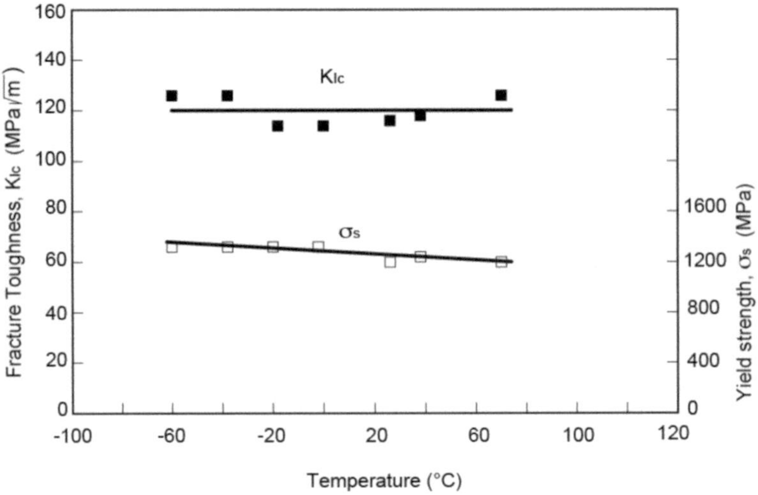

Fig. 9.47 Temperature dependence of the yield strength, σ_s, and plain strain toughness, K_{Ic}, of high toughness HP-9–4-25 high-alloy high strength martensitic steel. Also here there is no evidence of any brittle to ductile transition temperature NDT. The K_{Ic} curve is flat (data from [50])

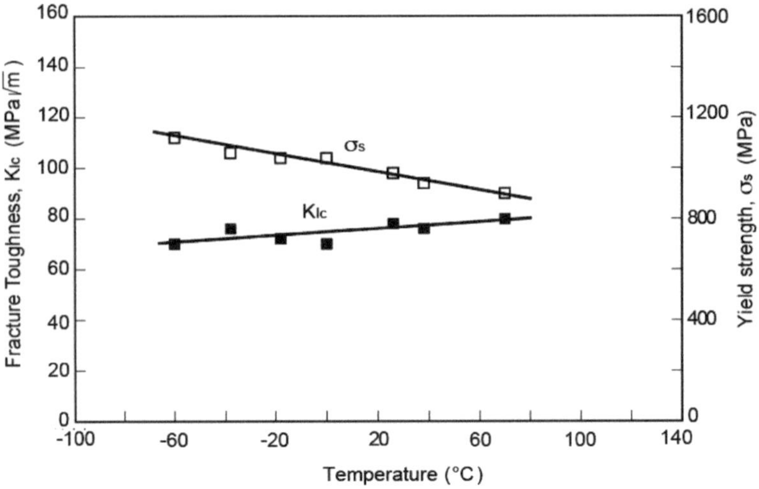

Fig. 9.48 Temperature dependence of the yield strength, σ_s, and plain strain fracture toughness, K_{Ic}, of high strength low toughness titanium alloy (data from [50])

9.15 Dynamic Load Effect on Fracture Toughness

At the time of issuance of the ASTM E399 in the early 1970s that introduced specimen dimensions requirements for valid static fracture toughness measurements, K_{Ic}, the influence of strain rate, $\dot{\varepsilon}$, on fracture toughness had not been adequately considered.

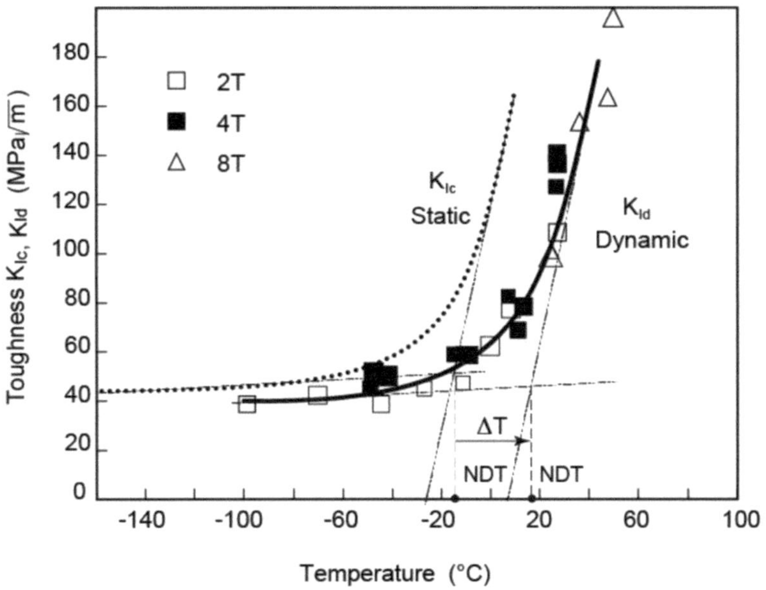

Fig. 9.49 Temperature shift, ΔT, of the dynamic toughness curve, K_{Id}, with respect to the static curve, K_{Ic} (data from [46] Curve 642,090-A)

It was generally understood that fracture toughness would decrease with increasing loading rate, in particular in the transition temperature region for BCC materials. We have seen in Sects. 1.13 and 5.5, how the increase in the strain rate, $\dot{\varepsilon}$, results in a parallel increase of the yield strength of the material and in a strong reduction of its ductility and, therefore, it is rather obvious to expect a consequent reduction of the fracture toughness, K_{Ic}, with increasing $\dot{\varepsilon}$ and, therefore, $dK/dt = \dot{K}$. To fully investigate the effects of the loading rate on the fracture toughness, work was sponsored, in the '70 s, at Westinghouse Electric Corporation to study the properties of medium strength carbon steels in use in the nuclear industry using CT compact specimens up to 8" thick. There were and still there are no standard testing procedures to cover this kind of testing. The result is that dynamic test data often cannot be judged accurately as to whether they represent true material behavior or represent results which have been seriously influenced by testing conditions. Dynamic fracture toughness were developed for loading times approaching 1 ms at temperatures ranging from – 50 °F (– 45.5 °C) to 150 °F (65.5 °C). The results obtained at Westinghouse have been consistent in trend. The corresponding fracture toughness was indicated as dynamic fracture toughness, K_{Id}. The results showed that the dynamic NDT of the material, which is also given by the Pellini Test, Sect. 3.7, increased by about 40 °C and a temperature dependence of the dynamic toughness, K_{Id}, similar to the static one, K_{Ic}, except that the trend curve was shifted towards higher temperatures, Fig. 9.49, by about 40 °C [46]. The same trend is observed in the low alloy steel A 517-F, as shown in Fig. 9.50 [46]. Also shown are the lines of constant $\beta = 0.4$, Eq. (4.95).

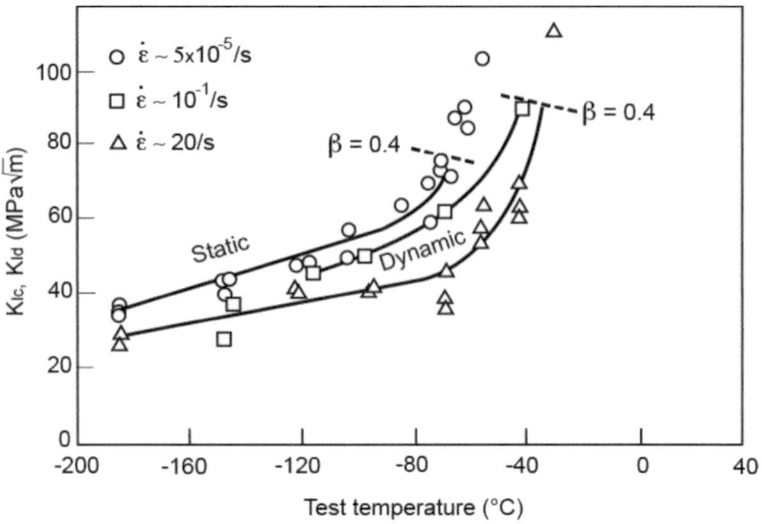

Fig. 9.50 Effect of temperature and strain rate on fracture toughness of A517-F steel (from [46] Curve 647,318-A). Also shown are the lines of constant $\beta = 0.4$ (see Sect. 9.12)

This shift by dynamic effect is not seen, as expected, in 18Ni (250) high strength steel, as shown in Fig. 9.51 [46]. This steel, in fact, has not got a transition temperature, NDT, since it does not have a BCC structure. Its low toughness, already under static loads, has not left any further embrittlement to dynamic loads. At the time, the results of this investigation were most important in that the application of linear-elastic-fracture mechanics to higher temperatures appeared to be demonstrated and the temperature dependence of dynamic toughness was quantified beyond previous data. The isothermal variation of fracture toughness as a function of the loading rate is shown in Fig. 9.52 for a carbon steel, type A 533 Gr. B, Class 1 [47].

9.16 Fracture Toughness Lower Bound Reference Curve

The ASTM size requirements, Eq. (9.47), creates not few a problem to technicians who have to run valid fracture toughness tests with BCC materials, in particular. These materials, as previously said, show a toughness rising high with temperature, in particular beyond the NDT, as Figs. 9.38, 9.39, 9.40 and 9.41 clearly show. To break an 8" or 12" (20 or 30 cm) specimen huge loading cells are needed in the range of 35 MN while common material testing machines have a 150 $\div$ 200 KN capability, besides that 8" $\div$ 12" specimens are no longer easy to obtain and handle. Design codes have then faced the question to provide a single lower bound design curve valid for all carbon steels, in particular.

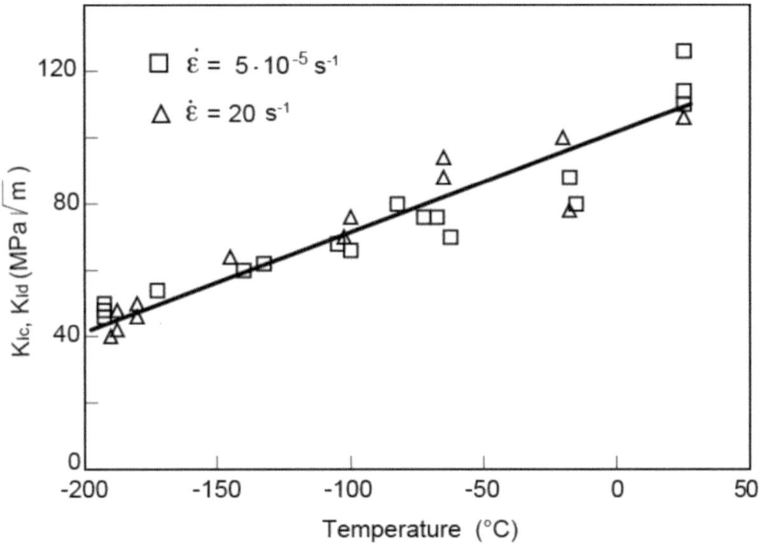

Fig. 9.51 Effect of temperature and strain rate on fracture toughness of 18 Ni (250) steel (from [46] Curve 647,320-A)

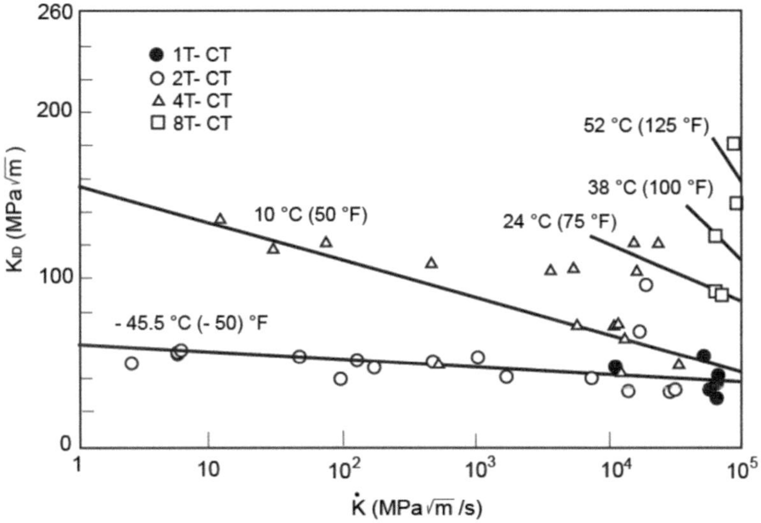

Fig. 9.52 Isothermal variation of fracture toughness with loading rate [47]

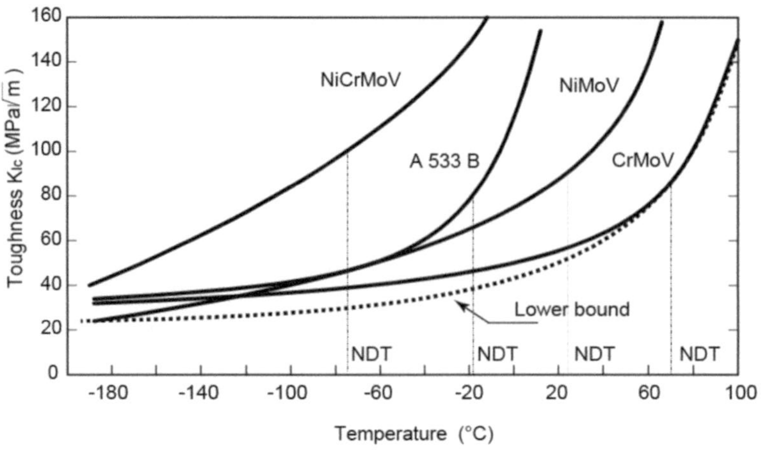

Fig. 9.53 Fracture toughness curves of different steels and relative lower bound

Again, the need was particularly felt in the nuclear industry where medium strength carbon steels, type SA 533, SA 508 or SA 302, were used in the construction of thick pressure vessels and pipes. However, to derive a single or reference lower bound toughness curve valid for all carbon steels for design purposes didn't appear to be an easy task. This can be clearly understood by drawing all curves shown in Figs. 9.38, 9.39, 9.40 and 9.41 in a single diagram, as shown in Fig. 9.53. The various curves seem not to have anything in common at all. At 60 °C, for example, CrMoV steel has an average toughness of about 75 MPa$\sqrt{}$m which is completely out of range for the other three steels that would be strongly penalized if that curve would be used as lower bound reference toughness.

Even within the same class of steel, large variation in fracture toughness, KIc, can be observed, which cannot be uniquely ascribed to the metallurgical variability. This can be observed in Fig. 9.54 [48] for different heats of A 533-B steel. The scatter bands for the different plates A, B and C of different production is as large as the entire variation among completely different steels shown in Fig. 9.53. The reason for that discrepancy lays in the NDT of the steels that is very different from each other and, often, also different from heat to heat of the same steel. Therefore, it appeared logic to index the fracture toughness, K_{Ic}, of the various steels to the relative NDT to solve the problem. This can be seen in Fig. 9.55. Effectively, the various curves, though not overlapping completely, seemed to come close to each other. That was the basis for inventing the concept of a reference curve K_{IR} valid for all carbon steels of medium to low yield strength not higher than 450 MPa at ambient temperature.

The first attempt to build such a reference curve of lower bound toughness, K_{IR}, was done in the early '70 s by the nuclear industry. All experimental K_{Ic} data available at that time were collected by Westinghouse and in January 1971 a Task group was formed by the PVRC (Pressure Vessel Research Committee) to review the state of the art and make recommendations on toughness requirements for ferritic materials

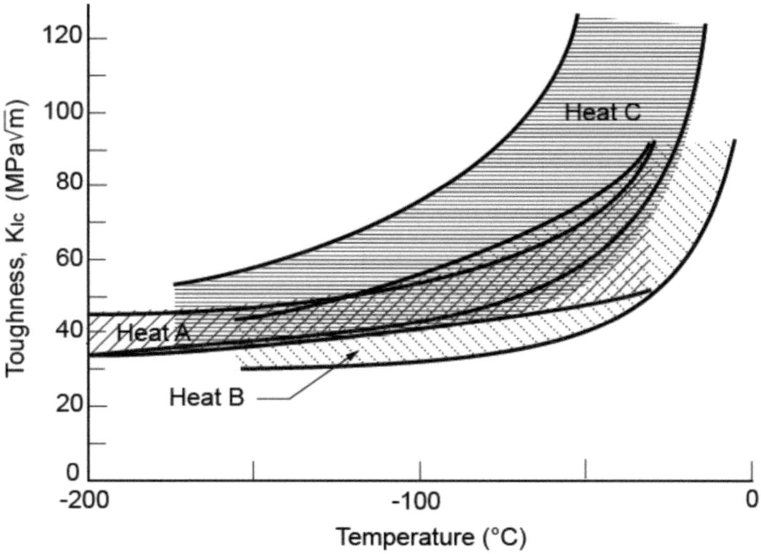

Fig. 9.54 Toughness scatter bands for three different heats of A 533-B steel [48]

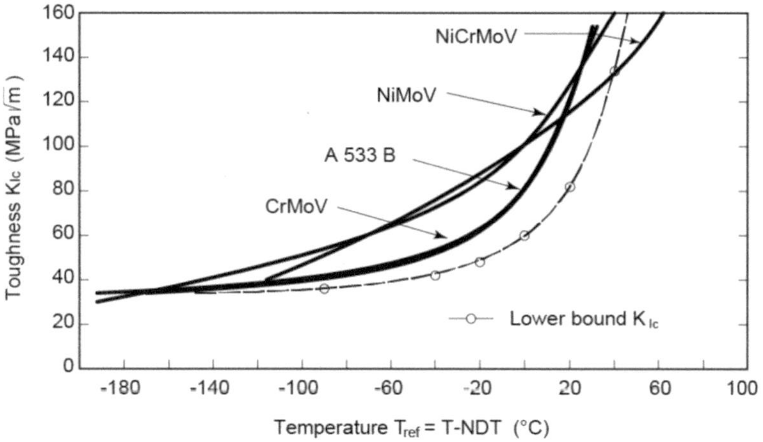

Fig. 9.55 Fracture toughness curves of various steels indexed to the reference temperature $T_{ref} = T - NDT$. Also shown is a possible lower bound curve valid for all steel

in nuclear power plant components. All data collected were put together in a diagram where the temperature, T_{ref}, indexed to the NDT

$$T_{ref} = T - NDT \tag{9.78}$$

At the beginning the experimental data collected were only relative to static K_{Ic} tests. The relative ASME lower-bound curve by digitizing and curve fitting was found to be

$$K_{Ic} = 28.9 + 34.3\exp[(T - RT_{NDT})/33.2] \tag{9.79}$$

Later, dynamic toughness data K_{Id} and crack arrest toughness data K_{Ia}, Chap. 12, were also used to draw a new lower bound curve known as K_{IR} reference curve. The curve, initially adopted in 1972 as a lower bound representation, was based on a set of 173 linear elastic fracture toughness, K_{Ic}, values and 50 linear elastic arrest toughness, K_{Ia}, values for 11 heats of RPV steel. The use of the RT_{NDT} to normalize temperature was intended to account for the heat-to-heat differences in fracture toughness transition temperature, thereby collapsing the fracture toughness data onto a single curve. It took almost all the toughness data ever produced on ferritic steels, static K_{Ic}, dynamic K_{Id} and crack arrest K_{Ia}, to build a single curve that was continuously adjourned. Because of the time and cost to derive it, the curve was nicknamed "*One million dollar curve*". The K_{IR} is then the lower bound curve drawn for all ferritic data, Fig. 9.56 [49]. In practice, the curve became a sort of material property curve that was incorporated in the Appendix G, Section III, of the ASME Boiler and Pressure Vessel Code, Nuclear Design Practice and later in the Section XI of the ASME Code, Inservice Inspections. The K_{IR} curve has been continuously upgraded with data coming from all the world on ferritic steels to be sure that no experimental point was ever falling below it. To-day, the concept of a lower bound curve based on collected experimental data is untenable.

Balkey and Furchi [50] found that material variability is by far the leading cause of uncertainty.

A similar conclusion was also reached by Kirkemo et al. [51]. Therefore, it became evident that a more advanced approach had to be adopted based on a probabilistic procedure to treat the subject of the possible lower bound reference curve.

The probabilistic approach, based on the Weibull distribution, Chap. 17, led to the introduction of a new lower-bound K_{Ic} and K_{Ia} curve

$$\begin{aligned} K_{Ic} &= 36.5 + 22.783 \cdot \exp[(0.036(T - RT_{NDT})] \\ K_{Ia} &= 29.4 + 13.675 \cdot \exp[(0.0261(T - RT_{NDT})] \end{aligned} \tag{9.80}$$

The new ASME probabilistic K_{IR} curve defined by the lower of Eqs. (9.80) lies just a little below the "One million curve" based on collected experimental measurements. The fundamental result of the FAD approach, Sect. 3.10, was that at 33 °C above the NDT a fully brittle fracture caused by cleavage was no longer possible. This temperature limit moved to 66 °C for very large sections. This finding actually means that a measured static fracture toughness over 160 $\mathrm{MPa}\sqrt{\mathrm{m}}$ does not belong to the linear elastic fracture mechanic field. Over that toughness value, LEFM theory is no longer a single parameter theory based on the K_I and new parameters are needed, though the dimension requirements, Eq. (9.52), are respected. This conclusion has

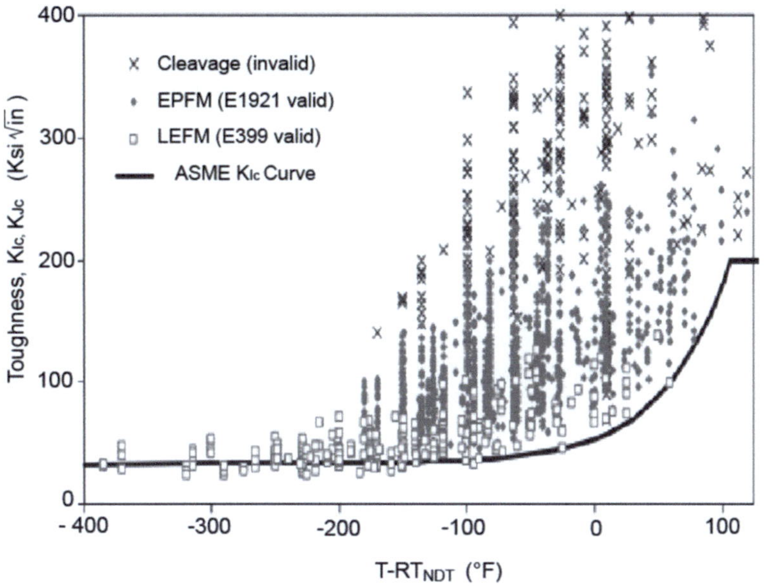

Fig. 9.56 Lower bound reference curve K_{IR} valid for ferritic steels [45] based on the overall population of cleavage fracture toughness data, K_{Ic}, for nuclear RPV steels. All values are plotted as measured and are normalized relative to the ASME value of RT_{NDT}

been experimentally confirmed by a set of five 6" CT specimens tested at different temperatures, as shown in Fig. 9.57 [52]. The material was A533 B Cl1 steel whose RT_{NDT} was −17.8 °C. It can be seen that at −3.9 °C, i.e., at about 14 °C above the RT_{NDT} of the material, the fracture is completely brittle, as it should be. However, already at 37.8 °C and, in particular, at 54 °C, i.e., at 55 °C and 72 °C over the RT_{NDT} of the material, respectively, the fracture surface clearly shows signs of fibrous fracture with shear lips that are indicative of a fracture not completely brittle, see Chap. 6. At 93 °C the fracture appears completely ductile, but we are already at 111 °C over the RT_{NDT} of the material. This is also the reason why the ASME code poses a limit to the reference toughness curve, K_{IR}, by truncating it at 220 MPa $\sqrt{m}$. On the upper shelf, the de facto K_{Ic} limit of applicability of 220 MPa $\sqrt{m}$ exceeds all available data. Over that value there is no sense to go with the K_{Ic} concept of brittle fracture toughness. Actually, that meaning is already lost well before that limit, despite scarce mention or defense of this value in the literature. For instance, Eq. (9.80) states that a toughness $K_{Ic} = 150$ MPa $\sqrt{m}$ is reached at about 45 °C over the transition temperature RT_{NDT}. Even assuming a low RT_{NDT} of about – 10 °C that would mean a temperature T of 35 °C at which the yield stress of a carbon steel may hardly be higher than 500–600 MPa. The corresponding plastic zone at the crack front has already reached the remarkable size of 3 to 5 mm, Eq. (8.9), which cannot be compatible with the hypothesis of a sharp crack at the base of LEFM and of the SSY condition as well. In the continuous challenge to find a lower bound curve, it soon became apparent

Fig. 9.57 Fracture surfaces of 6TCT specimens of A533 B Cl1 steel tested at different temperatures [52]

that the degree of conservatism provided by the K_{Ic}/RT_{NDT} description of fracture toughness was inconsistent from material to material. The curve, in fact, had a major drawback: the transition temperature indices used to determine RT_{NDT} are based on the Charpy V-Notch and nil-ductility (NDT) transition temperatures that have no direct correlation with the fracture toughness, K_{Ic}, see Sect. 3.2. Moreover, data so far collected demonstrates that conservatism cannot be guaranteed in all cases, especially in the lower shelf of the K_{Ic}/RT_{NDT} curve where a limiting asymptotic value is reached at about 35 MPa $\sqrt{m}$.

9.17 Fracture Testing of Welds

A workmanlike weld is certainly superior to the base metal on which it was made. However, a multipass weld leaves at its borders a region, no more than 2.5–3 mm thick, that has been subjected to the weld heat input and has undergone metallurgical transformations without melting, Fig. 9.58. In this area, called heat affected zone (HAZ), three main region can be distinguished.

What actually happens in these regions depends on the weld heat input of the first and subsequent weld passes and on the cooling rate. A long and complex story of heating and reheating affects the material changing its properties. From a general point of view, the material in the first region, region (1) of Fig. 9.59, has been heated at higher temperatures (>1100 °C) over the critical point (about 723 °C in steels) without

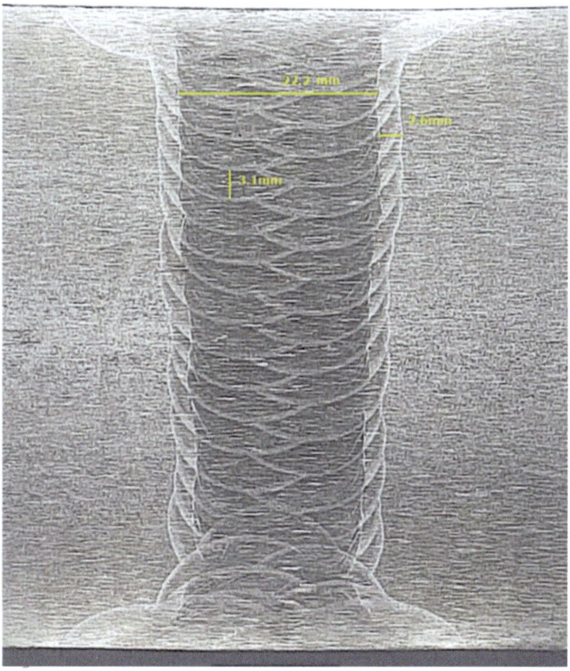

Fig. 9.58 Cross-sectional macrostructure of a multipass butt weld in a steel plate revealing the columnar grains (darker zones white bordered) and the adjacent HAZ (Courtesy of ATB RIVA CALZONI Spa, Brescia, Italy)

melting and then cooled down which produces coarse grains of low strength and low toughness. This is the most critical region of the HAZ that shows the lower toughness. In the second region, region (2) of Fig. 9.59, where the temperature was in 1000–900 °C range, grains hve been refined acquiring a superior strength and toughness, GRHAZ region. This is the high strength, high toughness zone. Finally, a third region appears at the interface with the base metal note as intercritically reheated coarse-grained heat affected zone (ICCGHAZ), region (3) of Fig. 9.59. The ICCGHAZ may have higher or lower toughness, depending on the temperature and cooling rate history, which is mainly due to the formation of necklace-type brittle martensitic-austenitic phases known as M-A constituents. The cooling and solidification of the weld pool leads to the shrinkage of the weld metal that is prevented by the adjacent base metal. The result is the build-up of an internal stress state known as welding residual stress state. The intensity of such a stress state depends on the volume ratio of melted pool to base metal. The most credited distribution of residual stresses in a butt weld is schematized in Fig. 9.60 [53, 54]. Longitudinal residual stresses,σ_l, are generally higher than transverse stresses,σ_t, whose maximum remains at about 30 to 40% of the longitudinal maximum. In the case of complete weld shrinkage prevention the resulting longitudinal residual stresses,σ_l, may reach the yield strength of the weld material. As it can be seen in Fig. 9.60, longitudinal stresses,σ_l, reach the maximum at the center of the weld, where the shrinkage of the molten pool is maximum during solidification. Since residual stresses are self-balancing stresses, these tensile stresses must be balanced by compressive stresses in the base material.

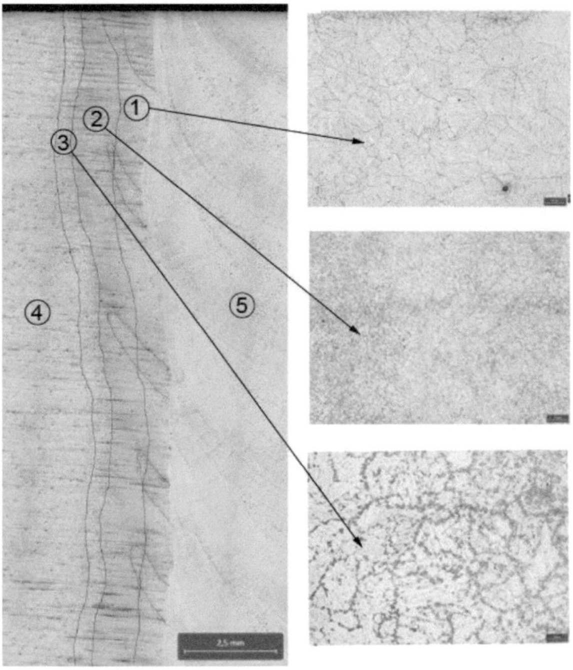

Fig. 9.59 Macro-micrographs of the HAZ with the three regions: (1) CGHAZ, (2) GRHAZ and (3) ICCGHAZ, formed between the weld deposit (5) and the base material (4) (Courtesy of ATB RIVA CALZONI Spa, Brescia, Italy)

1. coarse grain region (CGHAZ)
2. grain refined region (GRHAZ)
3. intercritical HAZ (ICCGHAZ)

According to Saint Venant principle residual stresses must extinguish within a distance two–three times the characteristic dimension of their cause, which is the weld size. Figure 9.61 presents an experimentally determined distribution of longitudinal residual stresses in a butt weld [55] and two possible interpolating curves proposed by Masubuchi and Martin [56]

$$\begin{aligned} &curve\ a\ \sigma_l = \sigma_o e^{-0.5(x/c)^2}\left[1-(x/c)^2\right] \\ &curve\ b\ \sigma_l = \sigma_o \tfrac{1-(x/c)^2}{1+(x/c)^4} \end{aligned} \tag{9.81}$$

in which σ_o is the maximum stress attained on the weld centerline and c the width of the tension zone, as shown in Fig. 9.60(b). These high-level stresses, though classified as secondary stresses, can jeopardize the integrity of the structure so that it is important to eliminate them. To remove residual stresses and recover the quality of the HAZ, welds are heat treated at a temperature of about 600–650 °C for 2–3 h. Not always this post weld heat treatment (PWHT) can be possible, as in the case of austenitic stainless steels where the PWHT would precipitate chromium at grain boundaries, as complex chromium-carbide, $Cr_{23}C_6$, and sensitizes the steel to stress

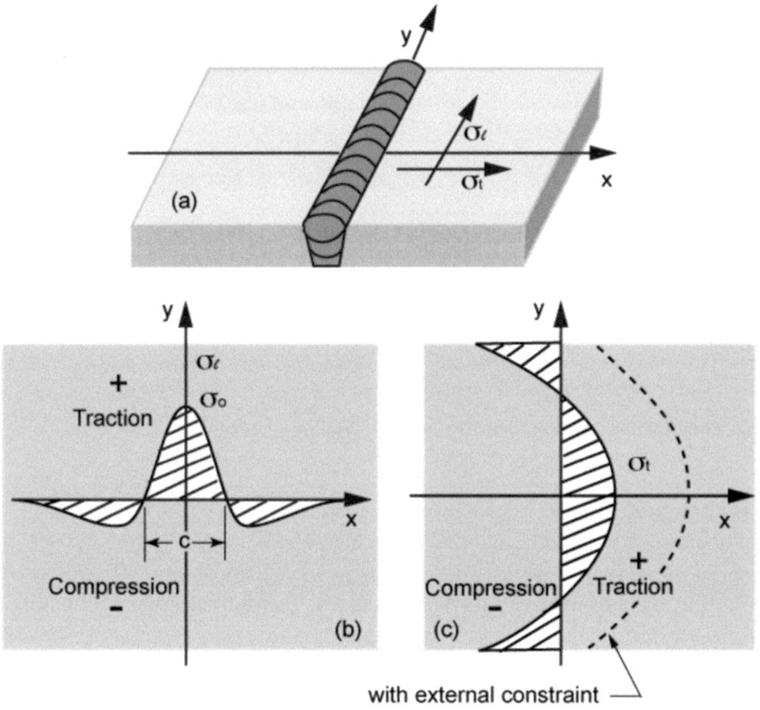

Fig. 9.60 Typical distribution of (**b**) longitudinal σ_l and (**c**) transverse σ_t residual stresses in (**a**) a butt weld

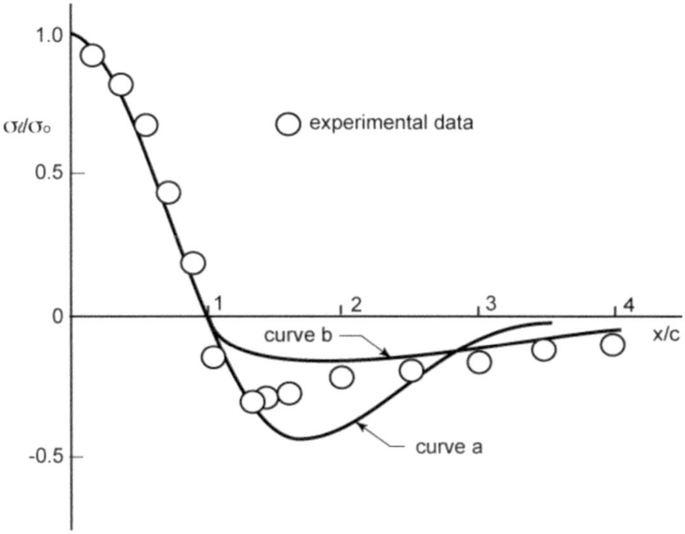

Fig. 9.61 Experimentally determined residual stress distribution in a butt weld (from [53])

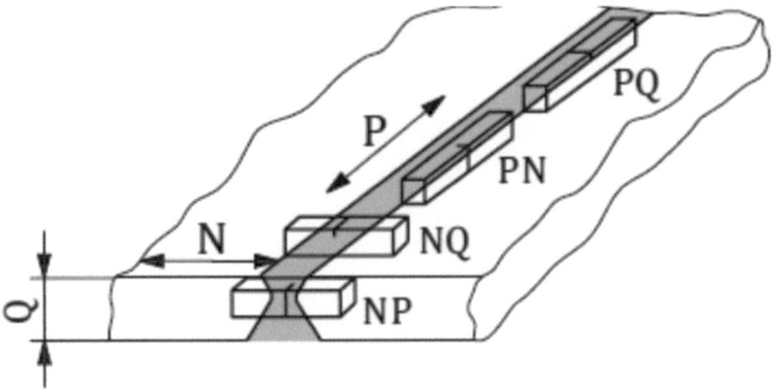

Fig. 9.62 Identification of specimen and crack orientation in a weld

corrosion cracking [57]. Also, in some precipitation hardened steels PWHT cannot be done because it would change the strength of the material. In these cases, a warm prestressing procedure may be effective, Chap. 13. The problem with fracture testing of a weld that has not received a PWHT and has residual stresses still acting is that by cutting a small specimen in the weld region residual stresses may be washed out. As the base metal, also welds show anisotropy depending on the orientation of the specimen used for testing. This anisotropy comes from the weld structure that has a columnar aspect, as shown in Fig. 9.58 and is schematically shown in Fig. 9.63. Orientation of both the specimens and the crack in the weld is defined by to letters as shown in Fig. 9.62.

The first letter N or P identifies the orientation of the specimen that can be normal or parallel to the weld seam, the second letter P or Q identifies the crack orientation that can be parallel (P) or normal (Q) to the weld seam.

The columnar structure of the weld will appear on the specimen surface after fracture occurred. Depending on the specimen orientation, columnar grains will be oriented on the crack plane or normal to the crack plane. This is shown in Fig. 9.64 for two SEN(B) 1" thick specimens of P 690 QL steel having NP and NQ orientation, respectively, and in Fig. 9.65 for two CT 1" thick specimens having PQ orientation, respectively, tested at –10 °C. It is clear that the CGHAZ, region (1) or, sometimes, region (3) of Fig. 9.59 are the weakest regions of the weld. Therefore, in testing a weld that has not received a PWHT care must be used to position the crack in that zone which is not so simple due to the extremely small dimension of the HAZ which is no larger than 0.5 mm, Figs. 9.58, 9.59 and 9.63. Furthermore, measuring the CTOD of the HAZ, Sect. 9.8, care must also be used to enter Eq. (9.40) with the proper yield stress, σ_s, that should be that of the HAZ and not of the weld metal. Since it is not possible to run a traction test with specimens machined from that zone, the yield stress, σ_s, must be deduced from the grain size and/or the hardness of the zone itself.

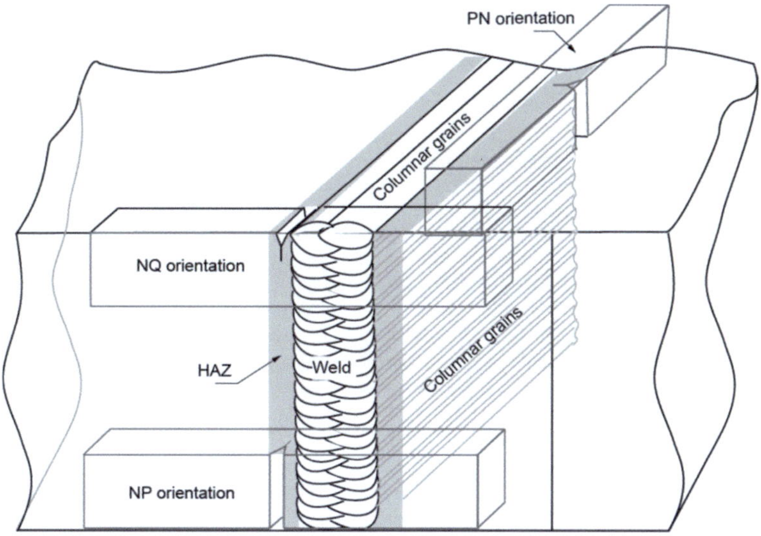

Fig. 9.63 Schematic of the columnar structure of a weld seam and specimen position in the HAZ. The HAZ width has been enlarged for convenience

9.18 Shallow Cracks

We have addressed the issue of shallow cracks in Sect. 4.4 from a phenomenological point of view observing that a short crack has too small unstressed material left behind to exert a strong reaction R to give rise to a significant stress component in the x direction, σ_x, Fig. 4.12b. This results in a low constraint at the crack tip which produces low biaxiality and, in thick specimens, also low triaxiality that would not generate conditions favorable to a completely brittle fracture event. Effectively, studies conducted at the University of Kansas [58–61], at the Edison Welding Institute in the US [62] and at the Welding Institute in the UK [63] have shown that the toughness measured with specimens containing short or shallow cracks is always significantly higher than that measured with specimens with deep cracks. The shallow-crack work at the University of Kansas was based on SE(B) specimen, Fig. 9.18, of A36 and A517 steel having different thickness, B, depth, W, and crack depth, *a*. This increase in toughness was found to take place in the transition region of the material becoming higher and higher with temperature, as shown in Fig. 9.66 [58, 59]. For the A517 steel the fracture toughness increase associated to shallow cracks was much higher, about 4 times in the upper shelf. The toughness is measured through a critical value of the CTOD, see Sect. 8.6. Finite element analysis indicated that the cause of the fracture toughness increase was geometry dependent and, therefore, could not be associated with a material property variability [64]. The plastic zone around the crack tip was found to be distinctly different depending on the *a/W* ratio. For the shallow-cracks, the plastic zone emanating from the crack tip expanded

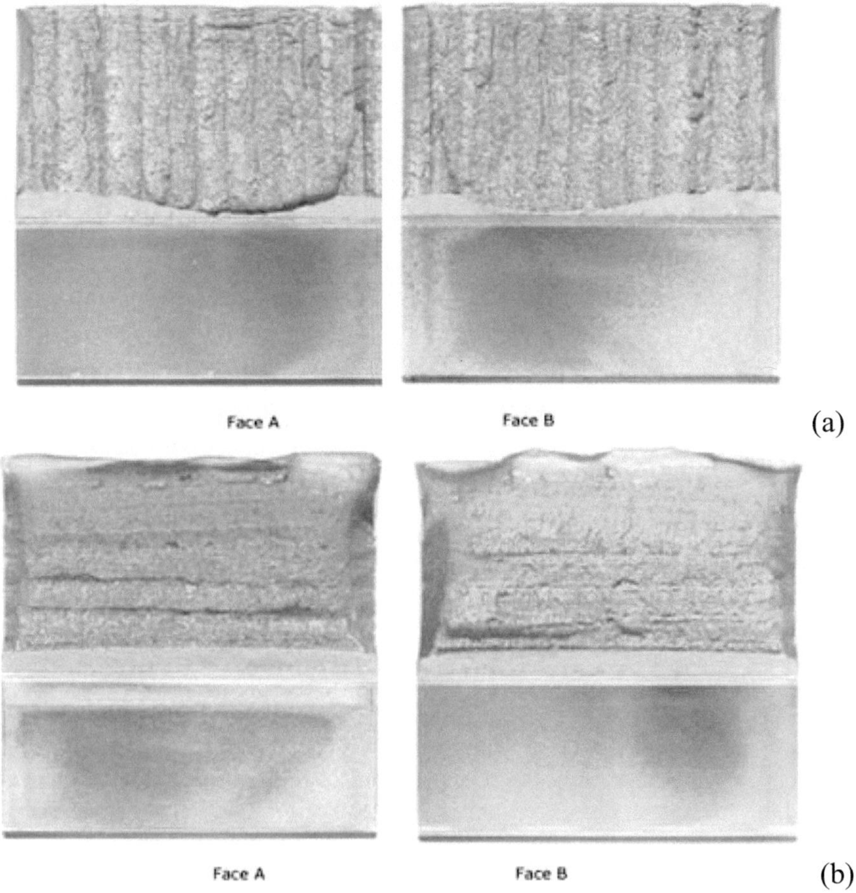

Fig. 9.64 Morphology of the fracture surface of SEN(B) 1" thick specimens having: **a** NP orientation and **b** NQ orientation, respectively, showing columnar grains perpendicular and parallel to the crack front, respectively (Courtesy of ATB RIVA CALZONI Spa, Brescia, Italy)

without boundaries interacting with the specimen surface leading to a sudden formation of the plastic hinge. For deep-cracks, the opposite was true: a plastic hinge was formed before the plastic zone interaction with the specimen surface. Figure 9.67 shows shallow and deep-cracked specimens plastic zone for comparison [65]. It is clear that with such an extended plastic zone with a/W ratios equal or smaller than 0.2 the constraint effect is completely lost turning a plain strain condition into a plain stress condition with the unavoidable sharp increase in the fracture toughness.

We have seen how the CTOD is related to the Irwin K_I stress intensity factor, Eq. (8.67) that can be put in the more general form

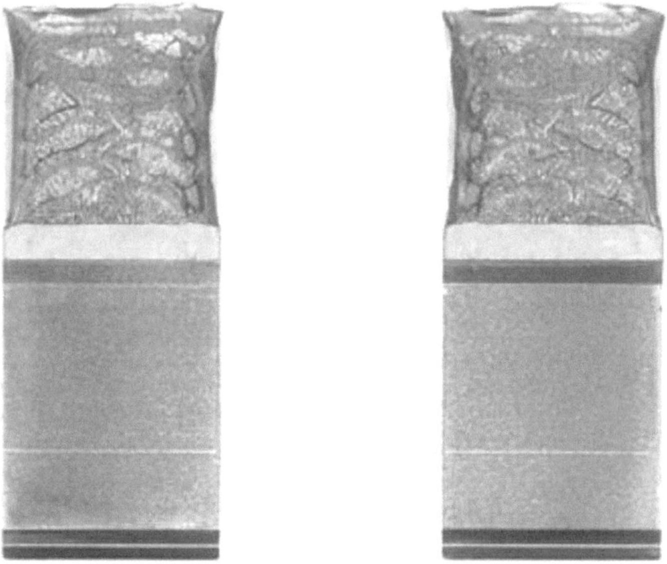

Fig. 9.65 Morphology of the fracture surface of a CT 1" thick specimen having: PQ orientation, showing columnar grains that appear as small fans (Courtesy of ATB RIVA CALZONI Spa, Brescia, Italy)

$$\begin{aligned} K_I &= \sqrt{m\sigma_f E'\delta_t} \\ CTOD &= \delta_t \end{aligned} \tag{9.82}$$

where σ_f is the flow stress of the material

$$\sigma_f = \frac{\sigma_s + \sigma_u}{2} \tag{9.83}$$

and E' is the Young's modulus in plain strain conditions, $E' = E/(1 - \nu^2)$ and m is the plastic constraint factor already seen in Fig. 8.2. When K_I reaches the critical value K_{Ic}, also CTOD does, which is shown in Figs. 9.66a and b. In the case of deep-cracks m has been found to be equal to $1.5 \div 1.8$. For shallow-cracks $m = 1.1$ and $E' = E$ without the increasing factor $1/(1 - \nu^2)$ typical of the plain strain conditions. Systematic studies were conducted at ORNL within the HSST program on A533 B Cl1 steel SE(B) specimens, having W = 100mm and different thickness B and length L with a/W ratios varying from 0.1 to 0.5. Results are shown in Fig. 9.68 [65]. Temperature scale is indexed to the specific RT_{NDT} since specimens were coming from two different heats of the same material having different RT_{NDT} equal to –35 and –15 °C, respectively. In all cases, the apparent fracture toughness K_c at 33 °C over the RT_{NDT} was in excess of 160 MPa$\sqrt{m}$.

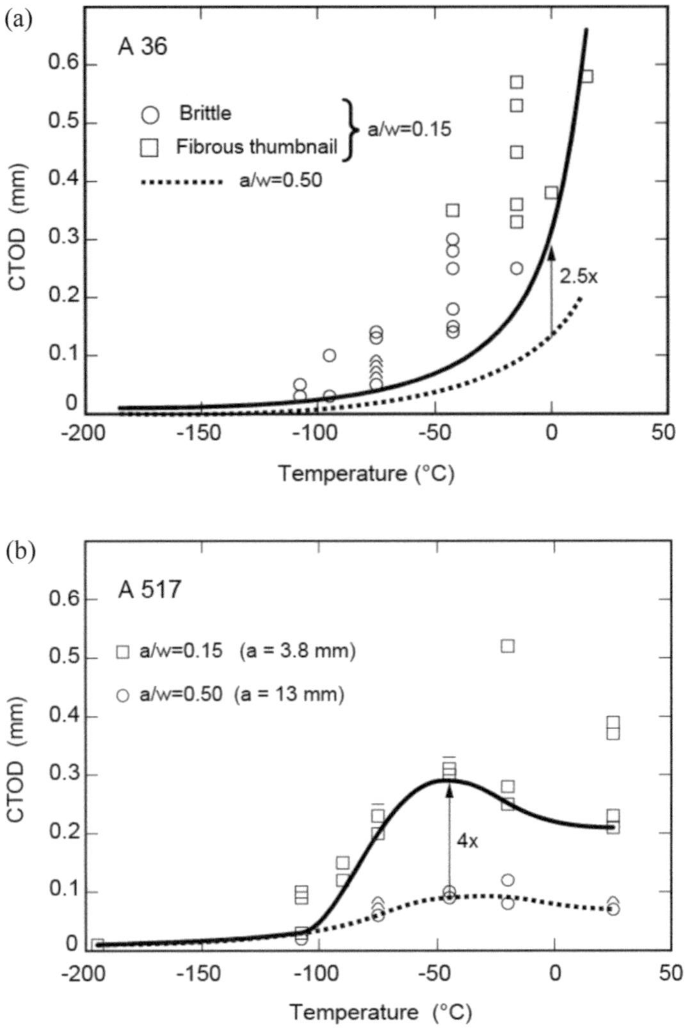

Fig. 9.66 **a** Critical CTOD versus temperature for A36 steel specimens with *a/W* ratios of 0.15 and 0.50, **b** for A517 steel specimens [58, 59]

It is evident the systematic increase of the toughness in the transition zone, as expected. However, the K_c curve for shallow-cracks always remains above the corresponding K_c curve for deep-cracks. Between the two curves there is a temperature shift ΔT of about 35 °C. Interesting enough, this is the same temperature shift between static and dynamic fracture toughness obtained on the same steel, Fig. 9.49.

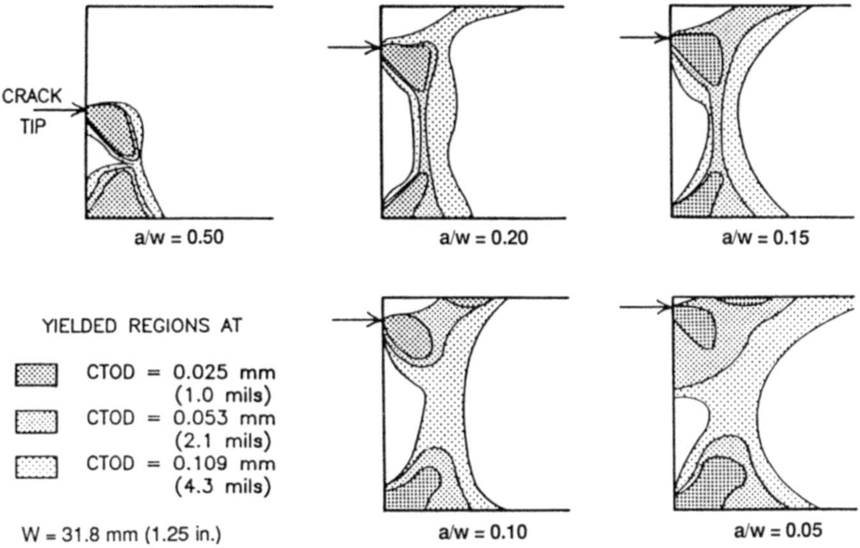

Fig. 9.67 Von Mises stress distribution for 2-D plain strain A36 steel specimens [64]

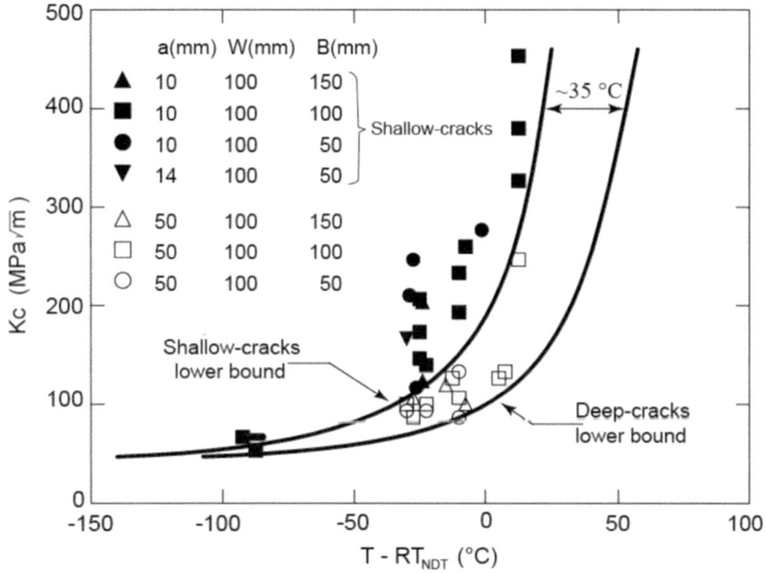

Fig. 9.68 Temperature dependence of the apparent toughness K_c of SE(B) specimens of A533 B Cl1 steel having shallow and deep-cracks [65]

9.19 Wide Plate Test

Beside its large use in crack arrest testing, Fig. 12.5, the wide plate test (WPT) represents the final verification that the toughness of a material as measured on small standard specimens may be really used to assess the response of a very large structure such as a pressure vessel or a large diameter, thick pipe under the operating conditions. The question arises as to whether or not a fracture toughness, K_{Ic}, measured with few small specimens, can be representative of a large volume from both the constraint and process volume. The problem of data scatter and process volume will be also addressed in Chap. 15. Using a specimen having a width W, Fig. 9.70a, of 40 cm, which is about 15 times larger than that of a 1 T-CT specimen, we may expect a process volume ahead of the crack tip of the wide plate 5 times larger than that of a single CT specimen and, therefore, we may also expect that failure occurs in the WPT at a lower value than that predicted by one or two 1 T-CT specimens. Particular attention must be used in the introduction of the surface crack in the wide plate. A machined notch, though with a very small crack tip radius, cannot represent a real sharp crack, as described in Sect. 9.1. However, the fatigue sharpening of a notch in a wide plate is hard to be done. The alternative is to deposit a brittle weld around the notch as a possible crack starter, as shown in Fig. 9.70b, and sharpen the notch tip by electro erosion, Fig. 9.70c. An electron beam weld would be preferred since its low heat input would not alter the characteristic of the base material. The crack size, $2c_f$, used for the final WPT verification is that corresponding to the largest size expected to develop in the real structure at its target final life. The initial crack size, $2c_o$, is equal to the minimum size that can be *seen* with a certain confidence using non-destructive methods, generally ultrasonic examination, with an aspect ratio length $2c$ to depth, a, normally assumed as: $2c/a = 5$ or even $2c/a = 10$. Generally, the initial crack is supposed to be a surface crack to which it is associated the largest applied K_I, *ceteris paribus*. The final length, $2c_f$, is obtained from the initial length, $2c_o$, using the Paris low of fatigue crack growth, of the type

$$\frac{dc}{dN} = C \cdot \Delta K_I^n \tag{9.84}$$

where C and n are material characteristics. It is good practice to adopt a safety factor of at least 2 on the initial crack length $c_i = 2c_o$. As shown in Fig. 9.71, the point of maximum depth, point A of Fig. 9.70c, is that in which the applied K_{IA} reaches its maximum. It is $K_{IA} > K_{IB}$, therefore during the fatigue life of the real component, point A of the initial crack is likely to growth at a faster rate than point B, continuously changing the aspect ratio of the crack. However, during fatigue crack growth calculations it may be convenient to assume that the two points, A and B, grow at the same rate, keeping the initial aspect ratio, $2c/a$. The procedure that leads to the final crack dimensions to be assumed and introduced in the WPT, $2c_f$, is schematically shown in Fig. 9.69.

An unexpected problem may arise with the WPT. Experience with very wide-plates of A 533 B Cl.1 steel used in the crack arrest tests, Fig. 12.5, has shown that

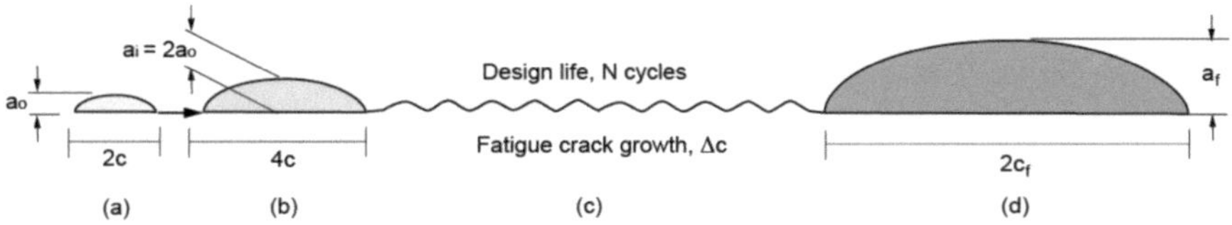

Fig. 9.69 Schematic of the basic logic process leading to the definition of the final crack size to be introduced in the WPT: **a** initial crack size equal to the minimum that can be detected by the nondestructive examination adopted, **b** use of a safety factor of 2 on that initial size, **c** calculation of fatigue crack during the operational life and **d** final crack size to be introduced in the wide plate for testing

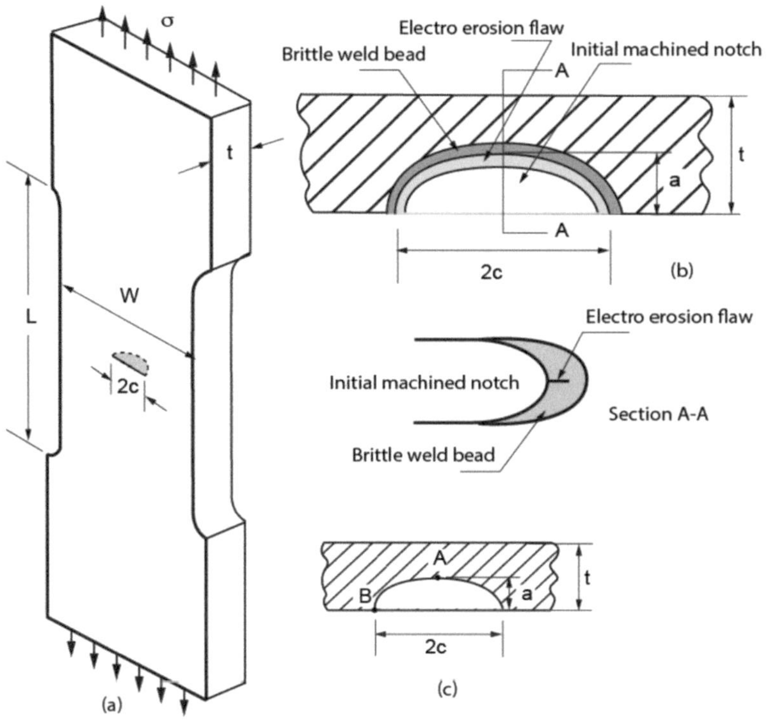

Fig. 9.70 **a** Wide plate, **b** schematic of the machined notch, weld bead and electro erosion flaw, **c** point A is the point of maximum applied K_I

cleavage crack initiation occurred at K_I values well above those, K_{Ic}, predicted with small specimens. In fact, the K_I/K_{Ic} ratios ranged from 2.0 to even 4.0. Surprising enough, the toughness of the wide plate was by far higher than that measured with small specimens. The answer to this unexpected discrepancy was finally found using the maximum principal stress criterion proposed by Anderson and Dodds, Eq. (6.3), based on achieving a critical area within a selected stress contour. The normalized area $A(\sigma_o/J)^2$ was by far higher in the CT specimens than in the wide plate.

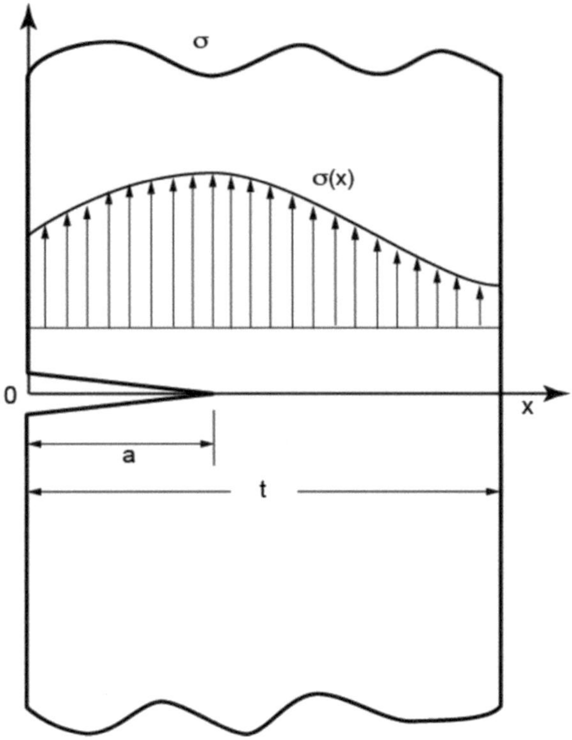

Fig. 9.71 Coordinate system for crack opening stress field

9.20 Influence Coefficients for Polynomial Stress Distributions

We have so far considered a uniform stress field of a given amplitude σ or an external force F applied to the body containing the crack. This may not be the case when, for instance, a thermal gradient is present in the beltline region of a pressure vessel during a transient that may also be a severe thermal shock. Mc Gowan [66] has treated the case of a continuous surface crack either longitudinal or circumferential in the beltline region of a pressure vessel with a radius to thickness ratio $R/t = 10$ subjected to the most general stress distribution that can be approximated by a third order polynomial of the form

$$\sigma = A_o + A_1 x + A_2 x^2 + A_3 x^3 \qquad (9.85)$$

as shown in Fig. 9.71.

When this is the case, the stress intensity factor at ahead of the crack can be determined by the expression

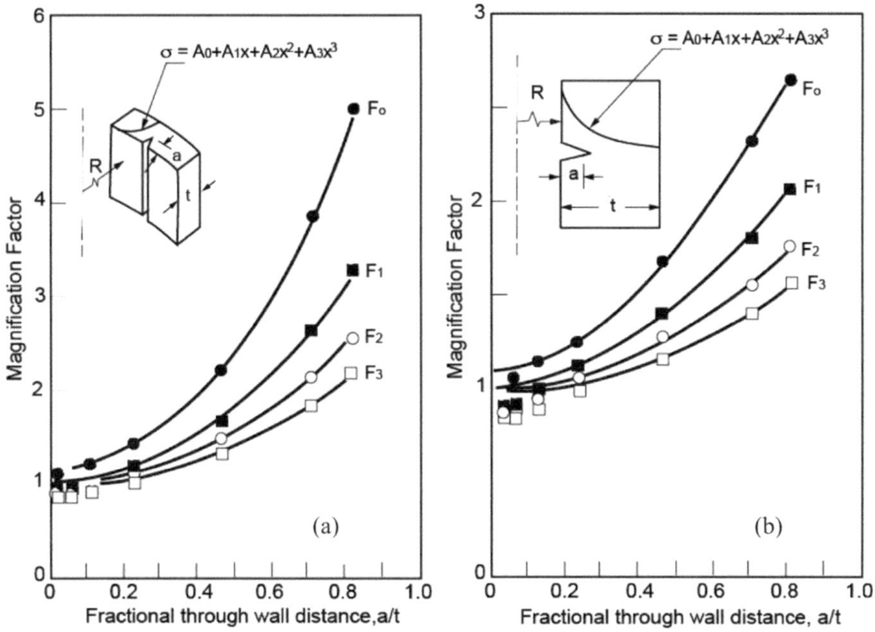

Fig. 9.72 **a** Magnification factors for a continuous longitudinal crack in a cylinder under non-uniform stress field, **b** magnification factors for a continuous circumferential crack in a cylinder under non-uniform stress field, as function of the fractional distance *a/t* [66]

$$K_I = \sqrt{\pi a}\left(A_o F_0 + \frac{2a}{\pi} A_1 F_1 + \frac{a^2}{2} A_2 F_2 + \frac{4a^3}{3\pi} A_3 F_3 \right) \tag{9.86}$$

where a is the crack depth, A_o through A_3 are the coefficients of the stress distribution defined in Eq. (9.85) and F_o though F_3 are influence coefficients or magnification factors defined in Figs. 9.72 [66].

Later, Milella and Pini [72] have addressed the same problem for a finite longitudinal and circumferential semielliptical crack in a vessel wall as function of the angle, θ, as shown in Fig. 9.73. For a finite crack under a non-uniform stress field $\sigma(x)$

$$\sigma(x) = A_0 + A_1 \frac{x}{t} + A_2 \left(\frac{x}{t}\right)^2 + A_3 \left(\frac{x}{t}\right)^3 \tag{9.87}$$

the stress intensity factor K_I at any point of the crack front defined by the angle θ can be written as

$$K_I = \sqrt{\frac{\pi a}{Q}} \left(\cos^2\theta + \frac{a^2}{c^2} \sin^2\theta \right)^{1/4}$$

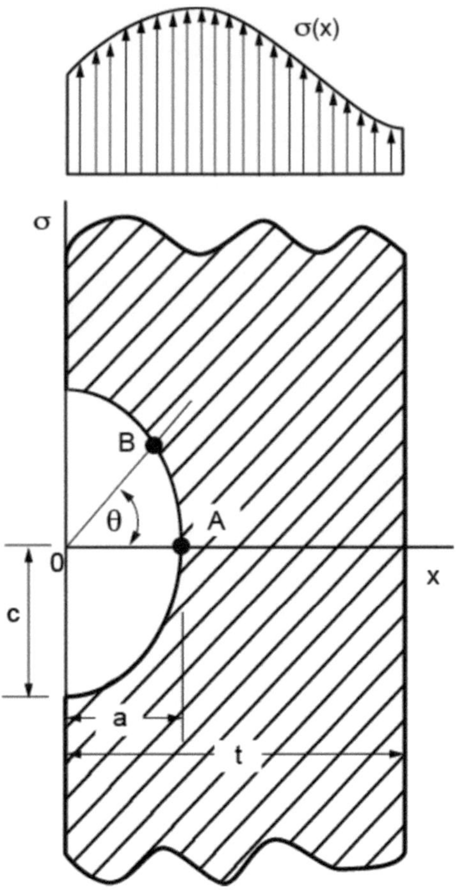

Fig. 9.73 Semielliptical surface flaw of aspect ratio *2c/a* in a vessel wall of thickness *t* under arbitrary stress field σ*(x)*

$$\left[A_0 H_0(\theta) + \frac{2}{\pi}\frac{a}{t} A_1 H_1(\theta) + \frac{1}{2}\left(\frac{a}{t}\right)^2 A_2 H_2(\theta) + \frac{4}{3\pi}\left(\frac{a}{t}\right)^3 A_3 H_3(\theta)\right] \tag{9.88}$$

where Q is a flaw-shape parameter, which is based on the solution of an elliptical integral of the second kind whose good approximation is given by the expression

$$Q = 1 + 1.464\left(\frac{a}{c}\right)^{1.65} \quad for\ a \leq c \tag{9.89}$$

and $H_o(\theta)$, $H_1(\theta)$, $H_2(\theta)$ and $H_3(\theta)$ are nondimensional magnification factors or influence coefficients relative to each of the four stress components, A_i with $(i = 0, 3)$, which the general stress field can be subdivided into by Eq. (9.87).

In practice, using the polynomial form of Eq. (9.87) the generic stress state acting on the vessel wall is broken down into four separate loading components, as represented in Fig. 9.74. To derive the $H_i(\theta)$ factors as function of any aspect ratio, *a/c*,

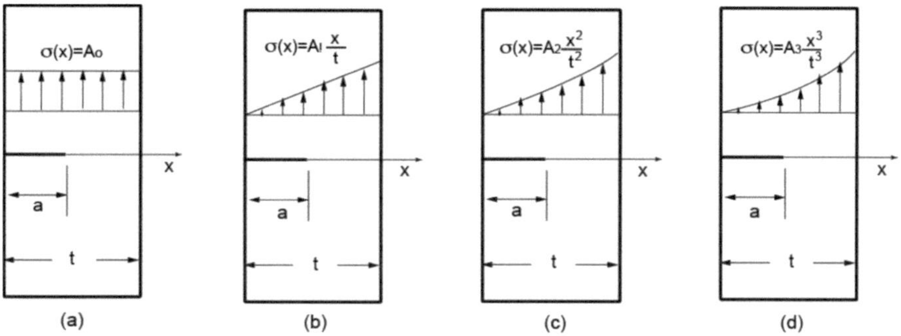

Fig. 9.74 Schematic of the possible subdivision of the most generic stress field, σ(*x*), into four components, of **a** uniform, **b** linear, **c** quadratic and **d** cubic trend, respectively, to each of which is associated a different influence coefficient $H_i(\theta)(i = 0, 3)$

crack depth to thickness ratio, *a/t*, and angle, θ, Milella and Pini used the technique to derive the best fitting function of three variable, *f(x,y,z)* using all data at the time available in the open literature for surface cracks in cylindrical geometries [76–78]. Table 9.5 lists the available solutions used.

In general, it is always possible to derive the function *f(x,y,z)* of three variables as a polynomial expression of a single variable *x,* in which the coefficients are function of the remaining two variables *y* and *z*

$$f(x, y, z) = \sum_{i=1}^{n+1} C_i \cdot x^{(i-1)} \tag{9.90}$$

where *n* is the degree of the polynomial of the *x* variable and

$$C_i = g(y, z) \tag{9.91}$$

The coefficients C_i can be, in turn, expressed as polynomial function of *y* with new coefficients A_{ij} function of the remaining variable *z*

$$C_i = \sum_{j=1}^{m+1} A_{i,j} \cdot y^{(j-1)} \tag{9.92}$$

with

$$A_{i,j} = h(z) \tag{9.93}$$

and

$$A_{i,j} = \sum_{k=1}^{p+1} B_{i,j,k} \cdot z^{(k+1)} \tag{9.94}$$

In this manner, the problem is broken down into three part each of which represents the derivation of a best-fitting function of a single variable, knowing only few values of it. The coefficients C_i are first found, keeping y and z constant, then $A_{i,j}$ with z constant and finally $B_{i,j,k}$; then the overall function $F(x,y,z)$ is assembled and the H_i polynomial expressions found. The degree chosen for each polynomial was 2 for the x and y variables, that correspond to *a/c* and *a/t*, respectively, and 3 for z, that corresponds to the angle θ. Table 9.2 lists the values of the coefficients $B_{i,j,k}$ for each magnification factor, H_1, H_2, H_3. Knowing the $B_{i,j,k}$ coefficients it is possible to write the numerical solution for any ratio *a/c, a/t* and any value of the angle θ as

$$\begin{aligned} f(x,y,z) = {} & B_{1,1,1} + B_{2,1,1} \cdot x + B_{3,1,1} \cdot x^2 + B_{1,2,1} \cdot y + B_{1,3,1} \cdot y^2 + B_{1,1,2} \cdot z \\ & + B_{1,1,3} \cdot z^2 + B_{1,1,4} \cdot z^3 + B_{2,2,1} \cdot xy + B_{2,1,2} \cdot xz + B_{1,2,2} yx + \ldots \end{aligned} \tag{9.95}$$

where $x = a/c$, $y = a/t$ and $z = \theta$. Figure 9.75a shows the influence coefficient H_o obtained using Eq. (9.95) for several *a/c* values and $a/t = 0.5$, as function of the angle θ. Also shown are the 3-D FE solutions used as reference [67]. The error between the polynomial expressions H_i and the 3-D Fe solutions is always limited to some percent. Figure 9.75b through Figure 9.75d show the other three influence factors H_1, H_2 and H_3 for several *a/c* values and $a/t = 0.5$ [67].

It is interesting to note how the influence coefficient H_0 always reaches its maximum at the deepest point A for $\theta = 90^\circ$, Fig. 9.73, with the exception of low *a/c* ratios, while the other magnification factors H_1, H_2 and H_3 reach their maxima at $\theta = 0^\circ$. This actually means that under a constant stress state, Fig. 9.74a, the crack is likely to advance on the sides more than in depth, becoming a continuous longitudinal flaw. The opposite is true under a stress gradient where the crack tends to grow in depth more than on the sides (Table 9.6).

Concluding Remarks

There is a large availability of specimens to measure the fracture toughness, K_{Ic}, of materials. However, within this wide availability, some specimens are more suitable to provide full constraint in the crack tip region or process volume and maintain SSY conditions. Constraint plays a fundamental role in fracture mechanics. Among these high constraint specimens, the compact type, C(T), behaves certainly better than plain type specimens such as CC(T), SEC(T) or DEC(T), since it provides a biaxial loading condition at the crack tip. The Round notch bar specimen, RNB(T), is probably the most effective of them all. SEN(B) specimen represents the best solution since it provides high constraint and reduces the applied load at fracture due to the amplifying effect exerted by the bending arms. However, to get a valid result specimen dimensions must comply with minimum size requirements. Stringent thickness and size requirements

Fig. 9.75 Trend of the Magnification Factors: **a** H_o, **b** H_1, **c** H_2 and **d** H_3 [67]

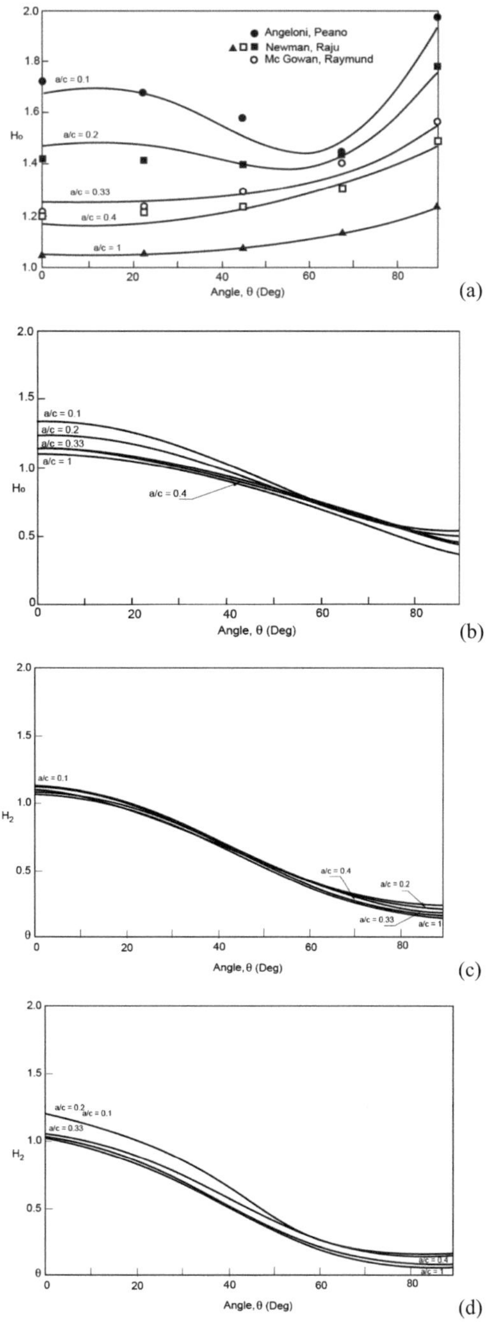

Table 9.6 Table of B(i,j,k) factors for each influence coefficient H_i

	H_o	H_1	H_2	H_3
B(1,1,1)	0.9850	1.010971	1.032670	1.228144
B(2,1,1)	0.0220	−0.051576	−0.173828	−0.722941
B(3,1,1)	−0.0114	0.164259	0.324154	0.713695
B(1,2,1)	0.4545	−0.131577	−0.518258	−1.430378
B(2,2,1)	−0.1054	1.057495	2.107098	4.644866
B(3,2,1)	−0.2620	−0.963560	−1.657703	−3.312582
B(1,3,1)	2.8515	2.049634	1.811909	2.232682
B(2,3,1)	−10.2202	−6.985,003	−5.969865	−7.000699
B(3,3,1)	7.4242	5.106468	4.335075	4.956183
B(1,1,2)	0.0037	0.005403	0.000637	−0.014035
B(2,1,2)	−0.0220	−0.034418	−0.022365	0.005030
B(3,1,2)	0.0181	0.029789	0.022614	0.005894
B(1,2,2)	0.0203	0.006388	0.014660	0.044969
B(2,2,2)	−0.0453	0.039409	0.030102	−0.040420
B(3,2,2)	0.0284	−0.040581	−0.042072	−0.000650
B(1,3,2)	−0.0019	−0.004997	−0.015866	−0.038671
B(2,3,2)	−0.0191	−0.043479	−0.026680	0.027429
B(3,3,2)	0.0154	0.040442	0.036738	0.004174
B(1,1,3)	−0.00004	−0.000320	−0.000359	−0.000217
B(2,1,3)	0.0004	0.000884	0.000613	0.000481
B(3,1,3)	−0.0003	−0.000781	−0.000674	−0.000704
B(1,2,3)	−0.0015	−0.000561	−0.000520	−0.000359
B(2,2,3)	0.0048	0.000733	−0.000040	−0.001182
B(3,2,3)	−0.0033	−0.000221	0.000583	0.001720
B(1,3,3)	0.0002	−0.000081	0.000070	0.000083
B(2,3,3)	0.0003	0.001746	0.001682	0.002469
B(3,3,3)	−0.0005	−0.001592	−0.001743	−0.002509
B(1,1,4)	0.0000008	0.000002	0.000002	0.000003
B(2,1,4)	−0.000004	−0.000005	−0.000003	−0.000005
B(3,1,4)	0.000003	0.000004	0.000004	0.000006
B(1,2,4)	0.00001	0.000006	0.000005	0.000002
B(2,2,4)	−0.00004	−0.000015	−0.000005	0.000014
B(3,2,4)	0.00003	0.000009	0.0000004	−0.000016
B(1,3,4)	−0.0000014	0.0000005	−0.0000003	0.0000002
B(2,3,4)	0.0000015	−0.000008	−0.000010	−0.000025
B(3,3,4)	0.0000003	0.000007	0.000010	0.000023

are placed by ASTM and ISO standard to limit the extension of the plastic zone ahead of the crack tip to ensure SSY conditions. These requirements often invalidate the results of fracture testing done with too small specimens. Fracture toughness in BCC materials is strongly dependent on temperature and loading conditions, whether static or dynamic. High scatter of experimental results is observed in the transition zone of ferritic steels or using small specimens therefore, a probabilistic approach is needed to obtain the toughness at the required level of reliability. The ASME Code provides a lower bound toughness reference curve for low-alloyed ferritic steels. LEFM remains a single parameter theory based on the Irwin stress intensity factor, K_I, as long as the fracture toughness remains substantially low, below 150 MPa$\sqrt{m}$ in ferritic steels (better applications are around 40–60 MPa$\sqrt{m}$), which for BCC materials are reached below RT_{NDT}. Shallow cracks tend to increase the apparent toughness of materials above the RT_{NDT} but not below the RT_{NDT}. Dynamic loads sharply reduce the toughness od materials in the transition region and above. They are less or not effective at all below the transition temperature where the material toughness is so poor that there is no room left for further reduction.

References

1. Robinson, E.L.: Bursting test of steam-turbine disc weels. Trans. ASME **66**, 373–386 (1944)
2. Sankey, G.O.: Spin fracture tests of nickel-molibdenum-vanadium rotor steels in the brittle fracture range. ASTM Proc. **60**, 721–732 (1960)
3. Winne, D.H., Wundt, B.M.: Application of the Griffith-Irwin Theory of crack propagation to the bursting behavior of discs, including analytical and experimental studies. Trans. ASME 80 (1958)
4. Nichols, R.W., Burdekin, F.M., Cowan, A., Elliot, D., Ingham, T.: The use of critical crack tip opening displacement techniques for the selection of fracture resistant materials. In: Proceeding of Symposium on Fracture Toughness, Risley, April 1969
5. Timoshenko, S.P., Goodier,J.N.: Theory of Elasticity, 3rd edn, p. 81. McGraw-Hill (1970)
6. Sankey, G.O.: Spin tests to determine brittle fracture under plain strain, experimental mechanics. J. SESA 529–538 (1968)
7. FALSIRE PHASE II, CSNI Project for Fracture Analyses of Large-Scale International Reference Experiments, OCDE/GD (96)187, November 1996
8. Tada, H.:2H the stress analysis of cracks handbook, Del Research C., Hellerton, Pennsylvania (1973)
9. Manjoine, M.J.: Biaxial brittle fracture test. J Basic Eng Trans ASME 293–298 (1965)
10. Bush, A.J., Wilson, W.K.: Experimental Determination of the Energy Release Rate of Biaxial Brittle Fracture Specimens, Westinghouse Research report for AEC, WERL-8844-2 (August 1964)
11. Wilson, W.K.: Analitycal Determination of Stress Intensity Factors for the Manjoine Brittle Fracture Test Specimen, Westinghouse Research Report for AEC, WERL 0029-3 (August 1965)
12. Wilson, W.K.: Optimization of WOL Brittle Fracture Test Specimen, Westinghouse Research Report 66-B40-BTLFR-R1 (February 1966)
13. Novak, S.R., Rolfe, S.T.: J. Mater. **4**(3) (1969)
14. Saxena, A., Hudak, S.J.: Review and extension of compliance information for common crack growth specimens. Int. J. Fract. **14**(5), 453–467 (1978)

15. E 399-90, Standard Test Method for Plain Strain Fracture Toughness of Metallic Materials, American Society for Testing and Materials, Philadelphia, PA, (1990)
16. Kapp, J.A.: Improved Wide Range Expressions for Displacements and Inverse Displacements for Standard Fracture Mechanics Specimens, JTEVA (1989)
17. ASTM E 1820-08a, Standard Test Method for Measurement of Fracture Toughness
18. Nisitani, H., Noguchi, H., Mori, K.: Analysis of single edge cracked specimen under three or four-point bending by body force doublet method. Trans. Jpn. Soc. Mech.Eng. **52**(474), 539–543 (1986)
19. Gross, B., Srawley, J.E.: Stress-intensity factor for three-point bend specimens by boundary collocation, NASA TN D-3092 (1965)
20. Srawley, J.E.: Wide range of stress intensity factors for ASTM E 399 standard fracture toughness specimens. Int. J. Fract. Mech. (12), 475–476 (1976)
21. Irwin, G.R., Kies, J.A., Smith, H.L.: Fracture strengths relative to onset and arrest of crack propagation. Proc. ASTM **58**, 640–660 (1958)
22. Wundt, B.M.: A Unified Interpretation of Room-Temperature Strength of Notched Specimens as Influenced by Their Size", paper 59-MET-9, ASME (1959)
23. Lubahn, J.D.: Experimental determination of energy release rates of notch bending and notch tension. Proc. ASTM **59**, 885–915 (1959)
24. Benthem, J.P., Koiter, W.T.: Asymptotic Approximation to Crack Problems, Method of Analysis and Solutions of Crack Problems, 3, p. 131, Noodhoff Int. Pub., Edited by G. C. Sih (1973)
25. Srawley, J.E., Brown, W.F.: Fracture Toughness Testing Methods, American Society for Testing and Materials, ASTM STP 381, pp.133–198 (1975)
26. Milella, P.P., Pini, A.: Calibrazione Sperimentale della Cedevolezza di un Provino Compact Crack Arrest Attraverso l'Uso del Metodo dell'Attrito, Rapporto Tecnico ENEA(1987)
27. Ripling, E.J., Crosley, P.B.: Contribution to the ORNL HSST Program, Quarterly Report, Nov. 1979
28. Newman, J.C., Jr.: Crack-Opening Displacement in the Center-Crack, Compact and Crack-Line-Wedge-Closed Specimens, NASA Langley Research Center, Hampton, Va., June 8, 1976 and NASA TN D-8268, Washington, DC, July, 1976
29. Freed, C.N., NRL Report 6654 (1967)
30. BS 5762, Methods for Crack Opening Displacement (COD) Testing. British Standard Institution, London (1979)
31. ASTM E813-81. Standard Test Method for Measurement for J_{Ic}, a Measure of Fracture Toughness. West Conshohocken (PA, USA): ASTM International (1981)
32. E1290-2, Standard Test Method for Crack Tip Opening Displacement Fracture Toughness Measurement, American Society for Testing and Materials, Philadelphia, PA (2002)
33. Fourney, W.F.: Investigation of Damping and Cleavage-Fibrous Transition in Reactor-Grade Steels, NUREG/CR-2751, ORNAL/TM-8369/V2, Whitmanand Bryan Eds. (1982)
34. Sanford, R.J.: Application of the least-square method of photoelastic analysis. Exper. Mech. **6**(20), 192–197 (1980)
35. Sanford, R.J., Fourney, W.L., Chona, R., Irwin, G.R.: Photoelastic Study of the Influence of Non-Singular stresss in Fracture Test Sprcimens, NUREG/CR 2179 (1981)
36. Irwin, G.R.: Use of Thickness Reduction to Estimate Values of K, NUREG/CR-5697 ORNL/Sub/79–7778/5, US NRC, November (1991)
37. US Steel Co. Seminar on missile steels (1960)
38. Irwin, G.R.: The Effect of Size Upon Fracturing, American Society for Testing and Materials, ASTM STP 158, pp. 176–194 (1954)
39. Tiffany, C.P., Masters ,J.N.: Applied Fracture Mechanics, ASTM-STP 381, pp.249–277 (Jan 1975)
40. Irwin, G.R.: Dimensional and Geometric Aspect of Fracture, Troy Conference on Fracture of Engineering Material, Eastern, N.Y., Cha. Of ASM, August 24–25 (1959)
41. Steigerwald, E.A.: Plain strain fracture toughness of high strength materials. Eng. Fract. Mech. **1**(3), 473–494 (1969)
42. Kaufman, J.G., Moore, R.L., Schilling, P.E.: Eng. Fract. Mech. 2(3):197–210 (1971)

43. Brown, W.F., Jr.: Review of Developments in Plain Strain Fracture Toughness Testing, American Society for Testing and Materials, ASTM STP-463, 1960.
44. Brown, W.F., Jr., Srawley, J.E., Plain Strain Crack Toughness testing of High Strength Metallic Materials, American Society for Testing and Materials, ASTM STP 410, Philadelphia (1966)
45. E399-90 Standard Test Method for Plain-Strain Fracture Toughness of Metallic Materials, ASTM, American Society for Testing and Materials, Philadelphia, PA (1970). (Reapproved in 1990)
46. Shabbits, W.O., Wessel, E.T.: Fracture Toughness of ASTM A 533 Grade B, Class 1 Heavy section submerged arc weldments. ASME J. Basic Eng. (1971)
47. Wessel, E.T.: Linear elastic fracture mechanics of thick-walled, welded steel pressure vessels. In: Material Property Considerations, Proceedings of Symposium on Fracture Toughness, Risley, pp. H1–H41 (April 1969)
48. Landes, J.D., Begley, J.A.: Test results from J-integral studies: an attempt to establish a Jic testing procedure. ASTM STP 560, American Society for Testing and Materials, pp. 170–186 (1974)
49. Greenberg, H.D., Wessel, E.T., Clark, W.G. And Pryle, W.H.: Critical flaw size for brittle fracture of large turbine genarator rotor forgings. In: Proceeding of 5th International Forgemasters Meeting, Terni, Italy, May (1970)
50. Wessel, E.T., Clark, W.G., Wilson, W.K.: Engineering method for the design and selection of materials against fracture. US Army Tank and Automotive Center Report, Final Technical Report, Control, DA 30-069 AMC 602(T), June 24 (1966).
51. Kirk, M.T.: The Technical Basis for Application of the Master Curve to the Assessment of Nuclear Reactor Pressure Vessel Integrity, Report Prepared for NRC (Oct 2002)
52. Shabbits, W.O.: Dynamic Fracture Toughness Properties of Heavy Section ASTM A 533 Grade B, Class 1 Steel Plate, ORNL Heavy Section Steel Technology Program, Technical Report No. 13, Westinghouse WCAP-7561 (December 1970)
53. Stonesifer, R.B., Rybiki, E.F.: Development of Models for Warm Prestressing, NUREG/CR-4491, MEA-2122, p. 15 (Jan 1987)
54. PVRC Ad Hoc Group on Toughness Requirements, "PVRC Recommendations on Toughness Requirements for ferritic Materials ", Welding Research Council, WRC Bulletin No. 175 (Aug 1972)
55. Balkey, K.R., Furchi, E.L.: Probabilistic fracture mechanics sensitivity study for plant specific evaluation of reactor vessel pressurized thermal shock. ASME PVP **92**, 71–85 (1984)
56. Kirkemo, F., Hessan, G., Lereim, J.: Probabilistic calibration of fracture mechanics design format applied in offshore structural design. In: Proceeding of OMAE-90, pp. 215–221 (1990)
57. Merkle, J.G.: An Examination of the Size Effects and Data Scatter Observed in Small-Specimen Cleavage Fracture Toughness Testing, NUREG/CR-3672, ORNL/TM-9088, p. 45 (1984)
58. Masubuchi, K.: Control of Distortion and Shrinkage in Welding, Welding Research Council Bulletin No. 149 (April 1970)
59. UKAEA, An Assessment of the Integrity of PWR Pressure Vessels, 2nd Report by Dr. W. Marshall Study Group (Mar 1982)
60. Kanazawa, T., Oba, H., Susei, S.: The effect of welding residual stress upon brittle fracture propagation, report II. Jpn. Soc. Naval Arch. **110**, 359–368 (1961)
61. Masubuchi, K., Martin, D.C.: Investigation of residual stresses in steel weldments. Battelle Memorial Institute, Columbus, Ohio, Defense Technical Information Center (1998)
62. Milella, P.P.: Fatigue and Corrosion in Metals, Chapter 18, 2nd edn. Springer (2024)
63. Sorem, W.A., Dodds, R.H. Jr., Rolfe, S.T.: The Effect of Crack Depth on Elastic-Plastic Fracture Toughness in Bend Bar Specimens. University of Kansas, SL Report 89–1, October (1989)
64. Smith, J.A., Rolfe,S.T.: The Effect of Crack Depth to Width Ratio on Elastic-Plastic Fracture Toughness of a High Strength Low-Strain Hardening Steel, Butler Manufactoring Co. & University of Kansas, WRC Bullettin 358, Welding Research Council, November (1990)
65. Smith, J.A., Rolfe,S.T.: The Effect of Crack Depth to Width Ratio on Elastic-Plastic Fracture Toughness of a High Strength Low-Strain Hardening Steel, Butler Manufactoring Co. & University of Kansas, February (1990)

66. Sorem, W.A., Dodds, R.H. Jr., Rolfe, S.T.: An Analitycal Comparison of Short Crack and Deep Crack CTOD Fracture Specimens of an A 36 Steel, WRC Bullettin 351, Welding Research Council, February (1990)
67. Edison Welding Institute, An International Research Project to Develop Shallow Crack Fracture Mechanics Tests, North American Contribution, 4th Progress Report, EWI Report No. J6098-12-88, October (1988)
68. The Welding Institute and Edison Welding Institute, An International Research Project to Develop Shallow Crack Fracture Mechanics Tests, Proposal, CP/MAN/3299-1, July (1987)
69. Theiss, T.J.: Recommendations for the Shallow-Crack Fracture Toughness Testing Task Within the HSST Program, Oak Ridge National Laboratory, US-NRC NUREG/CR-5554, September (1990)
70. Theiss, T.J., Shum, D.K., Rolfe, S.T.: Experimental and Analytical Investigation of the Shallow-Flaw Effect in Reactor Pressure Vessels, US-NRC NUREG/CR-5886, ORNL/TM-12115, July (1992)
71. McGowan, J.J.: An Assessment of the Beneficial Effects of Warm Prestressing on the Fracture Properties of Nuclear Reactor Vessels under Severe Thermal Shock, Westinghouse Elect. Co., WCAP-9178 (1978)
72. Milella, P.P., Pini, A.: Stress Intensity Factor Solutions for Semielliptical Surface Flaw of Any Aspect Ratio and Depth Longitudinally Oriented in Cylinders Subjected to Arbitrary Loadings, ENEA/DISP Report, ACO/ATEM DT 2 (1988)
73. McGowan, J.J., Raymund, M.: Stress intensity factor solutions for internal longitudinal semielliptical surface flaws in a cylinder under arbitrary loadings. Fract. Mech. ASTM-STP **677**, 365–380 (1979)
74. Labbens, R., Helliot, J.: Semielliptical cracks in the meridian plane of a cylinder subjected to stress gradient. STM-STP **677**, 341–364 (1979)
75. Newman, J.C., Raju, IS: Stress Intensity Factor Influence Coefficients for Internal and External Surface Cracks in Cylindrical Vessels, ASME-PVP, vol. 58, pp. 37–48 (1982)
76. Milella, P.P., Sampietri, C.: Influence Functions for Surface Flaws in Pressure Vessels, SMiRT 8th, Vol. G, Paper G 4/4, pp. 159–168, 19–23 August 1985
77. Angeloni, P., Peano, A.: Analisi di Difetti Superficiali Semiellittici ed Assialsimmetrici Disposti Assialmente in un Contenitore Cilindrico, Enea-Ismes Contract, Interim Report, ISMES 3399 (1987)
78. Angeloni, P., Peano, A.: Analisi di Difetti Superficiali Semiellittici ed Assialsimmetrici Disposti Circonferenzialmente in un Contenitore Cilindrico, Enea-Ismes Contract, Interim Report, ISMES 2579 (1986)

Chapter 10
Elastic-Plastic Fracture Mechanics

10.1 Introduction and Historical Sketch

The fundamental result of linear elastic fracture mechanics, LEFM, is that all the energy release processes that set the unstable propagation of a crack and those of energy absorption that represent, ultimately, the resistance R offered by the material, see Sect. 7.3, can be expressed through a single parameter: the K_c. In as long as LEFM applies, this parameter is independent of the stress state, the geometry of the system and the length or position, internal or superficial, of the crack and represents a characteristic of the material: the toughness or fracture resistance. The stability of a defect of given length under a certain stresses state is judged by the comparison between a calculated stress intensity factor, K, and the critical K_c. Since K represents the amplitude of the elastic singularity $1/\sqrt{r}$, i.e. the strength of the elastic stress field in front of the crack, it must be deduced that in practice, even if in real materials the resistance to fracture propagation is offered, in particular, by the small plastic enclave that develops at the front of a crack, what in fact governs the fracture itself is precisely the elastic stresses field that exists ahead of the crack: when it reaches a critical value, i.e. when K_I, K_{II} or K_{III} reach the respective K_{Ic}, K_{IIc} or K_{IIIc} values, the flaw becomes unstable. This apparent paradox is perfectly understandable if we argue that, within the field of validity of LEFM, there is a very precise biunivocal correspondence between the elastic stress field beyond the plastic zone and the plastic deformation field inside it: the elastic stress field fixes the boundary conditions for the plastic zone it contains. When this stress field reaches the limit value, K_c, also deformations in the plastic enclave reach an overall limit state and fracture occurs. All what happens in the very small plastic zone at the crack tip is controlled by the external elastic stress field governed by K, and vice versa. This is true whenever the dimensions of the plastic zone are small compared to those in which the elastic field Eqs. (7.111) hold. Since the latter are valid just in a neighborhood r of the crack tip in the singularity-dominated zone in which the terms in r/a of order higher than the first are negligible with respect to unity, point 2 of Sect. 7.10, it must be concluded that

P. P. Milella, *Fundamentals of Fracture Mechanics*,
https://doi.org/10.1007/978-3-032-23081-2_10

LEFM is valid whenever the radius of the plastic zone r_p is negligible with respect to the K-dominance zone. These are the conditions that we have already defined of SSY. Whenever the dimensions B, W and a, Fig. 9.8, decrease so much that the apparent toughness of the material, K_c, increases becoming geometry dependent, the radius r_p is no longer negligible. Under these conditions, the elastic stresses field ahead of the crack cannot be characterized by a single-valued parameter: the K. In these conditions, the interaction between the crack tip plastic area and the edge of the structure becomes so intense as to influence the elastic stress field which, eventually, may no longer exist. This is the limit case of *General Yielding* (GY) in which the entire section containing the defect is stressed beyond the elastic limit. A typical example is shown in Fig. 10.1 for a cracked body subjected to Mode III opening, out-of-plane shear, Fig. 7.14. The plastic enclave, which in an infinite plate is small, under the same situation of applied stress state stretches more and more as the width, W, of the plate is reduced until it reaches the outer edge, when W becomes twice the size of the crack [1]. Now, the elastic stress field does not exist anymore. Well before this last situation occurs, the measured toughness, K_c, will no longer be unique, but will depend on the geometry of the system: the case $\mathrm{W} = 2a$ of Fig. 10.1, for instance, has an apparent K_c greater than that relating to the case $\mathrm{W} = 4a$ which, in turn, will certainly be higher than that relative to the case $W = \infty$ that measures the real toughness K_{IIIc} of the material. Between these two limiting situations: the loss of a single-valued parameter elastic stress field, K, and the GY, one may argue whether the plastic stress field that takes over the elastic one presents a singularity of any kind and can ever fix, in analogy with the elastic case, the boundary conditions for something smaller inside it, but still much larger than the plastic zone of the elastic case, in which the deformations will be so high as to give rise to the formation and coalescence of the pores, typical of ductile fracture, Fig. 6.12. Unfortunately, historically this transfer of attention from the elastic stress field to the plastic field ahead of the crack was particularly difficult and, above all, discouraged by the fact that the only solutions available at the time were those concerning the cases of GY obtained with the slip-line field theory [2] which clearly indicated the non-uniqueness of the plastic field, which, on the contrary, strongly depended on the geometry of the system and the loading conditions, whether plane stress or plane strain. Fig. 10.2 illustrates some typical cases. The first case (a) is that of a plate containing a double surface crack under remote traction and plain strain conditions. It is shown the relative theoretical Prandtl field of constant shear stress, τ_s, equal to $\sigma_s/2$, the uniaxial yield stress of the material. This plastic field has been experimentally evidenced using dye liquid or liquid penetrant testing as shown in Fig. 10.3a [3]. It can be noted the good correspondence of the experimental plastic stress field with the predictions of the slip-line field theory. The stress σ_y at the crack tip, evaluated according to the Orowan model [4], is

$$\sigma_y = 2 \cdot \tau_s\left(1 + \tfrac{\pi}{2}\right) \tag{10.1}$$

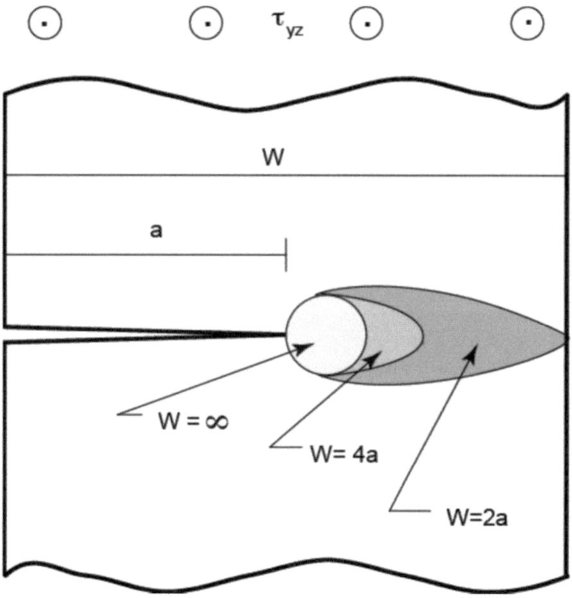

Fig. 10.1 Effect of width, W, reduction on the plastic zone size in a plate subjected to Mode III opening

as already seen, Eqs. (8.42), and the strain shows a singularity in the two central area where the slip lines merge, Eq. (8.43). The same case, but in plain stress conditions, Fig. 10.2b, results in a stress without any strain singularity

$$\sigma_y = 2 \cdot \tau_s \tag{10.2}$$

that is, without the hydrostatic elevation $(1 + \pi/2)$ and with a totally different deformation field that leads to the necking formation. If the crack is internal, Fig. 10.2c and d, under plain strain conditions the maximum stress changes again to become

$$\sigma_y = 2 \cdot \tau_s \tag{10.3}$$

The strain field will also change, which will be the one shown in Fig. 10.2c, evidenced in Fig. 10.3b [3], always with the technique of penetrating liquids. Under plain stress conditions, Fig. 10.2d, the stress σ_y will be

$$\sigma_y = \sqrt{3} \cdot \tau_s \tag{10.4}$$

Once again with different deformation field, singularity-free, while the ratio between the maximum stress σ_y relative to the external and internal crack, moves from $(1 + \pi/2) \approx 2.57$ to $2/\sqrt{\pi} \approx 1.15$ with the change of conditions from plane strain to plane stress, respectively, with a drastic reduction of hydrostatic elevation. This simple review indicates that under fully plastic conditions or GY, the flow field

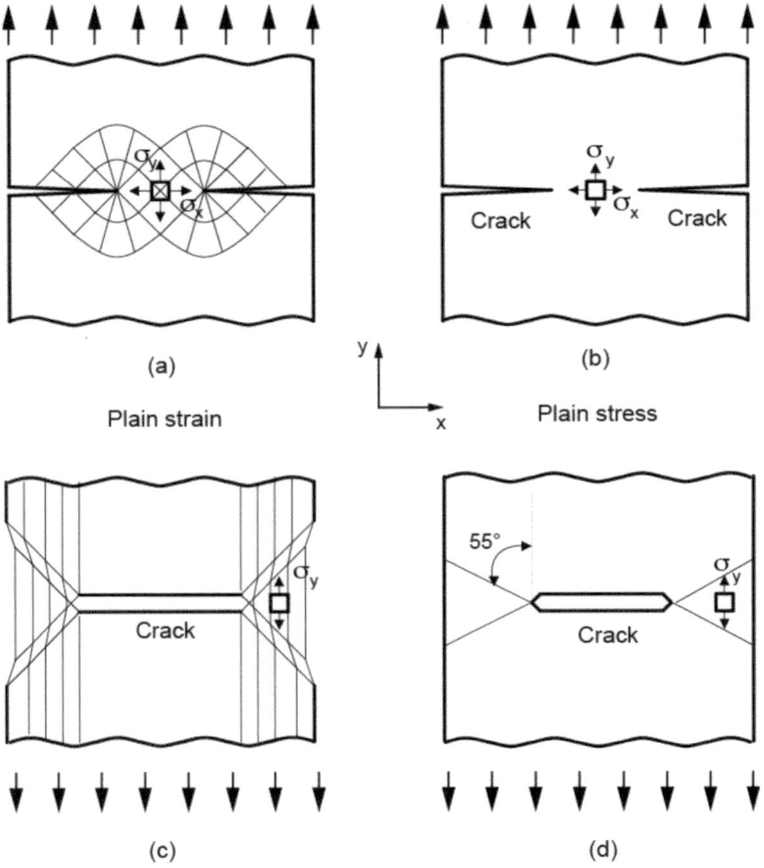

Fig. 10.2 Plastic strain fields accordingly to slip-line field theory

emanating from the crack or, better, from the notch, since the original crack became a notch, for nonhardening materials is not unique but depend on geometry.

This dependence introduces enormous difficulties on the practical ground since it warns that a fracture produced in small specimens under GY conditions could not occur in a real structure that may have a completely different notch tip stress field. The similitude principle that characterizes LEFM cannot be invoked anymore. However, it must be recalled that slip-line field theory applies only to ideally elastic-plastic bodies. In real materials even a modest strain hardening, can completely change the overall picture. The influence of the strain hardening and geometric conditions on the plastic stress field, evidenced by McClintock [1, 4], is enormous giving rise, under certain circumstances, to a well-defined crack tip stress field. The search for this singularity dominated plastic stress field was precisely the path followed in the US in the wake of the Irwin K stress intensity factor that was defining a self-similar elastic stress field with a singularity of the type $1/\sqrt{}$. In effect, between the validity

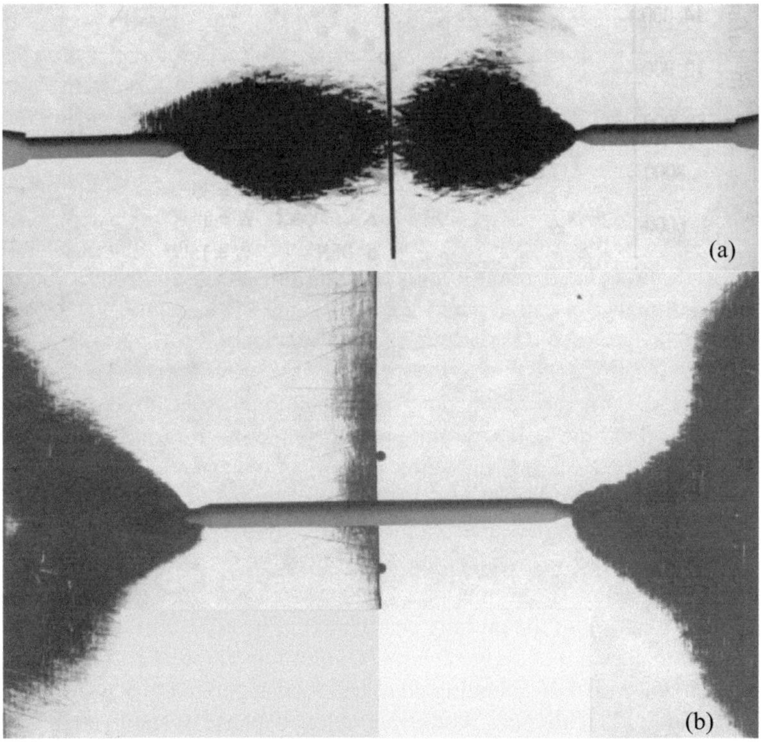

Fig. 10.3 Experimental evidence of plastic strain fields in a traction panel containing: **a** two edge cracks and **b** a central crack, using a penetrant liquids technique. Note the ~107° opening of the plastic delta for the central crack [3]

field of the LEFM and the GY an entire spectrum of intermediate situations exists in which the residual elastic stress field is no longer self-similar and a plastic stress field appears that gradually substitutes the elastic one becoming more and more relevant. In this substitution process that actually leads to the transition from LEFM to EPFM a new scenario appears in which the plastic stress field, grown beyond the limited border assigned by the LEFM, but well before being boundless, becomes dominant presenting a singularity of the type $(1/r)^{1/n+1}$ and always remaining self-similar, as it was the elastic one. It is the so-called HRR stress field or Hutchinson, Rice and Rosengren stress field, Sect. 10.7 and Eqs. 10.50 and 10.51, based on the J-integral. Note that the J-integral is based on the non-linear elasticity theory in which the material follows the Hooke's low with a strain hardening exponent $m < 1$, $n > 1$, Eq. (1.19), and any deformation is perfectly reversible.

In this condition, an EPFM exists as a single parameter theory, the J-integral, that appears to be an energy release rate, precisely the Griffith $\mathcal{G}$ parameter, Eq. (7.27). In analogy to the linear elastic case in which there was a K-dominance setting boundary conditions for a smaller plastic area inside, in this new situation there will be an area around the crack tip in which the stress field is totally plastic, defined and dominated by the J that sets the boundary conditions for smaller area inside in which the material behavior is no-longer non-linear elastic and therefore no longer described by any theory of plasticity, where a process of damage nucleation takes place based on pores formation, growth and coalescence that leads to the plastic crack propagation, see Sect. 6.3. Figure 10.4 schematizes this fundamental passage from LEFM to EPFM. As for the elastic case, it will now be possible to define a limiting value of J, the so-called J critical or J_c, at which the smaller zone of high plastic deformations inside the J-dominance region will fail. However, at variance with the K-dominance situation where the first failure of an inclusion at the border of the plastic enclave triggered the brittle fracture, see Section 6.4, in this new situation ductile failure is non-contextual to the first crack tip grain failure. Therefore, the J_c does not define the moment of plastic instability but marks the point of a sub-critical tearing process that always precedes the catastrophic plastic fracture to be reached by a further load increase. It is only after a small amount of stable tearing and crack growth, Δa, that instability may be reached that proceeds at a lower speed with respect to the brittle fracture. All this new situation is schematized in Fig. 10.5 that presents a load-subcritical crack grow diagram, F-Δa, of any test piece containing a crack. The initial straight line, up to the load F_c, marks an apparent growth of the crack caused by the blunting of the initially sharp crack, Fig. 6.22. The compliance of the testing machine does not distinguish between blunting and real crack growth.

It is only at the end of this blunting phase, after an apparent crack growth, Δa_c, that the notch tip starts to really tear. At this point, a new phase of stable tearing initiates ending at a maximum F_{max}-Δa_{max} point where the load-structure system decides how to behave. Three cases have been schematically shown in Fig. 10.6 as typical. The first, case (1), refers to a structure directly loaded in traction where the stable growth of the crack does not stop but continues catastrophically because the applied load does not decrease. At the opposite extreme, case (3), the structure is loaded under displacement-controlled conditions. The stable growth of the crack unloads the system and the crack eventually stops. In between these two limiting situations there is case (2) where the structure is loaded through a spring coupling, for instance the compliance of the loading system. Depending on the specimen to loading system compliance or stiffness, K_{sc}/K_{lsc}, the structure will fail or not. Figure 10.6 shows a piping containing a circumferential through wall crack loaded in three point bending under displacement controlled conditions. The crack first blunts. This can be seen because the parallel lines spacing is 1 cm. Then a slant tearing appears that proceeds steadily till the fast fracture occurs, but the fracture stops. While in the US the plastic stress field approach was followed, based on the J-integral, a different root was followed in the UK where the Griffith theory was born. It may appear strange and even difficult to believe but in the UK the US linear elastic approach based on the K concept initially was not considered on the basis that a single-valued parameter could not characterize the crack tip stress field when extensive yielding occurred and a national COD based criterion was used, which, in turn, always at the beginning, was not considered in the US. Two divergent roads were followed that, eventually,

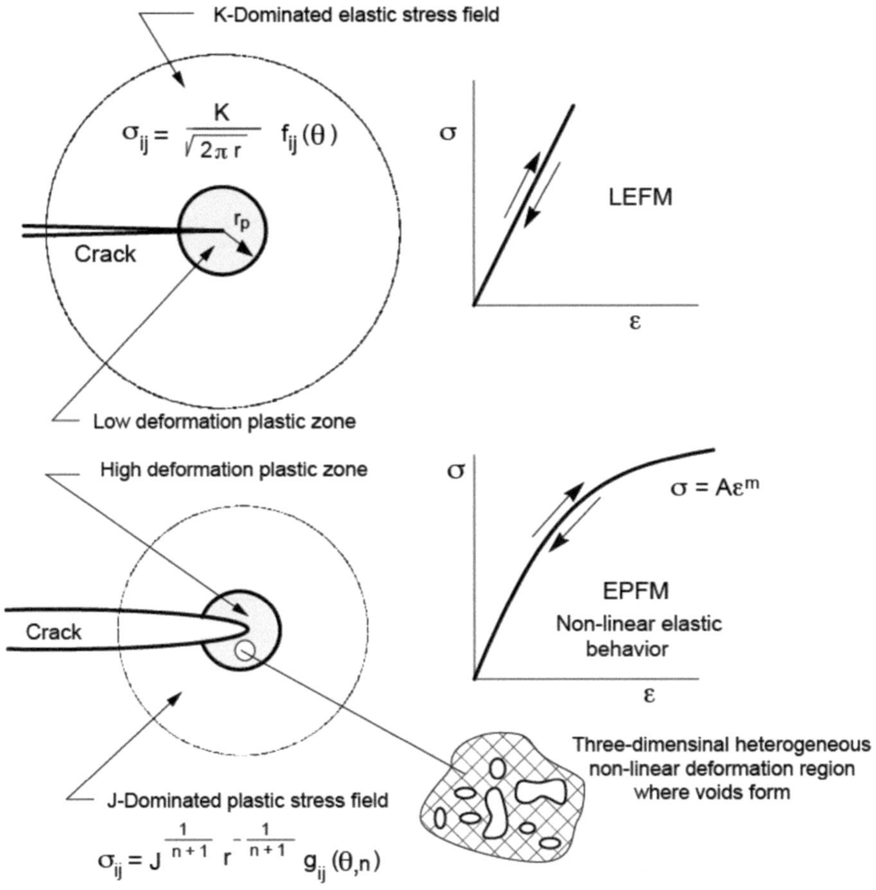

Fig. 10.4 Schematic view of the: **a** LEFM and **b** EPFM characteristics. Note that the exponent $m = 1/n$

would converge towards a common understanding. In the following sections these two approaches will be described, starting from the British one.

10.2 The CTOD Approach to Elastic-Plastic Fracture Mechanics

The first tentative to overcome the limits of LEFM was done in the UK by Cottrell [5] and Wells [6] who, independently, proposed in 1961a fracture parameter alternative to Irwin K_I. They focused the attention on the amount of crack tip opening displacement, CTOD, to characterize the crack tip stress and strain conditions also in cases where the material would be too tough to be treated with LEFM. It was recognized that

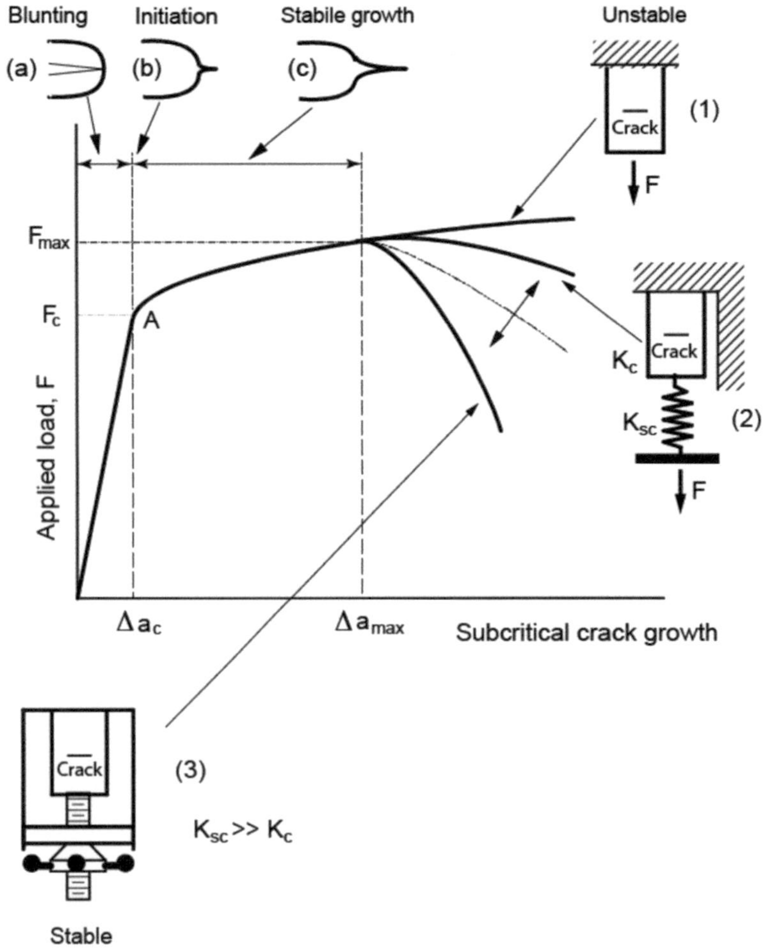

Fig. 10.5 schematic view of the possible behavior of a system containing a crack under elastic-plastic loading conditions. After the initial blunting of the crack that terminates in A at F_c, crack stable tearing initiates that terminates at F_{max} with: (1) unstable propagation, (3) crack arrest or (2) in-between situation

fracture would depend on the plastic strain field ahead of the crack tip that would open the tip more or less, depending on the material behavior. In a totally brittle fracture, the crack tip opening would remain confined so as to almost leave the initial sharpness of the crack, whereas in ductile or LSY (Large Scale Yielding) or GY (General Yield) condition the crack would blunt and become a large notch, as schematically shown in Fig. 10.7.

Thus, if fracture occurs at a critical value of K or $\mathcal{G}$ it is to be expected that this itself defines a critical value of CTOD. The concept of fracture at a critical value of CTOD or δ is relevant to materials outside the range of validity of linear

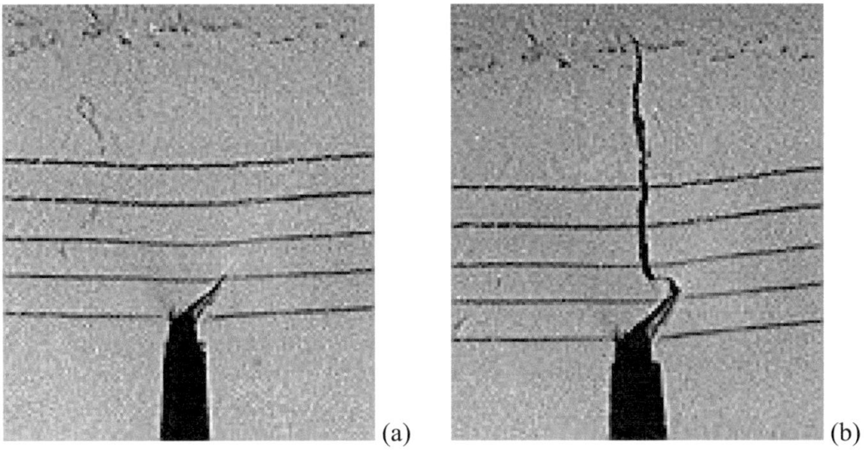

Fig. 10.6 **a** Blunting and stable tearing in a 4" A106 carbon steel pipe containing a 180° through wall crack under three-point bending; **b** after ~1.5 cm stable growth a sudden fracture occurred and then arrested because of the displacement-controlled loading conditions (Milella)

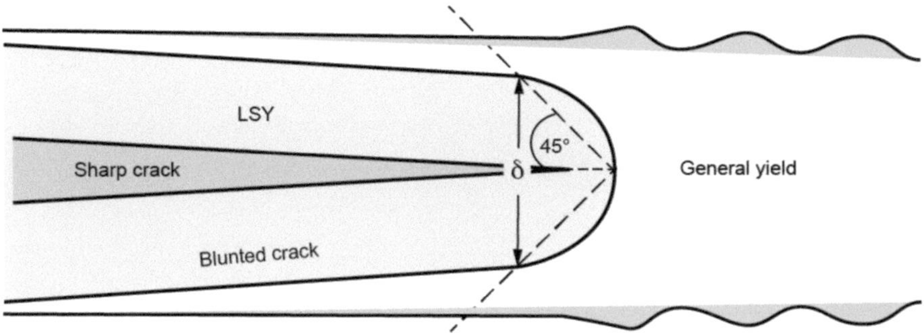

Fig. 10.7 In LEFM crack tip opening displacement (CTOD) is negligible and the crack does not lose its sharpness, remaining practically closed, in LSY it blunts more or less depending on material, temperature and opening load and under GY it spreads to the border. Also shown is a possible definition of CTOD = δ for LSY condition

elastic fracture mechanics. Figure 10.8 shows a set of Charpy-size specimens of mild steel having an initially sharp slit, loaded in slow bending at different displacements that resulted in different CTOD. After loading, each specimen was sectioned at mid-thickness and the CTOD optically determined [7]. The behavior is ductile, and the initial sharp crack blunts without fracturing showing the large-scale yielding (LSY) condition schematized in Fig. 10.7. Equal CTOD would mean equal crack tip conditions, independently of the specimen and crack sizes, shape or loading conditions and, therefore, a critical CTOD followed by an unstable fracture or stable tearing, depending on whether the behavior were brittle or ductile, would define the same critical event in any specimen or structure of the same material. Wells showed that the critical value of CTOD determined in bend specimens and wide plate specimens (representing structural components) of the same thickness were equivalent. Thus, he was able to demonstrate transferability of fracture toughness determined from test specimens to other structural geometries. This is shown to be true in Fig. 10.9 [8] for three specimens of the same material but different sizes containing identical cracks fractured at – 80 °C. In all cases the fracture conditions were brittle. The test piece having the smallest ligament underwent general yielding of the compression zone, the other two didn't but all failed at comparable crack tip deformation. Early experiments utilized a paddle-shaped gauge that was inserted into the slit as close as possible to the crack tip, the so-called *paddle COD-meter* (Fig. 10.9).

As the crack opened under an applied load, the paddle gage rotated, and this rotation was converted into an electrical signal by a transducer which could be calibrated

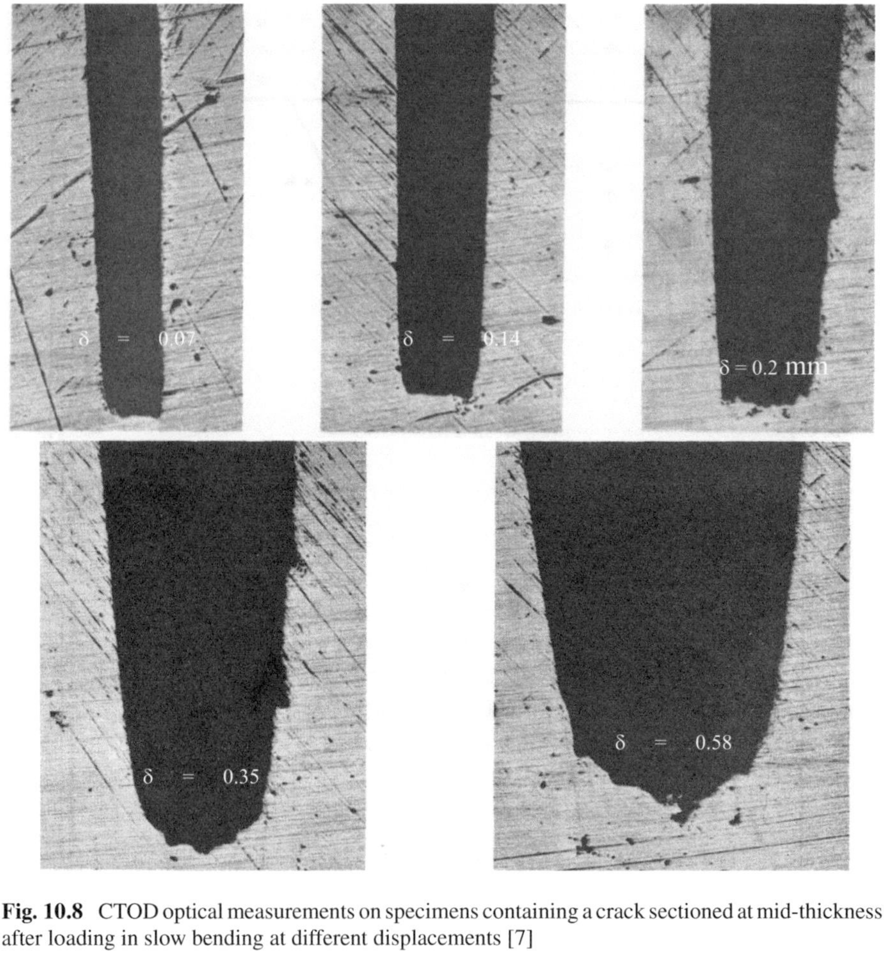

Fig. 10.8 CTOD optical measurements on specimens containing a crack sectioned at mid-thickness after loading in slow bending at different displacements [7]

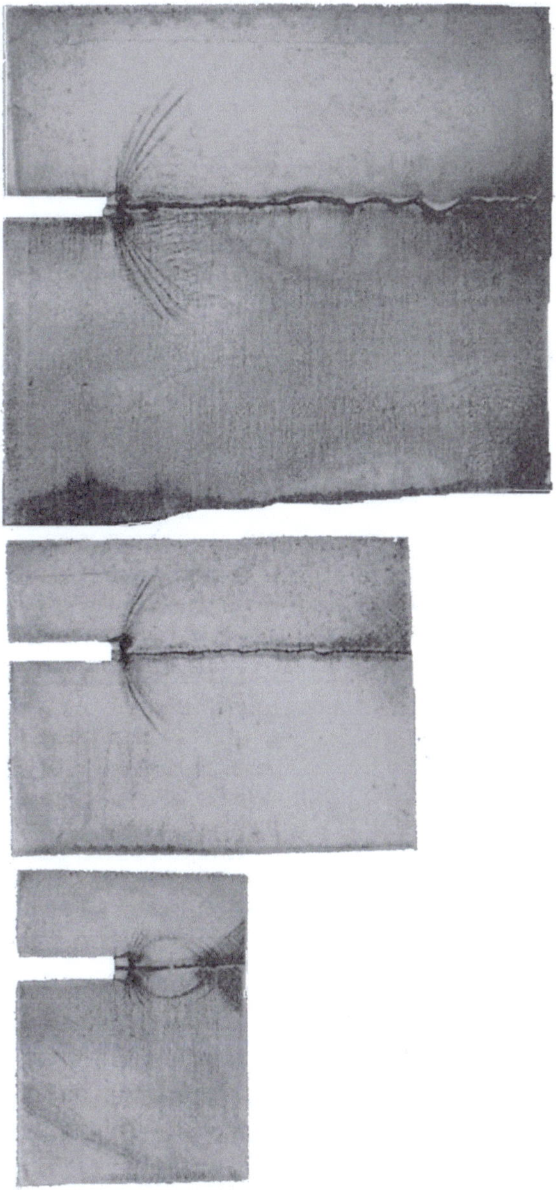

Fig. 10.9 Sections of mild steel specimens having the same notch geometry fractured at –80 °C. The plastic deformation preceding fracture has been etched using Fry's reagent [8]

in terms of CTOD. However, the method was inaccurate because it was difficult to reach the bottom of the crack, even impossible when the machined notch was fatigue pre-cracked. Therefore, the CTOD was later measured indirectly from a clip gage reading mounted on the external face as COD or CMOD (Crack Mouth Opening Displacement), as schematically shown in Fig. 10.10 for a SEN(B) specimen. From

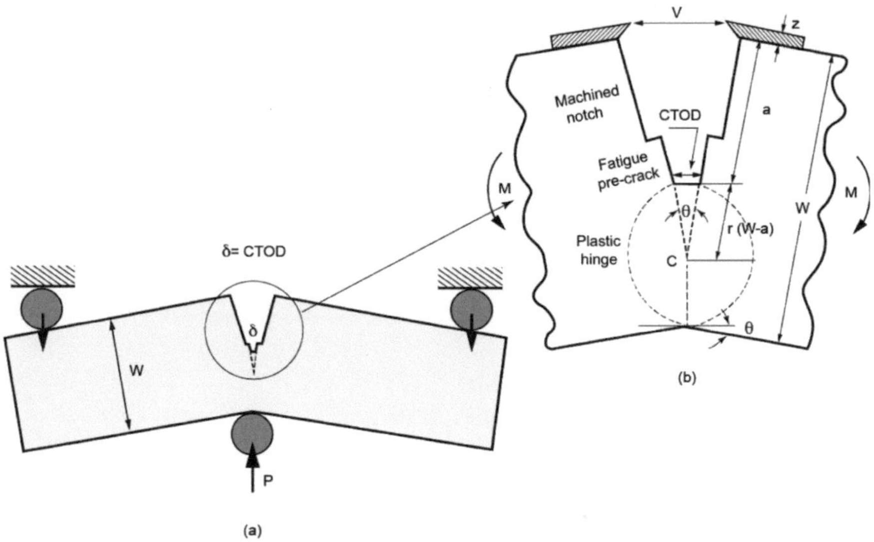

Fig. 10.10 **a** Three-point bend or SEN(B) specimen. CTOD is related to the crack mouth opening V through a rigid rotation of the specimen arms around point C following hinge formation **b** Detail of the notch area

the figure we can estimate CTOD from geometric considerations as

$$\frac{\delta}{r(W-a)} = \frac{V}{r(W-a)+a} \tag{10.5}$$

Therefore

$$\delta = CTOD = \frac{r(W-a)V}{r(W-a)+a} \tag{10.6}$$

where r is a dimensionless constant between 0 and 1.

However, also the hinge model was found to be inaccurate, because of the difficulty pinpointing the precise location of the hinge point C and the r distance, Fig. 10.10b. Therefore, today CTOD is measured indirectly from K or J, as described in Sect. 11.15.

10.3 Analytical Expression of CTOD

Using the Dugdale model, Figs. 8.3 and 10.11, Goodier and Field [9] and Stone and Burdekin [10] found the expression of CTOD for a central crack in a plate using a Westergaard stress function Z(z), see Sect. 7.10, for a central crack of length $2a_o$, opened by a pair of splitting forces, P, acting against the crack surface at the position

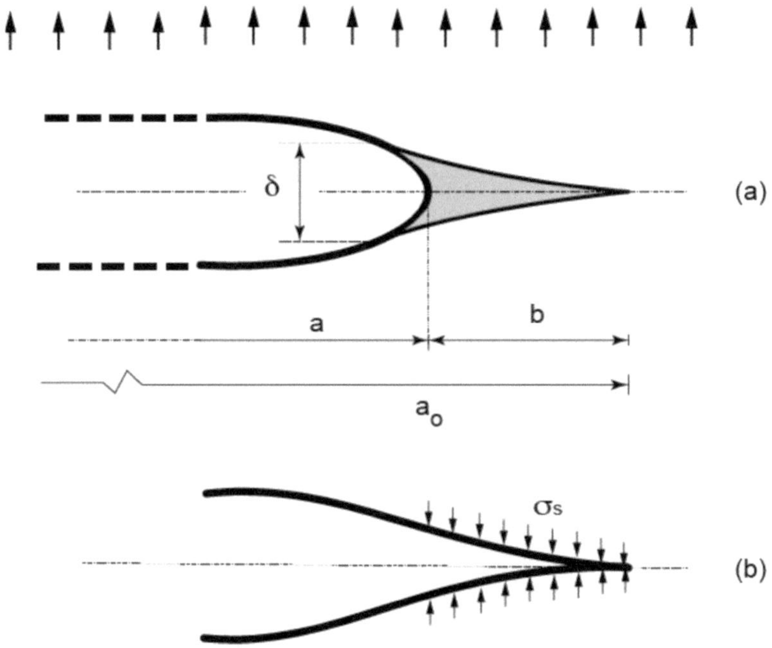

Fig. 10.11 Dugdale model of plastic zone

y = 0, x = b

$$Z(z) = \frac{2P_z\sqrt{a_0^2 - b^2}}{\pi\left(z^2 - b^2\right)\sqrt{z^2 - a_0^2}} \tag{10.7}$$

Assume that a uniform pressure, σ_s, is acting against a length *db* of the crack surface, $P = \sigma_s db$, as shown in Fig. 10.12 using the strip yield model, than the function can be derived from

$$\begin{aligned} Z(z) &= -\int_a^{a_o} \frac{2\sigma_s z\sqrt{a_o^2 - b^2}}{\pi\left(z^2 - b^2\right)\sqrt{z^2 - a_o^2}} db \\ &= -\frac{2\sigma_s}{\pi}\left[\frac{z}{\sqrt{z^2 - a_o^2}} cos^{-1}\left(\frac{a}{a_o}\right) - cot^{-1}\left(\frac{a}{z}\frac{z^2 - a_o^2}{a_o^2 - a^2}\right)\right] \end{aligned} \tag{10.8}$$

to which the external stress field, σ, must be added, so that it becomes

Fig. 10.12 Variation of the ψ function with the σ/σ_s ratio

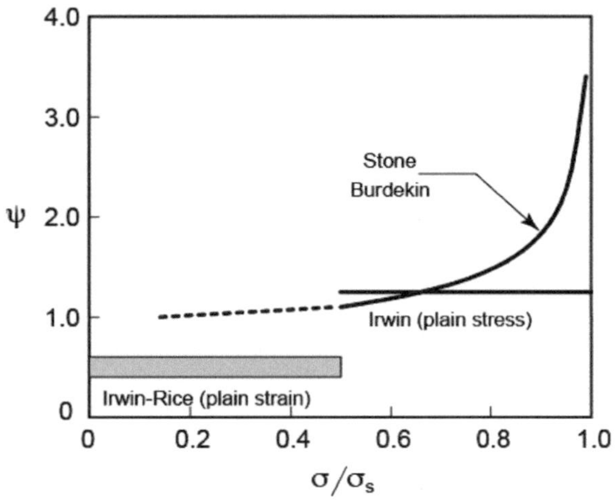

$$Z(z) = \frac{z\sigma}{\sqrt{z^2 - a_o^2}} - \frac{2\sigma_s}{\pi}\frac{z}{\sqrt{z^2 - a_o^2}} cos^{-1}\left(\frac{a}{a_o}\right) + \frac{2\sigma_s}{\pi} cot^{-1}\left(\frac{a}{z}\sqrt{\frac{z^2 - a_o^2}{a_o^2 - a^2}}\right) \tag{10.9}$$

If we want to avoid that for $x = a_o$ the function *Z(z)* assumes an infinite value (elastic-ideally plastic materials), the first two terms of Eq. (10.9) must be equal, so that posing

$$k = \frac{a}{a_o} = cos\left(\frac{\pi\sigma}{2\sigma_s}\right) \tag{10.10}$$

Eq. (10.9) becomes

$$Z(z) = \frac{2\sigma_s}{\pi} cot^{-1}\left(\frac{k}{z}\sqrt{\frac{z^2 - a_o^2}{1 - k^2}}\right) \tag{10.11}$$

and recalling the first of Eq. (7.117), we have that the displacement v at every point in front of the tip of the crack will be given by the only imaginary part of the integral of the function $\overline{Z}(z)$ which holds

$$Im\overline{Z}(z) = \frac{2\sigma_s}{\pi} \cdot \left[ka_o \cdot cot^{-1}\left(\frac{1}{a_o}\right)\sqrt{\frac{a_o^2 - z^2}{1 - k^2}} - z \cdot coth^{-1}\left(\frac{k}{z}\sqrt{\frac{a_o^2 - z^2}{1 - k^2}}\right)\right] \tag{10.12}$$

and, finally, we can find the value of the $CTOD = \delta_t = 2 \cdot v = 2 \cdot \text{Im } Z(z)/E$ at point $z \equiv (a_o, 0)$

$$CTOD = \delta_t = \frac{8a\sigma_s}{\pi E} ln\left[sec\left(\frac{\pi\sigma}{2\sigma_s}\right)\right]$$
$$= 2\left(\frac{2\sigma_s}{\pi\sigma}\right)^2 ln\left[sec\left(\frac{\pi\sigma}{2\sigma_s}\right)\right] G\frac{1}{\sigma_s} \tag{10.13}$$

expanding the *ln sec*

$$ln\left[sec\left(\frac{\pi\sigma}{2\sigma_s}\right)\right] = \left[\frac{1}{2}\left(\frac{\pi\sigma}{2\sigma_s}\right)^2 + \frac{1}{12}\left(\frac{\pi\sigma}{2\sigma_s}\right)^4 + \frac{1}{45}\left(\frac{\pi\sigma}{2\sigma_s}\right)^6 + \ldots\right] \tag{10.14}$$

With the assumption $\sigma \ll \sigma_s$, it is possible to arrest the expansion at the first term

$$CTOD = \delta_t = \frac{\pi a\sigma^2}{E\sigma_s}$$
$$= \mathcal{G}\cdot\frac{1}{\sigma_s} \tag{10.15}$$

In non-strain hardening materials we can express the *CTOD* in a more general form introducing a factor ψ, which is a function of the ratio σ/σ_s [11] and write

$$CTOD = \mathcal{G}\frac{1}{\sigma_s}\psi \tag{10.16}$$

For the Dugdale strip-yield model it is

$$\psi = 2\cdot\left(\frac{2\sigma_s}{\pi\sigma}\right)^2 ln\left[sec\left(\frac{\pi\sigma}{2\sigma_s}\right)\right] \tag{10.17}$$

The ψ factor can be considered the inverse of the plastic constraint factor *pcf*, $\psi = 1/pcf$, already presented in Sect. 8.2, Fig. 8.3, that measures the elevation of the yield strength of the material, σ_s, due to the degree of triaxiality occurring at the crack tip. According to Westergaard-Irwin, Eqs. (8.67) and (8.68), ψ varies from 0.57 to 0.73 depending on the factor α while according to Rice, in plain strain conditions ψ is equal to 0.50 or 0.58, whether considering the von Mises or the Tresca yield criterion, respectively. Table 10.1 reassumes the values of the ψ function so far considered.

Figure 10.12 shows the trend of the function ψ with the ratio σ/σ_s. Full-line curves indicate the field of validity. As can be seen, between $0.5\cdot\sigma/\sigma_s$ and $0.7\cdot\sigma/\sigma_s$ the Irwin function in plane stress and that of Stone and Burdekin assume values very close to each other. Above $0.7\cdot\sigma/\sigma_s$ the latter moves away leaving the former until it reaches the value of 3.893 for σ/σ_s equal to 0.995, see also Fig. 8.3. It is important to recall that the expression of plastic zone according to the Dugdale model is based on the hypothesis of elastic-ideally plastic behavior. This means that during the opening of the crack the material undergoes a very large deformation, but the stress cannot exceed the yield stress σ_s, except by triaxiality.

Table 10.1 Values of the (phi) function according to Westergard-Irwin, Rice, Stone-Burdekin, Dugdalel and ASTM 1820

CTOD	Westergaard-Irwin		Rice		Stone-Burdekin, Dugdale	ASTM 1820
$G/\sigma \cdot \psi$	Plain strain	Plain stress	von Mises	Tresca		
$\psi \rightarrow$	$\frac{(1-\nu^2)}{(1.75 \div 1.35)}$	$4/\pi$	$\frac{(1-\nu^2)}{1.48}$	$\frac{(1-\nu^2)}{1.29}$	$2\left(\frac{2\sigma_s}{\pi\sigma}\right)^2 \cdot ln\left[sec\left(\frac{\pi\sigma}{2\sigma_s}\right)\right]$	$\frac{(1-\nu^2)}{2}$

However, in a plain stress condition, which is typical of small thicknesses or large plastic slips, we cannot expect the yield stress to increase so much by triaxiality. The presence of work hardening in real materials will allow a higher reaction and, therefore, to develop a smaller apical plastic area ρ. The result is similar to that already seen in the derivation of the Irwin plastic zone whose size reduces because the yield strength σ_s increases by triaxial effect, Eq. (4.31) and Fig. 4.8. The consequence is that also the value of ψ will be reduced.

Hahn, Serrate and Rosenfield [11] evaluated the issue both in plain strain and in plane stress condition. They found that in the case of plain strain condition the function, ψ, in Eq. (10.16), which they called d_n, ranged from 0.8, for $1/n = 0$, to 0.1 for $1/n = 2$. At variance, in plain stress conditions, d_n ranged from 1.0 to 0.1. This is shown in Fig. 10.13. They also observed a dependence on the constant α in Ramberg's Osgood equation (1.41). From an operational point of view, we can recall the flow-stress σ_f, Eq. (9.58), equal to the average between the yield strength σ_s and the ultimate stress σ_u:

$$\sigma_f = \frac{\sigma_s + \sigma_u}{2} \tag{10.18}$$

which considers the work hardening through a fictitious increase in the fracture stress, compared to the yield stress, σ_s.

The flow stress can be conveniently used in GY situations in materials having a strain hardening, as it will be shown in Chap. 14 dealing with the problem of piping fracture. Similarly to what was done for the *CTOD*, Hahn and Rosenfield [12] based on an analysis by Goodier and Field [13] also derived an expression for COD, i.e. for the opening at the center of the crack, Fig. 8.22, which extends well beyond the elastic field of validity of Eq. (8.63)

$$\mathrm{COD} = \mathcal{G}\frac{1}{\sigma_s}\psi_2 \tag{10.19}$$

with ψ_2 given by

$$\psi_2 = ln\left[\frac{1+sin\left(\frac{\pi\sigma}{2\sigma_f}\right)}{1-sin\left(\frac{\pi\sigma}{2\sigma_f}\right)}\right] \tag{10.20}$$

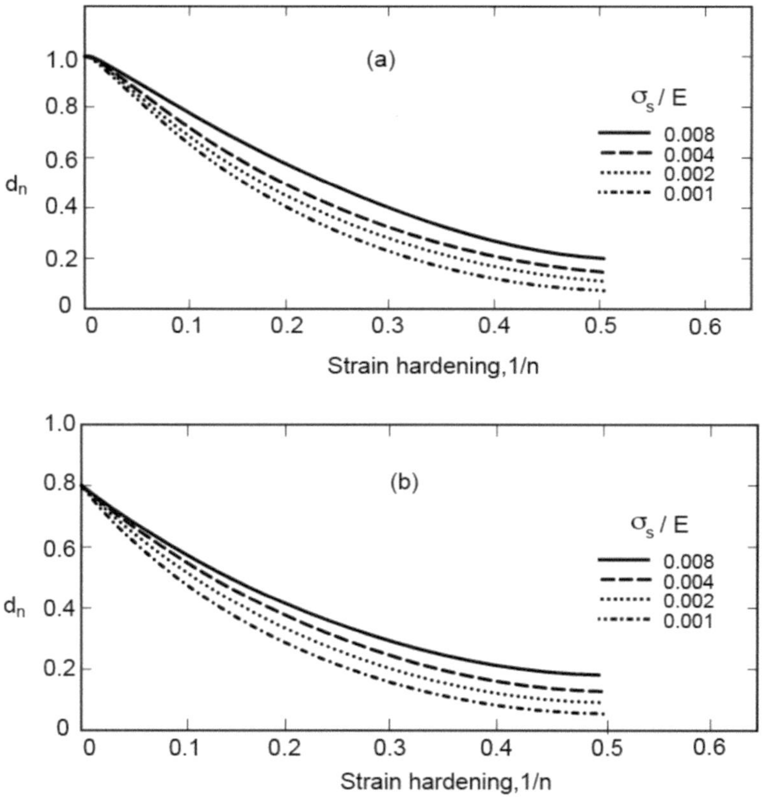

Fig. 10.13 Variation of d_n parameter with strain hardening $1/n$ for several values of the yield strain $\varepsilon_s = \sigma_s/\text{E}$ and $\alpha = 1$, **a** in plain stress condition and **b** in plain strain condition [11]

The two functions CTOD and COD, Eqs. (10.16) and (10.19), do not vary with the same rate, being $\dot{\delta}_t = \partial CTOD/\partial\sigma$, greater than that $\dot{\delta} = \partial COD/\partial\sigma$. This means that, even if the *CTOD* starts from about zero, sooner or later the two quantities will become equal or very close to each other.

Consider, in fact, the expression (10.16) of the *CTOD* with the correction factor ψ given by the Stone and Burdekin model, Eq. (10.17), and the *COD*, Eq. (10.19), normalized to $\sigma a/E$, being a, as usual, the half-length of a central flaw through an infinite plate, σ the remote stress and E the Young's modulus of the material. The two dimensionless functions are plotted in Fig. 10.14 for an ideal elastic-plastic material. As can be seen, in the linear elastic field, $\sigma/\sigma_s < 0.5$, CTOD remains well below the corresponding COD, as expected. However, the growth rate of the *CTOD* is larger than that of the *COD*. Therefore, as the applied stress σ increases, *COD* and *CTOD* becomes closer and closer. As general plastic flow approaches, their distance reduces till vanishing at GY. This general trend suggests that the shape of the crack will also change from linear elastic to elastic-plastic and, eventually, to GY conditions. The initially sharp flaw, in fact, will increasingly lose its original shape by blunting first

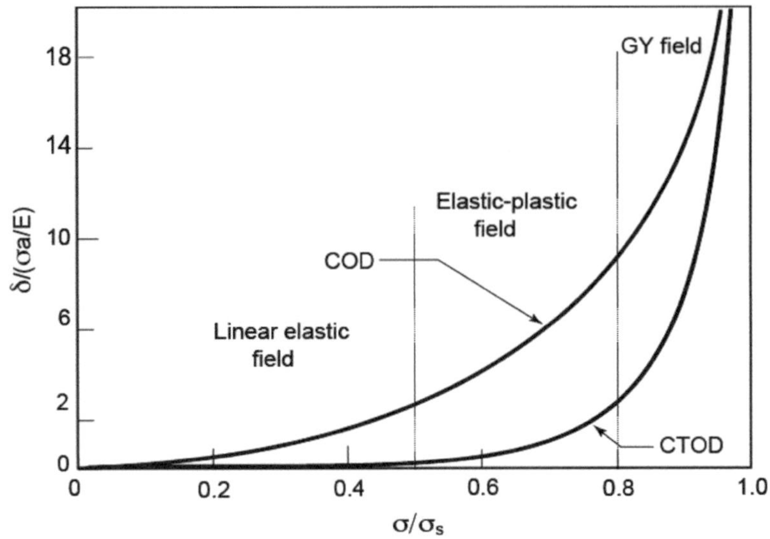

Fig. 10.14 Variation of normalized COD or CTOD with σ/σ_s for a non-hardening material

and then becoming flat where it will no longer be possible to distinguish between *COD* and *CTOD.* This is shown schematically in Fig. 10.7. In practice, in finite bodies the plastic zone will grow until it reaches the free edge opposite to the initial flaw. This is precisely predicted by the mathematic theory of plasticity [2] and is the consequence of the fact that the fundamental equations in ideally elastic-plastic materials are hyperbolic instead of elliptical, as in the linear elastic case.

10.4 Experimental Basis for CTOD

We have seen that CTOD can be theoretically assessed through the strip yield model as per Eq. (10.13)

$$CTOD = \frac{8a\sigma_s}{\pi E} ln\left[sec\left(\frac{\pi\sigma}{2\sigma_s}\right)\right] \tag{10.21}$$

Therefore, Eq. (10.21) can be conveniently used to check the validity of CTOD as a fracture mechanics criterion. This check has been done using pipes under pressure containing longitudinal cracks of length $2a$ opened by the hoop stress [14]. However, Eq. (10.21) is valid for a flat plate containing a central crack. Wells [6] suggested that the plastic zone size in cylinders compared with flat plate is increased in the same proportion as the crack extension force G at the crack tip. Folias [15] has shown that under elastic condition this increase is equal to

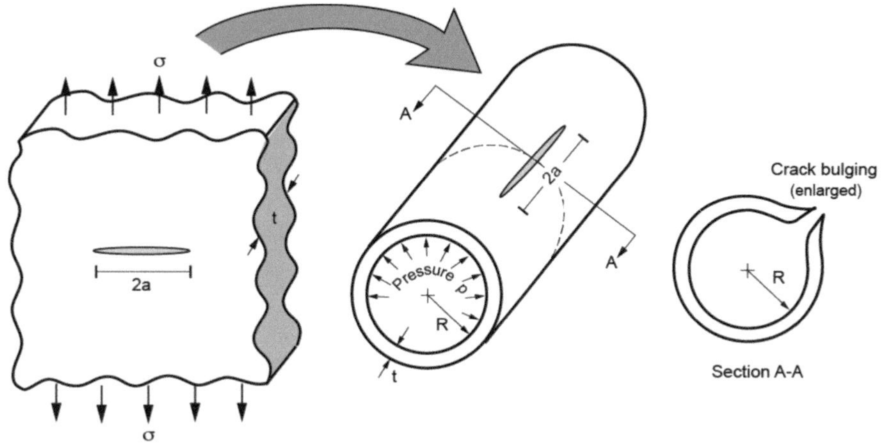

Fig. 10.15 Schematic showing the passage from a plate under remote tension containing a central through crack to a pressure pipe containing a longitudinal through crack. The equivalence is made taking into consideration also the bending effect of the bulging (enlarged for convenience) on the crack tip stress state

$$M^2 = 1 + 1.6\ a^2/Rt \tag{10.22}$$

where R is the pipe radius, t the thickness and a the half crack length. The Folias correction takes into account the bending effect of crack bulging, though very small, on the crack tip stress state, as schematically shown in Fig. 10.15. By applying M also to CTOD, as suggested by Wells, Eq. (10.21) becomes

$$CTOD = \frac{8a\sigma_s}{\pi E} ln\left[sec\left(\frac{\pi\sigma_H}{2\sigma_s}\right)\right]\left(1 + 1.6\frac{a^2}{Rt}\right) \tag{10.23}$$

where σ_H is the applied hoop stress.

Analysis of data obtained from tests on zirconium alloy pressure pipes [15] had shown that the Folias geometrical correction factor, Eq. (10.22), in the relationship between CTOD and hoop stress more closely defined the yielding behavior than any previously proposed model. Tests have been made at the UKAEA by Cowan and Kirby [14] on pipes of 5 ft (150 cm) inside diameter, 1 in. (25.4 mm) thick, made of three different steels. The steels were a mild steel BS1501-151, a 0.36% C normalized steel and a Mo-B low carbon steel, whose strength ranged from 230 MPa to 460 MPa. Pipes were carrying a longitudinal slot whose ends were sharpened with a jeweler's saw giving 0.008 in. wide notches (0.2 mm). The so obtained longitudinal cracks were sealed against leakage by applying thin sheets of mild steel, aluminum and neoprene to the inside surface of the vessel. The slits had a central enlargement to accommodate a jig holding a rotating paddle to measure CTOD during the test. Three different crack lengths of 24", 12" and 6", respectively, were introduced in pipes. The correspondence of measured and predicted values of CTOD for the three

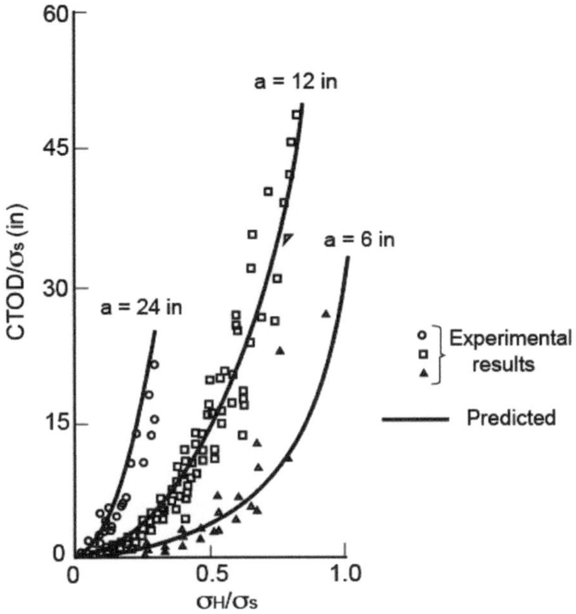

Fig. 10.16 Relationship between predicted CTOD and applied hoop stress for pipes with 6", 12 and 24" long cracks [14]

steels are shown in Fig. 10.16. CTODs values have been normalized to the yield strain, CTOD/ε_s, with $\varepsilon_s = \sigma_s/\mathrm{E}$, and the hoop stress, σ_H, to the yield stress, σ_s, σ_H/σ_s. The agreement is rather good, particularly at low stress levels, considering that the CTOD readings were made using of a cod-meter whose precision is not very high. At the higher value of CTOD, where the crack tip opening changes rapidly for a small change in stress, slow crack growth may be the dominant factor influencing the actual CTOD that would not be considered in Eq. (10.23).

Under progressively more ductile conditions the larger values of CTOD are believed to be associated with increasing amounts of slow crack growth. It must be also noted that the initial Folias factor M given by Eq. (10.22) overestimate the correction for larger crack size, as discussed in Sect. 14.2.1. Where a more accurate value of the M factor has been used, Eq. (14.4). Tests were also conducted at different temperatures. At temperatures where ductile tearing preceded fast fracture the value of the critical CTOD was not well defined. Notched bend specimens showed ductile fracture initiation at temperatures lower than that at which it was found in pipes. Under even higher ductile conditions, pipes gave higher values of CTOD at failure than bend specimens. When there was appreciable ductile tearing, there was little correspondence between pipes and specimens.

Hahn et al. [16] have shown that failure conditions may be better expressed in terms of flow stress, σ_f, rather than yield stress, σ_s, using the square root of the Folias correction factor M

$$\sigma_H = \sigma_f \left(1 + 1.61\frac{a^2}{Rt}\right)^{-\frac{1}{2}} \quad (10.24)$$

where the flow stress, σ_f, Eq. (10.18).

The use of the flow stress, σ_f, seems to be more realistic for strain hardening materials since the CTOD expression (10.21) is obtained for non-hardening materials.

10.5 COA and CTOA

Figure 10.17 shows an example of experimental detection of the shape of a central through crack in a magnesium plate of 41x10 cm, 1.6 mm thick, subjected to increasing levels of external traction [17]. The initial shape refers to incipient GY. It is very clear that the CTOD has reached the value of the corresponding COD and the flaw has completely lost its sharpness.

The experimental survey presented in the sequence of Fig. 10.16 also shows another important feature: the tearing of the material at the tip of the blunted notch. In an ideally elastic-plastic material as soon as the GY condition is reached, at $\sigma = \sigma_s$, the notch tip stress no longer increases, and the plastic zone extends to the edge in a physical continuum in which the deformation of the plastic zone equals the uniform crack opening, which is an empty space. In real materials with work hardening, this cannot be possible, and the material must tear to accommodate, in some way, this difference in deformability. With the tearing, the material ahead of the flat flaw will be able to accommodate the deformation request. This local tearing recreates a physical tip. Two new parameters are thus identified: the *COA*, or crack opening angle and the *CTOA*, or crack tip opening angle. The first, *COA*, represents the angle that the front tearing forms with the *COD*, Fig. 10.18. The second, *CTOA*, identifies, instead, the opening angle of the tearing itself. *COA* and *CTOA* are shown schematically in Fig. 10.18. De Koning [18] was the first to observe that under increasing loads, the *CTOA* after a first phase of growth stabilized remaining practically constant

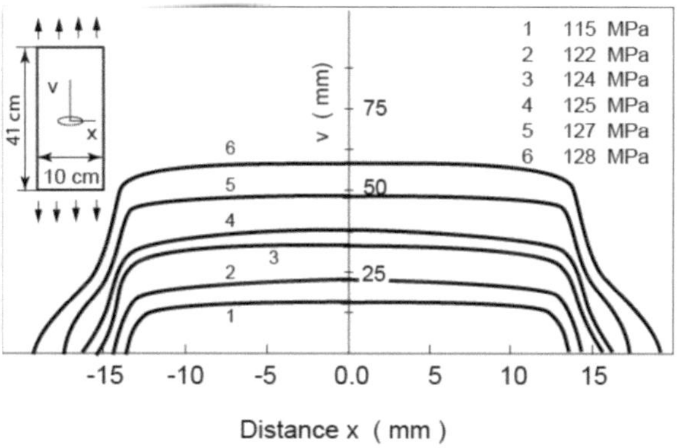

Fig. 10.17 Experimental survey of the crack shape, starting from initiation of GY, revealing the progressive widening of the notch as the external stress increases and the edge tearing of the material [17]

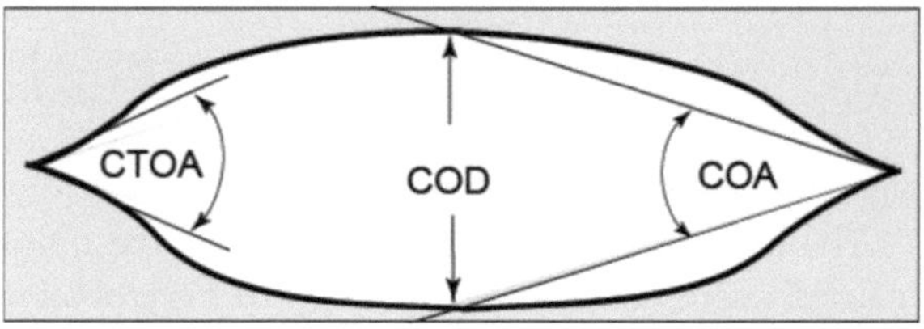

Fig. 10.18 Schematic representation of COA and CTOA

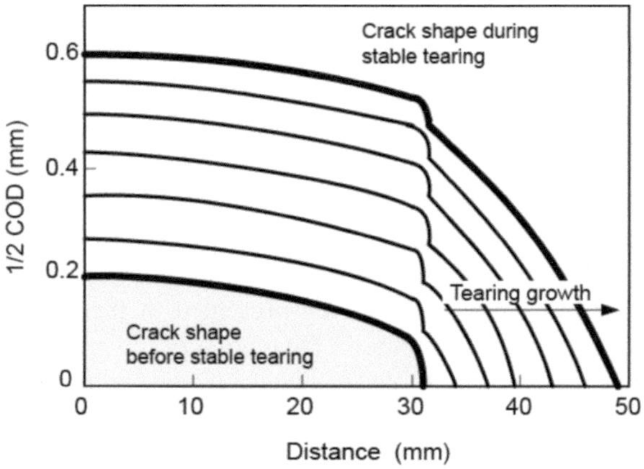

Fig. 10.19 Variation of the shape of a notch after first tearing in a 2024-T3 aluminum sheet under increasing lods. The CTOA doesn't change during the tearing progression [18]

until the moment of unstable propagation of the new crack with consequent final plastic collapse of the structure. The results obtained by de Koning on a thin panel of aluminum 2024-T3 are shown in Figs. 10.19 and 10.20. FE calculations simulating the crack growth process through the release of crack tip nodes, applied to fracture mechanics specimens such as SE(B) [19] and C(T) [20], have shown that after blunting, tearing initiates with decrescent CTOA. CTOA, then, stabilize after a subcritical growth Δa, that depends on the specimen dimensions that in a C(T) specimen is about one tenth of the ligament b, as shown in Fig. 10.19 [18] for a stainless steel type 304 at 288 °C. However, it must be recalled that the effective CTOA depends on the crack tip mesh dimensions (Fig. 10.21).

10.6 The *J* Contour Integral

In 1967 Rice [21] presented an integral, that he called *J*-integral, that would have opened new and surprising perspectives in fracture mechanics extending the field of validity of the stress analysis of cracks well beyond the elastic limits.

Its power derives from three fundamental properties (Fig. 10.22):

(1) it is capable of defining the elastic-plastic stress field generated by a crack in a body under external loads measuring its singularity at the crack tip,

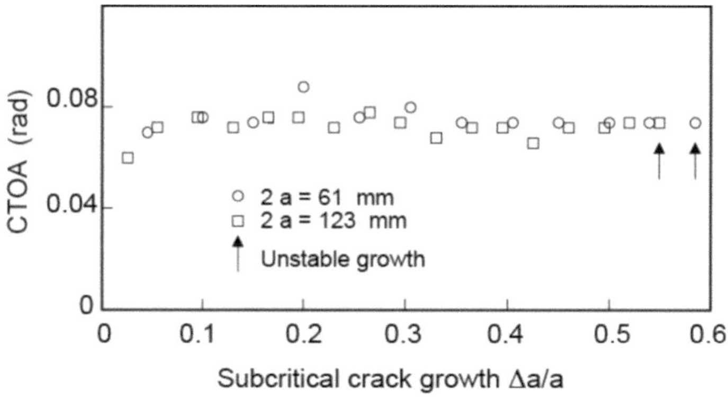

Fig. 10.20 Variation of CTOA with applied load, for two central crack lengths in a panel of aluminum 2024-T3. CTOA remains constant till final instability [18]

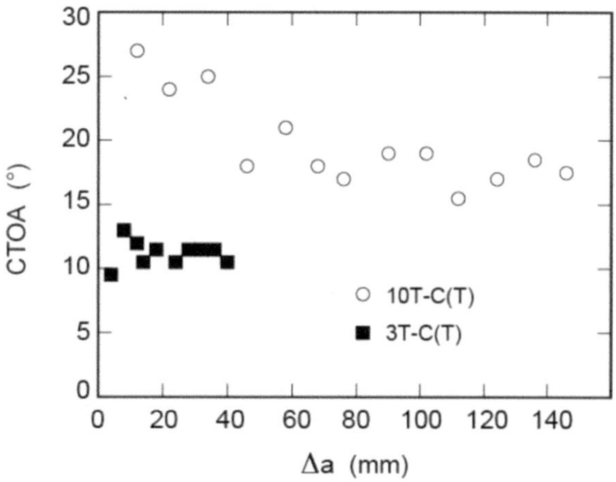

Fig. 10.21 Variation of CTOA with subcritical crack growth, Δa, obtained with a FE analysis of a 1TC(T) specimen of stainless steel, type 304, with planar dimensions of 3t and 10T at a temperature of 288 °C [18]

(2) it is of simple experimental measurement,
(3) it is of simple analytical calculation.

Assume to have a body of unit thickness containing a through flaw whose surfaces are parallel to the *x* axis, as shown in Fig. 10.22.

Also, assume that the system is subjected to an external plain strain state or generalized bidimensional stress state so as to describe the stress field by two stress components, σ_x and σ_y, the third component, σ_z, being given by $\sigma_z = \nu(\sigma_x + \sigma_y)$ or $\sigma_z = 0$ in plain stress. Suppose that the material is elastic or non-linear elastic. This last case refers to a material that follows a hardening low of the type

$$\sigma = K\varepsilon^n \tag{10.25}$$

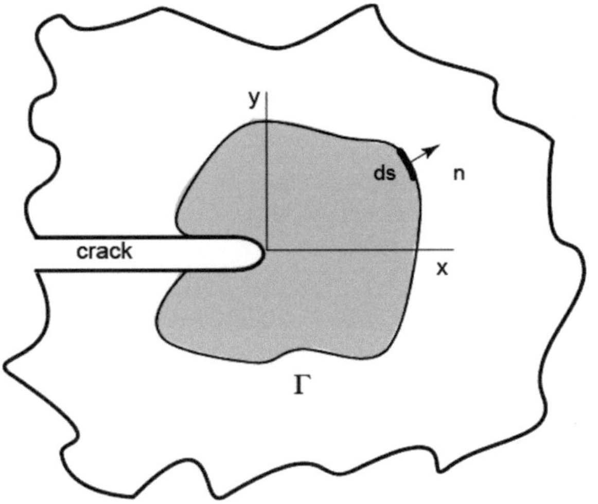

Fig. 10.22 Body of unit thickness containing a through wall crack under remote loading. Γ represents a generic curve that closes on both crack surfaces

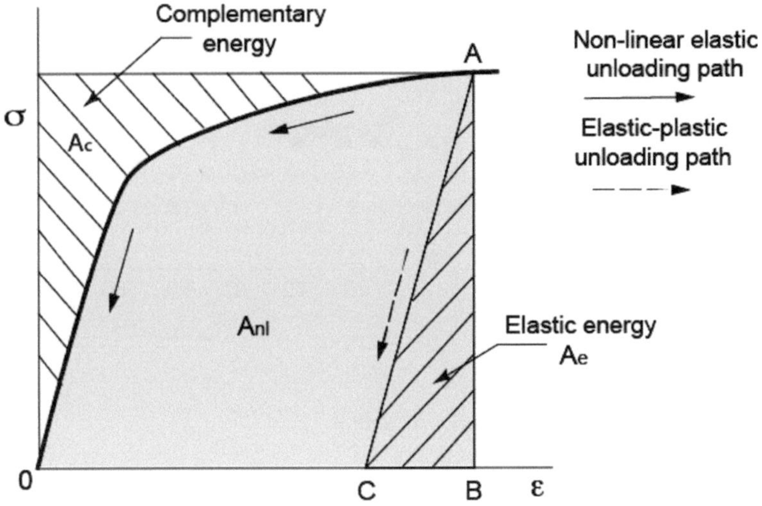

Fig. 10.23 Unloading paths followed by a material point, A, according to either the non-linear elasticity theory, path AO, or to the elastic-plastic theory, path AC. In the first case, upon unloading, there is a complete recovery of the energy stored, A_{nl}, in the second case only the elastic energy, A_e, is recovered

with $n < 1$. In non-linear elasticity it is assumed that any point, A, reached by the material during the loading phase, upon unloading follows the same path as it was loaded, Fig. 10.23. This is a material behavior completely different from that considered by the elastic-plastic theory where the same point would follow a linear unloading path with the slope equal to the Young's modulus, as shown in Fig. 10.22. Without any unloading the two theories are equivalent. The area A_{nl} under the $\sigma - \varepsilon$ curve is also called pseudo energy. It represent energy dissipated as heat, so almost none of it can be recovered, see Sect. 1.8.

This behavior is fundamental for the considerations that will be presented and that will define the J-integral as J-deformation because non-linear elasticity is based on the deformation theory of plasticity or Hencky's theory of plasticity. The theory assumes two more conditions:

(a) plastic deformations are independent of the hydrostatic component of stress $\sigma_m = (\sigma_1 + \sigma_2 + \sigma_3)/3$
(b) as the strain increases the stress components are always in fixed proportion to each other and this, in practice, requires the plastic strain components to be very small.

With such premises, let's be W the deformation energy density introduced in the system that, recalling the unit thickness assumption, will be given by the area under the stress-strain curve, $\sigma - \varepsilon$

$$W = W(x, y) = W(\vec{\varepsilon}) = \int_0^{\vec{\varepsilon}} \sigma_{ij} ds \tag{10.26}$$

$\vec{\varepsilon} = [\varepsilon_{ij}]$ being the deformation tensor. Consider, then, the integral, defined as J-integral or simply J, given by

$$J = \int_\Gamma \left(W dy - \vec{T} \cdot \frac{\partial \vec{u}}{\partial x} ds \right) \tag{10.27}$$

where Γ is any counter-clock line that surrounds the crack tip, T is the traction vector acting along the Γ curve whose components are

$$T_i = \sigma_{ij} n_j \tag{10.28}$$

with n being the unit out-word vector normal to the Γ curve, u the displacement vector and ds a length increment along the contour Γ. A remarkable characteristic of the J-integral thus defined, which makes its use fruitful in the field of fracture mechanics, derives from the fact that it is possible to prove, through Green's theorem, that its value vanishes, $J = 0$, if the path of integration Γ is any closed contour, which actually means that the value of J is independent of the particular path chosen whenever the contour starts at any point on the lower side of the crack and arrives to any point of the upper side. Consider, in fact, the flaw of Fig. 10.24 and the ABCDEFA contour that surrounds it, which, being closed, makes the value of $J = 0$.

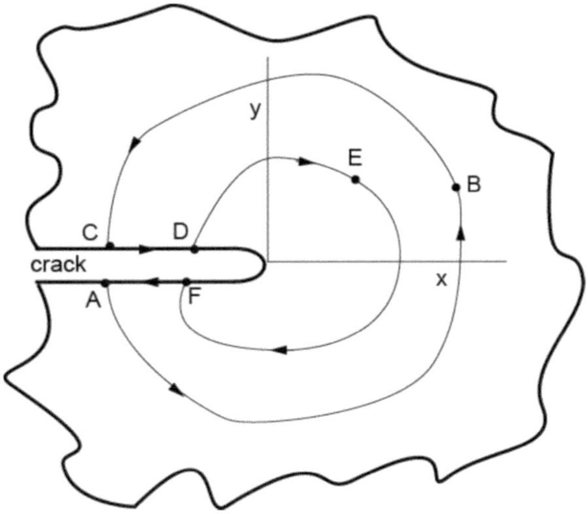

Fig. 10.24 Closed contour ABCDEFA for the calculation of the J-integral

On the AF and CD lines J is obviously zero because $T = 0$, being free surface sections on which the stress is zero and $dy = 0$ because the faces are parallel to the x-axis. Therefore, the value of integral calculated along the ABC contour must be equal, but with opposite sign, to that calculated along the DEF contour.The opposite sign derives only from the direction that in ABC is counterclockwise and in DEF clockwise. Consider, then, a virtual growth da of the crack of length a, Fig. 10.25, whose contour is Γ_p. The J-integral along this contour reduces to the first term only of Eq. (10.27) because the traction vector T is zero on any free surface

$$J = \int_{\Gamma_p} W dy \tag{10.29}$$

Let's now evaluate the deformation energy, dU, lost by the material as a result of this growth da. In a first approximation, this energy may be that contained in the plastic area, ΔA, of Fig. 10.25 and, therefore, recalling that the body has unit thickness, write

$$-dU = \int_{\Delta A} W da dy \tag{10.30}$$

where ΔA is the gray area of Fig. 10.25. Eq. (10.30) gives:

$$-\frac{dU}{da} = \int_{\Gamma_p} W dy \tag{10.31}$$

exactly equal to the integral (10.29) which gives the value of J. Therefore, the J-integral is the negative rate of change of the potential energy per unit thickness with respect to crack extension and is measured in Nm/m^2. For a brittle material this quantity, i.e., the J, can be physically interpreted as the energy that is available for

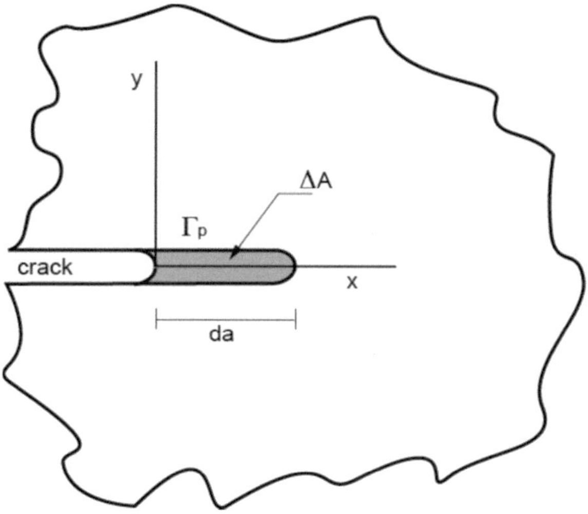

Fig. 10.25 Subcritical crack growth *da* with energy release *ΔA*

further crack extension; it is the well-known driving force $G = -dU/dA$ already introduced with Griffith's theory extended to real materials, Sect. 7.6.

This is hardly surprising. The potential energy of an elastic body, linear or non-linear, can, in fact, be expressed as

$$U = \int_V WdV - \int_S T_i u_i dS \tag{10.32}$$

where W is, as usual, the strain energy density (energy per unit volume), T_i and u_i are the forces acting on the outer surface S and their displacements, respectively. Assuming the unit thickness or referring to the generalized two-dimensional system already introduced, Eq. (10.32) can be written as

$$U = \int_S Wdxdy - \int_S T_i u_i dS \tag{10.33}$$

Therefore

$$\frac{dU}{dx} = \int \left(Wdy - T_i \frac{\partial u_i}{\partial x} dS \right) \tag{10.34}$$

which is nothing but the *J*-integral defined by Eq. (10.27). Note that Eq. (10.34) represents the complementary energy of the system or the strain energy minus the total work done by the system of external forces. This complementary energy is shown, in Fig. 10.23, by the dashed area A_c above the force-displacement curve which in a nonlinear material is less than that, A_{nl}, under the curve itself. As such, the quantity U of Eq. (10.34) is negative. Unfortunately, this physical interpretation of J cannot be valid for elastic-plastic materials, simply because plastic deformations are not reversible. For reversibility to apply, one must demonstrate that the same result is obtained both by loading a cracked body and then permitting an incremental crack extension *da* or extending the initial crack of a quantity *da* and then loading the body. This equality cannot be proved for plastic deformations. Consequently, the J-integral cannot be directly related to crack extension in a sense of a crack driving force G as in the case of linear elastic materials. For nonlinear elastic materials the physical significance of J is that of an energy difference between two similar bodies, loaded in the same manner and having incrementally differing crack sizes, as schematically shown in Fig. 10.26.

Note that this difference is precisely that of the respective complimentary energies, as it was shown for linear elastic materials, see Sect. 7.14. It should also be noted that Eq. (10.34) relates $-dU$ to the nonlinear elastic energy release, i.e., to the entire area A_{nl} under the stress-strain curve, $\varepsilon - \sigma$, and not to just the elastic component, A_e, as the elastic-plastic theory would predict. This is the consequence of the hypothesis made of nonlinear elasticity and is schematized in Fig. 10.23. This hypothesis will certainly pose a limit to the applicability of the J-integral to real elastic-plastic materials, which will be addressed in Sect. 11.8. In particular, in the plastic domain where fracture is always preceded by a subcritical growth, as mentioned in Sect. 10.1 and shown in

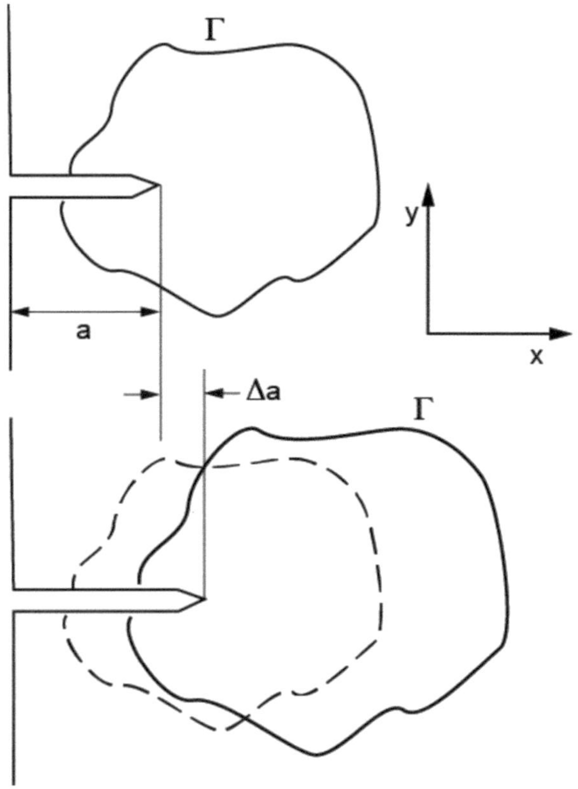

Fig. 10.26 The J-integral can be interpreted as the potential energy difference between two identical bodies having incrementally differing crack lengths, Δa

Figs. 10.5 and 10.6, the difference between the two energies can become significant and the J, as will be said, can also lose its meaning. Given the independence of the J from the integration contour, choosing an integration path alternative to that assumed before, for example a large circumference around the tip of the crack so that the plastic stress field with a singularity will not affect at all the calculation, the same result must be obtained (point 3. Above: simple analytical assessment). Rice has actually shown [22] that under plain strain conditions, *SSY* conditions, and, therefore, in the validity field of LEFM in which the stress field σ_{ij} is defined by Eqs. (7.114) and the plastic zone is of negligible size, the J holds

$$J = \left(1 - \nu^2\right) \frac{K_I^2}{E} \tag{10.35}$$

and recalling Eq. (7.141) which establishes the equivalence between $\mathcal{G}$ and K_I it must be concluded that under these conditions it is

$$J = \mathcal{G} = -dU/dA \tag{10.36}$$

10.7 Singularity of the Non-linear Elastic Stress Field. The HRR Stress Field

The independence of the value of the Rice integral from the chosen path of integration, as long as it contains the tip of the crack, lead us to believe that the J-integral must somehow characterize the only element common to all paths: the crack tip. From a morphological point of view, what characterizes the crack end is the *CTOD* or δ_t. Therefore, J must be equivalent to the δ_t, as we have already seen, which means that the latter, like the former, will have to express a criterion of criticality. Wells [6, 9, 23] had already formulated in the 60s this principle, today absolutely recognized and accepted, of the existence and independence of the δ_t critical from any geometric condition and load, δ_{tc}. The question is more complex if approached from a stress state point of view. In this respect, the front of a crack is characterized by a stress field, otherwise called proximity field. The J, therefore, will somehow have to measure the intensity of this field. We have already seen, Eq. (10.30), that in SSY conditions in which the plastic area ahead of the crack has negligible dimensions compared to any other and the elastic stress field is defined by Eqs. (7.114), there is a perfect equivalence between J and K. It is now appropriate to broaden the study by exploiting the nonlinear elastic base of the J that will allow us to go beyond the conditions of *SSY*. Rice and Rosengren [24] showed that the J-integral calculated on a circumferential path of radius r around the crack is written as

$$\frac{J}{r} = \int_{-\pi}^{+\pi} \left\{ w \cdot [\varepsilon(r,\theta)] \cdot cos\theta - T(r,\theta) \cdot \frac{du}{dx}(r,\theta) \right\} d\theta \tag{10.37}$$

This means that the integrand must show a singularity of the type $1/r$. Since the terms in the wright side of Eq. (10.37) are represented by products σ·ε, Rice and Rosengren were tempted to conclude that the singularity $1/r$ must exist in the product itself

$$\sigma_{ij}\varepsilon_{ij} \to \tfrac{1}{r} f_{ij}(\theta) \tag{10.38}$$

This position, which is evident in the linear elastic case, is also valid in the ideally plastic one in which a Prandtl field exists with a singularity, as in the case of Fig. 10.2a or 8.17. In the case of strain hardening materials, Eq. (10.38) implies that the crack tip stress-strain distribution must take the form

$$\sigma_{ij} \to r^{-\frac{1}{n+1}} \cdot F_{ij}(\theta, n);\ \varepsilon_{ij} \to r^{-\frac{n}{n+1}} \cdot G_{ij}(\theta, n) \tag{10.39}$$

with F_{ij} and G_{ij} functions of the work hardening coefficient n, with n being the exponent of the Ramberg-Osgood equation, Eq. (1.40), and of the anomaly θ. Hutchinson [25] always exploiting the criterion of independence of the J-integral from the path of integration, equated the value calculated on a contour far from the crack end and totally in the linear elastic remote field to that assessed with a proximity path around the nonlinear field, thus coming to find that the amplitude of the proximity field had to have the form

$$K = \left(\frac{\pi}{\alpha I_n}\right)^{\frac{n}{n+1}} \left(\frac{\sigma}{\sigma_s}\right)^{\frac{2n}{n+1}} \tag{10.40}$$

α and $n = 1/m$ always being the coefficients of the ROR, Eq. (1.40), σ_s the yield stress and I a normalization constant dependent on strain hardening n. Since the singularity of the proximity field, Eq. (10.38), exists and its amplitude is given by Eq. (10.39) we can write:

$$\sigma_{ij} \to \left(\frac{\pi\sigma^2}{\alpha I_n r\sigma_s^2}\right)^{\frac{1}{n+1}} \cdot F_{ij}(\theta, n) \tag{10.41}$$

Introducing the J-integral as a measure of this amplitude, we have, since the function F_{ij} is necessarily dimensionless:

$$\sigma_{ij} = \sigma_s\left(\frac{JE}{\alpha I_n r\sigma_s^2}\right)^{\frac{1}{n+1}} \cdot F_{ij}(\theta, n) \tag{10.42}$$

Similarly for deformations

$$\varepsilon_{ij} = \varepsilon_s\left(\frac{JE}{\alpha I_n r\sigma_s^2}\right)^{\frac{n}{n+1}} \cdot G_{ij}(\theta, n) \tag{10.43}$$

Equations (10.41) and (10.42) represent the so-called Hutchinson, Rice and Rosengren field or *HRR stress field*. The values of the coefficients I_n are tabulated in Table 10.2 for some values of n for both conditions of plain strain or plain stress [25].

The values of the coefficient I_n can be derived from the formulas:

$$I_n = -\frac{2.9}{n} + 7.2\sqrt{0.12 + \frac{1}{n}}$$

$$I_n = -\frac{4.6}{n} + 10.3\sqrt{0.13 + \frac{1}{n}} \tag{10.44}$$

valid under plain stress and plain strain conditions, respectively. The functions $F_{ij}(\theta, n)$ and $G_{ij}(\theta, n)$ of Eqs. (10.42) and (10.43) are tabulated in Tables 10.3 and 10.4, respectively, for some values of the work hardening coefficient n and for $\theta = 0$, in other words in the plane of the flaw. What makes the existence and uniqueness of J possible is precisely the presence of work hardening since the angular functions $F_{ij}(\theta, n)$ and $G_{ij}(\theta, n)$ of the stress field, Eqs. (10.42) and (10.43), are not uniquely defined for an ideal elastoplastic material *(n* $= 0$*)* and the dominant singularity is, as mentioned in Sect. 10.1, geometry-dependent. It will be precisely the hardening exponent n to determine, therefore, the "*force*"of the singularity. As n becomes extremely large, non-singular terms not considered in Eqs. (10.42) and (10.43) may become at least as important as singular terms.

Therefore, a certain finite value of n, even small, represents the necessary condition for elastoplastic fracture mechanics (EPFM) to be a one-parameter theory and will justify the use of J or the equivalent δ_t, as a measure of the singularity of the elastoplastic proximity field. Also for EPFM, different geometries under different load conditions, but resulting in the same J can be considered similar in terms of fracture strength, as they were for LEFM in terms of K.

Table 10.2 Values of the parameter I_n for some work hardening coefficients n under plain stress and plain strain conditions, respectively

	$n = 3$	$n = 5$	$n = 9$	$n = 12.5$
I_n plain stress	3.86	3.41	3.03	2.87
I_n plain strain	5.51	5.01	4.60	4.40

Table 10.3 Values of the functions $F_{ij}(\theta, n)$ and $G_{ij}(\theta, n)$ for $\theta = 0$ under plaint stress conditions

n	F_{xx}	F_{yy}	F_{xy}	G_{xx}	G_{yy}	G_{xy}
3	1,6454	1,9407	0,000	−0,0145	0,0145	0,000
5	1,6836	2,2172	0,000	−0,0183	0,0183	0,000
8	1,7118	2,4180	0,000	−0,0170	0,0170	0,000
10	1,7243	2,4969	0,000	−0,0156	0,0156	0,000

Table 10.4. Values of the function $F_{ij}(\theta, n)$ and $G_{ij}(\theta, n)$ for $\theta = 0$ in plain strain conditions

n	F_{xx}	F_{yy}	F_{xy}	G_{xx}	G_{yy}	G_{xy}
3	0,7699	1,1061	0,000	0,2092	0,6956	0,000
5	0,6907	1,1349	0,000	0,1186	0,7602	0,000
8	0,6433	1,1449	0,000	0,0679	0,7894	0,000
10	0,6281	1,1473	0,000	0,0521	0,7971	0,000

10.8 Proximity Field and Remote Field *J*-Integral

It has already been observed, in Sect. 10.6, that a characteristic of the J-integral is to be independent of the integration path chosen for its calculation, provided that it encloses the tip of the crack. This is certainly true for a stationary crack and can be easily verified by numerical calculation as done, early in the 70s, by Wilson [26] using a CT specimen and integration paths of increasing mean radius, r_m, to (W-a) ratio, ranging from a very close 0.025, which we will define as *proximity field* and develops just around the plastic HHR field, to 1, this last value corresponding to the outer edge of the specimen itself or specimen contour, which for this reason we will call *remote field*. It can be seen, Fig. 10.27, that, beside a sudden fall of the proximity field, the nondimensional ratio

$$\frac{K_J B \sqrt{\pi W}}{P} \tag{10.45}$$

where K_J is the K_I value derived from the J, remains rather constant. This implies that the large gradient of the plastic field and the effects of the elastic unloading in the proximity field that become significantly different from the nonlinear elastic ones have an effect on the value of the J-integral. For nonstationary cracks, Shih et al. [27] found that such independence was no longer respected in the proximity field integral. They applied the technique of numerical calculation with release of knots ahead of the crack, to a 4T-C(T) specimen of A 533 B using various integration paths ranging from the one immediately around the crack tip enclosing the plastic field, to the remote field J, which, evidently, develops in a totally elastic field. The results are shown in Fig. 10.28. It can be observed how the various curves J_R, calculated on 4 different paths, deviate from each other after just 4 mm that in a 4" C(T) specimen correspond to a 4% subcritical growth, $\Delta a = 0.04\ b$, apparently in agreement with the size requirement set by Eq. (11.58). In particular, the remote field integral, shown in Fig. 10.28 as J_6, coincides with that measured experimentally using Eq. (11.16). On the contrary, the curve relative to the proximity field, indicated by J_2, after about 4 mm of subcritical growth flattens out in contravention of the criterion expressed by Eq. (11.66), of $\omega \gg 0$.

This means that the effects of elastic unloading in the proximity field become significant and predominant compared to nonlinear elastic ones (see Fig. 11.23) and the *HRR* field of J dominance loses its meaning. More recent calculations conducted

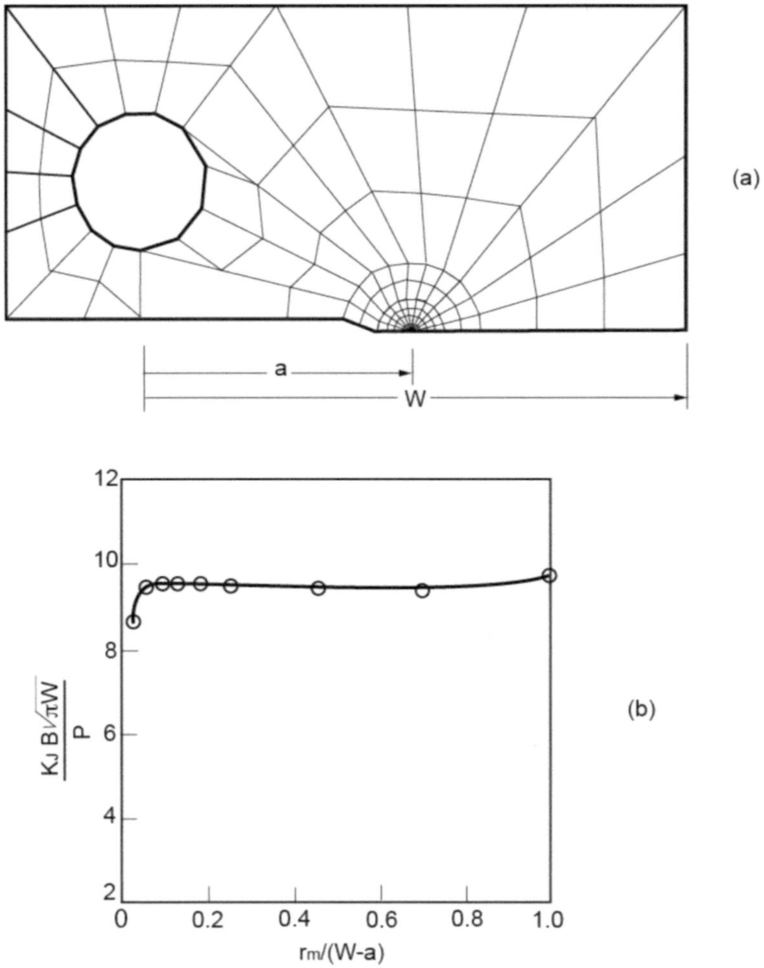

Fig. 10.27 **a** Finite element representation of one half of a CT specimen with a/W=0.5 and **b** K_J calculated from J-integral over paths of mean radius r_m from the crack tip [27]

by Papaspyropoulos, Marschall and Landow [20], always using a finite element technique with nodal release and incremental plasticity theory, have confirmed those of Shih *et al.*, enriching the scenario considering C(T) specimens of stainless steel type SA 304, of 1", 3" and 10" thick (25, 75 and 250 mm). In all three cases examined, after an initial subcritical growth that depended on the size of the specimen, the remote field integral and the proximity field integral moved away from each other. For 3" and 10" specimens this happened for growths Δa equal to 10 and 15% of the ligament b, respectively.

For the 1" specimen, independence from the contour was lost almost immediately. Dodds, Tang and Anderson [27], Fig. 10.29, applying the same technique to an SE(B)

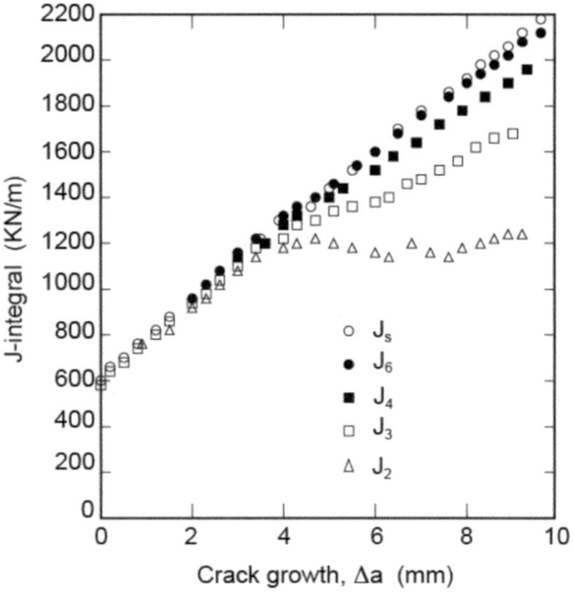

Fig. 10.28 Including the proximity field, J_2, and remote field integral, J_6, (based on [26])

specimen with $a/W = 0.5$ and $n = 0.1$ and 0.2, respectively, loaded at $J = J_{Ic}$, found for the stationary case almost the same result shown in Fig. 10.27, that is to say the independence of J from a path of diameter greater than about 10% of the ligament b, while in the case of subcritical growth the proximity field integral was almost zero, grew rapidly and became independent of the integration path for a radius R greater than 20 25÷% of b. Again, with subcritical growth, the calculated value of J coincided with that measured experimentally. We must then consider that for a non-stationary crack there can still be an *HRR* field dominated by J beyond the boundary placed by 11.54, but its extension is not easy to determine a priori and, in any case, to evaluate the intensity of the singularity through calculations, it is necessary to get away as far as possible from the proximity field toward the remote field. This is probably the reason why the J_R curve of experimental derivation seems to have that ability to extend, sometimes, beyond the limits defined by 11.54 and 11.50: it, in fact, arises precisely from measurements that collect the contribution of the remote field and that is why it coincides with that calculated one along the same field. However, this independence of the J-integral from the chosen contour only along remote paths, together with the observation that the J-proximity can also be zero, still raises serious questions. If, in fact, the *HRR* field, which is a proximity field, is uniquely defined by the value of the remote field J, it means that the proximity J must be in constant relation to the remote one, otherwise we will not have a one-parameter theory. Looking at Fig. 10.31, this does not appear to be the case for the specimen in question. The peculiarity of the remote field J-integral does not end here. Referring to a previous analysis by Rice and Sorensen [28], Rice, Drugan and Sham [29] found that under *SSY* conditions the CTOD velocity, $\dot{\delta}_t$, resulting from a

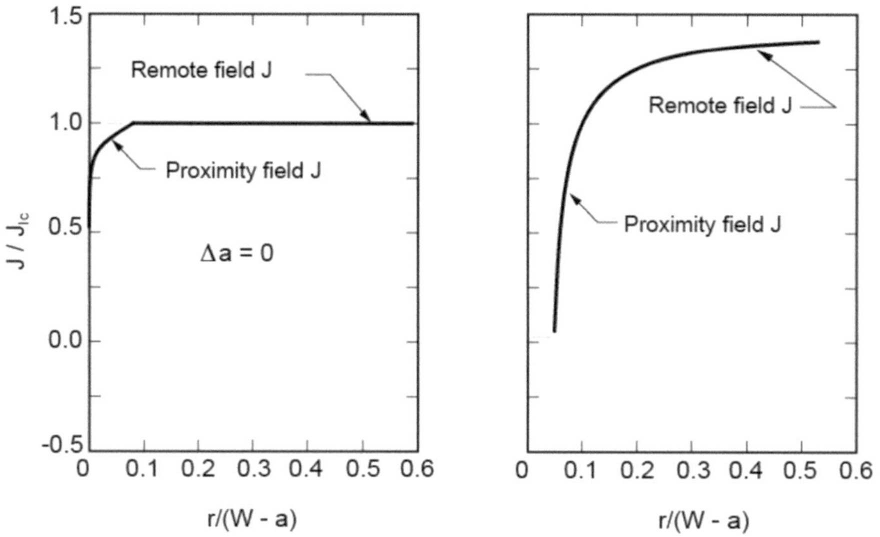

Fig. 10.29 FE calculated *J-R* curves for different integration paths of increasing radius, for a stationary crack, Δa=0, and (**b**) nonstationary crack, Δa=10·δ_{Ic}, in a SE(B) specimen with a/W=0.5 (based on [27])

stationary propagation at the same velocity, $\dot{a}$, was given, for $r \to 0$, by

$$\dot{\delta}_t = \beta \tfrac{\sigma_s}{E}\ \dot{a}\ ln\left(\tfrac{R}{r}\right) + \tfrac{\alpha}{\sigma_s}\ \dot{J} \tag{10.46}$$

where E is the Young's modulus of the material, σ_s the yield stress, R a parameter that in first approximation measures the maximum extension of the plastic zone and β and α two parameters that for *SSY* conditions hold: β =5.083 for ν =0.3 and 4.385 for ν =0.5 and α ~ 0.1·β. Note that for a stationary crack, $\dot{a}$ = 0, and Eq. (10.43) reduces to

$$\dot{\delta}_t = \tfrac{\alpha}{\sigma_s}\ \dot{J} \tag{10.47}$$

from which

$$\begin{aligned} \delta_t &= \int \frac{\alpha}{\sigma_s} dJ \\ &= \frac{\alpha}{\sigma_s} J \end{aligned} \tag{10.48}$$

which represents the well-known relationship between δ_t and J given by Eq. (10.16). Equation (10.46), again for asymptotic conditions $r \to 0$, should continue to hold even in situations of *LSY* or *LCG*, see Sect. 10.2, but certainly with β and α different from those relative to the *SSY* condition, and in the case of *GSP* if the geometry is such as to develop an adequate state of constraint as in a specimen in bending with deep crack. The separation of variables suggests expressing the $\dot{J}$ as

$$\dot{J} = \mu\dot{a} + \nu\dot{\delta} \tag{10.49}$$

where $\dot{\delta}$ is the rate of displacement along the points of application of the load with μ andν functions of a and δ. Substituting Eq. (10.49) into Eq. (10.46) yields

$$\dot{\delta}_t = \left[\beta\tfrac{\sigma_s}{E}ln\left(\tfrac{R}{r}\right) + \tfrac{\alpha}{\sigma_s}\mu\right]\dot{a} + \tfrac{\alpha}{\sigma_s}\nu\dot{\delta} \tag{10.50}$$

For an ideally rigid-plastic material, with $\sigma_s/E \to 0$, the opening rate $\dot{\delta}_t$ can only depend

exclusively on $\dot{\delta}$ both for the stationary and nonstationary crack. For this to happen it is necessary that the term in parentheses that multiplies $\dot{a}$ is zero with the first addend of finite value equal to $-\mu \cdot \alpha/\sigma_s$. This, however, is not possible because the term R, to balance the elimination of σ_s/E $\to$ 0, would have to contain a factor e elevated to the inverse of σ_s/E that would diverge leaving the extension of the plastic zone R unlimited. Therefore, the only possibility left is that it is $\mu = 0$ with which $\dot{J}$ and, therefore, $\dot{\delta}$ will no longer depend on $\dot{a}$

$$\dot{J} = \nu\dot{\delta}$$

$$\dot{J} \neq \dot{J}(\dot{a}) for \frac{\sigma_s}{E} \to 0 \tag{10.51}$$

10.9 Modified *J*

Let's start analyzing a SE(B) type specimen made of a rigid-plastic material. The moment M necessary to continue producing deformation is [30]

$$M = \tfrac{0,63}{\sqrt{3}}\ \sigma_s\ b^2 \tag{10.52}$$

and will produce a crack tip opening velocity equal to

$$\dot{\delta}_t = 0,37\ b\ \dot{\theta} \tag{10.53}$$

B being the ligament ahead of the crack and θ the angle of rotation of the arms of the specimen. Equations (10.51) and (10.52) apply to a stationary crack as well as to a subcritical growing crack. The relative J, again for the ideal rigid-plastic case, is given by (11.14)

$$J = \frac{2}{b} M \theta \tag{10.54}$$

Recalling Eq. (10.52)

$$J = 0,73 \, \sigma_s \, b \, \theta \tag{10.55}$$

Its derivative is

$$\begin{aligned} \dot{J} &= 0,73 \, \sigma_s \dot{b} \theta + 0,73 \, \sigma_s b \dot{\theta} \\ &= -\frac{J}{b} \dot{a} + 0,73 \, \sigma_s b \dot{\theta} \end{aligned} \tag{10.56}$$

being $b = (W - a)$. The J-integral does not have, therefore, the required property of independence from $\dot{a}$. Let's now evaluate it on a remote path $\Gamma_1 + \Gamma_2 + \Gamma_3$ that coincides with the edge of the specimen, as shown in Fig. 10.30.

Considering one half of the specimen, it will be

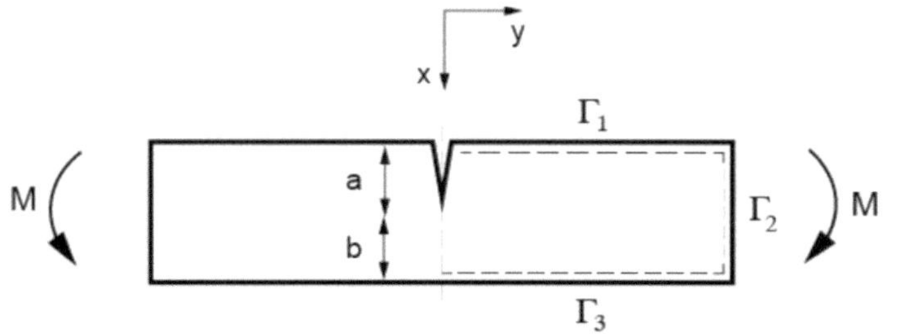

Fig. 10.30 Schematic of an SE(B) specimen showing the remote field integration path (dashed line)

Fig. 10.31 Phisical meaning of J_{Mpl} integral

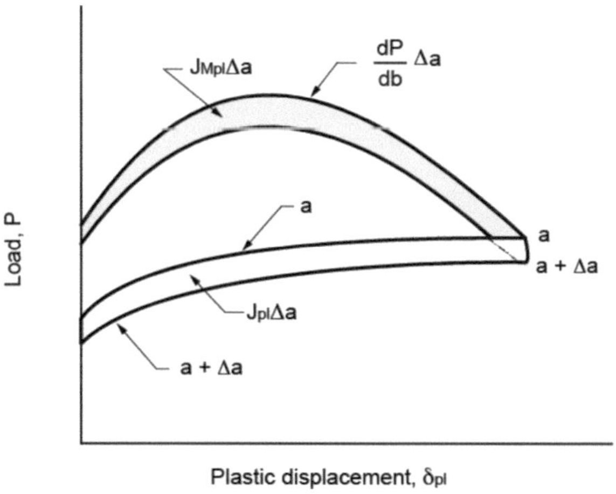

$$J_{rem} = 2 \int_{\Gamma_1+\Gamma_2+\Gamma_3} \left(Wdy - T_i \frac{\partial u_i}{\partial x} ds\right) \tag{10.57}$$

The term T_i cancels along the paths Γ_1 e Γ_3 and its integral is zero on Γ_2, a pure moment being applied. Also the strain energy density, W, becomes zero along Γ_1 e Γ_2. It remains, therefore, its value on Γ_3

$$J_{rem} = 2 \int_{\Gamma_3} Wdy = \left(2\frac{\sigma_s}{\sqrt{3}}\right)(-2u_y) \tag{10.58}$$

where u_y is the displacement in the y direction of the remote point of the rigid arm on which the moment M is applied. In a rigid-plastic specimen, this displacement is obtained by a rotation around a hinge positioned at a distance $x = 0.37{\cdot}b$ below the tip of the crack and is [30]

$$2u_y = -0,63 \int_0^\theta b\,d\theta \tag{10.59}$$

Equation (10.65) becomes

$$J_{rem} = 0,73\ \sigma_s \int_0^\theta bd\theta \tag{10.60}$$

whose derivative is

$$\dot{J} = 0,73\ \sigma_s\ b\ \dot{\theta} \tag{10.61}$$

therefore, independent of $\dot{a}$, as required by Eq. (10.51). We must conclude that, for an SE(B) geometry under *LCG* or even *SSY* conditions, it is the remote field integral and not others that must be chosen to ensure independence from the integration path, in accordance with the analyses of Dodds et al. shown in Fig. 10.30b and, therefore, for the assessment of the singularity of the *HRR* field. It can be shown that for a geometry DE(T) the opposite is true: it is the proximity integral that satisfies Eq. (10.59) and is, probably, path-independent, while for a CC(T) both integrals do satisfy the condition. All this favors the belief that it is not possible to obtain a resistance curve J_R that is valid for every geometry, dimension and subcritical growth using a single parameter. Ernst [31] tried to overcome this apparent limitation of J by introducing a modified version J_M defined as follows

$$J_M = J - \int_{a_0}^{a} \left.\frac{\partial(J - \mathcal{G})}{\partial a}\right|_{\delta_{pl}} da \tag{10.62}$$

where $\mathcal{G}$ represents the rate of release of elastic energy, Eq. (10.36), and, therefore, it is $J - \mathcal{G} = J_{pl}$ where J_{pl} represents the plastic component of the J-integral, see Eq. (11.42), a_o and a are the initial and final length of the crack, respectively and δ_{pl} the plastic component of the displacement of the load application points: $\delta_{pl} = \delta - \delta_{el}$. Having thus defined the J_M, it turns out that the growth rate $\dot{J}_M$ and the dJ_M/da are worth:

$$\dot{J}_M = \dot{J} - \left.\frac{\partial J_{pl}}{\partial a}\right|_{\delta_{pl}} \dot{a}$$
$$\frac{dJ_M}{da} = \frac{dJ}{da} - \left.\frac{dJ_{pl}}{da}\right|_{\delta_{pl}} \tag{10.63}$$

In the case of ideally rigid-plastic material, $\sigma_s/E \rightarrow 0$, in which it will also be $\mathcal{G} \rightarrow 0$and J_{pl} ~J, recalling the general expression (10.49), we have

$$\dot{J}_M = \mu\dot{a} + \nu\dot{\delta} - \mu\dot{a} = \nu\dot{\delta} \tag{10.64}$$

therefore, independent of $\dot{a}$ satisfying the Rice condition (10.51). Ernst suggested an operative formulation of J_M in incremental form similar to that expressed by Eq. (11.44) for the measurement of J by the unloading compliance method. Starting from the simplified general formulation of J_{pl} for a bending or tensile geometry [32], one can write

$$\left.\frac{\partial J_{pl}}{\partial a}\right|_{\delta_{pl}} = -\, m\, \frac{J_{pl}}{b} \tag{10.65}$$

where m is a factor that depends on the crack length a and on the work hardening n. Consequently, Eq. (10.62) of J_M becomes

$$J_M = J + \int_{a_0}^{a} \frac{m}{b} J_{pl} da \tag{10.66}$$

Equation (10.66) can be expressed as

$$J_{M,i+1} = J_{i+1} + \Delta J_i + \left(\frac{m}{b} J_{pl}\right)_i (a_{i+1} - a_i) \tag{10.67}$$

where ΔJ_i=0 for i=0. In particular, for a C(T) specimen m is

$$m = \left(1 + 0,76\frac{b}{w}\right) \tag{10.68}$$

For an SE(B) with deep crack m=1. An alternative form of Eq. (10.66) can be introduced by expressing J_M in the usual form

$$J_M = J_{el} + J_{M,pl} \tag{10.69}$$

and remembering the expression (11.30) for the J_{pl} we get

$$J_M = J_{el} + \frac{\mu}{b} \int_0^{\delta_{pl}} P\, d\delta_{pl} \tag{10.70}$$

P being the load per unit thickness, P=F/B.

Its physical significance is, therefore, that of a difference in plastic deformation energy, or irreversible work carried out by external forces, between two similar bodies, containing two cracks whose length differs by Δa, subject to the same deformation history, as schematically shown in Fig. 10.31.If, as we have just seen, the J_M integral appears to meet Rice's requirement of independence from the growth rate of the crack $\dot{a}$, still many doubts remain about its actual usefulness and ability to represent a curve -Δa independent of the geometry of the system, thickness and type of load. Some experimental evidence would indicate that is actually independent of the thickness of the specimen even well beyond the limits imposed on subcritical growth, others instead [33] show, as for A 533 steel at temperature above ambient temperature, even results less aggregated between them than those obtained with the J traditional. After all, J_M too is based on the theory of nonlinear elasticity to which it will always have to pay a tribute of non-representativeness of the behavior of real materials. In conclusion, further studies are still needed to give credit to J_M of qualities not recognized to the simple J.

10.10 *J-Q* Theory

Equation (10.42) that defines the crack tip stress field in the J-dominance region, when applied to a finite geometry is affected by the same approximation and limitation already discussed in Sect. 7.17 for the K_I dominance introducing the T-stress. Despite the influence on the near tip stress, the T-stress has no effect on J because T is a non-singular term. The issue of the inadequacy of the J to describe the crack tip stress field in finite geometries was first addressed by O'Dowd and Shih [34, 35] who proposed a two-parameter theory introducing the additional term to the HRR stress field given by Eq. (10.42)

$$\frac{\sigma_{ij}}{\sigma_0} = \left(\frac{J}{a\varepsilon_0\sigma_0 I_n r}\right)^{\frac{1}{n+1}} F_{ij}(\theta, n) + Q\left(\frac{r}{J/\sigma_0}\right)^q \hat{\sigma}_{ij}(\theta, n) \tag{10.71}$$

where J set the scale over which large deformations occur and Q quantifies the opening stress magnitude. The terms, σ_o and ε_o in Eq. (10.71) are the coefficients of the Ramberg-Osgood, RO, Eq. (1.41), often assumed to be equal to the yield stress σ_s and yield strain ε_s, respectively. At variance with the T-stress, the second-term of Eq. (10.72) operates not only on σ_x, but also on σ_y, reducing them if $Q < 0$. O'Dowd and Shih found, through a systematic FE study of the crack tip stress fields using modified boundary layer solution (MBL), that the power q of the second term in Eq. (10.71) is approximately zero for $n = 1/m \geq 0.25$ and that in the forward sector, $|\theta| \leq \pi/2$, the second order normal stresses $\hat{\sigma}_{yy}$ and $\hat{\sigma}_{xx}$ are approximately equal while the second-order shear stress $\hat{\sigma}_{xy}$ is approximately zero. Therefore, Q represents in practice a constant field whose amplitude is a hydrostatic or triaxiality term. The crack tip triaxiality cannot be described by the only J field because, adding a hydrostatic stress to the plastic zone, precisely that controlled by J, of non-compressible material, since plastic deformations develop without any volume change, no strain variation can be produced. Therefore, a *J-Q* combination is needed that defines both the singularity and the level of deformation of the crack tip and the degree of triaxiality. With these findings, Eq. (10.71) simplifies to

$$\frac{\sigma_{ij}}{\sigma_0} = \left(\frac{J}{a\varepsilon_0\sigma_0 I_n r}\right)^{\frac{1}{n+1}} F_{ij}(\theta, n) + Q\delta_{ij} \qquad (10.72)$$

δ_{ij} being the Kronecker delta. Equation (10.73) describes the stress field in a finite geometry as the HRR solution (first term) plus an additive constant term, the Q-stress. However, analysis of cracks in finite bodies [36] reveals that Q values referenced to crack tip stress fields in infinite bodies with the SSY, T=0 reference solution lose self-similarity much later than with the infinite bodies solution HRR, as shown in Fig. 10.32 for SE(B) specimen with a strain hardening exponent $n = 1/m = 0.1$. Also note that by using the nondimensional scale of distance $r/(J/\sigma_o)$ the Q-stress should remain constant, but it does not when referenced to the HRR stress field. Therefore Eq. (10.73) can be also put in the form

$$\sigma_{ij} = \left(\sigma_{ij}\right)_{SSY,T=0} + Q\sigma_0\delta_{ij} \qquad (10.73)$$

where the first term is the infinite body SSY K_I stress field for which the T-stress is $T = 0$.

Nevertheless, the self-similarity is lost, anyhow, with increasingly high load levels, probably because the ratio $a/W = 0.15$, being representative of a shallow crack, has a T<0, see Sect. 7.17, Table 7.3. It can be also seen how the Q-stress presents an insignificant dependence on the distance r and the angular position within a significant interval of the nondimensional distance, r, from the crack tip for $J/\sigma_o < r < 5J/\sigma_o$.

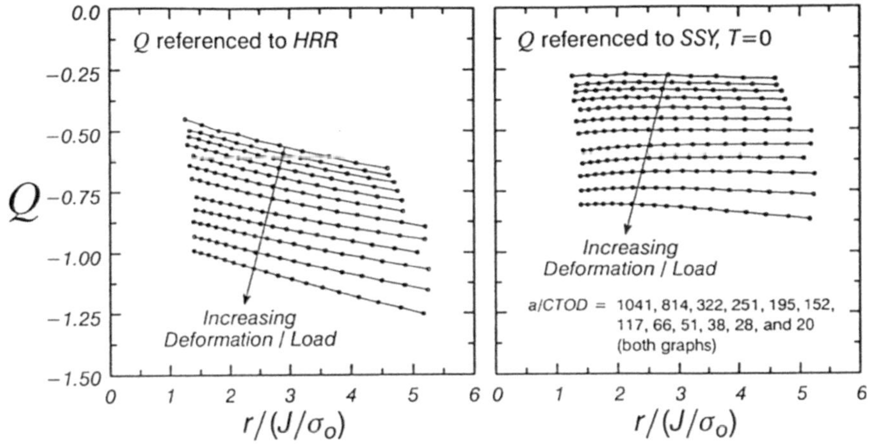

Fig. 10.32 Effect of infinite reference solution on radial dependence of Q for an $a/W = 0.15$ SE(B) specimen having a strain hardening exponent $n = 0.1$ [36]

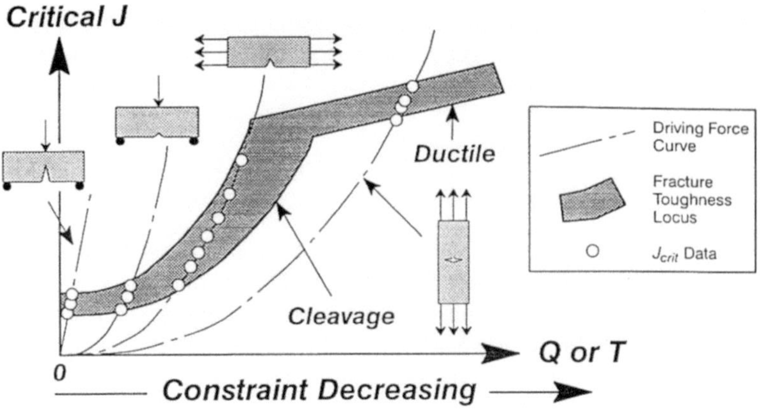

Fig. 10.33 Schematic of fracture toughness locus using various specimen types, with different a/W ratios, as function of loss of crack tip constraint represented by the T or Q parameter [36]

Moreover, also the Q-stress assumes negative values. Several investigators have tried to generate fracture loci associated to the T-stress and Q-stress values as indices od loss of constraint. This is schematically shown in Fig. 10.33 [36].

As Kirk and Dodds put it, considerable debate surrounds the application of the T-stress as a constraint parameter under LSY conditions rather than Q. The T approach has the advantage of simplicity relative to Q, requiring only linear-elastic analysis rather than elastic-plastic. It is a fact, however, that as we get more and more away from finite geometries under SSY conditions to more and more LSY conditions the T-stress may only approximate the elastic-plastic stress field of J dominance.

Concluding Remarks

The Irwin stress intensity factor, K_I, defines the amplitude of the stress field ahead of the crack tip as long as SSY condition prevails. As the crack front opens and blunts, becoming a notch, this condition is lost and the K_I parameter ceases to characterize as a single parameter the notch tip stress field. This happens, in particular, with very tough materials or ferritic steels used at or above the RT_{NDT}. At variance with the K_I, the CTOD, introduced in the UK, is not restrained by the operating conditions, i.e. whether the material is behaving in a ductile or brittle fashion, and applies also in cases where the crack tip opens and the SSY condition is lost. However, the experimental measurement of CTOD is not so straightforward and accurate. Therefore, fracture mechanics standards, both ASTM and ISO, provide methods to derive CTOD from a K_I or J_I measurement which actually raises the question as to whether or not this parameter is really useful. As the SSY condition is lost but well before General Yielding, GY, occurs, there is an area where the crack tip stress field, though no longer elastic, still remains self-similar and its amplitude is measured by a new parameter: the Rice J-integral. Strain hardening now plays a major role. J-integral provides a means of characterizing deformations in the crack tip vicinity without focusing attention directly at the crack tip. This is accomplished by expressing the average crack tip elastic-plastic fields in terms of deformation energy within an arbitrary region surrounding the crack tip. This allows us to choose a path containing the crack tip far from the apical zone where the stress field is extremely variable and difficult to assess. It may be even convenient to use as integration path the border of the specimen. However, at variance with the linear elastic behavior where the reaching of the critical value K_{Ic}, triggers brittle fracture immediately, in the elastic-plastic domain when the critical J_{Ic} value is reached fracture may not necessarily occurs, since it is preceded by the blunting of the crack tip and a local tearing in which the crack starts to grow steadily before unstable fracture occurs. When the Rice J-integral starts to lose its dominance in defining the crack tip stress field becoming a non-singular term, a new parameter may be added called Q-stress which represents a constant field whose amplitude is a hydrostatic or triaxiality term.

References

1. McClintock, F.A., Irwin,G.R.: Plasticity Aspects of Fracture Mechanics, American Society of Testing and Materials, 67th annual meeting, Chicago, June 1964, ASTM STP-381, pp. 84–113, Third Printing, January (1975)
2. Hill, R.: The Mathematical Theory of Plasticity, Oxford University Press (1950)
3. McCabe, D.E.: Summary of Elastic-Plastic R-Curve Studies, NUREG/CP-0064, CSNI Rep. No. 105, Proceedings of the 2nd CSNI Workshop on Ductile Fracture Test Methods, pp. 235–257, August (1985)
4. McClintock, F.A.: Plasticity Aspects of Fracture. In: Liebowitz, H. (ed.) Fracture, vol. 3, pp. 47–225. Academic Press, London (1971)
5. Cottrell, A.H.: Iron and Steel Institute Spec. Rep., vol. 69, p. 281 (1961)
6. Wells, A.A.: Proceedings of Crack Propagation Symposium, College of Aeronautics, Cranfield (1961)
7. Nichols, R.W., Burdekin, F.M., Cowan, A., Elliot, D., Ingham, T.: The use of critical crack opening techniques for the selection of fracture resistant materials. In: Proceeding of Symposium on Fracture Toughness Concepts of Weld. Structure Steel, pp. F1–F113, Risley (April 1969)
8. Knott, J.F.: Fundamentals of Fracture Mechanics, p. 153. London Butterworths (1973)
9. Burdekin, F.M., Dawes, M.G.: Practical use of linear elastic and yielding fracture mechanics with particular reference to pressure vessels. In: Proceedings of the Institute of Mechanical Engineers Conference, London, pp. 28–37 (May 1971)
10. Wells, A.A.: Application of fracture mechanics at and beyond general yielding. Br. Weld. J. **10**, 563–570 (1963)
11. Hahn, G.T., Sarrate, M., Rosenfield, A.R.: Criteria for crack extension in cylindrical pressure vessels. Int. J. Fract. Mech. **5**(3), 187–210, (1969)
12. Hahn, G.T., Rosenfield, A.R., Sarrate, M.: Observations of Yielding Accompanying Crack Growth, Inelastic Behavior of Solids, pp. 673–69. MacGrow-Hill, New York (1970)
13. Goodier, J.N., Field, F.A.: Plastic Energy Dissipation in Crack Propagation. In: Drucker, D.C., Gilman, JJ. (eds.) Fracture of Solids, p. 103. . New York (1963)
14. Cowan, A., Kirby, N., The Application of C.O.D. Measurements to Large Scale Test Behaviour, Proceedings of the Symposium on Fracture Toughness Concepts for Weldable Structural Steel, Risley, pp. D1-D27, April 1969

15. Folias, E.S.: The Effect of Initial Curvature on Cracked Sheets, Final report to NSF on Grant GK-1440, UTEC Do-68-070, August 31 (1968)
16. Hahn, G.T., Sarrate, M., Rosenfield, A.R.: Criteria for Crack Extension in Cylindrical Pressure Vessels. Int. J. Fract. Mech. **5**, 187–210 (1969)
17. Kobayashi, A.S., Engstrom, W.L., Simon, B.R.: Crack Opening Displacements at Normal Strains in Centrally Notched Plates, Off. of Naval Research, NR 064 478, T.R. No. 4, (January 1968)
18. De Koning, A.U.: A Contribution to the Analysis of Slow Stable Crack Growth, National Aerospace Laboratory Report No. NLR MP 7035U, The Netherlands (October 1975)
19. Dodds, R.H., Tang, M., Anderson, T.L.: Numerical Modelling of Ductile Tearing Effects on Cleavage Fracture Toughness, NRC NUREG/CR 6162 (June 1994)
20. Papaspyropoulos, V., Marschall, C., Landow, M.: Predictions of J-R Curves with Large Crack Growth from Small Specimen Data, NRC NUREG/CR-4575 (September 1986)
21. Rice, J.R.: A path Independent Integral and the Approximate Analysis of Strain Concentration by Notches and Cracks, Brown University, Division of Engineering, Providence, ARPA SD-86, Report E39 (May 1967)
22. Rice, J.R.: A path independent integral and the Approximate Analysis of Strain Concentration by Notches and Cracks, Journal of Applied Mechanics, pp. 379-386 (1968)
23. Wells, A.A.: IIW Houdrement Lecture. Brit. Weld. J. **12**(1), 2 (1965)
24. Rice, J.R., Rosengren, G.F.: Plane strain deformation near a crack tip in a power-low hardening material. J. Mech. Phys. Solids **16**, 1–12 (1968)
25. Hutchinson, J.W.: Singular behaviour at the end of a tensile crack in a hardening material. J. Mech. Phys. Solids **16**, 13–31 (1968)
26. Wilson, W.K.: Methods of K Determination, Westinghouse Nuclear Internal Document
27. Dodds, R.H., Tang, M., Anderson, T.L.: Numerical Modeling of Ductile Tearing Effects on Cleavage Fracture Toughness, US NRC NUREG/CR-6162 (June 1994)
28. Rice, J.R., Sorensen, E.P.: J. Mech. Phys. Solids **26**, 163–186 (1978)
29. Rice, J.R., Drugan, W.J., Sham, T.L.: Elastic-Plastic Analysis of Growing Cracks, ASTM STP-700, pp. 189–221 (1980)
30. Rice, J.R.: Proceedings, Mechanics and Mechanisms of Crack Growth, M. J. May Ed., British Steel Co., Physical Metallurgy Centre Publication, pp. 14–39 (1974)
31. Ernst, H.A.: Material Resistance and Instability Beyond J-Controlled Crack Groth, Westinghouse Scientific Paper 81-1D7-JINTF-P6 (3 Dec 1981)
32. Ernst, H.A.: ASTM STP **791**, 499–519 (1982)
33. Koppenhorfer, K.C., Dodds, R.H.: Size and Deformation Limits to Maintain Constraint in KIc and JIc Testing of Bending Specimens, NRC NUREG/CR-6191 (October, 1995)
34. O'Dowd, N.P., Shih, C.F.: Family of crack-tip fields characterized by a triaxiality parameter-I Structure of fileds. J. Mech. Phys. Solids **39**, 989–1015 (1991)
35. O'Dowd, N.P., Shih, C.F.: Family of crack-tip fields characterized by a triaxiality parameter-II. Fract. Appl. J. Mech. Phys. Solids **40**, 939–963 (1992)
36. Kirk, M.T., Dodds, R.H.: An Analytical and Experimental Comparison of J_I Values for Shallow Through and Part Through Surface Cracks, NUREG/CR-5970 (July 1993)

Chapter 11
Experimental Elastic–Plastic Fracture Mechanics

11.1 Experimental Verification of J as a Fracture Criterion

Begley and Landes [1] used Rice's definition of the J-integral, Eq. (10.33), as a convenient basis for the experimental measurement of J. Equations (10.31) and (10.33) can be used in the most general form for a body of thickness B.

$$J = -\frac{dU}{da}\frac{1}{B} \tag{11.1}$$

where—dU is the energy released during an elemental crack extension da. Consider, now, two similar specimens, for instance two SEC(T), Fig. 9.4a, having an edge crack of length a and $a + \Delta a$, respectively. It has been shown in Sect. 7.14 that for linear elastic materials the potential energy U of the system containing a crack, Eq. (7.131), is the complementary energy and any variation in that energy following a virtual crack extension da is given by the difference between the complementary energies relative to two equal systems of nearly similar crack lengths that differ only by da. Obviously, for nonlinear elastic materials the same assumption holds. Now, if we load two specimens having a cack of length a and $a + \Delta a$, respectively, with the same force F, case (a) of Fig. 11.1, the potential energy change $-\Delta U$ will be given by the difference between the respective complementary energies, dashed area of Fig. 11.1a which will approximate $-dU$ that appears in Eq. (11.1) the more the smaller the value of Δa. If the two bodies were subjected to the same fixed displacement Δ, Fig. 11.1b, the corresponding energy change $-\Delta U$ would always be given by the difference between the respective complementary energies, which is equal to the former $-\Delta U$ since the area ABC, Fig. 11.1a, is a diffential of the second order $1/2\Delta F d\Delta$ as Δa tends to da. It is possible to affirm that the point of arrival B of Fig. 11.1a will be reached under the action of the same force F either by the panel containing the crack of length $a + \Delta a$ or by that containing the shorter crack a allowing the growth $d\Delta$ at constant force F. Under the hypothesis of nonlinear elasticity, in fact, which provides

P. P. Milella, *Fundamentals of Fracture Mechanics*,
https://doi.org/10.1007/978-3-032-23081-2_11

for the complete reversibility of deformations, unloading the system with the longer crack $a + \Delta a$ will retrace exactly the loading curve returning to point 0 and since the system does not recognize if point B was reached directly with the crack $a + \Delta a$ or in the two phases F–Δ plus $d\Delta$ it is possible to conclude that point B of arrival will be common to the two paths. Landes and Begley [2] observed that rather than working with two systems containing cracks of different lengths, though very close, and measure the relative complementary energies it was more convenient, from the experimental point of view, to work with several specimens of the same size, type and material, at least four or five containing cracks of different sizes, a_i, very close to each other, obtained by fatigue-precracking the initial machined notch, see Sect. 9.2.

Assume, then, to load four specimens of the type shown in Fig. 11.2a to get an entire family of load vs. displacement curves, F-Δ, shown in Fig. 11.2b. The area under each curve is the so-called pseudopotential energy, U_{ij}, absorbed by the specimen with the crack a_i at the displacement, Δj. These energies are then plotted in a diagram energy vs. crack length, U_{ij}–a_i, obtaining another family of curves, as shown in Fig. 11.2c. The tangent to each curve at any point, $\mathrm{d}U/\mathrm{d}a$, provides the value of $J{\cdot}B$, where B is the specimen thickness. Finally, a new diagram is plotted with the calculated.

$$J_{ij} = \frac{1}{B}\frac{\partial U_{ij}}{\partial a_i} \tag{11.2}$$

As function of the displacement Δ_j, as shown in Fig. 11.2d. The plot provides the calibration curves for the specimens used, at the test temperature. It is clear

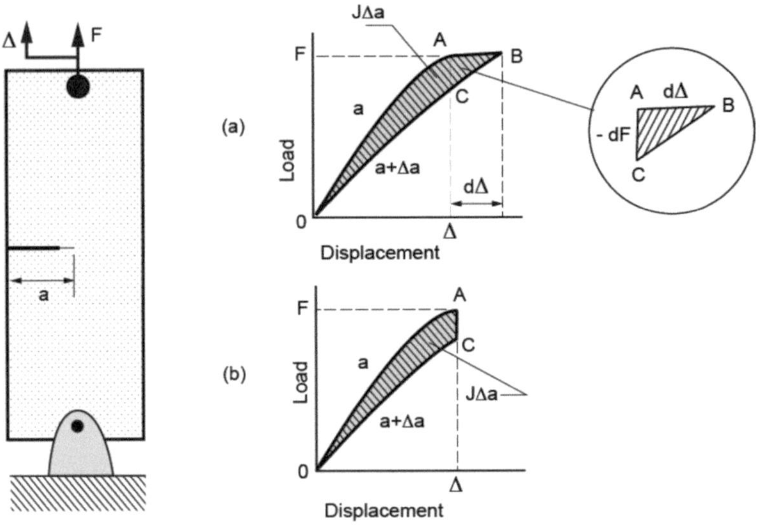

Fig. 11.1 Load diagrams for two SEC(T) panels having crack lengths a and $a + da$, respectively, loaded **a** in traction and **b** by a fixed displacement Δ

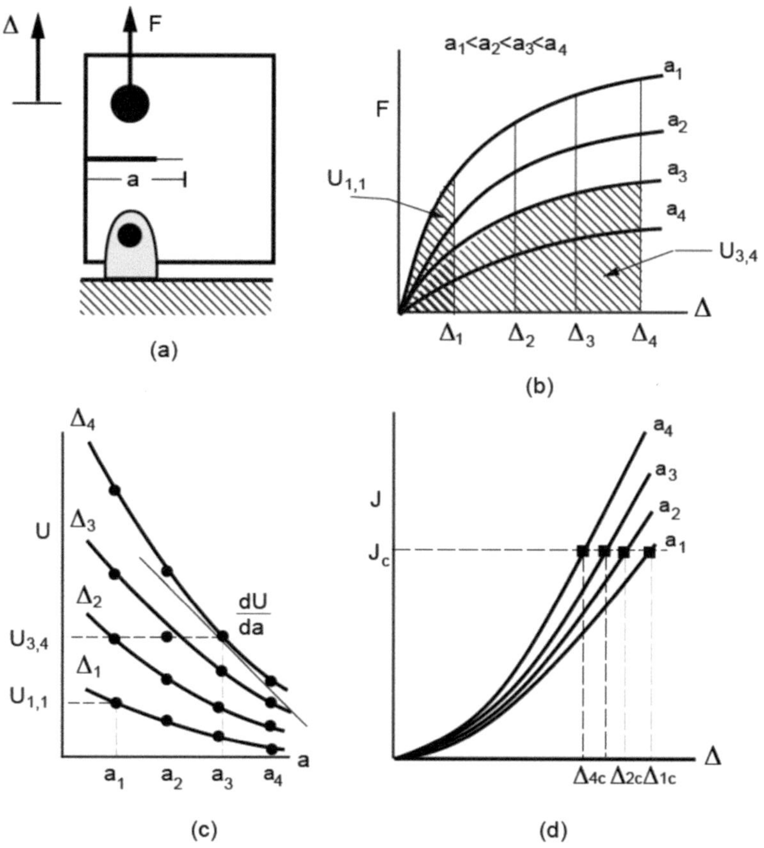

Fig. 11.2 Schematic of early experimental measurement of *J*-integral using multiple specimens procedure

that if the *J*-integral actually provides a criterion for criticality, then reporting on the calibration curve of each specimen the value of the displacement Δ_{ic} at which the first sign of stable propagation occurs, it must appear that, although different from each other, the various Δ_{ic} yield the same J_c, as shown in Fig. 11.2d. Landes and Begley, applying the methodology just described to three 1″ CC(T) specimens, Fig. 9.5, of a medium strength steel Ni–Cr–Mo–V for rotors containing a central crack of length equal to 8.9, 11.4 and 14.0 mm respectively, found that this was true [2]. At the testing temperature of 119 °C, the strain at which the three cracks began to grow was different, but the critical value J_c that corresponded to the critical deformations was practically the same, varying by 8% only from 168 kJ/m^2 to 184 kJ/m^2, as shown in Fig. 11.3 [2]. The steel had a yield strength $\sigma_s = 916$ MPa and ultimate tensile strength equal to 1020 MPa at room temperature and $\sigma_s = 855$ MPa and ultimate tensile strength equal to 957 MPa at the testing temperature of 119 °C.

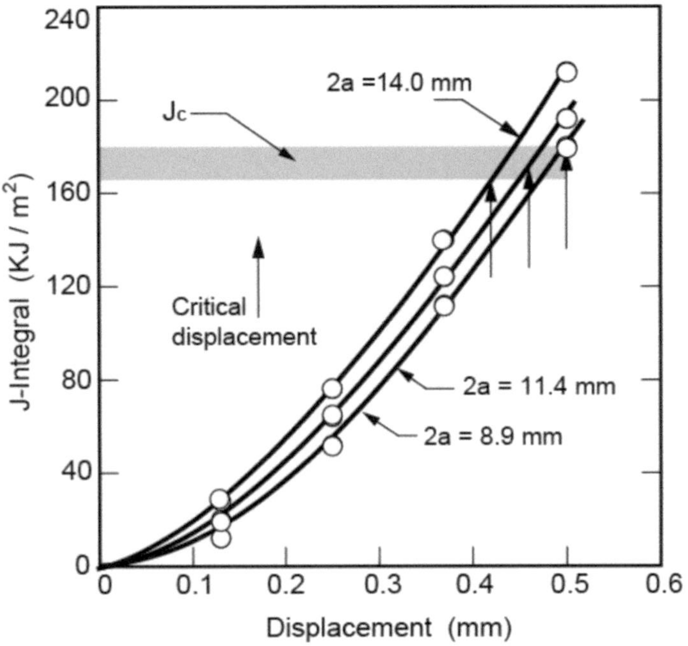

Fig. 11.3 *J*-integral versus displacement for 1″ CC(T) panels of NiCrMoV steel at 119 °C showing J_{Ic} [2]

In analogy to the linear elastic case in which the *K* relative to the first mode of crack opening was denoted by K_I, Fig. 7.14, and the critical one by K_{Ic}, also now we will denote J_{Ic} the critical value of the *J*-integral, recalling that, unlike the elastic case, this time this critical value, J_{Ic}, identifies the first occurrence of a stable tear at the crack tip, Fig. 10.5b, and not the possible unstable propagation. Using other specimens, type SE(B) Fig. 9.19, with cracks ranging from 3.6 mm to 7.1 mm Landes and Begley obtained for J_{Ic} an average value of 187 kJ/m^2. Similar results were found by Kobayashi et al. [3]. It is certainly worth noting that 187 kJ/m^2 is equivalent, on the basis of Eq. (10.40), to 202 MPa$\sqrt{}$m which would require a thickness of 6"(15.24 cm) if measured using a C(*T*) specimen with the LEFM K_{Ic} procedure, Table 9.5, besides that at that temperature the K_I cannot be applied any longer. Note the appearance of the curves of Fig. 11.3 that after a phase of exponential growth, linearize. This can be explained imagining to have two bodies containing cracks of very close length, *a* and $a + da$.

Their characteristics force–displacement, *F*–*Δ*, are shown in Fig. 11.4a for the two cases of elastic–plastic behavior with and without strain hardening, respectively. Consider now the elastic–plastic case without strain hardening, first. Under a generic load *F* the two bodies will show displacements *Δ* that differ from each other by $d\Delta$ in the linear field. Using Rice's interpretation, Eq. (11.1), assuming a unit thickness *B*, the area between the two load curves at the applied force *F* in the linear field will hold:

$$J = \frac{F}{2}\frac{\mathrm{d}\Delta}{\mathrm{d}a} \tag{11.3}$$

Consider, now, the compliance $C = \Delta/F$

$$J = \frac{F^2}{2}\frac{dC}{da} = \mathscr{G} \tag{11.4}$$

Equation (11.4) represents the fundamental relation already found, Eq. (9.20), that links the energy release rate G to load F and compliance C. Equation (11.4) can also be written as:

$$J = \frac{1}{2}\frac{\Delta^2}{C^2}\frac{\mathrm{d}C}{\mathrm{d}a} \tag{11.5}$$

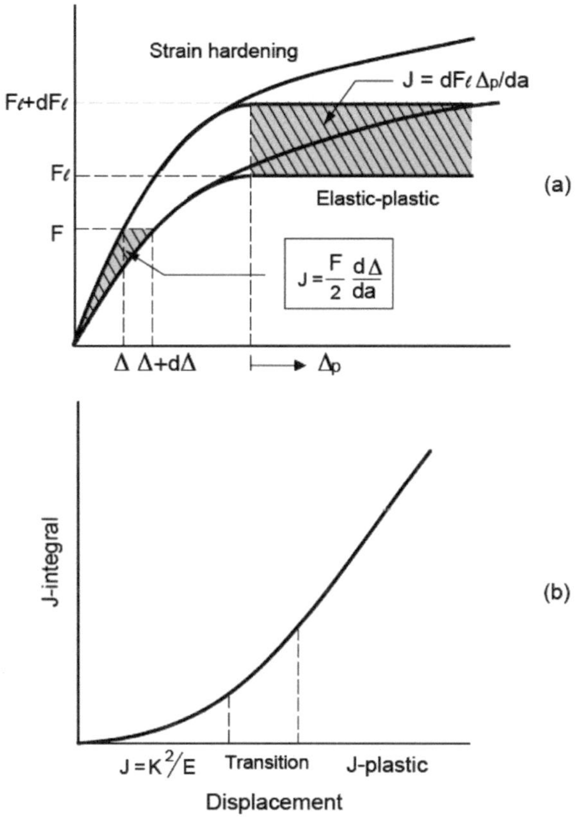

Fig. 11.4 **b** Schematic of J versus displacement obtained from (**a**) the areas under the load displacement curves for elastic–plastic and strain hardening materials. F_l is the limit load

from which it can be inferred that in the linear field $J = (1-\nu^2)K^2/E$ is proportional to the square of displacement Δ and will therefore have a parabolic trend, as shown in Fig. 11.4b. Increase the load acting on the two systems. As we enter the plastic domain by reaching the limit load, F_l, the displacement Δ will grow indefinitely, as shown in Fig. 11.4a. The area between the two load curves relative to the bodies containing a crack of length a and $a + da$, respectively, will provide the value of the J-integral.

$$J \cdot da = \Delta_p \cdot dF_l \tag{11.6}$$

Therefore, now J will be proportional to the displacement Δ assuming a linear trend. The approximation of the area between the two load curves with just the plastic one will be all the truer the larger Δ_p. If we move from the elastic–plastic material to one that works harden, we will have, in practice, the same linear behavior. Looking at Fig. 11.3 it will be noted that for Ni–Cr–Mo at 90 °C the calibration curves are effectively composed of a parabolic initial part, where the behavior of the material is predominantly elastic and the K_I concept holds, followed by a linear trend where the material representative of a plastic behavior and where the critical displacement occurred and K_I loses any meaning. Also note that, being CTOD proportional to G, Eq. (10.12), and, therefore, to J, also CTOD must have the same trend.

11.2 *J*-integral Measurement: Area Method and the η Factor

Though interesting, the method proposed by Begley and Landes to measure J_{Ic}, described in the previous section, is quite laborious. The simplest method for measuring the fracture toughness J_{Ic} is certainly the one proposed by Rice [4] in 1973 which makes use of only one single specimen. It can be observed that the area between the two load curves F–Δ relative to two specimens containing cracks of length a and $a + d\Delta$ very close to each other, Fig. 11.1a, is

$$U = \int\int df \cdot d\Delta \tag{11.7}$$

Recalling the definition of J given by Eq. (11.1) we get

$$\begin{aligned} J &= -\frac{1}{B}\int_0^{\Delta} \frac{dF}{da} \mathrm{d}\Delta \\ &= \frac{1}{B}\int_0^{F} \frac{d\Delta}{da} \mathrm{d}F \end{aligned} \tag{11.8}$$

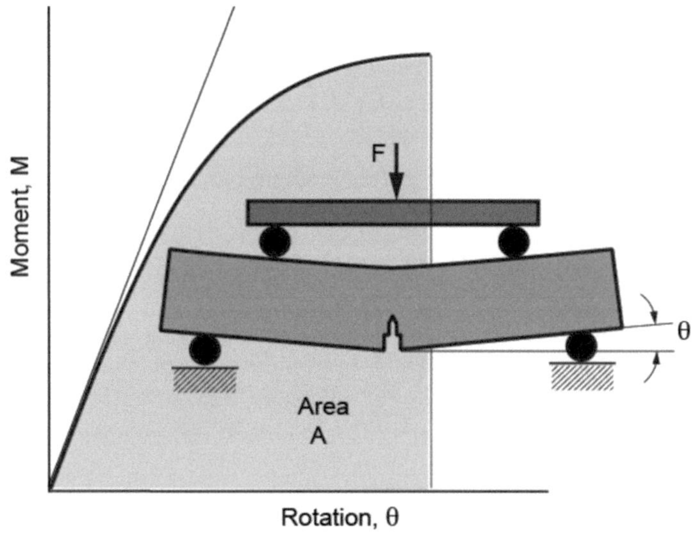

Fig. 11.5 Moment-rotation diagram in a four-point bend specimen. The area under the curve is the total energy absorbed during the rotation

Using a SEN(B) specimen subjected to a pure bending moment, M, (four-point bending), Fig. 11.5, we have

$$J = -\frac{1}{B}\int_0^M \frac{d}{\mathrm{d}b}\mathrm{d}M$$
$$= -\frac{1}{B}\int_0 \frac{dM}{\mathrm{d}b}\mathrm{d} \tag{11.9}$$

where θ is the angle of rotation of the arms of the specimen, which takes the place of the displacement, $\varDelta$, and b represents the length of the ligament in front of the crack, $b = W - a$, Fig. 9.19.

Since the angle of rotation θ is a function of the ligament b according to a relation of the type

$$\theta = f\left(\frac{M}{b^2}\right) \tag{11.10}$$

it is

$$\frac{d\theta}{db} = -\frac{2M}{b^2}f'\left(\frac{M}{b^2}\right) \tag{11.11}$$

It also is

$$\frac{d\theta}{dM} = \frac{1}{b^2} f'\left(\frac{M}{b^2}\right) \tag{11.12}$$

deriving $f'(M/b^2)$ from Eq. (11.11) and substituting into Eq. (11.12) yields

$$\frac{d\theta}{db} = -\frac{2M}{b}\frac{d\theta}{dM} \tag{11.13}$$

and Eq. (11.9) becomes

$$J = \frac{2}{Bb}\int_0^\theta M\,\mathrm{d}\theta \tag{11.14}$$

Note that the integral that appears in Eq. (11.14) is precisely the area *A* under the *M–θ* curve, Fig. 11.15. Thus, Eq. (11.14) takes the final form.

$$J = \frac{2A}{Bb} \tag{11.15}$$

which has a general validity. If we use a SEN(B) specimen under three-point bending, Fig. 9.19, the area, *A*, under the force versus load-line displacement curve, *F–Δ,* can be also used to calculate *J*, Fig. 11.6a. Two cases can be distinguished:

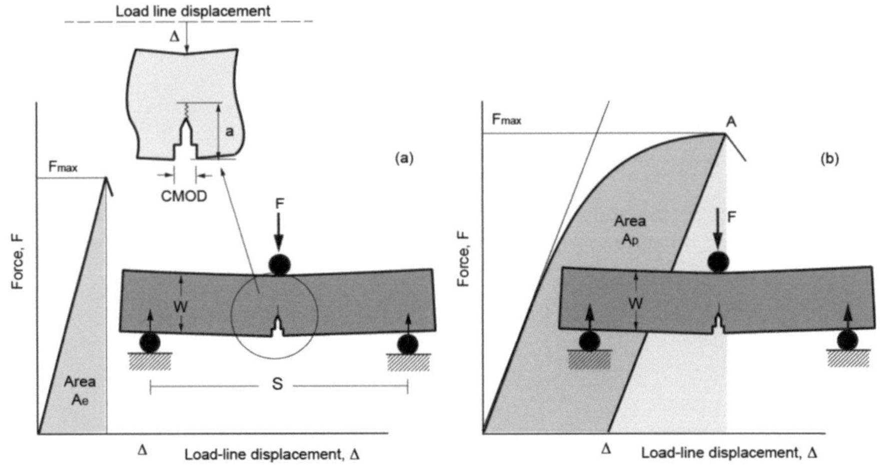

Fig. 11.6 Schematic of load–displacement, *F–Δ*, for a SE(B) specimen. If the behavior is linear, case (**a**), the area A_e gives a *J*-derived K_{Jc}, otherwise, case (**b**), it yields the plastic component of the *J*-integral J_p

Case (a): the behavior is linear, then the procedure described in Sect. 9.7 can be used for a SEN(B) specimen to calculate the corresponding K_{Ic}.

Case (b:) the behavior is not linear, then Eq. 11.15 provides the plastic component of J_p of the total J.

$$J = J_e + J_p \tag{11.16}$$

Following an approach originally offered by Merkle and Corten [5], later taken up by Sumpter and Turner [6], the plastic component, J_p given by Eq. (11.15) can be put in the most general form.

$$J_p = \frac{\eta_p A_p}{Bb} \tag{11.17}$$

where A_p is the plastic area under the force-load line displacement curve, *F–Δ*, as shown in Fig. 11.6(b). The elastic component J_e, Eq. (11.16), can be assessed through the usual relation (10.32).

$$J_e = \frac{K_I^2}{E'} \tag{11.18}$$

As for η_p, whose existence Ernst and Paris have proven [7], it is function of the ratio *a/W*, with $W = a + b$, that can be evaluated starting from Eq. (11.4) which expresses the relationship between *J*, *F* and the compliance $C = \Delta/F$ of the system [8].

The ASTM E 1820-08a Standard provides the following expressions for η_p.

$\eta_p = 1.9$ if Δ is the load-line displacement, otherwise.

$\eta_p = 3.667–2.199 + 0.437(a/W)^2$ if the crack mouth opening displacement CMOD, *V*, is measured instead of *Δ*, as shown in the detail of Fig. 11.6a.

Note that the term *B* that appears in Eq. (11.17) is substituted by B_N, the reduced section thickness, when the specimen is side grooved. As for the elastic component, J_e, given by Eq. (11.18), it is also possible to use an alternative expression similar to Eq. (11.17) using the area A_e under the curve *F–Δ*, or *F–V*, Fig. 11.16a.

$$J_e = \eta \frac{A_e}{Bb} \tag{11.19}$$

where the η-factor is always

$$\eta = \frac{b}{C} \frac{\partial C}{\partial a} \tag{11.20}$$

The η-factor for the elastic case is found to be 2 for a SEN(B) specimen for almost all crack size, *a*, larger than 40% of the width *W*.

Some adjustments will be necessary when using a *C(T)* specimen in which in addition to the bending stress also a traction component acts on the ligament in front of the crack, Fig. 9.8 and Eq. (9.12). Equation (11.17) always holds but with a different value of η_p [9].

$$\eta_p = 2\frac{(1+\alpha)}{(1+\alpha^2)} \tag{11.21}$$

with

$$\alpha = \left[\left(\frac{2a}{b}\right)^2 + 2\left(\frac{2a}{b}\right) + 2\right] - \left(\frac{2a}{b} + 1\right) \tag{11.22}$$

or, according to Logsdon [10]:

$$\eta_p = 2 + 0.522\frac{b}{W} \tag{11.23}$$

The real problem with all elastic–plastic fracture mechanics tests is the subcritical crack growth, Δa, that may occur during the test, Figs. 10.5 and 11.32. When the maximum load, F_{max}, is attained, point A of Fig. 11.6b, the original crack may have grown a distance Δa. This is shown in Fig. 11.7 for a SEN(B) specimen of hard steel tested at 0 °C. The consequence is that the measured J is overestimating the actual or critical value at initiation, J_{Ic}. To avoid that, it is possible to adopt a direct current potential drop technique, as shown in Sect. 11.4, to pinpoint the precise moment when the crack starts to grow steadily and pick up the actual force F_{max} and the corresponding area, A_p, under the F–Δ curve. Alternatively, a very simple and effective method can be used to detect the stable crack growth initiation point, F_i to which it is associated a plastic component $V_{i,p}$ of the total displacement V. At variance with the elastic regime where the initial crack does not blunt and remains sharp, in the plastic regime the crack starts to blunt before stable tearing initiates. Three different phases or Zones can be identified, as shown in Fig. 11.8. Initially, Zone I or early phase, the crack under load remains sharp, it doesn't open. The corresponding load–displacement, F–V, is a straight line, line 0A of Fig. 11.8a. During this phase of loading, the compliance, $C = dV/dF$, does not change and remains constant, line O'A' of Fig. 11.8b. Continuing to load, the crack starts to blunt, see Fig. 10.5. This is seen by the system as a spurious crack growth and the load–displacement line starts to bend, Zone II and curve AB of Fig. 11.8a.

During this phase, the compliance C starts to grow exponentially, curve A' B' of Fig. 11.8b. The blunting phase ends up with a real tearing of the material at the crack tip, Zone III of Fig. 11.8a, till the specimen fails, point C. When the first tearing occurs the compliance C changes its growth from exponential to linear, line B' C' of Fig. 11.8b. Therefore, the point of first subcritical crack growth, point B of Fig. 11.8a, can be pinpointed with great accuracy by looking at the compliance trend, point B' of Fig. 11.8b. The compliance curve can be drawn by points using the load–displacement, F–V, record. The compliance at any point is just given by:

Fig. 11.7 Fracture appearance of a SEN(B) specimen 60 mm thick of hard steel in which it is possible to recognize the fatigue precracking, light band that extends over the machined notch, and the subcritical crack growth Δa, darker band over the previous precracking growth (Courtesy of ATM RIVA CALZONI Spa, Brescia, Italy)

$$C = \frac{\mathrm{d}V}{\mathrm{d}F} \sim \frac{\Delta V}{\Delta F} \tag{11.24}$$

Figure 11.9 shows a test curve obtained with a SEN(B) specimen of high strength steel, 60 mm thick and 240 mm wide. Point F_i has been obtained with the compliance method just described and shown in Fig. 11.10. Once the displacement V_i at initiation is located on the compliance versus displacement curve, C–V, Fig. 11.10, the value of the corresponding force, F_i, can be assessed entering the diagram of Fig. 11.9 with that value V_i. From F_i the parallel at the initial linear elastic slope will yield the $V_{i,p}$ value and that of the plastic area component, A_p, to be used in Eq. (11.17).

Both ASTM E 1820-08a and EN ISO 12135 Standards provide detailed rules to measure the J-integral from SEN(B) and C(T) specimens. However, both standards just warn that by considering the area, A_p, at maximum load, F_{max}, in Eq. (11.17), as shown in Fig. 11.6, the resulting J is not corrected for subcritical crack extension, Δa that has surely occurred prior to that maximum load. Nevertheless, both standards do not provide any rule to derive the actual force, F_i, at which subcritical crack growth initiates.

11.3 Measurement of the *J–R* Curve: Multiple Specimens Method

In the ductile field the knowledge of the entire R curve is at least as important as the knowledge of J at initiation, J_{Ic}. Two materials having equal J_{Ic} may have different fracture behaviors if their respective resistance curves, R, are different. To build the entire R curve the area method, described in the previous Section, can be used with

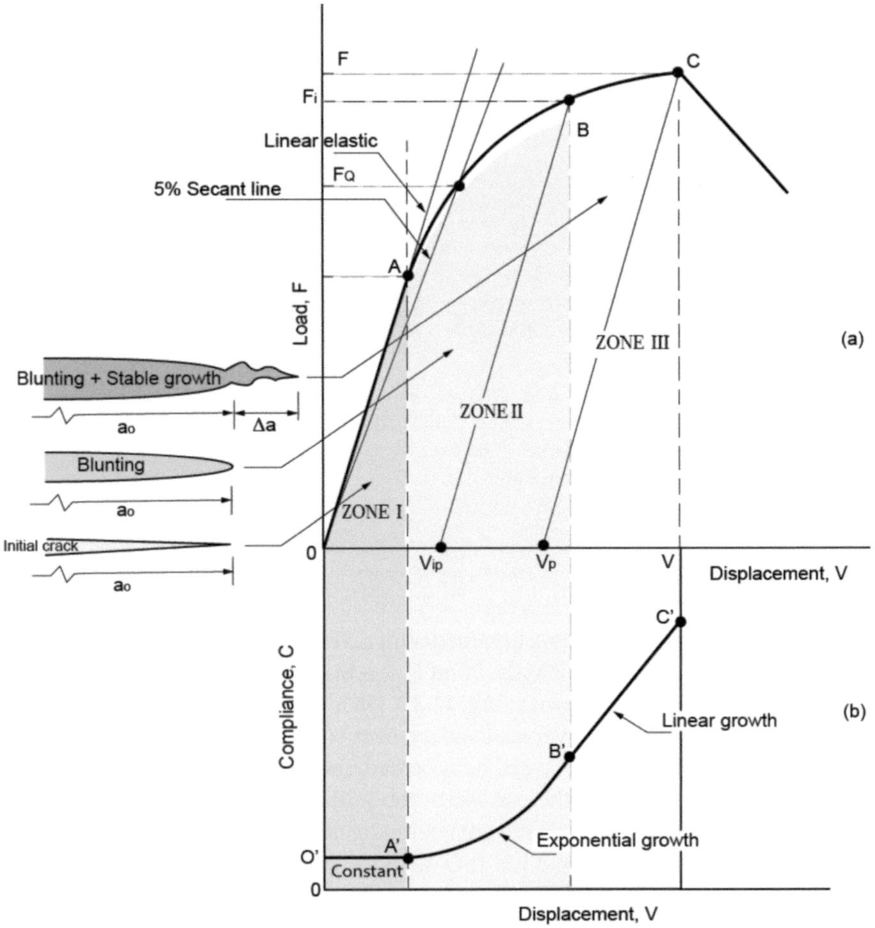

Fig. 11.8 **a** Load–displacement test record, *F–V*, in which three zones can be identified. Zone I refers to the initial linear behavior of the specimen when the crack remains close and sharp, line 0A. Zone II refers to the blunting of the crack where the *F–V* trend is no longer linear. Zone III starts with the first subcritical crack growth at the applied force F_i to which the plastic component of the displacement $V_{i,p}$ corresponds. **b** Compliance, *C*, where the three zones are characterized by a constant trend, line O'A', an exponential growth, line A'B', and a linear growth, B'C'. Also shown is the 5% secant that intercept the force F_Q on the *F–V* curve used for K_I calculation

a multiple specimen procedure. To this purpose, several specimens are tested, for example C(T) specimens.

Each specimen has a notch fatigue-precrack of the same initial length, *a*. The specimens are then loaded to different displacements, Δ_i, Fig. 11.11a.At each displacement, the plastic component of the area under the curve, $A_{p,i}$, is assessed, Fig. 11.6b, and the corresponding J_i calculated using Eqs. (11.17) with the proper η factor.

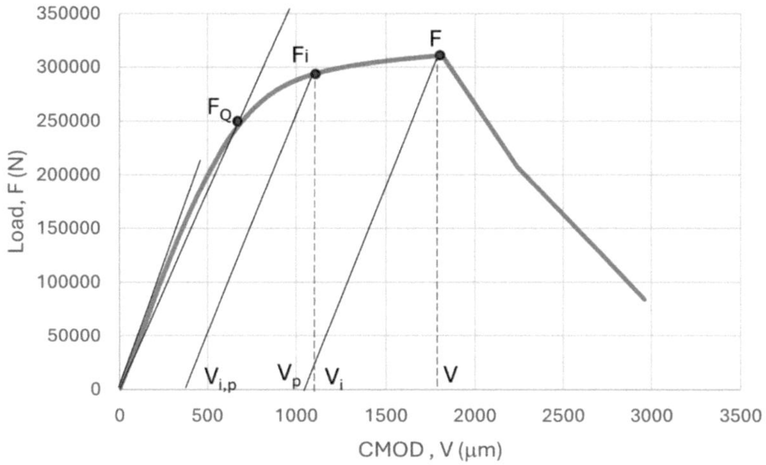

Fig. 11.9 Load-displacemet, F–V, diagram of a SEN(B) specimen of high strength steel, 60 mm thick and 240 mm wide, showing points F_Q, force at 5% offset line, F_i, force at subcritical crack growth initiation and F (maximum force) with the corresponding plastic component of displacement $V_{i,p}$ and V_p, respectively. Displacements were measured at the crack mouth (CMOD)

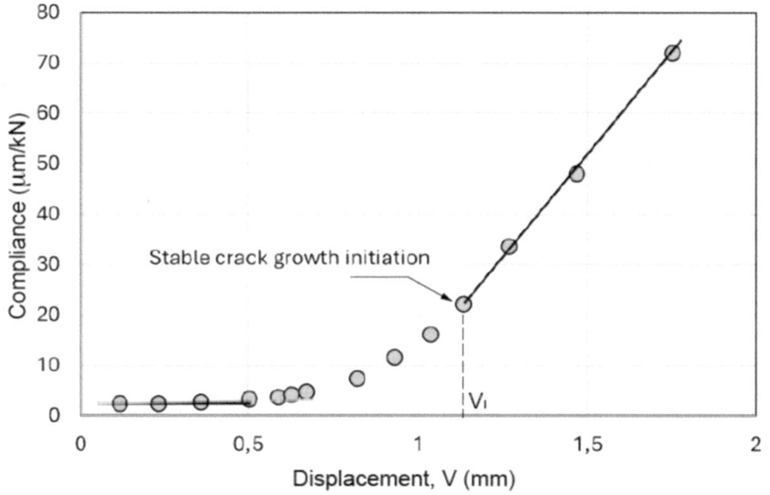

Fig. 11.10 Compliance measurement by points showing the three zones of Fig. 11.8b. Stable crack growth initiation is located at the beginning of the linear trend at a displacement V_i

To know each subcritical crack growth, Δa_i, associated with the given displacement Δi each specimen is put in a furnace at 300–400 °C for 15 min. During this permanence in the furnace, the new free surface produced by the crack growth, Δa_i, is heat tinted by oxidation, Fig. 11.12a. The specimen is then put in liquid nitrogen

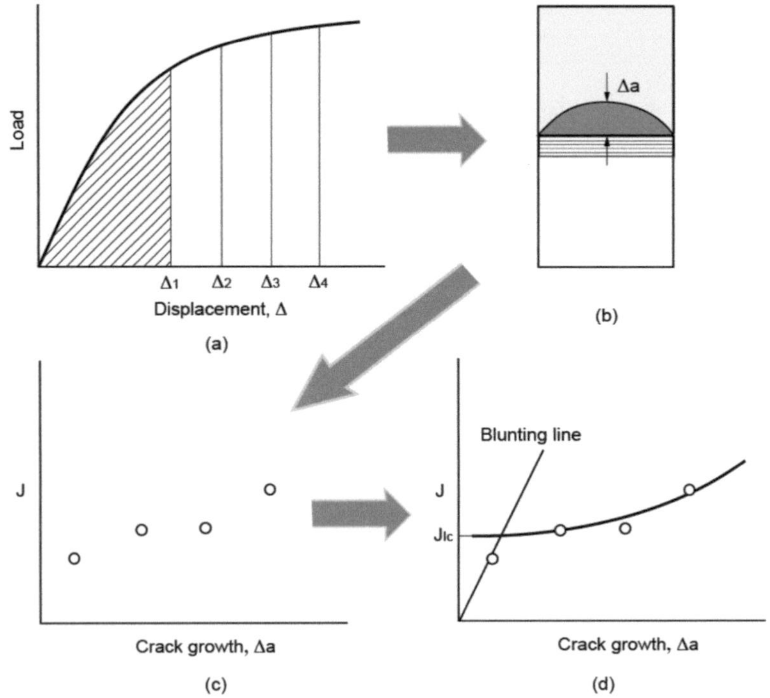

Fig. 11.11 Procedure for R curve construction and J_{Ic} measurement using a multiple specimen procedure

and broken open or just broken if it is made of high strength material. The subcritical crack growth, Δa_i, will then appear dark against the crystalline brittle fracture, Fig. 11.12b [11].

Note that the crack grew with a non-uniform front, with a maximum at the center leaving a kind of nail shaped appearance. The calculated J_i are then plotted against the heat-tinted Δa_i, Fig. 11.12c and a best fit line drawn, as schematically shown in Fig. 11.12d. This is the *J–R* curve of the material.

The intersection of this curve with the blunting line, see Sect. 11.7, yields the J_{Ic}, Fig. 11.11d. The blunting line is built with measurements of the stretched zone (SZ), Sect. 11.7, which is not a real growth, Δa, but the blunting of the initially sharp crack that precedes the tearing, Figs. 10.5 and 10.7. The extension of the stretched zone, Δa_{sz}, can be taken, as a first approximation, as

$$\Delta a_{sz} \approx \frac{1}{2} \quad \delta_t = \frac{1}{2} \; CTOD \tag{11.25}$$

and recalling the expression of *CTOD* given by Eq. (10.12) and the equality $j = \mathscr{G}$ we finally get the blunting line equation.

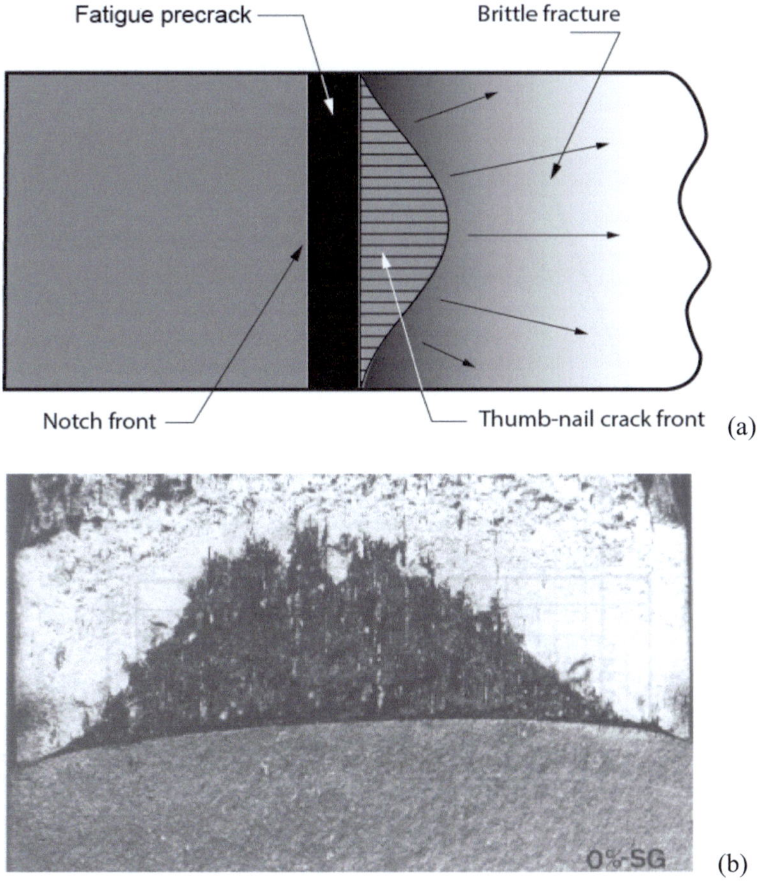

Fig. 11.12 **a** Schematic of the possible thumb-nail subcritical crack growth or tunnelling inside the specimen thickness that does not appear on the external surface; **b** subcritical crack growth observed by heat-tinting in a 1″ CT specimen of carbon steel, type A 542 [11]

$$\Delta a_{sz} = \frac{1}{2}\ \psi \frac{J}{\sigma_s} \tag{11.26}$$

where the values of the function ψ depend on the material and load conditions, as it will be better said in Sect. 11.7.

The R curve obtained using this procedure would appear with an upward trend after the initial almost linear part, as shown in Figs. 11.11d or 11.13. This is because the procedure overestimates the actual resistance of the material to crack extension.

In fact, the plastic area, $A_{p,i}$, under the curve that measures the energy absorbed by each specimen at increasing load levels always refers to the initial crack length, a, and not to the actual initial length, $a + \Delta a_i$. This is the same error that affects the experimental R curve found with DT tests, shown in Fig. 3.50. The real R-curve bends

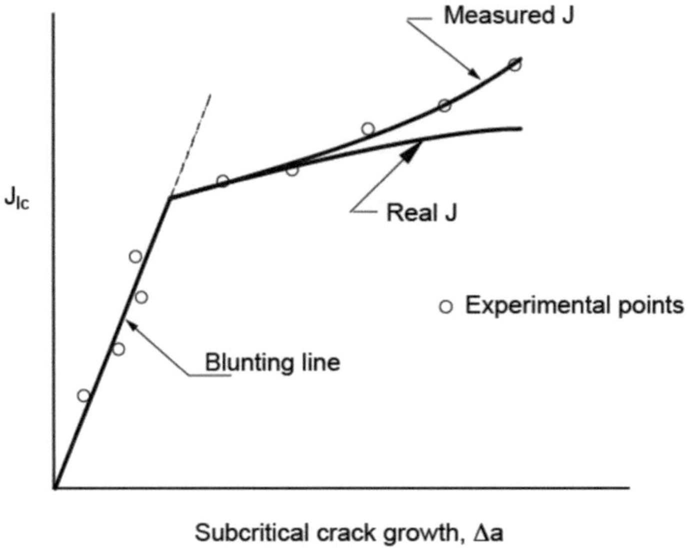

Fig. 11.13 Schematic of the J_{Ic}–Δa resistance curve bending up-ward obtained using specimens with the same initial crack length. The real *R*-curve has the opposite trend. The blunting line is obtained from the stretched zone measurements

in the opposite direction, as schematically shown in Fig. 11.13. This overestimation is shown in Fig. 11.14 for A216 steel [12].

The authors used precisely the multiple specimen technique just described to monitor crack growth by heat tinting each specimen and calculate the corresponding

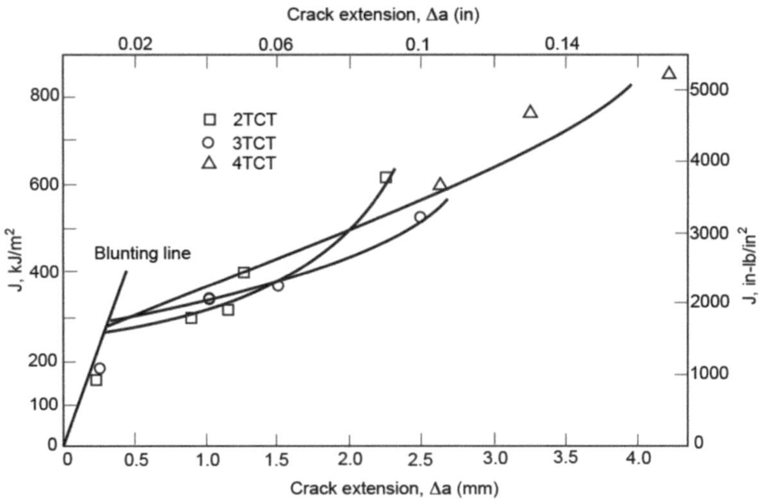

Fig. 11.14 J versus crack extension for CT specimens of A216 steel tested at 120 °C [12]

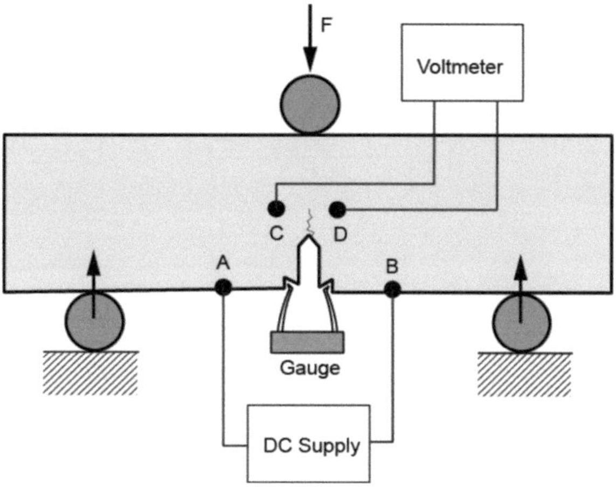

Fig. 11.15 Schematic of DCPD application to a SE(B) specimen to detect the moment of first stable growth of the notch crack

J. All curves show the upward trend. The blunting line, J–Δa_{sz}, of Fig. 11.14 is obtained using Eq. (11.26) with $\psi = 1.0$ and the flow stress σ_f instead of the yield stress $\sigma_{s,}$ Eq. (10.21).

Several CT specimens of different size, namely 2T CT, 3T CT and 4T CT were used to also assess the size effect. The value of J_{Ic} at initiation was obtained, as said, by the intersection of the J–Δa or R-curve with the straight line passing through the origin.

11.4 Measurement of the *J*-integral and *J–R* Curve: DCPD Method

The boresome heat tinting procedure to measure each crack length can be overcome by using non-visual methods such as, for example, the DCPD technique, or *direct current potential drop*, Fig. 11.15. A DC supply maintains an electric potential difference, V, between two points A and B on the surface of a SE(B) specimen on each side of the notch. The specimen is crossed by a constant direct current.

Two more electrodes, C and D, are placed at a certain distance on both sides of the crack tip, reading the local electric potential through a voltmeter. The crack growth results in the reduction of the cross-section area of the specimen which is directly associated with an increase in the electrical resistance. Therefore, as a stable tearing initiates, caused by the load increase, the voltmeter reads a drop in the electric potential between points C and D. The variation of the electric potential may be also read at the same A and B points. The amount of potential drop, ΔV, depends on both the specimen and the crack geometry. Once calibrated, the potential drop, ΔV, provide the actual crack length. This would allow to measure not only the critical value J_{Ic}, but also the amount of any further growth and build the entire J versus

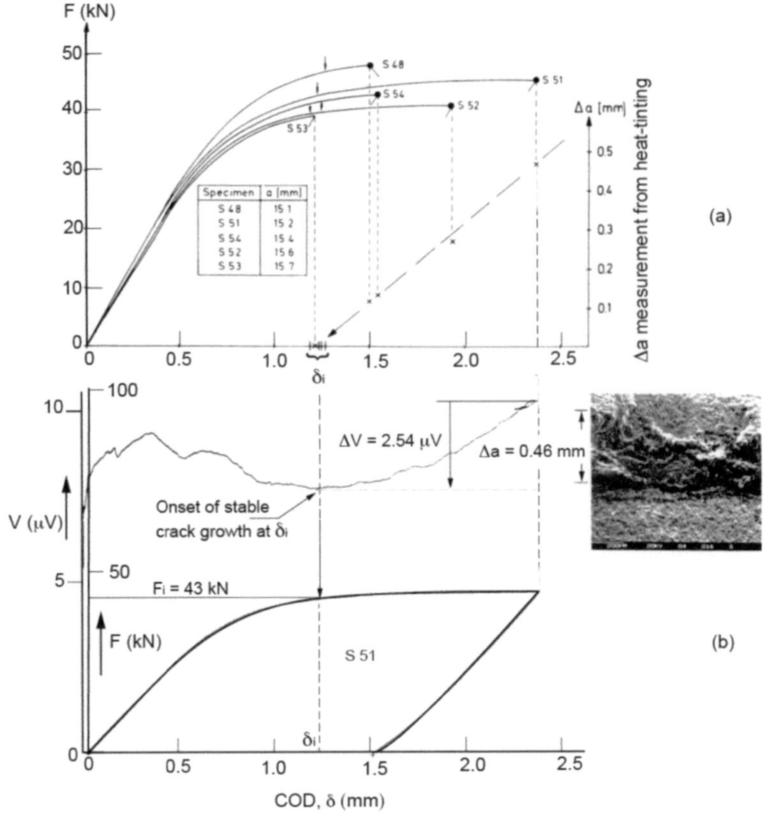

Fig. 11.16 Schematic of DCPD method. **a** Five WOL-X type specimens prepared with DCPD leads are loaded up to five different displacements close to each other. Subsequent heat-tinting marks the crack growth, Δa, in each specimen. **b** Load–displacement obtained for specimen S 51 where crack initiation occurs at δ_i corresponding to the minimum in the Potential-Displacement curve, *V*–*δ*. Also shown is a SEM of the stretched zone plus the stable crack growth of 460 μm in specimen S 51 (integrated from [13])

$\varDelta a$ curve, i.e., the *R* resistance curve using a single specimen through the procedure presented in Sect. 11.3.

Figure 11.16a [13] shows an example of the DCPD applied to WOL-X Type specimens, Fig. 9.9. Five WOL 25 X-Type specimens with a mean crack length $\overline{a}$ $= 15.4$ mm have been loaded to different levels of opening displacements COD od δ at room temperature. After unloading, the extent of stable crack growth $\varDelta a$ was read on each of the five specimens using a heat tinting technique, Fig. 11.16a. These five measurements were extrapolated back to a critical point at $\varDelta a = 0$ yielding a displacement at initiation $\delta_i = 1.2$ mm at an applied force $F_i = 43$ kN. Actually, this initial displacement,δ_i, represents a sort of mean value among the five displacements at initiation, different but very close to each other. This displacement corresponds to a minimum on the potential versus displacement curve, V–δ, of Fig. 11.16b. Note that from this minimum COD the potential curve starts to grow steadily indicating thee stable crack growth, Δa.

Figure 11.16b also reports the load–displacement curve of the S 51 specimen of Fig. 11.16a together with a SEM which shows a fully developed stretch zone at the crack tip with the stable crack growth of about 450 μm corresponding to the potential drop at the minimum of 2.54 μV. Specimen S 53 had a sudden instability soon after initiation. Once the system has been calibrated it can be possible to immediately detect both the occurrence of the first stable crack growth and the amount of subsequent growth Δa on any other specimen without using the heat tinting procedure [14–16]. C. Marschall [17], *et.al*, adopted an alternative method using a 3T-C(T) compact specimen of Type 304 austenitic stainless steel. The preparation of the specimen for testing is shown in Fig. 11.17. Leads to supply constant current were placed at points C and C', almost midway between the load line and the back face.

Potential leads were placed at four different locations: A–A', B–B', C–C' and D–D'. Testing of the compact specimen employed partial unloading at various stages of the test. Ten unloading were used. This was done to mark the crack front for later measurements of actual crack extension. Figure 11.18 [17] shows the electric potential, U, measured at A-A' versus load line displacement. The curve shows an initial region of rising potential that is approximately linear with displacement. This increase in potential is the result of crack blunting. After this linear rise the electric potential deviate from linearity, as shown in Fig. 11.15. This deviation, marked I in Fig. 11.18, is indicative of crack initiation followed by stable crack growth.

The corresponding load-line displacement was $\Delta = 14.2$ mm. The other couple of points, B–B', C–C' and D–D' yielded equal results. Knowing this displacement,

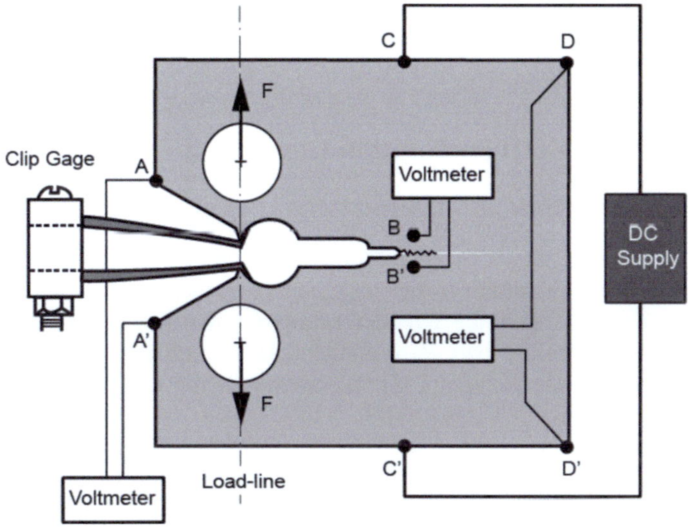

Fig. 11.17 Schematic of DCPD method for monitoring crack extension in a compact specimen

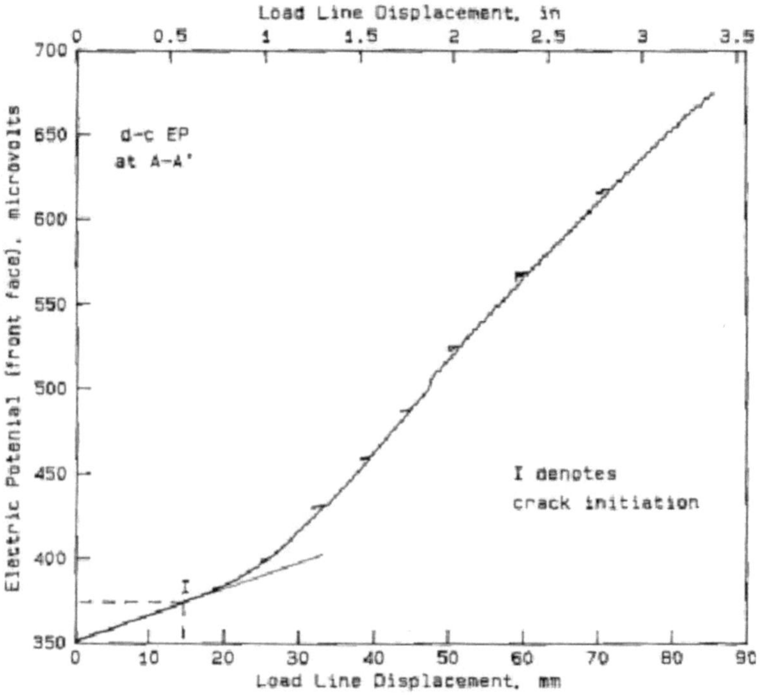

Fig. 11.18 Electric potential at A-A' versus displacement [17]

from the load–displacement curve it was possible to derive the load at which initiation occurred.

11.5 *J–R* Curve Overestimation Correction

The overestimation of the *J–R* curve described in Sect. 11.3 and shown in Figs. 11.13 and 11.14, can be corrected by expressing J as a function of the two independent variables: a, the current length of the crack, and Δ, the displacement at the load line. To this end, it can be used the Wilson's expression that links, in the elastic field, the K factor to the applied load, P, or moment, M, for unit thickness in a specimen containing a crack under bending [18]. In case of bending M.

$$K = 3.98\frac{M}{b^{3/2}} \approx 4.0\frac{M}{b^{3/2}} \tag{11.27}$$

where b is the length of the ligament in front of the crack. Remembering the general expression (9.30) we can write.

$$M = \frac{\Delta E b^{3/2}}{4\sqrt{W}} f(a/W) \tag{11.28}$$

E being the Young's modulus of the material, *W* the size of the specimen, $W = a + b$, and *f(a/w)* its calibration function. Putting as a first approximation $M = \mathrm{P}\cdot a$, being *P* the load applied per unit thickness, we will have

$$P = \frac{\Delta E b^{3/2}}{4a\sqrt{W}} f(a/W) = \frac{\Delta E b^{3/2}}{4(W-b)\sqrt{W}} f(a/W) \approx \frac{\Delta E b^2}{4W} f(a/W) \tag{11.29}$$

Equation (11.29) links the external load *P* to the load displacement *Δ* and, by means of the nondimensional function *f(a/W),* to the current crack length *a*. It can be put in the most convenient form:

$$P = \frac{b^2}{W} F\left(\frac{\Delta}{W}, \frac{a}{W}\right) \tag{11.30}$$

where the function *f(a/W)* can be expressed by two functions of Δ/W ed a/W.

$$F(\Delta/W, a/W) = G(a/W) \cdot H(\Delta/W) \tag{11.31}$$

Indeed, the separation of the function *F* into the product of two functions each dependent on a single variable is possible only if there exists a constant scale factor between the various trends *F*–Δ for any value of *a*/*w* [19]. With this Eq. (11.21) can be written as

$$J = \frac{\eta}{b} \int_0^{\Delta} P \mathrm{d}\Delta = \eta G \frac{b}{W} \int_0^{\Delta} H \mathrm{d}\Delta \tag{11.32}$$

The separation of variables *a* and *Δ* will allow to write the total differential of *J* as:

$$\mathrm{d}J = \left[-\left(\eta - 1 - \frac{b}{W} \frac{\eta'}{\eta} \right) \frac{\eta G}{W} \int_0^{\Delta} H \mathrm{d}\Delta \right] \mathrm{d}a + \left(\frac{\eta b}{W} G H \right) \mathrm{d}\Delta \tag{11.33}$$

therefore

$$J = \int_{a_o}^{a} \left(-\gamma \frac{\eta G}{W} \int_0^{\Delta} H \mathrm{d}\Delta \right) \mathrm{d}a + \int_0^{\Delta} \frac{\eta b G}{W} H \mathrm{d}\Delta$$

$$= -\int_0^{\Delta} \frac{J}{b}\gamma \mathrm{d}a = \int_0^{\Delta} \frac{\eta b G}{W} H \mathrm{d}\Delta \tag{11.34}$$

with

$$\begin{aligned} \eta' &= \frac{d\eta}{d(a/w)} \\ \gamma &= \left(\eta - 1 - \frac{b}{w}\frac{\eta'}{\eta}\right) \end{aligned} \tag{11.35}$$

Recalling Eq. (11.24).

$$\gamma = 1 + 0,76\frac{b}{w} \tag{11.36}$$

Equation (11.34) allows to calculate the J-integral along any path even in case of growth of the flaw. Consider, then, the graph in Fig. 11.19. The three curves indicate the paths, P–Δ, for three stationary crack of lengths a_o, a_1 and a_2, respectively. Assume to move along the curve relative to the flaw a_1.

Without any crack growth, the J-integral in A will be given by Eq. (11.34).

$$J_A = \frac{\eta b G}{W}\int_0^{\Delta_A} H \mathrm{d}\Delta = \frac{\eta}{b}\int_0^{\Delta_A} P \mathrm{d}\Delta \tag{11.37}$$

In B the value of J is

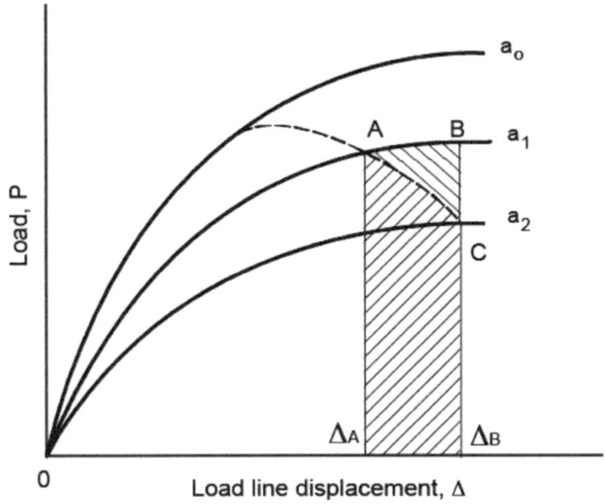

Fig. 11.19 Load–displacement diagram, P–Δ, for stationary cracks of length a_o, a_1 and a_2 and subcritical crack growth

$$J_B = \frac{\eta b G}{W} \int_0^{\Delta_B} H \mathrm{d}\Delta = \frac{\eta}{b} \int_0^{\Delta_B} P \mathrm{d}\Delta \tag{11.38}$$

Therefore, the two integral are related as

$$J_B = J_A + \frac{\eta}{b} \int_0^{\Delta_B} P \mathrm{d}\Delta \tag{11.39}$$

If we now imagine that in the transition from A to B the crack has grown and, therefore, the system has moved to C along the path A-C, then the integral J_C will hold, according to (11.34) and (11.38).

$$J_C = J_B - \int_{a1}^{a2} \frac{J}{b} \gamma \mathrm{d}a \tag{11.40}$$

and using Eq. (11.39)

$$J_C = J_A + \frac{\eta}{b} \int_{\Delta_A}^{\Delta_B} P \mathrm{d}\Delta - \int_{a1}^{a2} \frac{J}{b} \gamma \mathrm{d}a \tag{11.41}$$

If points A, B and C are very close to each other, we can approximate the differential *da* with the difference (a_2–a_1) and the area integral with the area A_{AB} enclosed between Δ_A and Δ_B of Fig. 11.16, so that Eq. (11.41) becomes

$$\begin{aligned} J_C &= J_A + \frac{\eta}{b} A_{AB} - \frac{\gamma}{b} J_B (a_2 - a_1) \\ &= \left(J_A + \frac{\eta}{b} A_{AB}\right)\left[1 - \frac{\gamma}{b}(a_2 - a_1)\right] \end{aligned} \tag{11.42}$$

and more generally

$$J_{i+1} = \left(J_i + \frac{\eta}{b} A_{i,i+1}\right)\left[1 - \frac{\gamma}{b}(a_{i+1} - a_i)\right] \tag{11.43}$$

where the subscripts i and $i + 1$ indicate the subsequent point of assessment of the various quantities in the generic loading phase that brings the system from i to $i + 1$ and $A_{i,i+1}$ is the area under the load curve between the displacements Δi and Δ_{i+1}. Equation (11.43) is the expression found by Ernst, Paris and Landes [19] that allows the calculation of the *J*-integral under non-steady crack conditions taking into account subcritical growths since the crack length changes continuously during the test, the *J*-integral must be calculated incrementally. Assuming that the first apparent subcritical growth occurs at the load step $j + 1$, in the previous one, j step, the system will still be in linear elastic regime, neglecting the blunting, where J is exactly equal to K^2/E', with E' equal to E/(1 − ν^2) or simply$E$, depending on whether there are plain strain or plain stress conditions. Therefore, the relative J_j can be indicated as J-elastic, J_{el}, to differentiate from the plastic one, J_{pl}, which will begin to be evaluated with the next step. Recalling the most general expression of J.

$$J = J_{el} + J_{pl} \tag{11.44}$$

with

$$\begin{aligned} J_{el} &= \frac{K^2(1-\nu^2)}{E} = \frac{\eta}{b}\int_0^{\Delta_j} P d\Delta \\ J_{pl(i+1)} &= \left[J_{pl(i)} + \frac{\eta}{b}A_{i,j+1}\right]\left[1 - \frac{\gamma}{b}(a_{i+1} - a_i)\right] \end{aligned} \tag{11.45}$$

Equation (11.44), and its explicit form (11.45), is used in the calculation procedure to get the corrected J-integral through the unloading compliance method described in the next section. Every single step $i, i + 1\ldots$ will correspond to each phase of loading–unloading. It should not be forgotten that load P is normalized to the thickness B of the specimen and, therefore

$$P = \frac{F}{B}; A = \frac{1}{B}\int F d\Delta \tag{11.46}$$

F being the force applied to the specimen. It is worth noting that the procedure for separating J into $J_{el} + J_{pl}$ should not necessarily lead to the assumption that is $\eta_{el} = \eta_{pl}$. The construction of the R curve or J_R can be done using a function of interpolation of experimental data of the type.

$$J_R = C(\Delta a)^N \tag{11.47}$$

with C and N parameters to be determined. Once drown the J-R curve through an interpolation equation of the type (11.47) it is possible to determine J_{Ic} directly from its intersection with the blunting line avoiding performing also the linear regression, as previously mentioned.

11.6 *J–R* Curve Obtained by the Unloading Compliance Method

The need for continuous monitoring of the crack growth, Δa, during the testing to derive the R curve and the wish to avoid the cumbersome procedure of the multiple-specimens approach has led to the introduction of the unloading compliance method

[20]. The method consists in the progressive loading of a single specimen through higher and higher load steps, each followed by a partial unloading and reloading till the final fracture that, in very high toughness material or at temperatures over the RT_{NDT} for BCC materials, may not occur during the test prior to a large displacement. The result is shown in Fig. 11.20 for a 2T-CT specimen of A 106 carbon steel, having a yield strength of 317 MPa and ultimate strength of 503 MPa at RT, tested at 135 °C. Steel is used in the construction of pipes.

The partial unloading during the test provides, with its slope, the value of the compliance $C = \Delta/F$ needed to assess the crack length, Δa, at each load step. To that purpose, Eq. (9.17) is used solved for a through the function $V_y(a/W)$.

$$V_y\left(\frac{a}{W}\right) = \frac{\Delta}{F}BE = CBE \tag{11.48}$$

Therefore, Δ must be the displacement read on the load line. As the crack growths, the specimen becomes more compliant (less stiff). The variation of the compliance from two successive unloading is generally very small and so is the corresponding J variation. However, the to-day use of an ad hoc software relaxes the previous attention of the researchers spent in enlarging each slope to assess compliance. The unloading is normally no larger than 10–20% of the load at which it occurs. This limitation was introduced to comply with the requirement imposed by the J-deformation definition based on the non-linear theory of elasticity that can represent real material behavior as long as no unloading occurs, see Sect. 10.6 and Fig. 10.23.

At the time it seemed that such partial unloading could be enough large to measure compliance C but not too large to violate the request of no unloading at all.

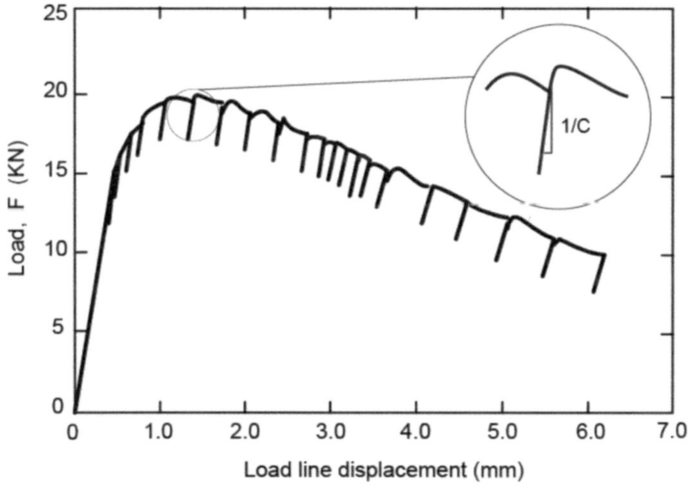

Fig. 11.20 Load–displacement diagram obtained at 135 °C for a carbon steel, type A 106, with a 2T-CT specimen, using the unloading compliance method

Today, this initial prudence has been validated revealing an even larger possibility of unloading, exceeding 20%. Generally, three types of loading–unloading patterns may be observed, as shown in Fig. 11.21. If the load-unload-reload phase is immediate, a load–displacement records type (a) is generally obtained. However, ductile metals tend to display relaxation softening or a Bauschinger effect, see Sect. 1.12, so that a type (b) record may be obtained, in particular when the reloading is not immediate. A third type of record, type (c) of Fig. 11.21, is obtained with metals subjected to dynamic strain aging at elevated temperature, 100 °C < T < 250 °C. This is evidenced by a slight overshoot in load immediately following the unloading slope. This last effect can be seen in the A 106 steel of Fig. 11.20 tested at 135 °C. The areas under the load–displacement curve of Fig. 11.17 at each load-unload step provides the value of the J associated with the corresponding Δa. The final result of the unloading compliance procedure is the J-R curve shown in Fig. 11.22 [21]. Data obtained with the unloading procedure are compared with those obtained with the multiple-specimens technique. The two methods yield consistent results. Also shown is the blunting line of Eq. (11.26) or Eq. (11.53) whose intersection with the R curve provides the value of the critical J_{Ic} that, in this case, varies from 236 to 309 kJ/m^2.

The data obtained with the multiple-specimen method are within the dispersion band of those relating to two C(T) specimens used with the unloading compliance method. It should be also considered that the unloading compliance procedure, like that of multiple specimens, reduces the effects of metallurgical variability on the assessment of J_c. This is because it increases the process volume by the progressive growth of the crack that continuously scans more and more material that is equal to that of the process volume associated with the single act of growth multiplied by the number n of unloading–reloading events. Therefore, a J–R curve obtained with a single specimen will have to compare well with the other obtained through a multiple specimen procedure. However, minimization does not mean elimination. In materials containing many non-metallic impurities, such as pipe steel A 106 (see Fig. 2.42), few

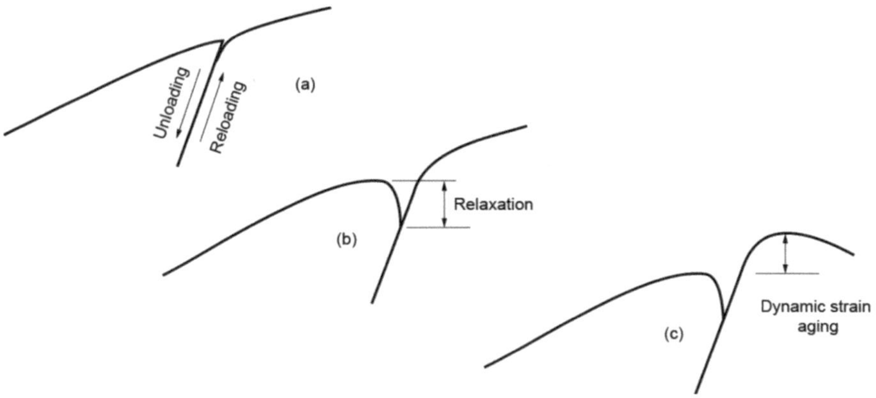

Fig. 11.21 Segments of load–displacement records showing time dependent effects

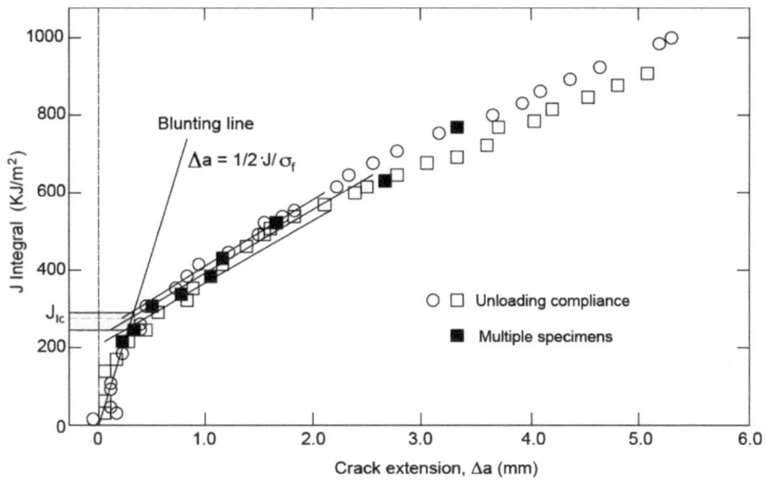

Fig. 11.22 J-R curve for A 106 steel at 135 °C obtained with the unloading compliance and multiple specimens technique [21]. Also shown is the blunting line whose intersection with the *R* curve yields the critical J_{Ic}

specimens will not be sufficient to fully capture the metallurgical variability. Note that the curvature of the *R* curves of Fig. 11.22 is upward as it should be because the data have been corrected using the procedure outlined in Sect. 11.5. Also note that in the initial linear phase of loading, Fig. 11.20, partial loading–unloading steps have almost the same slope indicating almost the same compliance. This is because, as said, in this initial phase the crack doesn't grow but is subjected to a pure blunting that results in an apparent change of compliance.

However, it is very difficult to assess this change through the consequent variation of the load-unload slopes. This is why the blunting line construction is not based on the experimental measurements of the stretched zone but is expressed through a pre-defined line equation or through calculation of *J*, as shown in Sect. 11.10. As the crack becomes very short, the compliance equation becomes less sensitive to crack length, but the specimen limit load also increases, which increases the length of the allowable elastic unloading, and the total effective crack length measurement resolution is marginally affected.

11.7 Stretched Zone and Blunting Line

The examination at the scanning electron microscope (SEM) of the fracture surface of a specimen containing a crack, reveals as already shown in Sect. 6.4, that even a completely brittle fracture is always preceded by a ductile tearing of the material, afterwards the fracture occurs. Examples of this have been reported and identified precisely as ductile tearing in Sect. 6.4 and shown in Figs. 6.23 and 6.24. However, within this initial plastic deformation ahead of the crack tip that precedes fracture or stable crack growth, more or less pronounced depending on the temperature and/or the ductility of the material, it is possible to observe a narrow featureless zone immediately adjacent to the fatigue precrack. This microscopic area that does not possess any known fractographic feature of either brittle or ductile fracture, was first observed by Wolf [22] and Spitzig [23] and has been called stretched zone. This part of the initial apparent growth was assumed to be associated with crack opening (blunting) and is primarily a crack front geometry change rather than actual material separation. An example of stretched zones observed in 7178-T651 aluminum alloy and type 403 stainless steel is shown in Fig. 11.23 a and b, respectively [24].

The width of the stretched zone depends on the temperature and/or the ductility of the material. Bates and Clarck [24, 25] reported stretched zone sizes in 1-inch-thick 1T-C(T) specimens of Ni–Mo–V ranging from $2.4 \cdot 10^{-4}$ in. ($6.1 \cdot 10^{-3}$ mm) to $7.7 \cdot 10^{-4}$ in. ($1.95 \cdot 10^{-2}$mm) and to $50 \cdot 10^{-4}$ in. (0.13 mm) at temperatures of –73 °C, –18 °C and 24 °C, respectively. Increasing the specimen size from 1-inch to 2-inch-thick caused a decrease in stretched zone width from $50 \cdot 10^{-4}$ in. (0.13 mm) to $30 \cdot 10^{-4}$ in. (0.76 mm), at RT. These observation showed that the more brittle the material (lower temperature or larger specimen size), the smaller the width of the stretched zone. By lowering the temperature to observe a totally brittle cleavage fracture, the ductile tearing may disappear almost at all, but the stretched zone always remains, as shown in Fig. 11.24a and b for 1T CT specimen of A508 Cl3 steel broken at –20 °C and – 100 °C, respectively [26]. The corresponding J_{Jc} values were 63.7 kJ/m^2 and 22 kJ/m^2, respectively. The two types of crack extension, one apparent and the second real, i.e., blunting and stable tearing, are schematically shown in Fig. 11.25. The stretched zone region (SZ) appears to be at an angle of about 45° whereas the stable tearing, Δa, seems to connect this region to the following unstable fracture that appears flat.

It must be said that the transition between the SZ and the material tearing that precedes fracture is not always clearly defined. Therefore, also the initial straight line or blunting line may be of difficult determination. Measurements of the stretch zone, Δa_{sz}, obtained by Clark and Bates [24, 25] on Ni–Mo–V steel, HP 9–4–25 steel, A533 steel, Type 403 Stainless steel, 7079-T6 aluminum and 5456-H321 Aluminum have been collected together with data obtained by Iwadate et al. [26] on A508 carbon steel and plotted in Fig. 11.26 as function of the K_Q/σ_s ratios.

K_Q denotes the provisional fracture toughness which is not necessarily a valid K_{Ic} because it may not comply with the size requirements expressed by Eq. (9.50).

The stretched zone data seem to be aligned along the straight line or blunting line of equation.

$$\Delta a_{sz} = 10^{-3} \cdot \left(\frac{K_Q}{\sigma_s} \right)^2 \tag{11.49}$$

Recalling Eq. (10.32).

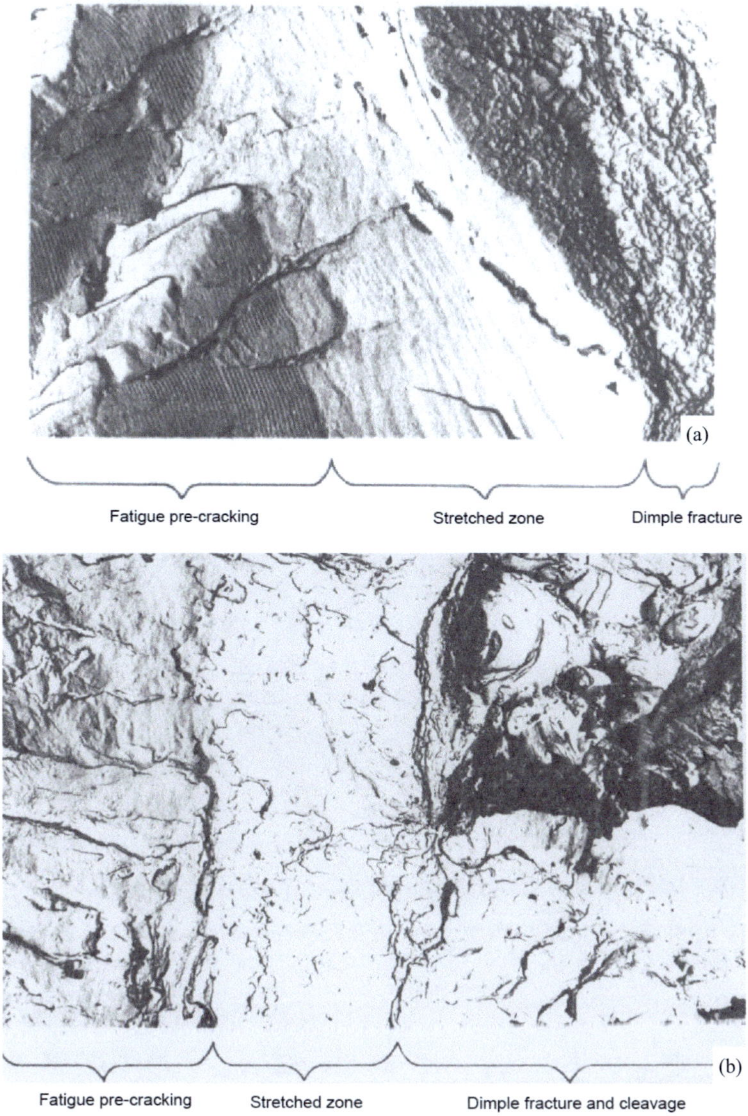

Fig. 11.23 Fractograph of the stretched zone observed in: **a** 7178-T651 aluminum and **b** in type 403 stainless steel [24]

$$\begin{aligned}\Delta a_{sz} &= 10^{-3} \cdot \frac{JE}{\sigma_s^2} \\ &= 10^{-3} \cdot \frac{J}{\sigma_s^2 \varepsilon_s}\end{aligned}$$

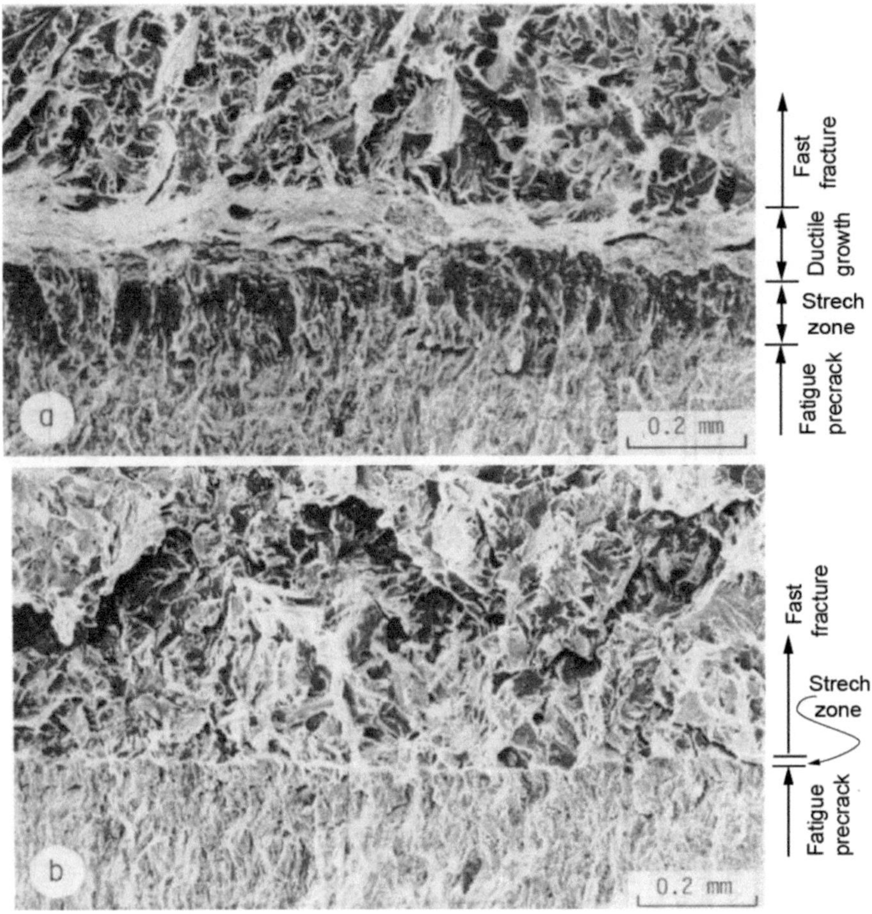

Fig. 11.24 SEM fractographs of 1T CT specimen of A508 GrB steel tested at: **a** –20 °C and **b** –100 °C, respectively [26]

$$\sim \frac{1}{2}\frac{J}{\sigma_s} \tag{11.50}$$

where ε_s is the strain at yield, $\varepsilon_s \sim 2{\cdot}10^{-3}$. Using the CTOD expression of the stretched zone as in Fig. 11.25, it is

$$\Delta a_{sz} = \frac{1}{2}\frac{CTOD}{\tan(45)} \tag{11.51}$$

Recalling Eqs. (10.20) and (10.40), yields

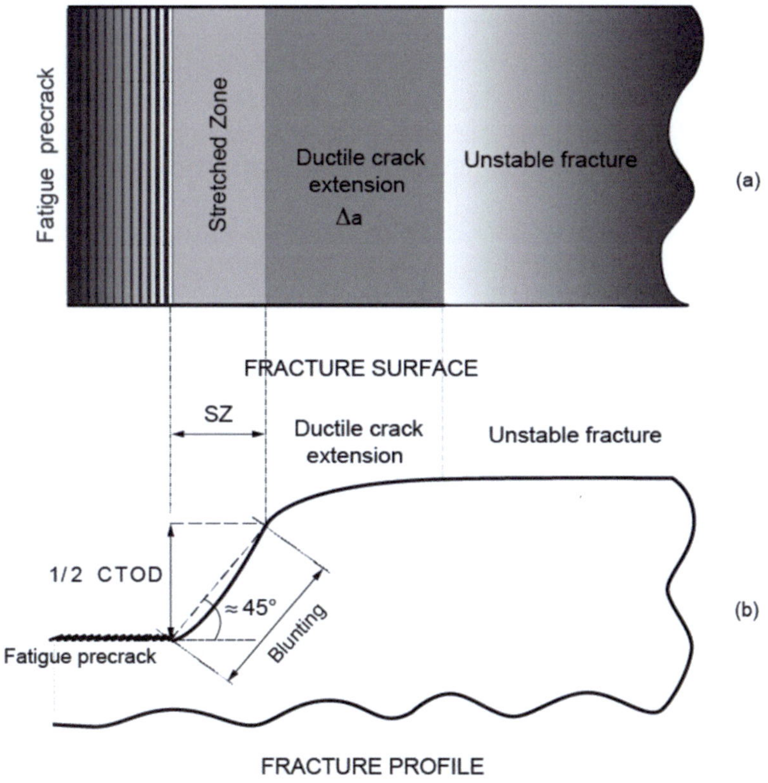

Fig. 11.25 Fracture appearance of a J_{Ic} specimen showing the SZ and the ductile crack extension zone that precede unstable fracture

$$\Delta a_{sz} = \frac{1}{2}\frac{J}{\sigma_s}\psi \tag{11.52}$$

where the factor ψ takes the values given in Table 10.1 depending on the load conditions. If we look for a stretch that precedes the ductile tearing, when the material works at the beginning or in the upper zone of the brittle-ductile transition, we have to consider a value of $\psi \sim 1$. A first massive research of the ψ factor was run by Clarke [27] using 1T, 2T and 4T-CT specimens of A 508 carbon steel having yield stress $\sigma_s = 517$ MPa. The notch tip was not sharp to avoid tearing for a better measuring of Δa_{sz}. Clarke used the unloading compliance procedure to evaluate Δa_{sz}. Results are shown in Fig. 11.27. As it can be seen, all experimental points fall above the blunting line of Eq. (11.52) with $\psi = 1$ indicating a $\psi = 0.4$–0.6. It must be noticed, however, that most data are extremely high, some over 1 mm. This was particularly true for 1T-CT specimens whose arms suffered excessive bending during the test and the corresponding data needed be corrected for rotation, see Sect. 11.15. The correction drew the experimental points closer to the blunting line, but always on

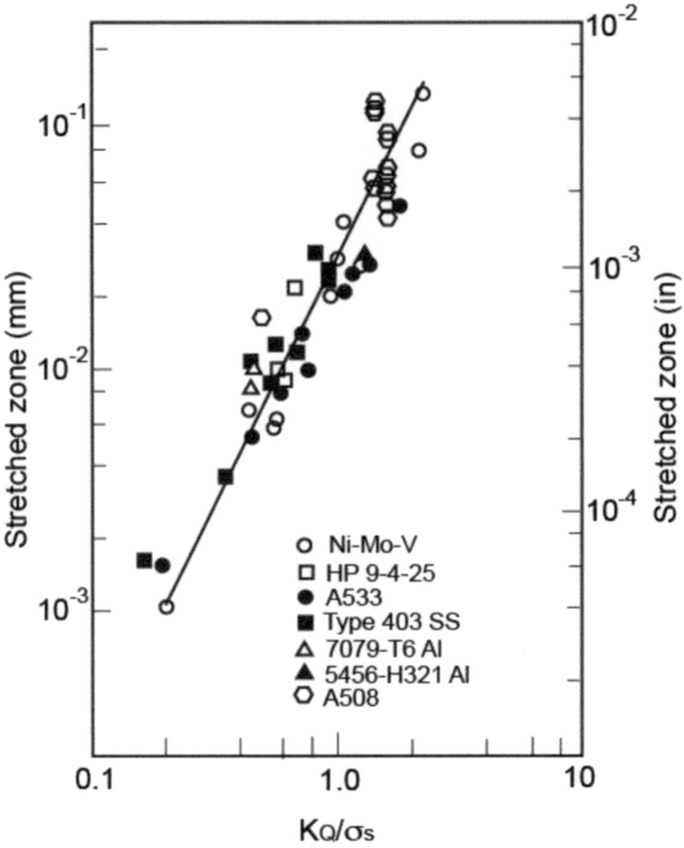

Fig. 11.26 Relationship of width of the stretched zone to relative toughness factor (data from [23–25])

the left. However, all data overestimate the real values by 60–100% because of the notch tip that was not sharp but blunted, as shown by Roos and Kockelmann [28] using specimen of various thickness of 20 MnMoNi 55 containing notches and sharp cracks. Moreover, it must be recalled that Eq. (11.52) and the values of the factor ψ of Table 10.1 are based on linear-perfectly plastic material behavior without hardening.

Therefore, the experimental points found by Clarke definitely suggest a value of $\psi = 0.5$. To overcome the limit posed by the elastic-perfectly plastic barrier, the flow stress, σ_f, was used instead of the yield stress, σ_s. Significant are the results obtained in a benchmark on stretched zone size measurement [29]. Eight different laboratories in different countries were given 0.5T-CT specimens from the same heat of A 533 B2 steel, having a flow stress, $\sigma_f = 632$ MPa, to test at 149 °C and measure the stretched zone, Δa_{sz}. Results are shown in Fig. 11.28. The scatter in the Δa_{sz} values obtained by the 8 laboratories is really impressive which was not seen in the subcritical crack growths Δa nor in the R curves that, on the contrary, were rather

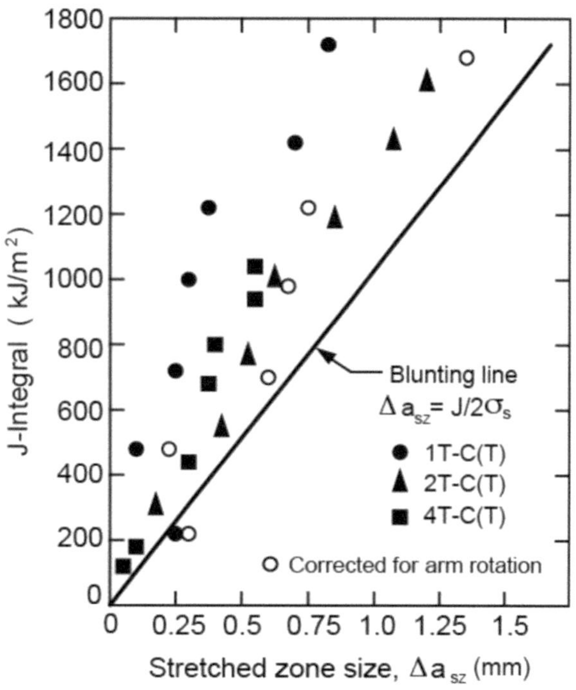

Fig. 11.27 Stretched zone measurements obtained on A 533 steel with CT specimens of various size. Data obtained with 1T-CT were corrected for arm rotation [27]

consistent. Discharging are results obtained by laboratories 4 and 6 that measured negative Δa_{sz}. Labs 6 and 7 has measured uncredible large Δa_{sz} values even at very low J. The trend lines or blunting lines shown have different slope indicated by ψ. The principal cause of data dispersion was found to lay in the combination lab-difficulty of measurement.

Study performed by Kobayashi et al. [30] for a large variety of metal alloys at room temperature, by Mills [31], Davis [32] for stainless steel type A 316 and by Heerens, Cornec and Schwalbe [29] for X 6CrNi1811 stainless steel, A 572 20 MnMoNi 55 carbon steel and Al 2024-FC aluminum gave the results reported in Table 11.1, together with the respective flow stress, σ_f. It is interesting to note the progressive increase of the ψ factor with σ_f, for both steels and aluminum alloys. The relative impact on the measurement of J_Q, which is a provisional J_{Ic}, is shown in Fig. 11.29.

In the high toughness stainless steel, type A 316, J_Q drops by 13% from 910 kJ/m^2 to 788 kJ/m^2, passing from $\psi = 1$ to $\psi = 0.5$, respectively. For the 2 ¼ Cr-1Mo the reduction is about the same while becomes practicaly negligible for a strength low toughness steel like NiCrMoV steel. Iwadate [26] has shown that the apparent crack growth Δa_{sz} continues also after the first local tearing the initiates subcritical crack propagation and then disappears. This evidence is an indirect confirmation of the crack profile that, after an initial phase of CTOA growth, remains constant, see Sect. 10.5 and Fig. 10.20.

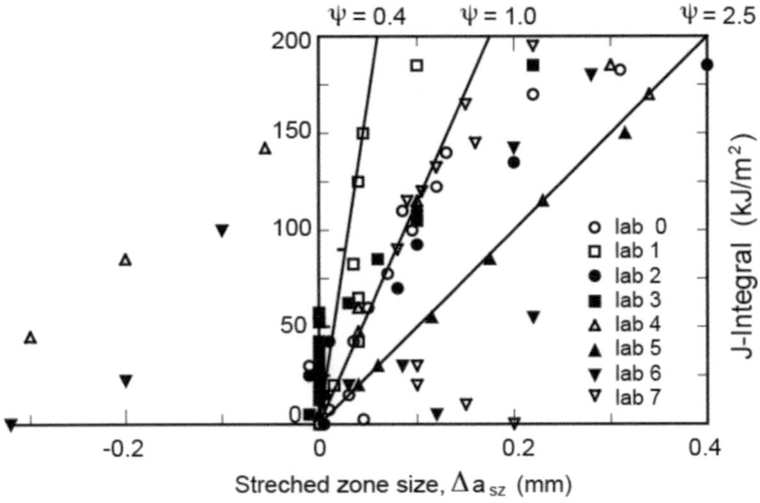

Fig. 11.28 Measurement of the stretched zone done by seven different laboratories around the world on 0.5T-CT specimen machined from the same heat of A 533 steel [29]

Table 11.1 Steels

Alloy	σ_f (MPa)	ψ	$\psi/2$
A 316	380	0,26	0,13
X6CrNi1811	431	0,43	0,215
A 304	496	0,46	0,23
A 572	496	0,44	0,22
20MnMoNi55	545	0,43	0,215
A 533-B1	588	0,5	0,25
HT 60	617	0,49	0,245
10B35 (600 °C)	755	0,5	0,25
HT 80	789	0,59	0,295
4340	1068	0,56	0,28
10B35 (400 °C)	1362	0,88	0,44
10B35 (200 °C)	1744	1,14	0,57
Aluminum alloys			
5083–0	217	0,46	0,23
2024	146	0,49	0,245
7NO1-T6	321	0,5	0,25
2017-T3	384	0,52	0,26
7075-T6	537	0,88	0,44
Titanium			
Ti-6Al-4 V	970	1,58	0,79

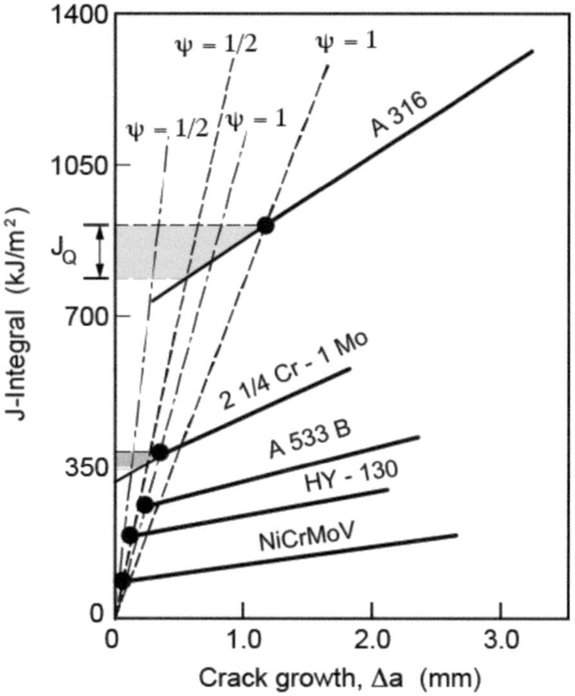

Fig. 11.29 Determination of provisional J_Q from *J-R* curves for different alloys. The blunting lines shown are relative to a ψ factor equal to 1 and 0.5, respectively, for the two separate class of steels: stainless steel and carbon steel (data collected from [29–31])

This randomness in the experimental determination of the stretched zone size, Δa_{sz}, together with the difficulty and the time-consuming process of detecting such small a dimension, of the order of tens of microns, discouraged from proceeding along this way favoring the definition of a blunting line equation, though not rejecting the option of measuring the stretched zone size. The ASTM E 1820 – 08 Standard Test Method for Measurements of Fracture Toughness has adopted a blunting line, called construction line, of slope $2 \cdot \sigma_f$ yielding the final blunting line equation.

$$\Delta a_{sz} = \frac{1}{2}\frac{J}{\sigma_f} \tag{11.53}$$

The ASTM E 1820–08 procedure defines a provisional J_{Ic}, called J_Q, obtained from intersection of the *R* curve with the line parallel to the blunting line of Eq. (11.53) but shifted by 0.15 mm, as schematically shown in Fig. 11.30. The provisional J_Q becomes a valid J_{Ic} after the dimensional verification, as described in the next Section.

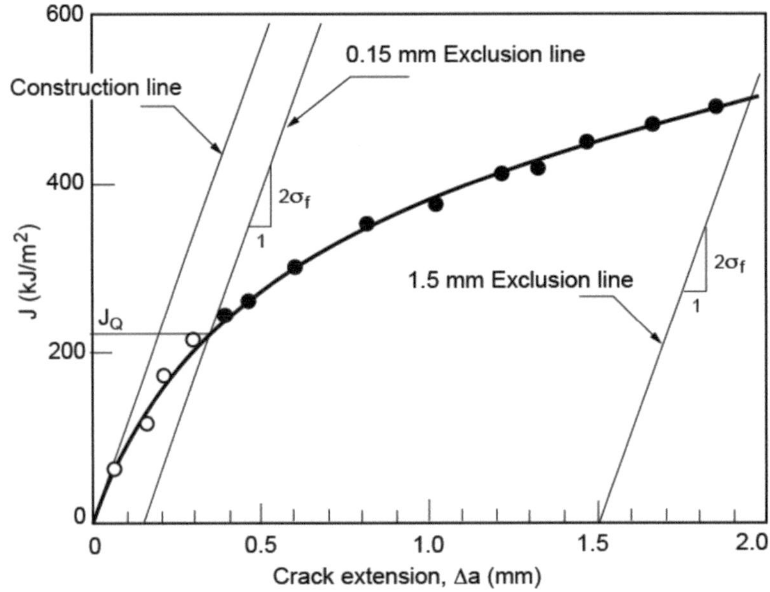

Fig. 11.30 Determination of provisional J_{Ic}, indicated as J_Q, from the intersection of the 0.15 mm exclusion line with the *R* curve

11.8 Validity Limit of *J*-Integral

The existence of the *HRR* stress field makes the transition from LEFM, where there is a *K* dominance stress field is around the crack tip, which remains sharp, to EPFM in which a nonlinear elastic *J* field dominates and the crack blunts and grows steadily till fracture eventually occurs. In both cases, the stress fields have a dominant singularity that, for the *HRR* field, is of the type $1/r^{(1/1+n)}$ with the *J*-integral measuring its amplitude, while for the former is of the type $1/\sqrt{r}$, whose amplitude is defined by the stress intensity factor *K*. We know that LEFM as a single parameter theory: the *K*, requires that the plastic enclave at the crack tip be very small, practically negligible, and this happens for $\beta < 0{,}4$ (see Sect. 9.8) or when the thickness *B* of the piece containing the crack and other dimensions comply with Eq. (9.50). The question that now arises is, obviously, about the limits within which the *J*-integral and its defined *HRR* stress field hold their validity, and, like *K* but in a smaller region containing a nonlinear field, Fig. 10.4, define a unique criterion of critical independence of the geometry or loading condition. The question, this time, is certainly more complex because of the double nature of the *J:* stationary and non-stationary, as schematized in Fig. 10.5 with the two essential moments of initiation or critical and propagation of stable tearing that ends with the final fracture. The *J* stationary is the J_{Ic} at the moment of stable tearing initiation, the non-stationary *J* provides the *R* curve associated with the crack growth. Therefore, we will have to define two conditions of validity: one for the stationary crack and another that regulates subcritical crack growth, Δa. At variance with LEFM where the only condition required was that of plain strain to restrain the crack tip plastic zone within negligible size, with EPFM the assumptions, conditions and limitations are more complicated:

1. plain-strain conditions are required at the crack tip fracture process zone, though not so stringent,
2. concentrated slips emanating from the crack tip to nearby boundaries in front or at 45 deg through the thickness have to be avoided that would cancel the HRR field,
3. the crack growth should be confined to some extent to avoid that energy release from real elastic–plastic material behavior may be too different from that coming from the non-linear elastic theory on which J is based,
4. cleavage would not intercede on the J–R curve.

At variance with LEFM, conditions 1. to 4. cannot be found simply by testing, as discussed in Sect. 9.8. Begley and Landes [1, 2, 33] presented experimental data for a wide range of specimen sizes and several crack geometries which showed that the critical J_{Ic} is independent of the extent of plastic deformation and specimen geometry as long as some minimal size requirements, as suggested by Paris [34, 35] were met.

$$b \geq 25\frac{J}{\sigma_s} \tag{11.54}$$

Subsequent work by Begley and Landes [36] reported that the J-resistance curves for a CC(T) panel and a C(T) specimen were quite different though the size requirements exceeded Eq. (11.55). Nevertheless, the correspondent J_{Ic} was the same. McMeeking and Parks [37], following a numerical approach, found that, for low hardening materials the J crack tip stress fields for SEC(B) and SE(T) specimens, Figs. 9.5 and 9.19, provided the same SSY stress field if condition (11.55) were verified. Even more surprising was the fact that this equivalence also existed for $n = 0$, apparently contrary to what was said in Sect. 10.7 about the need for even a small but finite value of work hardening n to sustain the singularity of the apical field expressed by Eq. (10.47) and, therefore, to support the very existence of J. It should be borne in mind that condition (11.55) has a more general meaning than just referring to the thickness alone. It includes also other dimensions that characterize the geometry considered such as the crack length, a, and the ligament b, the smallest of which must comply with the condition of validity for J. Equation (11.54) will, then, be written in the most general form.

$$a, b, B \geq 25\frac{J}{\sigma_s} \tag{11.55}$$

The same calculations done by McMeeking and Parks [37] showed that passing from a geometry SEN(B) in bending to a CC(T) in traction the condition (11.54) became more severe.

$$b \geq 200\frac{J}{\sigma_s} \tag{11.56}$$

and, this time, it depended on n that in this case had been assumed equal to 0.1. This is not surprising since we know that panel geometries and in particular CC(T), have a negative *T-stress*, see Sect. 7.17, and, therefore, will require a thickness and a ligament much greater than those required under bending to ensure the dominance of the singular term. Question is about the radius R_{HHR} of the domain around the crack tip in which the small nonlinear deformation field is precisely predicted by the *HRR* field. If we refer to Fig. 8.20, we can assert that the nonlinear elastic field, having to contain the inelastic apical wedge of larger deformations, must be at least greater than $2{\cdot}\delta_t$. Indeed, McMeeking [38], again using a numerical approach, found that in the cases of *SSY* and crack opening Mode I the effects of large deformations were significant within a distance R_{HHR} from the crack tip equal to 2–3 times δ_t and that the independence of the J-integral from the path of integration was ensured only if.

$$R_{HHR} = 3\delta_t \tag{11.57}$$

Since in cases of plain stain δ_t is approximately 0,5–0.25 times J/σ_s, see Eq. (10.20), with the highest value likely to be representative of the highest work hardening materials, condition (11.55) implies.

$$b > 50 \div 100\delta_t \tag{11.58}$$

which combined with Eq. (11.57) yields.

$$R_{HHR,\max} = 0.06 \div 0.12b \tag{11.59}$$

Thus, in an SEN(B) specimen the nonlinear stress field of J-dominance, the HRR field, will extend to a maximum of $6 \div 12\%$ of the ligament b. The same relation derived for a CC(T) for which Eq. (11.56) applies, gives

$$R_{HHR,\max} = 0.01b \tag{11.60}$$

and this actually means that the *HRR* field dominates a region equal to only one percent of the ligament when $n = 0.1$. For a 1″ thick specimen, $R_{HHR,max}$ will not exceed 0.2 mm in depth and it will be difficult, under these conditions, to consider the *HRR* field dominant in all those situations that tend to give fracture according to a mechanism not exclusively of cleavage, but with ductile participation from pore coalescence which, we know, involves areas even a few millimeters in front of the of the crack tip. Once again, the non-singular terms, not present in Eq. (10.47), will take over. Having established the first condition (11.54) and (11.56) for a stationary crack, we move on to deal with the most challenging question of the validity of the J-integral in situations of subcritical growth Δa which, ultimately, is to understand under what conditions and limitations the entire J_R curve found experimentally is actually a characteristic of the material and, like J_{Ic}, independent of the geometry of the system and load conditions. The *J*-integral, being based on the theory of nonlinear elasticity, is not able to represent those situations that involve non-linear deformations and unloading (11.31).

The former will certainly exist in the plastic wedge in direct contact with the crack tip, of which we have already said, the latter will accompany the subcritical growth Δa. With reference to Fig. 11.31, while according to nonlinear elasticity the *J*-integral expressed in its most general form as $J = J_{el} + J_{pl}$, is proportional to the difference between the complementary energies for two systems with cracks of length a and $a + \Delta a$, respectively, area OABO of Fig. 11.31, the energy that would instead be released by a real system with elastoplastic behavior following a growth Δa is associated with the elastic unloading CABC of FigIt is clear that the two areas OABO and CABC will be closer to each other the smaller the subcritical growth Δa. Therefore, a limit must be placed to Δa not just to the single act of growth as to the total growth to which the entire *R* curve refers. The more the crack grows, the more the crack tip plastic area will increase in proportions. Furthermore, we have also to consider the so-called elastic wake in which the elastic unloading takes place, which contains the anelastic wake associated with the large apical deformations, as schematized in Fig. 11.32. If the area of this region grew to the point of competing with that dominated by *J*, then the *HRR* field may cease to characterize with its singularity the fracture behavior of the system in which the crack is present. Since this area has an extension comparable to Δa, this last must be limited so that

$$\Delta a << R_{\mathrm{HHB,\,max}} \tag{11.61}$$

and recalling Eq. (11.61) for a system under bending.

$$\Delta a < 0.06 \div 0.12\, b \tag{11.62}$$

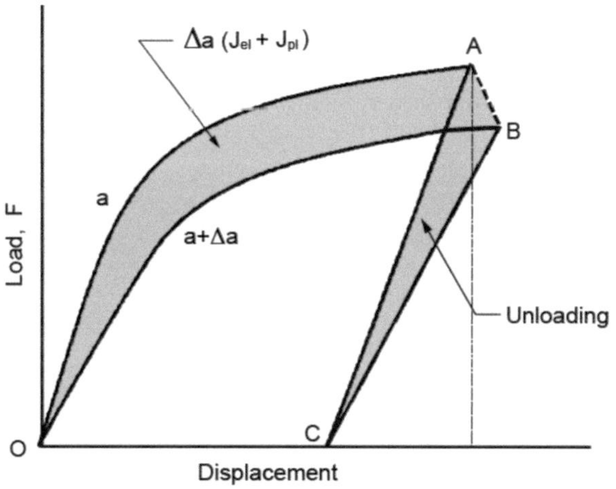

Fig. 11.31 Schematic of the energy release following a stable growth Δa according to the nonlinear theory of elasticity and the elastic–plastic theory

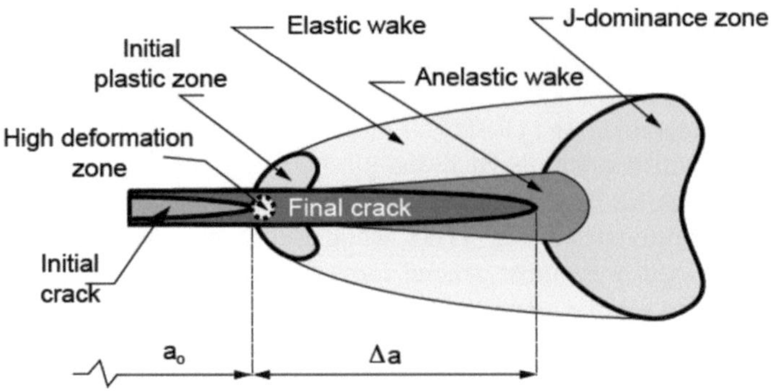

Fig. 11.32 Schematic of the elastic wake following a subcritical crack growth

Hutchinson and Paris [39], again following the theory of nonlinear elasticity, found that for a material having a work hardening n the variation of the deformation field, $d\varepsilon_{ij}$, resulting from an elemental crack growth da was equal to

$$d\varepsilon_{ij} = K_n J^{\frac{1}{n+1}} r^{-\frac{1}{n+1}} \left[\frac{1}{n+1} \frac{dJ}{J} f_{ij}(\theta) + \frac{da}{r} \beta_{ij}(\theta) \right] \tag{11.63}$$

with K_n proportionality factor and $f_{\mathrm{ij}}(\theta)$ and $\theta_{ij}(\beta)$ angular factors in a polar system r,θ of comparable value close to one. The first term in brackets in Eq. (11.63) represents the increase in the field of proportional deformations, as a result of the increase of the applied external load necessary for the continuation of subcritical growth, $d\varepsilon_{\mathrm{ij}} \sim \varepsilon_{ij}$, the second is instead that of non-proportional deformations, resulting from the subcritical growth itself, da. Equation (11.63) clearly indicates that if we want J to be the only parameter to characterize the apical stress and deformation fields or that the dominant field with a singularity expressed by J has mainly proportional stresses, of the two terms in parentheses the first must be predominant with respect to the second.

$$\frac{dJ}{J} >> \frac{da}{r} \rightarrow \frac{dJ}{da} >> \frac{J}{r} \tag{11.64}$$

Introducing a unit of length L

$$L = \frac{J}{dJ/da} \tag{11.65}$$

defined as the distance Δa at which the value of J measured on the tangent to the curve R in J_{Ic} doubles from J_{Ic} to $2{\cdot}J_{Ic}$, as shown in Fig. 11.33. We will have that Eq. (11.64) can be written as

$$L << r \tag{11.66}$$

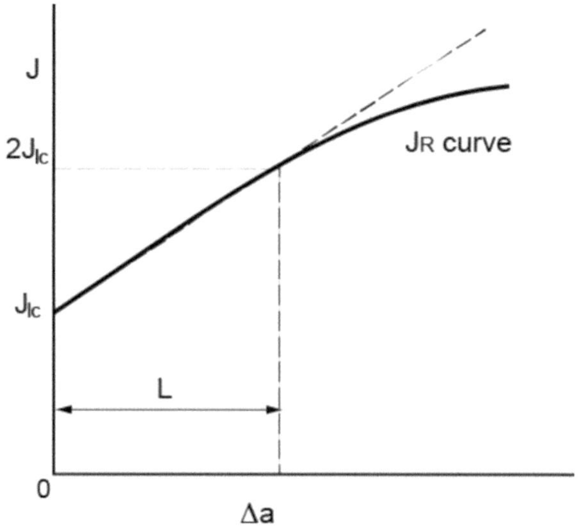

Fig. 11.33 Operative definition of the length L at which the value of J, measured on the tangent in J_{Ic} to the R curve, double from J_{Ic} to 2 J_{Ic}

If, therefore, it is also.

$$L << \Delta \tag{11.67}$$

then there will exist a region between L and R_{HHR}, around the crack tip

$$L << r < R_{HHR} \tag{11.68}$$

in which we can expect a minimal difference between the crack tip plastic deformations field predicted by the theory of nonlinear elasticity and that deriving from the theory of plasticity, so that in this region the J will actually uniquely define the crack tip singularity.

Once again it must be remembered that all what has been said is based on the assumption of the existence of some work hardening n, albeit small, but necessarily non-zero, so that the dominant singularity measured by J does not depend on the geometry of the system and J itself is unique. We now introduce a new dimensionless parameter ω thus defined.

$$\omega = \frac{b}{J_{Ic}}\left(\frac{dJ}{da}\right) >> 1 \tag{11.69}$$

Remembering Eqs. (11.60) and (11.67) we will have

$$\omega >> 8 \div 16 \tag{11.70}$$

This condition was first proposed by Hutchinson and Paris [39] and satisfies point 3. request. Under the given condition the applicability of the nonlinear theory or deformation theory is appropriate, the condition for the so called "J-controlled crack growth" are met, and J may be defined either by its contour integral, Eq. (10.32), or compliance counterparts, Eq. (11.8). The meaning of Eq. (11.69) is that in the presence of subcritical growths, the *HRR* field and, therefore, the J must grow sufficiently rapidly, as per condition expressed by Eq. (11.70) to ensure that the region R_{HHR} is always dominant over the inelastic one. It is not yet clear what the minimum value ω must take. It will certainly depend on geometry and work hardening n. An important consequence of the validity conditions for the growth of a crack controlled by J is that the entire resistance curve J–R obtained in a LGC condition be equal to that relative to a *SCG* situation or of SSY, both under plane strain and plane stress conditions. This means that we can use the J_{Ic} to derive a value of K_{Ic} not only in Region I of Fig. 11.46, as we already knew, see Eq. (10.40) in which the fracture is brittle and occurs along the blunting line, but also in the transitional region, Region II, with the obvious advantage that derives from conditions in Eq. (11.57) compared to that of Eq. (9.50) already seen for the validity of K_{Ic}. From condition (11.55), in fact, it can be deduced that the minimum thickness B for the validity of a $J_{Ic} \rightarrow K_{Ic}$ test will be

$$\begin{aligned} B &= 25\frac{J_{Ic}}{\sigma_s} = 25\frac{K_{Ic}^2}{E\sigma_s} \\ &= 25\varepsilon_s\left(\frac{K_{Ic}}{\sigma_s}\right)^2 \end{aligned} \tag{11.71}$$

ε_s being the strain at the elastic limit, equal to σ_s/E, which we can consider equal to about $2 \cdot 10^{-3}$, and Eq. (11.71) becomes

$$B \geq 0.05\left(\frac{K_{Ic}}{\sigma_s}\right)^2 \tag{11.72}$$

i.e. 50 times less restrictive than Eq. (9.50). The K_{Ic} thus obtained is normally referred to as K_{Jc} to indicate its derivation from a measure of J_{Ic}. Recently, Koppenhorfer and Dodds [40] found that, in truth, condition (11.54) would not be entirely independent of the value of work hardening coefficient n. For $n > 0.2$ it can be considered true, but for values of n smaller, even much less than 0.1, the multiplying factor of J/σ_s must pass from 25 to 200 with which Eq. (11.72) becomes

$$B \geq 0.5\left(\frac{K_{Ic}}{\sigma_s}\right)^2 \tag{11.73}$$

again expressing a restriction 6 times less limiting than (9.50).

11.9 Verification of Validity Limits for J

An extensive experimental and analytical campaign begin in order to verify the adequacy of the limits set by Eqs. (11.55), (11.62) and (11.71). Joyce et al. [41] and Joyce and Hackett [42] have long tested C(T) specimens of four different alloys having various sizes, arriving to impose subcritical growths, Δa, up to even 60% of the length of the ligament b, in clear contrast with the requirement set by Eq. (11.62). The method adopted for the construction of the R curve was that of unloading compliance. The alloys used were A 533 B2 and A 533 B H13 steel, 448 MPa yield strength, σ_s, and 534 and 520 MPa flow stress, σ_f, respectively, A 710 steel of 510 MPa yield strength and 555 MPa flow stress and 3-Ni having $\sigma_s = 613$ MPa and $\sigma_f = 672$ MPa, all tested at room temperature except A 533 B steel which had been tested at 150 °C. All specimens had a 0.6W depth initial crack size. Figure 11.34 shows the J–Δa results thus obtained. It is certainly difficult to assess either the presence or absence of a field singularity just looking at the experimental curves, besides that, regardless of the size of the specimen and the material used, all tests show a surprising uniform and continuous trend also well beyond the limits imposed by size requirements, from which nothing could be deduced about a possible loss of singularity of the *HRR* field. Figure 11.34 shows, for each specimen size, the validity range J-Δa defined by Eqs. (11.55) and (11.62). It can be observed that tests went well outside this field, reaching also, for steel A 533 B-02, about 60% of subcritical growth Δa and, for A 710 steel, about three times the maximum value allowed for J ($J_{max} = b{\cdot}\sigma_s/25$). The only exception to the regular trend was observed in one of the 1/2 T–CT specimens of 3-Ni steel that showed a flex at about 6 mm of crack extension, Δa, with a sudden surge in the final phase. A 533 B steel specimens, in particular, do not seem to be affected by le large crack extension, Δa, well beyond the theoretical allowance area for J, presenting an amazing continuity and overlapping of data coming from different specimen sizes. It must also be considered that in all cases examined the unloading compliance method led to subcritical growth provisions confirmed, with great precision, by the measurements carried out on the surfaces of the specimen broken open and this up to 50% of subcritical growth.

The only exception was observed on A 710 steel for subcritical growth from 40 to 60% where the value measured after the test was of about 20% higher, but in this case the specimen showed very clear signs of extensive deformation with marked contraction and reduction of thickness ahead of the notch.

On the contrary, when the same experimental data were plot in terms of *J-modified*, J_M, see Sect. 10.9, it was found a different course of the resistance curves, *R*, not only as a general trend, but also as a behavior between different alloys, as can be seen in Fig. 11.35 [41].

Now, three distinct regions can be distinguished: the initial and well-known region of crack tip blunting followed by that of real tearing resistance, the *R* curve, which would confirm the existence of the field singularity, and finally a third and new phase of exponential growth.

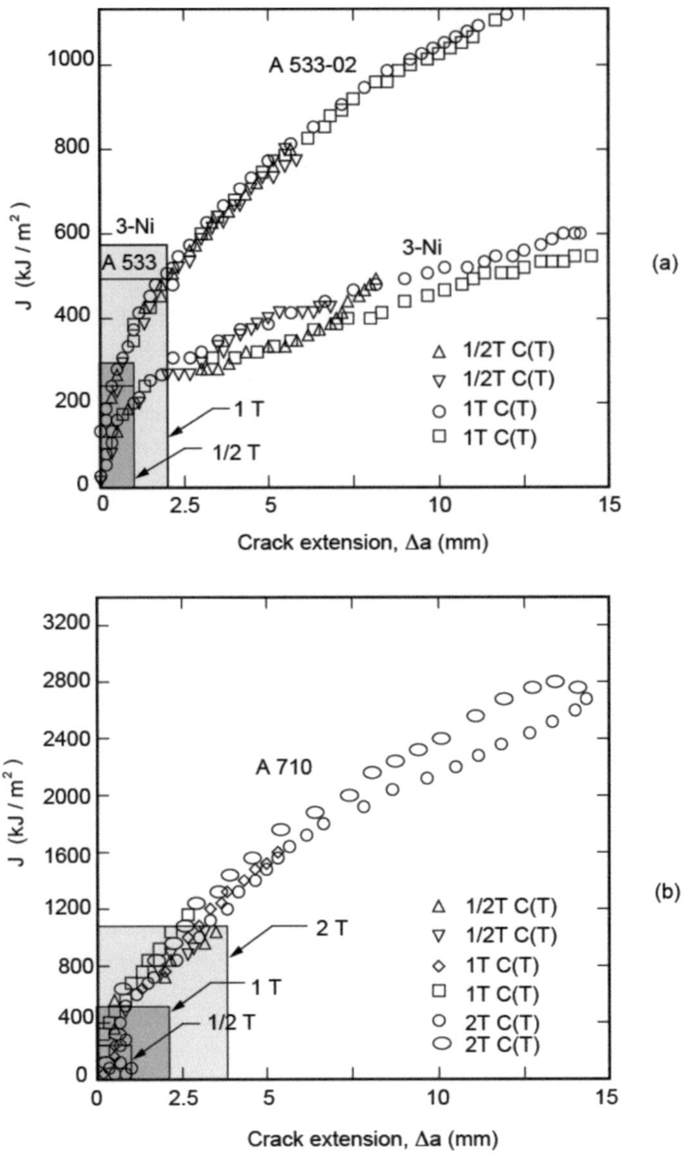

Fig. 11.34 Experimental *J-R* curves for three different steels obtained by the unloading compliance method. Also shown are the validity limits imposed to *J* indicated, for each steel and size, by the correspondent tinted area [41]

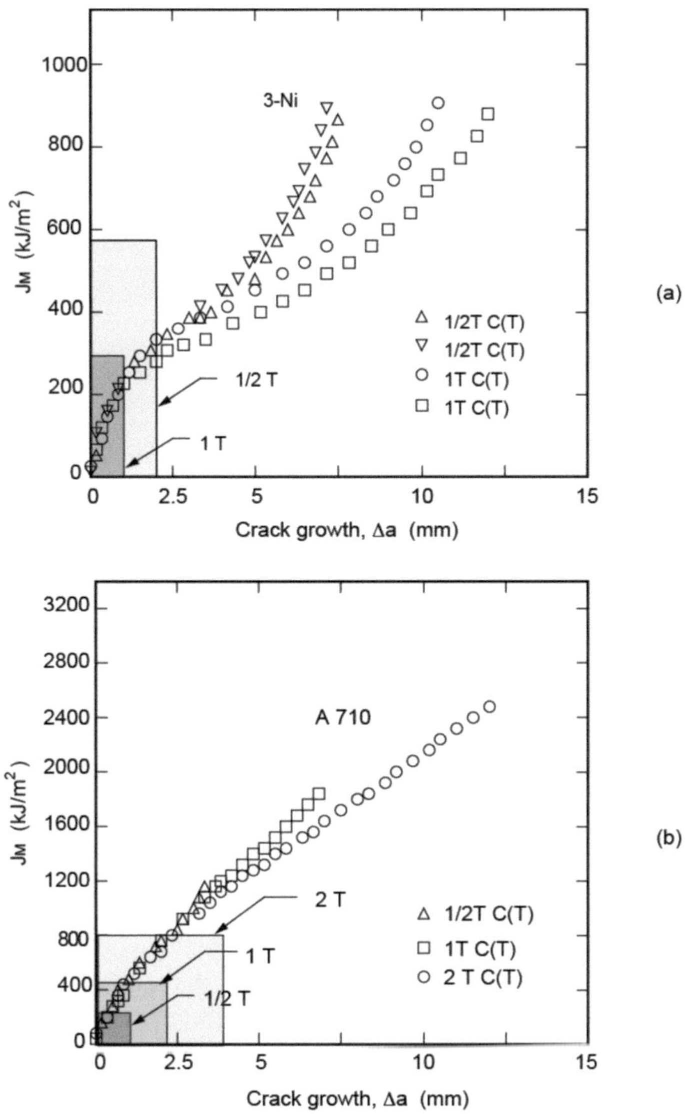

Fig. 11.35 Experimental J_M-R curves for: **a** 3-Ni steel and **b** A 710 steel. The former presents an upward bent well beyond the validity area that is lacking in the latter steel [41]

This time it is possible to recognize, at least for 3-Ni steel, what could be the moment of loss of the *HRR* field as the dominant field and, therefore, of validity of J.

This moment is probably represented by the inflection point of the J_M curves that also show a dependence on the thickness of the specimen, which, however, starts well beyond the region of validity of the J, allowing subcritical growths up to 20–30% of ligament b. What seems to be a simple hypothesis acquires more validity looking at

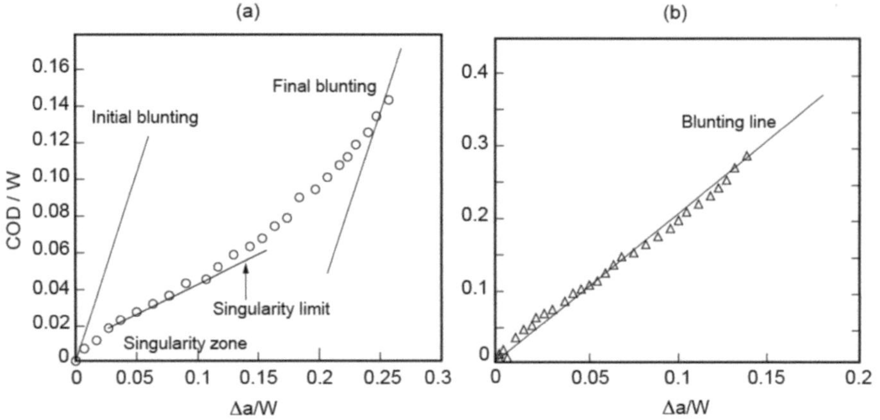

Fig. 11.36 COD variation with crack growth in: **a** 3-ni steel and **b** A 710 steel. Note in **a** the final re-blunting of the crack [41]

the trend of the *COD* as a function of the growth Δa. This is shown in Fig. 11.36 [42]. We can observe, for 3-Ni steel, the same trend in the *COD* obtained on half-inch C(T) specimens which confirms the onset, after a phase of real tearing, of a new very consistent crack tip opening, therefore of loss of triaxiality and of strong prevalence of large non-proportional deformations. It is particularly indicative that the upward trend identified on the J_M curves occurs for the same subcritical growth value Δa on the *COD* curve: the same phenomenon is confirmed by two distinct indicators. For A 710 steel, Fig. 11.36b, the experimental points are placed almost exactly on a straight line that has just the slope of the initial blunting line. As for A 533 B steel, it was observed the same trend of 3-Ni reflecting the possible loss of singularity of the HRR field, after about 30% of subcritical crack growth. There is therefore a certain capacity of steels, but not in all or in equal measure, to preserve a crack tip singularity field even in the presence of large subcritical growths.

This surprising capability of *J* to continue to be valid as unique parameter that characterizes the plastic stress field, even beyond the theoretical limits, is now interpreted with the fact that after the initial stretching and blunting of the crack tip that becomes a notch, a phase of tearing takes place which, in fact, resharpens the notch, as shown in Fig. 11.37 [28]. Recent studies conducted through finite element calculation, by Varias and Shih [43] on the crack tip fields under *SSY* and stationary growth conditions and by O'Dowd et al. [44] for the transient in which the CTOA value increases, see Sect. 10.5, have shown that the *re-sharpening* of the notch leads to the development of a state of high apical triaxiality as soon as the tip of the crack extends inside the material previously left in the elastic state and this even when growth starts from conditions of low triaxiality with T-Stress < 0.

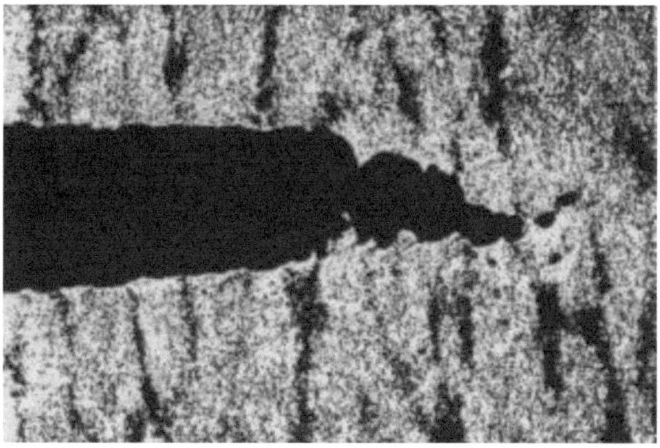

Fig. 11.37 Crack tip blunting followed by re-sharpening of the crack tip in a C(T) specimen of 20 MnMoNi55 at 80 °C [28]

11.10 Measurement of the J-integral and R Curve: The Pseudo Energy Method

The most general expression for J is

$$J = J_e + J_p \tag{11.74}$$

where J_e is the elastic component of J

$$J_e = \frac{K^2}{E'} \tag{11.75}$$

with $E'' = E/(1-\nu^2)$ being the Young's modulus in plain strain conditions. The plastic component of J can be calculated from the plastic pseudo energy, i.e., the area under the curve force-plastic displacement, F–Δ, or moment–angle of rotation, M–θ, relative to the plastic component, as shown in Sect. 11.2. It is evident that the same procedure can be applied to derive the entire R curve by measuring the pseudo energy at different load step. To this purpose the ASTM Standard E1820-08a describes a procedure to the plastic component J_{pi} at each load increment i can be assessed using

$$J_{p,i} = J_{p,(i-1)} + \frac{\eta_p}{b_{(i-1)}}\left[\frac{A_{p,\mathrm{i}} - A_{p,(i-1)}}{B_N}\right]\left[1 - \frac{\gamma_p(a_i - a_{i-1})}{b_{(i-1)}}\right] \tag{11.76}$$

were $A_{p,i}$ is the pseudo energy or area under the load versus plastic load line displacement curve at increment i, B_N the net specimen thickness at the side groove roots, η_i the plastic η factor at crack length a_i, b_i the remaining ligament $b_i = W$–a_i and W the specimen width.

The γ factor is given by

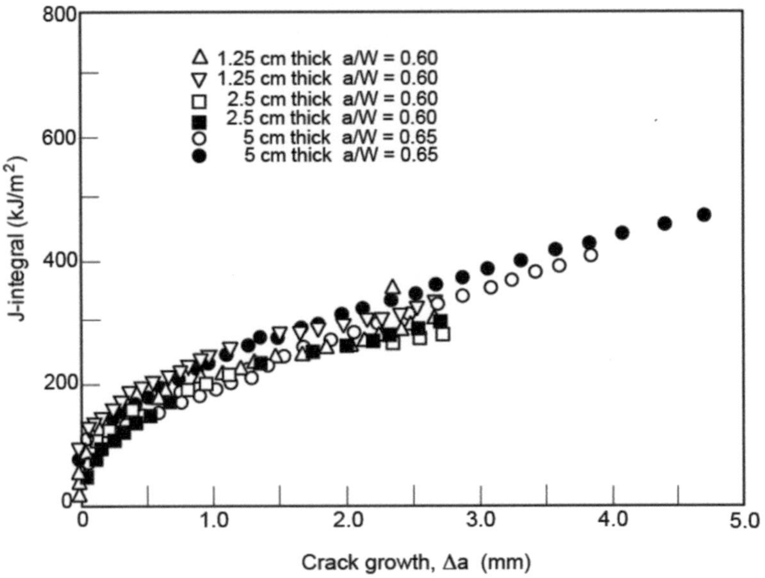

Fig. 11.38 *J–R* curves for HY-100 steel deep-cracked SE(B) specimens of various sizes [45]

$$\gamma_i = \left[\eta_i - 1 - \frac{b_i}{W}\frac{\eta_i'}{\eta_i}\right] \tag{11.77}$$

evaluated at each crack length a_i, and

$$\eta_i' = \frac{\partial \eta_i}{\mathrm{d}(a_i, W)} \tag{11.78}$$

For the deep cracked SE(B) specimen it has already been found, see Sect. 11.2, $\eta =$ 2 for the elastic case, and

$\eta_p = 1.9$ if Δ is the load-line displacement, otherwise.

$\eta_p = 3.667–2.199 + 0.437(a/W)^2$ if the crack mouth opening displacement CMOD, V, is measured instead of Δ, as shown in Fig. 11.6a, while the γ factor is equal to 1.0.

The final *J–R* curve is shown in Fig. 11.38 for HY100 hard steel [45].

11.11 Tearing Modulus

The construction of the entire *J–R* curve by any of the methods described in the previous sections (unloading compliance or pseudo energy) allows better treatment of material behavior of cracked bodies shown in Fig. 10.5 in terms of force-crack growth, $F–\Delta a$, and now converted into a $J–\Delta a$, as shown in Fig. 11.39 [46]. Here we assume that brittle fracture has not occurred because the material presents an intrinsically ductile behavior being an FCC material or, if it is a BCC material, because we are well above the NDT. Note that, in this latter case, it has been shown by Paris and Clarke that plotting the R curve using J/σ_f, instead of J as the ordinate, the trend is nearly independent of the temperature under plain strain condition [47]. After the initial blunting of the crack th*da*at is seen by the system as an apparent crack growth, initiates the stable tearing of the crack that proceeds with an initial slope $dJ/$, almost constant. If size requirements are met, Eq. (11.71), it may be assumed that the J–R curve is size independent, besides material variability always affecting test results. It is clear that the slope of the R curve represents a material characteristic to stable tearing. This characteristic may be expressed by the so-called tearing modulus T.

$$T = \frac{\mathrm{d}J}{\mathrm{d}a} \cdot \frac{E}{\sigma_f^2} = C \tag{11.79}$$

where E is the Young's modulus and σ_f the flow stress of the material. Eq. (11.79) defines a tearing modulus, also called T-material or T_{mat}, as a constant property of the material which means that the R curve of Fig. 11.39 is a straight line. At variance, many experimentally determined R curves appear to be curved concave downwards as tearing progress extensively. This implies that dJ/da, i.e., the tearing slope, decreases with crack extension, Δa. This can be seen in Fig. 11.34a for A 533 B steel.

However, although that portion of the curve is termed "stable tearing" it may not be stable turning into instability except in those configurations where there is substantial bending imposed on the remaining ligament, such as in deeply cracked C(T) or SEN(B) specimens. Within the limits of validity of the J–R curve, the assumption of a linear trend may be realistic. This can be seen in Fig. 11.40 for 5083–0 aluminum alloy [48]. The entire R curve has a variable slope and, therefore, a variable tearing modulus, T, continuously decreasing. However, the initial part of the curve, Fig. 11.40b, has a constant slope. Note the random dispersion of the experimental points in the blunting region, which correspond to the stretched zone sizes. This dispersion was already discussed and shown in Fig. 11.27 for A 533 steel.

11.12 Geometry Effect on *J* and R Curves

The considerations made in Sects. 11.8 and 11.9 on the validity limits of the J-integral and on the unknowns that still remain about its extension to the field of subcritical crack growth, in particular that of the large growths Δa that characterize totally plastic failures, in which the representativeness of a nonlinear elastic theory can no longer be argued, allow to outline some interesting elements. First, if size requirements (11.54) and (11.55) were met, which is also a constraint placed on thickness B.

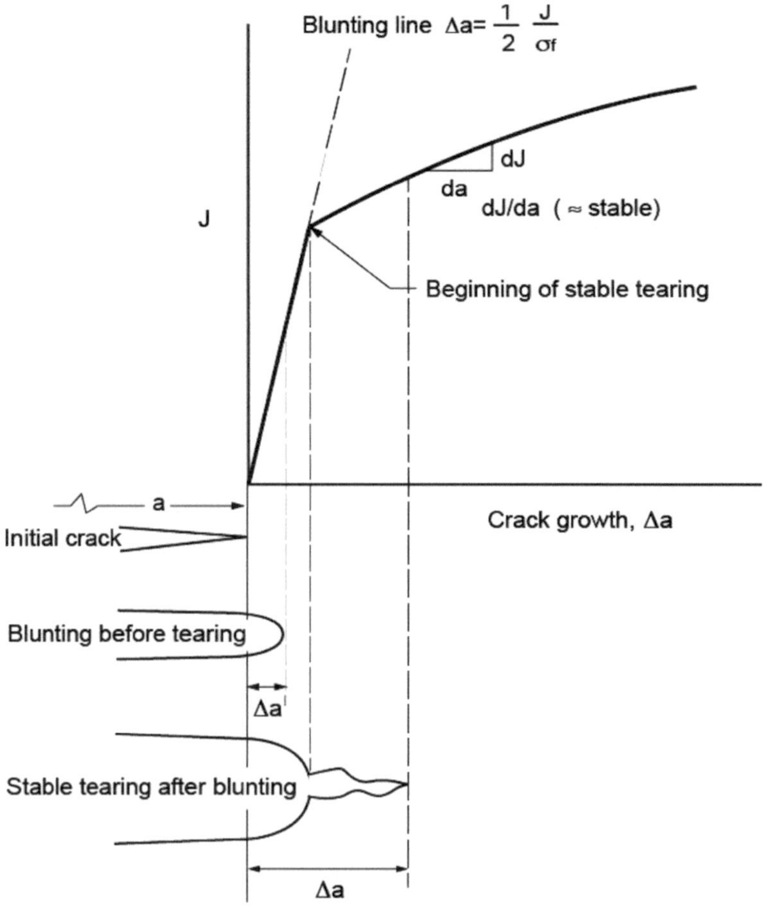

Fig. 11.39 *J–R* curve following Rice [46]

$$B \geq b \geq 25\frac{J}{\sigma_s} \tag{11.80}$$

we can consider that any possible limit to the J-integral does not affect the stationary crack, $\Delta a = 0$, and therefore, J_{Ic} whose assessment will probably be independent or very little dependent

on the geometry of the system on which, instead, depends the trend of the curve J_R and, therefore, the dJ/da and the modulus ω as well, Eq. (11.69).

Figure 11.41 shows precisely this circumstance, namely the independence of J_{Ic} and the dependence of the resistance curve R on the geometry of the system. The J_R curve is obtained at 120 °C using both a 25.4 mm thick panel with a central crack in traction, CC(T), and a 1T C(T) specimen of NiCrMoV steel [49]. The J_{Ic} is worth about 120 kJ/m^2 which means that with a yield stress of ~700 MPa, Eq. (11.55)

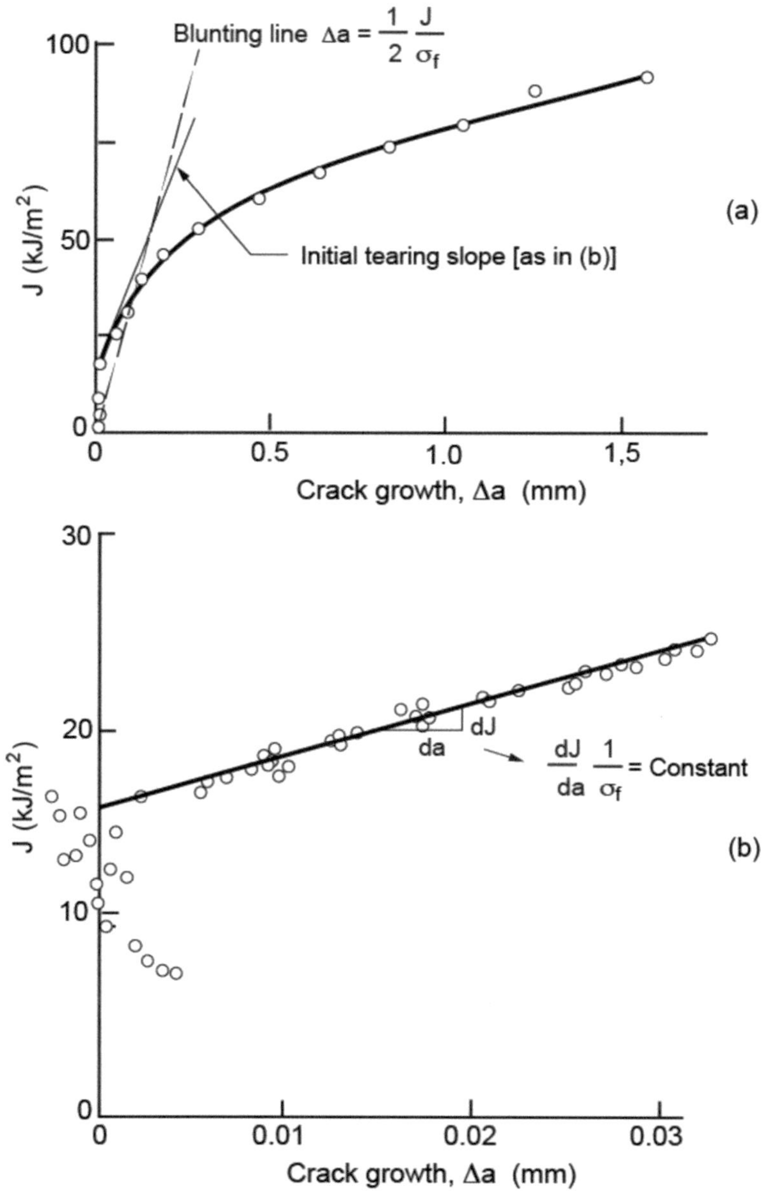

Fig. 11.40 **a** The *R* curve for 5083–0 aluminum alloy, **b** detail of the initial slope of the *R* curve (conv. in S.I. Unit from [48])

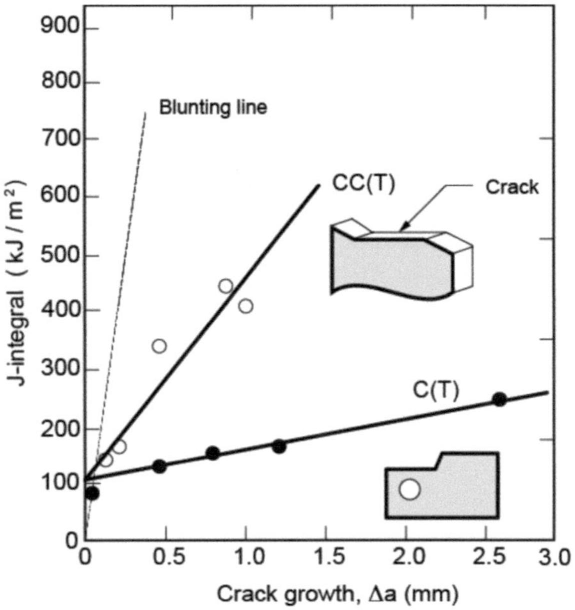

Fig. 11.41 *J–R* curve for NiCrMoV steel at 120 °C using 1T CC(T) and C(T) specimens [49]

requires for the C(T) specimen a minimum thickness *B* of only 4 mm and for the panel CC(T) about 34 mm, therefore slightly larger than that actually used in the tests. In the meantime, it should be noted that the use of *J* for the measurement of toughness is much more convenient from the point of view of the dimensional requirement. For the same type of steel, in fact, at –20 °C it is already necessary at least a 4" (100 mm) specimen, Fig. 9.38 and Table 9.5, to measure a valid K_{Ic} while for a toughness of 155 MPa$\sqrt{m}$, which corresponds to those 120 kJ/m^2 obtained at 120 °C, it takes a 5"–6" C(T) specimen. Returning to Fig. 11.41, it is evident that the value of J_{Ic} is practically the same for both specimens.

On the contrary, the slope of the interpolation line of the experimental data relative to the panel is much larger than that of the compact specimen, revealing an apparently higher tearing resistance and, therefore, a clear effect of the specimen geometry and load conditions. In both cases ω satisfies condition (11.69) being equal to 75 for the CC(T) and 10 for C(T). The blunting line shown in Fig. 11.41 is that with $\psi = 1$ in Eq. (11.52). For this type of steel, it would be more appropriate to assign $\psi = 0.5$ as in Table 11.1 which would double the slope of the blunting line, further reducing the gap between the J_{Ic} measured with the two specimens leaving the possible difference to the metallurgical variability.

The same effect can be observed in Fig. 11.42 for type A 304 stainless steel at room temperature [49], which has much higher toughness and is known to have no transition temperature due to its CFC structure, see Chap. 6 Concluding Remarques. The enormous difference in the behavior of the two geometries beyond the J_{Ic} can be explained through the considerations already made in Sect. 9.3 about the loss of constraint of panel specimens with respect to biaxial specimens and in Sect. 10.10

about the T stress and the Q stress. If, in fact, a geometry has a high value of the T stress, in particular negative, we must expect a loss of dominance of K_I and J_I in the elastic and plastic field, respectively. When this happens, we must also expect a change in the resistance curves, R, during subcritical propagation Δa. This is shown in Fig. 11.43a that shows the toughness J_{Ic} of A 533 B carbon steel at room temperature obtained with different types of specimens [45]. Besides a certain and expected dispersion of experimental data do to material variability, once again there is no significant variation of J_{Ic} with the C(T) and bend type specimens. Also shown, Fig. 11.43b, is the tearing modulus, T, or T_{mat}, which indicates the slope of the R curve, see Sect. 11.10 and Eq. (11.79). We observe e general upward trend reflecting the ability of the geometry to exert a restrain on the apical plastic zone. The key to reading can be, therefore, that of T-stress. In fact, reporting the numerically calculated value of the T-stress at the moment of ductile tearing for the various types of specimens we obtain the diagram of Fig. 11.44 for a HY-100 steel [45]. There is a definite downward trend passing from a SE(B) with a shallow crack ($a/W = 0.15$) or SE(T) and DE(T) to the more constraint SE(B) specimen.

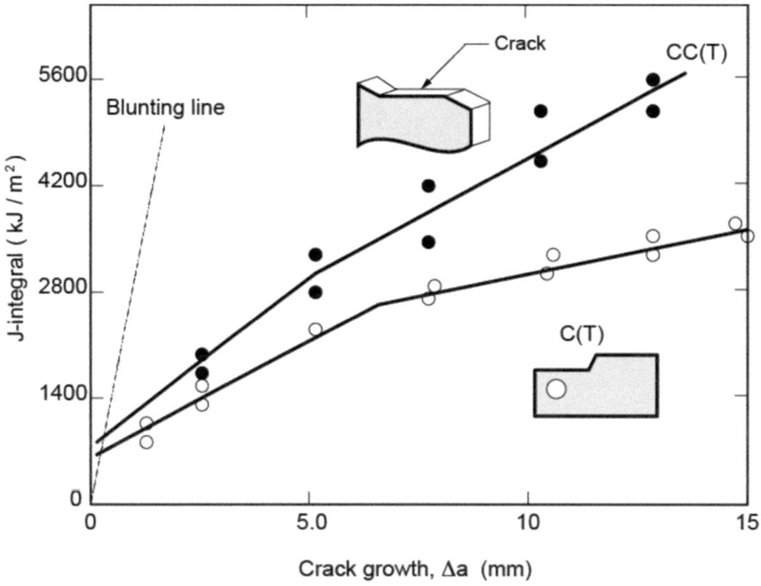

Fig. 11.42 J–R for type A304 stainless steel at room temperature using 10 mm thick CC(T) and C(T) specimens (adapted from [49])

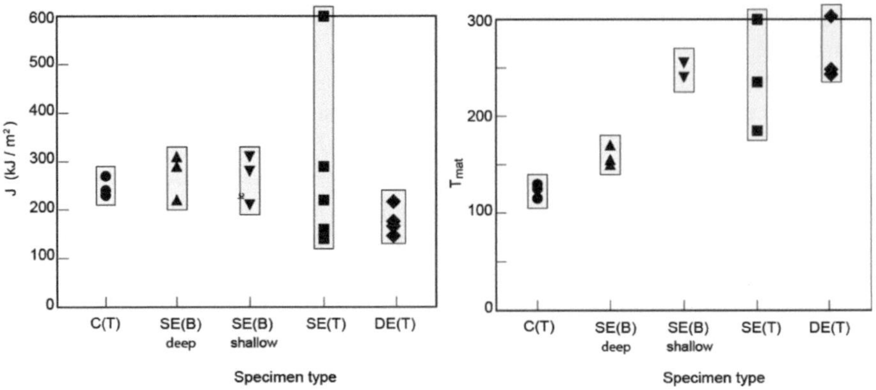

Fig. 11.43 J_{Ic} measured on A 533 B carbon steel using different test specimens. For the SE(B) it has been considered both the deep ($a/W = 0.6$) and shallow crack ($a/W = 0.15$) case. (data from [45])

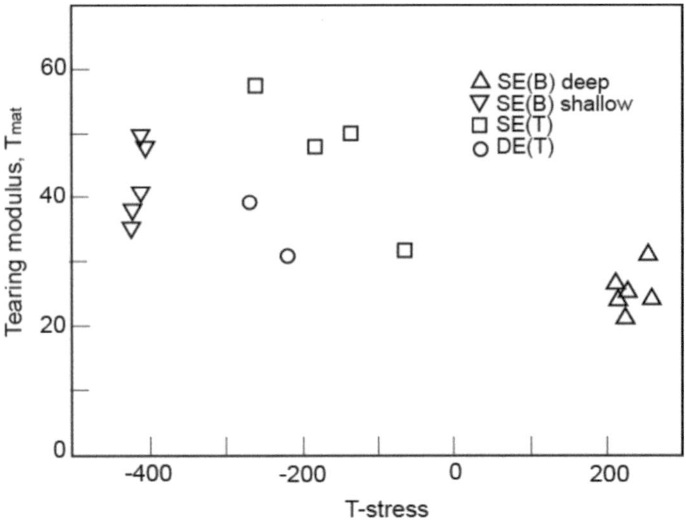

Fig. 11.44 T_{mat} versus T-stress for HY-100 steel [45]

11.13 Thickness Effect on *J*-integral and R Curve

It has been shown how the thickness, *B*, of the test piece introduces a strong constraint on the crack tip plastic zone size particularly with elevated a/W ratios that characterize deep flaws. This hardening effect will also have an impact on the shape of the *J-R* curve, changing the tearing modulus, T_{mat}. Figure 11.45 [50] shows the resistance curves *R* determined for a low-alloy carbon steel type A 302 B, with a yield strength of approximately 450 MPa and an ultimate stress close to 580 MPa, at a temperature of 82 °C, using C(T) specimens of various thicknesses, B, with a/w ratio of 0.5 and with side grooves equal to 20% of the thickness.

It can be observed how the slope dJ/da changes, i.e., the tearing modulus, T_{mat}, of the material, as the thickness increases, while the J_{Ic} remains practically constant, Fig. 11.46a [50], with the only exception of the 0.5" CT specimen that yielded J_{Ic} values varying from 60 kJ/m^2 to 120 kJ/m^2. This scatter must be ascribed to the statistical metallurgical variability associated with the lower process volume of this very small specimen more than to the constraint effect exerted by the thickness itself,

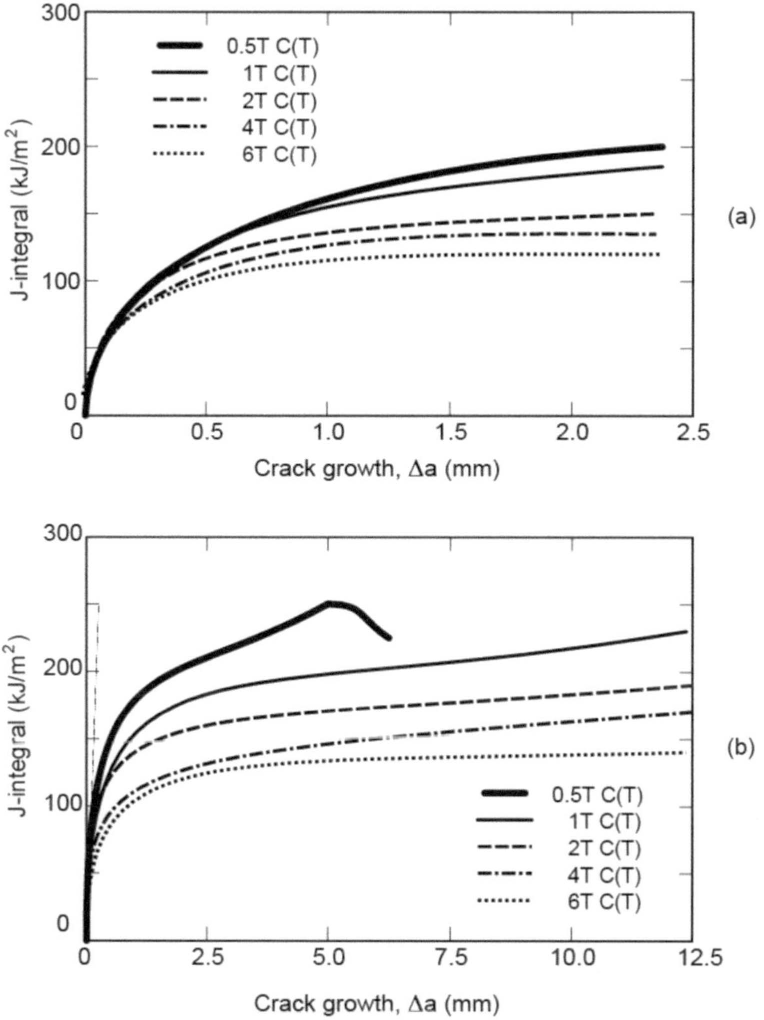

Fig. 11.45 **b** J–R curves obtained for A 302B steel at 82 °C with C(T) side grooved specimens of various thickness, B, **a** detail of the first 2.5 mm of crack growth (from [50])

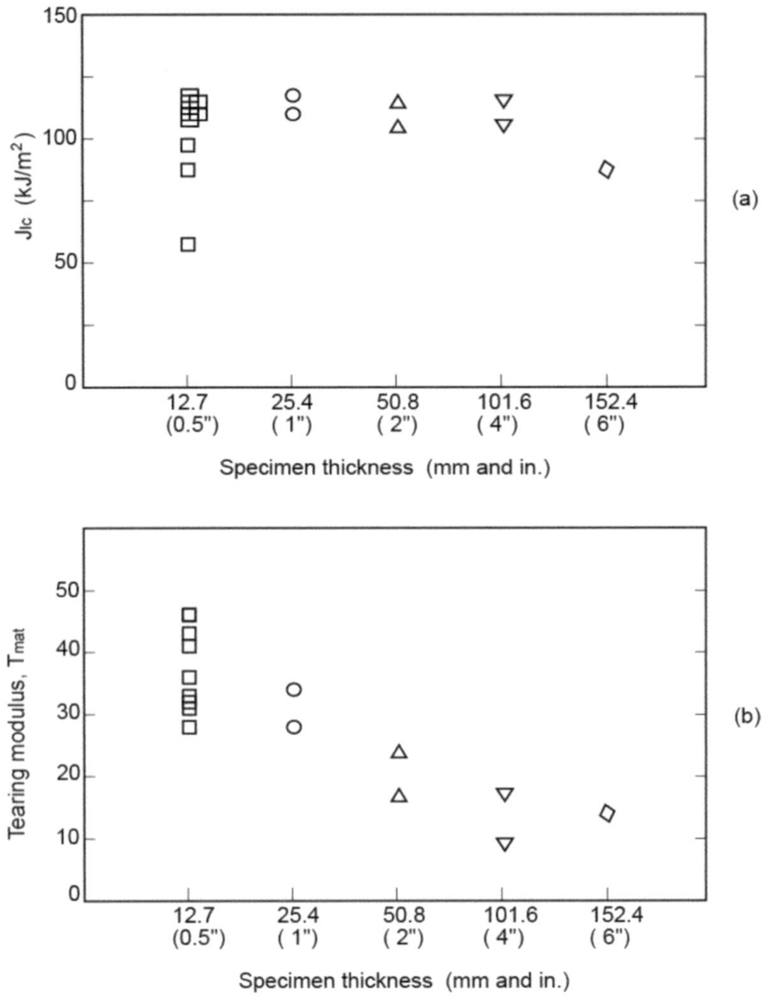

Fig. 11.46 **a** J_{Ic} and **b** tearing modulus dependence on the specimen thickness for A 302B carbon steel at 82°B [50]

which makes many small specimens be equal, from the metallurgical variability standpoint, to a single large specimen. At variance with the J_{Ic}, the tearing modulus, T_{mat}, does change with the thickness because of a constraint effect, as shown in Fig. 11.46b.

As already observed with K_{Ic} measurements, see Sect. 9.13, and for the very same reason also the *a*/*W* ratio must have an impact on the *J*–*R* curve, as shown in Fig. 11.47 [45] for HY-100 high strength steel of 747 MPa yield strength and 877 MPa ultimate strength already considered in Fig. 11.38.

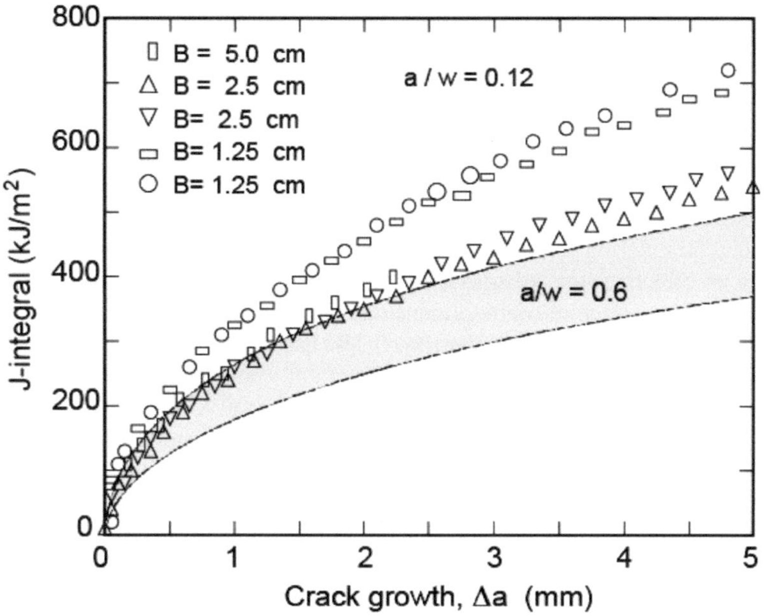

Fig. 11.47 *J-R* curves at ambient temperature obtained for HY-100 high strength steel using SE(B) specimens having different thickness, *B*, and *a*/*W* ratios. The gray band refers to specimens having $a/W = 0.6$ (from [45])

The *J-R* curves have been obtained making use of the pseudo energy method described in Sect. 11.10. Since SE(B) specimens had shallow cracks, a/*W* < 0.28, the η and γ factors have been adjourned [51].

$$\eta = 0,32 + 12 \cdot \frac{a}{w} - 49,5 \cdot \left(\frac{a}{w}\right)^2 + 99,8 \cdot \left(\frac{a}{w}\right)^3 \tag{11.81}$$

Consequently, the factor γ will no longer be that given by Eq. (11.36) but a new

$$\gamma = \frac{-12,22 + 106,7(\frac{a}{w}) - 236,6(\frac{a}{w})^2 - 924,6(\frac{a}{w})^3 + 4845,4(\frac{a}{w})^4 - 9880(\frac{a}{w})^5 + 9960(\frac{a}{w})^6}{0,32 + 12(\frac{a}{w}) - 49,5(\frac{a}{w})^2 + 99,8(\frac{a}{w})^3} \tag{11.82}$$

11.14 Effect of Temperature and Dynamic Load on *J*-integral

The mechanisms by which temperature and dynamic loads operate on the toughness, K_{Ic}, of materials have already been discussed in Sects. 9.10 and 9.11, respectively. In this Section we will show the same effects in relation to J-integral. We have seen that for ductile steels of the FCC family and, in particular, for BCC materials above the NDT it is no longer possible to measure fracture toughness K_{Ic} complying with size requirements because this would require prohibitive test piece thickness and, though assuming to be able to do so, with questionable results. With size limits (11.51) posed to the J_{Ic} the thickness of the specimen no longer appears to be a problem. However, with ductile materials, which for BCC materials would mean increasing the temperature, it is the very K_{Ic} concept that falls. Losing the SSY condition end entering the ductile field, even a piece of infinite dimension, though assuring an effective state of triaxiality, would not be sufficient to trigger a brittle fracture event since the absolute dimensions of the crack tip plastic zone would no longer be negligible becoming enormous. The use of the J-integral may appear to fill this gap, both dimensional and, more important, conceptual. Figure 11.48 shows a series of curves obtained with 1T C(T) specimens of A 508 Cl. 3 forging steel at different temperatures using the unloading compliance method [52]. As can be seen, at very low temperatures, −180 and − 100 °C, the load–displacement record is linear revealing brittle fracture activated by a cleavage mechanism. Toughness can be measured in terms of K_{Ic} with the usual MFLE procedures or derived from J_{Ic} using the equivalence expressed by Eq. (11.75). As the temperature increases, at –20 °C some ductility, though small, seems to affect the material response, but already at 25 °C a large deviation from linearity occurs with a reduction of the maximum load. At 100 °C the material enters a full ductile condition showing a clear decrease of the maximum force with respect to the –20 °C test. This is because as the temperature increases the strength of the material decreases penalizing its fracture resistance. The balance between an increasing toughness due to more ductility and a decreasing strength of the material, σ_s and σ_u, at a certain moment is lost when the strength reduction prevails over the ductility increase. Approaching the GY condition, in fact, it is the flow stress that determines the fracture resistance. This can be seen plotting the fracture toughness K_{Ic} or K_{Jc} versus temperature, as shown in Fig. 11.49 [52]. Already over 0 °C, after a maximum is reached, the toughness decreases. The very high values of toughness, over 200 MPa$\sqrt{}$m, would require very thick specimens to get a valid test, according to the size requirements, Eq. (9.50), well over 14"C(T). However, the toughness K_{Jc} was derived from a J_c measurement using Eqs. (11.71) and (11.72) taking benefit from the more favorable size requirements offered to J. The overall behavior can be described in a J versus temperature diagram, as shown in Fig. 11. In the low temperature-low toughness domain, Region I, J readings are on the blunting line, i.e., on the linear part of the J-R curve. As temperature increases, and so does toughness, fracture occurs on the initial part, almost linear, of the J-R curve beyond J_c, Region II.

The material behaves in a ductile fashion. By further increasing the temperature, the unloading compliance procedure may fail as we enter Region III where fracture occurs on the non-linear part of the J-R curve outside the validity limits set for J, see Sect. 11.9.

This is the region where toughness drops abruptly over 100 °C in Fig. 11.49. This is a behavior common to all steels and alloys having a BCC lattice, like ferritic steels, when used at temperatures beyond the NDT. At variance with that, in the upper shelf

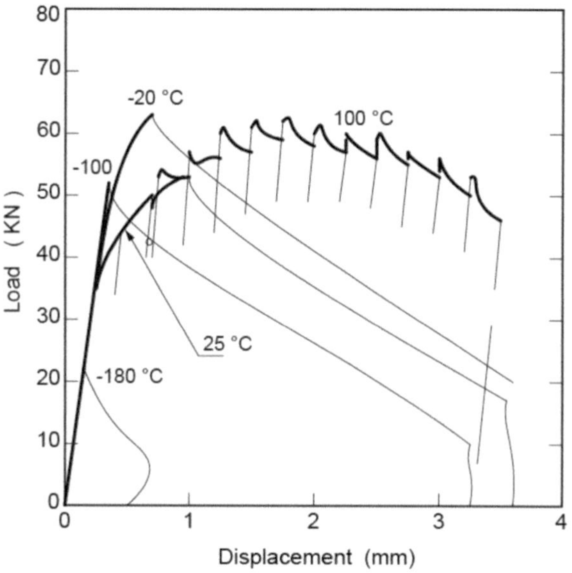

Fig. 11.48 Unloading compliance *J-R* curves obtained at various temperatures using 1T-C(T) specimens of A 508 Cl3 steel. At 25 °C and 100 °C fracture occurs after large displacements whilst at –180 °C and –100 °C it happens without any plastic deformation. Note that at 25 °C and 100 °C, the maxim force is lower than that reached at –20 °C [52]

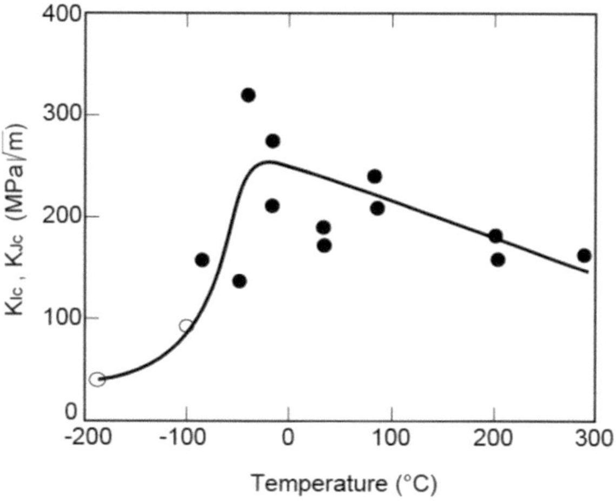

Fig. 11.49 K_{Ic} and K_{JQ} versus temperature diagram obtained using the unloading compliance procedure with 1T-C(T) specimens of A508 steel [52]

region any factor that raises the yield stress results in an improvement of fracture resistance. This is the case of dynamic loadings that do increase the yield strength of materials, see Sects. 5.4.2 and 5.4.3. This latter phenomenon will be observed not only in BCC materials, but also in FCC lattice materials that are equally strain rate sensitive. The resulting effect in the upper shelf region is diametrically opposed to that occurring in the brittle regime, Region I of Fig. 11.50, where any increase in yield strength results in an early fracture, since the fracture process is controlled by stress and not by deformations. Figure 11.51 precisely shows this phenomenon of increased toughness in the upper shelf for a 2.25 Cr 1-Mo steel for pressure vessels obtained with dynamic tests of 1TC(T) specimens with a $\dot{K}$ rate of $2 \cdot 10^4$ MPa$\sqrt{}$m/s [53]. Dynamic loading moves the entire fracture curve toward higher temperatures, as already seen in Fig. 9.44. In terms of *J-R* resistance curve, we will see an increase not only in J_{Ic}, but also in the slope *dJ/da* as the loading rate increases, as shown in Fig. 11.52 for a 3 Ni steel of 750 MPa ultimate strength and 645 MPa yield strength [53].

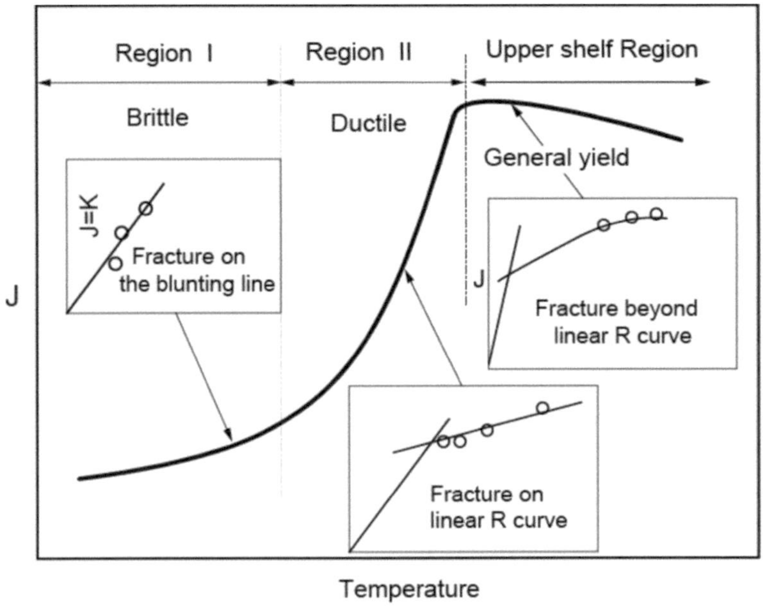

Fig. 11.50 The three Regions of fracture occurrence versus temperature on the *J–R* curve

Fig. 11.51 K_{Ic} and K_{JQ} versus temperature diagram obtained with 1T-C(T) specimens 2.25 Cr 1-Mo steel statically and dynamically loaded. K_{JQ} indicates non valid *K* values obtained from J_Q measurements (from [53])

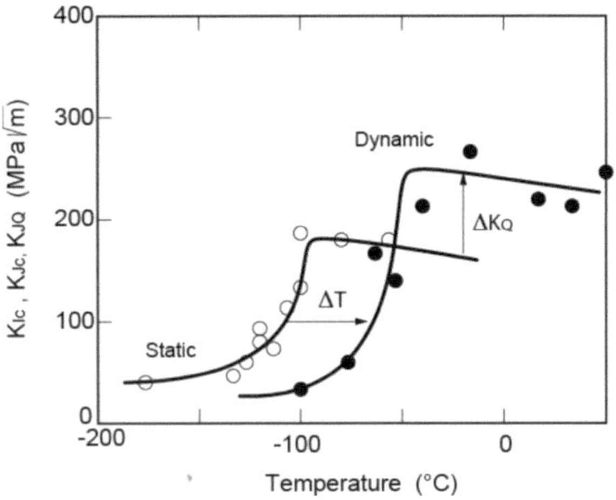

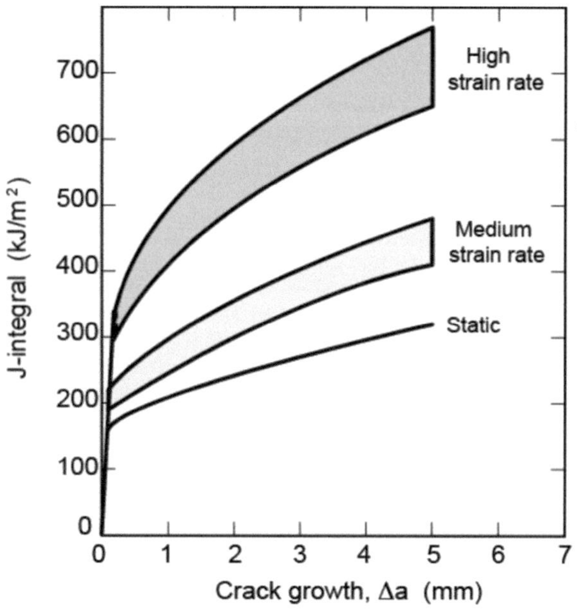

Fig. 11.52 *J-R* curves for a 3 Ni steel for three different loading conditions: static, medium strain rate and high strain rate (from [53])

Austenitic stainless steel and aluminum alloys are typical examples of FCC lattice material that have not got a transition temperature, NTD, but are, anyhow, sensitive to strain rate. Figure 11.53 shows the *J-R* curves obtained at 288 °C with C(T) specimen of A 304 stainless steel under static and dynamic conditions [54]. The beneficial effect of dynamic loading is evident.

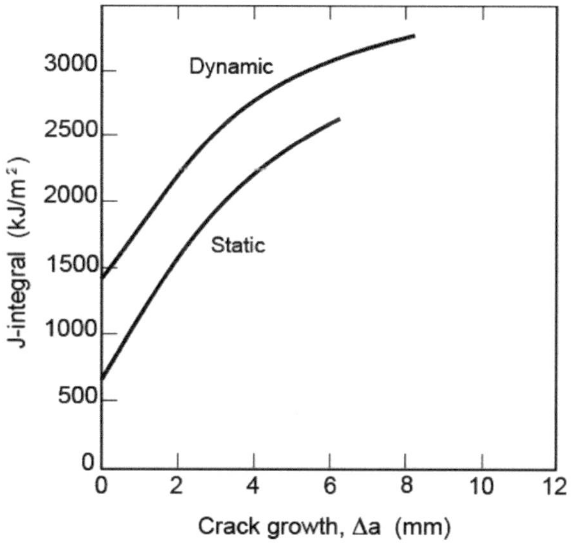

Fig. 11.53 *J-R* curves for A 304 stainless steel at 288 °C under static and dynamic loading (from [54])

11.15 CTOD Measurement

Born to be directly measured on the test piece using a cod-meter, see Sect. 10.2, CTOD or δ is now indirectly measured through the knowledge of K or J, using Eq. (10.12) where G = J.

$$\delta = \psi \frac{J}{\sigma_s} \tag{11.83}$$

with ψ ~ ½, see Sect. 11.5 and Table 11.1. Recalling Eq. (10.32) that links the stress intensity factor K to the J-integral, it also is

$$\delta = \frac{1}{2} \frac{K_I^2}{\sigma_s E\prime} \tag{11.84}$$

Equation (11.84) can be written in a more general form using the plastic constraint factor, m, following Eq. (8.62).

$$\delta = \frac{K_I^2}{m \sigma_s E\prime} \tag{11.85}$$

with the plastic constraint factor, m, that depends on the yield to ultimate stress ratio, σ_s/σ_u, given by Eqs. (8.63) or (8.64).

This indirect approach is used because the original cod-meter didn't yield reliable results. This also raises the question of whether CTOD is still a useful fracture mechanics parameter since the knowledge of either the K factor or J-integral already solves the issue of fracture integrity assessment. Equation (11.85) holds in the linear elastic domain under SSY condition.

Once the material enters the plastic regime and the SSY condition is lost, a plastic component of the CTOD, δ_p, develops. Therefore, experimental CTOD measurement is made by separating the CTOD into an elastic component, δ_e, and a plastic component, δ_p:

$$\text{CTOD} = \delta = \delta_e + \delta_p \tag{11.86}$$

As to elastic component, δ_e, Eq. (11.85) holds. The plastic component, δ_p, can be assessed through the specimen arms rotation about the plastic hinge, as shown in Fig. 11.54.

$$\delta_p = \frac{r(W-a)}{r(W-a)+a+z} V_p \tag{11.87}$$

where $a + z$ is the crack length a plus the knife-edge height z, r is a plastic rotational factor that defines the relative position of the hinge point, C, whose value ranges from 0 and 1 and V_p is the plastic component of the crack mouth opening

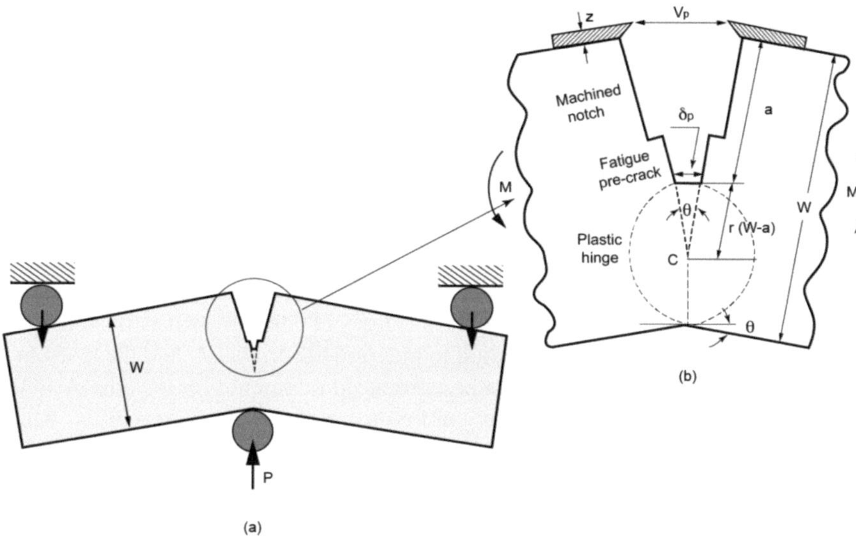

Fig. 11.54 Hing model for determining the plastic component of CTOD, δ_p

displacement or CMOD, read on the load–displacement record by constructing a line parallel to the elastic loading line, as shown in Fig. 11.8a. Caution must be taken about the possible subcritical crack growth that may occur during the loading phase, as already discussed in Sect. 11.2 where it was also shown how to pinpoint the actual load F_i of first tearing and, therefore, the displacement $V_{i,p}$ that must be used in Eq. (11.87), through the compliance curve record, Figs. 11.8b, 11.9 and 11.10.

for a SEN(B) specimen the plastic rotational factor is given by

$$r_p = 0.44 \tag{11.88}$$

for the compact specimen C(T) it is

$$r_p = 0.4\left\{1 + 2\left[\left(\frac{a}{b}\right)^2 + \left(\frac{a}{b}\right) + 0.5\right]^{\frac{1}{2}} - 2\left[\frac{a}{b} + 0.5\right]\right\} \tag{11.89}$$

where b is the length of the ligament ahead of the crack.

The measurement of CTOD is described in the ASTM standard E 1820 [55] and ISO 12135 [56].

11.16 Correction for Rigid Rotation

In Sect. 9.3 the WOL X-Type specimen was introduced, which was modified in the WOL T-type, Fig. 9.9. The reason for such a modification stemmed from the geometry of the specimen, which had too long an arm with respect to the eight. The specimen, therefore, was too compliant. In LEFM excessive deformations due to bending are not tolerated, which may be not the case in EPFM. However, large rotations of the specimen arms may introduce an error in the assessment of the applied moment due the shortening of the arm of the applied load P and in the deflection measurement thus reducing the accuracy of crack growth estimates, when using the unloading compliance method to measure the critical J_c and the resistance curve R. Therefore, a correction must be introduced to account for the change in the specimen compliance. Loss [57] was the first to provide such a correction, which was introduced in the ASTM E 813 Standard for J_c fracture toughness measurement with C(T) specimens. With reference to Fig. 11.47b we define two compliances, C_m and C_c as

$$C_m = \frac{d_m}{P_m}; \, C_c = \frac{2d_{mc}}{P_c} \tag{11.90}$$

where d_m is the measured crack opening displacement read on the load line, and d_c is the corrected value of d_m, P_m is the measured load, and P_c is the corrected value of P_m.

Assuming that the two crack surfaces rigidly rotate around a rotational center O, Fig. 11.55, to account for crack opening displacement in C(T) specimens, the crack size estimation should be corrected as

$$C_c = \frac{C_m}{\left[\frac{H}{R}\sin\theta - \cos\theta\right]\left[\frac{D}{R}\sin\theta - \cos\theta\right]} \tag{11.91}$$

However, as Cao and Bai pointed out [58], no detailed derivation of Eq. (11.91) can be found in the publication by Loss [57]. Referring to Fig. 11.55b, it is

$$\tan(\theta + \varphi) = \frac{\mathrm{d}_C + D}{R} \tag{11.92}$$

then the corrected d_c can be expressed as

$$d_c = \frac{\left[1 + (D/R)^2\right]\sin\theta}{\cos\theta - D/R \cdot \sin\theta} R \tag{11.93}$$

and

$$\frac{\mathrm{d}_m}{2} = D(\cos\theta - 1) + R \cdot \sin\vartheta \tag{11.94}$$

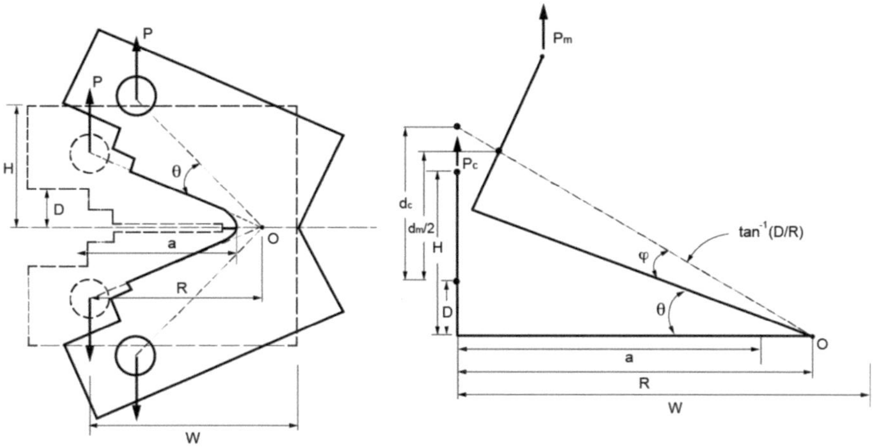

Fig. 11.55 Schematic of elastic compliance correction due to rigid rotation of specimen arms

The corrected load P_c can be obtained by equating the two applied moments about the rotational center O for both the deformed and undeformed configurations, Fig. 11.55b.

$$P_c \cdot R = P_m \cdot R_m \tag{11.95}$$

where

$$R_m = R \cdot \cos \vartheta - H \cdot \sin \vartheta \tag{11.96}$$

Equations (11.95) and (11.96) Equations (11.95) and (11.96) yield.

$$\frac{P_c}{P_m} = \cos\theta - \frac{H}{R} \cdot \sin\theta \tag{11.97}$$

From Eqs. (11.90), (11.93), (11.94) and (11.97) it is

$$\frac{C_c}{C_m} = \frac{\sin}{[\cos - D/R \sin\theta][D/R(\cos\theta - 1) + \sin\theta][\cos\theta - H/R\sin\theta]} \tag{11.98}$$

which is not exactly equal to Eq. (11.91) reported by the ASTM E 813. Using the Joyce et al. [59] approximation.

$$\frac{\mathrm{d}_m/2}{\mathrm{d}_C} = \cos\theta - \frac{D}{R}\sin\theta \tag{11.99}$$

Combining Eq. (11.97) with Eq. (11.99) we finally get

$$C_c = \frac{C_m}{\left[\frac{H}{R}\sin\theta - \cos\theta\right]\left[\frac{D}{R}\sin\theta - \cos\theta\right]} \tag{11.100}$$

which is exactly equal to Eq. (11.81) given by the ASTM standard.

References

1. Begley, J.A., Landes, J.D.: The J integral as a fracture criterion, fracture toughness. In: Proceedings of the 1971 Nat Symp On Frac Mech, Part II, ASTM STP 514, pp. 1–20 (1972)
2. Landes, J.D., Begley, J.A.: The effect of specimen geometry on J_{Ic}. In: Proceedings of the 1971 Nat Symp on Frac Mech, Part II, ASTM STP 514, pp. 24–39 (1972)
3. Kobayashi, A.S., Chiu, S.T., Beenwkes, R.A.: A numerical investigation on the use of J-integral. Eng. Fract. Mech.**5**, 293–305 (1973)
4. Rice, J.R., Paris, P.C., Merkle, J.G.: Some further results of J-integral analysis and estimates. ASTM STP **536**, 231–245 (1973)
5. Merkle, J.G., Corten, H.T.: A J-integral analysis for the compact specimen, considering axial forces as well as bending effects. J. Press. Vessel. Technol., Trans. ASME, 286–292 (1974)
6. Sumpter, J.D.G., Turner, C.E.: Cracks and fracture. ASTM STP **601**, 3–18 (1976)
7. Ernst, H.A., Paris, P.C.: Techniques of analysis of load-displacement records by J integral methods. NRC NUREG/CR 1222, (1980)
8. Clarke, G.A., Andrews, W.R., Paris, P.C., Schmidt, D.W.: ASTM STP **590**, 27–42 (1976)
9. Clarke, G.A., Landes, J.D.: J. Test. Eval. **7**(5), 264–269 (1979)
10. Logsdon, W.A.: Elasic plastic (J_{Ic}) fracture toughness values: Their..., ASTM STP 590, pp. 43–60 (1976)
11. De Vries, M.I. Experimental aspect of unloading-compliance J-tests, ductile fracture test methods. In: Proceedings CSNI workshop, p. 166, Paris, (1983)
12. Landes, J.D., Begley, J.A.: Test results from J integral studies—an attempt to establish a J_{Ic} Testing Procedure, Westinghouse Internal Document (1972)
13. Blauel, J.B., Hollstein, T.: On the determination of elasto-plastic materials parameters-results and problems, CSNI Specs. In: Meeting on Plastic Tearing Instability, NUREG/CP-0010, CSNI Rep. No. 39, Addendum, St. Luis, Missouri, USA (1979)
14. Johnson, H.H.: Mater. Res. Stand. **5**, 442–445 (1965)
15. Lowes, J.M., Fearnebough, G.D.: Eng. Fract. Mech. **3**, 103–108 (1971)
16. Vassilaros, M.G., Hackett, E.M.: ASTM STP **833**, 535–552 (1984)
17. Marschall, C.W., Held, P.R., Landow, M.P., Mincer, P.N.: Use of direct-current electric potential method to monitor large amount of crack growth in highly ductile metals. American Society for Testing and Materials, ASTM-STP 1074 (1990)
18. Wilson, W.K.: Stress intensity factors for deep cracks in bending and compact tension specimens. J. Eng. Fract. Mech., **2**(2), (1970)
19. Ernst, H.A., Paris, P.C., Landes, J.D.: Estimation on J-integral and tearing modulus T from a single specimen test record. ASTM STP **743**, 476–502 (1981)
20. Hiser, A.L., Loss, F.J.: Alternative procedure for *J-R* curve determination. NRC NUREG/CR-3402, (1983)
21. Sutton, G.E., Vassilaros, M.G.: Study of the effects of elastic unloading on the J-R curves from compact specimens. NRC NUREG/CR 4283, (1985)
22. Wolf, U.E.: Microfrachtographic of a fracture toughness specimen, reactor primary coolant system rupture study, AEC R&D Report GEAP-5474, Quart. Progr. Rep. No. 8, (1967)
23. Spitzig, A.: A fractographic feature of plain strain fracture in 0.4 C-Ni-Cr-Mo steel. Trans. ASM., **61**, 344 (1968)
24. Clark, W.G. Jr., Bates, R.C.: Microscopic aspect of fracture toughness. Westinghouse Scientific Paper 69–1E7-RDAFC-P1, (1969)

25. Bates, R.C., Clark, W.G., Jr.: Fractography and fracture mechanics. Trans. ASM **62**, 380–388 (1969)
26. Iwadate, T., Tanaka, Y., Ono, S.: An analysis of elastic-plastic fracture toughness behavior for J_{Ic} measurement in the transition region, ASTM, STP 803, vol. 2, pp. 531–561 (1983)
27. Clarke, G.A.: An evaluation of the unloading compliance procedure for J-integral testing in the hot cell. US NRC NUREG/CR-1070 (1980)
28. Roos, E., Kockelmann, H.: A survey of different experimental methods for the determination of crack initiation parameters of small and large specimens. In: Proceedings of a OECD/CSNI workshop on ductile fracture test methods, pp. 331–362. Paris (1983)
29. Heerens, J., Cornec, A., Schwalbe, K.H.: Determination of J_{Ic} using new blunting line construction. In: Proceedings of the second OECD/CSNI workshop on ductile fracture test methods, pp. 199–211. Paris, (1985). Published in 1988
30. Kobayashi, H, Nakamura, H., Nakazawa, H.: Evaluation of blunting line and elastic-plastic fracture toughness, ASTM-STP 803, vol. 2, pp. 420–438 (1983)
31. Mills, W.J.: J. Test. Eval. **9**, 56 (1981)
32. Davis, P.H.: An elastic-plastic fracture mechanics study of crack initiation in 316 stainless steel. ASTM-STP 803, vol. 2, pp. 611–631 (1983)
33. Landes, J.D., Begley, J.A.: Fracture analysis. ASTM STP **560**, 170–186 (1974)
34. Paris, P.C.: Written discussion of Ref. [1]
35. Paris, P.C.: Fracture mechanics in the elastic plastic regime. ASTM STP **631**, 13 (1977)
36. Begley, J.A., Landes, J.D.: Int. J. Fract. **12**, 764–766 (1976)
37. McMeeking, R.M., Parks, D.M.. On criteria of J-dominance of crack tip fields in large scale yielding. ASTM STP 668, (1979)
38. McMeeking, R.M.: J. Mech. Phys. Solids **25**, 357 (1977)
39. Hutchinson, J.W., Paris, P.C.: Stability analysis of J-controlled crack growth. ASTM STP **668**, 37–64 (1979)
40. Koppenhorfer, K.C., Dodds, R.H.: Size and deformation limits to maintain constraint in K_{Ic} and J_{Ic} testing of bending specimens, NRC NUREG/CR-6191, (1995)
41. Joyce, J.A., Davis, D.A., Hackett, E.M., Hays, R.A.: Application of J-integral and the modified J-integral to cases of large crack extension. US NRC NUREG/CR-5143, (1989)
42. Joyce, J.A., Hackett, E.M.: Development of an engineering definition of the extent of J singularity controlled crack growth, US NRC NUREG/CR-5238, (1989)
43. Varias, A.G., Shih, C.F.: Quasi-static crack advance under a range of constraints-steady-state fields based on characteristic length. J. Mech. Phys. Solids **41**(5), 835–861 (1993)
44. O'Dowd, N.P., Shih, C.F., Dodds, R.H.: The role of geometry and crack growth on constraint and implication for Ductile/Brittle Fracture, ASTM STP 1224, Philadelphia (1994)
45. Joyce, J.A., Link, R.E.: Effects of tensile loading on upper shelf fracture toughness, NRC NUREG/CR-6051, (1994)
46. Rice, J.R.: Elastic plastic fracture mechanics, the mechanics of fracture. ASME, F. Erdogan (1976)
47. Paris, P.C., Clarke, G.A.: Observations of Variations in fracture characteristics with temperature using a J-integral approach. In: Symp on elasic-plastic fracture, ASTM (1977)
48. Argy, G., Paris, P.C., Shaw, F.: Fatigue crack growth and fracture toughness of 5083–0 Aluminum Alloy, ASTM, STP 579 (1975)
49. Hiser, A.L., Terrel, J.B.: Size effects on J-R curves for A 302-B plate. US NRC NUREG/CR-5265, (1989)
50. Landes, J.D.: Size and geometry effects on elastic-plastic fracture characterization, pp. 194–225. CSNI Specialists Meeting on Plastic Tearing Instability, St. Luis, Missouri (1979)
51. Sumpter, J.D.G.: J_{Ic} determination for shallow notch welded bend specimens. Fatigue Fract. Eng. Mater. Struct. **10**(6), 479–493 (1987)
52. Brioschi, F.: Caratterizzazione della Virola Prototipo. Parte II—Meccanica della Frattura, ANSALDO-CISE, Relazione Tecnica, ING-MATE-86 R226 (1986)
53. Joyce, J.A., Hackett, E.M.: Development of an analytic key curve approach to drop tower J-R curve measurement, US NRC NUREG/CR-4782, (1986)

54. Marschall, C.W., Landow, M.P., Wilkowski, G.M.: Loading rate effects on strength and fracture toughness of pipe steels used in task 1 of the IPIRG program. US NRC NUREG/CR-6098, BMI-2175, (1993)
55. ASTM E 1820–08a.: Standard test method for measurement of fracture toughness
56. International Standard ISO 12135.: Metallic materials—Unified method of test for the determination of quasistatic fracture toughness, 3rd edn (2021)
57. Loss, F.J.: Structural integrity of water reactor pressure boundary components. Progress Resport Ending 30 Nov 1977, NRL Memorandum Report 3782 (1978)
58. Bao, C., Cai, L.: Investigation on compliance rotation correction for compact tensile specimen in unloading compliance method. Acta Mechanica Solida Sinica., **24**(2), (2011)
59. Joyce, J.A., Link, R.E.: Effects of constraint on upper shelf fracture toughness. In: Fracture mechanics, 26th Volume, ASTM STP1256, American Society for Testing and Materials, Philadelphia (1995)

Chapter 12
Dynamic Propagation and Arrest of a Crack

12.1 Introduction and Historical Remark

The finding that a running crack could arrest surprised researchers at NRL already in the mid-40s. Many Liberty ships, in fact, were found to be affected by long cracks, even meters long, that arrested without catastrophic consequences. At the time, LEFM with the Irwin K_I factor had still a long way to come and material toughness was more a general concept than a measurable material property. Therefore, researchers related the arrest of running brittle fractures to a local stress reduction or to some change in material properties, in particular ductility. This led to the formulation of the Robertson crack arrest test where a temperature gradient maintained along a test piece was used to change the ductility (today we can say the toughness) of the material and arrest a fracture triggered in the starter cold region by a dynamic load, Fig. 3.50. Results clearly indicated that for any combination of crack length-applied stress, a–σ, a temperature T_a, existed at which a running crack would eventually arrest. The formulation of this relationship was of empirical nature and entered in the FAD diagram of Fig. 3.51 as the CAT lower bound curve. Indeed, if today we were to revisit the CAT curve in terms of fracture toughness, K_{Ic}, we would obtain a curve much like the crack arrest toughness curve, K_{Ia}, which we will be discussing in the following Sections. The arrival, in 1988, to the formulation of the crack arrest toughness, K_{Ia}, measurable with small specimens was a long lasting and controversial event that still maintains several unresolved doubts. In 1965 Krafft and Irwin [1] already evidenced that there were two and not just one fundamental moment characterizing any fracture process: the first was the already known initiation event associated with the toughness of the material K_{Ic}, and the other, at that time still to be unveiled, the final event in which a fast-running crack arrested. Irwin and Wells [2] went on hypothesizing that crack arrest could be viewed as the reversal of crack initiation. By this reasoning the energy released rate after crack arrest, K_{Ia}, becomes a material property in much the same way the energy released rate at the initiation, K_{Ic}, is a material property. Few years later, in 1969, while investigating

P. P. Milella, *Fundamentals of Fracture Mechanics*,
https://doi.org/10.1007/978-3-032-23081-2_12

the effect of strain rate on the fracture toughness of a pressure vessel steel, Crosley and Ripling [3] at MRL noted that they were measuring a new quantity that seemed to show less variability than crack initiation toughness, K_{Ic}, measured in the same set of experiments. When the crack initiation K_Q was high, well over the K_{Ic} of the material, the crack would jump a distance before arresting, but the applied load at arrest was reasonably the same. The K_Q is a formally calculated K value equal to the stress intensity factor K_I which would exist if there were an initial crack with a length corresponding to the machined notch depth. In other words, if the initial sharp crack of length a blunts or is already a blunted notch, the corresponding K_I is not a valid K_{Ic} and is indicated as $K_Q > K_{Ic}$. K_Q, then, retains the concept of a potential stress intensity factor affecting the initiation of a notch once the tip starts to tear and becomes a crack. It measures the excess of energy release rate over that corresponding to K_{Ic}. Since Crosley and Ripling were using a contour or tapered double cantilever beam (TDCB), Fig. 12.1, in which the applied K_I does not change with crack length, a, this new quantity estimated with the crack length measured at arrest, which they called *Crack Arrest Toughness* and denoted by the symbol K_{Ia}, was practically the same which also implied its reproducibility. The measurement of the crack arrest toughness, K_{Ia}, as a material property was just born by chance. However, the theoretical anticipation of the fracture toughness, K_{Ia}, measurable through a static test was not free from criticism and doubts. At variance with fracture initiation, crack propagation and arrest is, in fact, a dynamic phenomenon that does not allow to be interpreted by mean of a static parameter just measuring the final crack length, a_a.

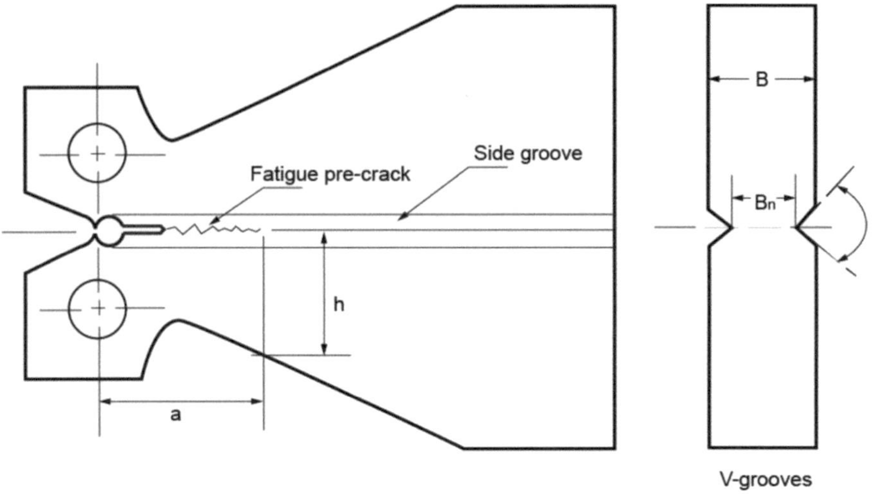

Fig. 12.1 Contoured or tapered Double Cantilever Beam (TDCB) used by Crosley and Ripling in their tests that led to the first formulation of the crack arrest toughness, K_{Ia}. This type of specimen has a K_I independent of the crack length a, at least in the tapered section

Therefore, in the Seventies, in particular, two different schools of fought faced each other in a long lasting and interesting scientific dispute: the one based on the Krafft-Irwin-Wells-Crosley-Ripling idea of K_{Ia} existence as material characteristic, and that initially sustained by Hahn, Kanninen, Hogland and Rosenfield at Battelle Memorial Institute, BMI, to-day Battelle Columbus Laboratories, BCL, who, recalling the theoretical treatments of Mott [4], Yoffe [5], Broberg [6], Eshelby [7] and Freund [8], argued that the phenomenon of crack propagation and arrest had to be treated within the context of a dynamic fracture mechanics theory which included:

(1) kinetic energy contribution,
(2) inertia effects and reflection of stress waves,
(3) dependence of dynamic fracture toughness, K_D, on crack speed
(4) viscous effects.

Viscous effects were soon recognized to be of minor importance and were neglected. In an infinite plate there is a one-to-one correspondence between the stress intensity factor at a moving crack tip and the static value for the same crack tip position [8]. Conversely, in a real, finite structure stress waves reflect off boundaries and reenter the process zone at the crack tip making K_{Ia} be an oversimplification. Moreover, the energy release rate during propagation, G_D or K_{ID}, depends on crack velocity which, in turn, depends on the initial energy available to propagation, K_Q. Probably, the constancy of K_{Ia} measurements obtained by Crosley and Ripling, they argued, was just a particular case due to the size of specimens used, an 8 inch long specimen, that did not allow large crack jumps before arresting and, therefore, did not give enough time to stress waves to reenter into the process zone feeding propagation. Crosley and Ripling, anyhow, continued their tests using the contoured or tapered DCB specimen of Fig. 12.1, and Hahn, Kanninen, Hogland and Rosenfield [9] proceeded on their idea of non-existence of a crack arrest toughness as a material characteristic using a different specimen, a wedge-loaded double cantilever beam, shown in Fig. 12.2 and described in the next section, long enough, even 500 mm, to allow long crack jumps.

If K_{Ia} were a material property as the arrest toughness theory demands, the crack lengths at arrest would be identical also with this new long specimen independently of the initial K_Q. The fact that experiments indicated that those jumps were not identical but depended on K_Q, they argued, was demonstrating the dependence of the arrest event on the energy available during propagation. They also stressed the dependence of the crack arrest on the dynamic fracture toughness K_{ID}, that indicates the variation of the stress intensity factor at the tip of a propagating crack on the crack velocity, V: $K_{ID} = K_{ID}(V)$. Studies by Kalthoff et al. [10, 11] were supporting the need for a complete dynamic analysis that would consider the K_{ID} and the reflected stress waves. They measured the dynamic stress intensity factor at the tip of a propagating and subsequently arresting crack, $K_{ID}(V)$, including that at arrest, K_{IA}, using the method of shadow patterns applicable to transparent materials like Araldite B [10] or that of caustics [11]. These measurements made with wedge loaded-DCB, of BCL type, specimens of Araldite B were compared with the static stress intensity factor, K_I, and the static crack arrest K_{Ia} obtained on TDCD specimens of MRL

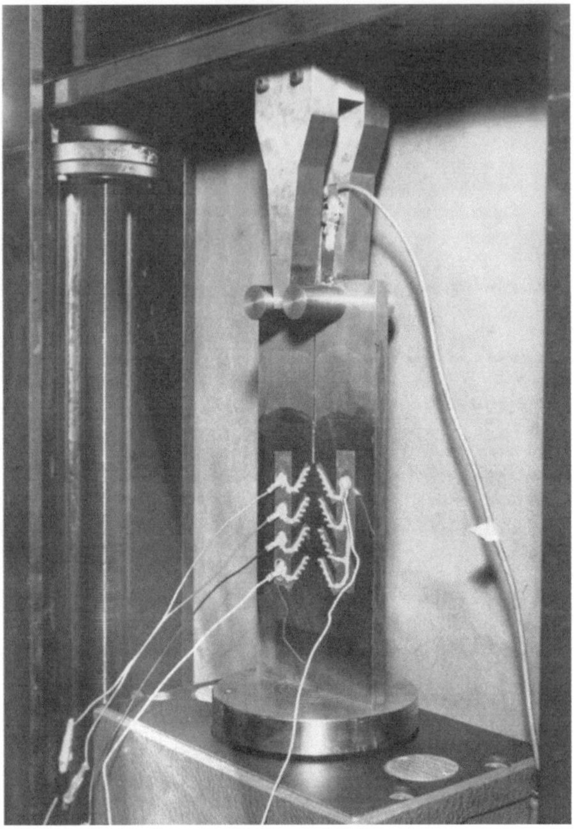

Fig. 12.2 Testing arrangement of wedge loaded-DCB specimen used at BCL [9]

type. Results are shown in Fig. 12.3 [11]. As it can be seen, the starting point is the same K_Q, indicated in Fig. 12.3 as K_{Iq}, higher than the K_{Ic} of the material to have an excess of driving force that accelerates the crack. The toughness K_{Ic} is indicated in Fig. 12.3 as K_I^{stat}. However, after an initial sudden drops and during propagation the K_{ID} maintains lower than the corresponding static value K_I till an inversion point where it prevails over the static stress intensity factor that continues to decrease, till the crack arrests at a value K_{IA} higher than the corresponding static K_{Ia}. This K_{IA} value is indicated in Fig. 12.3 as K_{Ia}^{dyn} or K_{Im}, while K_{Ia} is indicated as K_{Ia}^{stat}. The difference between these two values, K_{IA}–K_{Ia}, increased with increasing crack jumps, $\Delta a = a_a – a_o$, revealing that K_{IA} was not a material characteristic. After arrest, the dynamic stress intensity factor, K_{IA}, approaches the corresponding static factor, K_{Ia}, through oscillation of decreasing amplitude (time scale of Fig. 12.3) till the two stress intensity factors reached a final equal value. The initial amplitude of the dynamic stress intensity factor oscillation soon after arrest

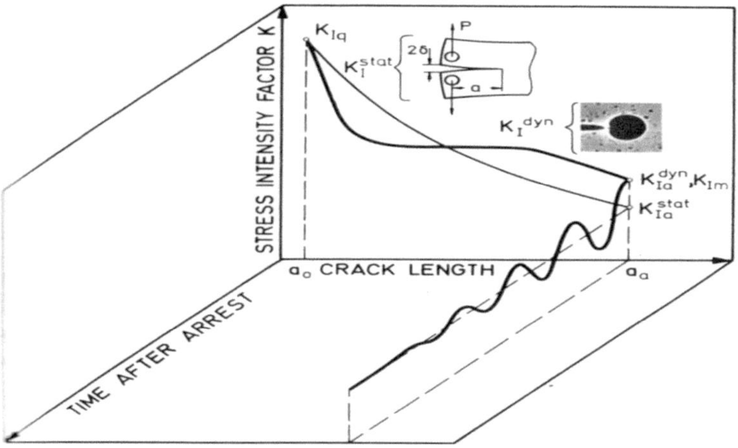

Fig. 12.3 Schematic of stress intensity factors for propagating and arresting crack [11]

$$\frac{K_{IA} - K_{Ia}}{K_{Ia}} \tag{12.1}$$

is a measure of the influence of the dynamic effects on the crack arrest event. Differences between the dynamic, K_{IA}, and the static stress intensity factor, K_{Ia}, curves became smaller for cracks propagating at lower maximum velocity indicating that dynamic effects decrease with decreasing velocity, as one may expect. A large number of tests made with standard duplex wedge loaded DCB specimens, see Sect. 12.3, definitely indicated that in long test sections the kinetic energy would return to the advancing crack tip process zone contributing to the fracture energy, changing the measured K_{Ia}. Numerical and experimental investigation revealed that the recovery of kinetic energy contributes to driving the running crack further and that dynamic fracture toughness K_{ID} can be defined as a function of crack velocity as

$$\frac{K_{ID}}{K_{Ic}} = 1 + A_1\left(\frac{\dot{a}}{C_0}\right) + A_2\left(\frac{\dot{a}}{C_0}\right)^2 \tag{12.2}$$

were K_{Ic} is the static initiation toughness, C_o is the sound wave velocity in the metal and A_1 and A_2 are material constants. However, the scientific debate did not end because the same experiments run using long test sections also contained arguments supporting the opposite thesis, precisely the original idea of Crosley and Ripling of a K_{Ia} existence and characteristic of the material. The paradox can be explained observing that the long specimens used at BCL, even 500 mm long had close borders, over and below the running crack, in the order of 70 mm. It was precisely this short boundaries distance coupled to a long crack jump and not just the long jump, responsible for the returning of the stress waves into the process zone. Should the boundaries be far enough the energy waves would not reenter. If those 500 mm were representing a very thick real structure, like a thick pressure vessel,

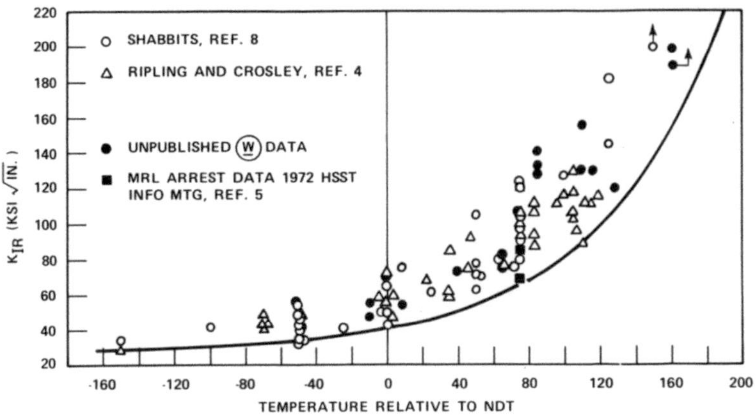

Fig. 12.4 Comparison of reference stress intensity factor curve, K_{IR}, with experimental data available in 1972 [12]

the other dimension, the height, would have been even larger. Probably, in such a real structure there would be no time left to the reflected stress wave to reenter the process zone. The crack would arrest before. However, the dependence of K_{ID} on the crack velocity, $\dot{a}$, still remained an argument of discussion, which was not just a heuristic question. The crack arrest issue, in fact, soon became particularly important in the nuclear industry that started to investigate the possibility that a crack running through the nuclear vessel wall would arrest without jeopardizing its integrity and to use a material characteristic, precisely the K_{Ia}, to calculate that jump. Therefore, in January 1971, the Pressure Vessel Research Committee (PVRC) formed a Task Group to review current knowledge on crack arrest and prepare recommendations on toughness requirements for ferritic steels in nuclear power plant components. It was then that the PVRC Task Group [12] developed the reference stress intensity factor curve, K_{IR}, shown in Fig. 12.4, using the drop weight NDT as a basic fracture toughness temperature index, see Sects. 3.7 and 9.12.

The K_{IR} reference curve was taken as the lower bound for all known dynamic, K_{Id}, and crack arrest, K_{Ia}, fracture toughness data. It may appear that PVRC supported the idea of an arrest toughness K_{Ia} measurable with small specimens under static conditions, without any dynamic analysis. Actually, the PVRC Task Group was only using the K_{Ia} available data to draw a reference curve, K_{IR}, as the lower bound for ferritic steels of all known toughness data, among which the arrest toughness seemed to provide the lowest values, without taking a position on the existence of a static toughness K_{Ia} as a material characteristic. The debate whether or not a static K_{Ia} toughness represented a material characteristic went on. One significant outgrowth of the resulting research and concern was the ASTM sponsored Symposium on Fast Fracture and Crack Arrest of June 28–30, 1976 [13], which started a cooperative ASTM program in 1978. In addition, the US NRC founded a large research program, within the framework of the HSST program, aimed at assessing the possibility that a running crack triggered in a pressure vessel by a thermal shock event could eventually arrest. The program, called PTS (Pressurized Thermal Shock) also made use of wide-test plates (1 × 1 × 0.15 m) welded to two pull plates for a total length of 9.6 m. The wide-plate test section had an edge crack 0.2 m deep.

The test rigs were put in traction in the NIST (National Institute for Standard and Technology, former National Bureau of Standards NBS) testing machine in Gaithersburg, MD, as shown in Fig. 12.5 [14]. The thermal gradient was achieved by cooling the edge of the plate containing the crack with liquid nitrogen while the opposite border was maintained at a temperature of about 168 °C. Figure 12.6 shows a section of test WP-1.7 where two cleavage crack run-arrest events occurred, which can be distinguished in the figure [14]. Under the ASTM cooperative program, tests were also run to ascertain the independence of K_{Ia} from the initial K_Q. The use of a notch instead of a sharp crack made possible the loading at nominal K_Q levels higher than the K_{Ic} of the material. The K_Q, then, became a convenient test parameter to reference. If the K_{Ia} were not a material characteristic, the crack would jump further, the higher the value of the initial K_Q yielding a lower K_{Ia}. Figure 12.7 [15] shows the variation of K_Q with the notch root radius measured using CCA (Compact Crack Arrest) specimens of A 533 steel at various temperatures. The cooperative testing program started using wedge-loaded CT type and wedge loaded-DCB specimen of SA 533, grade B, Class 1 steel that were given to 30 different laboratories. Results are shown in Fig. 12.8. WCA data (ENEA data) of Fig. 12.8 were obtained in Italy with WCA type specimens, Fig. 12.17b. Results clearly indicate a dependence of K_{Ia} on K_Q. However, though all data were obtained at room temperature, caution must be taken since the relative RT_{NDT} was not the same for all specimens, ranging from –25 to –10 °C. The idea of a K_{Ia} characteristic of the material seems to be vanishing. In 1988 ASTM arrived to define a standard specimen for testing K_{Ia}, which is the wedge-loaded CT specimen or CCA specimen discussed in Sect. 12.4. After several decade of arguing, the story of crack arrest is not yet over, though all researchers to-day seem to agree on some well-established points: the K_{Ia}, is the static crack arrest toughness that can be experimentally determined and represents a conservative estimate of K_A, the actual arrest toughness that seems to be dependent on K_{ID}, the dynamic toughness, see next section. It is also lower than both the static K_{Ic} and the dynamic toughness, K_{Id}, obtained under dynamic loads, Fig. 12.10, and, therefore, is presently used as a lower bound reference curve K_R, Fig. 9.50. Today, the initial conflict started in the Seventies between the *dynamic fracture mechanics* approach and the *quasi-static* interpretation of a run arrest event still remains, though its is generally accepted that the latter yields an underestimation of the real arrest toughness that get closer and closer as the crack jump decreases,

12.2 The Kinetic Energy Role

To address the issue of crack arrest it is convenient to distinguish between three different loading conditions: (a) energy release rate, $\mathcal{G}$, increasing with increasing crack extension, (b) energy release rate, $\mathcal{G}$ constant and independent of the crack

Fig. 12.5 HSST wide-plate crack-arrest test specimen installed in NIST testing machine [14]

growth and (c) energy release rate, $\mathcal{G}$ decreasing with crack extension. The first condition, case (a), corresponds to a system containing a crack loaded by a force F that is maintained always at the same level so that $\mathcal{G}$ increases as the crack length, a, increases, Fig. 12.9a. This is the case of fracture mechanics specimens containing a crack, central or edge crack, loaded in traction or bending, where the load F or moment M is imposed and held constant. The second case, case (b), refers to a system containing a crack where the applied force F results in a constant K_I or J_I independent of the crack length, a. This is the case of the tapered DCB specimen used by Crosley and Ripling, shown in Fig. 12.1. The last conditions, case (c), pertain to all systems containing a crack wedge-loaded, as the DCB used at BCL and shown in Fig. 12.2 and the wedge-loaded CCA specimen of Fig. 12.8, where the propagation of the crack results in a parallel increase of the specimen compliance and decrease of the applied load. These three basic cases are shown in Fig. 12.9 together with the resistance R curve of the material. However, both the J-R curve and the $\mathcal{G}$ energy release rate or J shown are relative to the static case. Therefore, they can be used to assess the moment of instability, case (a) or initial crack extension, cases (b) and

Fig. 12.6 Fracture surface of specimen WP-1.7: **a** overall fracture surface and **b** close-up of initial crack, cleavage run-arrest events [14]

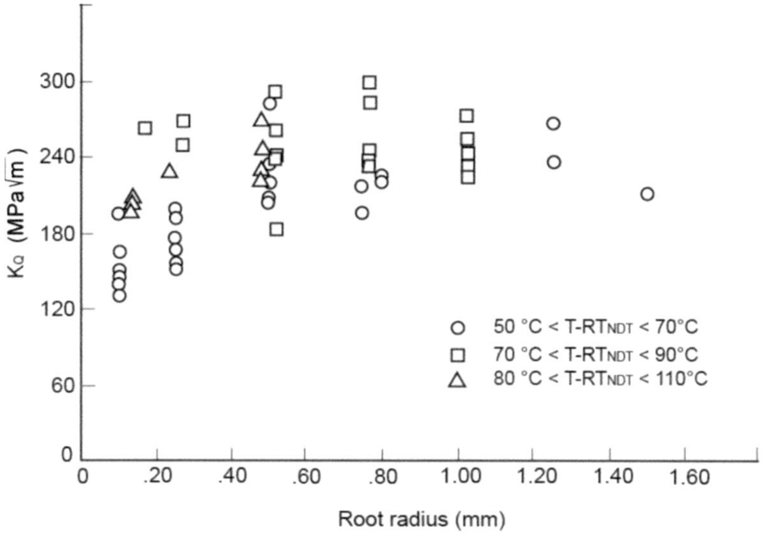

Fig. 12.7 Dependence of K_Q on the root radius of the starter notch in A 533 steel [15]

(c) but cannot be used to describe conditions existing after the initiation moment. These parameters, in fact, may change during crack propagation leading to crack arrest. New $\mathcal{G}$ and R formulations are needed. There is a general agreement that the fracture resistance and crack arrest capability of a material is controlled by the fracture energy, R_D, defined as the energy consumed by plastic deformation and

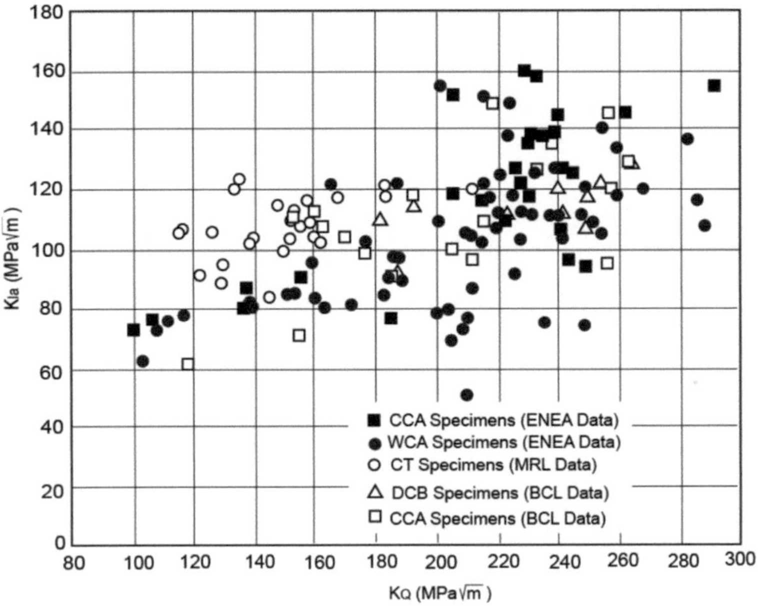

Fig. 12.8 Dependence of K_{Ia} on K_Q for SA533 Gr.B, Cl.3 steel. ASTM Cooperative Program test data (data from [15–19])

fracture process at the tip of a propagating crack per unit area of crack advance or the corresponding dynamic toughness K_{ID} [21, 22]. All these researchers believe that this fracture energy or dynamic toughness is a function of crack velocity, V, i.e., $R_D(V)$ or $K_{ID}(V)$ with a minimum $R_{D,min}$ or $K_{ID,min}$ either at zero or finite velocity. Therefore, criteria for propagation and crack arrest based on energy conservation can be expressed as

$$\mathrm{G} = R_D(V) \tag{12.3a}$$

Criteria for crack propagation:

$$K_I = K_{ID}(V) \tag{12.3b}$$

$$\mathrm{G} < R_{D,min} \tag{12.3c}$$

Criteria for crack arrest:

$$K_I < K_{ID,min} \tag{12.3d}$$

where $\mathcal{G}$ is the crack tip energy release rate and K_I the corresponding stress intensity factor (or J_I). Note that $R_D(V) \rightarrow J_{Ic}$ as $V \rightarrow 0$. The crack must arrest when the

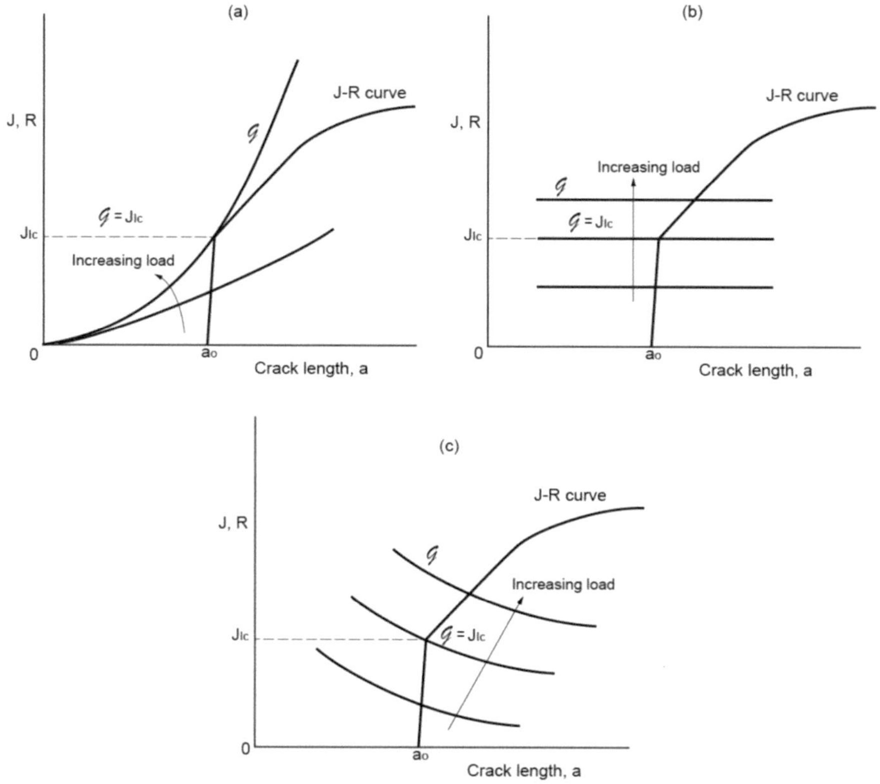

Fig. 12.9 Schematic of the three different cases of loading conditions of a system containing a crack: **a** by applying a force, F, or moment, M, held constant, **b** by applying a force, F, or moment, M, to a tapered specimen having a K_I or J_I independent of crack length, ***a*** and **c** by applying a constant displacement, wedge-loading. Also shown is the resistance *J-R* curve of the material

minimum in the fracture energy-crack velocity dependence or toughness exceeds $\mathscr{G}$ or K_I. However, Eq. (12.3) do not take into consideration the role of the kinetic energy that contributes to recovered strain energy.

When kinetic energy associated with fracture propagation returns to the crack tip process zone, prior to the crack arrest, it converts into fracture energy and must be taken into consideration in the energy balance. The initial energy release rate $\mathscr{G}_1$ is given by

$$G_I = \frac{1}{B}\left(\frac{\partial W}{\partial a} - \frac{\partial U}{\partial a}\right) \tag{12.4}$$

with W being the work done by the external forces acting on the system, U the potential energy (elastic deformation energy) and B the thickness of the test piece. Shortly after initiation the crack attains a velocity V that builds up kinetic energy T

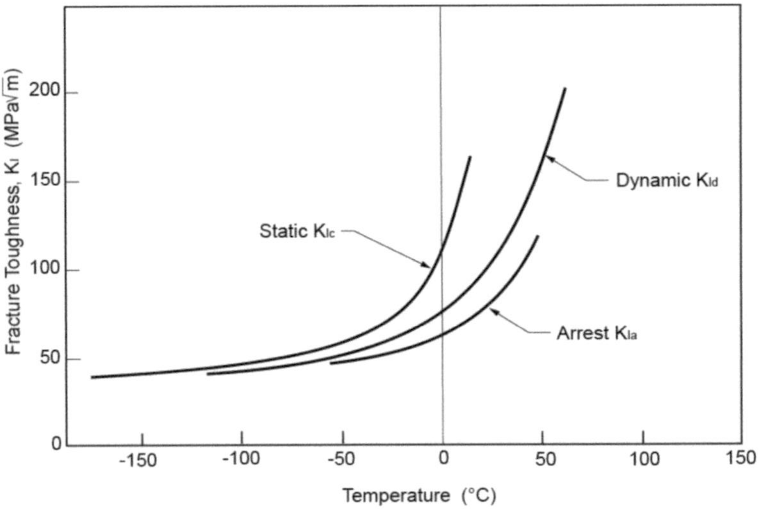

Fig. 12.10 Variation of static K_{Ic}, dynamic K_{Id} and arrest K_{Ia} toughness with temperature in A 533 Gr B Cl 1 carbon steel

so that we can write

$$\mathcal{G}_{ID} = \frac{1}{B}\left(\frac{\partial W}{\partial a} - \frac{\partial U_D}{\partial a} - \frac{\partial T}{\partial a}\right) \tag{12.5}$$

where U_D is the potential energy (strain energy) under the dynamic conditions of propagation and T the kinetic energy. Physically, $\mathcal{G}$ is the energy per unit area of crack extension that is available to sustain crack propagation and R is the energy dissipation rate per unit area of crack growth, hence propagation will occur when, and only when

$$\mathrm{G}(t, V) = R(V) \tag{12.5a}$$

Equation (12.5a) shows the dependency of G on both time, t, and crack velocity, V, while the resistance depends on the crack velocity, V. If it is

$$\mathrm{G}(t, V) < R(V) \tag{12.5b}$$

the crack cannot continue its propagation and arrests. The new term, U_D, can be much different from the static U energy term that appears in Eq. (12.4) because the strain rates ahead of a running crack are very high, even much larger than those attainable in a dynamic test for measuring the dynamic crack initiation toughness, K_{Id}. These very high strain rates result in a sudden ponderous increase of the crack tip yield stress, σ_s, see Sect. 5.4.3, and consequent large reduction of both the crack tip plastic zone radius and the dynamic toughness, K_{ID}, of the running crack and is particularly felt at temperatures near or above the NDT where the material is very sensitive to strain rate, see Fig. 9.47. Accordingly, one might anticipate the behavior shown in Fig. 12.10 that refers to A 533 Gr B Cl 1 carbon steel. At very low temperatures where the static toughness reaches the lowest values, static K_{Ic}, dynamic K_{Id} and crack arrest K_{Ia} toughness are not greatly different. At variance, where K_{Ic} increases more rapidly with temperature, the dynamic toughness, K_{Id}, is significantly lower than the static one and the arrest toughness, K_{Ia}, which is associated with the maximum possible strain rates, is significantly lower than the previous two toughness. Equation (12.5b) can be expressed [23] through a minimum fracture resistance either in terms of energy, $R(V)_{min}$, or stress intensity, K_{Im}, Eq. (12.3d) as

$$K_I < K_{Im} \tag{12.5c}$$

Examples of possible K_{ID} crack velocity curves are shown in Fig. 12.11. The dynamic fracture toughness K_{ID} may have a constant value with crack velocity, Fig. 12.11a or a minimum at zero velocity, Fig. 12.11b or at high crack velocity, as shown in Fig. 12.11d. In this last case the material property governing crack arrest length is not simple K_{Im}, but the portion of K_{ID} crack velocity curve sampled during the run arrest event (just before and soon after the K_{Im}). Figure 12.12 [24] is an examples of K_{ID} crack velocity curve for a VCN-100 steel (UNI 38 NiCrMo4) in two heat-treated conditions (annealed, $\sigma_s = 390$ MPa and Q&T, $\sigma_s = 1360$ MPa) and 4340 steel which resemble the Fig. 12.11d trend.

Figure 12.13 [15] presents a crack velocity profile measured with a 2″ CCA specimen, see Sect. 12.4, of A 533 Gr. B, Cl. 2 533 steel, that after the initial sudden jump to a very high value in the hardened zone at the notch tip shows a drop to a rather constant value of about 600 m/s in the base material with a consequent constant dynamic fracture toughness $K_{ID} = K_{Im}$, as in Fig. 12.11a. Hahn et al. [20]

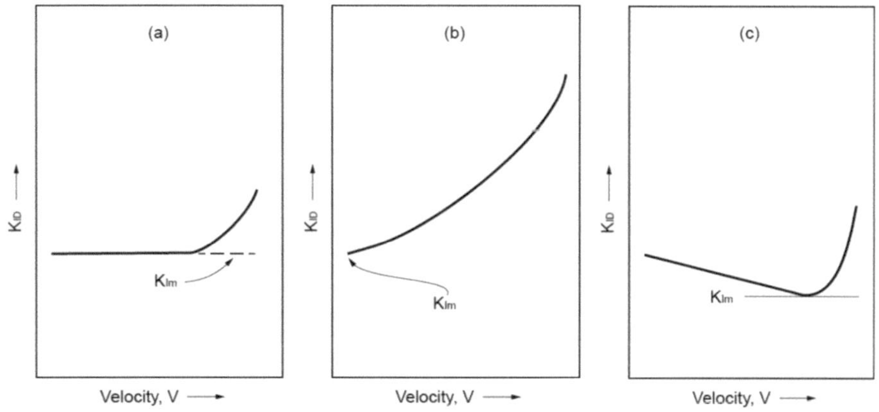

Fig. 12.11 Possible fracture resistance crack velocity curves: **a** and **b** are examples where the minimum K_{Im} corresponds with zero velocity, and **c** where K_{Im} corresponds with a high velocity

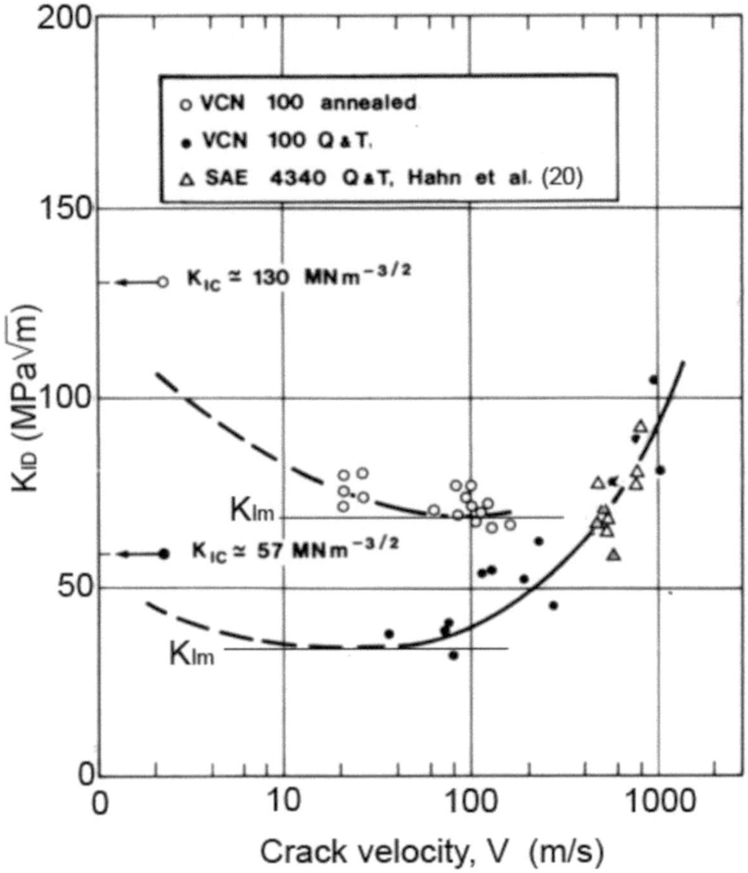

Fig. 12.12 K_{ID} crack velocity curves for VCN-100 and 4340 steels (integrated from [24])

have shown that the crack arrest toughness, K_{Ia}, approaches K_{Im} in the limit of an infinitesimal crack jump when the minimum of velocity curves occurs at zero crack velocity. In this cases Eq. (12.5c) may be written as

$$K_{Ia} = K_{Im} \tag{12.5d}$$

where K_{Ia} is a useful approximation for small crack jumps.

A small crack jump may seem a non-operative definition. Calculations made by Hahn et al. [20] for rectangular DCB specimens, see Sect. 12.3, show that K_{Ia} underestimates K_{Im} by 10 percent for a 20-mm jump, and 20 percent for a 30-mm jump. Kalthoff et al. [25], in remarkable experimental research with DCB specimens of Araldite B, have shown that effectively differences between staticK_{Ia} and dynamic $K_{IA} = K_{Im}$ become smaller and smaller as the crack jump reduces which actually

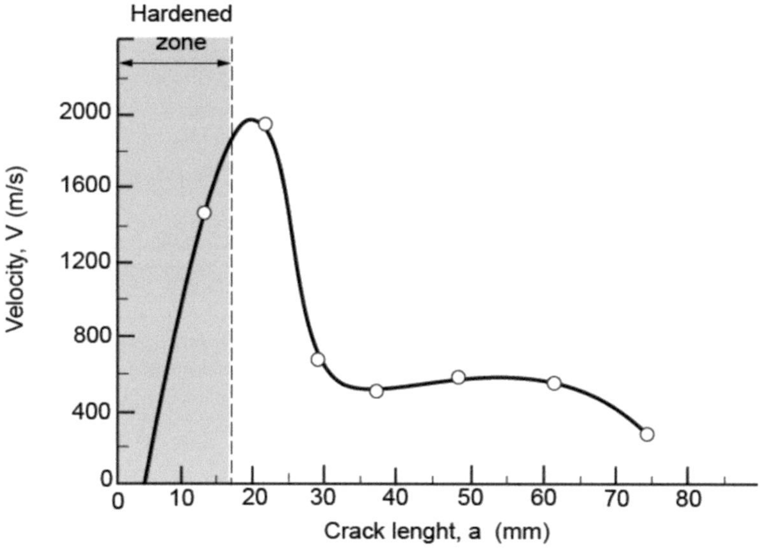

Fig. 12.13 Crack velocity measured at ENEA on an instrumented 2″ CCA specimen of A 533 Gr. B, Cl. 2 material at RT [15]

means when the crack velocity reduces, as shown in Fig. 12.14. The static K_{Ia} consistently underestimates K_{Im} and cannot be used when K_{Im} corresponds with a high crack velocity.

It clearly appears that early in the run-arrest event, dynamic effects are so large that static calculations do not describe the specimen behavior. At a certain point into the transient, dynamic response overwhelms the static one till equality is reached between K_{Ia} and K_{Im} at the final zero speed. The option that the entire kinetic energy reenters into the process zone is considered by putting equal zero the integral of the difference between Eqs. (12.4) and (12.5). Recalling the relationship between G and K_I, Eq. (7.141), and considering that there is no external work W done on the system, we have

$$\int_{a_0}^{a_0+\Delta a} (\mathcal{G}_I - \mathcal{G}_{ID})da = \frac{1-\nu^2}{E}\int_{a_0}^{a_0+\Delta a} \left(K_I^2 - K_{ID}^2\right)da = 0 \tag{12.6}$$

The limits of integration are the initial crack length, a_o, and the arrest one, $a_o + \Delta a$. Equation (12.6) states that all kinetic energy made available as fracture energy reenter into the crack tip process zone and supports crack propagation. This is certainly an approximation of what may happen and clearly requires a large crack jump that takes enough time to let the kinetic energy reenter into the process zone as reflected waves, not just one but several times to be completely accounted for.

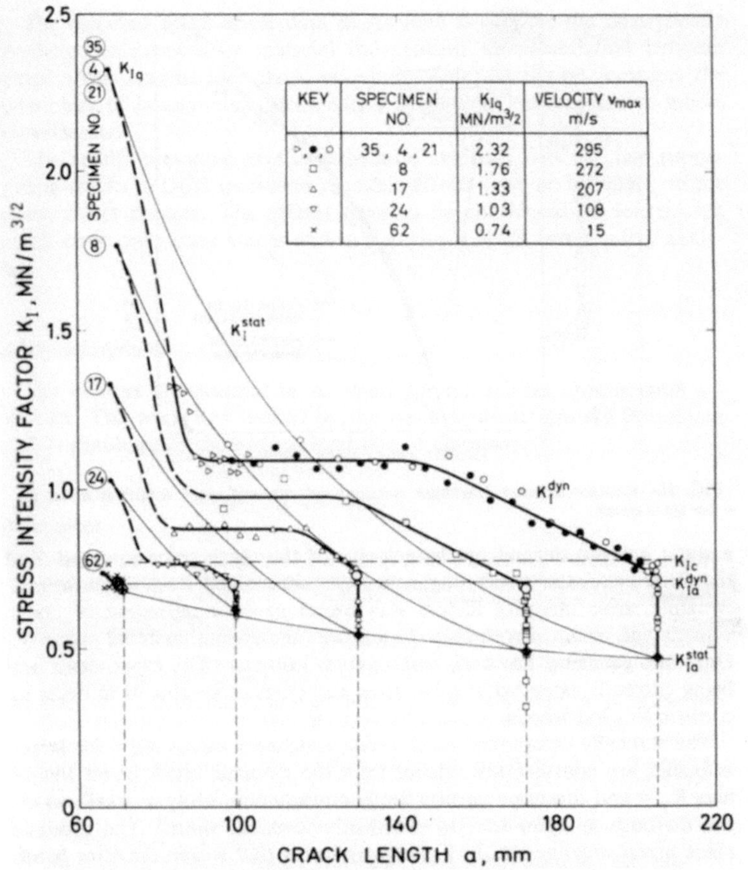

Fig. 12.14 Crack arrest values determined in the experimental program by Kalthoff et al. using Araldite B DCB specimens. It can be seen how the difference between (K_{Ia}^{stat}) and K_{IA} (K_{Ia}^{dyn}) decreases as the crack jump reduces, specimen (62), and so does the crack velocity [25] (Symbols are the same used in Fig. 12.3)

In a small specimen this may not happen, and this is probably the reason why the specimens used by Crosley and Rippling, see previous section, where yielding almost constant K_{Ia} values. Equation (12.6) can also be used to assess crack arrest graphically, as shown in Fig. 12.15. Crack arrest is predicted when the two shaded areas, A and B, are equal. Using a technique similar to that of Broberg [6], Freund [8] found

$$K_I = \sqrt{A(V)} \cdot \sqrt{\frac{E\mathcal{G}_I}{1-\nu^2}} \tag{12.7}$$

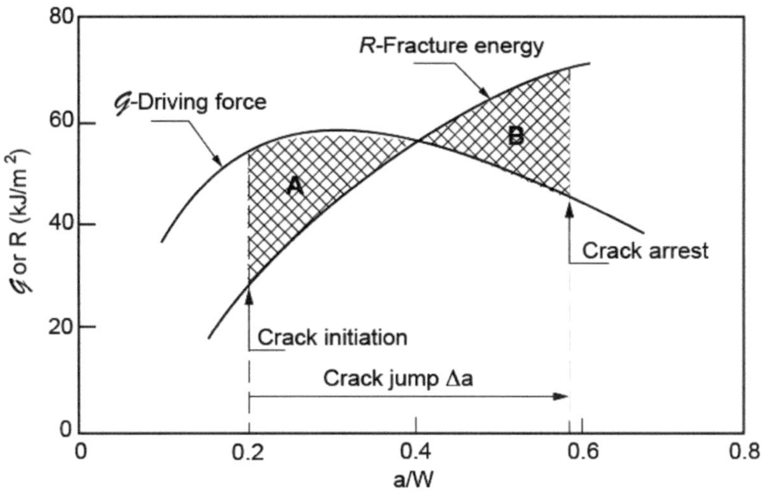

Fig. 12.15 Schematic of approximate analysis using the method of kinetic energy reuse: the area A, which is the excess of driving force over the fracture energy absorbed, must be equal to the area B, which is the excess of fracture energy over driving force

and

$$K_{ID} = \sqrt{A(V)} \cdot \sqrt{\frac{ER_{ID}}{1 - \nu^2}} \tag{12.8}$$

where the function $A(V)$ depends on the crack velocity, V. The value of $\sqrt{A(V)}$ for steel is close to unity for relatively low crack speeds.

$$1 \leq \sqrt{A(V)} \leq 1.1 \tag{12.9}$$

for

$$0 \leq V \leq 1500 \text{ m/s } (0.3C_o) \tag{12.10}$$

where C_o is the velocity of sound in steel. Measurements of crack velocity, V, were made at ENEA with a 2"CCA specimen of A 533 Gr. B, Cl. 2 material, Fig. 12.12 [15]. The crack velocity, very high in the hardened zone of the notch, decreases to remain almost constant at about 600 m/s. This independence of the crack velocity with crack length would result in a parallel independence of the dynamic $K_{ID}(V)$ from crack length.

12.3 The DCB Specimen

The main competitor in the academic dispute on whether or not the crack arrest toughness had to be a material characteristic, was the wedge-loaded Double Cantilever Beam specimen (DCB) designed at Battelle Columbus (BCL). Its main features were the wedge-loading system, Fig. 12.2, that provided an energy release rate $\mathcal{G}$ decreasing with crack length, a, Fig. 12.9c, that led to the arrest of a running crack, and its size, W, that was wide enough to provide a crack jump sufficiently long to last the time needed to an almost complete return of reflected stress waves into the crack tip process zone. Its shape and dimensions are shown in Fig. 12.13. The notch tip was not fatigue pre-cracked. The radius ρ was chosen to load the system to a required K_Q level above the K_{Ic} toughness of the test material before triggering the fracture, Figs. 6.26 and 6.27. This expedient allowed to store more energy in the specimen at crack initiation than possible with a sharp crack. Notch tip radii were varying from 0.25 mm to about 2 mm allowing to reach a ratio of 2 over the K_{Ic}. Figure 12.16 shows two type of specimens. Type(a) is the standard specimen used with brittle steels like 4340 high strength steel or low carbon steels at low temperatures (below the NDT). When the resistance to crack initiation is high, as in tough, low strength materials at temperature over the NDT, type (b) duplex specimen is used in which a starter section of brittle steel is welded, by electron beam weld, to the test section. The unstable crack propagation is easily achieved in the brittle section and enters the test section at high velocity, where arrests.

Typical measures for steels used at BCL, both ordinary and duplex specimen, were: width, 250 mm $\leq$ W $\leq$ 500 mm, 50 mm $\leq$ H $\leq$ 75 mm, 0.9 $\leq a_o$/H $\leq$ 1.5. The static stress intensity factor K_I at initiation is [26].

$$K_I = \frac{P}{B\sqrt{H}} \cdot f\left(\frac{a}{W}\right) \tag{12.11}$$

$$f\left(\frac{a}{W}\right) = 2\sqrt{3}\left(\frac{a}{H} + 0.64\right) \tag{12.12}$$

with the displacement δ on the load line, COD, being [22].

$$\delta = \frac{P(1-\nu^2)}{EB} \cdot h\left(\frac{a}{W}\right) \tag{12.13}$$

$$h\left(\frac{a}{W}\right) = 4\left(\frac{a}{H}\right)^3\left(1 + 0.64\frac{H}{a}\right)^3 \tag{12.14}$$

The load P opening the specimen can be obtained from the force F applied to the wedge using the friction methodology explained in Sect. 12.6. In the duplex specimen the starter section width has an impact on the arrest toughness, as schematically shown in Fig. 12.17 [20] for a starter section of very hard 0–1 steel and a test section of high strength 4340 steel. According to the kinetic energy reuse or balance, in

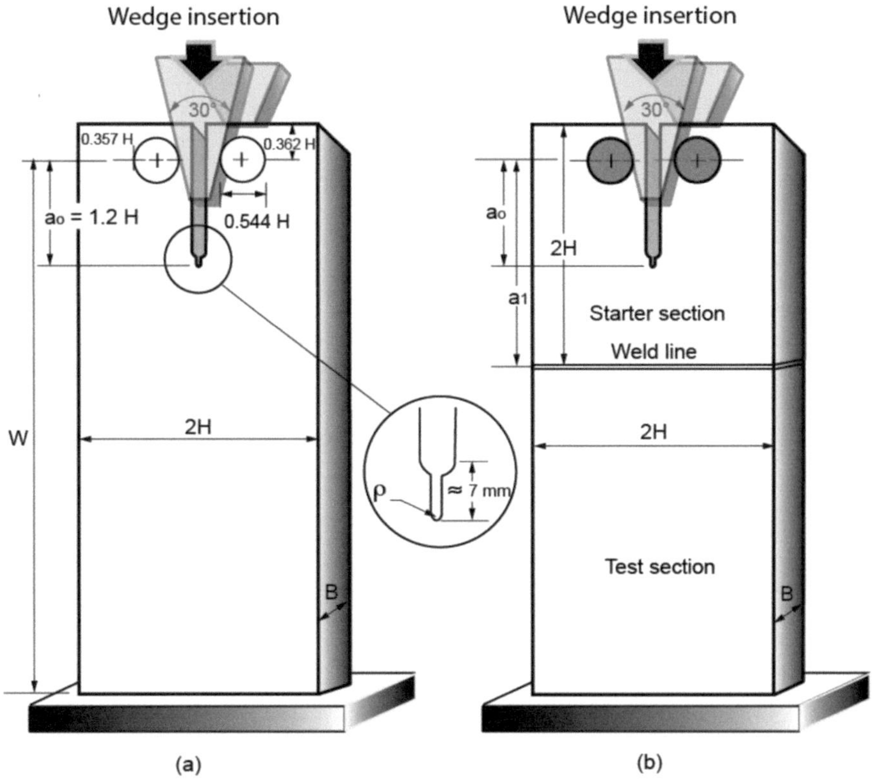

Fig. 12.16 Wedge-loaded DCB specimen: **a** ordinary, **b** duplex

the short brittle starter, Fig. 12.17a, the crack, of length a_I, absorbs less energy per unit extension made available by release of strain energy. This energy, region A of Fig. 12.17b, is converted into kinetic energy that reenters the process zone in the test section feeding the strain energy in driving the crack. Since it must be absorbed by the material before arresting, region B of Fig. 12.17b, it takes a smaller crack jump to make the balance. The crack will arrest when the net loss in strain energy is equal to the net gain in fracture energy, i.e., the area A must be equal to the area B. With a longer starter section of low toughness, Fig. 12.17c, loaded to the same K_Q level, more kinetic energy is generated that will return in the process zone in the test section contributing to the driving force. A longer crack propagation occurs, Fig. 12.17d resulting in a lower crack arrest toughness, K_{Ia}. The compliance, C, of the Battelle DCM specimen is shown in Fig. 12.15 [19]. Its equation is

$$C = \frac{4a^3}{EBH^3}\left[1 + 0.811\left(\frac{H}{a}\right)^3\right] \tag{12.15}$$

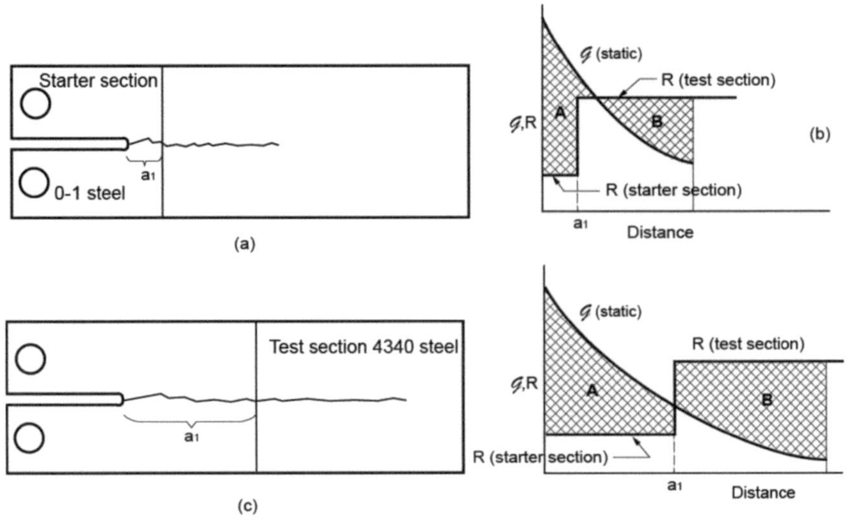

Fig. 12.17 Schematic illustration of the reutilization of kinetic energy in the test section of a duplex DCB specimen of very hard O–1 steel coupled to high strength 4340 steel (from [20])

where a is the crack length, B the thickness and $2H$ the beam height.

12.4 Transverse-Wedge Loading Compact Crack-Arrest Specimen

By the end of the Seventies the original contoured specimen initially developed at MRL, Fig. 12.1, had been substituted by the new transverse-wedge loaded compact specimen shown in Fig. 12.19. The specimen, called CCA (Compact Crack-Arrest) is loaded by the insertion of a wedge between two split pins that enter and fit the central hole. Specimen dimensions are shown in Fig. 12.20a. Note the clip cage applied on the surface to measure the CMOD. At the time the force opening the specimen was not known because of the friction angle between the wedge and the split pins that remained undetermined. In 1980 Milella developed a simple method to measure this friction angle and assess with great accuracy the actual force opening the specimen, as discussed in Sect. 12.6. To allow for a longer crack jump without impinging the ASTM requirements, Milella [15] introduced a compact specimen similar to the CCA but a little longer, which was called wedge-loaded compact crack arrest specimen or WCA. The WCA specimen is shown in Fig. 12.17b along with its dimensions. We know, however, that larger crack jump results in larger differences between staticK_{Ia} and dynamic $K_{IA} = K_{Im}$, as discussed in Sect. 12.2 and Fig. 12.18.

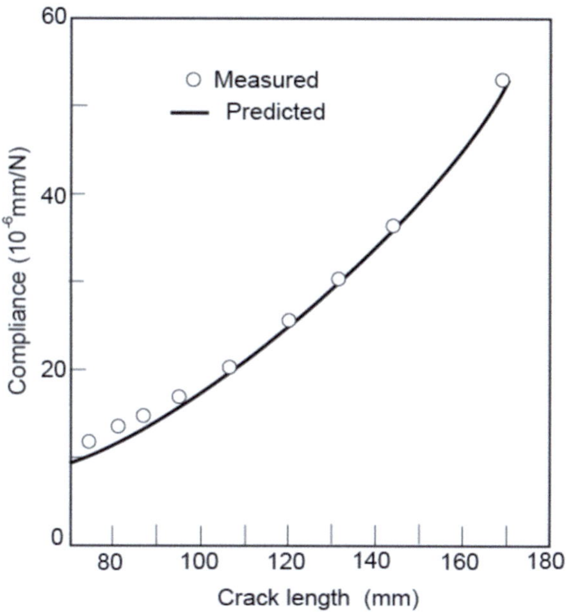

Fig. 12.18 Comparison of measured and predicted compliance [19]

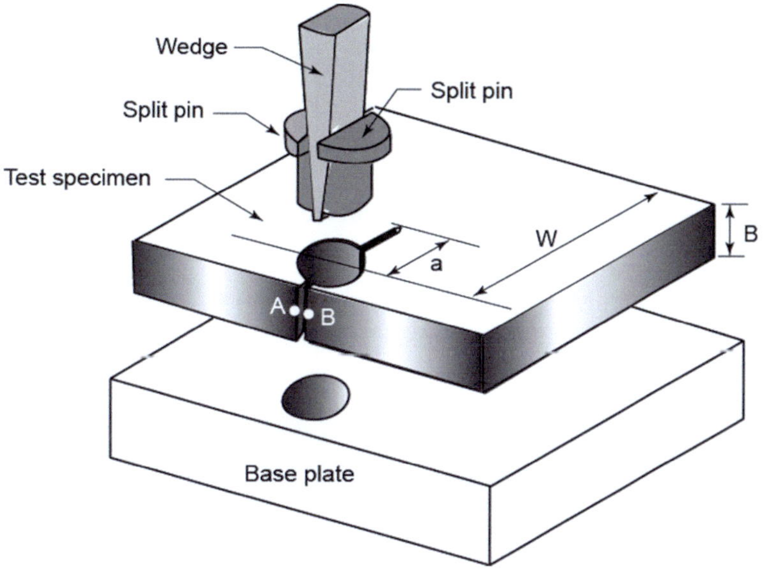

Fig. 12.19 Transverse-wedge loading arrangement of a compact-type specimen CCA

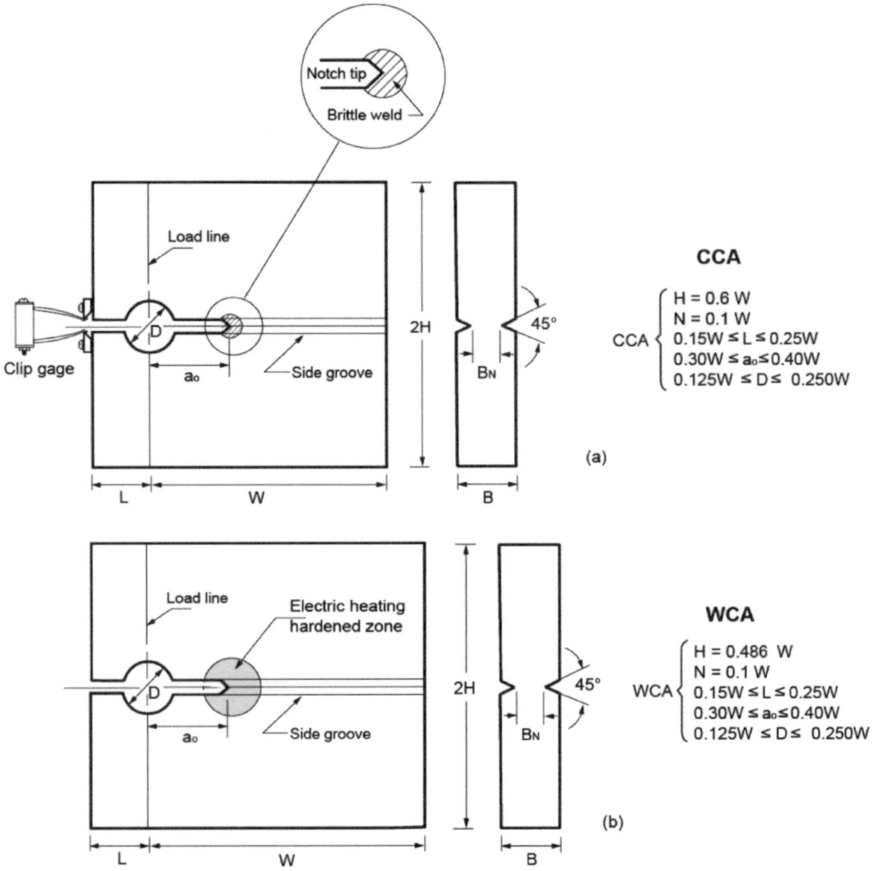

Fig. 12.20 The **a** wedge-loaded CCA specimen with COD gage and **b** WCA specimen, respectively

As shown in Fig. 12.20b, the surface clip gage was no longer needed since the force F opening the wedge was directly measured from the test record through the assessment of the friction angle.

Crack arrest toughness is particularly difficult to measure at temperatures in the range and above the NDT where the material starts to present a certain ductility and then becomes ductile enough to avoid unstable brittle fracture propagation, in particular under wedge-loading conditions. In this field of temperatures, initiation of unstable fracture under static loading may become a problem for a CCA specimen. Therefore, a brittle weld bead was deposited ahead of the initial blunt notch where a sharp crack tip was machined, as schematically shown in Fig. 12.20a. However, this solution was not always effective at temperatures over the NDT at which the weld bead would even detach from the base material stopping the crack from entering the test metal. To overcome this problem, Milella [15, 16] used to harden the metal by electric current heating, applying two electrodes on both sides of the specimen in the area of the crack tip, Fig. 12.20b. The heating produced by the current flow, which brings the metal over the transformation temperature, is followed by a rapid self-quenching that results in non-tempered hard martensite formation. To this purpose, it was used a spot weld apparatus of 260–350 kVA, with applied contact force of 30 kN and a contact area of 490–1020 mm^2 [16]. Though better than local brittle weld, the solution, however, was not really effective as expected because the electric current was not equally heating the entire thickness which concentrated only on two small surface areas no more depth than 1/3 of the thickness, B, as schematically shown in Fig. 12.21a. This happens because, independently of the electric power used, the electric flow through the thickness widens with the depth, Fig. 12.21b, reducing the power intensity in the central part of the specimen that will never reach a temperature high enough to allow a metallurgical transformation. Therefore, the technique was improved by using first a cylindrical electrode and then a prismatic one [16] positioned on the bottom of a notch, shaped to fit the notch profile and a second electrode on the opposite face, as shown in Fig. 12.21c. The power used was just 20 kVA for 90 s with a contact area of 33 mm^2.This second approach revealed very effective, allowing to run tests even at 75 °C over the reference temperature, RT_{NDT}, the highest temperature ever reached in a K_{Ia} crack arrest test with a CCA specimen.

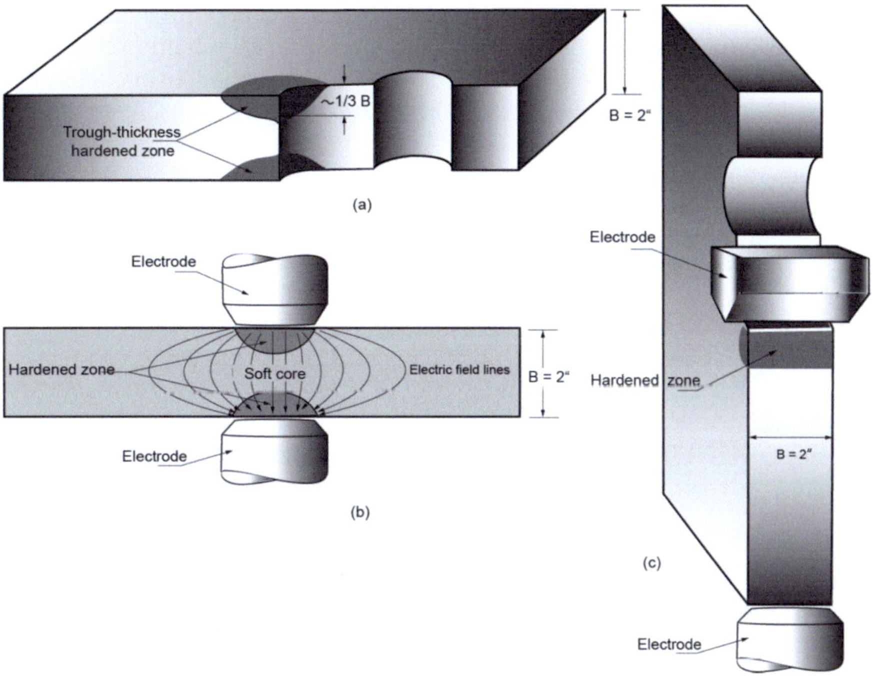

Fig. 12.21 Schematic of local hardening through electric heating at a temperature over the metallurgical transformation temperature of the steel followed by rapid self-quenching

12.5 Calibration of the CCA and WCA Specimens

Newman [27] using a boundary collocation method to calculate the displacement δ at the load-line of a CCA specimen, gave an analytical solution for the functions $g(a/W) = \text{CEB}$, $f(a/W)$ and $h(a/W)$, see Eqs. (9.25), (9.29) and (9.30). Later, Crosley and Ripling [17], using a grasping system capable of opening the specimen like a wedge, through which it was possible to measure directly the opening force F, found experimentally the calibration function g(a/W) with C = 0.77 y/F, where 0.77 is the load line displacement, δ, to clip gage displacement, y, ratio, δ/y. The equation that interpolates the experimental results obtained by Crosley and Ripling is

$$g(x) = \frac{9{,}85 - 0{,}17x + 11{,}0x^2}{(1-x)^2} \tag{12.16}$$

Figure 12.22 presents a comparison between the theoretical g(a/W) function of Newman and the experimental function obtained by Crosley and Ripling, where the respective functions g(a/W) are multiplied by $(1-x)^2 \cdot \delta/y$. In 1981 Milella and Pini [28], using the methodology outlined in Sect. 12.6 to measure the opening force F exerted by the wedge on the split pin, obtained an experimental curve using three CCA specimens of 2 ¼ Cr-1 Mo Ni steel. After each test, the crack length a of each specimen was increased for a total of 21 different crack lengths. The results are shown in Fig. 12.22, compared with the Newman and the Crosley and Ripling results. The agreement is rather excellent, also confirming the validity of the friction angle method to measure the force exerted by the wedge on the split pins, described in Sect. 12.6. The ratio δ/y measured by Milella and Pini for the CCA specimen is shown in Table 12.1.

Figure 12.23 presents the experimental results of the function g(a/W) = yEB/F obtained by Milella and Pini compared with the results obtained by Crosley and Ripling for the same function. The equation of the curve interpolating the experimental results is

$$g(x) = \frac{3.906 + 12.758x + 14.778x^2 - 60.871x^3 + 49.039x^4}{(1-x)^2} \tag{12.17}$$

Using Eq. (12.17) in Eqs. (9.29) and (9.31) it is possible to obtain the two functions $f(x)$ and $h(x)$, respectively, whose equations are

$$f(x) = \sqrt{\frac{\left(20.58 + 42.32x - 182.61x^2 + 257.03x^3 - 98.08x^4\right)(1-x)}{2\left(1-\nu^2\right)\left(3.91 + 12.76x + 14.78x^2 - 60.87x^3 + 49.09x^4\right)^2}} \tag{12.18}$$

and

Fig. 12.22 Comparison of the theoretical calibration function $g(x)\,(1{-}x^2)\,\delta/y$ obtained by Newman and the experimental calibration functions obtained by Ripling and Crosley and Milella and Pini, respectively, for the CCA specimen

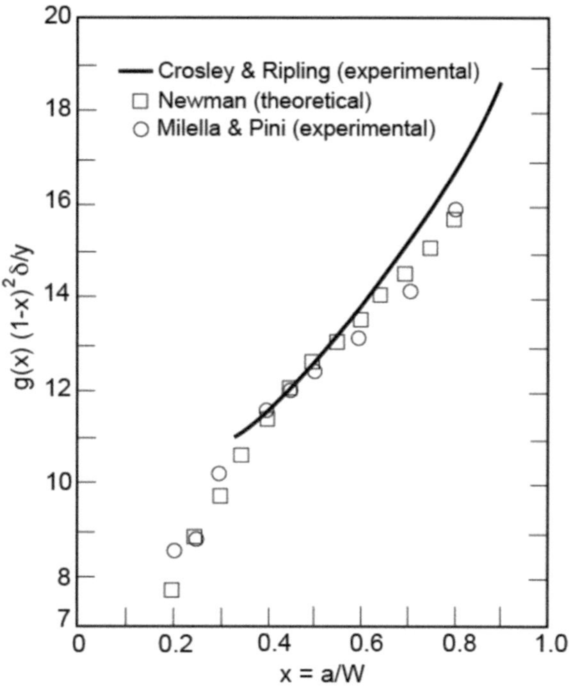

Table 12.1 Values of the δ/y ratio as function of the crack length a for a CCA specimen

a/W	0.2	0.3	0.4	0.5	0.6	0.7	0.8
δ/y	0.77	0.77	0.763	0.763	0.766	0.773	0.782

$$h(x) = \sqrt{\frac{\left(20.58 + 42.32x - 182.61x^2 + 257.03x^3 - 98.08x^4\right)}{2\left(1-\nu^2\right)(1-x)^3}} \tag{12.19}$$

The three functions $g(x)$, $f(x)$ and $h(x)$ are shown in Fig. 12.24. The same procedure has been applied to assess the calibration function $g(x)$, $f(x)$ and $h(x)$ for the longer WCA specimen of Fig. 12.20b which are

$$g(x) = \frac{5{,}073+16{,}568+19{,}192x^2-79{,}053x^3+63{,}687x^4}{(1-x)^2} \tag{12.20}$$

$$f(x) = \sqrt{\frac{\left(26{,}714+54{,}952x-237{,}15x^2+333{,}801x^3-127{,}374x^4\right)(1-x)}{2\left(1-\nu^2\right)\left(5{,}073+16{,}568x+19{,}192x^2-79{,}053x^3+63{,}687x^4\right)^2}} \tag{12.21}$$

$$h(x) = \sqrt{\frac{26{,}714+54{,}952x-237{,}15x^2+333{,}801x^3-127{,}374x^{44}}{2\left(1-\nu^2\right)(1-x)^3}} \tag{12.22}$$

Equations (12.20), (12.21) and (12.22) are in Table 12.2. Equation (9.28).

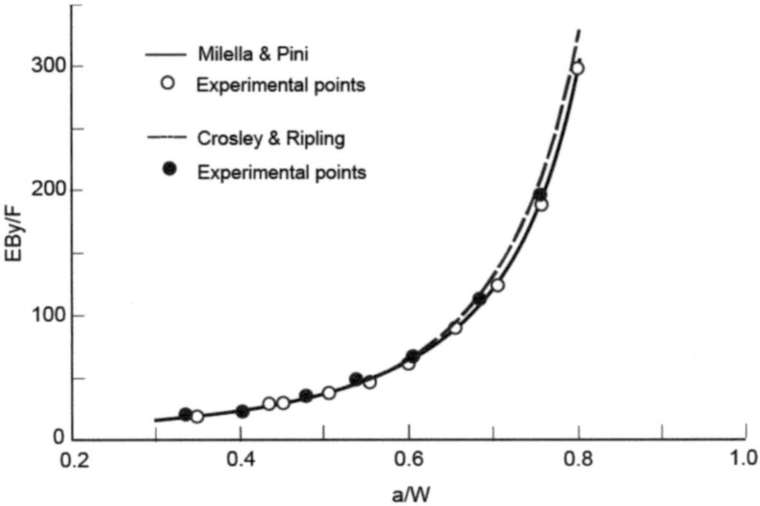

Fig. 12.23 Comparison of experimental results obtained by Milella and Pini with the calibration curve obtained by Ripling and Crowsley for the CCA specimen

Fig. 12.24 Plot of the calibration functions $g(x)$, $f(x)$ and $h(x)$ for the CCA specimen

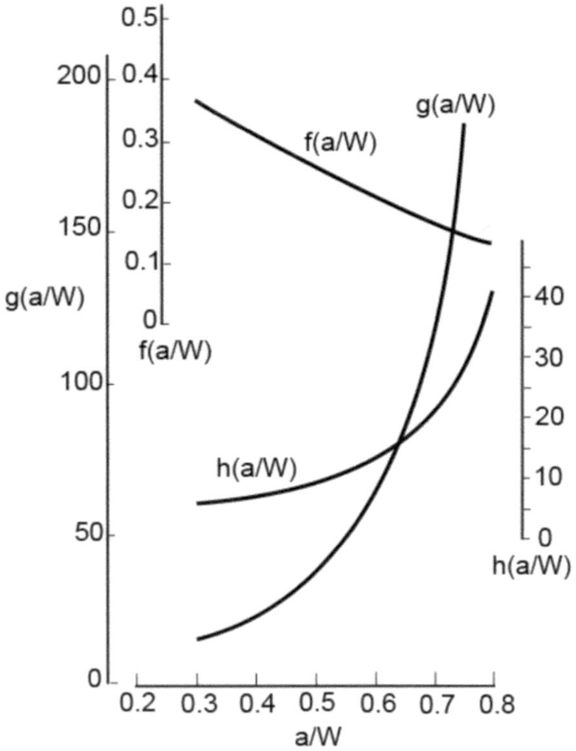

Table 12.2 Values of the experimental data, g_{ex}(a/w) of their interpolation function g(a/w), of the functions h(a/w) and f(a/w) obtained by Milella and Pini and of the theoretical $g_{N\delta}$ (a/w) e g_{Ny}(a/w) for several values of the ratio a/W for a WCA specimen

a/W	$g_{ex}(a/w)$	*g(a/w)*	*h(a/w)*	*f(a/w)*	$g_{N\delta}(a/w)$	$g_{Ny}(a/w)$	*y/δ*
0.25	16.88	16.76	5.26	0.40	15.64	12.30	0.786
0.30	20.12	20.72	5.79	0.36	19.90	15.51	0.779
0.35	26.36	25.53	6.42	0.32	25.12	19.37	0.771
0.45	39.58	39.08	8.15	0.27	39.73	30.34	0.763
0.50	50.19	49.01	9.53	0.25	50.33	38.42	0.763
0.55	64.17	62.55	11.1	0.23	64.59	49.36	0.764
0.60	82.76	81.88	13.5	0.21	84.63	64.85	0.766
0.65	112.7	111.09	16.9	0.19	114.3	87.91	0.769
0.70	159.7	158.34	21.8	0.18	161.0	124.0	0.770
0.75	240.0	241.52	29.4	0.16	240.8	186.5	0.774
0.80	390.0	405.53	42.6	0.14	392.4	305.7	0.779

$$KI = yE\sqrt{\frac{B}{B_n W}}f(x) \tag{12.23}$$

or (9.30).

$$KI = \frac{F}{\sqrt{BB_n W}}h(x) \tag{12.24}$$

can be used to calculate the K_Q at initiation and the K_a at arrest knowing the initial crack size, *a*, and the crack length at arrest.

12.6 Crack Opening Force in Transverse-Wedge Loading

With transverse-wedge loading the actual value of the force, N, opening the specimen is not known because of the friction between the wedge and the split pins or between the wedge and the rods, as in the case of the DCB, so that the loading conditions must be deduced from the COD readings done by a clip gage placed on the specimen surface, Fig. 12.20. There is, however, a very simple method of measuring the friction acting against the wedge and, therefore, measuring the actual force, N, opening the specimen [29]. In the absence of friction, N is related to P, the load applied to the wedge, Fig. 12.25, by the relation.

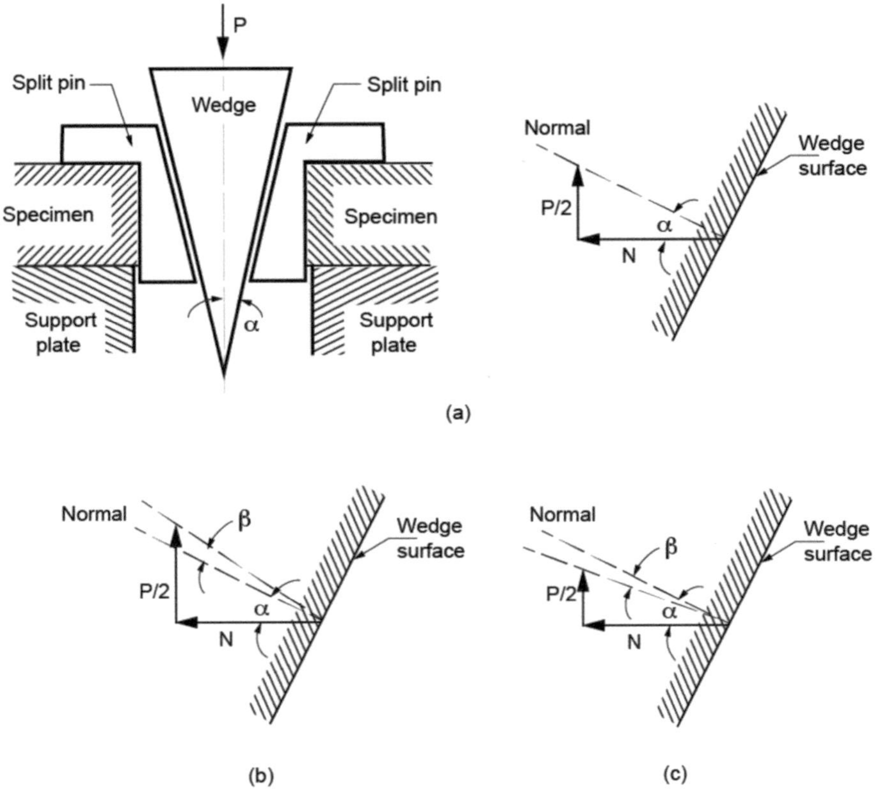

Fig. 12.25 Schematic of the force P acting on the wedge and force N opening the specimen: **a** in absence of friction the two components P/2 and N depend on the wedge semi angle α, **b** wedge insertion with friction, the β friction angle is added, and **c** wedge withdrawal, reversal of motion the β angle is subtracted [29]

$$N = \frac{P}{2}\frac{1}{tan\alpha} \tag{12.25}$$

where α is the wedge half angle. When friction is present, Fig. 12.25b, the appropriate relation is

$$N = \frac{P}{2}\frac{1}{tan(\alpha + \beta)} \tag{12.26}$$

where β is the friction angle. Hence, if β can be determined, N can be assessed from the known value of P and α. All what must be done is the momentarily reverse the direction of travel of the crosshead test-machine. When this is done, the friction that was previously acting against wedge insertion now acts against wedge withdrawal, Fig. 12.25c. The result is a sudden measurable drop in load with no change in displacement, as schematically shown in Fig. 12.26. The magnitude of the load

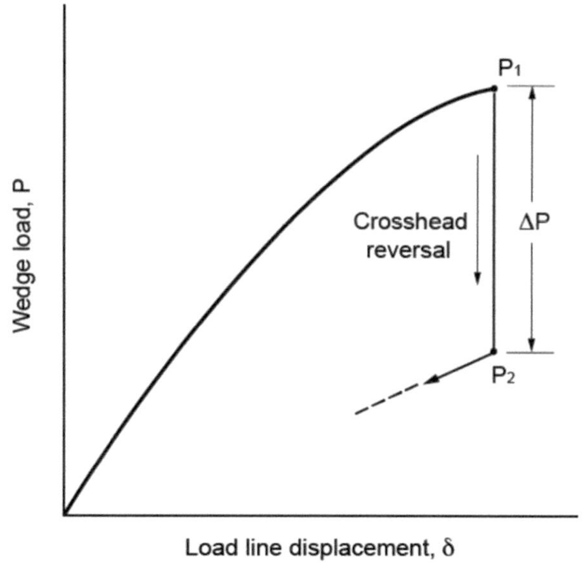

Fig. 12.26 Schematic of load versus displacement record for transverse wedge insertion and withdrawal

drop relative to the load that existed just prior to crosshead reversal can be used to assess β, as follow: let N_1 and P_1denote values at the instant of crosshead reversal and N_2 and P_2 denote values at the end of the abrupt load drop.

Then, from Fig. 12.25b and c it is

$$N_1 = \frac{P_2}{2 \cdot tan(\alpha - \beta)} \tag{12.27}$$

$$N_2 = \frac{P_2}{2 \cdot tan(\alpha - \beta)} \tag{12.28}$$

Inasmuch as the displacement does not change in this interval, N_1, can be set equal to N_2. Hence, from Eqs. (12.27) and (12.28).

$$\frac{tan(\alpha - \beta)}{tan(\alpha + \beta)} = 1 - \frac{\Delta P}{P_1} \tag{12.29}$$

where $\Delta P = P_1 – P_2$. If both α and β are small, Eq. (12.29) can be simplified to

$$\frac{a}{\beta} = \frac{2P_1}{\Delta P} - 1 \tag{12.30}$$

To confirm the adequacy of Eqs. (12.29) and (12.30) to yield the correct value of N, an indirect method can be used: the load line displacement, δ, of transverse wedge-loaded compact specimen was measured and the specimen compliance.

$$C = \frac{\delta}{N} = \frac{2\delta}{P} tan(\alpha + \beta) \tag{12.31}$$

was used to calculate the crack length, which could be compared to the actual crack length. Values of C from Eq. (12.31) were converted to crack length, a, through the relation.

$$C = \frac{g(a/W)}{EB} \tag{12.32}$$

where E is the Young's modulus, B the specimen thickness and $g(a/W)$ is the calibration function from an analysis by Newman [30], shown Fig. 12.15. This check has been done extensively with a result of 1% to 5% accuracy [31].

12.7 Crack-Arrest Testing

In 1988, ASTM published a standard for crack arrest testing, E 1221 [30]. It prescribes the use of a side grooved crack-arrest compact, CCA, specimen, the type described in Sect. 12.4. The standard procedure has not yet recognized that the normal load, N, can be easily assessed through the friction procedure previously described, Eq. (12.13), therefore is still based on the COD measurement obtained by means of a clip gage, as shown in Fig. 12.19, that must be multiplied by 0.77 to obtain the load-line corresponding displacement. This procedure has to face the problem of spurious displacements caused by the wedge-pin assembly and the local yielding occurring in the starter notch, which must be somehow recognized and not considered in the test analysis. Therefore, to identify these displacements a cyclic loading procedure is adopted making the assessment of K boresome long lasting procedure. Figure 12.27 is a test record of 1″ CCA specimen of SA 533 Gr.B, Cl.1 steel at RT. The first cycle can be considered the settlement cycle where residual displacements of the test rig complex occur. In the case of Fig. 12.27 it is worth about 0.25 mm and must be subtracted to the final displacement.

The specimen is then reloaded and unloaded and this procedure continues till fracture occurs. This happens in the 5th cycle for the specimen of Fig. 12.20 where a sudden load drop occurs followed by a small jump and a final arrest. The question of whether or not the relative subsequent zero-load offset displacements preceding the last cycle must be subtracted or not to the final displacement at crack instability is not completely clear, at the moment. They may include plastic strains in the notch tip hardened zone.

The ASTM standard requires that half of these plastic offsets be included in the stress intensity K_{Ia} calculation. At final arrest the wedge is extracted, and load goes to zero. At this point, the specimen is heat tinted at 250–350 °C for some 10–20 min to mark the crack propagation and the specimen broken open to measure the final crack length. This is shown in Fig. 12.28. Calculation performed according to the ASTM procedure for the test whose graph is reported in Fig. 12.27 yielded a value

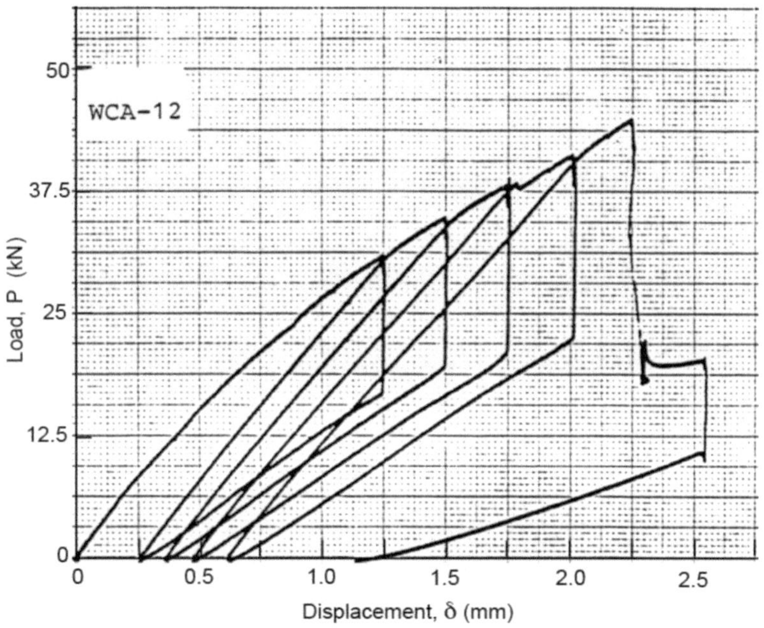

Fig. 12.27 Test diagram obtained at RT with a CCA specimen of SA 533 steel as per ASTM cyclic test procedure for crack-arrest measurement. Fracture occurred at the 5th cycle

of $K_{Ia} = 139$ MPa$\sqrt{}$m. The friction angle β assessed through Eq. (12.17) was for each cycle: $\beta_1 = 1.49$, $\beta_3 = 1.49$ and $\beta_5 = 1.40$. Using Eq. (12.13), the normal force, N, opening the specimen at arrest was: N = 178 kN and the corresponding crack arrest toughness was: $K_{Ia} = 138$ MPa$\sqrt{}$m, practically equal to the ASTM value. The corresponding initiation toughness was $K_Q = 227$ MPa$\sqrt{}$m. At this point, the ASTM standard requires a validity check to ensure that the plastic zone is small compared to specimen dimensions

$$\mathrm{W} - a_a \geq 0.15\ \mathrm{W} \tag{12.33}$$

$$\mathrm{W} - a_a \geq 1.25\left(\frac{K_a}{\sigma_{sd}}\right)^2 \tag{12.34}$$

$$B \geq 1.0\left(\frac{K_a}{\sigma_{sd}}\right)^2 \tag{12.35}$$

$$a_a - a_o \geq \frac{1}{2\pi}\left(\frac{K_a}{\sigma_{sd}}\right)^2 \tag{12.36}$$

Fig. 12.28 View of the heat-tinted final fracture crack length obtained in a CCA specimen of SA 588 steel at RT

where a_a is the crack length at arrest, a_o the initial crack length and σ_{sd} the dynamic yield strength since unstable crack propagation produces very high strain rates capable of raising the yield stress of the material much over its quasi-static value, σ_s, see Sect. 5.4.3. The ASTM standard specifies the dynamic yield stress, σ_{sd}, to be 205 MPa over the quasi-static yield stress, σ_s, which is rather conservative. Once the requirements (12.33), (12.34), (12.35) and (12.36) are met, the calculated, K_{Ia}, can be considered a valid K_{Ia}.

References

1. Kraft, J.M., Irwin, G.R.: Crack-velocity considerations, fracture toughness testing and its applications, ASTM, STP-381, pp. 114–129 (1965)
2. Irwin, G.R., Wells, A.A.: Metallurgical review **10**, 223 (1965)
3. Crosley, P.B., Ripling, E.J.: J Basic Eng, Trans, ASME, 91, Series D, 3 (1969)
4. Mott, N.F.: Engineering **165**, 16 (1948)
5. Yoffe, E.H.: The moving griffith crack. Phil. Mag. **42**, 739–750 (1951)
6. Broberg, K.B.: J. Appl. Mech., ASME, 546 (1964). Also: Recent Progress in, Broberg, et al. (eds), p. 125 (1967)
7. Eshelby, J.D.: The force on an elastic singularity. Phil. Trans. **244**, 87–111 (1951)
8. Freund, L.B.: Journal of Journal of Applied Mechanics, ASME the Mechanics and Physics of Solids, vol. 20, Part 1, p. 129 and Part. 2, p. 141 (1972). Also: J. Mechanics and Physics of Solids, vol. 21, p. 47 (1973)
9. Hahn, G.T, Gehlen, P.C., Hoagland, R.G., Kanninen, M.F., Popelar, C., Rosenfield, A.R.: Critical experiments, measurements and analysis to estabilish a crack arrest methodology for nuclear pressure vessel steels, BMI-1934, 1937, 1939,1951, 1959 (1975–1976)
10. Kaltoff, J.F, Beinert, J., Winkler, S.: Measurements of dynamic stress intensity factors for fast running and arresting cracks in DCB specimens. In: ASTM symposium on fast fracture and crack arrest, STP-627, pp. 161–176 (1976)

11. Kaltoff, J.F, Beinert, J., Winkler, S., Klemm, W.: Experimental analysis of dynamic effects on different crack arrest test specimens. In: ASTM symposium on crack arrest methodology and applications, STP-711, p. 111 (1978)
12. PVRC Recommendations on toughness requirements for ferritic materials, PVRC Ad Hoc groupson toughness requirements. Welding Research Council Bulletin, WRC-175, (1972)
13. Hahn, G.T., Kanninen, M.F., (eds.): Fast fracture and arrest, ASTM STP-627. American Society for Testing and Materials, (1977)
14. Naus, D.J., Keeney-Walker, J., Bass, B.R., Bolt, S.E., Fields, R.J., deWit, R., Low III, S.R.: High-temperature crack-arrest behavior in 152-mm-thick SEN wide plates of quenced and tempered A 533 Grade B Class 1 Steel, NUREG/CR-5330, ORNL/TM-11083, p. 31 (1989)
15. Milella, P.P., Squilloni, A., Traficante, M.: Crack arrest study, testing methodology, procedure and results in Italy, Presentation to the ASTM Round Robin on, K_{Ia}, Testing Workshop, MPA, Stuttgart (1985)
16. Milella, P.P., Pini, A. Squilloni, A.: ENEA program on pressure vessel integrity, HSST meeting on dynamic fracture and crack arrest program, Gaithersburg, MD, (1987)
17. Crosley, P.B., Ripling, E.J.: Crack arrest studies, EPRI NP-1225, Final report, p. I-21 (1979)
18. Hahn, G.T, Gehlen, P.C., Hoagland, R.G., Kanninen, M.F., Popelar, C., Rosenfield, A.R.: Critical experiments, measurements and analysis to estabilish a crack arrest methodology for nuclear pressure vessel steels, NURER/CR-1887 (1981)
19. Hahn, G.T, Corten, H.T., Gehlen, P.C., Hoagland, R.G., Kanninen, M.F., Popelar, C., Rosenfield, A.R., et al.: Critical experiments, measurements and analysis to estabilish a crack arrest methodology for nuclear pressure vessel steels, NURER/CR-0057, BMI-1995, Progress Rep. October 1976-September 1977
20. Hahn, G.T, Gehlen, P.C., Hoagland, R.G., Kanninen, M.F., Popelar, C., Rosenfield, A.R., deCampof, V.S.: Critical experiments, measurements and analysis to estabilish a crack arrest methodology for nuclear pressure vessel steels, BMI-1937, p. 6.3 (1975)
21. Hahn, G.T, Hoagland, R.G., Kanninen, M.F., Rosenfield, A.R.: Pressure vessel technology, Part II, ASME, p. 981–994 and Part III, p. 121 (1974)
22. Pressure vessel technology, Part III, ASME, p.126 (1974)
23. Hoagland, R.G., Rosenfield, A.R. Gehlen, P.C., Hahn, G.T.: A crack arrest measuring procedure for K_{Im}, K_{ID} and K_{Ia} properties
24. Angelino, G.C.: Influence of the geometry on unstable crack extension and determination of dynamic fracture mechanics parameters. ASTM-STP **627**, 392–407 (1976)
25. Kalthoff, J.F., Beinert, J., Winkler, S.: Measurements of dynamic stress intensity factors for fast running and arresting cracks in double-cantilever-beam specimens. ASTM-STP **627**, 161–176 (1976)
26. Kanninen, M.F.: An augmented double cantilever beam model for studying crack propagation and arrest. Int. Journal of Frac. **9**(1), 83–91 (1973)
27. Newman, J.C.: NASA Report TND-8268 (1976)
28. Milella, P.P., Pini, A., Barra Caracciolo, M.: Assessment of crack arrest properties of nuclear pressure vessel steels by means of small specimens. In: SMiRT 10 conference, pp. 125–130. Anaheim, Ca, USA, (1989)
29. Milella, P.P., Marschall, C.W., Rosenfield, A.R.: A method to obtain the crack-opening force in transverse-wedge loading. Int. J. Fract. **17**(4), R77–R81 (1981)
30. E1221–96.: Standard method for determining plain-strain crack-arrest toughness, KIa, of Ferritic Steels, American Society for Testing and Materials, Philadelphia, PA. (1996) (Reapproved 2002)
31. Newman, J.C., Jr.: Crack-opening displacement in the center-crack, compact and crack-line-wedge-closed specimens, NASA Langley Research Center, Hampton, Va., June 8, 1976 and NASA TN D-8268, Washington, DC, (1976)

Chapter 13
Warm Prestressing

13.1 Introduction and Historical Remarks

As Nichols pointed out very clearly in his historical review of the issue of overstressing, written in 1967 [1], the use of loading a structure under controlled conditions to a load level higher than that to be sustained in service as a means of proving that the structure is unlike to fail in service, is a procedure as old as the engineering practice itself. It is, however, only in recent years that it is also discovered that overstressing a structure at temperatures at which it does not fail by brittle fracture, may result beneficial from the point of view of fracture resistance. It is a fact, at the time resulting completely unexplainable, that none of the Liberty ships launched during warm seasons or having the first cruise in warmer seas, underwent a brittle fracture event even when, later, were sent to icy-cold arctic seas of the Nord where similar vessels failed by brittle fracture. The ships launched during warm weather or sealing first in warmer seas precisely underwent a warm prestressing event that preserved them from subsequent brittle fracture. There is indeed considerable evidence that under some circumstances overstressing can improve the toughness of materials. These circumstances are, in general, the temperature at which the overstressing is practiced and the load history following the prestressing. Warm prestressing (WPS) is an experimentally observed phenomenon in which a material containing a flaw exhibits an apparent increase in fracture toughness due to previous loading at a higher temperature. This is schematically shown in Fig. 13.1. A material loaded at a temperature, T_1, well over the NDT, loading path O-A of Fig. 13.1, is then unloaded along the reverse course A-O at the same temperature. The material is then cooled at the operating temperature, T_2, this time below the NDT where its behavior is brittle, path O-B, and reloaded. Arrived in C, which corresponds to the fracture toughness K_{Ic} of the material at that temperature, it should fail, and it would certainly fail, beside the metallurgical variability scatter, if it were tested directly at that load and temperature without WPS. But experience shows that it doesn't fail. Load can be further increased to point D before it eventually fractures experiencing a fracture toughness,

P. P. Milella, *Fundamentals of Fracture Mechanics*,
https://doi.org/10.1007/978-3-032-23081-2_13

K_{If}, increase, well in excess of K_{Ic}: $K_{If} > K_{Ic}$. WPS has been effective. Caution must be taken not to embrittle the material by strain-aging, either static or dynamic. Static strain aging may occur in the plastic zone ahead of the crack tip, whether a crack already is present in the material, after an overstressing particularly large. Strain aging embrittled may result dangerous during reloading at lower temperature. However, static strain-aging requires some time to take place, days or even months. Therefore, its possible detrimental effect will not be felt at all if reloading at lower temperature occurs soon after unloading, without any delay or after a short waiting time. As we shall see, the maximum benefit of warm prestressing occurs precisely during a loading-cooling-reloading at failure sequence, without any waiting time. At variance, dynamic strain-aging (DSA) occurs precisely during plastic deformation. In carbon steels DSA occurs, under quasi-static loading, in a temperature range of 120 to 250 °C which is right in the upper shelf toughness range where warm prestressing may be practiced. Historically, the first beneficial effects of overstressing were discovered in relation with residual stresses. If a structure had not been thermally stress relieved, residual stresses from previous fabrication processes, a welding for instance, can sum up to the service stresses, which can significantly increase the risk of fracture. Overstressing can reduce or fully eliminate residual stresses. In this context the overstressing technique as been referred to as mechanical stress relief. Where notches are present, overstressing can, by causing localized plastic strain to exceed yield strain, leave a compressive residual stress state and thus reduce the risk of failure.

Weck [2], already in 1965, stated that a thermal stress relief can be omitted even in the presence of large crack-like defects and even in materials that are brittle at the minimum service temperature, provided that the structure has been subjected to a mechanical stress relief at a higher stress than the service stress and at a temperature higher than the NDT. Nichols [2], reports that beneficial results arise from overstressing even of structure with a notch which did not contain significant residual stresses from a welding processes, for example structures which had previously been thermally stress relieved. The argument was that such overstressing would produce what he called "notch nullification" by causing large yielding at the tip of any dangerous notch and so, on removal of the stress, leaving beneficial residual stresses which will reduce the effect of subsequent applied loads, increasing the stress required to cause failure. The following sections will deal with all possible beneficial effects of warm overstressing or prestressing (WPS).

13.2 Effect of Overstressing on Welds

One of the first application of the overstressing technique was in the field of welding. The cooling and solidification of the weld pool is accompanied by a parallel shrinkage that is prevented by the adjacent base metal which remains solid and fix. The result is the build-up of an internal stress state known as welding residual stress state. The intensity of this stress field depends on the volume of the melted pool with respect to that of the two pieces of base metal to be welded. The higher this latter volume the

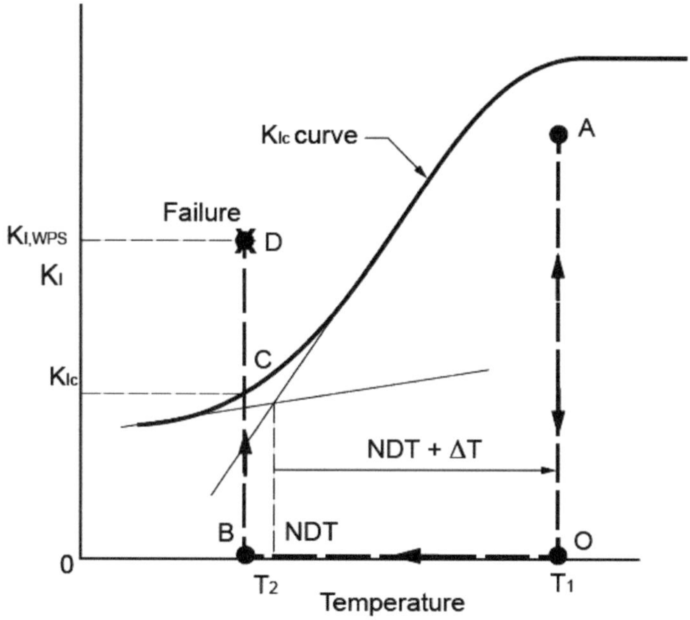

Fig. 13.1 The material loaded along the path O-A-O-B-C-D experience WPS so that, upon reloading at a temperature below the NDT, it does not fracture at C, as it should, but at D, i.e., at $K_{I,WPS} >> K_{Ic}$

higher the intensity of the residual stress field which may reach the yield stress. The most credited distribution of residual stresses in a butt weld is schematized in Fig. 13.2 [3, 4]. Longitudinal residual stresses, σ_l, are generally higher than transverse stresses, σ_t, unless the workpiece has an external constraint than can prevent, fully or in part, the weld shrinkage. In the case of complete weld shrinkage prevention, the resulting residual stresses, σ_l and σ_t, may reach the yielding strength of the material. As it can be seen in Fig. 13.2b, longitudinal stresses, σ_l, reach the maximum on the centerline of the weld. These tensile stresses must be balanced by compressive stresses in the base material. Since residual stresses are self-balancing, tensile stresses must be balanced by a surrounding compressive stress state. According to Saint Venant principle residual stresses, treated as a local disturbance, must extinguish within a distance two–three times the characteristic dimension of their cause, which is the weld size.

Figure 13.3 presents an experimentally determined distribution of longitudinal residual stresses in a butt weld [5] and two possible interpolating curves proposed by Masubuchi and Martin [6] of equation

$$
\begin{aligned}
&curve\ a\ \ \sigma_l = \sigma_o e^{-0.5(x/c)^2}\left[1-(x/c)^2\right] \\
&curve\ b\ \ \sigma_l = \sigma_o \frac{1-(x/c)^2}{1+(x/c)^4}
\end{aligned}
\quad (13.1)
$$

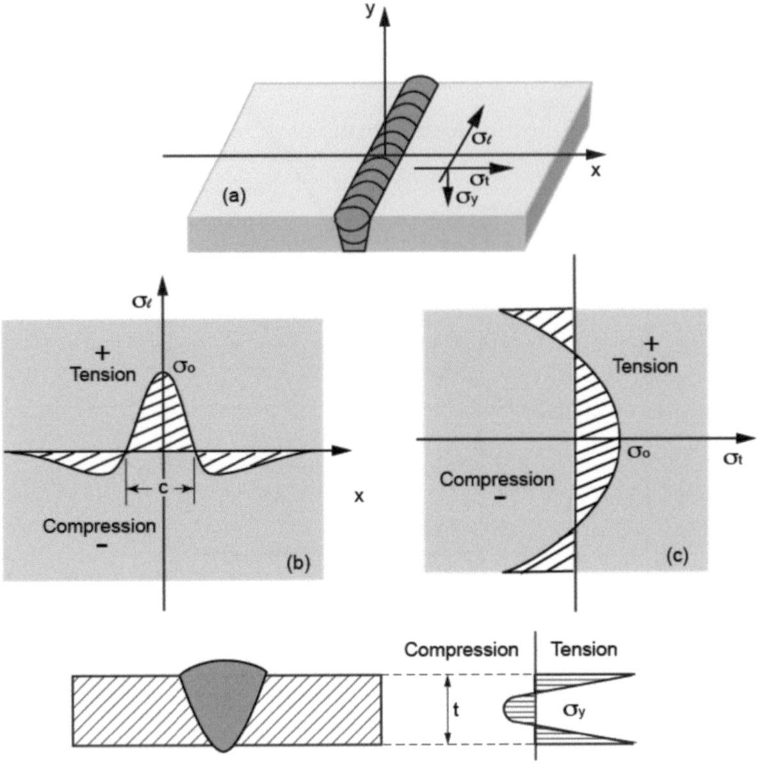

Fig. 13.2 Schematic of residual stress state in a butt-weld (**a**): **b** longitudinal stresses σ_l, **c** transverse stresses σ_t and through the thickness stresses

where σ_o is the maximum stress attained on the weld centerline and c the width of the tensile zone, shown in Fig. 13.2b, which is generally larger than the weld seam. The maximum tensile stress σ_o is often of the order of the yield stress σ_s of weld metal for low carbon steel weldments. For high strength steels σ_o can be considerably lower than the yield stress σ_s, as shown in Fig. 13.4 [7].

The transverse stress σ_t, Fig. 13.2c, is normally smaller than 20% of the longitudinal stress σ_l. The area surrounding the weld, where residual stresses arise is the so-called Heat affected zone or HAZ. Generally, this area is characterized by a coarse-grained structure with poor metallurgical and mechanical properties. Hence the need to restore lost metal properties and eliminate residual stresses. This is normally done by a post-weld heat treatment (PWHT) heating the metal beyond 730 °C at a controlled heating rate and then cooling-down under controlled cooling rate. Unfortunately, PWHT cannot be done on thin structures that would deform permanently nor on austenitic stainless steels because austenite sensitizes to stress corrosion cracking when maintained over 600 °C (see Section.). Already in 1937 Liljeblad [8] published an experimental work on the effect of overstressing on residual welding stresses. He concluded that any applied load, no matter how small, will cause residual stresses

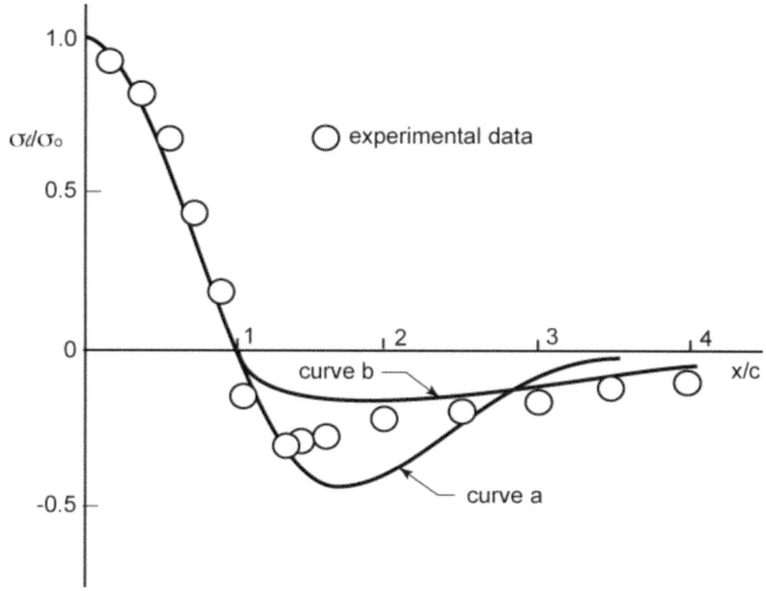

Fig. 13.3 Experimentally determined residual stress distribution in a butt weld and interpolation curves (modified from [5])

Fig. 13.4 Relationship between σ_o and σ_s (experimental results) [7]

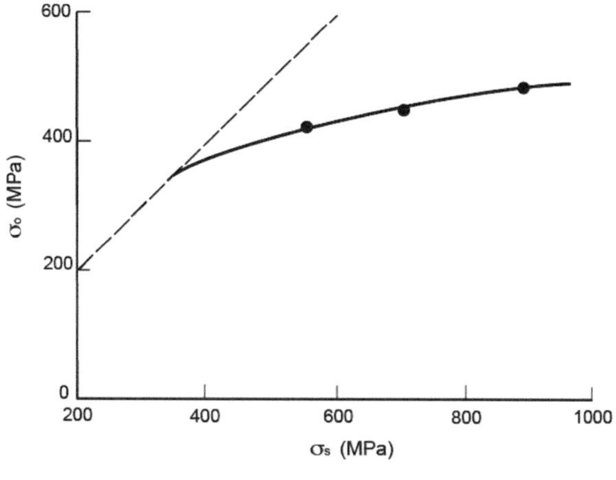

close to yield to pass the yield point, thus reducing the residual stress state. Liljeblad considered this method very effective in reducing welding residual stresses stating that at ASEA had used it for fabricating beams, steam boilers and welded frames. An experimental work by Kihara et al. [9] claimed that welding residual stresses were reduced from the as-welded value by an amount equal to the pretensioning stress. His results are shown in Fig. 13.5 [9, 10]. However, caution must be taken to avoid fracture during overstressing, in particular when a crack is present or supposed to exist.

Nichols reports experiments done at the University of Illinois [10] on ASTM A212 carbon-manganese steel showing that the effect of prestressing on welded wide plates containing a variety of notches were equivalent to those obtained with a post weld heat treatment. The mechanism of residual stresses reduction can be explained as follows. Consider a test piece containing a butt weld that produces close to yielding longitudinal residual stresses only on some limited region on the weld center line, as schematized in Fig. 13.2b. The corresponding point on the stress–strain curve is A in Fig. 13.6. Imagine that the application of an overstress would bring this material point beyond the yield stress, σ_s, of the material, point B of Fig. 13.6. Upon unloading, point B will follow line BC parallel to the loading line 0A arresting somewhere between D and C. The final point depends on the ratio of the process volume V_{res}, which is the volume of the weld residual stresses, to V_{uns}, which is the total volume of

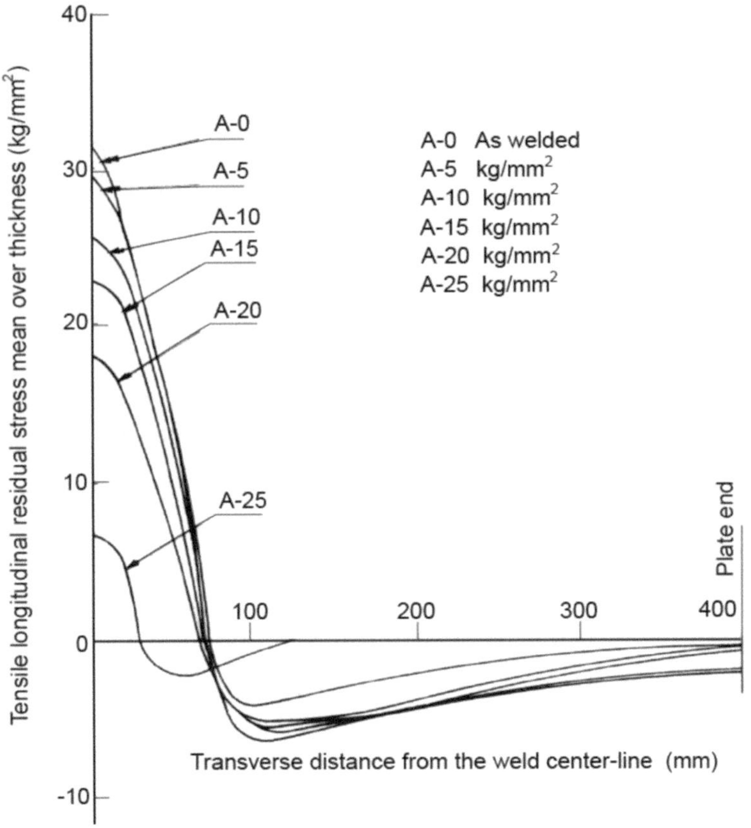

Fig. 13.5 Distribution of weld residual longitudinal stresses in as welded and mechanically stress-relieved conditions [9, 10]

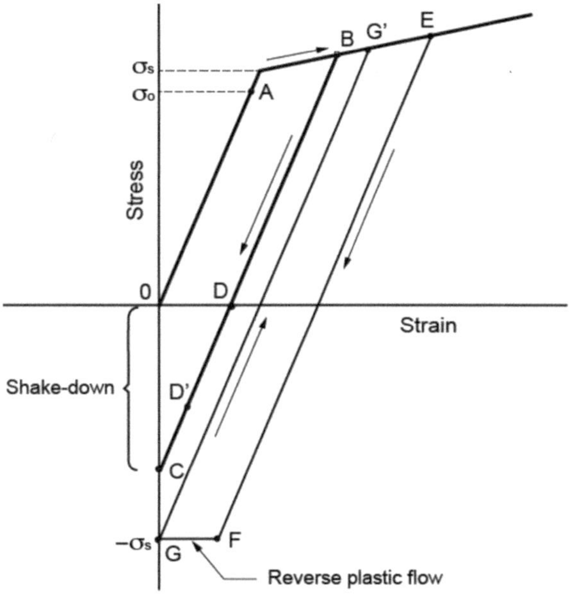

Fig. 13.6 Schematic of shake-down effect. Test piece subjected to bending. **a** outermost metal fibers from neutral axis exceed yield strength, **b** upon unloading fibers that exceeded yield strength remain in either compressive or traction stress state, depending on whether during the loading phase they exceeded yield in traction or compression

the unstressed and practically free of weld residual stresses, Vres/V_{uns}. The smaller this ratio, the more point B get closer to C. In a wide thick section, where residual stresses interest just a very small fraction of the entire specimen volume, $V_{res}/V_{uns} <$ < 1, point B falls in C thus remaining in a residual compression stress state forced to be at zero strain by the surrounding much larger elastic volume that, upon unloading, wonts to get back to zero stress and zero strain. This spring back effect is called shake down.

Note that the material interested by the shake-down effect has virtually increased its yield stress which is now $\sigma_s' = \sigma_s + \sigma_{sd}$, σ_{sd} being the shake-down stress. The effect holds as long as the unloading starts from a point on the stress strain curve not larger than G' in Fig. 13.6. Beyond that point, unloading produces a reverse plastic flow with following hysteresis loop. Localized yielding produced by overstressing a specimen containing a non-heat-treated weld also occurs in any specimen or structure containing a sharp notch. This is shown in Fig. 13.7. In this case a cylindrical specimen carrying a circumferential notch is subjected to an overload F_1 that causes a small volume of material at the notch tip to reach yielding, as shown in Fig. 13.7b where the material has been assumed to be elastic-perfectly-plastic. If the volume of the plastic enclave so generated is very small, negligible with respect to the surrounding volume of elastic material, shake-down will take place leaving, upon unloading, the residual stress state shown in Fig. 13.7c with a compression peak stress $-\sigma_{nt}$ at the notch tip (point 2 of the notch tip load history of Fig. 13.7. It is from this residual stress state $-\sigma_{nt}$ that the following loading at a stress amplitude σ_a will operate. Accordingly, the resulting cycles 2, 3, 4, 5, 6 …will produce a much lower peak stress σ_{max} and a negative mean stress σ_m, as shown by the notch tip stress history of Fig. 13.7. Note

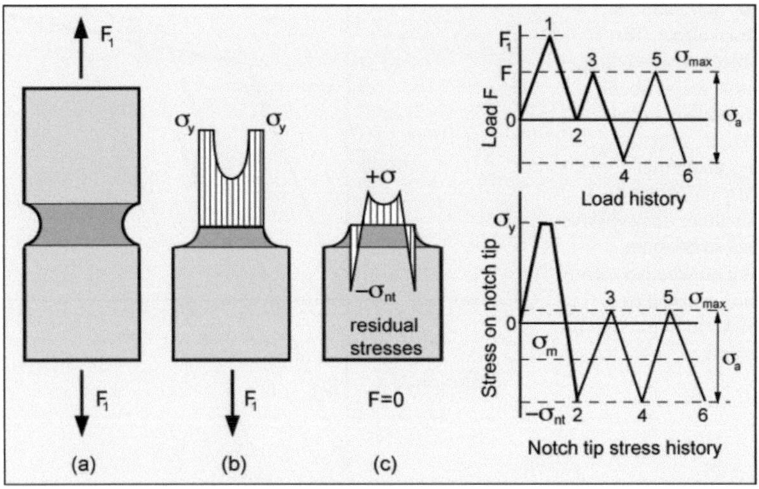

Fig. 13.7 Cylindrical notched specimen subjected (**a**) to a prestressing overload F_1 that (**c**) leaves a thin surface layer around the notch in a compression stress state. The following cycling will be applied with a beneficial compressive mean stress

that prestressing is not influencing the stress amplitude σ_a of the following cycles, if fatigue were acting, that remains the same, but only the peak stress σ_{max} and the mean stress σ_m that becomes negative. The overall residual stress state is necessarily self-equilibrating since it is composed of internal stresses, therefore

$$\int_{-w/2}^{w/2} \sigma_l \, dx = 0 \tag{13.2}$$

with W being the width of the test piece, which actually means that the two areas, that of positive tensile stresses and that of negative compression stresses, must be equal. The moment also must be zero

$$\int_{-w/2}^{w/2} x\sigma_l \, dx = 0 \tag{13.3}$$

Hence, the distribution of residual stresses σ_{res} following the initial prestressing will be that depicted in Fig. 13.5b. When the material is later stressed in the elastic field after this initial prestressing, residual stresses σ_{res} will sum with their sign to the applied stress σ

$$\sigma_{tot} = \sigma + \sigma_{res} \tag{13.4}$$

reducing the total stress state σ_{tot} where they are compressive and increasing it where they are of the same sign of the applied stress σ.

13.3 Experimental Results for Mechanical Stress Relief

Probably, one of the most extensive study on the effects of warm prestressing was done in Japan by Kihara et al. [11, 12] already in 1959, and reported by Nichols in his monumental review report [1]. Their results on tensile tests of notched, welded specimens of 18 mm thick mild steel plate are shown in Fig. 13.8 [12]. In these tests, specimens, which did not receive any post weld heat treatment for stress relieve, were tested at –25 and –50 °C after they had been loaded to various degree at 20 °C. Two reference curve are shown. The first, solid curve in the upper part of the diagram, indicates the fracture stress required to cause failure at general yield in specimens that received a PWET and were free from residual stresses. The curve is divided in two sectors by the transition temperature at 0 °C. In the first sector below the transition temperature, where the material has poor ductility, the fracture stress increases with decreasing temperature because the yield strength of the material increases.

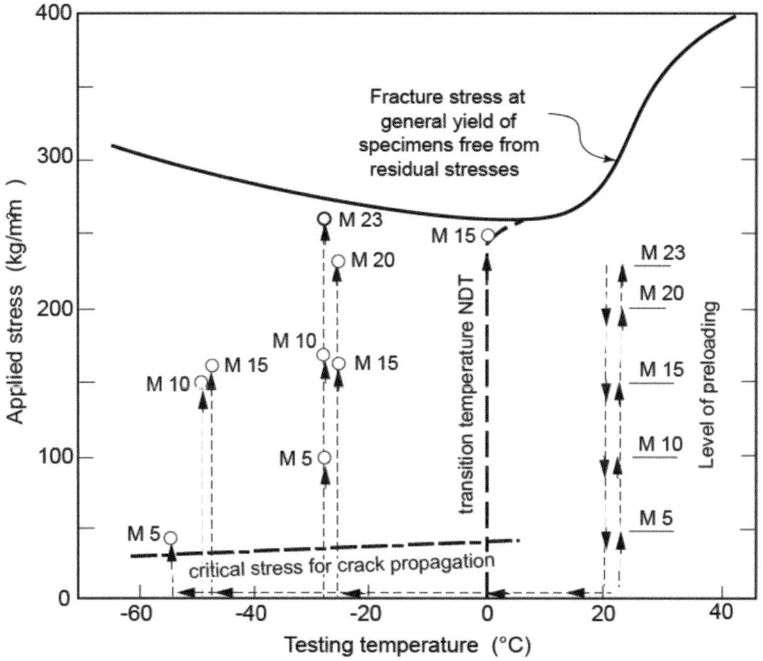

Fig. 13.8 Summary results of tensile tests on notched, welded specimens of mild steel showing the effect of warm prestressing on fracture strength (from [1])

In this sector, since we are dealing with notches and not sharp cracks, failure may not occur by sudden cleavage below the yield strength and the notch creates a triaxial stress state or plain strain condition that increases the material strength, see Chap. 4 and Fig. 4.8. In the second sector, beyond the transition temperature, the ductility of the material sharply increases as to require higher and higher stresses to cause general yield. It is worth noting that the material behavior is described in terms of stress at general yield and not toughness, K_{Ic} or K_Q, simply because the testing dates back to 1959 and, therefore, precedes the introduction of the toughness K_{Ic}. The second curve, dashed line on the bottom, indicates the critical stress in specimen that did not receive any PWHT and fully retain residual stresses close to yielding. All specimens tested and shown in the diagram were warm prestressed at 20 °C at various load levels. This temperature is well over the critical temperature for low stress initiation in this material being about 0 °C. The letter M indicates the level of prestress. All specimens only failed in subsequent low temperature test when the applied stress was nearly equal or somewhat higher than the level of stress applied in the preloading. The higher the preloading, the higher the failure stress in low temperature test until it approached the stress for general yield in specimens that were thermally stress relieved. The authors claimed that the beneficial effect of warm prestressing in reducing or even eliminating weld residual stresses was evident and that mechanical stress relieving was equivalent to a PWHT and recommended this use in pressure vessels. We may retain some doubts about this last sentence because the mechanical treatment does not restore the metallurgical properties of the metal jeopardized by the welding process, but the benefit, at least as long as the notch remains a notch and does not turn into a sharp crack, is really evident.

13.4 Loading Sequence Effects on Warm Prestressing

One of the most serious accident a nuclear pressure vessel may face is the so-called thermal shock. In a thermal shock accident, the nuclear pressure vessel operating at 288 °C loses the primary cooling water system because of a complete severance of a main water piping and must be refrigerated by the emergency water taken from a tank located outside the plant containment system. Therefore, this water may even be taken at 0 °C, if the accident occurs in winter. The consequence of the emergency icy-cold water injection into the vessel is the thermal shock. In a *pure* thermal shock there is no pressure acting and the only stresses in the vessel beltline arise from the thermal gradient. Assuming that a longitudinal flaw is present on the internal surface of the vessel, the injection of very cold water may trigger a brittle fracture. With a moderate deep longitudinal flaw having a depth, *a*, to thickness, *t*, ratio: $a/t = 0.12$, that in a 15 cm thick wall means 1.8 cm, calculations indicate that the applied stress intensity factor, K_I, varies during the transient as shown in Fig. 13.9. Meanwhile, the toughness, K_{Ic}, of the material at the crack tip drops during the same transient, as the cold wave moves deeper and deeper into the vessel wall, as also shown in Fig. 13.9, so that a moment occurs when the applied K_I exceeds the material toughness, K_{Ic}, point B of Fig. 13.9, and unstable fracture should happen. However, it must be noted that the K_I reaches its peak, $K_{I,max}$, early into the transient, point A of Fig. 13.9, when the vessel wall is still hot and the toughness, K_{Ic}, still high, and drops monotonically afterwards. Therefore, if warm prestressing (WPS) occurs the toughness of the metal should remain frozen around that warm prestressing high value, $K_{I,max}$, dashed line of Fig. 13.9 for $\alpha = 1$, and fracture will not happen. Unstable brittle fracture should not happen even for $\alpha = 0.25$, indicating a partial WPS effect. Suffice it to say the enormous importance of warm prestressing (WPS) in the nuclear power industry, so that McGowan [13], at Westinghouse Nuclear, reanalyzed the data reported by Nichols [1] together with those obtained by Yukava [14]. These data are summarized in Table 13.1 and shown in Fig. 13.10 [13] as function of the parameter α.

$$\alpha = \frac{\left(\sigma_f' - \sigma_f\right)}{\left(\sigma_{ps} - \sigma_f\right)} \tag{13.5}$$

where σ_f' is the prestressed fracture strength, σ_f the unprestressed fracture strength and σ_{ps} the prestress. The stress and strength are nominal net section value.

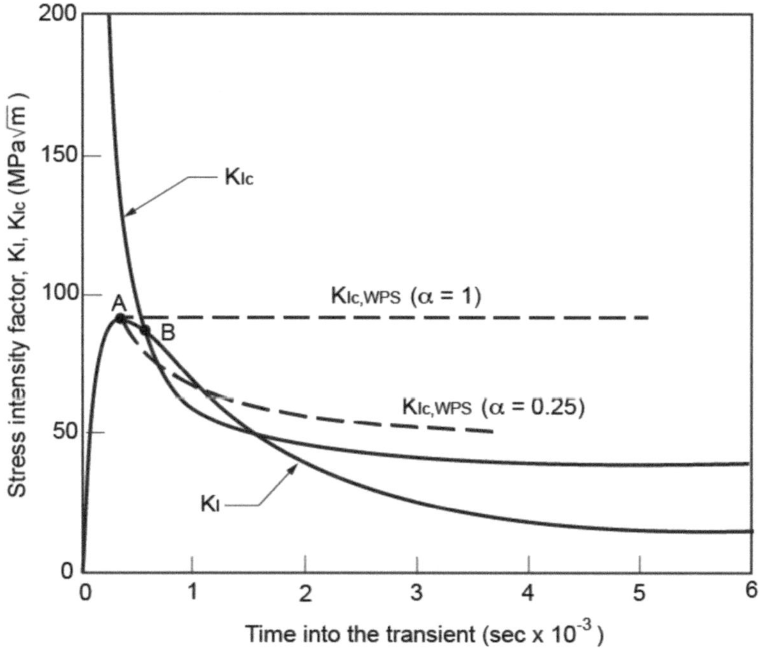

Fig. 13.9 Applied stress intensity factor, *KI*, and fracture toughness,, variation with time at the tip of a longitudinal crack, 12% of the vessel wall deep ($a/t = 0.12$) during a severe thermal shock (modified from [13])

Table 13.1 Summary of Warm prestressing studies on low-alloy steels

References	Material	Test specimens	Remarks
Brothers and Youkawa [15]	NiMoV rotor steel 586–759 MPa Yield	Notched bend and tensile specimens, fatigue cracked or with sharp notches	
Sankey [16]	NiMoV rotor steel 517–689 MPa Yield	Notched spin test	
Wessel and Baudry [17]	NiMoV rotor steel	Precracked plate tensile	
Barson and Watson [a]	MoV steel, 689 MPa UTS	Notched tensile, 0.001″ root radius	Study effect of prestr. temperature and time
Andrews [18]	A508 CL 2 steel 345–485 Mpa Yield	Notched bend fatigue cracked	Study of load history following prestress

[a] Unpublished work quoted in Nichols [1]

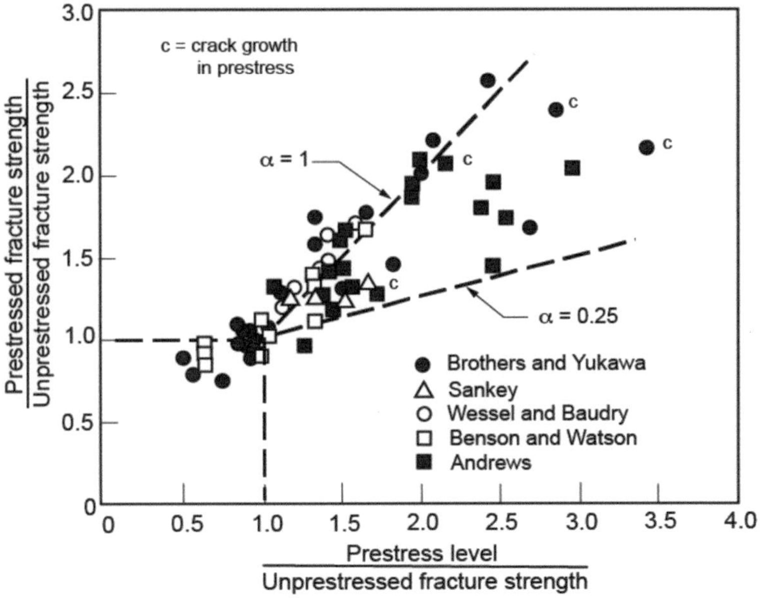

Fig. 13.10 Compilation of warm prestressed test results for low-alloy steels [13]

A value of $\alpha = 1$ means that the fracture strength is increased by the amount of applied prestress while a value of $a < 0$ means that the fracture strength is lowered by prestressing. Figure 13.9 shows that much of the data falls at $\alpha = 1$, with $\alpha = 0$, 25 being a lower bound.

The data, except for that attributable to experimental scatter, show that prestressing is effectively operating and that no harmful effects are produced by the prestressing ($\alpha < 0$).

The data falling near the lower bound limit may be due to one of the following reasons:

(1) the crack had a stable tearing,
(2) plain stain conditions occurred, as opposite to plain stress conditions,
(3) strain aging occurred following prestressing or dynamic strain aging during prestressing,
(4) unloading between the application of prestressing and subsequent fracture test at low temperature, as explained in the next Section.

As to stable tearing, it may be prevented making sure that the prestressing overload, though large enough to produce a visible effect, is small enough to avoid tearing. As to the second cause, Shankey [16] noted that in testing rotor steels the prestressing was not as effective for large, thick specimens as it was for smaller, thinner ones. This was due to the plain strain conditions existing in the thicker sections as opposite to the plain stress conditions for the smaller ones that resulted in a smaller plastic zone ahead of the crack tip. This is the case of a large section pressure vessel under severe thermal shock as shown in Fig. 13.8 by the dashed line labelled $K_{I,WPS}$ ($\alpha = 0.25$). As to the third point, Hawthorne and Loss [19] and Succop et al. [20] have reported the effect of strain aging. The fourth point is of great importance. It has to do with the load history between prestressing and fracture testing, as discussed in the next section.

13.5 Effect of Loading Sequence

Harrison and Farnehough [21] investigated the effect of the loading sequence on the warm prestress of carbon-manganese steel. They studied the first and the second loading sequence of Fig. 13.11. The first loading path, LCF, is relative to a specimen first loaded (L) at the prestress temperature in the upper shelf region of fracture toughness, cooled (C) to the operating temperature and then brought to fracture (F), without any unloading. In the second loading sequence, LUCF, the specimen is first loaded at the prestress temperature, always in the upper shelf region of fracture toughness, unloaded and then cooled to the service temperature before reloading till fracture occurs.

For all specimens, the prestressed temperature was the room temperature, while all specimens were fractured at –190 °C. The results of the experiments are shown in Fig. 13.12 [21]. They show that.

(1) a threshold warm prestressing level of 75% of K_{Ic} must be reached before any benefit is gained,

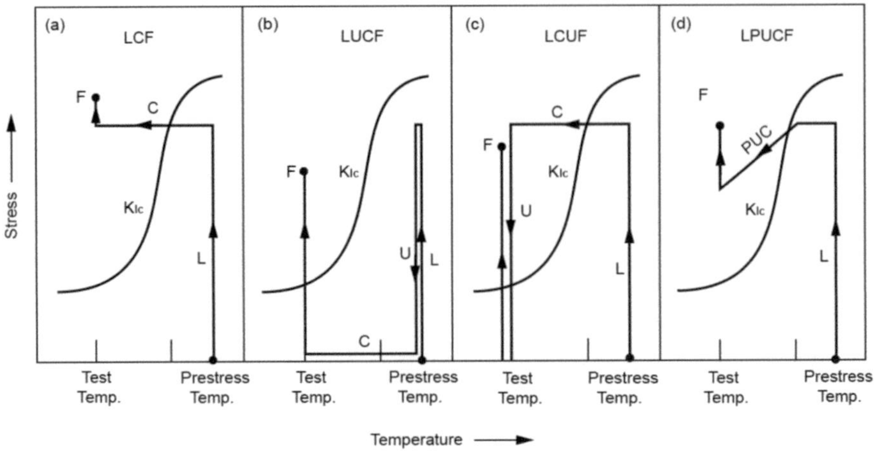

Fig. 13.11 Schematic of loading sequences

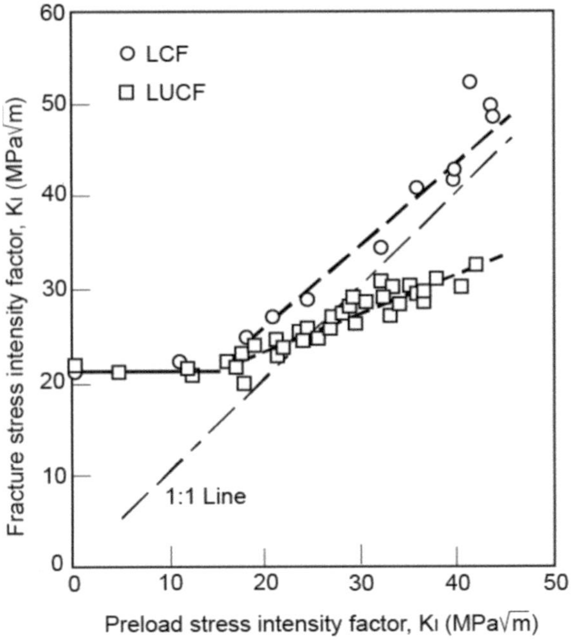

Fig. 13.12 Relation between prestress and fracture stress intensity factors, K_I, for low-alloy carbon-manganese Steel [21]

(2) for the LCF sequence, Fig. 13.11a, the measured fracture toughness $K_{I,F}$ is always above the warm prestressing toughness, $K_{I,WPS}$

(3) for the LUCF sequence, Fig. 13.11b, the measured fracture toughness $K_{I,F}$ is always below that of the LCF for the same prestress load, but higher than the K_{Ic} at the testing temperature,

(4) the fracture toughness $K_{I,F}$ varies linearly with the prestress $K_{I,WPS}$ and the slope $K_{I,F}/K_{I,WPS}$ is less than 1.

McGowan [13] reports a series of tests performed by Andrews [18] on A508 Cl2 steel also considering a third loading sequence in which the specimen was loaded, cooled, unloaded, and then fractured, cycle LCUF of Fig. 13.11c. Results are shown in Fig. 13.12 [13, 21]. Again, the warm prestressing temperature was the ambient temperature, and the fracturing was done at –196 °C.

Andrews results confirm McGowan results as far as the LCF and LCUF sequences are concerned and show that the LUCF cycle is the less effective of them all. Loss et al. [22] used 38 mm and 76 mm thick four-point bend specimens of A508 Cl2 steel preloaded at RT at 59 MPa$\sqrt{m}$ and 84 MPa$\sqrt{m}$, respectively. These values of prestressing were the maximum allowable by the ASTM E399 procedure for those thicknesses, see Sect. 9.8 and Table 9.5. However, they also tested some specimens at $K_{I,WPS}$ values higher the ASTM maximum allowable. The aim was to check PWS under plain stress conditions. The test failure temperature was varying from –73 °C to –196 °C. Loss et al. also considered a new loading sequence that was: load, partially unloading-cooling, and fracture, LPUCF, as shown in Fig. 13.11d, which is, in many respects, close to the LCF cycle, Fig. 13.11a (Fig. 13.13).

This new loading cycle was used because it resembles the trend during a LOCA accident, as shown in Fig. 13.9. Results are shown in Figs. 13.14 and 13.15.

Figure 13.14 indicates that the WPS benefit in terms of the $K_F/K_{I,WPS}$ ratio is obtained for the case of LPUCF cycle, small ΔT with respect to the WPS temperature and low $K_{I,WPS}$. The same data presented in Fig. 13.15 clearly shows that the elevation of K_{Ic}, as reflected by K_F, increases with WPS level.

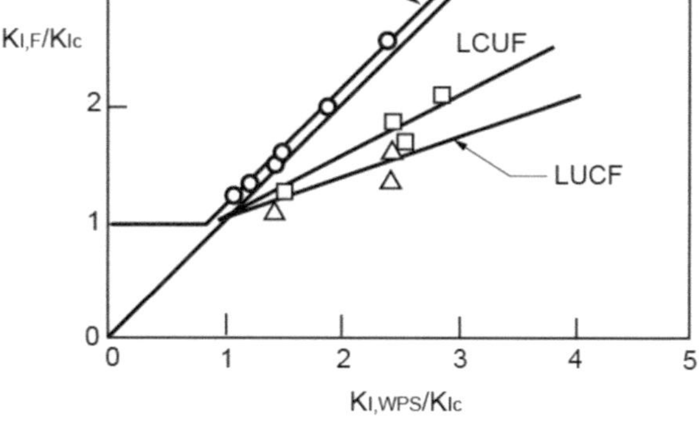

Fig. 13.13 Effect of different loading sequences on warm prestressing for A508 Cl 2 Steel [13, 18]

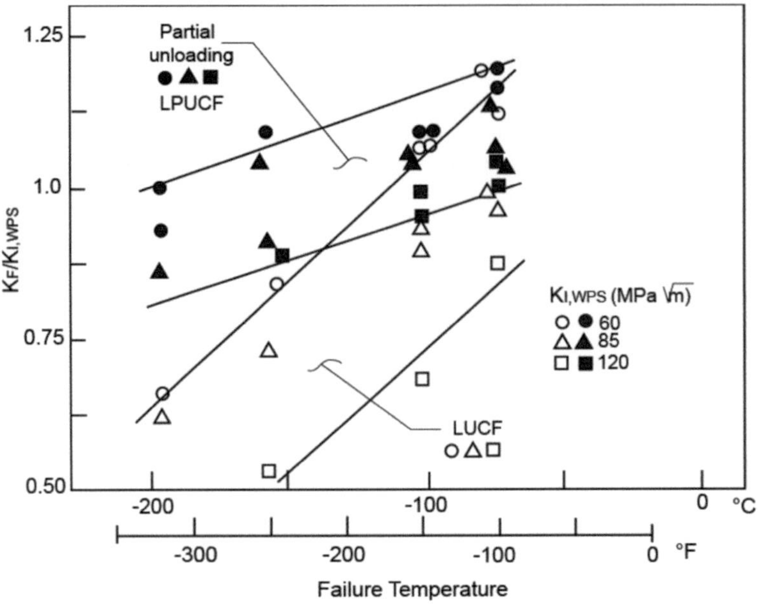

Fig. 13.14 Relationship between the failure level K_F and the failure temperature, for different warm prestress level $K_{I,WPS}$, and loading sequence, for four-point bend specimens of A508 Cl2 steel [22]

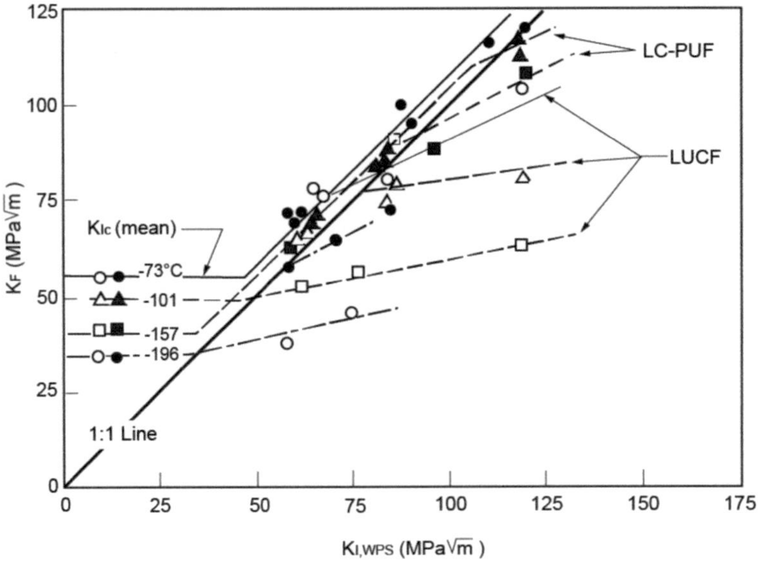

Fig. 13.15 Comparison of the fracture level K_F and the warm prestress level $K_{I,WPS}$ for two different loading sequence in four-point bend specimens of A508 Cl2 steel. Also indicated are the mean fracture toughness K_{Ic} at the failure temperature for the virgin material [22]

The magnitude of K_{Ic} elevation can be observed by comparing K_F with the mean values K_{Ic} of the virgin material for the appropriate failure temperature. Loss concluded that for a $K_{I,WPS}$ up to 120 MPa$\sqrt{}$m, the WPS phenomenon will (a) provide an effective elevation of K_{Ic}, and (b) assure that the failure level will equal or exceed the level of WPS. For example, a K_{Ic} elevation in excess of 100% is exhibited by the extreme of the partial unloading data at –73 °C, whereas essentially no elevation in K_{Ic} is exhibited by the LUCF data at –196 °C. Similar results were obtained by Stonesifer, et al. [23] on A533 GrB low alloy carbon steel. Two different loading sequence were used: the LUCF and the LPUCF. All tests were run on 4" CT specimens that were warm prestressed in the upper shelf region of the material at 177 °C. For this steel the NDT was equal to 0 °C, therefore the WPS temperature was 177 °C above the NDT, well above the FTP temperature, see Chap. 3, Fig. 3.47. At this temperature a large crack tip blunting occurs and, probably, also a stable tearing within the plastic zone. Temperature chosen for loading to fracture were –23 °C and –95.5 °C for the LUCF sequence, while for the LPUCF sequence the test temperature was –95.5 °C, well into the low shelf behavior of the material. In the LUPCF sequence the initial loading was followed by 1/3 partial unloading till the final temperature of –95.5 °C and then loaded to fracture. The mean fracture toughness, K_{Ic}, of the virgin material was $K_{Ic} = 110$ MPa$\sqrt{}$m at –25 °C with a standard deviation $\Delta = 21.4$, and $K_{Ic} =$ 55 MPa$\sqrt{}$m with a standard deviation $\Delta = 10.4$ at –95.5 °C, respectively. For all loading sequences the specimens were prestressed at = 192 MPa$\sqrt{}$m. Results are shown in Fig. 13.16 [23]. The beneficial effect of WPS is rather clear for the LPUCF loading sequence in which the K_F was always higher, by 13%, than the $K_{I,WPS}$. This occurrence can be better seen in the Weibull two parameter statistical distribution of Fig. 13.17. Interesting enough, the low probability tail of fracture toughness K_F equals the mean $K_{I,WPS}$. This actually means that the WPS evident benefit of this sequence is not always expected or not for all specimens. However, the tendency is that K_F equals or passes the WPS $K_{I,WPS}$. For the LUCF loading sequence, the probability (frequency) distribution is shown in Figs. 13.18 and 13.19 for the two cases of failure temperature of –23 °C and –95.5 °C, respectively [23].

The net, high gain obtained in the toughness K_F at –23 °C with respect to that of the virgin material at the same temperature is evident and quantifiable in about 145 MPa$\sqrt{}$m. Also the toughness K_F increase with respect to the WPS value, $K_{I,WPS}$ = 192 MPa$\sqrt{}$m, can be observed, though the tail of this distribution is falling within that of the WPS toughness distribution. At –95.5 °C the failure toughness gain still remains with respect to that of the virgin material at the same temperature, $\Delta K =$ 105.7 °C, but there is no increase, but rather a decrease, with respect to the WPS value, $K_F = 174$ versus = 192.8 MPa$\sqrt{}$m. The conclusions that can be drawn are that WPS shows that:

(1) WPS always has a beneficial effect increasing the failure fracture toughness, K_F with respect to that, K_{Ic}, of the virgin material at the same temperature,
(2) WPS temperature and the initial $K_{I,WPS}$ level is important,
(3) the maximum beneficial effect is obtained with a loading sequence that does not consider any unloading, LCF cycle, or just a partial unloading, LPUCF,

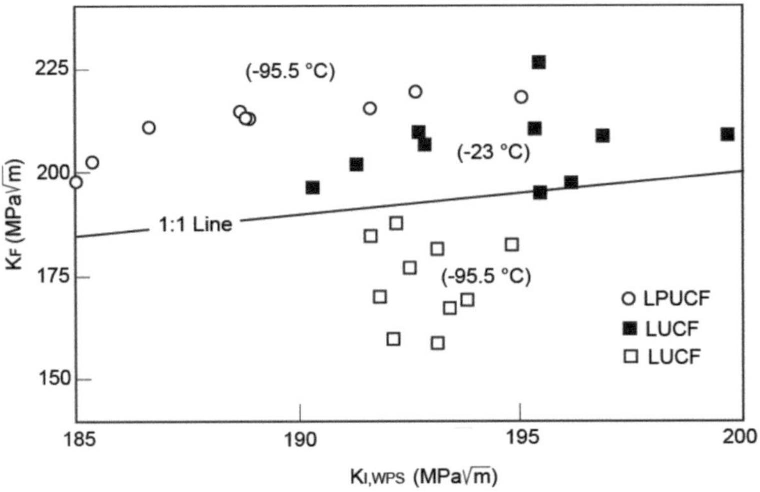

Fig. 13.16 Comparison of the fracture level K_F and the warm prestress level $K_{I,WPS}$ for two different loading sequence in A508 Cl2 carbon steel and two different testing temperatures (plot of data from [23])

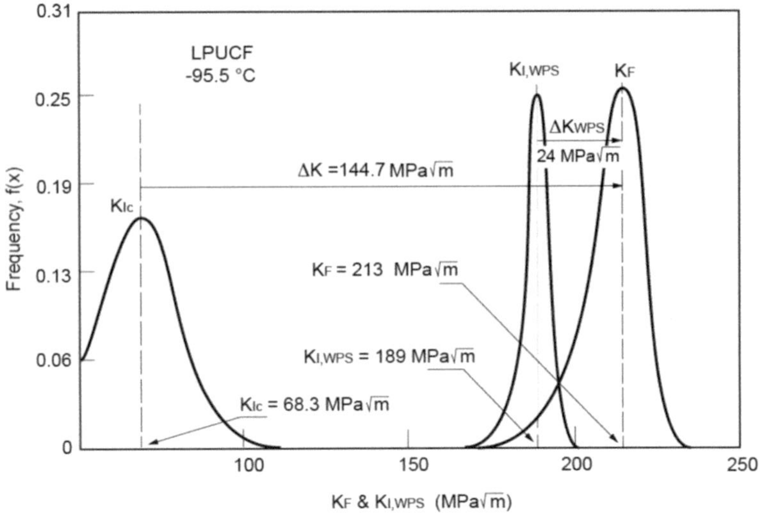

Fig. 13.17 Two-parameter Weibull distribution of K_F and $K_{I,WPS}$ following a LPUCF loading sequence with failure at 95.5 °C for A 533 GrB carbon steel. Also shown is the K_{Ic} distribution at –95.5 °C (integrated from [23])

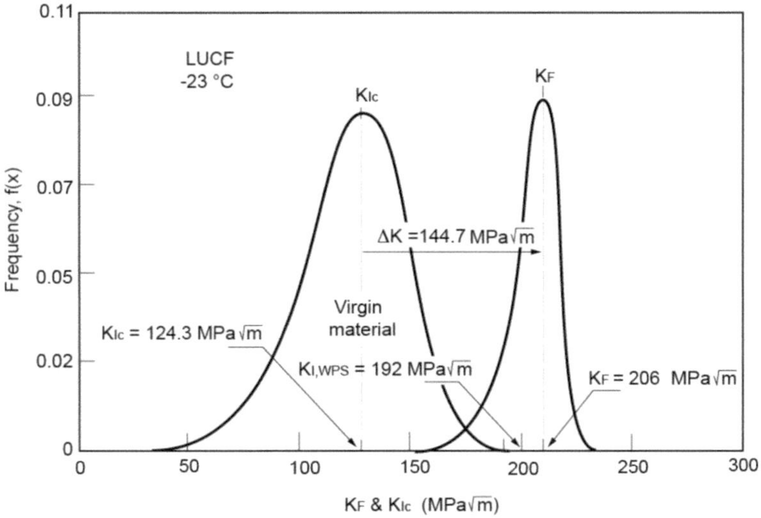

Fig. 13.18 Two-parameter Weibull distribution of K_{Ic} at –23 °C and K_F following a LPUCF loading sequence with failure at –23 °C for A 533 GrB carbon steel (from [23])

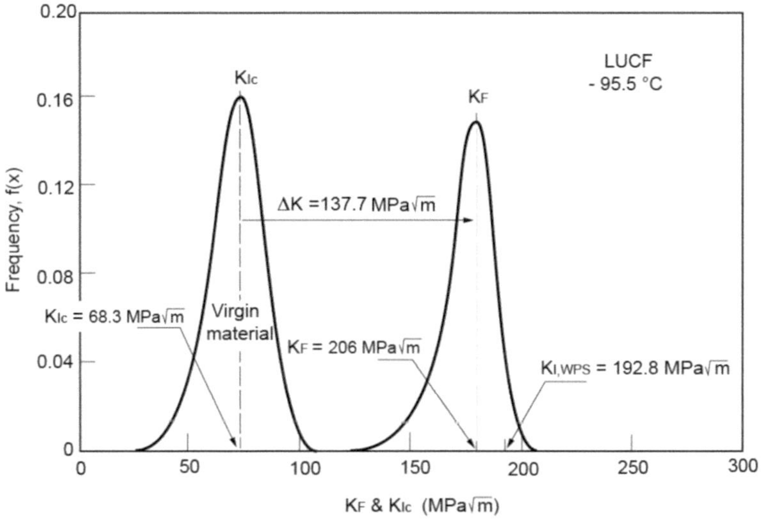

Fig. 13.19 Two-parameter Weibull distribution of K_{Ic} at –95.5 °C and K_F following a LUCF loading sequence with failure at –95.5 °C for A 533 GrB carbon steel (from [23])

(4) the loading sequence LCUF performs better than the LUCF one
(5) the beneficial effect of WPS reduces with decreasing failure temperature, but still remains.

13.6 Theory of WPS

Nichols [1] recognized three mechanisms which account for the benefits of warm prestressing:

(1) introduction of residual stresses on unloading,
(2) work hardening in the plastic zone (general modification of the unstressed microstructure),
(3) crack tip blunting by plastic flow.

The first mechanism, Fig. 13.6, may not be so important as it is for the reduction of residual stresses since the most efficient loading sequence is the LCF that does not consider any unloading, Fig. 13.11a. What seems to be clear is that during the cooling phase, either with or without unloading, the crack remains stable though the applied $K_{I,WPS}$ level is higher than that at which unstable crack propagation is expected, based on the K_{Ic} of the virgin material, $K_{I,WPS} >> K_{Ic}$. Ahead of the crack tip a plastic zone is formed when a specimen containing a crack is loaded at the WPS temperature, T_{WPS}, to the initial $K_{I,WPS}$ level, Fig. 13.20a. When the specimen temperature is reduced to the final value, T_f, keeping constant the load, which is to say both the initial $K_{I,WPS}$ level and the entire stress field, K_I, the yield strength, $\sigma_{s,WPS}$, increases to $\sigma_{s,Tf}$, which results in a plastic zone contraction, as shown in Fig. 13.20a. Question arises as to why this reduced plastic zone remains dormant while at the temperature T_f the virgin material would fail well before the applied load? A possible explanation is that during the WPS the crack blunts so much as to turn into a notch. The blunting changes the stress field that is no longer described by the K_I, as shown in Fig. 13.20b, but by a new stress field without singularity. The corresponding plastic zone is wider and so remains, after cooling. The new toughness is no longer the K_{Ic} of the virgin material but a new one, K_Q, much higher than that, even two or three times higher, typical of a blunted notch, as shown in Figs. 6.26 and 6.27. Therefore, the notch is stable. It takes a further increase, though small, of the applied load to initiate a tearing to resharpen the notch tip and trigger a brittle fracture because the temperature is so low that cannot sustain a crack. This kind of blunting that results in a LGC after loading in the ductile regime of the material is shown in Fig. 13.21 [24] for the crack tip of a specimen of A533 GrB carbon steel. The initially sharp crack obtained by fatigue precracking has turned into a notch having a CTOD of about 0.3 mm. Note the metallurgical structure that has been evidenced by heating the specimen at 700 °C for three hours followed by furnace cooling that has produced a recrystallization of the grains subjected to high strains. Grains up to 0.1 mm ahead of the notch, after the thermal treatment turned their bainitic structure into ferrite, that appears white with a volume increase that let them to be recognized at the microscopy observation.

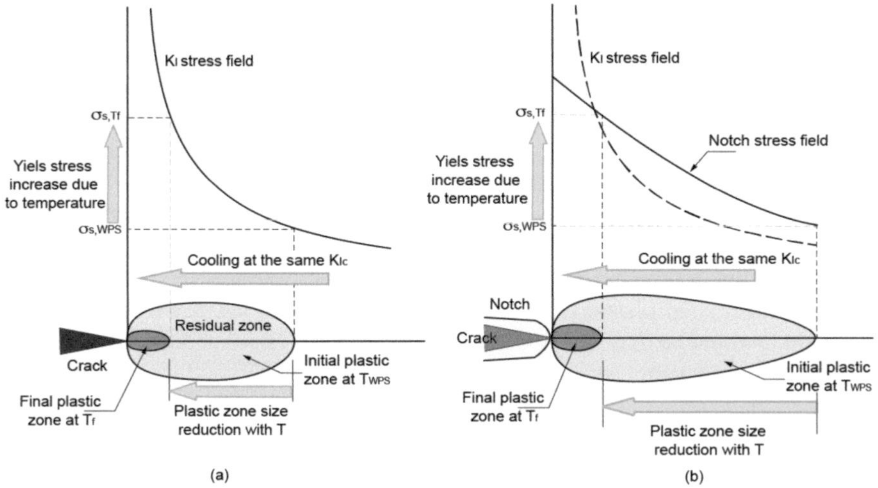

Fig. 13.20 Schematic of the reduction of the plastic zone: **a** ahead of the crack tip due to a cooling at fixed load, **b** ahead of a blunt notch

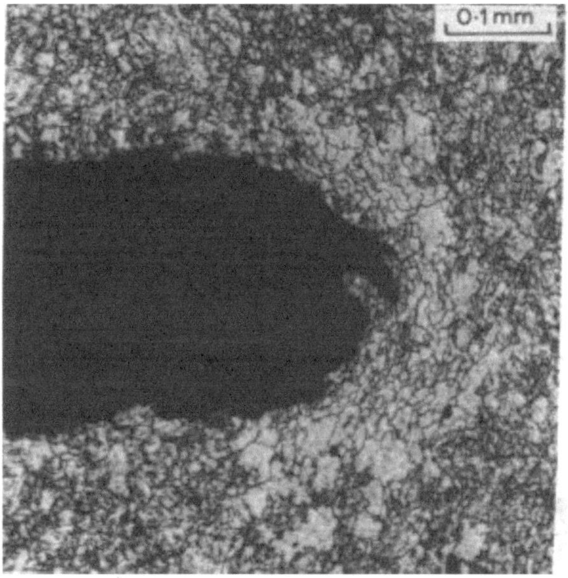

Fig. 13.21 Recrystallization of heavily deformed grains ahead of a crack tip turned into a notch that appears wither and larger than the surrounding matrix. This area represents the stretched zone [24]

This area represents the stretched zone, see Sect. 11.7. These metallurgical transformations and the possible work hardening may play a role in the loading sequence where unloading is present. The reason why in these loading sequences failure occurs prior to reach the WPS toughness level, though much over the toughness of the virgin material, may be due to the resharpening or to the partial closure of the notch tip operated by the load reversal that squeezes the apical plastic zone. Chell and Vitek and Chell [25, 26] differentiated between the initial plastic zone of extension a_{res}, Fig. 13.22a, which was formed during the WPS, and the final plastic zone, of extension a_f, set up at lower temperature with higher yield stress. They argued that the initial plastic zone, which they called *residual zone*, remained practically frozen during the cooling phase, maintaining an internal plastic strain field also locked that could not be further increased because dislocations activated and multiplied during WPS upon cooling and/or unloading became immobile. Therefore, it was not a true plastic zone but, rather, the memory of a previously active plastic zone. At variance, the final and smaller plastic zone, internal to the residual zone, had mobile dislocations and it would be there that plastic flow could occur upon increasing the applied load by an infinitesimal amount. Consequently, Chell and Vitek [25] proposed as a failure criterion the attainment of a critical force on all mobile dislocations of the smaller plastic zone set up at cooling. This force can be inferred by evaluating the J that includes only elastic strains, denoted J_e, at variance with the usual J where total strains and distortions are used. Therefore, J_e is path dependent and must be calculated choosing a contour Γ_f that includes only mobile dislocations, i.e., the integration path must be the boundary between the final plastic zone at the failure temperature and the residual plastic zone, as schematically shown in Fig. 13.22a. It is postulated that failure occurs when the force on these mobile dislocations attains a critical value that equals the friction stress opposing dislocation motion that is assumed to be the flow stress $\sigma_{f,Tf}$ of the material at the final temperature T_f, Eq. (10.22), and this force must be used to assess J_{Ic}. As per Eq. (10.40), it is $J_{Ic} = K_{Ic}^2$/E', then the failure condition becomes

$$J_e = J_{Ic} = \frac{K_{Ic}^2}{E'} \tag{13.6}$$

Therefore, if we were able to calculate J_e from the loading sequence, which remains constant during the LCF and varies for the other sequences, Eq. (13.6) would be used to predict the K_{Ic} at the failure temperature. It is clear that such an integral cannot be computed via normal integration along the contour Γ_f so that Chell et al. used the Bilby-Cottrell-Swinden (BCS) dislocation model [27] to describe the plastic zones with mobile and immobile dislocations with the strip-yield model of Fig. 10.12, as shown in Fig. 13.22b. In the BCS model, dislocations are continuously distributed within the plastic zone with a density characteristic of a uniformly stressed solid containing a crack. In the strip-yield model this density D(x) is linked to the function, $\phi(x)$, expressing the displacement v along the x axis, in the mobile dislocation zone by the relation

$$\phi(x) = b\int_{x}^{a_f} D(x)dx \tag{13.7}$$

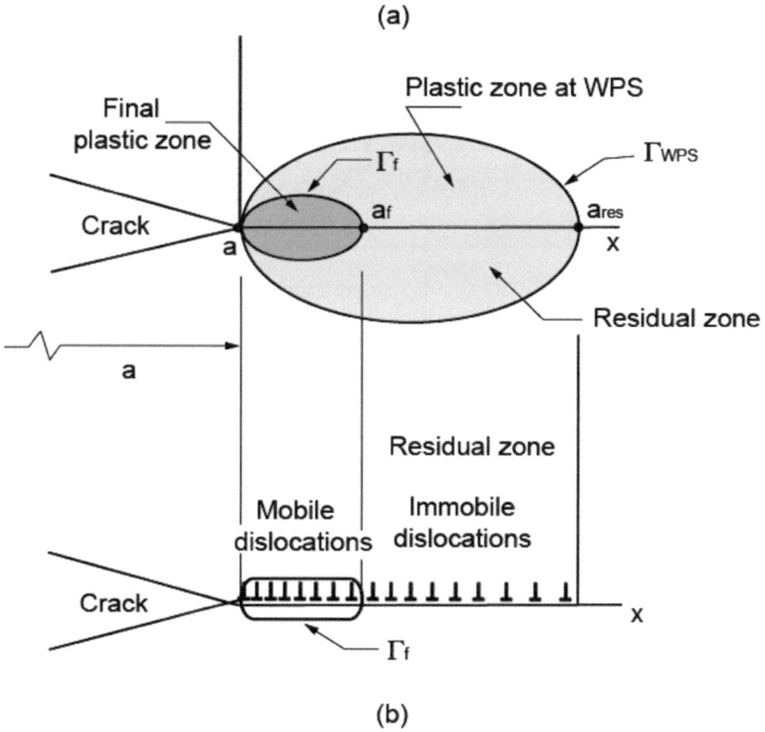

Fig. 13.22 Schematic representation of the situation ahead of the crack tip upon cooling and **b** its simulation by the yield-strip model

where b is the Burger's vector, see Sect. 2.8. J_e is then given for the contour Γ_f by the expression

$$J_e = \sigma_{f,T_f}\left[\phi(a) - \phi\left(a_f\right)\right] \tag{13.8}$$

where $\phi(a)$ is the displacement at the crack tip (CTOD) and $\phi(a_f)$ is the displacement at the mobile dislocation border following the loading sequence. Note that for a LCF sequence Eq. (13.8) predicts the same toughness reached during WPS. Equation (13.8) actually provides the way J_e can be calculated. For any other loading sequence is required to assess the displacement at the crack tip (CTOD) and that at the border of the mobile dislocation zone, whose extension depends on the yield stress of the material at the cooling temperature. Besides, the Chell et al. J_e based model, there is also the theory proposed by Curry [28] of a critical stress reached at a specified distance ahead of the crack tip. With this model, unstable fracture is attained when the maximum principal stress normal to the crack plane, σ_{yy}, exceeds a fracture stress σ_f over a characteristic distance.

Curry model was inspired by the Ritchie-Knott-Rice theory [29] of a critical fracture stress, σ_f, reached at a certain distance ahead of the crack tip that would

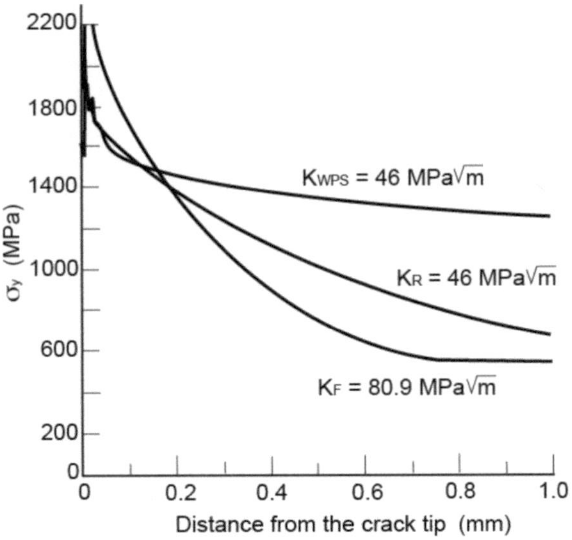

Fig. 13.23 Comparison of crack plane normal stresses, σ_y, for the WPS solution, the reference solution and the failure solution [22]

initiate a brittle fracture by breaking some inclusion and that they evaluated in a couple of grains, for the high-nitrogen mild steel they used. To check the validity of the Curry critical distance, Stonesifer et al. [30] performed an accurate FE calculation for the LUCF type loading sequence. The FE simulation considered a bend specimen 76 mm thick (3.0 in) of A 508 Cl2 steel. The finite-element grid had elements in the vicinity of the crack tip as small as 0.005 mm and contained about 1,000 degrees of freedom. For these specimens Loss et al. [22] had run WPS tests loading at 24 °C at $K_{I,WPS} = 119.1$ MPa$\sqrt{m}$ with unloading at –101 °C and failure at $K_F = 80.9$ MPa$\sqrt{m}$. In addition, they also performed the same FE calculations on a reference specimen which had not been given a WPS load history and fractured at the same temperature of –101 °C to compare the stress field at failure with those for the WPS specimen. If the stress field of the WPS specimen and that of the reference specimen were accurate representations of the distribution at fracture initiation, then it would be possible to use those distributions to determine the critical stress and distance, if any, for the Curry WPS theory. This comparison is shown in Fig. 13.23 and resulted in a critical stress of about 1,410 MPa with an associated critical distance of about 0.180 mm. Note that this distance is rather consistent with those found experimentally at ORNL on SA 533 carbon steel specimens broken from 71 MPa$\sqrt{m}$ to 109 MPa$\sqrt{m}$ (see Table 6.1). The critical stress and distance should be those at the intersection of the failure curve, at $K_F = 80.9$ MPa$\sqrt{m}$ and the reference curve, at $K_R = 46$ MPa$\sqrt{m}$ and are equal to 1,410 MPa and 0.180 mm, respectively. Also interesting is the fact that the curve corresponding to the maximum applied preload (WPS) tends to intersect other curves at about the same stress and distance.

References

1. Nichols, R.W.: The Use of Overstressing Techniques to reduce the Rick of Subsequent Brittle Fracture, Review Prep. For Commission X, Int. Inst. of Welding (1967)
2. Weck, R.: Australian National Welding Convention. melburne (1965)
3. Masubuchi, K.: Control of Distortion and Shrinkage in Welding. Welding Research Council Bulletin No. 149, April, (1970).
4. UKAEA.: An Assessment of the Integrity of PWR Pressure Vessels. 2nd Report by Dr. W. Marshall Study Group (1982)
5. Kanazawa, T., Oba, H., Susei, S.: The effect of welding residual stress upon brittle fracture propagation, Report II. Japan Soc. Naval Arch. **110**, 359–368 (1961)
6. Masubuchi, K., Martin, D.C.: Investigation of Residual Stresses in Steel Weldments. Battelle Memorial Inst., Columbus, Ohio, Defense Technical Information Center (1998)
7. Wu, X.R., Carsson, J.: Welding residual stress intensity factors for semielliptical surface cracks in thin and thick plates. Eng. Fract. Mech. **19**, 407–427 (1984)
8. Liljeblad, R.: ASEA J. 23 (1937)
9. Kihara, H., Masubuchi, K., Kusuda, T., Iida, K.: IIW-X-219 (1959)
10. Hall, W.J., Joshi, J.R., Munse, W.H.: Welding j. Res. Supp., p 182S (1965)
11. Kihara, H., Masubuchi, K., Kusuda, T, Iida, K.: IIW-X-219 (1959)
12. Kihara, H., Masubuchi, K., Iida, K., Oba, H.: Effect of Stress Relieving on Brittle Fracture Strength of Welded Steel Plate, IIW-X-218 (1959)
13. McGowan, J.J.: An Assessment of the Beneficial Effects of Warm Prestressing on the Fracture Properties of Nuclear Reactor Vessels Under Severe Thermal Shoch, Westinghouse Electric Co., WCAP-9178 (1978)
14. Yukava, S.: Evaluation of Periodic Proof Testing and Warm Prestressing Procedure for Nuclear Reactor Vessels, HSSTP-TR-1 (1969)
15. Brothers, A.J, Yukawa, S.: The Effect of Warm Prestressing on Notch Fracture Strength, Trans. ASME, V. 85, Series D, pp. 97–104 (1963)
16. Sankey, G.O.: Spin-fracture tests of nickel-molibdenum-vanadium rotor steels in the brittle fracture range. Proc. ASTM **60**, 721–732 (1960)
17. Wessel, E.T, Baudry, R.A.: Discussion of Chapter 1-Brittle Fracture of Metals and Chapter 2–Engineering Aspects of Evaluating Notch Toughness, ASM, Metals Park, Ohio, pp. 37–40 (1964)
18. Andrews, W.R.: Effect of Loading Sequence on Notch strength of Warm prestressed Alloy Steel, Trans. ASME, vol. 92, Series b, pp. 785–791 (1970)
19. Hawthorne, G.R., Loss, F.J.: The Effect of Coupling Nuclear Radiation with Static and Cyclic Service Stresses and Periodic Proof Testing, NRL-6620, July (1976)
20. Succop, L.N., et al.: Effect of warm overstressing on pressure vessel steel properties. Weld. J. **49**, 351s–361s (1970)
21. Harrison, T.C., Farnerhough, G.D.: The influence of warm prestressing on the brittle fracture of structure containing sharp defect. Trans. ASME **94**, 373–386 (1972)
22. Loss, F.J., Gray, R.A., Hawthorne, J.R.: Significance of Warm Prestress to Crack Initiation During Thermal Shoch, NRL/NUREG Report 8165, September 29 (1977)
23. Stonesifer, R.B., Rybicki, E.F., McCabe, D.E.: Worm Prestress Modeling: Comparison of Models and Experimental Results, NUREG/CR-5208, MEA-2305, April (1989)
24. Heerens, J., Cornec, A., Schwalbe, K.H.: Determination of J_{Ic} Using New Blunting Line Construction, Proceedings of the Second OECD/CSNI Worckshop on Ductile Fracture Test Methods, pp.199–211, Paris (1987)
25. Chell, G.G., Haigh, J.R., Vitek, V.: A theory of warm prestressing: experimental validation and the implication for elastic plastic failure criteria. Int. J. Fract. **17**(1), 61–81 (1981)
26. Chell, G.G.: Some Fracture Mechanics Applications of Warm Prestressing to Pressure Vessels, 4th Intern. Conference on Pressure Vessel Technology, London, pp. 117–124, 19–22 May (1980)

27. Bilby, B.A., Cottrell, A.H., Swinden, K.H.: Proc. Roy. Soc. London **A272**, 304 (1963)
28. Curry, D.A.: A Micromechanistic Approach to Warm Prestressing of Ferritic Steels, CERL Lab Note RD/L/N103/79, September (1979)
29. Ritchie, R.O., Knott, J.F., Rice, J.R.: On the relationship between critical tensile stress and fracture toughness in mild steel. J. Mech. Phys. Solids **21**(6), 395–410 (1973)
30. Stonesifer, R.B., Rybicki, E.F., McCabe, D.E.: Development of Models for Warm Prestressing, NUREG/CR-4491, MEA-2122, January (1987)

Chapter 14
Pressure Vessels and Piping Integrity Assessment

14.1 Introduction and Historical Remarks

Pipes are certainly the most common and convenient means of transportation of liquid and gas media. Pipelines run all over the earth, onshore and offshore, for millions of kilometers. They normally have a longitudinal seam and a girth weld, which result in millions of kilometers of welds that are the most probable site of defects formation. Figure 14.1 presents some typical examples of defects in girth welds observed in a 10" A106B carbon steel pipe.

Pre-service pressure test has always been the usual tool to prove the system integrity arguing that if failure were not occurring under a pressure higher than the service pressure *a fortiori* it would not occur during service life. One of the earliest pre-service piping test procedures made use of gas under pressure. Gas seemed to be the most convenient and straightforward means of pressurization of a pipe. However, it became soon evident the devastating effect that such tests could have on piping integrity. Gas decompression rate is usually lower than brittle crack propagation rate in steel can reach 2,000 m/s. A gas under pressure retains enough energy to sustain the unstable propagation of a crack through the entire length of a pipeline, opening it up like a zip, Fig. 14.4. Therefore, in the pre-service testing, gas pressure was confined to no more than 10 percent of the service pressure and the entire pipeline divided into sections of shorter length, or the gas substituted by an uncompressible liquid so that the first leak would result in a sudden pressure drop, arresting crack propagation without jeopardizing the integrity of the entire pipeline. However, hydrostatic tests may be hazardous when residual water remaining inside the system, also as simple moisture can damage the piping. The world economy relies on pipelines. Interesting enough, some of the first experimental programs, were run in the Sixties at General Electric, California, and Battelle Memorial Institute (BMI), Columbus Laboratories, Ohio, for the US Atomic Energy Commission [1–4] and were concerned with carbon and stainless-steel pipes of the primary cooling system of nuclear power plants. That initial demand for safety expressed by the nuclear industry grew in the Seventies

P. P. Milella, *Fundamentals of Fracture Mechanics*,
https://doi.org/10.1007/978-3-032-23081-2_14

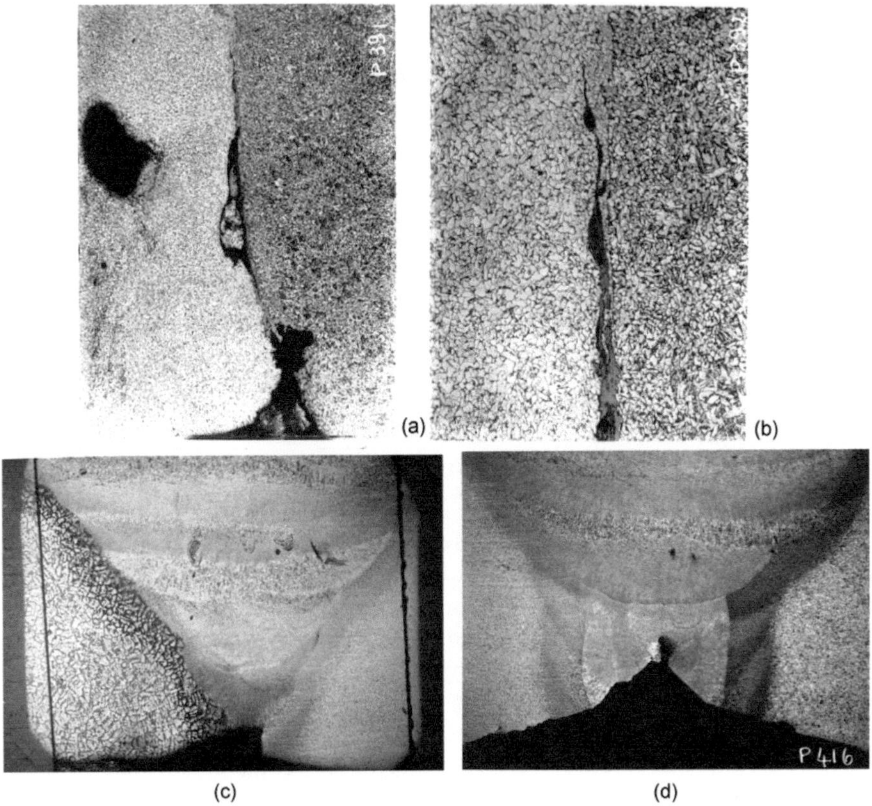

Fig. 14.1 Defects in a 10" A106B carbon steel pipe girth weld: **a** lack of side fusion, **b** detail of (**a**) revealing that the flaw is a sharp transgranular crack, **c** undercut and **d** lack of penetration

and was the driving force for both the NRC "Degraded Piping Program" and the IPIRG (International Piping Integrity Research Group) program [5]. The IPIRG program was funded by a consortium of organizations from nine nuclear nations: Canada, France, Italy, Japan, Sweden, Switzerland, Taiwan, United Kingdom, and the United States of America. The program task was to develop data needed to verify engineering methods for assessing the integrity of cracked piping used in nuclear power plants. The five-year program, conducted at Battelle Columbus, Ohio, was completed in July 1991. Parallel and in addition to these programs, in 1981 also the Italian Nuclear Agency ENEA (Ente Nazionale per l'Energia Nucleare) started a large research program on pipe fracture behavior [6]. Same research was done in the UK where the CEGB (now Nuclear Electric) developed an advanced new procedure known as R-6 method of failure avoidance [7]. The following sections will deal with the fundamentals and results of those research programs on piping integrity analysis, that used a fracture mechanics approach, both linear-elastic and elastic–plastic, or a general yield criterion, which led to a comprehensive knowledge of pipe fracture behavior and turned into design standards. In any piping or pressure vessel fracture analysis some fundamental concepts must be kept. First, thin sections piping generally behave in a ductile manner and are not subjected to cleavage fracture. This is particularly true with FCC materials. With BCC materials this is also true at temperatures at or over RT_{NDT}. The concept of thin section is relative and generally associated with the *t/R* ratio where *t* is the thickness of the pipe and *R* its radius. A *t/R* ratio ≤ 10 is considered to describe small sections or shells structures. Secondly, dealing with shells, LEFM and the K_I concept do not apply and EPFM or GY should be used. Generally, as already pointed out, a toughness in excess of 130–150 MPa $\sqrt{m}$ is considered representative of loss of SSY conditions.

14.2 Pipe Flaws Analysis

14.2.1 Through-Wall Longitudinal Cracks

Pipes and shells are normally loaded by internal pressure, sometimes as high as 150 bars or even higher, so that the main concern was initially focused on longitudinal flaws possibly existing in longitudinal welds, since the hoop stress, σh, is twice the longitudinal or axial stress, σl, acting on girth welds and circumferential defects. The first semi-empirical solution to the problem was conducted using a fracture mechanics stress intensity approach much like the CTOD approach described in Sect. 10.4. A K_{Ic} formulation was developed with the premise that a cylindrical vessel loaded by an internal pressure, containing a longitudinal slit, could be treated like a flat panel of the same thickness, and containing the same through-crack as the vessel, as schematically shown in Fig. 14.2, provided the nominal stress on the panel is taken equal to the hoop stress multiplied by a magnification factor M, using the approximate solution proposed by Folias, Eq. (10.18), and shown in Fig. 14.3 by the upper curve, which is valid for linear elastic-perfectly plastic nonhardening materials.

This correction factor for the plane plate is needed in order to consider bending stresses due to the local crack tip bulging, as schematically presented in Fig. 10.16, particularly high in thin shell of tough materials. Figure 14.4 [2] shows three failure example of a local bending stress, ranging from a rather brittle behavior, Fig. 14.4a, almost without any apparent local bulging and resulting in a simple leak, to a complete opening of the flaw that literally leaves the pipe flat open, Fig. 14.4c.

The last picture, Fig. 14.4c, has been chosen to better visualize, through a dramatic example, the bulging effect which we are referring to. A further modification is needed to account for the plasticity developed at the tip of the notch using the Dugdale plastic zone factor based on the strip-yield model, see Eq. (8.13). With these corrections, and with reference to Fig. 14.2, the first semiempirical equation was proposed by Swedlow and Williams [8] as:

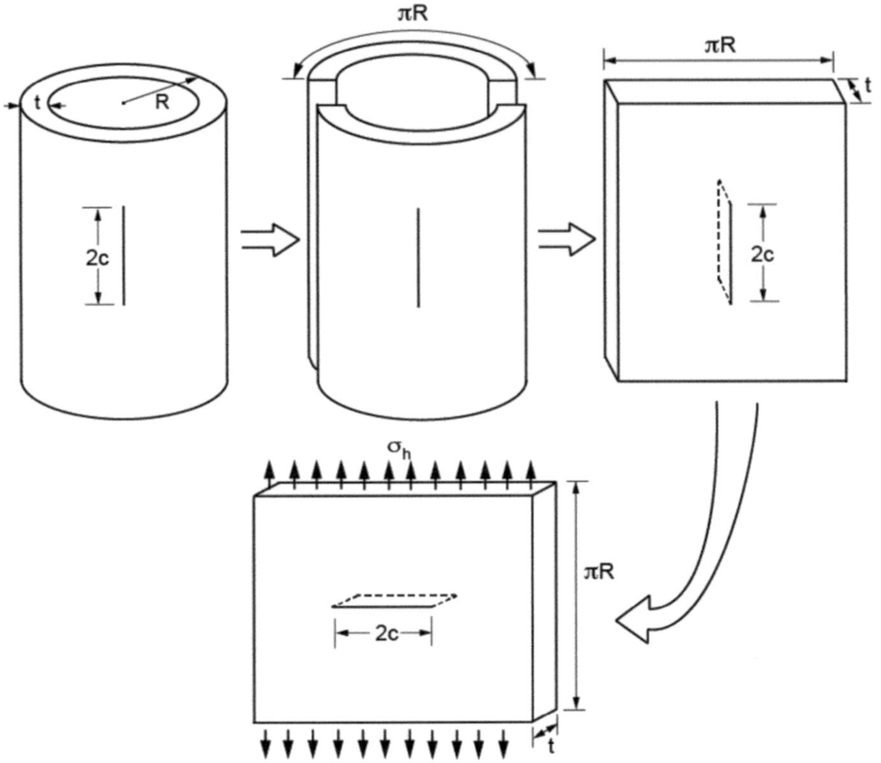

Fig. 14.2 Schematic of a trough-wall longitudinal flaw in a pipe and its correspondence with a flat panel

$$K_C^2 = \frac{\sigma_h^2 \pi c}{\cos \theta} M^2 \left(\frac{4 - K}{2} \right) \tag{14.1}$$

where

K_c = critical stress intensity factor
c = through-wall half crack length
R = average vessel radius
t = wall thickness
σh = nominal hoop stress

σo = failure stress of unflawed vessel
K = $(3 - 4\nu)$ plain strain
K = $(3 - 4\nu)/(1 + \nu)$ plain stress

$$M^2 = \left(1 + 1.61 \frac{c^2}{Rt} \right) \text{and}\, \theta = \frac{\pi}{2} \frac{\sigma_h}{\sigma_o} \tag{14.2}$$

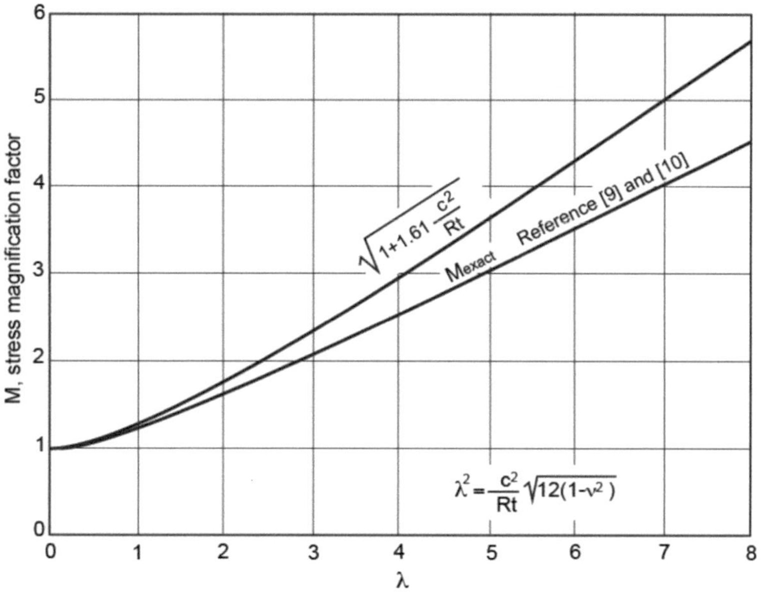

Fig. 14.3 Relationship between λ and stress magnification factor M [9]

Equation (14.1) was modified substituting the original M factor given by Eq. (14.2), with a new and more exact magnification factor M_{exact}

$$K_C^2 = \frac{\sigma_h^2 \pi c}{\cos\theta}\left(M_{exact}^2\right)\left(\frac{4-K}{2}\right) \tag{14.3}$$

where M_{exact} is the solution later obtained by Folias [9] and by Erdogan and Kibler [10], and shown in Fig. 14.3 by the curve labeled Reference [9] and [10] as function of

$$\lambda^2 - \frac{c^2}{Rt}\sqrt{12\left(1 - \nu^2\right)} \tag{14.3b}$$

Twenty-one full scale experiments on pipe of A106B pipe, having various diameter, ranging from 4 to 12 inch, and different schedule (thickness) having different through-wall longitudinal flaw sizes were tested by Reynolds at RT [1]. The material is a low cost 0.30% carbon steel representative of routine manufacturing practice. The pipes were hydrostatically loaded with water or water-ethylene glycol mixture under pressure. The longitudinal through-wall cracks were sealed from inside with a1/16-in-thick type 304 stainless steel patch to avoid leakage and depressurization. Values of the critical stress intensity factor, K_c, representing the fracture toughness, were calculated using either Eq. (14.1) or (14.3), this last with [1].

Fig. 14.4 Modes of crack extension observed in 24 × 1.70-inch A106 pipe [2] producing more or less severe local bulging

$$M_{exact}^2 = \left(1 + 0.4\frac{c^2}{Rt}\right) \tag{14.4}$$

Results are shown in Fig. 14.5 [1]. Those obtained using Eq. (14.1) show a continuously growing K_c with crack half-length, c, ranging from about 80 ksi$\sqrt{\text{in}}$ (~88 MPa$\sqrt{\text{m}}$) to about 250 ksi$\sqrt{\text{in}}$ (~245 MPa$\sqrt{\text{m}}$).

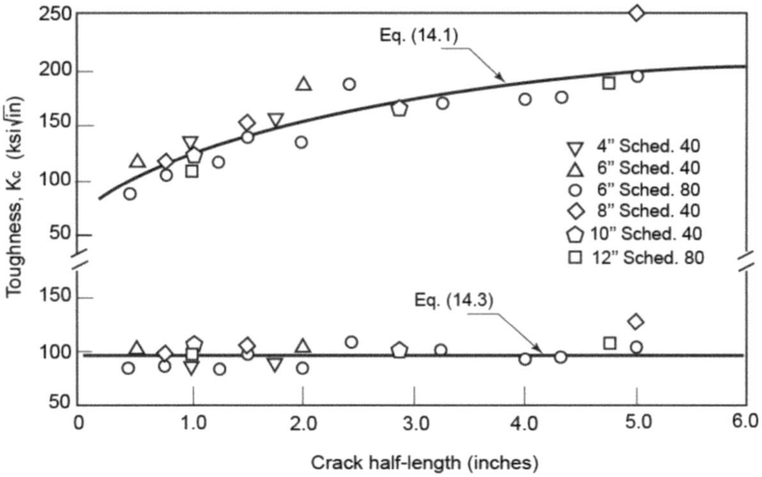

Fig. 14.5 Critical stress intensity factor, K_c, for ASTM A106B pipes containing axial through-wall flaws. Equation (14.3) is predicting a constant toughness (redesigned from [1])

This apparent dependency of the measured toughness on crack length is not only infringing the concept of toughness as a material characteristic but is also violating the common idea that short cracks show a more ductile behavior than longer cracks, see Sect. 9.17.

The increase of K_c with increasing flaw length is the result of the first Folias M factor given by Eq. (14.2) that grows too much with crack half-length with respect to M_{exact}, Fig. 14.3. Using Eq. (14.3) with M given by Eq. (14.4), the K_c, becomes practically constant, as it should be, at about 95 ksi$\sqrt{\text{in}}$ (~104 MPa$\sqrt{\text{m}}$), as shown in Fig. 14.5. However, a question still remains in that a measured toughness of about 104 MPa$\sqrt{\text{m}}$ is too low for a structure, the pipe, that should behave and fail in a ductile manner, at least for the very high R/t ratio considered.

Compact CT specimens ½ inch thick taken from pipes of A106B steel pipes were yielding toughness K_{Jc} values, derived from J measurements, ranging from 190 MPa$\sqrt{\text{m}}$ to 250 MPa$\sqrt{\text{m}}$, depending on specimen orientation [12]. Figure 3.26 presents the results of metallographic examination of a longitudinal section of a pipe of A106B carbon steel [12] showing a considerably high anisotropy introduced by the rolling process. The basic metallurgical appearance is a fine-grained, pearlite-ferrite structure with pearlite (iron carbide) elongated in longitudinal bands oriented parallel to the longitudinal axis of the pipe. This anisotropy appears in the relative Charpy-V energy (resilience), Fig. 3.27 [1], for specimen taken from ASTM A106B 6" Schedule 80 pipe. The lowest resilience, and therefore toughness, is that yielded by specimens A and B with a T-L and T-S orientation having the crack as the through-wall crack of Fig. 14.2. The specimens used by Reynolds were not valid according to the ASTM size requirements for linear elastic fracture mechanics measurements, see Sect. 9.8, however, ½ inch was the only size possible from the pipe thickness available and the expected behavior was not really elastic but fully plastic. This raises the question of validity of Eq. (14.3), which can be solved for σh knowing the toughness of the material, K_{Ic}, obtained with valid fracture mechanics specimens or K_{Jc} derived from J measurements with smaller specimen. The problem with Eq. (14.3) is that it contains two conflicting element, namely the linear elastic K_c and the elastic–plastic Folias-Erdogan-Kibler correction factor M. Moreover, the M factor is valid for linear elastic-perfectly plastic nonhardening materials that cannot be representative of an elastic–plastic behavior of a real strain hardening material. To solve the puzzle, it may be convenient to go back to the original CTOD $= \delta_t$ formulation given by Eq. (10.17) corrected with the Folias M or M_{exact} factor of Eq. (14.4) for the hoop stress σh, and recalling the relationship given by Eq. (10.19) between δ_t and G and the equality (10.40), it is:

$$\begin{aligned} \delta_t &= \frac{8\sigma_s c}{\pi E} \ln\left[\sec\left(\frac{\pi M \sigma_h}{2\sigma_s}\right)\right] \\ \delta_t &\approx \frac{G}{\sigma_s} = \frac{K^2}{E} \end{aligned} \tag{14.5}$$

where σ_S is the yield stress of the material. Hahn et al. [13] introduced a new stress σ^* that must be interpreted as an average flow stress acting in the plastic zone, that cannot be precisely defined. At low nominal stresses the plastic strains generated in the crack tip plastic zone are small, and $\sigma^* = \sigma s$ is probably a good estimate. At higher stress levels the influence of strain hardening is felt, with $\sigma^* = \sigma u$, the ultimate tensile strength being an approximate upper limit. Therefore, the second of Eq. (14.5) solved for K becomes:

$$K_c^2 = (\sigma^*)^2\left[\frac{8c}{\pi}\, ln\, sec\left(\frac{\pi M \sigma_h}{2\sigma^*}\right)\right] \tag{14.6}$$

For very high toughness materials or relatively thin pipes which fail by large-scale yielding (LSY) or plastic instability, it is simply:

$$\sigma^* = \sigma_h M \tag{14.6a}$$

Equation (14.6a) states that failure occurs when the stress in the region of the crack tip reaches a critical stress or flow stress σ^* whose value in Hahn et al. experiments was

$$\sigma^* = \sigma_h \mathrm{M} = 0.9\,\sigma_s \tag{14.6b}$$

i.e., close to the yield stress. The LSY hypothesis reveals itself to be very effective for relatively thin vessels of though materials for which $\sigma^* = \sigma f$, i.e., the flow stress equal to half the sum of yield stress and ultimate stress, Eq. (10.14). Hahn et al. were also suggesting that the flaw stress criterion should apply when $(Kc/\sigma s)^2/c \geq 7$, and

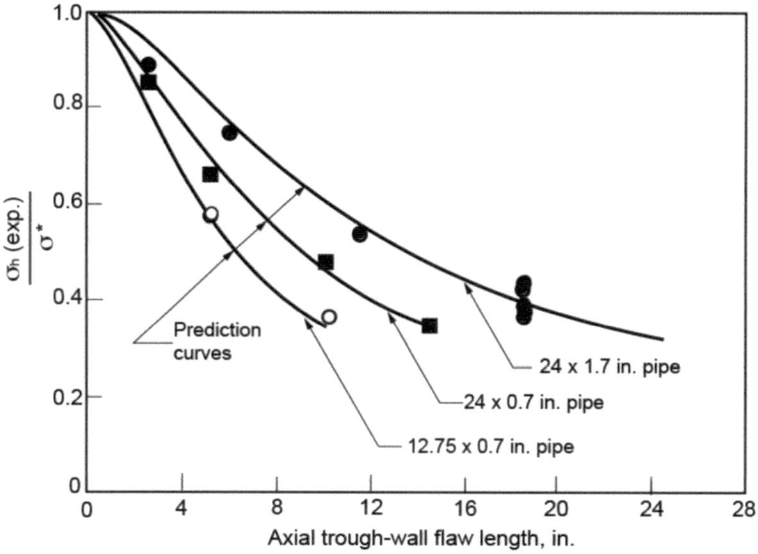

Fig. 14.6 Comparison of experimental critical flaw size data with predicted behavior for three A106B carbon steel pipe sizes [4]

for values less than 7 the K_c criterion should apply with $\sigma^* = \sigma s$. Solving Eq. (14.6) for the hoop stress at failure, σh, the K_c becomes a J-derived K_{Jc}. Equations (14.6) and (14.6a) cover the complete range of toughness.

On this issue see also Sect. 10.3. Figure 14.6 [4] presents a plot of experimental axial through-wall critical flaw data obtained with 24 in. A106B carbon steel pipes of two different thicknesses: 1.7 in. and 0.7 in., respectively, and a 12.75 in. diameter, 0.7 in. thick pipe of the same steel, along with calculated flaw size curves using Eq. (14.6a) and

$$\sigma^* = \frac{\sigma_s + \sigma_u}{2.4} \tag{14.6c}$$

14.2.2 *Part-Through Crack*

A more realistic defect configuration than the through-wall crack from the viewpoint of occurrence in service is the part-through or surface flaw, Fig. 14.7. An empirical equation has been developed by Hahn et al. [13] from experimental results on thin-wall pipes to predict the failure pressure of axial surface flaws

$$\sigma_h = \sigma^* \left[\frac{A_0 - A}{A_0 - A(1/M)} \right] = \sigma^* \left[\frac{t/d - 1}{t/d - 1/M} \right] \tag{14.7}$$

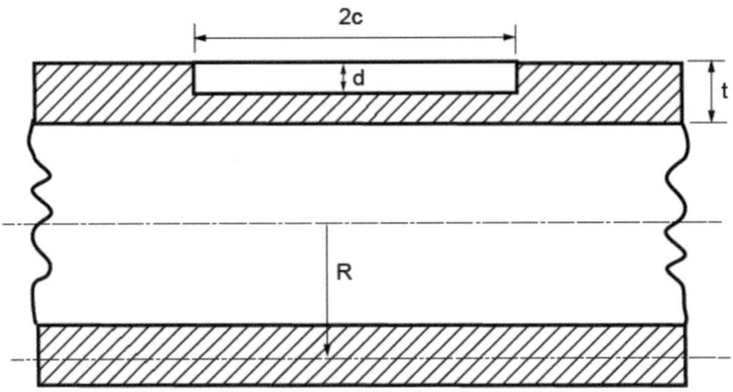

Fig. 14.7 Part-trough wall or surface longitudinal flaw in a pipe

where

c = through-wall half crack length
R = average vessel radius
t = wall thickness
d = surface flaw depth
A_o= $2ct$, area of through-wall of same length as the surface flaw
A = $2cd$, area of through wall flaw
σh = nominal hoop stress.
σ^* = flow stress.

This equation indicates that very long flaws reduce the failure pressure in proportion to the depth of a flaw as M, the magnification factor for a through-wall flaw given by Eq. (14.4), increases, as with long flaws. Figure 14.8 presents surface flaw failure curves, calculated using Eq. (14.7), and experimental results for 0.6, 0.7 and 0.9 in.-deep flaws in 24 in. diameter, 1.5 in. thick, type 316 stainless steel pipe [2]. The experimental points correspond quite well with the predicted failure curves based on Eq. (14.7) indicating the good agreement between the prediction and the experimental data.

14.3 Circumferential Flaws. NSC Criterion

14.3.1 Through-Wall Crack

We have shown in Fig. 14.2 how a cylindrical vessel can be treated like a flat panel of the same thickness containing the same through-crack. The same premise holds for a circumferential flaw, as schematically shown in Fig. 14.9: a pipe carrying a circumferential crack of length $2a$ can be considered as a flat plate with a surface

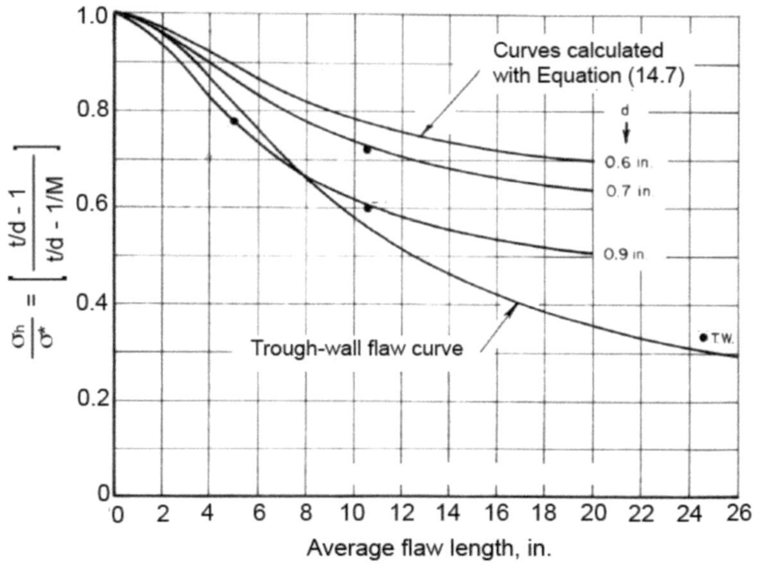

Fig. 14.8 Surface flaw behavior in 24 in. diameter, 1.50 in. thick type 316 stainless steel pipe [2]

crack of depth a. The traction stress acting on the panel is the longitudinal stress, σl, generated in the pipe by the pressure, which is one-half the circumferential or hoop stress, σh.

As a consequence, the pressure burst strength of a pipe should be rather insensitive to circumferential flaws. Pressure burst tests on circumferentially flawed A106B carbon steel or stainless-steel pipes show that, effectively, the crack develops a lateral tearing at about 45°, as shown in Fig. 14.15, rather than continuing to grow in the

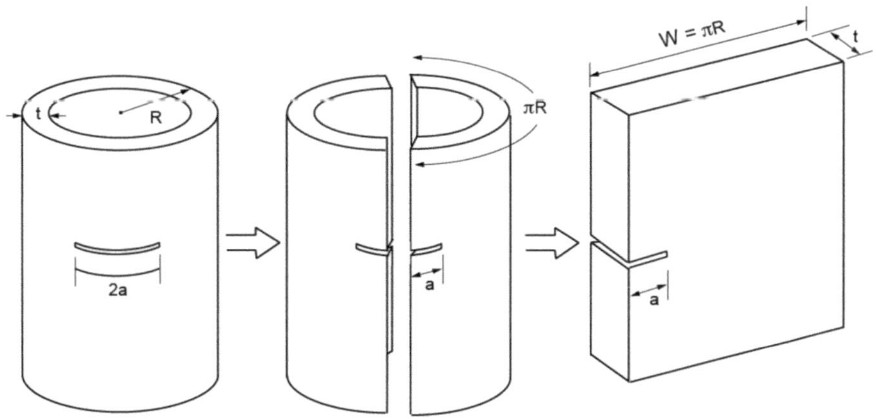

Fig. 14.9 Circumferential flaw in a pipe and its correspondence with a flat panel loaded in traction

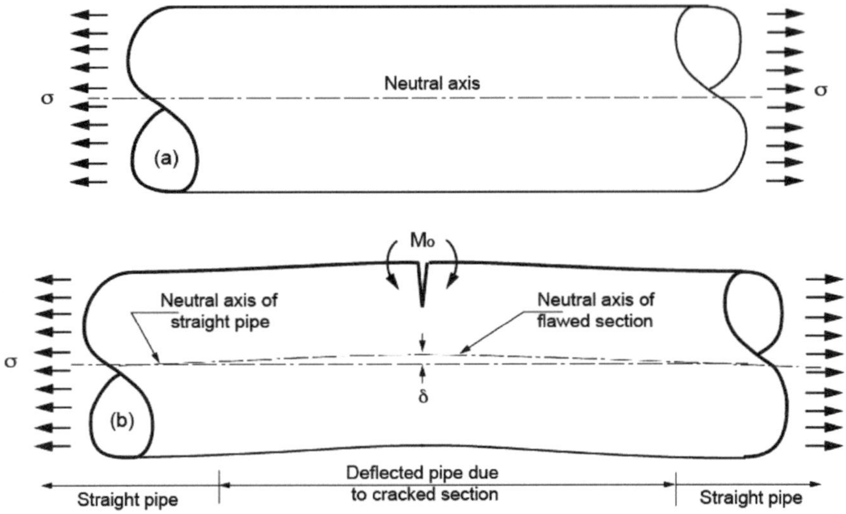

Fig. 14.10 **a** Neutral axis of a straight pipe without crack under traction σ, **b** schematic of the induced bending moment M_o at the flaw section in a straight pipe carrying a circumferential flaw under axial traction σ

circumferential direction, unless the pipe is small and the crack is rather long, having an initial extension of 180°. A straight pipe of diameter D and thickness t carrying a circumferential crack under axial traction, σ, develops a local bending moment, M_o, as shown in Fig. 14.10. The induced moment, M_o, is equal to the axial force times the distance, δ, between the neutral axis of the straight pipe and that of the cracked pipe.

$$M_o = (\pi D t)\delta \tag{14.8}$$

However, the real danger of a circumferential crack in a pipe does not come from the pressure but, rather, from the bending moment due to an earthquake or associated with misalignments introduced during installation and differential displacements between anchor points of the pipeline, which can jeopardize the integrity of the system more than the internal pressure. The main feature of a pipe is to be a thin-wall structure. The importance of thickness in the fracture behavior of a body of ductile material has been deeply discussed in Sect. 3.10, 4.4, 4.5, 4.6 and 9.8. This feature prevents the occurrence of a brittle fracture, at least in carbon steel at temperatures higher than the transition temperature, RT_{NDT}, see Sect. 3.10 and 9.12, and in FCC structure materials of high ductility, like stainless steel. Therefore, for ductile materials failure is likely to be controlled by Large Scale Yielding (LSY), as already assumed and verified for surface flaws. Under this basic assumption, a criteria for pipes with through-wall circumferential crack was developed by Eiber et al. [4] and Kanninen et al. at BMI [15] assuming that failure occurs when the stress state on the net cross section carrying the crack reaches a limit value, σf, which is the flow stress of the material. The fracture criterion was called Net Section Collapse load criterion (NSC). This method is of particular interest since it is currently used in several internationally accepted criteria as a means of evaluating postulated cracks in nuclear piping systems. One major assumption of this method is that the material toughness is high enough that failure is governed by the flow stress or collapse stress rather than by the linear-elastic K_{Ic} which, indeed, loses its meaning at all.

Also crack growth is not taken into account. In addition, and sustainment of this hypothesis, the NSC criterion embodies the assumption that piping with very high *R/t* ratios has unlimited ductility, and thus the only effect of a crack or defect is to reduce the cross-sectional load-carrying capacity of the pipe. With reference to Fig. 14.11, for an unpressurized pipe with a circumferential through-wall crack subjected to pure bending, an idealized fully plastic bending stress, σ, is assumed in the plane of the crack, see Fig. 3.10. A free-body analysis shows that the point of stress inversion, defined by the β angle in Fig. 14.11, is

$$\beta = (\pi - \gamma)/2 \tag{14.9}$$

The net-section-collapse moment balancing the external applied bending moment M is given by:

$$M = 2\sigma_f R^2 t(2\sin\beta - \sin\gamma) \tag{14.10}$$

It is, then, clear that Eq. (14.10) represents a valid failure criterion if, solved for σ

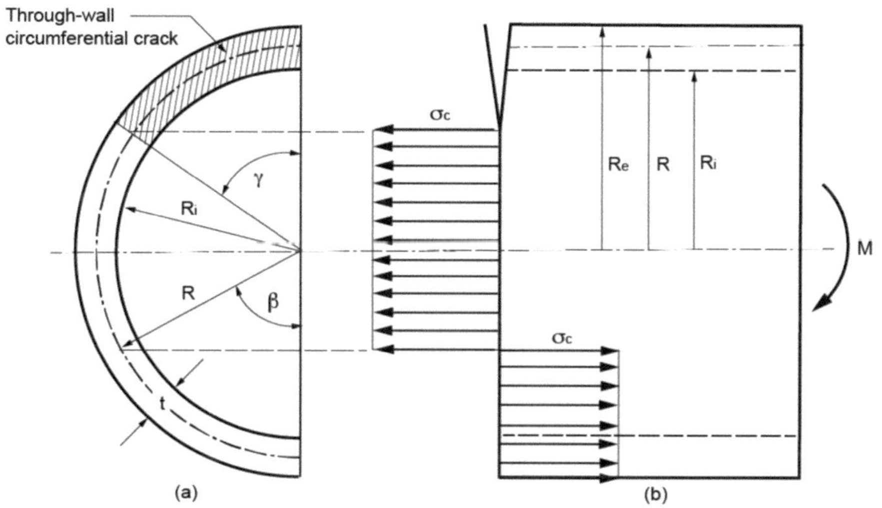

Fig. 14.11 Net section collapse load criterion for through-wall circumferential crack

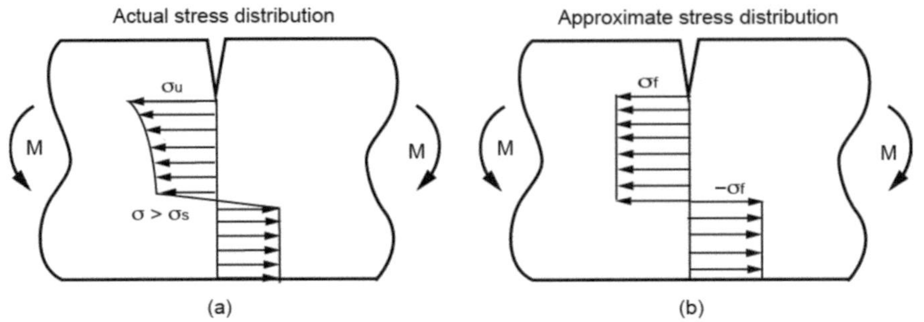

Fig. 14.12 NSC load criterion. Stress distribution on the cracked section: **a** actual and **b** approximate

$$\sigma_f = \frac{M}{2R^2t(2\sin\beta - \sin\gamma)} \tag{14.11}$$

it will always yield the same critical value, σf, independently of the crack size, γ angle, and pipe diameter or thickness. It must be understood, however, that the schematization of the stress state shown in Fig. 14.11 is a simplification of what is likely to be the real stress distribution on the cracked section, as schematically shown in Fig. 14.12. It is rater evident that stresses near the crack tip must reach the ultimate stress value, σu, at failure. To verify the validity of the NSC load criteria, a large experimental program was run in the U.S. [16] and in Italy [6, 17] on A106B carbon steel pipes and 316L stainless steel pipes of 4", 6" and 8" size, loaded under four point bending (pure moment), having different schedule, carrying circumferential through-wall and part-through wall cracks, at RT and 280 °C. Results for the A106B pipes with circumferential through-wall cracks loaded at RT are shown in Fig. 14.13 [17, 18].

Here, the maximum NSC stress σf calculated from Eq. (14.11), knowing the experimental maximum moment M, is plotted versus the initial half-crack angle γ. It can be seen that most of the experimental results clearly indicate that both small and large cracks, either in small or large pipes, fall within the flow stress $\sigma f = (\sigma s + \sigma u)/2$ scatter band measured on traction specimens that was varying from a minimum of 355 MPa to a maximum of 420 MPa with an average of 380 MPa. Same results were obtained for SA 316 stainless steel pipes of 6 in. and 8 in. diameter.

However, it must be said, that the experimental moment, M, is not the maximum moment at failure but that applied to the pipe by the loading cell compatible with the maximum pipe bending allowed by the testing apparatus, as shown in Fig. 14.14 for a 6 in. stainless steel 316L pipe tested at 280 °C [20] that reached the maximum stroke possible. In almost all cases pipes did not break. After crack tip blunting, the initial flaw started to grow steadily at an angle at about 45° with the crack plane, also bifurcating and the test was stopped. This behavior common to all pipes is evidenced in Fig. 14.15 [12] that shows a loading sequence in a 6 in. A106B carbon steel pipe

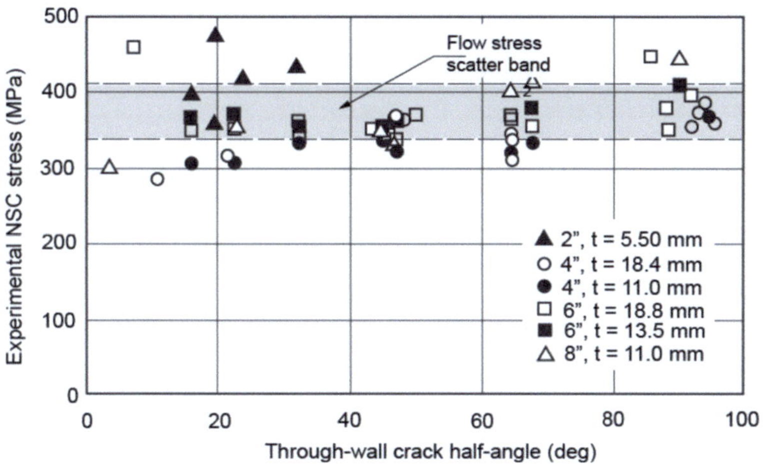

Fig. 14.13 Results of the Italian research program on 2″, 4″, 6″ and 8″ A106B pipe of different schedule, having through-wall circumferential cracks tested at RT. The experimental NSC stress is plotted as function of the crack half-angle [17, 18]. Also shown is the scatter band (brown band) of the flow stress σf

with a 90° through-wall crack, tested at RT. The initial sharp crack tip, Fig. 14.15a, blunts as the applied load increases becoming a notch larger and larger, Fig. 14.15b and c.

At a certain moment in the loading sequence, a slant tearing appears at 45° that grows till the end of the test, Fig. 14.15d. This last occurrence was confirmed by ½ CT specimens machined from a pipe sample, having the crack oriented as the crack in the pipe that developed the same tearing, Fig. 14.15e, with bifurcation but without unstable fracture, confirming the full ductile behavior of the pipe. Figure 14.15f is a detail of the specimen crack tip blunting and 45° tearing with bifurcation. Note that the crack grew about 1 cm before tearing initiated and about 4 cm at the end of the test. In very few cases, just three out of 80 tests at RT on A106B, the pipe failed catastrophically with a 180° through-wall crack, as shown in Fig. 14.16.

All experimental results shown in Fig. 14.13 are replotted in Fig. 14.17 as function of the applied bending moment, Eq. (14.10), with $\sigma f = 380$ MPa. It is interesting to note how the trend relative to smaller diameter pipes flattens which is indicative of potential instability since a small increase of the applied stress or of the crack length due, for instance, to subcritical crack growth from tearing, Fig. 14.16, or service fatigue may result in pipe fracture. Moreover, ovalization of the cross section, which is much higher in smaller diameter pipes, is a contributing factor to instability.

Fig. 14.14 Four-point bend testing apparatus showing the maximum bending achievable in an AISI 316L, 6 in. stainless steel pipe, carrying a through-wall circumferential crack, at 280 °C [20]

14.3.2 Surface Cracks

Surface cracks can be loaded by internal pressure, *p*, and external moment *M*. The NSC criterion applied to surface cracks assumes that also the remaining ligament, *t-d*, in the cracked section reaches fully plastic conditions at failure, as schematically shown in Fig. 14.18. A simple beam theory analysis developed in Reference [15] wherein static equilibrium conditions are enforced shows that the maximum moment-carrying capacity is given by

$$M_{NSC} = 2\sigma_f R^2 t\left(2\sin\beta - \frac{d}{t}\sin\gamma\right) \tag{14.12}$$

where

$$\beta = \frac{\pi - \gamma(d/t)}{2} - \frac{\pi R_i^2 p}{4R\sigma_f t} \tag{14.13}$$

with σf being the flow stress of the material, the average of the yield and ultimate stress, $\sigma_f = (\sigma_s + \sigma_u)/2$, *p* the internal pressure, R_i the inside radius of the pipe and *R* the mean radius of the pipe. the only difference Thus, the only difference between

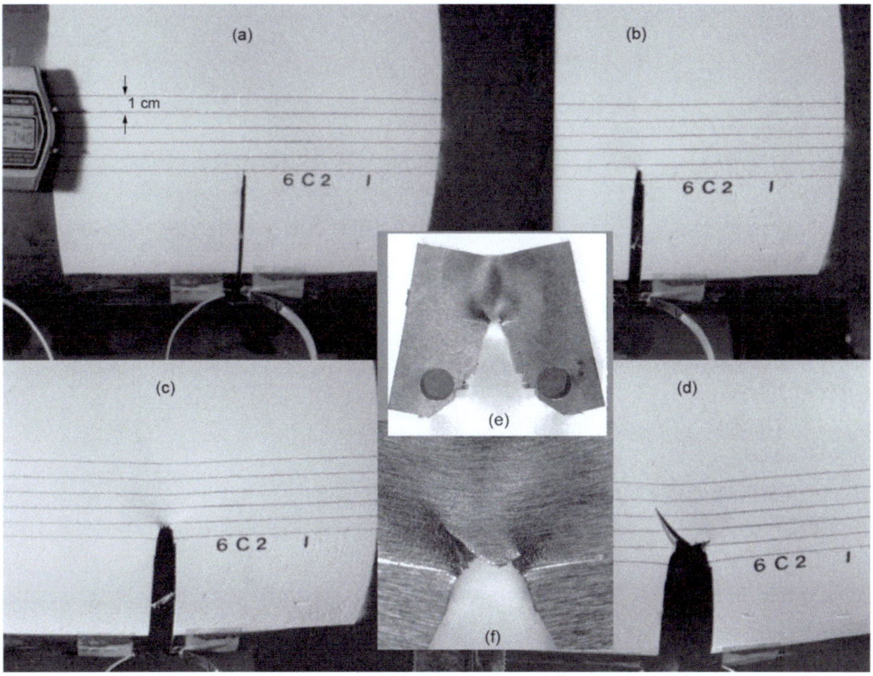

Fig. 14.15 Four-point bend loading sequence in A106B 6 in. pipe carrying a circumferential through-wall crack 90° wide (Milella). Also shown for comparison: **e** the 1/2CT specimen machined from a pipe sample that developed the same tearing at 45° to the crack plane, and **f** detail of crack tip specimen tearing and bifurcation that reproduces exactly that observed on the pipe [12]

surface cracks NSC formulation for the pure bending case and the combined pressure and bending case is the determination of the stress inversion angle β. The effect of pressure is to shift the neutral axis down, thus lowering the predicted net-section collapse moment. For the case of a pipe subjected only to pressure loads, i.e., pure pressure, the NSC formulation can be done by rearranging Eq. (14.12). For this to be the case, it is

$$2\sin\beta = (d/t)\sin\gamma) \tag{14.14}$$

Substituting Eq. (14.13) into Eq. (14.14) yields

$$\sigma_{NSC} = \sigma_f\left[1 - \frac{\gamma}{\pi}(d/t) - \frac{2}{\pi}\arcsin\left(\frac{d}{2t}\sin\gamma\right)\right] \tag{14.15}$$

The above analysis assumes that the ratio R/t is large (> 10) and, therefore, thin-wall pipe approximations are valid. Experimental results obtained on 304 stainless steel pipes at RT and 288 °C [21] are shown in Fig. 14.19 as ratio of experimental maximum moment, M, to the Net Section Collapse Moment, M_{NSC}, given

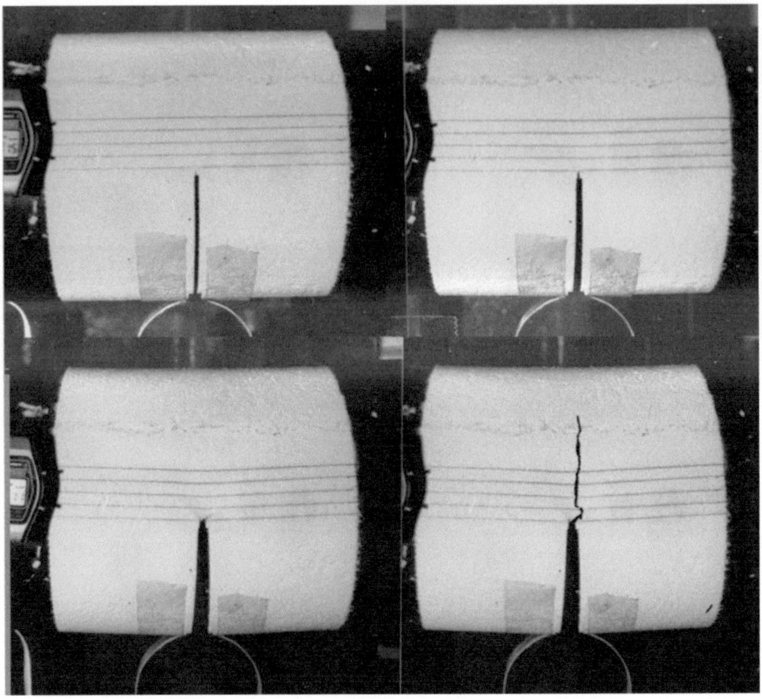

Fig. 14.16 Loading sequence of A106B 4" pipe carrying a 180° circumferential through-wall crack that failed after an initial blunting and 45° tearing

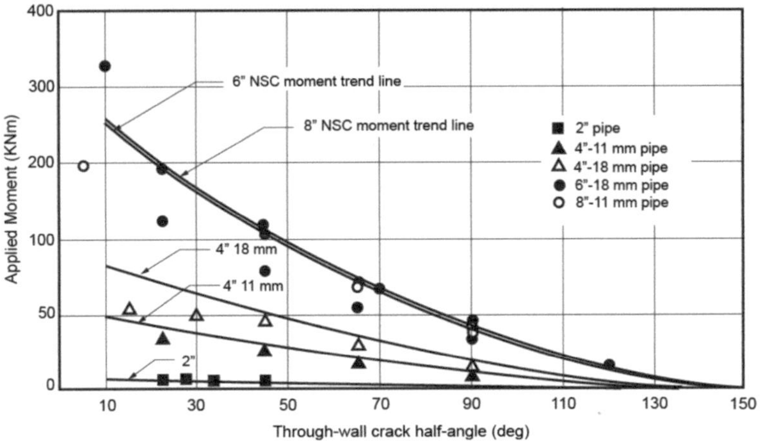

Fig. 14.17 Experimental results obtained on 6" 316L stainless steel pipes of different thickness at RT [6, 20]. The 6 in. 14 mm pipe had the through-wall flaw in a TIG weld. Solid curves are theoretically calculated using Eq. (14.10) with the flow stress, $\sigma f = 400$ MPa

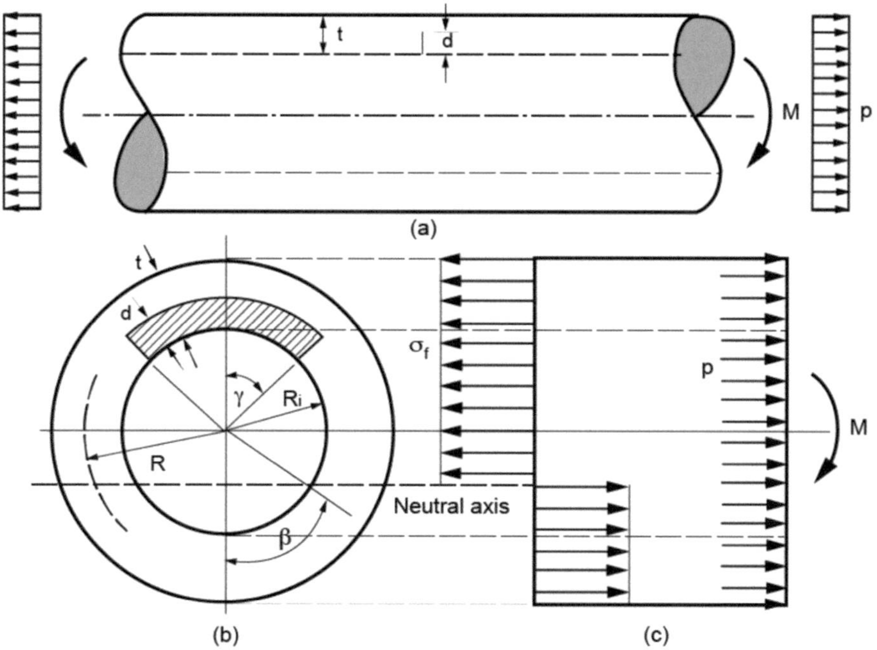

Fig. 14.18 Schematic of idealized surface-cracked pipe geometry and loading conditions

by Eq. (14.12). Note that pipes were tested under pure bending moment without any pressure, $p = 0$.

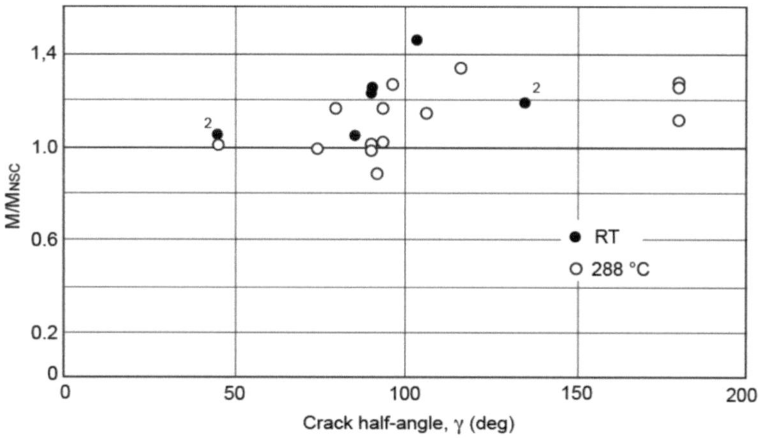

Fig. 14.19 Results of tests run at RT and 288 °C on TP304 stainless steel pipes carrying a surface crack under four-point bending. [data taken from Ref. [21])

Surprising enough, results indicate a clear underprediction of the Net Section Collapse Moment, M_{NSC}, with respect to experimental applied moment, despite the fact that smaller pipes with higher *a/t* ratios showed some ovalization with reduction of the vertical diameter. Figure 14.20 shows the same data of Fig. 14.19 as function of pipe diameter and, Fig. 14.21, of the *a/t* ratio. Smaller pipes seem to beave better than larger pipes. In one case the applied moment was exceeding the NSC moment by more than 40%.

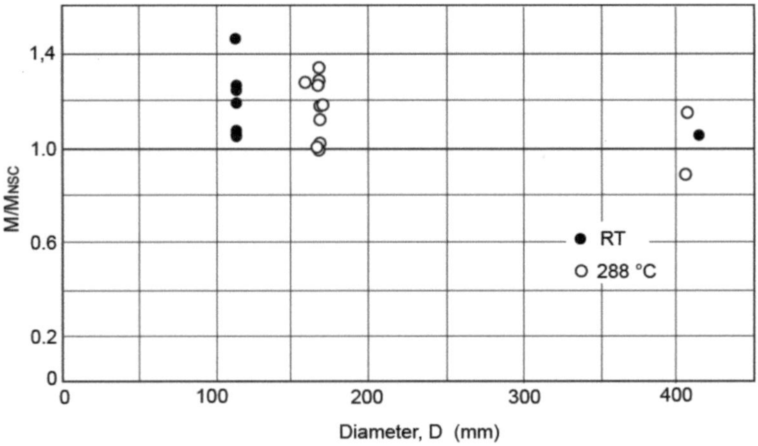

Fig. 14.20 Results of tests run at RT and 288 °C on TP304 stainless steel pipes carrying a surface crack [data taken from Ref. [21])

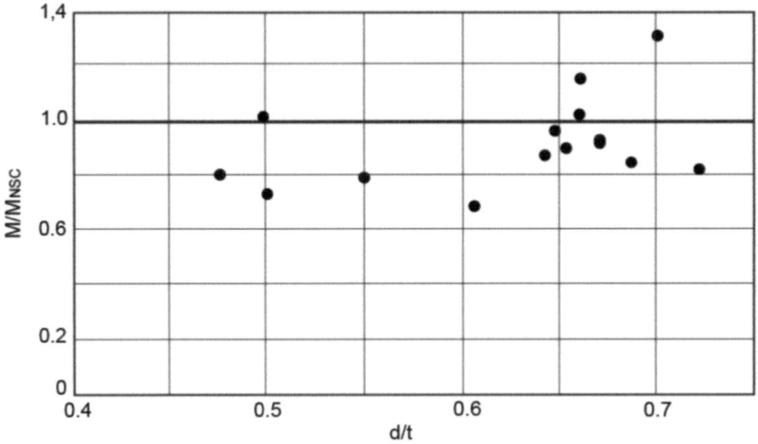

Fig. 14.21 Results of tests run at RT and 288 °C on TP304 stainless steel pipes carrying a surface crack as function of the *d/t* ratio [data taken from Ref. [21])

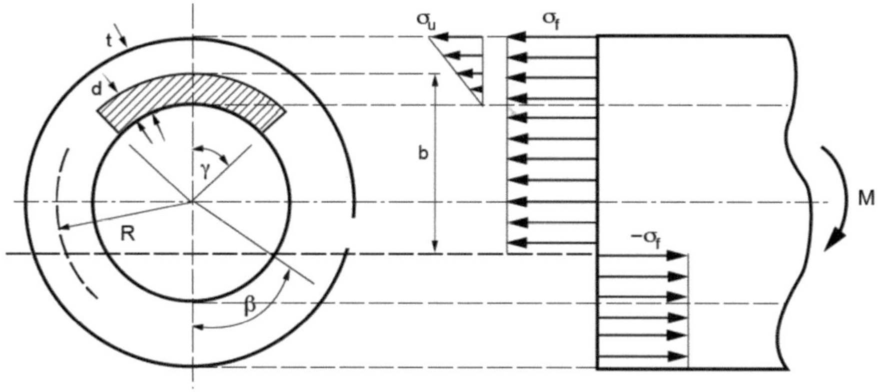

Fig. 14.22 With surface cracks the stress in the ligament can reach the ultimate strength σ_u increasing the M_{NSC} moment

This occurrence is better understood by looking at the dependence of the M/M_{NSC} ratio on the *d/t* ratio, as shown in Fig. 14.21. Experimental data clearly indicate that larger crack angles associated with small *d/t* ratios increase the pipe resistance over the M_{NSC} prediction, which is practically impossible since the NSC criterion assumes that the pipe section is completely under general yielding condition. This particular finding suggests that the real stress state in the uncracked ligament may exceed the flow stress σ_f reaching the ultimate strength value, as shown in Fig. 14.22.

The additional ΔM_{NSC} can be expressed as [22]

$$M = 2\sigma_f R^2 t(2\sin\beta - \sin\gamma) + \sigma_c \lambda Rt\left[\cos\left(\gamma + \frac{\lambda}{3R}\right) + \cos\beta\right] \tag{14.16}$$

with

$$\begin{aligned} \sigma_c &= \sigma_u - \sigma_f \\ \lambda &= \sqrt{\gamma Rt} \end{aligned} \tag{14.16b}$$

or simply

$$\Delta M_{NSC} = 2\gamma R(t-d)\frac{(\sigma_u - \sigma_f)}{2}b \tag{14.16c}$$

with b given by:

$$b = R\left(\frac{1}{2} + \cos\frac{\pi - (d/t)\gamma}{2}\right) \tag{14.17}$$

One may argue that the flow stress criterion by taking half of the yield stress plus the ultimate stress is penalizing the stress state with respect to the actual stress distribution. However, by flattening the stress state at σf over the entire cross section it is added additional stress in the areas closer to the neutral axis that have lower capability of reaction. On the contrary, by adding the stress triangle of Fig. 14.22 in the ligament more distant from the neutral axis, pipe reaction is increased. It is clear that the increment ΔM_{NSC} given by Eqs. (14.16) and (14.16c) must be added to the conventional M_{NSC} moment only in case the applied moment M is higher than M_{NSC}: $M > M_{NSC}$. Introducing the correction expressed by Eq. (14.16), the scattered data of Fig. 14.19 appear to compact along the $M/M_{NSC} = 1$ line, as shown in Fig. 14.23. To decide whether or not the correction is needed it can be used the Plastic-Zone Screening Criterion described in the following section. By reaching the ultimate stress in the outer region of the ligament, as supposed by the additional stress triangle of Fig. 14.22, also means that before overall failure occurs the ligament must fail, and the surface crack becomes a trough-wall crack. When this occurs, a leak takes place depressurizing the pipe if pressure is given through an uncompressible fluid such as water or oil. Therefore, the loading conditions change from pressure plus bending moment, $p + M$, to simple bending moment, M. This change results in a sudden drop of the stress state with an increase of the stress inversion angle, β, and that moves the neutral axis upwards. Consequently, the pipe may not fail at all leading to what is referred to as *leak before break*.

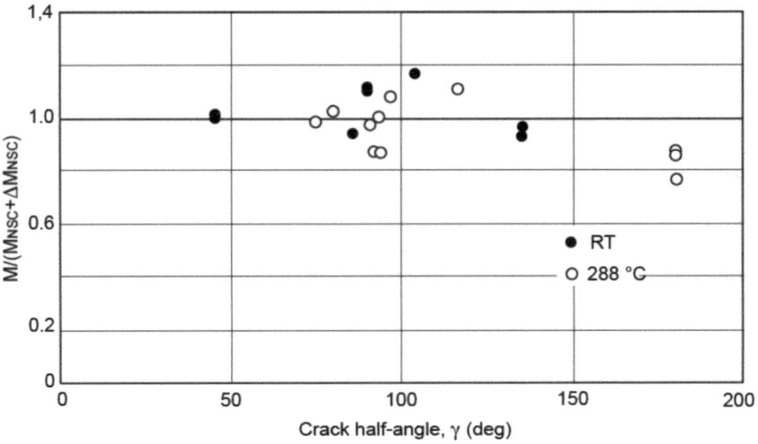

Fig. 14.23 Results of tests run at RT and 288 °C on TP304 stainless steel pipes carrying a surface crack shown in Fig. 14.19 reanalyzed using the additional moment expressed by Eq. (14.16)

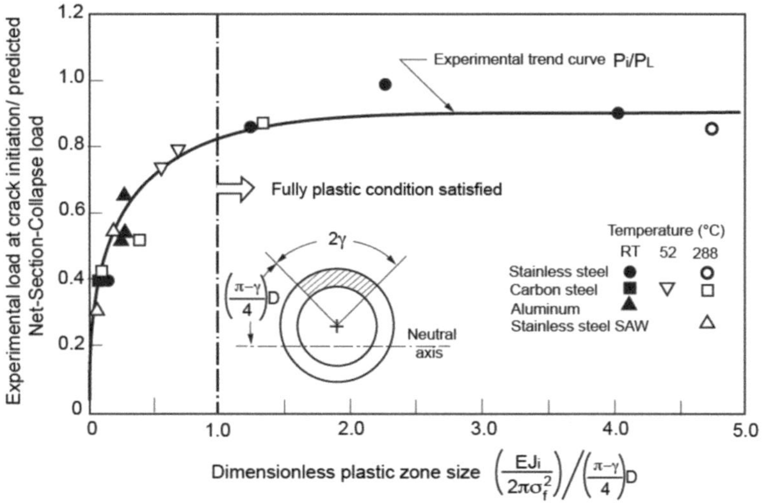

Fig. 14.24 Ratio of load at crack initiation to predicted net-section-collapse load as a function of the dimensionless plastic-zone size parameter (from [23])

14.4 Plastic-Zone Screening Criterion

The use of the NSC method is based on the assumption that the pipe material is tough enough that gross yielding of the net section occurs prior to failure regardless of pipe size, pipe strength, or crack size, as results shown in Figs. 14.13 and 14.17 indicate. Not always this condition is really met, and this can be causing an overestimation of the M_{NSC}, Eq. (14.10), with respect to experimental results.

To differentiate cases where net-section-collapse conditions are met from those in which they are not, a relatively simple screening criterion was developed, called Plastic-Zone Screening Criterion (PZSC) [23]. A plastic-zone model similar to the Irwin plastic-zone model was used to calculate the plastic-zone size. When the plastic-zone size is equal to the distance from the crack tip to the neutral axis it is assumed that a fully plastic condition is reached. Therefore, the formulation of the dimensionless plastic-zone parameter for through-wall crack takes the form:

$$\frac{EJ_i}{2\pi\sigma_f^2} \geq \frac{\pi - \gamma}{4} \cdot D \tag{14.18}$$

where

E = elastic modulus.
Ji = J value at crack initiation
σf = flow stress.
D = mean pipe diameter.
γ = half the initial crack angle in radians.

The term on the left of Eq. (14.18), $\left(EJ_i/2\pi\sigma_f^2\right)$, represents the plane stress plastic zone size at crack initiation, according to the Irwin hypothesis, see Sect. 8.2, Eq. (8.6), where J has been substituted to K_I using the usual relationship given by Eq. (10.40). Figure 14.24 [23] shows the experimental load at crack initiation relative to the predicted net-section-collapse load as a function of the dimensionless plastic-zone parameter for through-wall circumferentially cracked pipe in bending for several alloys and test temperatures.

These data show a well-defined experimental trend curve. If the dimensionless plastic-zone parameter is greater than 1.0, the load at crack initiation and the maximum load were close to the predicted net-section-collapse load. However, if the dimensionless plastic-zone parameter is less than 1.0, the net-section-collapse load underpredicts the experimental results. Note that the experimental data were calculated using a flow stress $\sigma f = 1.5(\sigma s + \sigma u)/2$. It can be seen that, out of seventeen experimental data, only five fall in the zone of validity of NSC load criterion.

The same screening criterion is used for surface flaws. When the Irwin plastic-zone size is equal to the remaining tensile ligament, it is assumed that a fully plastic condition is reached. This occurs for a circumferentially surface-cracked pipe when the dimensionless quantity given by Eq. (14.19) is satisfied:

$$\frac{EJ_i}{2\pi\sigma_f^2}/\left(\frac{\pi + (d/t)\gamma}{2} + \frac{\pi R_i^2 p}{4Rt\sigma_f}\right)\pi R \geq 1 \tag{14.19}$$

where

d = depth of the surface crack.
t = wall thickness of the pipe.
R = mean radius of the pipe.
R_i = inside radius of the pipe.
p = internal pressure (if any).

Again, the numerator of the left side of Eq. (14.19) $\left(EJi/2\pi\sigma f^2\right)$ represents the plane stress plastic zone size at crack initiation, according to the Irwin hypothesis. Since we want to verify whether the plastic zone has just reached the neutral axis to start a fully plastic condition, it would be more reasonable to use the yield stress, σs, rather than the flow stress, σf, therefore it is better to write Eq. (14.19) as:

$$\frac{EJ_i}{2\pi\sigma_s^2}/\left(\frac{\pi + (d/t)\gamma}{2} + \frac{\pi R_i^2 p}{4Rt\sigma_f}\right)\pi R \tag{14.19b}$$

The denominator of Eq. (14.19) or Eq. (14.19b) represents the distance to the neutral axis. The first term of the denominator is a crack geometry term. Embodied in this term is the crack depth to wall thickness ratio (*d/t*) and half the crack angle γ. The second term is a pressure term which accounts for the shift in the neutral axis due to the internal pipe pressure. Since the value of J at crack initiation is used in defining the plastic-zone size, the stress at crack initiation is compared with the

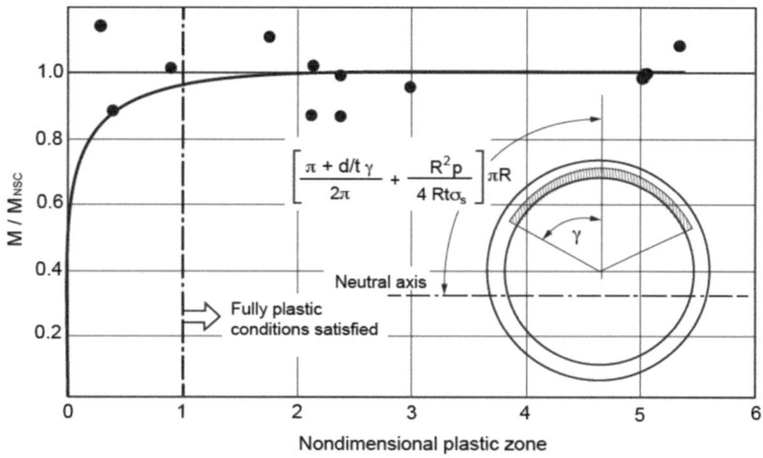

Fig. 14.25 Ratio of applied moment M to the NSC moment, M/M_{NSC}, as function of the nondimensional plastic zone according to the PZSC for the experimental points of Fig. 14.23

calculated net-section-collapse stress. Figure 14.25 shows the PZSC applied to the experimental results shown in Fig. 14.23. It can be seen that, with the exception of three data points, one of which is actually very close to the border of the fully plastic condition, all other points fall well into this required condition to make the NSC analysis work. The only point falling below the two borders set by the two lines, horizontal and vertical, of value 1 is relative to a large pipe of 406 mm diameter, 26.2 mm thick, carrying a surface crack of 106 degree, with $a/t = 0.66$ in which the fully plastic condition did not occur.

14.5 GE-EPRI Engineering Approach

As already discussed in Chap. 10, fracture mechanics analysis in the elastic–plastic regime involves a number of complications resulting from nonlinearities in deformations. These nonlinearities and large geometry changes also result in a significant amount of crack tip blunting prior to initiation and crack extension possibly leading to the final fracture. The various stages of ductile fracture process are schematically illustrated in Fig. 10.5 in terms of applied load P versus subcritical crack growth, Δa, and in Fig. 14.26 as J versus the applied load, P. Before blunting occurs, the elastic stage is essentially governed by the stress intensity factor K_I that can be easily calculated for any geometry under traction or bending moment and converted into a J-integral formulation, J_e, through the well-known relationship (10.32), where the e-subscript stands for elastic

$$J_e = \frac{K_I^2}{E\prime} \tag{14.20}$$

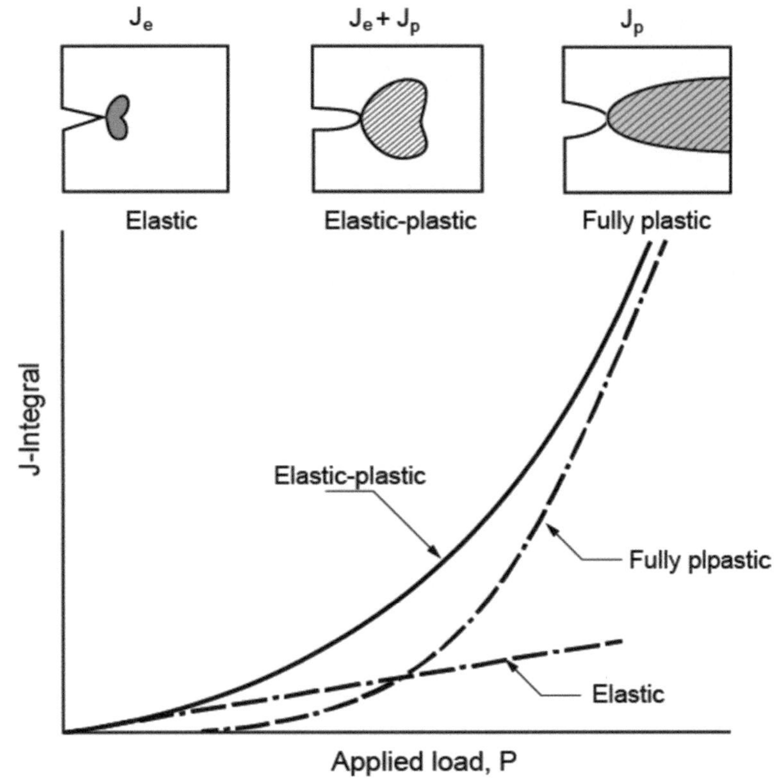

Fig. 14.26 The J-integral variation with applied load can be decomposed into the elastic and plastic components

with $E' = E/(1 - \nu^2)$ for plain strain condition or $E' = E$ for plain stress condition, while the fully plastic stage is governed by the J_p whose formulation requires same attention.

The peculiarity of the GE-EPRI engineering approach to pipe failure [24–26], or briefly Engineering Approach, rests in the ability to provide a simple J_p expression in which some constants appear whose values have been assessed through FE calculations and tabulated for a broad range of representative structural configurations. Once the two components are obtained, the elastic–plastic solution, shown in Fig. 14.26, is simply given by the summation:

$$J_{tot} = J_e + J_p \tag{14.21}$$

The J_{tot} is the driving force or energy release rate $\boldsymbol{G} = J_e + J_p$, that will be used with the material resistance curve R, see Sect. 11.6, to assess the critical condition shown in Fig. 14.27, as the tangency point between G and R, also shown in Fig. 7.8. To evaluate the fully plastic component of J we can start from the HRR field formulation,

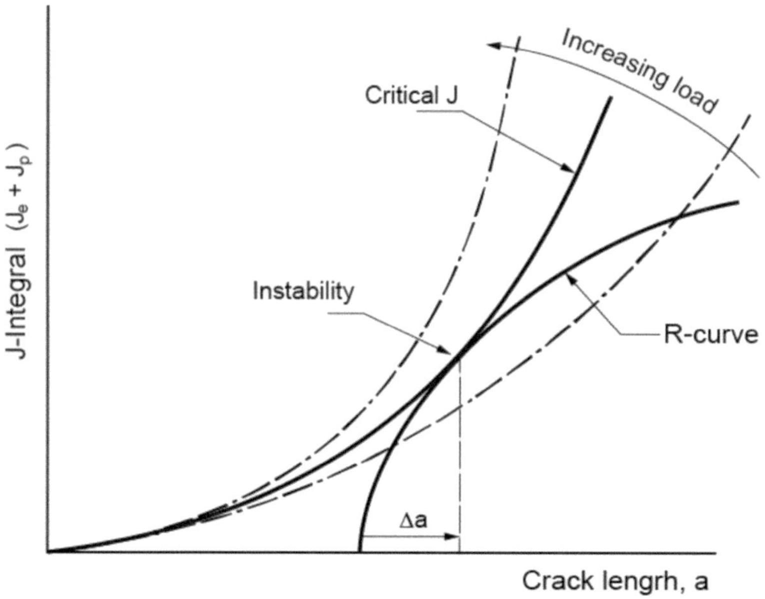

Fig. 14.27 Schematic representation of the *J-R* fracture assessment procedure

Eq. (10.46s), solved for J

$$J = \alpha \varepsilon_o \sigma_o I_n \left(\frac{\sigma_{ij}}{\sigma_o} \right)^{n+1} F_{ij}^{n+1} \tag{14.22}$$

with, ε_o, σ_o and n constants of the Ramberg-Hosgood expression, Eq. (1.41). For J-controlled conditions, the loading must be proportional as described in Sect. 10.6, i.e., the local stresses must increase in proportion to the applied remote load P or moment M.

Therefore, Eq. (14.22) can be written as:

$$\begin{aligned} J &= \alpha \varepsilon_o \sigma_o h l \left(\frac{P}{P_o} \right)^{n+1} \\ J &= \alpha \varepsilon_o \sigma_o h l \left(\frac{M}{M_o} \right)^{n+1} \end{aligned} \tag{14.23}$$

or

where.

h = dimensionless function of geometry and strain hardening.
l = characteristic length dimension that can be the crack length.
P_o = reference load.

M_o = reference moment.

The concept that fully plastic solutions are scalable with respect to applied load originated from the work of Goldman and Hutchinson [27]. The fully plastic equations for J, crack-mouth-opening displacement δ_p, and load line displacement Δ_p, assume the following form for most geometries

$$\begin{aligned} J_p &= \alpha\varepsilon_o\sigma_o b h_1\left(\frac{a}{W}, n\right)\left(\frac{P}{P_o}\right)^{n+1} \\ \delta_p &= \alpha\varepsilon_o a h_2\left(\frac{a}{W}, n\right)\left(\frac{P}{P_o}\right)^{n} \\ \Delta_p &= \alpha\varepsilon_o a h_3\left(\frac{a}{W}, n\right)\left(\frac{P}{P_o}\right)^{n} \end{aligned} \tag{14.24}$$

where a = crack length, and h_1, h_2, and h_3 = dimensionless parameters that depend on geometry and a/b ratio (b = uncracked ligament) and hardening exponent n.

The numerical values of parameters h_1, h_2, and h_3 and the limit loads P_o can be found tabulated in the open literature [25–29] and listed in Chap. 16 for several simple geometries. The elastic J shall be assessed by making use of the effective crack length, a_{eff}, that considers the correction for the plastic zone radius, r_p, that must depend on the strain hardening, n, of the material. Kumar and Shih [28] have proposed the following expression

$$a_{\text{eff}} = a_o + \phi r_p \tag{14.25}$$

where <

$$r_p = \frac{1}{\beta\pi}\left(\frac{n-1}{n+1}\right)\left(\frac{K_I}{\sigma_o}\right)^2 \tag{14.25a}$$

The ϕ factor in Eq. (14.25) attempts to reduce the overadjustment under condition of contained plasticity, while at the same time restraints the plasticity adjustment in the fully plastic regime. Kumar and Shih [28] choose to define ϕ as

$$\phi = \frac{1}{1 + (P/P_o)^2} \tag{14.25b}$$

Therefore, Eq. (14.25) becomes

$$a_{eff} = a + \frac{1}{1 + (P/P_o)^2}\frac{1}{\beta\pi}\left(\frac{n-1}{n+1}\right)\left(\frac{K_I}{\sigma_o}\right)^2 \tag{14.25c}$$

Equation (14.25c)) is based on a modified Irwin plastic zone correction [24] that considers strain hardening, with $\beta = 2$ for plain stress and $\beta = 6$ for plain strain conditions, see Eq. (8.9). Equation (14.25c) is a first-order correction.

14.6 Experimental Evidence

It will be shown, in the following, the analysis of a cracked pipe according to the GE-EPRI Engineering Approach. The pipe is a 6″ A106B carbon steel pipe carrying a through-wall 180° circumferential crack tested in four-point bending conditions at RT under the Italian program [6]. The J_{tot} used has the following expression

$$\begin{aligned} J &= J_e + J_P \\ &= \frac{P^2}{E'} F_1\left(a_{eff}, R\right) + \alpha \varepsilon_o \sigma_o c h_1 \left(\frac{P}{P_o}\right)^{n+1} \end{aligned} \quad (14.26)$$

where

P = applied load.
P_o = limit load.
F_1 = nondimensional factor for the pipe carrying a circumferential flaw.
h_1 = nondimensional function for the fully plastic solution for a pipe carrying a circumferential flaw
a_{eff} = effective crack length = $a + r_p$, given by Eq. (14.25c)).
r_p = plastic zone.
$\alpha, \varepsilon_o, \sigma_o$ = Ramberg–Osgood constants.
c = b – a.
b = πR.

The limit load P_o is given by

$$P_o = 0.536 \quad \sigma_o c^2 / b \quad (14.28)$$

Figure 14.28 shows the Ramberg–Osgood curve for the A106B material measured on a traction specimen taken from the pipe. Figure 14.29 shows the unloading compliance and the COD measurement obtained directly on the 6 in. pipe, see Sect. 11.6. Finally, the *J-R* estimation scheme is shown in Fig. 14.30a and b for the pipe carrying a circumferential 180° through-wall crack under both plain-stress and plain strain condition, respectively [18].

It can be seen how for the plain stress condition the tangency condition is reached at an applied load of bout 15 ton (between curves relative to 14 and 16 ton).

For the plain strain condition, Fig. 14.30b, this occurs at about 18.5 ton. The experimental critical load was about 15,580 kg suggesting a plain stress condition.

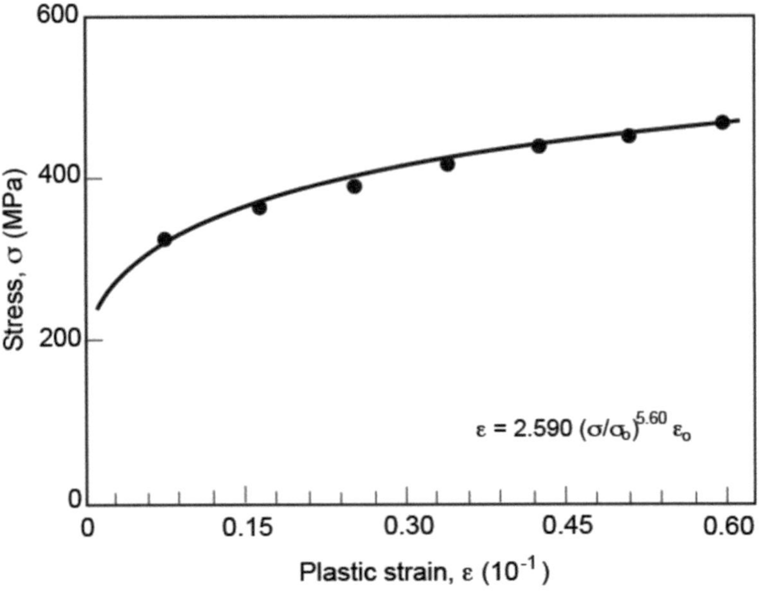

Fig. 14.28 Ramberg–Osgood representation of the stress-plastic strain for A106B steel at room temperature [18]

Fig. 14.29 Unloading compliance and COD measurement versus loading cell displacement measured on A106B steel pipe with 180 deg circumferential through-wall crack at RT [18]

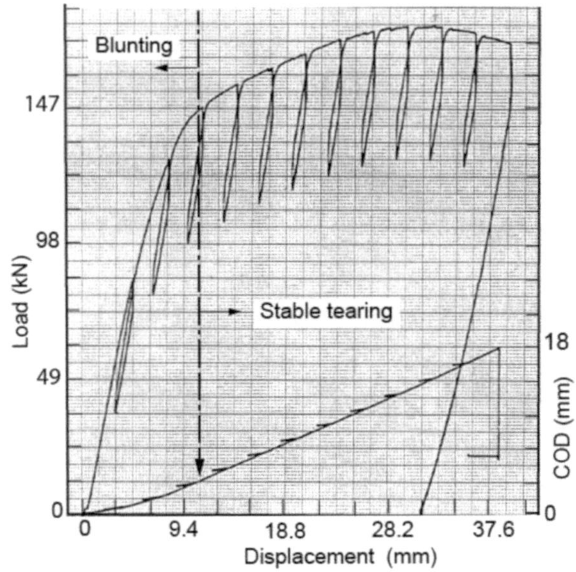

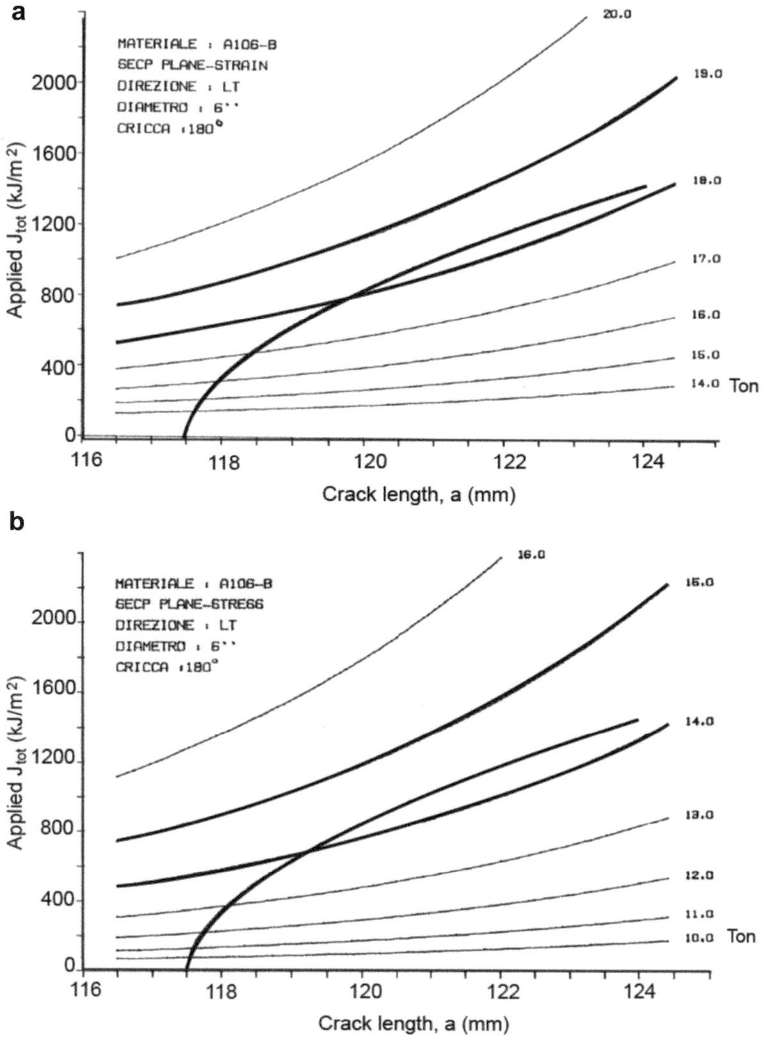

Fig. 14.30 **a** Applied *J-R* curve Estimation Scheme for A106B steel 6 in. pipe with a circumferential through-wall 180 deg crack under plain stress condition [18] **b** Applied *J-R* curve Estimation Scheme for A106B steel 6 in. pipe with a circumferential through-wall 180 deg crack under plain strain condition [18]

14.7 R-6 Failure Assessment Diagram Approach

In the previous sections of this chapter we have analyzed pipe fracture behavior both under fully plastic conditions (FP) and elastic–plastic conditions (EP), using the NSC criterion and the GE-EPRI engineering approach, respectively. The two methodologies are specific to those particular conditions as the K_I-based approach is for linear elastic conditions. In the mid-Seventies a new approach was developed in the UK, called R6 Method, that would cover the entire spectrum of possible failure mode caused by the presence of a crack, spanning from a linear elastic to a fully plastic condition. The basis of the R6 Method is that it is possible to identify two criteria which limit the load-carrying capability of a flawed structure. First, the applied linear elastic stress intensity factor, K_I, must not be greater than the fracture toughness of the material, K_{Ic}, secondly, the applied load, P, must not be greater than the plastic collapse load of the flawed structure, P_o. These two criteria represent two distinctive failure mechanisms. However, some interaction between the two mechanisms is to be expected which is precisely what the so-called R-6 Failure Assessment Diagram is intended to provide. The procedure, that was originally developed by the Central Electric Generating Board (CEGB) [30, 31] is based on the *two criteria* approach of Dowling and Townley [32]. Under this method, a FAD or Failure Assessment Diagram is introduced which represents the analytical link between the two distinct and separate mechanisms of failure that define a fail-safe region of operation of a flawed structure. The original approach, published in 1980 by the British Standards Institution as PD6493:1980 [33], was based on the CTOD idea described in Sect. 10.3. This last methodology, though sound in its basic concept, suffered from a number of shortcomings, having mostly empirical equations and variable level of conservativism. Therefore, a new edition of that procedure appeared in 1991 as PD6493:1991 Level 1. For the FAD procedure stresses acting on a section must be separated into primary stresses, σI, secondary stresses, σQ. A primary stress, or load controlled stress, is related to mechanical loading directly and satisfies force and moment equilibrium. Primary stresses are further categorized into membrane stresses, σm, that have a uniform distribution across a cross-section, like pressure stresses, and bending stresses, σb, that vary linearly. Primary stresses are not self-limiting and contribute to plastic collapse. Therefore, if a primary stress exceeds the yield strength of the material through the entire cross section of the piping, then failure can be prevented only by strain hardening and when the flow stress is reached over the entire cross section of the piping failure can occur by plastic collapse. Secondary stresses are developed by the constraint of displacements of a structure. These displacements can be caused either by thermal gradients and welding processes or by outwardly imposed restraint and anchor point movements. Under this loading conditions, the piping system must satisfy an imposed strain pattern rather than be in-equilibrium with imposed forces. Local yielding and minor distortions of the piping system tend to relieve these stresses. At variance with primary stresses, secondary stresses are self-limiting, i.e., the net force and bending moment must be zero, and do not contribute to plastic collapse. The FAD procedure defines a CTOD or δr ratio as:

$$\delta_r = \frac{\delta_I}{\delta_{mat}} \tag{14.29}$$

where δ_I is the applied elastic CTOD obtained from Eq. (10.19) and is related to J or K_I by Eq. (10.12) or Eq. (10.32):

$$\delta_I = \frac{K_I^2}{\sigma_s E'} for\, \sigma_I/\sigma_s \le 0.5 \tag{14.30}$$

with σ_I being the primary stress and

$$\delta_I = \frac{K_I^2}{\sigma_s E}\left(\frac{\sigma_s}{\sigma_1}\right)^2\left(\frac{\sigma_1}{\sigma_s} - 0.25\right) for\, \sigma_1/\sigma_s \ge 0.5 \tag{14.31}$$

with σ_l being the maximum stress, defined as the sum of the primary and secondary stresses, $\sigma_I + \sigma_Q$, assumed to act uniformly across the cross section. The CTOD option can be replaced by

$$K_r = \frac{K_I}{K_{mat}} \tag{14.32}$$

where K_I is the applied K and K_{mat} is the K_{Ic} of the material.

The second term considered in the Level 1 approach is the S_r ratio defined as:

$$S_r = \frac{\sigma}{\sigma_f} \tag{14.33}$$

where σ is the primary net section stress and σf the flow stress of the material. In a revised version a safety factor of $1/\sqrt{2} = 0.707$ is applied to the δr term whereas the S_r scale is restricted to $0.8{\cdot}S_r$ so that the resulting safety diagram is that of Fig. 14.31 named "Level 1".

Any loading condition resulting in a $\delta r - Sr$ combination that falls inside the safety domain is safe, otherwise it is unsafe. The evolution of Level 1 is Level 2 of the FAD curve, obtained using the Dugdale and Stone solution for the strip-yield model given by Eq. (8.14b):

$$K_{eff} = \sigma_s\sqrt{\pi a}\left[\frac{8}{\pi^2} lnsec\left(\frac{\pi\sigma}{2\sigma_s}\right)\right]^{1/2} \tag{14.34}$$

where the yield strength σS is considered as limiting value since it applies to ideally elastic–plastic materials without any strain hardening. In real materials Eq. (14.34) shall be modified substituting the yield strength σS with the more realistic flow stress σf. With this premise, the coordinates employed in the FAD transition diagram that defines the domain of safety operation of a flawed structure arbitrarily loaded are the ratios K_r and S_r, which are defined as follows:

$$\begin{aligned} K_r &= \frac{K_I}{K_{eff}} \\ S_r &= \frac{P}{P_o} or \frac{\sigma}{\sigma_f} \end{aligned} \tag{14.35}$$

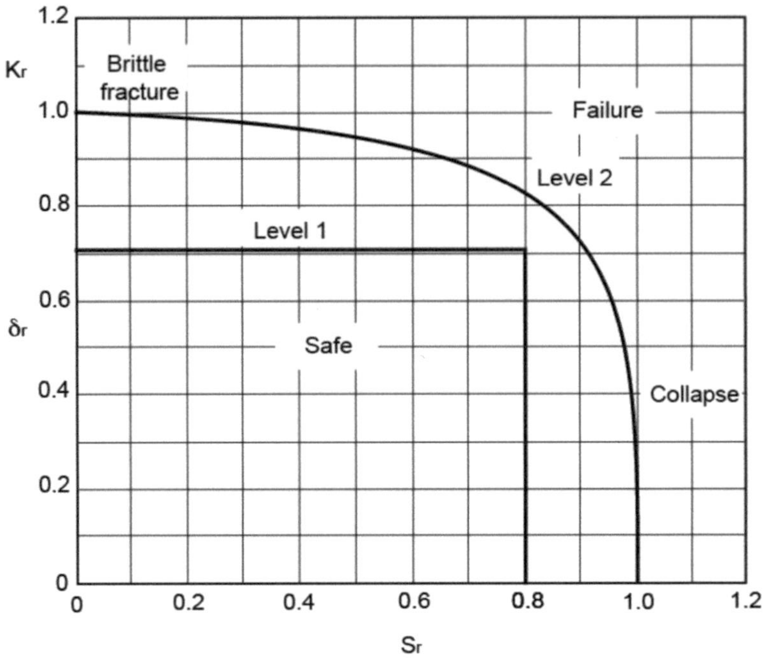

Fig. 14.31 Schematic of two-criteria approach R-6 Failure Assessment Diagram initially developed by Dowling and Townley [32] and adopted by the British Standards as Level 1 and Level 2

as in Eq. (14.33). As has been pointed out in Sect. 8.3, the strip-yield model has a built-in limit load criterion for perfectly plastic nonhardening materials. Extending its applicability to real materials require replacing the yield stress with the flow stress of the material. Introducing Eq. (14.35) into Eq. (14.34), changing the yield strength σS with the flow stress σf, and taking the reciprocal yields:

$$K_r = S_r\left[\frac{8}{\pi^2} lnsec\left(\frac{\pi}{2}S_r\right)\right]^{-1/2} \tag{14.36}$$

Equation (14.36) is plotted in Fig. 14.31 as Level 2 curve. The curve represents the locus of predicted failure points. As can be seen, the safety domain enlarges with respect to the Level 1 approach. If the applied conditions of the structure analyzed place the assessed operating point (S_r, K_r) inside the FAD Level 2 curve, the structure is deemed safe. Outside the safety domain, failure occurs which can be either prevalently brittle or ductile, depending on whether the operational point approaches the extreme left or right of the diagram. Completely brittle fracture is predicted when $K_r = 1$. If the toughness is very large, the structure will fail under fully plastic conditions by collapse, $S_r = 1$. For intermediate cases, collapse and fracture balance each other and both K_r and S_r are less than 1.0 at failure. The present form of the R-6 diagram is due to Harrison, Loosemore and Milne [34].

14.8 *J*-Controlled Failure Assessment Diagram

The FAD analysis discussed in the previous Section as R-6 methodology is the British approach to assess structure stability that, though using the Dugdale strip-yield model, recognized and *borrowed* the American Irwin's stress intensity factor K_I. Meanwhile, on the other side of the ocean, it was recognized that the R-6 methodology was a valid estimation scheme noting, however, that the Dugdale solution assumed the material to be elastic-perfectly plastic and was intended for a finite crack in a thin infinite sheet subjected to remote traction. Therefore, as a return of the courtesy, the Failure Assessment Diagram procedure was *borrowed* in the US and the estimation scheme of the GE-EPRI Engineering Approach, already discussed in Sect. 14.5, implanted [35, 36]. The procedure led to a new formulation of the failure assessment diagram, which showed that the failure line was not universal, as proposed by the R-6 procedure, but depended on the geometry of the cracked body, the type of loading and the material properties, specifically the strain hardening [37]. The following description of the procedure to derive the new set of curves is fully described in [25, 38] from which it is taken. The fundamental point was the use of the *J*-integral in the K_r term. Under *J*-controlled crack growth conditions, the equilibrium of a crack requires the crack driving force to equal the resisting force, i.e.,

$$J(a, P) = J_R(\Delta a) \tag{14.37}$$

where J_R is the material resistance or *R* curve, which depends on the amount of crack growth, Δa, Fig. 14.27, and *J* is the applied driving force. In the estimation procedure, the applied *J* is expressed as the sum of an elastic component, J_e, and a plastic component, J_p, as in Eq. (14.21) and shown in Fig. 14.26

$$J = J_e(a_e, P) + J_p(a, P, n) \tag{14.38}$$

where a_e is the effective crack length or adjusted crack length that takes into account also the crack tip plastic zone, and *n* is the strain hardening exponent of the material. The modified crack length a_e is given by Eq. (14.25). This modification essentially provides a plastic zone correction which is arranged to have a reduced effect at higher loads when plasticity contributions are given by the second term in Eq. (14.38).

Note that the effective crack length, a_e, appears only in the elastic component, J_e, of the applied *J*. Using Eq. (14.24) for the definition of *J*, Eq. 14.37 can be written as:

$$\hat{J}(a_e)\left[P/P_o\right]^2 + \hat{J}(a, n)\left[P/P_o\right]^{n+1} = J_R(\Delta a) \tag{14.39}$$

where the nondimensional functions topped by (^) are independent of the applied stress and are given by [27]

$$\hat{J}(a_e) = J_e/(\varepsilon_o \sigma_o a_e)$$
$$\hat{J}(a,n) = J_p/(\alpha \varepsilon_o \sigma_o a h_1) \tag{14.40}$$

Where α, ε_o, σo are the coefficients of the Ramberg–Osgood equation, h_I is the nondimensional function of Eq. (14.24) and P_o is the limit load. Note that the elastic J component, J_e, can be given an expression similar to the plastic J component, J_p, by setting $n = 1$ and $\alpha = 1$, i.e., elastic behavior without strain hardening, so that

$$J_e = \varepsilon_0 \sigma_0 a \left(\frac{P}{P_0} \right)^2 \hat{J}(a_e) \tag{14.40a}$$

with the exponent of the load ratio P/P_o obviously equal to 2, with $\hat{J}(a_e)$ being the proper nondimensional function for the geometry considered. Equation (14.39) can be rearranged to yield:

$$\frac{J_e(a,P)}{\hat{J}(a_e)[P/P_o]^2 + \hat{J}(a,n)[P/P_o]^{n+1}} = \frac{J_e(a,P)}{J_R(\Delta a)} \tag{14.41}$$

Recalling Eq. (10.32)

$$J_e = \frac{K^2}{E'} \tag{14.42}$$

where $E' = E$ for plain stress conditions and $E' = E/\left(1 - v^2\right)$ for plain strain problems, the nondimensional term $\hat{J}(a)$ can be determined directly from the available solutions for K_I.

A consequence of J-controlled crack growth is

$$E' J_R(\Delta a) = K_R^2(\Delta a) \tag{14.43}$$

where the K_R curve is obtained under small-scale yielding conditions (SSY). Now introduce a reference stress, σref,

$$\sigma_{ref} = (P/P_o)/\sigma_o \approx (P/P_o)/\sigma_s \tag{14.44}$$

so that the load ratio Sr becomes

$$S_r = \frac{\sigma_{ref}}{\sigma_s} \sim \frac{P}{P_o} \tag{14.45}$$

practically the same as in Eq. (14.35) but with a more general meaning, as it will be discussed in the following section. Also introduce the following nondimensional ratios for stress and crack driving force

$$K_r = K_I(a, P)/K_R(\Delta a)$$
$$J_r = J_e(a, P)/J_R(\Delta a) \tag{14.45a}$$

It follows from Eqs. (14.42) and (14.43) that:

$$K_r^2(a, P, \Delta a) = J_r(a, P, \Delta a) \tag{14.46}$$

Substituing Eqs. (14.44), (14.45) and (14.45a) into Eq. (14.41) yields:

$$\frac{S_r^2}{H_e S_r^2 + H_n S_r^{n+1}} = J_r = K_r^2 \tag{14.47}$$

where

$$H_e = \hat{J}(a_e)/\hat{J}(a)$$
$$H_n = \hat{J}(a, n)/\hat{J}(a)$$

Equation (14.47) defines a curve in the space K_r and S_r where the crack driving force (under dead-load condition or load controlled situation) is in equilibrium with the material resistance. The shape of the equilibrium curve depends on the crack configuration and material properties including the hardening exponent n since the nondimensional functions H_e and H_n depend on these quantities. The curve defines the equilibrium crack growth state in terms of the applied load, P, and crack length, a, for a given amount of crack extension. For example, if crack initiation is defined as failure, which actually means $\Delta a = 0$, then $K_r = K_{Ic}$ and $J_r = J_{Ic}$ and the elastic driving force term in Eq. (14.44) is normalized by the respective crack initiation resistance. In this case Eq. (14.46) defines the state of equilibrium for a given crack configuration just prior to crack extension. In place of the more appropriate elastic–plastic solutions given by Eq. (14.24), the Dugdale solution for J:

$$J\left(a, \sigma^{\infty}\right) = \frac{8}{\pi}\sigma_o \varepsilon_o a \ln\left[\sec\left(\frac{\pi}{2}\frac{\sigma^{\infty}}{\sigma_o}\right)\right] \tag{14.48}$$

may be used as the crack driving force in Eq. (14.37), and the equilibrium curve given by Eq. (14.46) becomes the failure line of Harrison, Loosemore and Milne [34] for crack growth:

$$\frac{S_r^2}{\frac{8}{\pi^2}\ln\left[\sec\left(\frac{\pi}{2}S_r\right)\right]} = J_r = K_r^2 \tag{14.49}$$

Again, the $K_r = f(S_r)$ curve becomes independent of the material strain hardening. It is worth noting that Eq. (14.46) is equal to Eq. (14.36) of the original R-6 FAD level 2 approach as long as failure occurs in the linear elastic region where K dominates, as it should be expected. However, as soon as that elastic region is left and some, though small, plasticity starts to appear in the fracture mode, invalidating the SSY condition, the two curves, i.e., the two approaches start to diverge. This is because, as the J-integral plastic component, J_p, of the J_{tot}, Eq. (14.21), starts to be felt and then prevails over the elastic $J_e = K^2/E'$ component, which actually means that fracture is no longer fully brittle, the effect of the strain hardening of materials, n, also starts to be felt. Note that, at variance with Level 2 FAD curve of Fig. 14.31 that defines a safety area within the limits $K_r \leq 1$ and $S_r \leq 1$, the new safety domain defined by Eq. 14.46 extend over the limit $S_r = 1$ because of the strain hardening of the material that affects the plastic component of J. Note that the strain hardening also affects the shape of the fracture assessment line as soon as some crack tip blunting appears and then some tearing develops moving the fracture appearance from totally brittle to ductile. This departure from the SSY condition increases with decreasing strain hardening values.

The sequence of Fig. 14.15 shows the blunting and tearing of an initially sharp crack in a ductile pipe with low hardening exponent. An analogous effect will be observed with respect to the crack depth, since short cracks develop larger plastic zones and behave in a more ductile fashion than long cracks, as explained in Sect. 9.13.

Figures 14.32 and 14.33 show the influence of both n exponent and a/b ratio for a center-cracked panel in traction, CC(T), Fig. 9.5a, in plain stress conditions, respectively, compared to the original R-6 Level 2 curve. It may surprise the progressive reduction of the fail-safe area with decreasing a/b ratio, i.e., for shorter cracks.

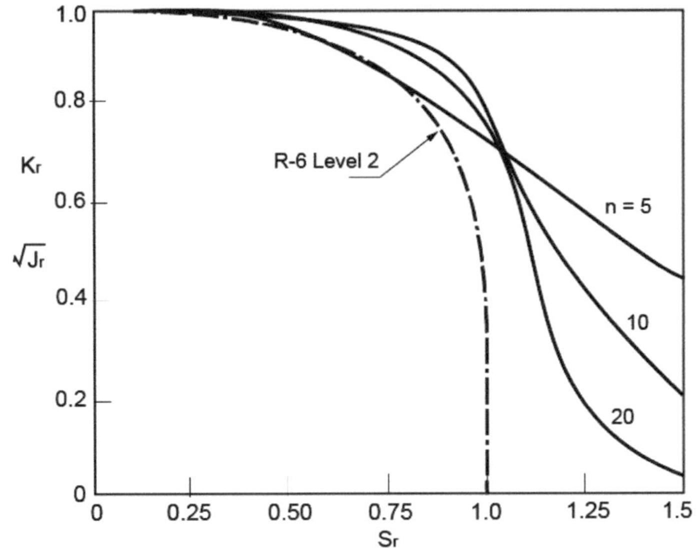

Fig. 14.32 Plain stress failure curves for center-cracked plate for several values of the strain hardening exponent (from [25])

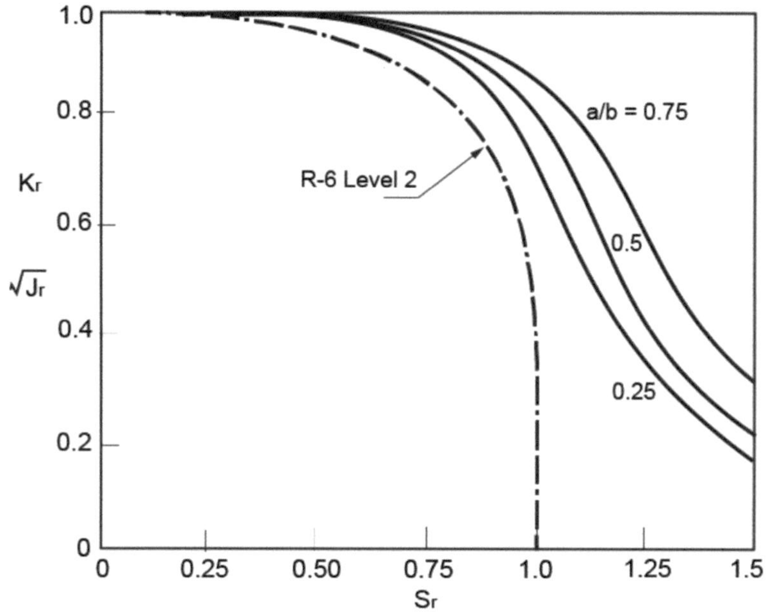

Fig. 14.33 Plain strain failure curves for Center-cracked plate for several crack length-to-width ratios (from [25])

The reason rests in the fact that by reducing the crack length, the apparent toughness, K_C, or J_c of the material increases and, therefore, the K_r or J_r corresponding to a given applied load decreases. But if we want to increase the applied load to regain in K_r then also S_r increases with a corresponding worsening of the working conditions.

Figure 14.34 shows the FAD for axially and circumferentially cracked cylinders of 304 stainless steel and A533-B carbon steel, respectively.

The crack length-to-wall thickness ratio was chosen to match the deepest relative crack depth for the maximum postulated flaw size in the ASME Pressure Vessel and Piping Code. Also shown is the R-6 Level 2 curve.

14.9 Approximation of the FAD Curve. Options 1, 2 and 3

Though Eq. (14.46) extends the FAD safety field of the previous R-6 Level 2 curve beyond point $S_r = 1$ by considering the strain hardening behavior of materials, it fails to provide a unique FAD curve valid for all materials. Figures 14.32 through 14.34 clearly indicate that the shape and position of the failure line depends on the geometry of the cracked body, the type of loading and material properties, strain hardening in particular. This last dependence is through the nondimensional function

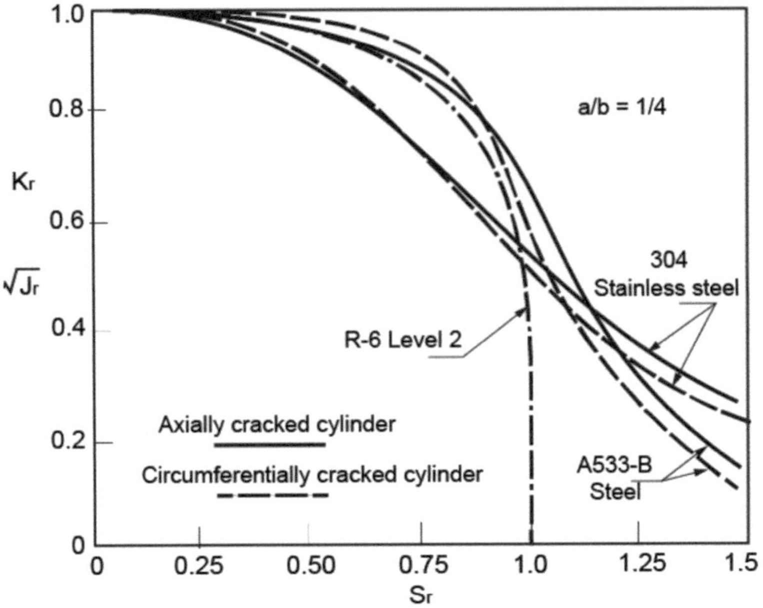

Fig. 14.34 Failure curves for axially and circumferentially cracked cylinders (from [25])

H_n in Eq. (14.47) that contains the nondimensional h_l function that defines the plastic component of J, J_p, Eq. (14.24). Moreover, Ainsworth [39] also observed that although Eqs. (14.21) and (14.24) were straightforward to use given a geometry for which h_1 is defined, there were some difficulties in fitting the stress–strain curves of work hardening materials with a unique Ramberg–Osgood Eq. (1.41) since the constants α and n which give a good fit at low strains may be significantly different from those which give a good fit at high strains which, in turn, may be much different from the entire range covered by a linear regression, as shown in Fig. 14.35 [23] for 304 Stainless steel at 288 °C. Because of the power-law dependence of Eq. (14.24) on n, large differences in calculated values of J can arise from different fits of Eq. (1.41) to the stress–strain curve.

Ainsworth [39] noticed that the factor h_l became relatively insensitive to material properties by adopting a proper reference stress, σref. Accordingly, the loas ratio S_r was substituted by a new term, L_r, defined as

$$L_r = \frac{\sigma_{ref}}{\sigma_s} \tag{14.50}$$

where σ_{ref} has been introduced with Eq. (14.44) so that

$$L_r = \frac{P}{P_0} \tag{14.51}$$

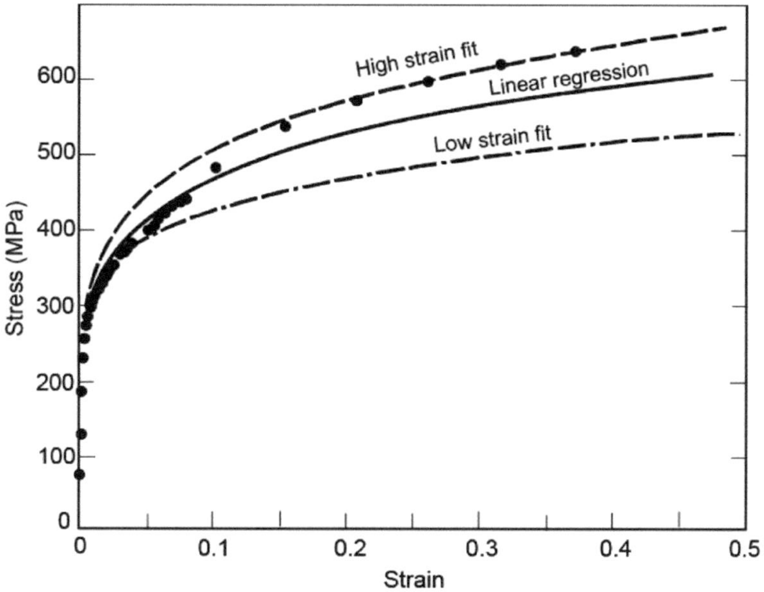

Fig. 14.35 Engineering stress–strain curve for 304 stainless steel at 288 °C using a high strain, low strain and total linear regression fit (redrawn from [23])

At variance with R-6 Level 2 curve that ends at $S_r = 1$, this new term can be larger than 1 since the limit load P_o is calculated using $\sigma o \sim$ as limit strength the applied load P is evaluated assuming that the applied stress can exceeds σs.

In any case, the load ratio L_r cannot exceed $\sigma f/\sigma s$ with σf being the flow stress of the material, Eq. (10.14), which poses a cut-off to the FAD curve. To cope with these new formulation, Ainsworth [39] reformulate the equations of the GE-EPRI Engineering approach already seen in Sect. 14.5, so that the actual stress–strain curve of the material could be used rather than the Ramberg–Osgood representation. Approximations were introduced to make the formulae geometrically independent, so removing the need for fully plastic power-low solutions for each geometry. The evaluation of the $J_{tot} = J_e + J_p$ then becomes:

$$J = \frac{K_I^2}{E\prime}\left(\frac{E\varepsilon_{ref}}{\sigma_{ref}} + \frac{1}{2}\frac{\sigma_{ref}^3}{E\varepsilon_{ref}\sigma_s^3}\right) = \frac{K_I^2}{E\prime}\left(\frac{E\varepsilon_{ref}}{\sigma_{ref}} + \frac{1}{2}\frac{L_r^3}{E\varepsilon_{ref}}\right) \qquad (14.52)$$

where the quantity ε_{ref} is a reference strain obtained from the true stress-true strain curve for the material, see Sect. 1.6, at a reference stress level $\sigma ref = Lr\sigma s$. This reformulation led to a new assessment curve

$$K_r = \left(\frac{E\varepsilon_{ref}}{\sigma_{ref}} + \frac{L_r^2}{2(1 + L_r^2)} \right)^{-\frac{1}{2}} \tag{14.53}$$

The first term $(E\varepsilon_{ref}/\sigma_{ref})$ in Eq. (14.53) describes both elastic and fully plastic behavior. However, it is not capable of describing the intermediate SSY region where the bulk behavior is elastic, $E\varepsilon_{ref}/\sigma_{ref} = 1$ but J exceeds its elastic value. In this region the required minor plasticity corrections are provided by the second term of Eq. (14.53). The form of the denominator $(1 + L_r^2)$ in the second term was based on the GE-EPRI scheme of Sect. 14.5 and is designed to phase out the minor corrections at large loads as the fully plastic term dominates. However, the phasing out should only occur when the fully plastic term does indeed become large $(E\varepsilon_{ref}/\sigma_{ref} > 1)$ rather than simply on the basis of the value of L_r. Therefore in R6 the denominator was made directly dependent on the magnitude of the first term by replacing $(1 + L_r^2)$ with $(E\varepsilon_{ref}/\sigma_{ref})$. Making this change in Eq. (14.53) and substituting $\sigma_{ref} = L_r\sigma_s$ leads to the Option 2 curve

$$K_r = \left(\frac{E\varepsilon_{ref}}{L_r\sigma_s} + \frac{L_r^3\sigma_s}{2E\varepsilon_{ref}} \right)^{-\frac{1}{2}} \text{ for } L_r \leq L_{r(max)} \tag{14.54}$$

where L_r is given by Eq. (14.50), ε_{ref} is the reference strain inferred from the true stress-true strain curve at various σ_{ref} and $L_{r(max)}$ is a cut off value

$$L_{r(max)} = \frac{\sigma_f}{\sigma_s} \tag{14.55}$$

However, it may be convenient, from a practical point of view which is also the point of view of standard practices, to provide a standard FAD curve equation that does not requires the analytical evaluation of Eq. (14.54). Three options, Option 1, Option 2 and Option 3 are available for generating such curves for the R-6 method. In each case the FAD curve has a cut off given by Eq. (14.55). The accuracy of the analysis and information needed to generate the respective curves grows with option number.

Option 1. The FAD curve is given by [40]

$$K_r = (1 - 0.14L_r^2)[0.3 + 0.7\exp(-0.65L_r^6)] \text{ for } L_r \leq L_{r(max)} \tag{14.56}$$

$$K_r = 0 \text{ for } L_r > L_{r(max)} \tag{14.57}$$

This FAD curve was obtained empirically as the lower bound of curves generated with Eq. (14.54) for a large variation of stress strain curves and is recommended for low-medium strength materials. The information needed is just the yield strength, σ_s, and the reference stress, ε_{ref}. Figure 14.36 represents the Option 1 R-6 FAD curve.

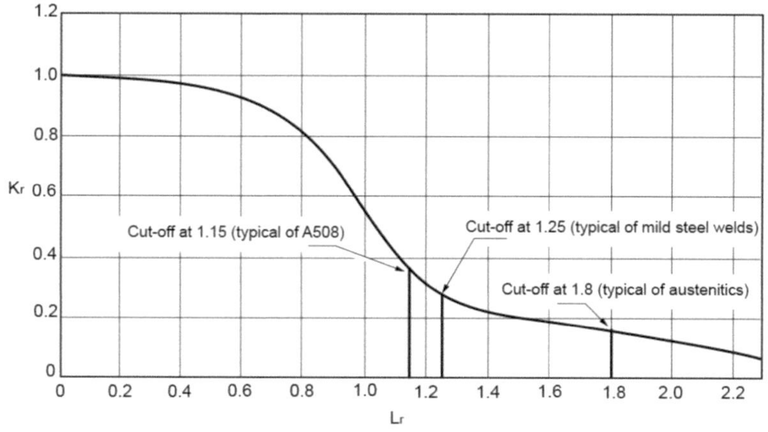

Fig. 14.36 Option 1 R-6 FAD curve with cut-off for several materials [40]

Recently, a modification of Eq. (14.56) has been proposed [41] which gives a better fit of Eq. (14.54), within 3%. Its equation is.

$$K_r = \left(1 + 0.5 \cdot L_r^2\right)^{-\frac{1}{2}}\left[0.3 + 0.7\exp\left(0.65L_r^6\right)\right] \text{ for } L_r \le L_{r(max)} \tag{14.58}$$

with the same cut-off of Eq. (14.57). Curves described by Eq. (14.56) or (14.58) do not take into consideration the effect that initial strain hardening rate has on the amount of strain needed to reach the flow stress, as shown in Fig. 14.37 where some examples of typical engineering stress–strain curves are shown. The data have been plotted with the stress axis normalized by the flow stress.

Option 2. The Option 2 FAD was developed by Ainsworth [39] from expressions for the J-integral taken from the GE-EPRI handbook [25]. The FAD curve is given by Eq. (14.54)with the same cut-off of Eq. (14.58) [40]. Its use is recommended for all materials regardless of the stress–strain behavior. The information needed is the full stress–strain curve up to a strain of 1%, σ_s, and the reference stress, ε_{ref}. Figure 14.38 [40] shows plots of the FAD derived from Eq. (14.54) for the same materials whose stress–strain curves are shown in Fig. 14.37.

Option 3. Option 3 Fad provides the most accurate analysis. The FAD curve is obtained using the J-integral analysis

$$K_r = \left(\frac{J_e}{J}\right)^{\frac{1}{2}} \text{ for } L_r \le L_{r(max)} \tag{14.59}$$

$$K_r = 0 \text{ for } \quad L_r > L_{r(max)} \tag{14.60}$$

where Je and J are the values of the J-integral obtained for an elastic analysis and an elastic-analysis, respectively, for the same load corresponding to the value L_r.

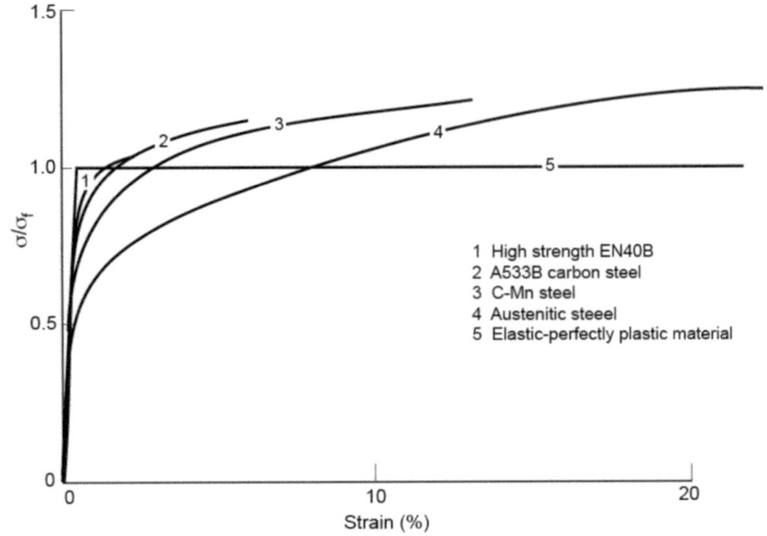

Fig. 14.37 Typical engineering stress–strain curves for several materials [40]

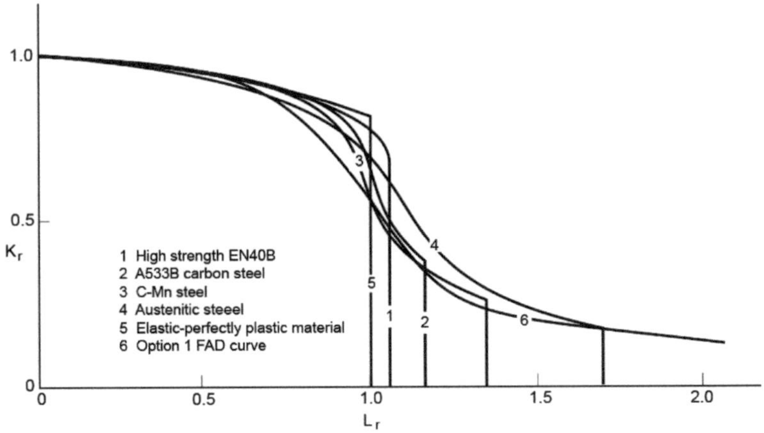

Fig. 14.38 Comparison of FAD curves obtained for the same materials shown in Fig. 14.37. also shown is the Option 1 FAD curve

option 3 curve is for use when a J value of the cracked structure is available. It has the potential for greater accuracy than the approximate curves of Option 1 and 2.

Its use is recommended for complex loading systems. The information needed is the full stress strain curve of the material, J and J_e, σ_s, and the reference stress, ε_{ref}. The three options presented result in three different FAD curves.

Figure 14.39 compares the general FAD curve given by Eq. (14.56) with those derived from use of Eq. (14.55) that is the base for material-specific FAD curves,

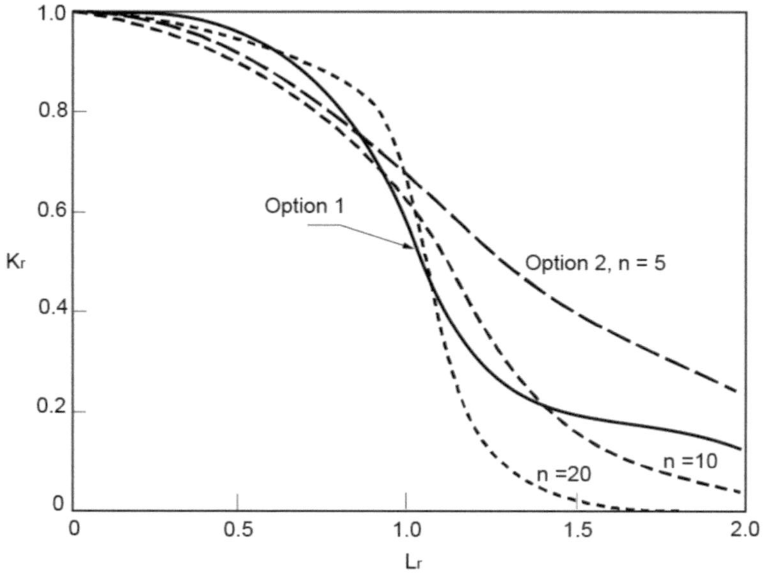

Fig. 14.39 Comparison of Option 1 FAD curve and material-specific FAD curves (Option 2)

Options 2 and 3. As can be seen, the major difference between FAD expressions occurs above net section yield, where $L_r > 1$, while for $L_r < 1$ there is little difference and the curves run reasonably close to each other. Options 2 and 3 FAD curves depend on the strain hardening exponent n. It may come as a surprise, but the Option 2 curve obtained for $n = 20$ lies below the Option 1 curve which is supposed to be a lower bound curve.

Observation of FAD curve clearly indicates the existence of three domains. In general, if the fracture parameter, K_r, is large compared with the load ratio, L_r, which actually means that it is operating in the upper left portion of the FAD curve of Fig. 14.40, then LEFM would be the most proper mean of analysis. Conversely, if the fracture parameter, K_r, is small compared with the load ratio term, L_r, then it is operating in the lower portion of the FAD curve where limit load condition should govern and, therefore, the Option 1 of the FAD procedure may be too conservative. In between these two extremes, elastic plastic conditions would govern and Option 1 should be valid, unless the operating point would be too close to the borderline of the FAD. These two extremes are roughly defined by $K_r/L_r \geq 1.8$ and $K_r/L_r < 0.3$, respectively, as schematically shown in Fig. 14.40.

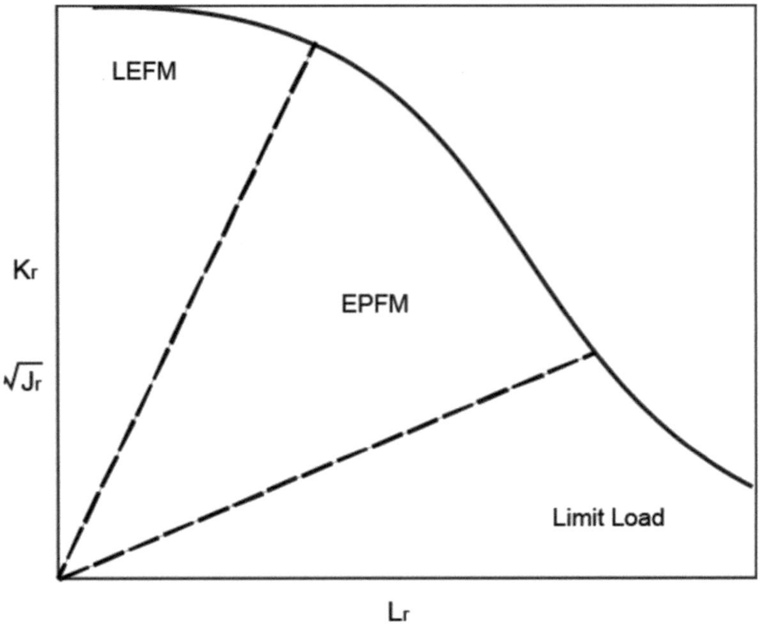

Fig. 14.40 Schematic of application fields of linear elastic fracture mechanics, elastic–plastic fracture mechanics and limit load analyses

14.10 Application of R6 FAD Methodology

Consider a large diameter pipeline of high strength steel type P690QL1, 190 cm radius and 60 mm thick, used in the construction of a hydroelectric plant. The steel has a yield strength of 670 MPa and ultimate strength of 770 MPa. The pipeline is carrying water at a maximum pressure of 11.15 MPa at a minimum operating temperature of 0 °C. The pipeline is supposed to have an operative life of 100 years. The first step is to assume an initial crack size and shape possibly existing in the pipe and then calculate its dimensions at the end of the 100 years life due to the operating fatigue cycles. To this purpose, an initial embedded elliptical flaw in a circumferential weld at 1/3 depth from the internal surface is assumed to exist, Fig. 14.41, whose size is $2a_o = 4$ mm and $2c_o = 40$ mm. This is the maximum defect size allowable according to ISO 15626 [41]. A fatigue crack growth analysis, see Sect. 16.7, has been performed using the Paris-Erdogan postulate, Eq. (16.5), with $C = 4{\cdot}10^{-8}$ and $n = 3$. The analysis has shown that the final crack size after 100 years life is $2a_f = 18$ mm and $2\ c_f = 42$ mm. Note that the choice of an embedded crack implies that the internal water never enters the crack so that its growth by stress corrosion is prevented.

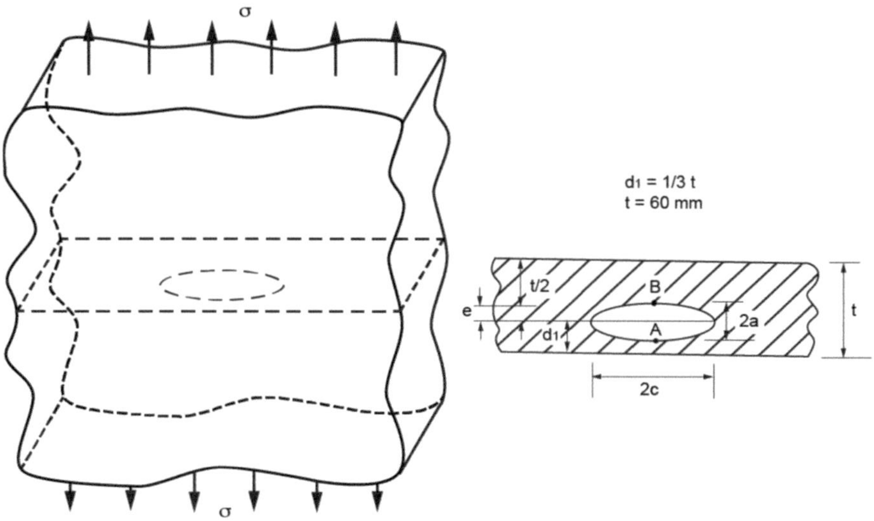

Fig. 14.41 Schematic of the initial embedded crack assumed to exist at 1/3t from the internal surface of the pipe. Because of the large diameter, the pipe wall has been approximated as a flat surface loaded by the hoop stress, as in Fig. 14.2

Twenty SEN(B) specimens, 240 × 60 mm, have been taken from the weld region and tested at 0 °C. The minimum value of toughness measured was $K_{Ic} = 145$ MPa $\sqrt{m}$ which has been assumed as reference fracture toughness. According to what has been said in Sect. 14.1, this toughness is representative of a loss of SSY conditions, and elastic–plastic behavior of the pipe is expected.

An R-6 FAD Option 1 analysis has been chosen whose curve is given by Eq. (14.58). This choice is based on the fact that the toughness measured, though rather high, is not high enough to be representative of a GY condition. Therefore, the operating region of the FAD is expected to be quite far from the tail region of the curve where the strain hardening of the material, n, can have some impact, see Fig. 14.39. The R-6 FAD Option 1 curve is, then, the one shown in Fig. 14.36 with a cut-off at $L_r = 1.5$. For the geometry considered, Fig. 14.41, the applied K_I at points A and B, respectively, is given by [41]

$$
\begin{aligned}
K_{IA} &= \sigma \frac{\sqrt{\pi a}}{\phi} M_A \\
K_{IB} &= \sigma \frac{\sqrt{\pi a}}{\phi} M_B
\end{aligned}
\tag{14.60}
$$

Where ϕ is the elliptical integral of the second kind, whose value is given by Eq. (9.64). The values of the two constant M_A and M_B for the initial crack size, $2c_o$, are worth [42] $M_{Ao} = 1.076$ and $M_{Bo} = 1.041$ and for the final one, 2*c*, $M_{Ao} = 1.035$ and $M_{Bo} = 1.015$. As expected, the maximum is reached at the deepest point A, though the two values are rater close to each other. As to stress, σ, in Eq. (14.60), since the crack is assumed to exist in a circumferential weld, to the primary hoop stress, σh, due to the internal pressure it must be added also the secondary residual stress, σres, resulting from welding. Therefore, the total applied stress is

$$\sigma_{tot} = \frac{P \cdot R}{t} + \sigma_{res} \tag{14.61}$$

Welding residual stresses, σres, are normally very high, of the order of the yield strength, σs of the material $\sigma res \sim \sigma s$. However, considering that the weld has been given a post weld heat treatment (PWHT) we may well assume that this residual stress state is about $\sigma res = 0.2\sigma s = 134$ MPa. Therefore, the total applied stress (14.61) turns out to be $\sigma_{tot} = 499.2$ MPa. With the initial and final crack sizes the applied K_I at point A is $K_{IA,o} = 44$ MPa $\sqrt{m}$ and $K_{IA,f} = 95.7$ MPa $\sqrt{m}$, respectively. With these values, the K_R given by Eq. (14.32) are

$$K_{r,o} = \frac{K_{I,o}}{K_{Ic}} = \frac{44}{145} = 0.3$$

$$K_{r,f} = \frac{K_{I,f}}{K_{Ic}} = \frac{95.7}{145} = 0.66$$

As to the L_r ratio, Eq. (14.51), it can be used the expression given by Miller [43]

$$L_r = \frac{p \cdot R}{t\sigma_s}\left[1 + \frac{0.4(2c)^2}{Rt}\right]^{1/2} \tag{14.62}$$

where *p* is the pressure, *R* is pipe radius and *t* is the pipe thickness. Using Eq. (14.62), the values of L_r for the initial and final crack are, $L_{r,o} = 0.54$ and $L_{r,f} = 0.55$, respectively. Therefore, the operating points for the initial and final crack are $A_o \equiv (0.54;0.3)$ and $A_f \equiv (0.55;0.66)$, respectively, as shown in Fig. 14.42. The points are well within the R-6 Option 1 curve.

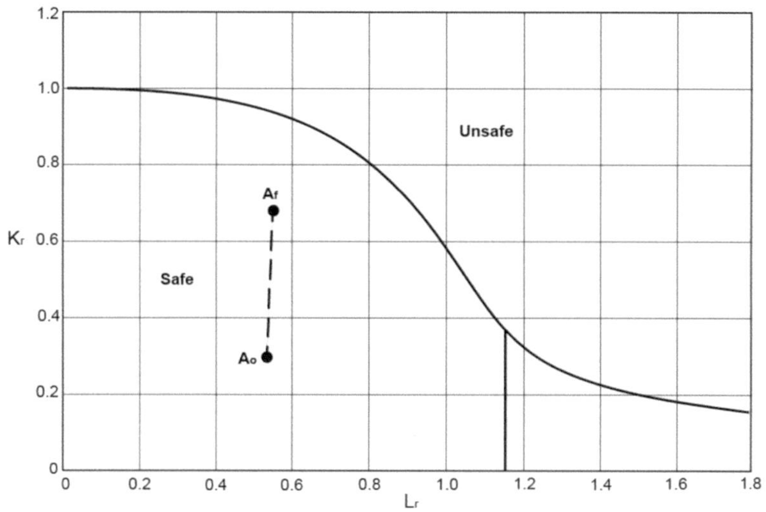

Fig. 14.42 R-6 FAD Option 1 analysis of the embedded crack of Fig. 14.42, in a pipeline of steel type P690QL1 under pressure. A_o and A_f represent the initial state and the final one after 100 years life, respectively. They are well inside the safe FAD region. Therefore, the assumed crack can be tolerated without jeopardizing the integrity of the pipe

References

1. Reynolds, M.B.: Failure Behavior in ASTM A106 Pipes Containing Axial Through-Wall Flaws, GEAP-5620 (1968)
2. Eiber, R.J., Maxey, W.A., Duffy, A.R., Atterbury, T.J.: Investigation of the Initiation and Extent of Ductile Pipe Rupture, BMI-1866, Phase I Final Report (1969)
3. Reynolds, M.B.: Failure Behavior of Flawed Carbon Steel Pipes and Fittings, GEAP-10236 (1970)
4. Eiber, R.J, Maxey, W.A., Duffy, A.R., Atterbury, T.J.: Investigation of the Initiation and Extent of Ductile Pipe Rupture, BMI-1908, Final Report (1971)
5. Wolkowyski, G.M., Scbmidt, R., Scott, P., Olson, S., Mahall, C.W., Kramer, G., Paul, D.: International Piping Integrity Research Group (IPIRG) Program, NUREG/CR-6233, Vol. 4, BMI-2177 (1997)
6. Milella, P.P.: Outline of nuclear piping research conducted in Italy. Nucl. Eng. Des. **98**, 219–229 (1987)
7. Milne, I., Ainsworth, R.A., Dowling, A.R., Stewart, A.T.: Assessment of the Integrity of Structures Containing Defects, R/H7/R6-Rev. 3, published by Central Electric Generating Board, England (1986)
8. Swedlow, J.L., Williams, M.L.: A Review of Recent Investigations into Fracture Initiation at Cacit, ARL 64–175 (1964)
9. Folias, E. S.: On the Effect of Initial Curvature on Cracked Flat Sheets, UTEC CE-69–002 (1969)
10. Erdogan, F., Kibler, J.J.: Cylindrical and spherical shells with cracks. Int. J. Fract. Mech. **5**(3), 229–323 (1969)
11. Swedlow, J.L., Williams, M.L.: A Review of Recent Investigation into Fracture Initiation at Galcit, ARL 64–175 (1964)
12. Lucon, E., Lazzati, A.: Caratterizzazione di Tubazioni in Acciaio Ferritico ed Inossidabile, Rapporto Finale, CISE 7229, Milano (1992) (Available from Milella)

13. Hahn, G.T., Sarrate, M., Rosenfield, A.R.: Criteria for crack extension in cylindrical pressure vessels. Int. J. Fract. Mech. **5**(3), 187–210 (1969)
14. Eiber, R.J, Maxey, W.A., Duffy, A.R., Atterbury, T.J.: Investigation of the Initiation and Extent of Ductile Pipe Rupture, BMI-1793 (1967)
15. Kanninen, M.F., Broek, D., Marschall, C.W., Rybichi, E.F., Sampath, S.G., Simonen, F.A., Wilkowski, G.M.: Failure Prediction for Sensitized Stainless Steel piping with Circumferential Cracks, Electric Power Research Institute, Report EPRI NP-192 (1976)
16. "Degraded Piping Program - Phase II", Semiannual Report March 1984 - September 1984 by Battelle-Columbus Laboratories, NUREG/CR-4082 (1985)
17. Galbusera, U., Squilloni, A.: Prove di Collasso su Tubazioni in Acciaio A106B da 2" e 8" e in Acciaio 316L da 6", CISE 288 73 (1986) (Available from Milella)
18. Galbusera, U., Squilloni, A.: Prove di Collasso su Tubazioni in Acciaio A106B e su Pannelli in Anticoroidal 100TA16, CISE 2252 (1984) (Available from Milella)
19. Galbusera, U., Squilloni, A.: Prove di Collasso su Tubazioni da 6" in Acciaio AISI 316 L alla Temperatura di 280 °C, Rapporto Topico CISE 4214 (1988) (Available from Milella)
20. Galbusera, U., Squilloni, A.: Prove di Collasso su Tubazioni in Acciaio AISI 316 L da 6", Rapporto Topico CISE 2884 (1988) (Available from Milella)
21. Krishnaswarny, P., Scott, P., Mohan, R., Rahman, S., Choi, Y. H., Brust, E., Kilinski, T., Francini, R., Ghadiali, N.C., Marschall, C.W., Wilkowski, G.: Fracture Behavior of Short Circumferentially Surface-Cracked Pipe, NUREG/CR-6298, BMI-2183 (1995)
22. Milella, P.P.: Recent Results and Future Programs on Pipe Fracture Tests in Italy, Leak-Before-Break: Progress in Regulatory Policies, NUREG/CP-0092, Tokio, May 14–15 (1987)
23. Wilkowski, G.M., and others.: Degraded Piping Program - Phase II", Semiannual Report, April 1985-September 1985, by Battelle Columbus Division, NUREG/CR-4082, vol. 3 (1986)
24. Shih, C.F., German, M.D., Kumar, V.: An engineering approach for examining crack growth and stability in flawed structures. Int. J. Pres. Ves. Piping **9**, 159–196 (1981)
25. Kumar, V., German, M.D., Shih, C.F.: An Engineering Approach for Elastic-Plastic Fracture Analysis, EPRI Report NP-1931, Electric Power Research Institute, Palo Alto, California (1981)
26. Kumar, V., German, M.D.: Elastic-Plastic Fracture Analysis of Through-Wall and Surface Flaws in Cylinders, EPRI Report NP-5596, Electric Power Research Institute, Palo Alto, California (1988)
27. Goldman, N.L., Hutchinson, J.W.: Fully plastic crack problems: the center cracked strip model under plain strain. Int. J. Solids Struct. **11**, 575–591 (1975)
28. Kumar, V., Shih, C.F.: Fully plastic crack solutions, estimation scheme and stability analysis for compact specimens, fracture mechanics. ASTM Spec. Tech. Publ. **700**, 406–438 (1980)
29. Yagawa, G., Ueda, H.: A Finite Element Analysis for Inelastic Behavior of Surface crack, Proceedings of Seminar on Leak-Before-Break: Progress in Regulatory Polices and Supporting Research, NUREG/CP-0092, Tokyo, pp. 427–461, May 14–15 (1987)
30. Milne, I., Ainsworth, R.A., Dowling, A.R., Stewart, A.T.:'Assessment of the Integrity of Structures Containing Defects, CEGB Report R/H/R6-Revision 3 (1986)
31. Milne, I., Ainsworth, R.A., Dowling, A.R., Stewart, A.T.: Background to and Validation of CEGB Report R/H/R6-Revision 3 (1987)
32. Dowling, A.R., Townley, H.A.: The effect of defects on structural failure: a two-criteria approach. Int. J. Press. Vessel Pip. **3**, 77–137 (1975)
33. PD6493:1980: Guidance on Some Methods for the Derivation of Acceptance Levels for Defect in Fusion Welded Joints. British Standards Institutions (1980)
34. Harrison, R.P., Loosemore, K., Milne, I.: Assessment of the Integrity of a Structure Containing Defects, CEGB Report No. R/H/R6 (1976)
35. Shih, C.F., German, M.D., Kumar, V.: An Engineering Approach for Examining Crack Growth and Stability in Flawed Structures, GE Co. TIS Report No. 80CRD205, September 1980, also in Int. J. Pres. Ves. Pip. pp. 159–196 (1981)
36. Shih, C.F., Kumar, V., German, M.D.: Studies on the Fracture Assessment Diagram Using the Estimation Scheme and *J*-Controlled Crack Growth Approach, GE Co. TIS Report (1981) also in ASTM STP36770S, 01 January 1983

37. Bloom, J.M.: Prediction of ductile tearing using a proposed strain hardening failure assessment diagram. Int. J. Fract. **6**, R63–R67 (1980)
38. Shih, C.F., Kumar, V. German, M.D.: Studies on the Fracture Assessment Diagram Using the Estimation Scheme and *J*-Controlled Crack Growth Approach, ASTM STP 803 (1981)
39. Ainsworth, R.A.: The assessment of defects in structures of strain hardening materials. Eng. Fract. Mech. **19**, 663 (1984)
40. Milne, I., Ainsworth, R.A., Dowling, A.R., Stewart, A.T., Background to and validation of CEGB report R/H/R6-Rev. 3. Int. J. Press. Vess. Pip. **32**, 105–196 (1988)
41. ISO 15626, Non-destructive testing of welds—Time-of-flight diffraction technique (TOFD)—Acceptance levels, Second ed., 2018–07
42. Murakami, Y.: Stress Intensity Factors Handbook, Committee on Fracture Mechanics, The Society of Material Science, Vol. 2, pp. 734–737. Japan, Pergamon Press (1988)
43. Miller, A.G.: Review of Limit Loads of Structures Containing Defects, 2nd edn, CEGB Report TPRD/B/OO93/N82--Rev. 1 (1985)
44. Newman. J.C., Raju. I.S.: Stress Intensity Factor Equations for Cracks in Three-dimensional Finite Bodies Subjected to Tension and Bending Loads, NASA Technical Memorandum 85793, NASA Langley Research Center, Hampton, VA (1984)

Chapter 15
Metallurgical Variability and Data Scatter

15.1 Introduction

We have seen that to obtain a valid fracture toughness measurement, K_{Ic}, is necessary to comply with the thickness requirements provided by Eq. (9.73). Any thickness, B, smaller than

$$B = 2.5\left(\frac{K_{Ic}}{\sigma_s}\right)^2 \tag{15.1}$$

yields a fracture toughness value, indicated as K_c or K_Q, generally higher than the true fracture toughness of the material K_{Ic}. This is shown in Figs. 9.30 through 9.32. Moreover, when small specimens are tested a large data scatter is observed that seems not to occur with larger and valid ones. This can be seen in Fig. 15.1 for 508B Cl.3 forging tested at – 20 °C [1]. The valid toughness, K_{Ic}, of this material at –20 °C is 97 MPa$\sqrt{}$m which requires a minimum test thickness of 9 cm, according to Eq. 15.1. Note that in Fig. 15.1 toughness values for smaller specimens, 1/2T, 1T and 5T, were obtained from J_{Ic} measurements using Eq. (10.32) and are indicated as K_J since the thickness requirements for measurements are much less restrictive, Eq. (11.73), therefore a small 1/2T or 1T specimen that does not comply with size requirements of Eq. (15.1) may yield a valid J_{Ic} value from which the toughness K_{Ic} can be derived. Two important aspects of small-specimen fracture toughness testing emerge from Fig. 15.1. First, data scatter increases as the specimen size decreases. Secondly, small specimens eventually yield the same fracture toughness value obtained with a single large and valid 10T CT specimen.

The same behavior can be observed for A533B Cl.1 steel tested at −20 °C, Fig. 15.2 [2]. These results indicate that the size requirement expressed by Eq. (15.1) may not be totally related to a size-induced constraint effect needed to maintain a plain strain condition since, as it was clearly expressed by Landes et al. [2] : "It would be hard to explain how some of the 1T specimens lose constraint while others

P. P. Milella, *Fundamentals of Fracture Mechanics*,
https://doi.org/10.1007/978-3-032-23081-2_15

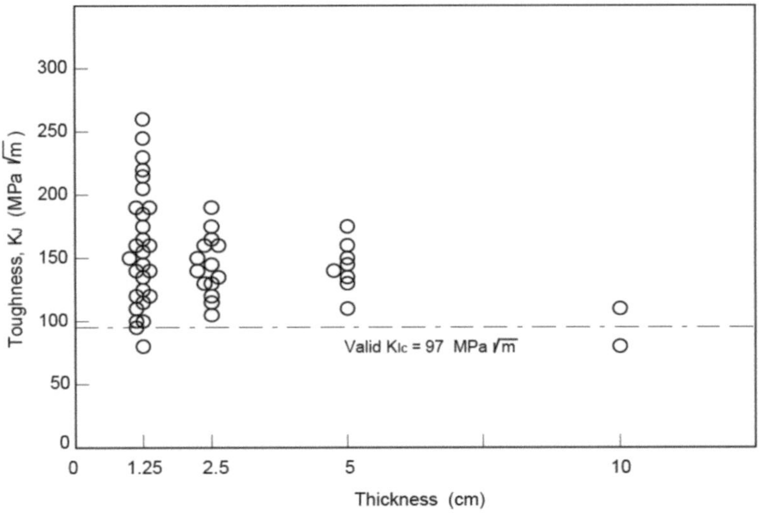

Fig. 15.1 Data scatter obtained in fracture toughness testing for A508 Cl.3 carbon steel specimens of different sizes at –20 °C [1]

do not because the specimens are all identical". This statement recognized the fact that size constraint is a mechanical effect and therefore it must exist to the same degree in all identical specimens. Therefore, it must be concluded that data scatter is related to the other parameter that changes with size which is the process volume. By increasing the thickness of the specimen, we also increase the volume of material initially interested by the fracture process that we call process volume, which is the very small volume ahead of the crack tip plastic zone where fracture initiates, see Sect. 6.4. By increasing the process volume, V_{pv}, we also increase the probability of finding the worst microscopic brittle region or ICR that triggers the fracture. The process volume effect is then related to metallurgical variability, which is a characteristic of the material. However, the process volume can be effectively increased either by increasing the thickness, B, of a single specimen or, which is perfectly equivalent, by increasing the number of small specimens that may not even comply with Eq. (15.1) size requirements. Moreover, there is also another feature that must be caught in Fig. 15.2, in particular. Small 1/2T specimens not only reach the valid value of toughness $K_{Ic} = 151$ MPa$\sqrt{}$m obtained with larger and valid specimens, but some even went below that value yielding an incredible low 90 MPa$\sqrt{}$m. The same happened with the 1T specimen that reached, one out of eight, 140 MPa$\sqrt{}$m that were not obtained with none of the two 2T specimens. This fact clearly indicates that the process volume of 8 1T specimens must be higher than that of two 2T specimens. This also means that by running additional tests with 2T valid specimens that lower value obtained with the 1T specimen series or the 1/2T series will be seen again: it is just a matter of process volume. Equal process volumes yield the same minimum fracture toughness value. Its effect will continue to be seen till the minimum value possible

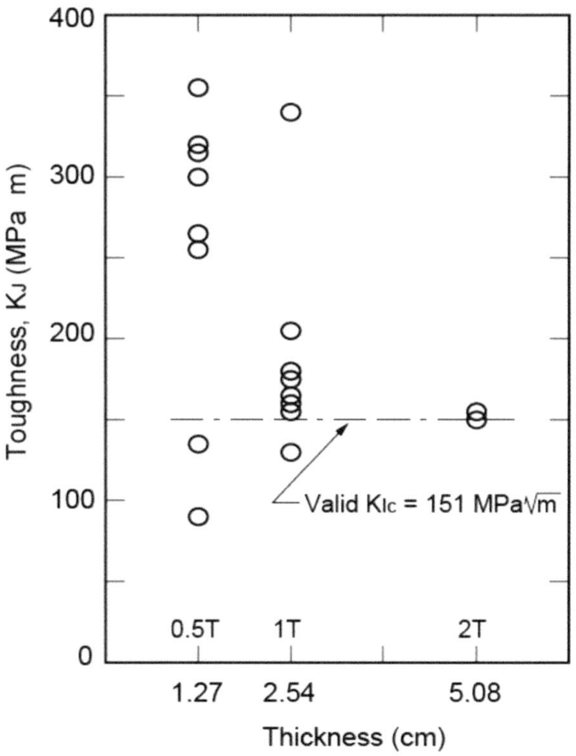

Fig. 15.2 Data scatter obtained in fracture toughness testing for A533B Cl.1 carbon steel specimens of different sizes at –20 °C [2]

will be reached, $K_{Ic,min}$, that indicates that also the worst metallurgical condition possible for that type of material has been reached. By further increasing the process volume there will not be any further fracture toughness reduction.

This minimum value can be assessed using a probabilistic analysis, as it will be shown in the following paragraphs. However, it is also clear that by reducing the size, the specimen is allowed to deform more and more which increases crack tip blunting thus excluding the possibility of a totally brittle fracture. This is particularly true for BCC materials in the temperature range over the RT_{NDT} where small non valid specimens will never reach the lower bound K_{Ic} toughness values obtained with large valid specimens, no matter how many they are. This is shown in Fig. 15.3 for A533B Cl.1 steel [3]. The 0.4T and 1T specimens at temperatures higher than 0 °C cannot reach the lower bound toughness values that are seen with the 4T specimens. Therefore, the constraint effect must also take a role in Eq. (15.1).

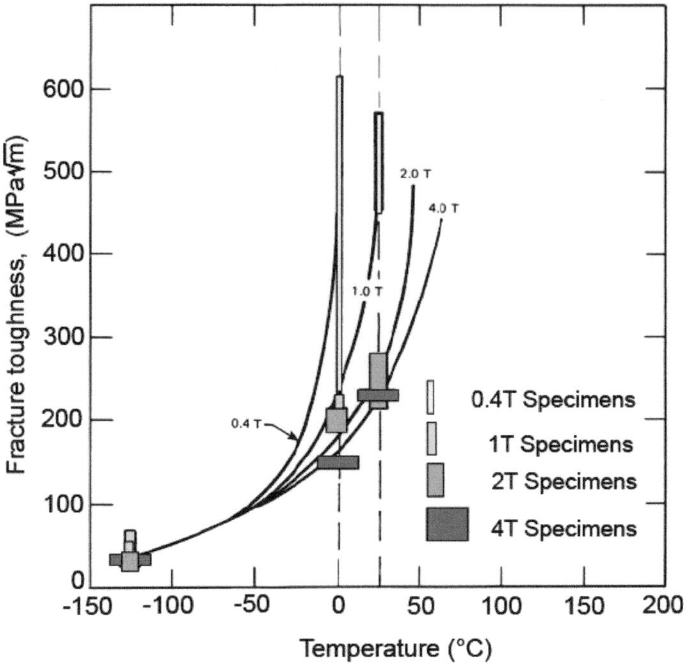

Fig. 15.3 Scatter ranges of K_c at cleavage instability versus specimen thickness and test temperature for A533B Cl.1 steel. Solid lines indicate the average value obtained for each specimen thickness (adjusted from [3])

15.2 Process Volume Size

It ha been shown in Sect. 6.4 that brittle fracture initiates at a certain distance, X, from the ductile growth, Δa, that always precedes cleavage, see Fig. 6.20. This distance is of the same order of magnitude of Δa, generally no larger than 2 or 3 times Δa, as shown in Table 6.1, which is of the order of some tens of micron in the totally brittle fracture regime. Within this distance X the high triaxiality elastic stress field is dominated by K. Ductile growth that precedes brittle fracture is related to the crack tip blunting equal to 1/2 CTOD. Therefore, using Eq. (11.84), we can say that the distance X is proportional to K_I^2:

$$X \propto \frac{K_I^2}{m\sigma_s E'} \tag{15.2}$$

where σ_s is the yield stress, m is a constraint factor that indicates the yield stress elevation due to triaxiality and E' the Young's modulus in plain strain condition. Assuming, as a first approximation, that the critical area ahead of the crack tip where fracture initiates has a circular shape as for the Irwin plastic zone, see Sect. 8.2, we can argue that the volume beyond the ductile growth Δa where brittle fracture initiates, which will be called process volume, V_{pv}, can be written as

$$V_{pv} = C \cdot B \cdot X^2 \tag{15.3}$$

where B is the thickness of the specimen and C is a constant. Recalling Eq. (15.2), it is:

$$V_{pv} = C \cdot B \cdot \left(\frac{K_I^2}{m\sigma_s E'} \right)^2 \tag{15.4}$$

which means that the process zone has a volume, V_{pv}, proportional to the 4th power of K_I:

$$V_{pv} = C' B K_I^4 \tag{15.5}$$

This actually means that for a measured and valid value K_{Ic} or K_{Jc} the only difference in the process volume between specimens of different sizes B is the thickness. Therefore, to obtain with a ½T CT specimen a K_{Ic} fracture toughness value as measured with a valid 4T CT specimen we need at least 8 small specimens. This can be seen in Fig. 15.4 for A508B Cl.3 steel forging [1].

Small specimens broken in a large number of tests yield the same low toughness value obtained with a single 4T specimen up to a temperature of –20 °C. From that temperature on small specimens cannot reach the lower bound value obtained with larger specimens because of loss of constraint. This is seen through the stable crack growth measurements that reveal that at –20 °C the crack starts to blunt and to growth subcritically, see Fig. 10.5.

15.3 The Irwin β_{Ic} Correction Factor

The first to address the issue of data scatter in fracture toughness measurements using, in particular, small specimens was George Irwin already in 1959 [4, 5]. He started precisely from the graph shown in Fig. 9.29 that refers to 7075T6 aluminum alloy. He readjusted the data, normalizing the toughness K_c to the true value K_{Ic}, which he supposed to know, and reporting on the x-axis instead of the thickness B its value normalized to the radius of the plastic zone ahead of the crack tip, r_p, in other words the inverse of the β factor given by Eq. (9.48) that will be called β_c. In this specific case, the true K_{Ic} toughness is 36 MPa$\sqrt{}$m obtained with the 30 mm specimen and the yield strength is $\sigma_s = 450$ MPa. The result is shown in Fig. 15.5. The solid line curve refers to experimental data relating to the aluminum alloy mentioned, but it is also valid for A508B Cl.3 forging as it wiil be seen. The curve can be conveniently used to adjust the fracture toughness data obtained with specimens that are undersized according to Eq. (15.1).

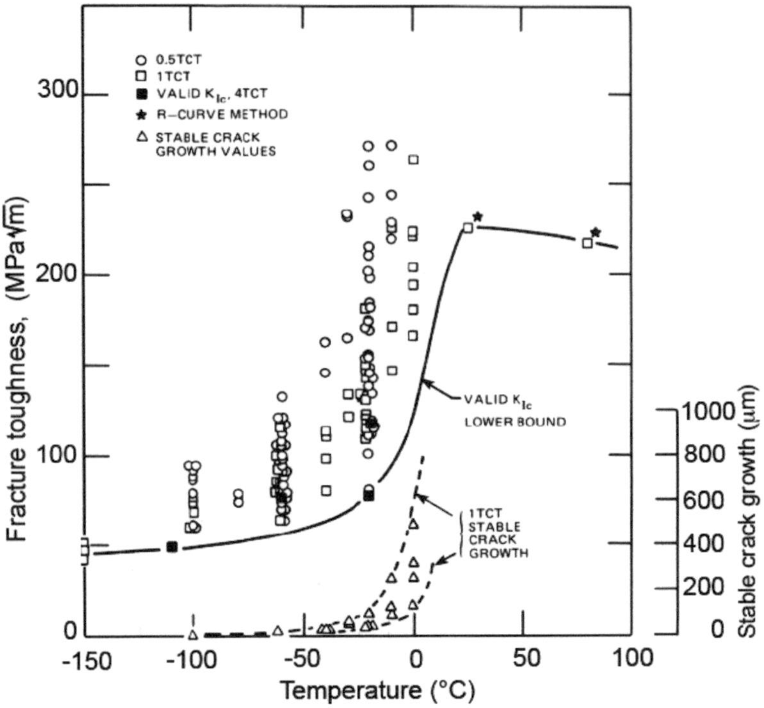

Fig. 15.4 K_J calculated from J at fracture versus temperature for A508B Cl.3 forging. Small specimens broken in a large number of tests yield the same low toughness value obtained with a single 4T specimen up to a temperature of −20 °C. Also shown is the measured stable crack growth [1]

In fact, if K_c is a toughness value obtained with a specimen of insufficient thickness B which corresponds to a certain value of $\beta_c = (K_c/\sigma_s)^2/B$, it will be sufficient to enter the curve of Fig. 15.5 at the abscissa $1/\beta_c$ and read the corresponding value of the ratio K_c/K_{Ic} from which the true value K_{Ic} will be obtained. This is an empirical procedure that bases its reason for use on the fact, evidently to be verified, that the trend of the curve $K_c/K_{Ic} = f(1/\beta_c)$ obtained through so few data is representative of a general behavior and is therefore characteristic not only of the alloy under consideration, but of several different alloys if not of any alloy, thus becoming an invariant. It should be noted that the ratio $K_c/K_{Ic} = 1$ is reached only asymptotically. For $\beta_c = 0.4$, which marks the dimensional limit for the validity of a K_{Ic} test, see Sect. 9.10, the K_c/K_{Ic} ratio is 1.079, which means that the corresponding K_{Ic}, which is the one we consider valid, is only about 8% higher than what would be obtained with an infinite thickness. The relationship between the ratio K_c/K_{Ic} and $1/\beta_c$ also implicitly identifies a correspondence between β_c and $\beta_{Ic} = (K_c/\sigma_s)^2/B$. In fact, the K_c/K_{Ic} ratio can also be written:

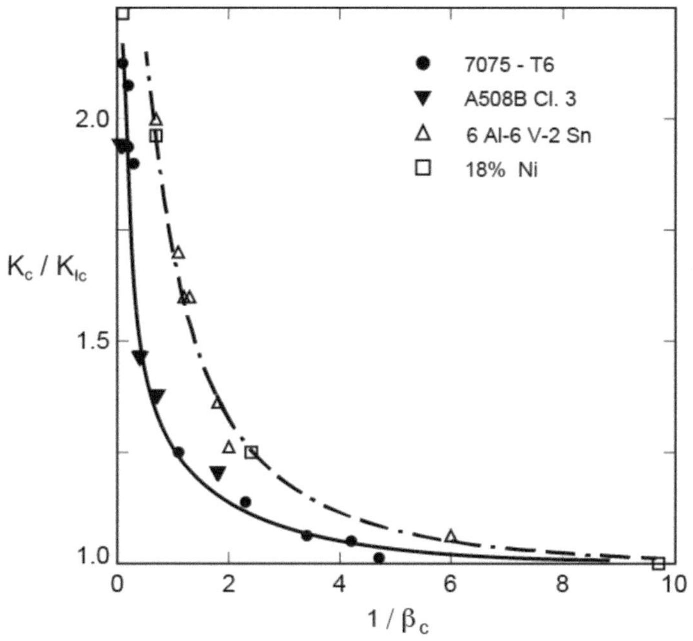

Fig. 15.5 Trend of the K_c/K_{Ic} parameter versus $1/\beta_c$ for four different alloys, two of which are medium strength and two high strength alloys

$$\frac{K_c}{K_{Ic}} = \sqrt{\frac{\beta_c}{\beta_{Ic}}} \tag{15.6}$$

and it can be verified that for the curve in question, shown in Fig. 15.5, the relation holds:

$$\beta_c = \beta_{Ic} + 1.4 \cdot \beta_{Ic}^3 \tag{15.7}$$

which, converted into K_c/K_{Ic}, provides:

$$\frac{K_c}{K_{Ic}} = \sqrt{1 + 1.4 \cdot \beta_{Ic}^2} \tag{15.8}$$

Equation (15.8) is precisely the empirical expression of adjustment of the experimental K_c data proposed by Irwin and known as Irwin's β_{Ic} factor equation. In practice, the following procedure is used. A fracture toughness test is carried out with a specimen of insufficient thickness B, which gives an invalid toughness value K_c, the corresponding value of β_c is calculated. At this point Eq. (15.7) is resolved with respect to β_{Ic}:

$$\beta_{Ic}^{3} + \frac{5}{7} \cdot \beta_{Ic} - \frac{5}{7} \cdot \beta_{c} = 0 \tag{15.9}$$

Equation (15.9) is a cubic equation with no quadratic term that can be solved by introducing the auxiliary parameters p, C_1, and C_2:

$$\begin{aligned} p &= \frac{5}{14}\beta_c \\ C_1 &= \sqrt{p^2 + \frac{(5/7)^3}{27}} + p \\ C_2 &= \sqrt{p^2 + \frac{(5/7)^3}{27}} - p \end{aligned} \tag{15.10}$$

by which β_{Ic} becomes equal to:

$$\beta_{Ic} = C_1^{1/3} - C_2^{1/3} \tag{15.11}$$

and finally the adjusted value K_{Ic} is obtained through Eq. (15.6). The solid curve in Fig. 15.5 is tabulated in Table 15.1 together with the values of $1/\beta_c$ obtained through relations (15.7) and (15.11). The interpolation curve of toughness data for type 7075T6 aluminum alloy in Fig. 11.5 also interpolates very well those obtained for a type A508B Cl. 3 forging. This means that the procedures shown can also be applied to low-alloy, low-yield carbon steels. Not all steels or alloys, however, follow the same trend. For instance, 18%Ni maraging steel with $\sigma_s = 1700$ MPa and high strength 6Al-6V-2Sn Titanium, $\sigma_s = 1218$ MPa, shown in Figs. 9.28 and 9.30 seem to follow a similar but different curve, indicated by the dotted line in Fig. 15.5, which interpolates experimental data obtained with the same procedure. Irwin Eq. (15.8) was written in 1959 but it was only in 1984 that Merkle proceeded with an extensive verification of its validity [6]. This is shown in Fig. 15.6 that refers to the data shown in Fig. 15.4 for A508B Cl. 3 forging.

The experimental data coming from small 0.5T and 1T specimens have been adjusted using the β_{Ic} approach. As can be seen, the original large scatter band shrinks down to values compatible with those obtained with valid 4T specimens. Also shown is the lower bound ASME Sec. XI reference K_{Ic} curve of Fig. 9.51, based on $RT_{NDT} = -25$ °C.

15.4 Weibull Analysis

Metallurgical variability and its effect on fracture toughness measured with small specimens, in particular, can be addressed using a statistical analysis. In this study the Weibull distribution will be used [7]. The Weibull probability density function, $p(x)$, in its simplest form (two parameters function) is defined as

Table 15.1 Values of $1/\beta_c$, β_{Ic} and corresponding K_c/K_{Ic} for 7075-T6 Aluminum alloy

$1/\beta_c$	β_{Ic}	K_c/K_{Ic}	$1/\beta_c$	β_{Ic}	K_c/K_{Ic}
0.01	4.08	4.94	0.55	0.876	1.441
0.02	3.216	3.94	0.06	0.839	1.409
0.03	2.790	3.455	0.65	0.805	1.382
0.04	2.520	3.149	0.70	0.775	1.357
0.05	2.326	2.932	0.75	0.747	1.335
0.06	2.177	2.767	0.80	0.722	1.316
0.07	2.057	2.635	0.85	0.698	1.298
0.08	1.958	2.526	0.90	0.676	1.281
0.09	1.874	2.435	0.95	0.656	1.266
0.10	1.80	2.356	1.0	0.637	1.253
0.15	1.540	2.080	1.2	0.571	1.207
0.20	1.372	1.909	1.4	0.518	1.173
0.25	1.251	1.788	1.6	0.475	1.147
0.30	1.157	1.697	1.8	0.438	1.126
0.35	1.081	1.625	2.0	0.406	1.109
0.40	1.018	1.567	2.2	0.378	1.096
0.45	0.964	1.518	2.4	0.354	1.084
0.50	0.917	1.476	2.6	0.332	1.075

$$p(x) = m\frac{x^{m-1}}{x_o^m}\exp\left[-\left(\frac{x}{x_o}\right)^m\right] \tag{15.12}$$

in which m is a shape parameter, also known as Weibull exponent or slope, and x_o a scale factor to which, as it will be shown, it is associated a probability of failure of 63.2%. Equation (15.12) is actually defining a family of functions since its shape depends on the value of the Weibull exponent m. Figure 15.7 presents several Weibull probability density functions $p(x)$ related to different m values. As it can be seen, the main difference between the Gauss probability density function and the Weibull function is in the fact that the former assumes that experimental data always follow the bell shaped distribution while the latter is providing a curve that follows the actual experimental distribution of data, which is the same difference between a tailor that always sews the same dress asking the client to adapt and another that sews a dress that fits the client.

From the probability density function it is derived the cumulative probability function $P(x)$ that is the probability that an element of the population assumes a value less or equal to a given x. The cumulative probability function $P(x)$ is

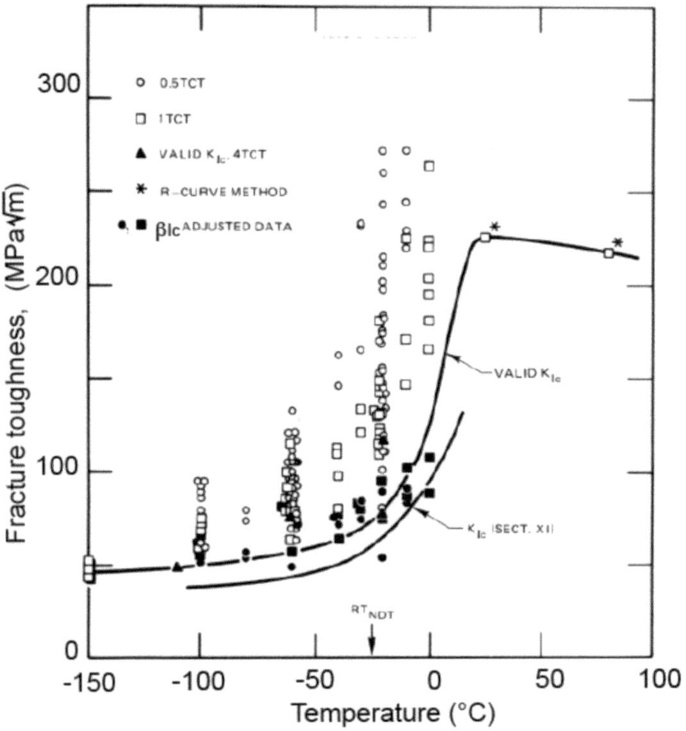

Fig. 15.6 Illustration of agreement between β_{Ic} adjusted 0.5T and 1T small-specimen fracture toughness values and 4T large-specimens for A508B Cl. 3 forging [6]

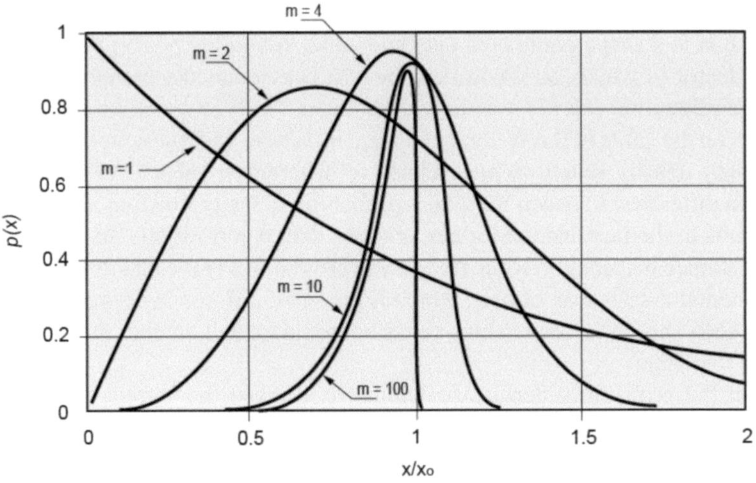

Fig. 15.7 Weibull probability density function *p(x)* plotted for different *m* values

$$P(x) = \int_{-\infty}^{x} p(z)dz = 1 - \exp\left[-\left(\frac{x}{x_o}\right)^m\right] \quad (15.13)$$

Having to do with fracture toughness measurements, K, Eq. (15.13) can be written as

$$P(x) = \int_{-\infty}^{x} p(z)dz = 1 - \exp\left[-\left(\frac{K}{K_o}\right)^m\right] \quad (15.14)$$

The cumulative probability function is plotted in Fig. 15.8 for some selected values of the Weibull exponent m equal to 1, 4, 10, 20 and 100, respectively. It shares some interesting features. First, all curves have the same inflection point of coordinates $x/x_o = K/K_o = 1$ and $P(x) = 0.632$ so that x_o represents the value below which 63.2% of the population falls. Secondly, it is interesting to note how data dispersion reduces progressively as the Weibull exponent m increases. For $0 < m \leq 1$ any value is possible from zero to infinity. For $m = 4$ the interval of possible values shrinks significantly, in practice it is given by $0.1 \leq x/x_o \leq 1.8$, while for $m = 100$ the cumulative probability function $P(x)$ approximate a step function that is to say a function with one single value with no dispersion. This actually means that all samples yield almost the same value. This, moreover, is confirmed by the same density probability function $p(x)$ that, as shown in Fig. 15.7, for m equal 1 or 2 indicates that any value is possible from 0 to infinity while for $m = 100$ the dispersion reduces to 1/2. Weibull exponent, therefore, may be thought as an index of the metallurgical variability in the data scatter. The values of the Weibull exponent m and of the shape factor K_o are easily assessed. By taking twice the logarithm of Eq. (15.13) it yields

$$\ln[1 - P(K_c)] = -\left(\frac{K}{K_o}\right)^m;$$

$$\ln\ln\frac{1}{1 - P(K)} = m\ln(K) - m\ln(K_o). \quad (15.15)$$

In a bi-logarithmic scale Eq. (15.15) is represented by a line whose slope is m and the known term is $m\ln(K_o)$. In practice, the steps to determine the Weibull parameters m and K_o are the following: first, list the experimental results x_i $(i = 1,2,3,\ldots,n)$ in ascending order; second, assign a rank to each results. The rank indicates the failure probability in percent associated to each result, i.e., the frequency. Since the entire population is never available, we can only estimate the rank of the sample population. In this respect, several options are available. One of the most common estimate is the median rank, i.e., that rank that is as likely to err on the high side, with respect

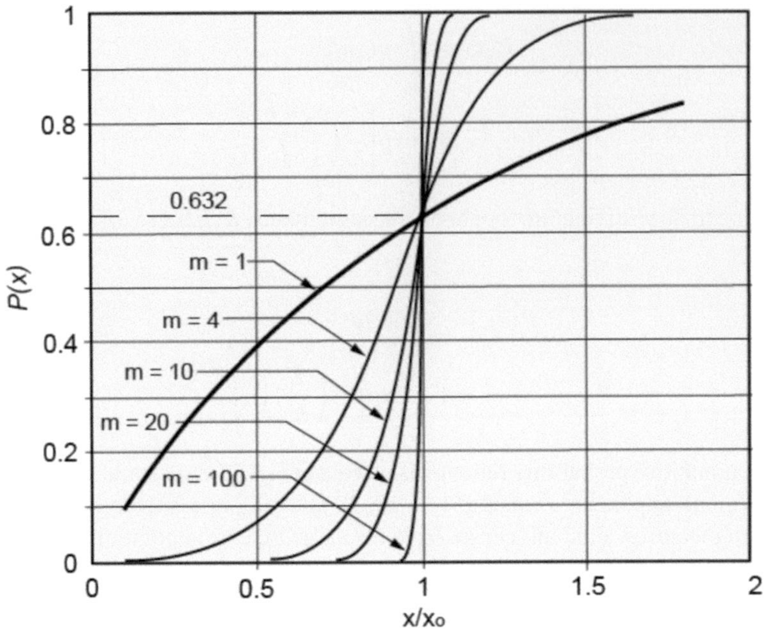

Fig. 15.8 Weibull cumulative probability function P for some values of the m exponent

to the median value, as on the low side. The following equation gives a very good approximation of the estimated median ranks for any sample size n

$$\text{Estimated median rank} = \frac{0.7}{n+0.4}, \frac{1.7}{n+0.4}, \frac{2.7}{n+0.4}, \ldots, \frac{n-0.3}{n+0.4} \tag{15.16}$$

A valid alternative to the median rank is the mean rank, which is given by the equation

$$\text{Estimated mean rank} = \frac{1}{n+1}, \frac{2}{n+1}, \frac{3}{n+1}, \ldots, \frac{n}{n+1} \tag{15.17}$$

which may be written as

$$P(K) = \frac{N}{n+1} \tag{15.18}$$

where N is the serial number of the experimental results listed in the ascending order and n the sample size or sample population. The number n of experimental data plus 1 is used to avoid the last value of the population has a probability equal 1 ($n = N$), i.e., certainty. Another method to assign a probability of occurrence or rank to each result is

$$P(K) = \frac{N - 0.5}{n} \tag{15.19}$$

Once a rank or probability of occurrence *P(x)* is assigned to each value of the sample population, the experimental results x_i are plotted in a bi-logarithmic scale *lnln[1/1-P(K)]-ln(K).* The best-fitting line will have a slope equal to *m*. Once *m* is inferred the scale factor K_o is obtained from the known term, *NT*, of the line equation

$$K_o = exp\left[-\frac{NT}{m}\right] \tag{15.20}$$

Once *m* and K_o are known, Eq. (15.13) can be used to assess the value of the variable *x* with the wanted probability *P(x)* or, which is the same, with the wanted probability of survival $R = 1 - P(K)$, which in case of fracture toughness becomes

$$K = K_0\{-\ln[1 - P(K)]\}^{1/m} \tag{15.21}$$

Figure 15.9 shows result of the Weibull analysis done by Milella for A508B Cl.3 1TCT specimens tested at Oak Ridge under the HSST Program [3] at −32 °C, −5 °C, 20 °C and 50 °C, respectively. As it can be seen, at −32 °C, −5 °C and 20 °C, the value of the Weibull exponent, *m*, is equal to 4.59, 3.84 and 4.97, respectively, which is very close to the theoretical value of 4 found for the process volume dependence on K_c, Eq. (15.5), confirming that data scatter in the experimental toughness results, K_c, depends on the probability to find an inclusion or ICR in the process volume, V_{pv}, ahead of the crack tip plastic zone. These results comply with Landes et al. results obtained for A471 NiCrMoV steel at 20 °C where *m* was about 5 [2] and with Milella and Bonora [8] results for A533B Cl.1 steel at −32 °C and −5 °C where *m* was equal to 4.7 and 5.6, respectively. For the A508B Cl.3 forging, the Weibull exponent, *m*, swells to $m = 13.42$ at 50 °C, Fig. 15.9, indicating that at this high temperature, 40 °C over the RT_{NDT}, SSY conditions are lost, and fracture is no longer produced by cleavage.

The K_I dominated region ahead of the crack tip does not exist any longer because the crack itself does not exist any longer being turned into a notch by the large blunting. Now the notch tip stress field is dominated by *J* which shows a large plastic component, J_p, in addition to the elastic one, J_e, see Sect. 11.2 and Eq. (11.16). Under this conditions the process volume swells and is no longer proportional to the 4th power of *K*. Milella et al. [8] analyzing data relative to CT specimens, Charpy specimens and RNB(T) specimens of different notch radii equal to1.2, 2, 2.4, 4, 6, 10, 20 mm and ∞, i.e., RB(T), respectively, have found that the *m* exponent was growing from 4 to 40, as shown in Fig. 15.10. They also shown that the *m* exponent depends on the triaxial factor *FT*, Eq. (4.33) that depends on the stress field σ_x, σ_y, σ_z associated with the particular type of crack or notch, according to the relationship [8]

$$m = 55.4 - 22.4 \cdot FT \tag{15.22}$$

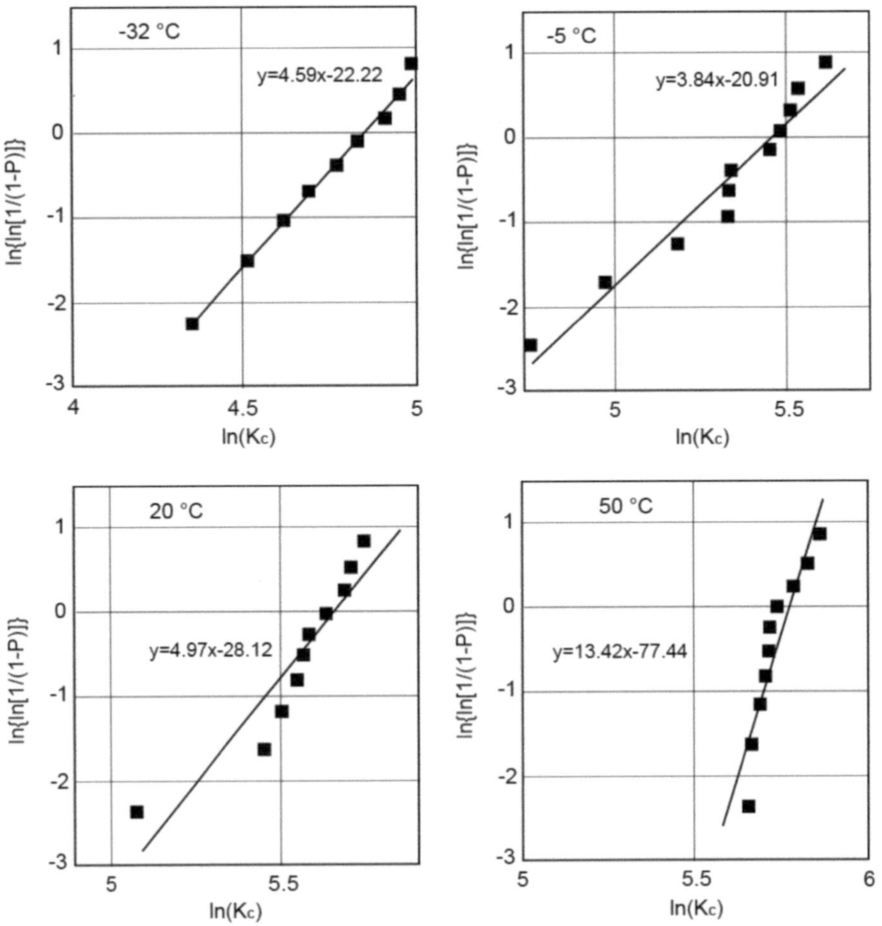

Fig. 15.9 Weibull distribution $ln\{ln[1/(1-P)]\}$ versus $ln(K_c)$ for A508B Cl.3 forging obtained by Milella analyzing 1TCT data obtained at ORNL at four different temperatures

For a smooth specimen, without notches of any kind, with a FT of 0.3, Eq. (15.22) yields the maximum value $m \approx 49$, while for a sharp crack that has an $FT = 2.28$ the minimum value is obtained, which is exactly 4.

15.5 Process Volume Effect

The real power of the Weibull analysis is to assess the process volume effect. We have seen how this volume, V_{pv}, in which brittle fracture first occurs is a very small region ahead of the crack tip, whose size is roughly given by Eq. (15.5). Figure 15.10 shows the crack front of a large size 4TCT specimen in which 8 1/2TCTspecimens

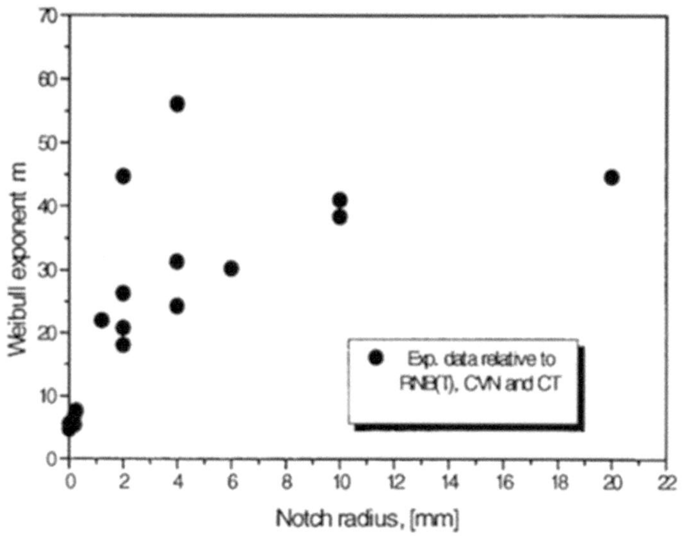

Fig. 15.10 Variation of the Weibull exponent m with the notch root radius of various types of specimens [8]

are inserted to show the difference in fracture toughness sampling size between large and small specimens. Small ICRs (isolated cleavage regions) are almost equally distributed along the crack front of large and small specimens. In BCC materials at very low temperatures, much below the RT_{NDT}, these small inclusions are responsible for cleavage fracture, therefore at very, very low temperature, for instance –150 °C, there is almost no difference in toughness, K_{Ic}, using small or large specimens, as it can be seen in Fig. 15.6 for A508B Cl.3 forging. This circumstance can be also seen in Fig. 9.51 where the spread of experimental fracture toughness results sharply reduces as the temperature decreases and below $T\text{-}RT_{NDT} = -200$ °F there is practically no spread.

However, as the temperature rises, larger and larger ICRs are needed to trigger brittle fracture. A large ICR has a much higher probability of being found in a large specimen than in a small one, therefore larger specimens will fail at lower K_c. Accordingly, also the spread of experimental values coming from small specimens will increase. The process volume of a large specimen, $V_{L,pv}$, can be seen as made of s statistically independent elemental volumes V_{pv}: $V_{L,pv} = s{\cdot}V_{pv}$. For example, for the case shown in Fig. 15.10 the process volume of the large 4TCT specimen, $V_{L,pv}$, is: $V_{L,pv} = 8V_{pv}$. From a probabilistic stand point, applying the weakest link criterion, it is possible to assert that the entire volume of a large specimen, $V_{L,pv}$, fails if at least one of its elemental constituent parts s of volume V_{pv} fails. Alternatively, the probability of non-failure or reliability $R_s = 1 - P_s(K)$ of the system formed by s unit volumes V_{pv} is given by the probability of non-failure of all s parts. Using the cumulative probability low

$$[1 - P_s(K)] = [1 - P(K)]^s. \tag{15.23}$$

What Eq. (15.23) is actually saying is that the probability of failure of a system formed by s pieces is higher than the probability of failure of each single piece. Using Eqs. (15.14) and (15.23) can be written as (Fig. 15.11)

$$\exp\left[-\left(\frac{K}{K_{so}}\right)^m\right] = \left\{\exp\left[-\left(\frac{K}{K_o}\right)^m\right]\right\}^s \tag{15.24}$$

or

$$\left(\frac{K}{K_{so}}\right)^m = s\left(\frac{K}{K_s}\right)^m \tag{15.25}$$

Recalling that $s = V_{L,pv}/V_{pv}$, Eq. (15.24) becomes

$$\frac{K_o}{K_{so}} = \left(\frac{V_{L,pv}}{V_{pv}}\right)^{1/m} \tag{15.26}$$

or more generally

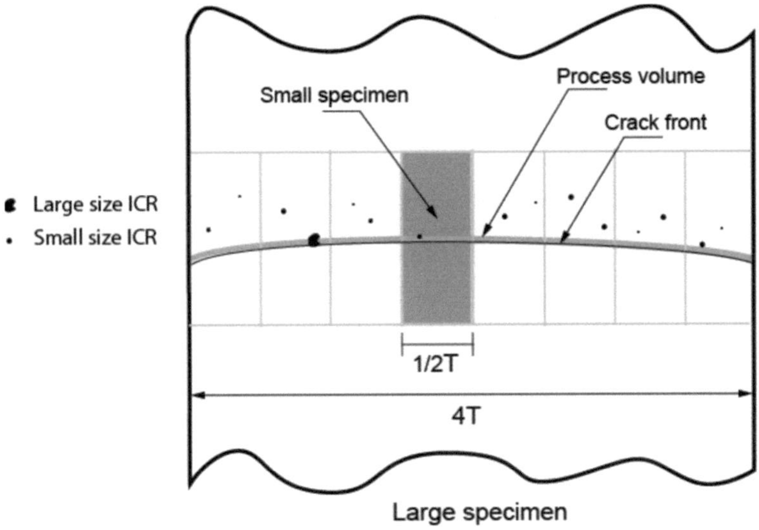

Fig. 15.11 Schematic view of the difference in fracture toughness sampling size between large and small specimens. Large inclusions or ICRs are more likely to be found ahead of a large crack front line in the process volume of a large thickness specimen than in a small one where small inclusions are more common

$$\frac{K_L}{K} = \left(\frac{V_{pv}}{V_{L,pv}}\right)^{\frac{1}{m}} \qquad (15.27)$$

Equation (15.27) holds the key to metallurgical variability through the process volume effect. It states that the fracture toughness measured with a large specimen, K_L, is equal to that measured with a single small specimen, K, times the inverse ratio of their process volumes elevated to $1/m$. since $m = 4$, at least as long as the K_I field dominates, by increasing the process volume eight times, $V_{L,pv}/V_{pv} = 8$, would decrease the fracture toughness measurement by 0.59, almost 50%.

It is clear that by further increasing the size and, therefore, the process volume, we may eventually arrive to predict a fracture toughness extremely low, incompatible with real experience. This is because Eq. (15.13) poses no lower limit to a measurable K that may even approach zero. Therefore, Eq. (15.13) must necessarily contain a limit lower value, K_{min}, below which it is not possible to go

$$P(x) = 1 - \exp\left[-\left(\frac{K - K_{min}}{K_o}\right)^m\right] \qquad (15.28)$$

Equation (15.28) introduces a more realistic version of the Weibull cumulative probability function known as three parameter Weibull distribution, K_o, K_{min} and m. The value of K_{min} at any temperature must be known in advance or guessed by experience. For instance, at very low temperature it has never been obtained an experimental fracture toughness value below 35 MPa$\sqrt{\text{m}}$ for a carbon steel. Knowing K_{min}, the Weibull analysis is done following exactly the same procedure presented for the two parameter Weibull approach.

References

1. Iwadate, T., Tanaka, Y., Ono, S., Watanabe, J.: An analysis of elastic plastic fracture toughness behaviour for JIC measurements in the transition region. In: Elastic Plastic Fracture: Second Symposium, Vol. II, ASTM STP 803, pp. 531–561 (1983)
2. Landes, J.D., Shaffer, D.H.: Statistical characterization of fracture in the transition region. Fract. Mech. ASTM STP **700**, 368–382 (1979)
3. Merkle, J.G.: An examination of the size effects and data scatter observed in small-specimen cleavage fracture toughness testing, NUREG/CR-3672, ORNL/TMM-9088 (1984)
4. Irwin, G.R.: Fracture mode transition for a crack traversing a plate. J. Basic Eng. **82**(2), 417–425 (1960)
5. Irwin, G.R.: Plastic zone near a crack and fracture toughness. In: Proceedings of the 7th Sagamore Ordnance Materials Research Conference, Syracuse Research Institute, 16–19 Aug 1960
6. Merkle, J.G.: An examination of size effects and data scatter observed in small specimen cleavage fracture toughness testing. Oak Ridge National Laboratory, NUREG/CR 3672, April 1984
7. Weibull, W.: A Statistical Theory of Strength of Materials. Royal Swedish Institute for Engineering Research, No. 151, Stockholm (1939)
8. Milella, P.P., Bonora, N.: On the dependence of the Weibull exponent on geometry and loading conditions and its implications on the fracture toughness probability curve using a local approach criterion. Int. J. Fract. **104**, 71–87 (2000)

Chapter 16
Stress Intensity Factor, *J*-Integral and Fully Plastic Solutions for Selected Configurations

16.1 Introduction

Linear elastic fracture mechanics and elastic–plastic fracture mechanics apply to two different situations where materials and structures behave in a brittle fashion without any apparent plastic deformation at the crack tip or with some more or less extended plastic deformation preceding fracture. In the case of fully brittle behavior, the crack tip stress field is fundamentally elastic and dominated by the Irwin stress intensity factor K, while the fully plastic behavior is dominated by the Rice J-integral. The most general situation is represented by a combination of the two cases and can be expressed, see Sect. 14.5 and Fig. 14.26, with a J_{tot} given by the sum of two terms

$$J_{tot} = J_e + J_p \tag{16.1}$$

withe the J elastic, J_e, is given by Eq. (14.20)

$$J_e = \frac{K_I^2}{E'} \tag{16.2}$$

When the plastic zone becomes so large to swell boundless over the entire section containing the crack, as schematically shown in Fig. 10.7, both LEFM and EPFM fail to describe the material behavior, defined as general yield, and new solutions are needed based on the limit load approach. Some of the linear elastic expressions for the stress intensity factor K have already been provided for the most common specimen geometries in Chap. 9. In the following sections, solutions will be given for the linear elastic stress intensity factor, K, for some three-dimensional geometries containing cracks. In addition, fully plastic solutions will be also provided for the most common fracture mechanics specimens and some cylindrical geometries representative of pipes and pressure vessels containing cracks.

P. P. Milella, *Fundamentals of Fracture Mechanics*,
https://doi.org/10.1007/978-3-032-23081-2_16

16.2 Linear Elastic Solutions for Some Three-Dimensional Geometries with Cracks

The following tables provide elastic solutions, K_I, for sone tridimensional geometries containing cracks. The K_I expressions are given in the most general form

$$K_I = (\sigma_m + H\sigma_b)\sqrt{\frac{\pi a}{Q}}f \tag{16.3}$$

where σ_m and σ_b are the tensile or membrane stress and the bending stress respectively, Q is the flaw shape parameter already seen, Eq. (9.64), and f a shape factor. Table 16.1 provides the linear elastic stress intensity factor solutions for a plate with a surface semielliptical cracks with depth a ≤ c, Fig. 16.1, while Table 16.2 [1] refers to the same geometry with a > c, Fig. 16.2. The case of an elliptical central crack, Fig. 16.3, is treated in Table 16.3 while the corner crack case is treated in Table 16.4, Fig. 16.4, for a ≤ c and in Table 16.5, Fig. 16.5, for $a > c$ [1].

16.3 Fully Plastic Solutions for Selected Specimens

We will consider, first, the compact specimen, Fig. 16.6, for which the linear elastic stress intensity factor solution, K_I, has already been given in Sect. 9.4 along with the load line displacement Δ and the COD or δ. Here, we will provide the J_p plastic solutions given in the most general form already seen in Sect. 14.5 and Eq. (14.24)

$$\begin{aligned} J_p &= \alpha\varepsilon_o\sigma_o b h_1\left(\frac{a}{W}, n\right)\left(\frac{P}{P_o}\right)^{n+1} \\ \delta_p &= \alpha\varepsilon_o a h_2\left(\frac{a}{W}, n\right)\left(\frac{P}{P_o}\right)^{n} \\ \Delta_p &= \alpha\varepsilon_o a h_3\left(\frac{a}{W}, n\right)\left(\frac{P}{P_o}\right)^{n} \end{aligned} \tag{16.4}$$

where α, $\varepsilon_o$andσ_o are the Ramberg-Hosgood constants, Eq. (1.41), a is the crack length, and h_1, h_2, and h_3 are dimensionless parameters that depend on geometry, a/W ratio, with W being specimen width and n is the hardening exponent. The load P_o that appears in Eq. (16.4) is a reference load or limit load. For the compact specimen, its value is given by [2]

$$P_o = 1.455\eta B b\sigma_o \tag{16.5}$$

under plain stain conditions and [2]

Table 16.1 Stress intensity factor solution, K_I, for semi-elliptical surface crack with $a \le c$

See Fig. 16.1 σ_m = membrane stress σ_b = bending stress M = applied moment $\sigma_b = \dfrac{Ms}{2I}$ $I = \dfrac{wb^3}{6}$	$K_I = (\sigma_m + H\sigma_b)\sqrt{\dfrac{\pi a}{Q}} f\left(\dfrac{a}{s}, \dfrac{a}{c}, \dfrac{c}{w}, \Phi\right)$ $Q = 1 + 1.464\left(\dfrac{a}{c}\right)^{1.65}$ $f = \left[M_1 + M_2\left(\dfrac{a}{t}\right)^2 + M_3\left(\dfrac{a}{t}\right)^4\right] g_\Phi g_w h$ $M_1 = 1.13 - 0.09 \cdot \left(\dfrac{a}{c}\right)$ $M_2 = -0.54 + \dfrac{0.89}{0.2 + \left(\frac{a}{c}\right)}$ $M_3 = 0.5 - \dfrac{1}{0.65 + \left(\frac{a}{c}\right)} + 14 \cdot \left(1.0 - \dfrac{a}{c}\right)^{24}$ $g_\Phi = \left[\left(\dfrac{a}{c}\right)^2 \cos^2\Phi + \sin^2\Phi\right]^{1/4}$ $g_w = \left[\sec\left(\dfrac{\pi c}{2w}\sqrt{\dfrac{a}{t}}\right)\right]^{1/2}$ $h = 1 + \left[0.1 + 0.35 \cdot \left(\dfrac{a}{t}\right)^2\right](1 - \sin\Phi)^2$ $H = H_1 + (H_2 - H_1)(\sin\Phi)^p$ $p = 0.2 + \dfrac{a}{c} + 0.6 \cdot \left(\dfrac{a}{t}\right)$ $H_1 = 1 - 0.34 \cdot \left(\dfrac{a}{t}\right) - 0.11 \cdot \left(\dfrac{a}{c}\right)\left(\dfrac{a}{t}\right)$ $H_2 = 1 - F_1\left(\dfrac{a}{t}\right) + F_2\left(\dfrac{a}{t}\right)^2$ $F_1 = -1.22 - 0.12 \cdot \left(\dfrac{a}{c}\right)$ $F_2 = 0.55 - 1.05 \cdot \left(\dfrac{a}{c}\right)^{0.75} + 0.47 \cdot \left(\dfrac{a}{c}\right)^{1.5}$

$$P_o = 1.072\eta Bb\sigma_o \tag{16.6}$$

for plain stress conditions, respectively, where

$$\eta = \sqrt{\left(\frac{2a}{b}\right)^2 + \frac{4a}{b} + 2} - \left(\frac{2a}{b} + 1\right) \tag{16.7}$$

The values of the dimensionless parameters h_1, h_2, and h_3 are listed in Table 16.6 for plain strain conditions and in Table 16.7 for plain stress conditions, for several values of the ratio *a/W* and strain hardening exponent, *n*.

For a Single-Edge Notch in Bending, SEN(B) or Three-point Bending specimen, Fig. 16.7, whose linear elastic solutions have been provided in Sect. 9.5, the plastic solutions are always given by Eq. (16.4) with load P_o under plain stain conditions

Table 16.2 Stress intensity factor solution, K_I, for a plate with a Semi-Elliptical Surface Crack with a > c [1]

See Fig. 16.2 σ_m = membrane stress σ_b = bending stress M = applied moment $\sigma_b = \frac{Ms}{2I}$ $I = \frac{ws^3}{6}$	$K_I = (\sigma_m + H\sigma_b)\sqrt{\frac{\pi a}{Q}} f\left(\frac{a}{s}, \frac{a}{c}, \frac{c}{w}, \varphi\right)$ $Q = 1 + 1.464\left(\frac{c}{a}\right)^{1.65}$ $f = \left[M_1 + M_2\left(\frac{a}{t}\right)^2 + M_3\left(\frac{a}{t}\right)^4\right] g_\varphi g_w h$ $M_1 = \sqrt{\frac{c}{a}}\left[1.0 + 0.04 \cdot \left(\frac{c}{a}\right)\right]$ $M_2 = 0.2 \cdot \left(\frac{c}{a}\right)^4$ $M_3 = -0.11 \cdot \left(\frac{c}{a}\right)^4$ $g_\Phi = \left[\left(\frac{c}{a}\right)^2 \sin^2\varphi + \cos^2\varphi\right]^{1/4}$ $g_w = \left[\sec\left(\frac{\pi c}{2w}\sqrt{\frac{a}{t}}\right)\right]^{1/2}$ $h = 1 + \left[0.1 + 0.35 \cdot \left(\frac{c}{a}\right)\left(\frac{a}{t}\right)^2\right](1 - \sin\varphi)^2$ $H = H_1 + (H_2 - H_1)(\sin\varphi)^p$ $p = 0.2 + \frac{c}{a} + 0.6 \cdot \left(\frac{a}{t}\right)$ $H_1 = 1 + F_{11} \cdot \left(\frac{a}{t}\right) + F_{12} \cdot \left(\frac{a}{t}\right)^2$ $H_2 = 1 + F_{21}\left(\frac{a}{t}\right) + F_{22}\left(\frac{a}{t}\right)^2$ $F_{11} = -0.04 - 0.41 \cdot \left(\frac{c}{a}\right)$ $F_{12} = 0.55 - 1.93 \cdot \left(\frac{c}{a}\right)^{0.75} + 1.38 \cdot \left(\frac{c}{a}\right)^{1.5}$ $F_{21} = -2.11 - 0.77 \cdot \left(\frac{c}{a}\right)$ $F_{22} = 0.55 - 0.72 \cdot \left(\frac{c}{a}\right)^{0.75} + 0.14 \cdot \left(\frac{c}{a}\right)^{1.5}$

given by [2]

$$P_o = \frac{1.455Bb^2}{S}\sigma_o \tag{16.8}$$

and for plain stress conditions by [2]

$$P_o = \frac{1.072Bb^2}{S}\sigma_o \tag{16.9}$$

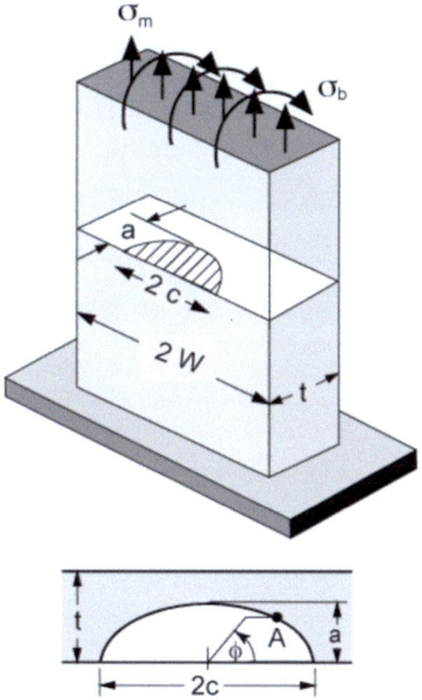

Fig. 16.1 Plate with a surface semielliptical cracks with a $\leq$ c under a traction stress or membrane stress, σ_m, and a bending stress, σ_b

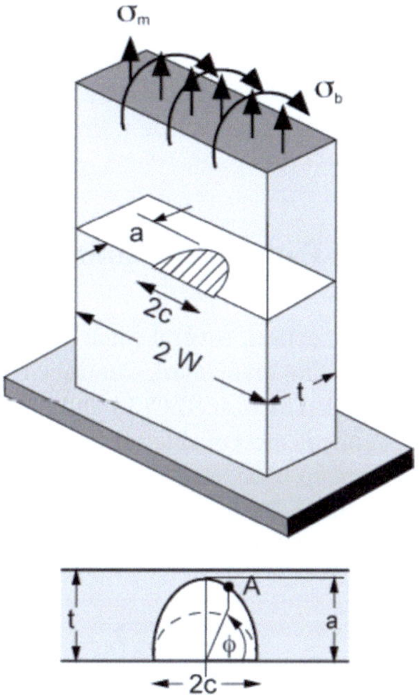

Fig. 16.2 Plate with a semielliptical surface crack with $a > c$ under a traction stress or membrane stress, σ_m, and a bending stress, σ_b

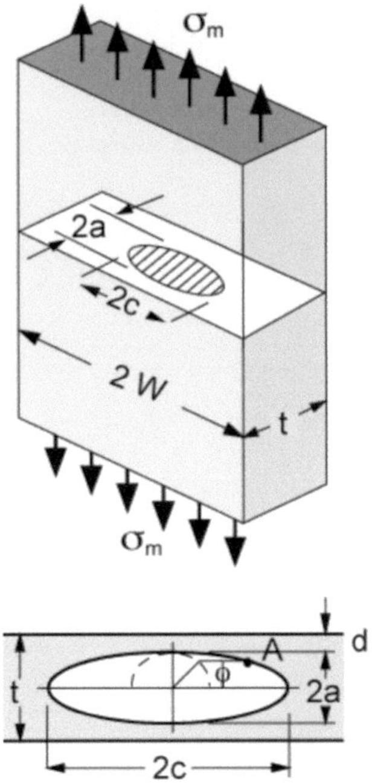

Fig. 16.3 Plate with an elliptical central crack under remote traction or membrane stress

The values of the dimensionless parameters h_1, h_2, and h_3 are listed in Table 16.8 for plain strain conditions and in Table 16.9 for plain stress conditions, respectively, for several values of the ratio a/W and strain hardening exponent, n.

16.4 Panels Containing Cracks in Traction

In this section, several plain geometries will be considered as they respond in the plastic domain that are: Central Crack Panel in Traction CCP(T), Single-Edge Crack Panel in Traction SEC(T) and Double-Edge Crack Panel in Traction DEC(T). All specimens are considered under plain strain and plain stress conditions. The same specimens have been already analyzed under linear elastic conditions and the relative stress intensity factor solutions, K_I, have been provided in Sects. 9.2. As to the Central Crack Panel in Traction CCP(T), Fig. 16.8, it is [2]

$$
\begin{aligned}
J_p &= \alpha\varepsilon_o\sigma_o\frac{ba}{W}h_1\left(\frac{a}{W},n\right)\left(\frac{P}{P_o}\right)^{n+1}\\
\delta_p &= \alpha\varepsilon_o a h_2\left(\frac{a}{W},n\right)\left(\frac{P}{P_o}\right)^{n}\\
\Delta_p &= \alpha\varepsilon_o a h_3\left(\frac{a}{W},n\right)\left(\frac{P}{P_o}\right)^{n}
\end{aligned}
\tag{16.10}
$$

Table 16.3 Stress intensity factor solution, K_I, for a plate with an Elliptical Central Crack [1]

See Fig. 16.3 σ_m = membrane stress	$K_I = \sigma_m \sqrt{\frac{\pi a}{Q}} f\left(\frac{a}{s}, \frac{a}{c}, \frac{c}{w}, \Phi\right)$ $\lambda = \frac{a}{a+d}$ $f = \left(M_1 + M_2\lambda^2 + M_3\lambda^4\right) g_\Phi g_w h$ $M_2 = \frac{0.05}{0.11 + \left(\frac{a}{c}\right)^{1.5}}$ $M_3 = \frac{0.29}{0.23 + \left(\frac{a}{c}\right)^{1.5}}$ $g_w = \left[\sec\left(\frac{\pi c}{2w}\sqrt{\frac{2a}{t}}\right)\right]^{1/2}$ $h = 1 - \frac{\lambda^4\sqrt{2.6 - 2\lambda}}{1 + 4\lambda}\lvert\cos\Phi\rvert$ *per* $a/c \leq 1$: $Q = 1 + 1.464\left(\frac{a}{c}\right)^{1.65}$ $g_\Phi = \left[\left(\frac{a}{c}\right)^2 \cos^2\Phi + \sin^2\Phi\right]^{1/4}$ $M_1 = 1$ *per* $a/c > 1$: $Q = 1 + 1.464\left(\frac{c}{a}\right)^{1.65}$ $g_\Phi = \left[\left(\frac{c}{a}\right)^2 \sin^2\Phi + \cos^2\Phi\right]^{1/4}$ $M_1 = \sqrt{\frac{c}{a}}$

where α, $\varepsilon_o$andσ_o are the Ramberg-Hosgood constants, Eq. (1.41), a is the crack length, and h_1, h_2, and h_3 are dimensionless parameters that depend on geometry and *a/b* ratio, with b being the uncracked ligament, W is the panel width as shown in Fig. 16.8, and n the hardening exponent. The load P_o that appears in Eq. (16.10) is a reference load or limit load. For the Central Crack Panel in Traction CCP(T) under plain stain conditions its value is given by [2]

$$P_o = \frac{4}{\sqrt{3}} B b \sigma_o \tag{16.11}$$

and for plain stress conditions [2] by

$$P_o = 2 B b \sigma_o \tag{16.12}$$

Table 16.4 Stress intensity factor solution, K_I, for a plate with Quarter-Elliptical Corner Crack with a ≤ c [1]

See Fig. 16.4 σ_m = membrane stress σ_b = bending stress M = applied moment	$K_I = (\sigma_m + H\sigma_b)\sqrt{\frac{\pi a}{Q}} f\left(\frac{a}{s}, \frac{a}{c}, \frac{c}{w}, \varphi\right)$ $Q = 1 + 1.464\left(\frac{a}{c}\right)^{1.65}$ $f = \left[M_1 + M_2\left(\frac{a}{t}\right)^2 + M_3\left(\frac{a}{t}\right)^4\right] h_1 h_2 g_\varphi g_w$ $M_1 = 1.08 - 0.03 \cdot \left(\frac{a}{c}\right)^{1.65}$ $M_2 = \frac{-0.44 + 1.06}{0.3 + \left(\frac{a}{c}\right)}$ $M_3 = -0.5 - 0.25 \cdot \left(\frac{a}{c}\right) + 14.8 \cdot \left(1 - \frac{a}{c}\right)^{15}$ $g_\varphi = \left[\left(\frac{a}{c}\right)^2 \cos^2\varphi + \sin^2\varphi\right]^{1/4}$ $g_w = \left[\sec\left(\frac{\pi c}{2w}\sqrt{\frac{a}{t}}\right)\right]^{1/2}$ $h_1 = 1 + \left[0.08 + 0.4 \cdot \left(\frac{a}{t}\right)^2\right](1 - \sin\varphi)^3$ $h_2 = 1 + \left[0.08 + 0.15 \cdot \left(\frac{a}{t}\right)^2\right](1 - \cos\varphi)^3$ $H = H_1 + (H_2 - H_1)(\sin\varphi)^p$ $p = 0.2 + \frac{a}{c} + 0.6 \cdot \left(\frac{a}{t}\right)$ $H_1 = 1 - 0.34 \cdot \left(\frac{a}{t}\right) - 0.11 \cdot \left(\frac{a}{c}\right)\left(\frac{a}{t}\right)$ $H_2 = 1 - F_1\left(\frac{a}{t}\right) + F_2\left(\frac{a}{t}\right)^2$ $F_1 = -1.22 - 0.12 \cdot \left(\frac{a}{c}\right)$ $F_2 = 0.55 - 1.05 \cdot \left(\frac{a}{c}\right)^{0.75} + 0.47 \cdot \left(\frac{a}{c}\right)^{1.5}$

The values of the dimensionless parameters h_1, h_2, and h_3 are listed in Table 16.10 for plain strain conditions and in Table 16.11 for plain stress conditions, respectively, for several values of the ratio *a/W* and strain hardening exponent, *n*.

For the Single Edge Crack Panel in Traction, SEC(T), Fig. 16.9, Eq. (16.10) olds. In plain strain condition it is [2]

$$P_o = 1.455\eta B b \sigma_o \tag{16.13}$$

where

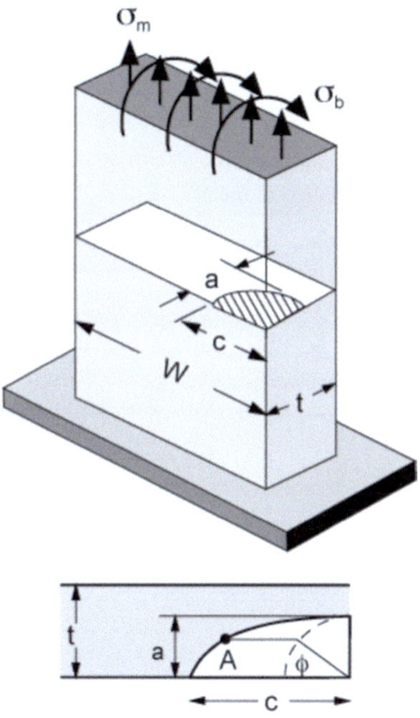

Fig. 16.4 Plate with a Quarter-Elliptical Corner Crack with a ≤ c

$$\eta = \sqrt{1 + \left(\frac{a}{b}\right)^2} - \frac{a}{b} \tag{16.14}$$

In plain stress condition it is [2]

$$P_o = 1.072\eta Bb\sigma_o \tag{16.15}$$

whith η given by Eq. (16.14). The values of the dimensionless parameters h_1, h_2, and h_3 are listed in Table 16.12 for plain strain conditions and in Table 16.13 for plain stress conditions, respectively, for several values of the ratio *a/W* and strain hardening exponent, *n*.

For the Double-Edge Crack Panel in Traction DEC(T), Fig. 16.10, Eq. (16.10) still old. In plain strain conditions it is [2]

$$P_o = \left(0.72 + 1.82\frac{b}{W}\right)Bb\sigma_o \tag{16.16}$$

while in plain stress conditions it is [2]

$$P_o = \frac{4}{\sqrt{3}}Bb\sigma_o \tag{16.17}$$

Table 16.5 Stress intensity factor solutions, K_I, for a plate with Quarter-Elliptical Corner Crack with a > c [1]

See Fig. 16.5 σ_m = membrane stress σ_b = bending stress M = applied moment $\sigma_b = \frac{Ms}{2I}$ $I = \frac{ws^3}{6}$	$K_I = (\sigma_m + H\sigma_b)\sqrt{\frac{\pi a}{Q}} f\left(\frac{a}{s}, \frac{a}{c}, \frac{c}{w}, \varphi\right)$ $Q = 1 + 1.464\left(\frac{c}{a}\right)^{1.65}$ $f = \left[M_1 + M_2\left(\frac{a}{t}\right)^2 + M_3\left(\frac{a}{t}\right)^4\right] g_\varphi g_w h_1 h_2$ $M_1 = \sqrt{\frac{c}{a}}\left[1.08 + 0.03\cdot\left(\frac{c}{a}\right)\right]$ $M_2 = 0.375\cdot\left(\frac{c}{a}\right)^2 \quad M_3 = -0.25\cdot\left(\frac{c}{a}\right)^2$ $g_\Phi = \left[\left(\frac{c}{a}\right)^2 \sin^2\varphi + \cos^2\varphi\right]^{1/4}$ $g_w = \left[\sec\left(\frac{\pi c}{2w}\sqrt{\frac{a}{t}}\right)\right]^{1/2}$ $h_1 = 1 + \left[0.08 + 0.4\cdot\left(\frac{c}{t}\right)^2\right](1 - \sin\varphi)^3$ $h_2 = 1 + \left[0.08 + 0.15\cdot\left(\frac{c}{t}\right)^2\right](1 - \cos\varphi)^3$ $H = H_1 + (H_2 - H_1)(\sin\varphi)^p$ $p = 0.2 + \frac{c}{a} + 0.6\cdot\left(\frac{a}{t}\right)$ $H_1 = 1 + F_{11}\cdot\left(\frac{a}{t}\right) + F_{12}\cdot\left(\frac{a}{t}\right)^2$ $H_2 = 1 + F_{21}\left(\frac{a}{t}\right) + F_{22}\left(\frac{a}{t}\right)^2$ $F_{11} = -0.04 - 0.41\cdot\left(\frac{c}{a}\right)$ $F_{12} = 0.55 - 1.93\cdot\left(\frac{c}{a}\right)^{0.75} + 1.38\cdot\left(\frac{c}{a}\right)^{1.5}$ $F_{21} = -2.11 - 0.77\cdot\left(\frac{c}{a}\right)$ $F_{22} = 0.64 - 0.72\cdot\left(\frac{c}{a}\right)^{0.75} + 0.14\cdot\left(\frac{c}{a}\right)^{1.5}$

The values of the dimensionless parameters h_1, h_2, and h_3 are listed in Table 16.14 for plain strain conditions and in Table 16.15 for plain stress conditions, respectively, for several values of the ratio *a/W* and strain hardening exponent, *n*.

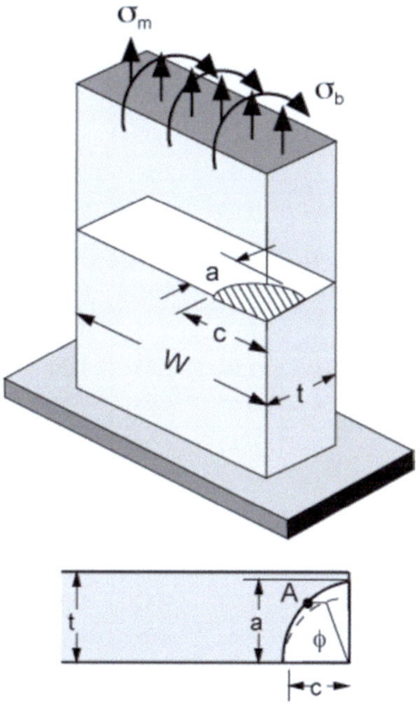

Fig. 16.5 Plate with a Quarter-Elliptical Corner Crack with a > c

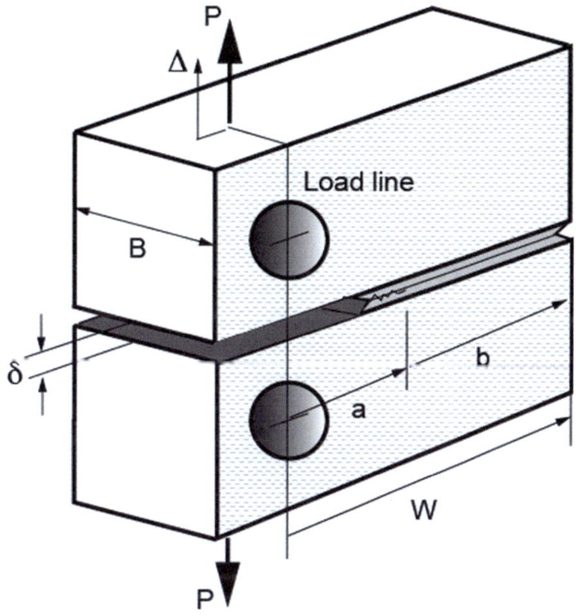

Fig. 16.6 Compact specimen. Symbols and dimensions

Table 16.6 Values of dimensionless h_1, h_2, and h_3 constants for compact specimen in plain strain [2]

a/W		n = 1	n = 2	n = 3	n = 5	n = 7	n = 10	n = 13	n = 16	n = 20
0.250	h_1	2.23	2.05	1.78	1.48	1.33	1.26	1.25	1.32	1.57
	h_2	17.9	12.5	11.7	10.8	10.5	10.7	11.5	12.6	14.6
	h_3	9.85	8.51	8.17	7.77	7.71	7.92	8.52	9.31	10.9
0.375	h_1	2.15	1.72	1.39	0.970	0.693	0.443	0.276	0.176	0.098
	h_2	12.6	8.18	6.52	4.32	2.97	1.79	1.10	0.686	0.370
	h_3	7.94	5.76	4.64	3.10	2.14	1.29	0.793	0.494	0.266
0.500	h_1	1.94	1.51	1.24	0.919	0.685	0.461	0.314	0.216	0.132
	h_2	9.33	5.85	4.30	2.75	1.91	1.20	0.788	0.530	0.317
	h_3	6.41	4.27	3.16	2.02	1.41	0.888	0.585	0.393	0.236
0.625	h_1	1.76	1.45	1.24	0.974	0.752	0.602	0.459	0.347	0.248
	h_2	7.61	4.57	3.42	2.36	0.81	1.32	0.983	0.749	0.485
	h_3	5.52	3.43	2.58	1.79	1.37	1.00	0.746	0.568	0.368
0.750	h_1	1.71	1.42	1.26	1.033	0.864	0.717	0.575	0.448	0.345
	h_2	6.37	3.95	3.18	2.34	1.88	1.44	1.12	0.887	0.665
	h_3	4.86	3.05	2.46	1.81	1.45	1.11	0.869	0.686	0.514
1	h_1	1.57	1.45	1.35	1.18	1.08	0.950	0.850	0.730	0.630
	h_2	5.39	3.74	3.09	2.43	2.12	1.80	1.57	1.33	1.14
	h_3	4.31	2.99	2.47	1.95	1.79	1.44	1.26	1.07	0.909

16.5 Fully Plastic *J* Solutions for Flawed Cylinders Under Internal Pressure

We have treated the subject of pipes and pressure vessels containing cracks in Sect. 14.5. In this section some values of the fully plastic solution J_p will be provided for cylinders under internal pressure, p. The linear Elastic Stress Intensity Factor solutions, K_I, have been given in Sect. 9.19. The general expression for J_p and takes the form

$$
\begin{aligned}
J_p &= \alpha\sigma_o\varepsilon_o b\left(\frac{a}{t}\right)h_1\left(\frac{a}{t}, n; \frac{R_i}{R_o}\right)\left(\frac{p}{p_o}\right)^{n+1} \\
\delta_p &= \alpha\varepsilon_o a h_2\left(\frac{a}{t}, n; \frac{R_i}{R_o}\right)\left(\frac{p}{p_o}\right)^{n}
\end{aligned}
\tag{16.18}
$$

where, with reference to Fig. 16.11, R_o is the external radius, R_i the internal radius, a is the crack depth, b the ligament, t the thickness of the cylinder, δ_p the COD and; α, $\sigma_o$andε_o are the usual Ramberg-Hosgood constants, Eq. (1.41), and n is the strain hardening exponent. Two cases will be considered. The first refers to a pipe or

Table 16.7 Values of dimensionless h_1, h_2, and h_3 constants for compact specimen in plain stress [2]

a/W		n = 1	n = 2	n = 3	n = 5	n = 7	n = 10	n = 13	n = 16	n = 20
0.250	h_1	1.61	1.46	1.28	1.06	0.903	0.729	0.601	0.511	0.395
	h_2	17.6	12.0	10.7	8.74	7.32	5.74	4.63	3.75	2.92
	h_3	9.67	8.00	7.21	5.94	5.00	3.95	3.19	2.59	2.023
0.375	h_1	1.55	1.25	1.05	0.801	0.647	0.484	0.377	0.284	0.220
	h_2	12.4	8.20	6.54	4.56	3.45	2.44	1.83	1.36	1.02
	h_3	7.80	5.73	4.62	3.25	2.48	1.77	1.33	0.990	0.746
0.500	h_1	1.40	1.08	0.901	0.686	0.558	0.436	0.356	0.298	0.238
	h_2	9.16	5.67	4.21	2.80	2.12	1.57	1.25	1.03	0.814
	h_3	6.29	4.15	3.11	2.09	1.59	1.18	0.938	0.774	0.614
0.625	h_1	1.27	1.03	0.875	0.695	0.593	0.494	0.423	0.370	0.310
	h_2	7.47	4.48	3.35	2.37	1.92	1.54	1.29	1.12	0.928
	h_3	5.42	3.38	2.54	1.80	1.47	1.18	0.988	0.853	0.710
0.750	h_1	1.23	0.977	0.833	0.683	0.598	0.506	0.431	0.373	0.314
	h_2	6.25	3.78	2.89	2.14	1.78	1.44	1.20	1.03	0.857
	h_3	4.77	2.92	2.24	1.66	1.38	1.12	0.936	0.800	0.666
1	h_1	1.13	1.01	0.775	0.680	0.650	0.620	0.490	0.470	0.420
	h_2	5.29	3.54	2.41	1.91	1.73	1.59	1.23	1.17	1.03
	h_3	4.23	2.83	1.93	1.52	1.39	1.27	0.985	0.933	0.824

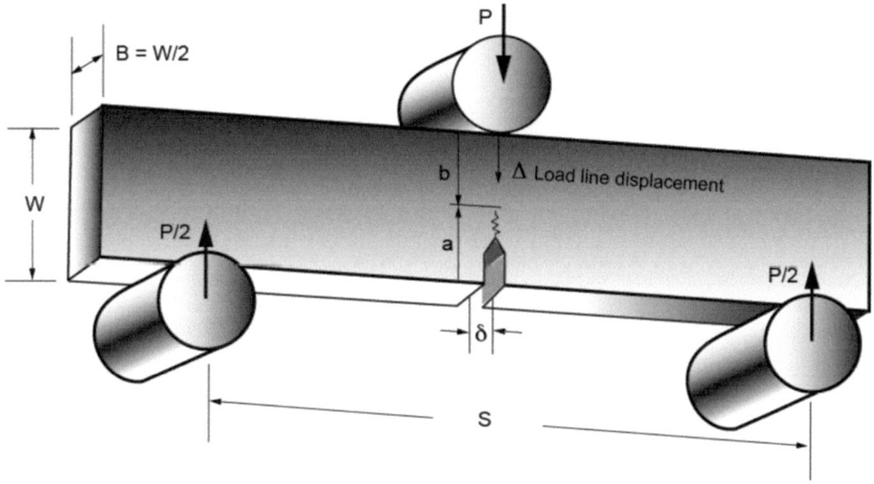

Fig. 16.7 Single-Edge Notch specimen in Bending, SEN(B). Symbols and dimensions

Table 16.8 Values of dimensionless h_1, h_2, and h_3 constants for a SEN(B) specimen in Plain Strain [2]

a/W		n = 1	n = 2	n = 3	n = 5	n = 7	n = 10	n = 13	n = 16	n = 20
0.125	h_1	0.936	0.869	0.805	0.687	0.580	0.437	0.329	0.245	0.165
	h_2	6.97	6.77	6.29	5.29	4.38	3.24	2.40	1.78	1.19
	h_3	3.00	22.1	20.0	15.0	11.7	8.39	6.14	4.54	3.01
0.250	h_1	1.20	1.034	0.930	0.762	0.633	0.523	0.396	0.303	0.215
	h_2	5.80	4.67	4.01	3.08	2.45	1.93	1.45	1.09	0.758
	h_3	4.08	9.72	8.36	5.86	4.47	3.42	2.54	1.90	1.32
0.375	h_1	1.33	1.15	1.02	0.084	0.695	0.556	0.442	0.360	0.265
	h_2	5.18	3.93	3.20	2.38	1.93	1.47	1.15	0.928	0.684
	h_3	4.51	6.01	5.03	3.74	3.02	2.30	1.80	1.45	1.07
0.500	h_1	1.41	1.09	0.922	0.675	0.495	0.331	0.211	0.135	0.0741
	h_2	4.87	3.28	2.53	1.69	1.19	0.773	0.480	0.304	0.165
	h_3	4.69	4.33	3.49	2.35	1.66	1.08	0.669	0.424	0.230
0.625	h_1	1.46	1.07	0.896	0.631	0.436	0.255	0.142	0.084	0.0411
	h_2	4.64	2.86	2.16	1.37	0.907	0.518	0.287	0.166	0.0806
	h_3	4.71	3.49	2.70	1.72	1.14	0.652	0.361	0.209	0.102
0.750	h_1	1.48	1.15	0.974	0.693	0.500	0.348	0.223	0.140	0.0745
	h_2	4.47	2.75	2.10	1.36	0.936	0.618	0.388	0.239	0.127
	h_3	4.49	3.14	2.40	1.56	1.07	0.704	0.441	0.272	0.144
0.875	h_1	1.50	1.35	1.20	1.02	0.855	0.690	0.551	0.440	0.321
	h_2	4.36	2.90	2.31	1.70	1.33	1.00	0.782	0.613	0.459
	h_3	4.15	3.08	2.45	1.81	1.41	1.06	0.828	0.646	0.486

vessel under internal pressure p carrying a continuous longitudinal crack, as shown in Fig. 16.11.

The hoop stress is opening the crack. For this case the limit pressure p_o is [2]

$$p_o = \frac{2}{\sqrt{3}} \frac{b\sigma_o}{R_c}; \; R_c = R_i + a \tag{16.19}$$

Table 16.16 provides the values of constants h_1, h_2 and h_3 for several values of the thickness to internal radius ratio, *t/R*, and *a/t* ratio, and for several values of the strain hardening exponent, *n*.

Table 16.9 Values of dimensionless h_1, h_2, and h_3 constants for a SEN(B) specimen in Plain Stress [2]

a/W		n = 1	n = 2	n = 3	n = 5	n = 7	n = 10	n = 13	n = 16	n = 20
0.125	h_1	0.676	0.600	0.548	0.459	0.383	0.297	0.238	0.192	0.148
	h_2	6.84	6.30	5.66	4.53	3.64	2.72	2.12	1.67	1.26
	h_3	2.95	20.1	14.6	12.2	9.12	6.75	5.20	4.09	3.07
0.250	h_1	0.869	0.731	0.629	0.479	0.370	0.246	0.174	0.117	0.0593
	h_2	5.69	4.50	3.68	2.61	1.95	1.29	0.897	0.603	0.307
	h_3	4.01	8.81	7.19	4.73	3.39	2.20	1.52	1.01	0.508
0.375	h_1	0.963	0.797	0.680	0.527	0.418	0.307	0.232	0.174	0.105
	h_2	5.09	3.73	2.93	2.07	1.58	1.13	0.841	0.626	0.381
	h_3	4.42	5.53	4.48	3.17	2.41	1.73	1.28	0.948	0.575
0.500	h_1	1.02	0.767	0.621	0.453	0.324	0.202	0.128	0.00813	0.0298
	h_2	4.77	3.12	2.32	1.55	1.08	0.655	0.410	0.259	0.0974
	h_3	4.60	4.09	3.09	2.08	1.44	0.874	0.545	0.344	0.129
0.625	h_1	1.05	0.786	0.649	0.494	0.357	0.235	0.173	0.105	0.0471
	h_2	4.55	2.83	2.12	1.46	1.02	0.656	0.472	0.286	0.130
	h_3	4.62	3.43	2.60	1.79	1.26	0.803	0.577	0.349	0.158
0.750	h_1	1.07	0.786	0.643	0.474	0.343	0.230	0.167	0.110	0.0442
	h_2	4.39	2.66	1.97	1.33	0.928	0.601	0.427	0.280	0.114
	h_3	4.39	3.01	2.24	1.51	1.05	0.680	0.483	0.316	0.129
0.875	h_1	1.086	0.928	0.810	6.46	0.538	0.423	0.332	0.242	0.205
	h_2	4.28	2.76	2.16	1.56	1.23	0.922	0.702	0.561	0.428
	h_3	4.07	2.93	2.29	1.65	1.30	0.975	0.742	0.592	0.452

The second case refers to a pipe or vessel under internal pressure p carrying a circumferential crack, as shown in Fig. 16.12. The hoop stress is opening the rack. Table 16.16 provides the values of constants h_1, h_2 and h_3 for several values of the thickness to internal radius ratio, *t/R*, and *a/t* ratio, and for several values of the strain hardening exponent, *n*.

For this case, the limit pressure p_o is [2]

$$p_o = \frac{2}{\sqrt{3}}\sigma_o\pi\left(R_0^2 - R_c^2\right) \tag{16.20}$$

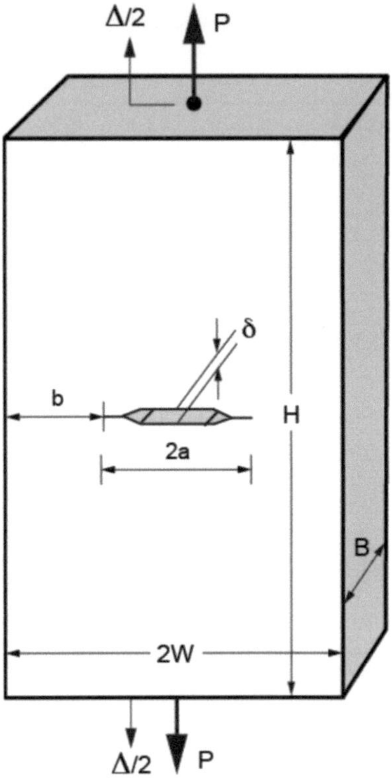

Fig. 16.8 Central crack panel in traction CCP(T)

Table 16.17 provides the values of constants h_1, h_2 and h_3 for several values of the thickness to internal radius ratio, t/R_i, and a/t ratio, and for several values of the strain hardening exponent, n.

Table 16.10 Values of dimensionless h_1, h_2, and h_3 constants for Central Crack Panel in Traction CCP(T) in Plain Strain [2]

a/W		n = 1	n = 2	n = 3	n = 5	n = 7	n = 10	n = 13	n = 16	n = 20
0.125	h_1	2.80	3.61	4.06	4.35	4.33	4.02	3.56	3.06	2.46
	h_2	3.05	3.62	3.91	4.06	3.93	3.54	3.07	2.60	2.06
	h_3	0.303	0.574	0.840	1.30	1.63	1.95	2.03	1.96	1.77
0.250	h_1	2.54	3.01	3.21	3.29	3.18	2.92	2.63	2.34	2.03
	h_2	2.68	2.99	3.01	2.85	2.61	2.30	1.97	1.71	1.45
	h_3	0.536	0.911	1.22	1.64	1.84	1.85	1.80	1.64	1.43
0.375	h_1	2.34	2.62	2.65	2.51	2.28	1.97	1.71	1.46	1.19
	h_2	2.35	2.39	2.23	1.88	1.58	1.28	1.07	0.890	0.715
	h_3	0.699	1.06	1.28	1.44	1.40	1.23	1.05	0.888	0.719
0.500	h_1	2.21	2.29	2.20	1.97	1.76	1.52	1.32	1.16	0.978
	h_2	2.03	1.86	1.60	1.23	1.00	0.799	0.664	0.564	0.466
	h_3	0.803	1.07	1.16	1.10	0.968	0.796	0.665	0.565	0.469
0.625	h_1	2.12	1.96	1.76	1.43	1.17	0.863	0.628	0.458	0.300
	h_2	1.71	1.32	1.04	0.707	0.524	0.358	0.250	0.178	0.114
	h_3	0.844	0.937	0.879	0.701	0.522	0.361	0.251	0.178	0.115
0.750	h_1	2.07	1.73	1.47	1.11	0.895	0.642	0.461	0.337	0.216
	h_2	1.35	0.857	0.596	0.361	0.254	0.167	0.114	0.0810	0.0511
	h_3	0.805	0.700	0.555	0.359	0.254	0.168	0.114	0.0813	0.0516
0.875	h_1	2.08	1.64	1.40	1.14	0.987	0.814	0.688	0.573	0.461
	h_2	0.889	0.428	0.287	0.181	0.139	0.105	0.0837	0.0682	0.0533
	h_3	0.632	0.400	0.291	0.182	0.140	0.106	0.0839	0.0683	0.0535

Table 16.11 Values of dimensionless h_1, h_2, and h_3 constants for Central Crack Panel in Traction CCP(T) in Plain Stress [2]

a/W		n = 1	n = 2	n = 3	n = 5	n = 7	n = 10	n = 13	n = 16	n = 20
0.125	h_1	2.80	3.57	4.01	4.47	4.65	4.62	4.41	4.13	3.72
	h_2	3.53	4.09	4.43	4.74	4.79	4.63	4.33	4.00	3.55
	h_3	0.350	0.661	0.997	1.55	2.05	2.56	2.83	2.95	2.92
0.250	h_1	2.54	2.97	3.14	3.20	3.11	2.86	2.65	2.47	2.20
	h_2	3.10	3.29	3.30	3.15	2.93	2.56	2.29	2.08	1.81
	h_3	0.619	1.01	1.35	1.83	2.08	2.19	2.12	2.01	1.79
0.375	h_1	2.34	2.53	2.52	2.35	2.17	1.95	1.77	1.61	1.43
	h_2	2.71	2.62	2.41	2.03	1.75	1.47	1.28	1.13	0.988
	h_3	0.807	1.20	1.43	1.59	1.57	1.43	1.27	1.13	0.994

(continued)

Table 16.11 (continued)

a/W		n = 1	n = 2	n = 3	n = 5	n = 7	n = 10	n = 13	n = 16	n = 20
0.500	h_1	2.21	2.20	2.06	1.81	1.63	1.43	1.30	1.17	1.00
	h_2	2.34	2.01	1.70	1.30	1.07	0.871	0.757	0.666	0.557
	h_3	0.927	1.19	1.26	1.18	1.04	0.867	0.758	0.668	0.560
0.625	h_1	2.12	1.91	1.69	1.41	1.22	1.01	0.853	0.712	0.573
	h_2	1.97	1.46	1.13	0.785	0.617	0.474	0.383	0.313	0.256
	h_3	0.975	1.05	0.970	0.763	0.620	0.478	0.386	0.318	0.273
0.750	h_1	2.07	1.71	1.46	1.21	1.08	0.867	0.745	0.646	0.532
	h_2	1.55	0.970	0.685	0.452	0.361	0.262	0.216	0.183	0.148
	h_3	0.929	0.802	0.642	0.450	0.361	0.263	0.216	0.183	0.149
0.875	h_1	2.08	1.57	1.31	1.08	0.972	0.862	0.778	0.715	0.630
	h_2	1.03	0.485	0.310	0.196	0.157	0.127	0.109	0.0971	0.0842
	h_3	0.730	0.452	0.313	0.198	0.157	0.127	0.109	0.0973	0.0842

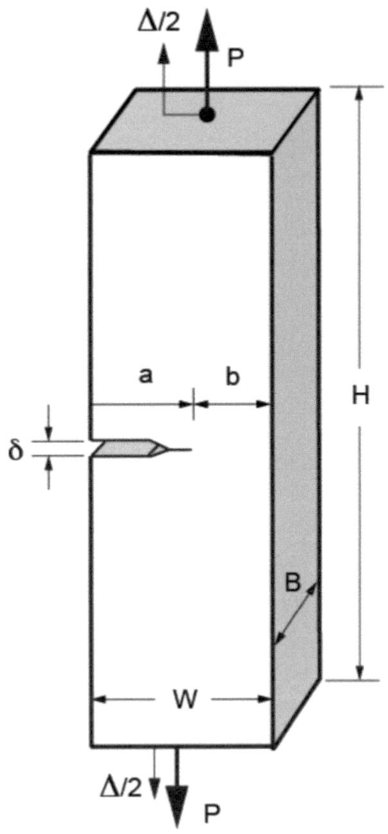

Fig. 16.9 Single edge crack panel in traction, SEC(T)

Table 16.12 Values of dimensionless h_1, h_2, and h_3 constants for Single Edge Crack Panel in Traction, SEC(T), in Plain Strain [2]

a/W		n = 1	n = 2	n = 3	n = 5	n = 7	n = 10	n = 13	n = 16	n = 20
0.125	h_1	4.95	6.93	8.57	11.5	13.5	16.1	18.1	19.9	21.2
	h_2	5.25	6.47	7.56	9.46	11.1	12.9	14.4	15.7	16.8
	h_3	26.6	25.8	25.2	24.2	23.6	23.2	23.2	23.5	23.7
0.250	h_1	4.34	4.77	4.64	3.82	3.06	2.17	1.55	1.11	0.712
	h_2	4.76	4.56	4.28	3.39	2.64	1.81	1.25	0.875	0.552
	h_3	10.3	7.64	5.87	3.70	2.48	1.50	0.970	0.654	0.404
0.375	h_1	3.88	3.25	2.63	1.68	1.06	0.539	0.276	0.142	0.0595
	h_2	4.54	3.49	2.67	1.57	0.946	0.458	0.229	0.116	0.048
	h_3	5.14	2.99	1.90	0.923	0.515	0.240	0.119	0.060	0.0246
0.500	h_1	3.40	2.30	1.69	0.928	0.514	0.213	0.0902	0.0385	0.0119
	h_2	4.45	2.77	1.89	0.954	0.507	0.204	0.0854	0.0356	0.0110
	h_3	3.15	1.54	0.912	0.417	0.215	0.085	0.0358	0.0147	0.0045
0.625	h_1	2.86	1.80	1.30	0.697	0.378	0.153	0.0625	0.0256	0.0078
	h_2	4.37	2.44	1.62	0.081	0.423	0.167	0.0671	0.0272	0.0082
	h_3	2.31	1.08	0.681	0.329	0.171	0.067	0.0268	0.0108	0.0033
0.750	h_1	2.34	1.61	1.25	0.769	0.477	0.233	0.116	0.059	0.0215
	h_2	4.32	2.52	1.79	1.03	0.619	0.296	0.146	0.0735	0.0267
	h_3	2.2	1.10	0.765	0.435	0.262	0.125	0.0617	0.0312	0.0113
0.875	h_1	1.91	1.57	1.37	1.10	0.925	0.702			
	h_2	4.29	2.75	2.14	1.55	1.23	0.921			
	h_3	2.01	1.27	0.988	0.713	0.564	0.424			

Table 16.13 Values of dimensionless h_1, h_2, and h_3 constants for Single-Edge Crack Panel in Traction SEC(T) in plain stress [2]

a/W		n = 1	n = 2	n = 3	n = 5	n = 7	n = 10	n = 13	n = 16	n = 20
0.125	h_1	3.58	4.55	5.06	5.30	4.96	4.14	3.29	2.60	1.92
	h_2	5.15	5.43	6.05	6.01	5.47	4.46	3.48	2.74	2.02
	h_3	26.1	21.6	18.0	12.7	9.24	5.98	3.94	2.72	2.0
0.250	h_1	3.14	3.26	2.92	2.12	1.53	0.960	0.615	0.400	0.230
	h_2	4.67	4.30	3.70	2.53	1.76	1.05	0.656	0.410	0.237
	h_3	10.1	6.49	4.36	2.19	1.24	0.630	0.362	0.224	0.123
0.375	h_1	2.88	2.37	1.94	1.37	1.01	0.677	0.474	0.342	0.226
	h_2	4.47	3.43	2.63	1.69	1.18	0.762	0.534	0.372	0.244
	h_3	5.05	2.65	1.60	0.812	0.525	0.328	0.223	0.157	0.102

(continued)

Table 16.13 (continued)

a/W		n = 1	n = 2	n = 3	n = 5	n = 7	n = 10	n = 13	n = 16	n = 20
0.500	h_1	2.46	1.67	1.25	0.776	0.510	0.286	0.164	0.0956	0.0469
	h_2	4.37	2.73	1.91	1.09	0.694	0.380	0.216	0.124	0.0607
	h_3	3.10	1.43	0.871	0.461	0.286	0.155	0.088	0.0506	0.0247
0.625	h_1	2.07	1.41	1.105	0.755	0.551	0.363	0.248	0.172	0.107
	h_2	4.30	2.55	1.84	1.16	0.816	0.523	0.353	2.42	0.150
	h_3	2.27	1.13	0.771	0.478	0.336	0.215	0.146	0.100	0.0616
0.750	h_1	1.70	1.14	0.910	0.624	0.447	0.280	0.181	0.118	0.0670
	h_2	4.24	2.47	1.81	1,15	0.798	0.490	0.314	0.203	0.115
	h_3	1.98	1.09	0.784	0.494	0.344	0.211	0.136	0.0581	0.0496
0.875	h_1	1.38	1.11	0.962	0.792	0.677	0.574			
	h_2	4.22	2.68	2.08	1.54	1.27	1.04			
	h_3	1.97	1.25	0.969	0.716	0.591	0.483			

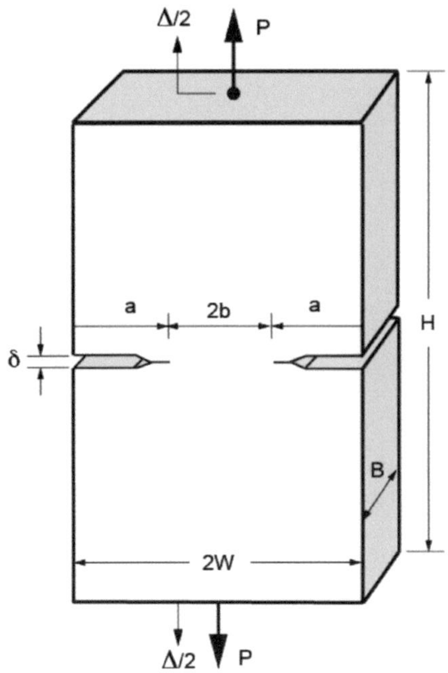

Fig. 16.10 Double-edge crack panel in traction DEC(T)

Table 16.14 Values of dimensionless h_1, h_2, and h_3 constants for Double-Edge Crack Panel in Traction DEC(T), in Plain Strain [2]

a/W		n = 1	n = 2	n = 3	n = 5	n = 7	n = 10	n = 13	n = 16	n = 20
0.125	h_1	0.572	0.772	0.922	1.13	1.35	1.61	1.86	2.08	2.44
	h_2	0.732	0.852	0.961	1.14	1.29	1.50	1.70	1.94	2.17
	h_3	0.063	0.126	0.200	0.372	0.571	0.911	1.30	1.74	2.29
0.250	h_1	1.10	1.32	1.38	1.65	1.75	1.82	1.86	1.89	1.92
	h_2	1.56	1.63	1.70	1.78	1.80	1.81	1.79	1.78	1.76
	h_3	0.267	0.479	0.698	1.11	1.47	1.92	2.25	2.49	2.73
0.375	h_1	1.61	1.83	1.92	1.92	1.84	1.68	1.49	1.32	1.12
	h_2	2.51	2.41	2.35	2.15	1.94	1.68	1.44	1.25	1.05
	h_3	0.637	1.05	1.40	1.87	2.11	2.20	2.09	1.92	1.67
0.500	h_1	2.22	2.43	2.48	2.43	2.32	2.12	1.91	1.60	1.51
	h_2	3.73	3.40	3.15	2.71	2.37	2.01	1.72	1.40	1.38
	h_3	1.26	1.92	2.37	2.79	2.85	2.68	2.40	1.99	1.94
0.625	h_1	3.16	3.38	3.45	3.42	3.28	3.00	2.54	2.36	2.27
	h_2	5.57	4.76	4.23	3.46	2.97	2.48	2.02	1.82	1.66
	h_3	2.36	3.29	3.74	3.90	3.68	3.23	2.66	2.40	2.19
0.750	h_1	5.24	6.29	7.17	8.44	9.46	10.9	11.9	11.3	17.4
	h_2	9.10	7.76	7.14	6.64	6.83	7.48	7.79	7.14	11.1
	h_3	4.73	6.26	7.03	7.63	8.14	9.04	9.40	8.58	13.5
0.875	h_1	14.2	24.8	39.0	78.4	140.0	341.0	777.0	1570.0	3820.0
	h_2	20.1	19.4	22.7	36.1	58.9	133.0	294.0	585.0	1400.0
	h_3	12.7	18.2	24.1	40.4	65.8	149.0	327.0	650.0	1560.0

Table 16.15 Values of dimensionless h_1, h_2, and h_3 constants for Double-Edge Crack Panel in Traction DEC(T), in Plain Stress [2]

a/W		n = 1	n = 2	n = 3	n = 5	n = 7	n = 10	n = 13	n = 16	n = 20
0.125	h_1	0.583	0.825	1.02	1.37	1.71	2.24	2.84	3.54	4.62
	h_2	0.853	1.05	1.23	1.55	1.87	2.38	2.96	3.65	4.70
	h_3	0.0729	0.159	0.26	0.504	0.821	1.41	2.18	3.16	4.73
0.250	h_1	1.01	1.23	1.36	1.48	1.54	1.58	1.59	1.59	1.59
	h_2	1.73	1.82	1.89	1.92	1.91	1.85	1.80	1.75	1.70
	h_3	0.296	0.537	0.770	1.17	1.49	1.82	2.02	2.12	2.20
0.375	h_1	1.29	1.42	1.43	1.34	1.24	1.09	0.970	0.873	0.674
	h_2	2.59	2.39	2.22	1.86	1.59	1.28	1.07	0.922	0.709
	h_3	0.658	1.04	1.30	1.52	1.55	1.41	1.23	1.07	0.830

(continued)

Table 16.15 (continued)

a/W		n = 1	n = 2	n = 3	n = 5	n = 7	n = 10	n = 13	n = 16	n = 20
0.500	h_1	1.48	1.47	1.38	1.17	1.01	0.845	0.732	0.625	0.208
	h_2	3.51	2.82	2.34	1.67	1.28	0.944	0.762	0.630	0.232
	h_3	1.18	1.58	1.69	1.56	1.32	1.01	0.809	0.662	0.266
0.625	h_1	1.59	1.45	1.29	1.04	0.882	0.737	0.649	0.466	0.0202
	h_2	4.56	3.15	2.32	1.45	1.06	0.790	0.657	0.473	0.0277
	h_3	1.93	2.14	1.95	1.44	1.09	0.809	0.665	0.487	0.0317
0.750	h_1	1.65	1.43	1.22	0.979	0.834	0.701	0.630	0.297	
	h_2	5.90	3.37	2.22	1.30	0.966	0.741	0.636	0.312	
	h_3	3.06	2.67	2.06	1.31	0.978	0.747	0.638	0.318	
0.875	h_1	1.69	1.43	1.22	0.979	0.845	0.738	0.664	0.614	0.562
	h_2	8.02	3.51	2.14	1.27	0.971	0.775	0.663	0.596	0.535
	h_3	5.07	3.18	2.16	1.30	0.980	0.779	0.665	0.597	0.538

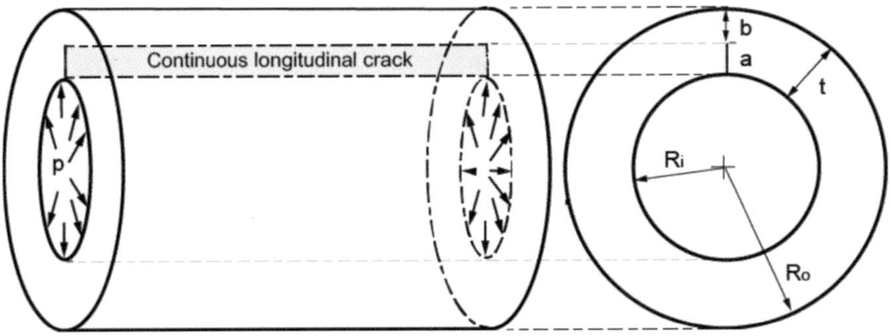

Fig. 16.11 Cylinder under internal pressure p with an internal continuous longitudinal crack

Table 16.16 Values of the functions h_1, h_2 and h_3 for a cylinder under internal pressure p carrying an internal longitudinal crack for several values of the ratio, t/R_i and a/t ratio, and for several values of the strain hardening exponent, n [2]

a/t		n = 1	n = 2	n = 3	n = 5	n = 7	n = 10
$t/R_i = 0.2$							
0.125	h_1	6.32	7.93	9.32	11.5	13.12	14.94
	h_2	5.83	7.01	7.96	9.49	10.67	11.96
0.25	h_1	7.00	8.34	9.03	9.59	9.71	9.45
	h_2	5.92	8.72	7.07	7.26	7.14	6.71
0.5	h_1	9.79	10.37	9.07	5.61	3.52	2.11
	h_2	7.05	6.97	6.01	3.70	2.28	1.25
0.75	h_1	11.0	5.54	2.84	1.24	0.83	0.493
	h_2	7.35	3.86	1.86	0.556	0.261	0.129
$t/R_i = 0.1$							
0.125	h_1	5.22	6.64	7.59	8.76	9.34	9.55
	h2	5.31	6.25	6.88	7.65	8.02	8.09
0.25	h_1	6.16	7.49	7.96	8.08	7.78	6.98
	h2	5.56	6.31	6.52	6.40	6.01	5.27
0.50	h_1	10.5	11.6	10.7	6.47	3.95	2.27
	h_2	7.48	7.72	7.01	4.29	2.58	1.37
0.75	h_1	16.1	8.19	3.87	1.46	1.05	0.787
	h_2	9.57	5.40	2.57	0.706	0.370	0.232
$t/R_i = 0.05$							
1/8	h_1	4.50	5.79	6.62	7.65	8.07	7.75
	h_2	4.96	5.71	6.20	6.82	7.02	6.66
1/4	h_1	5.57	6.91	7.37	7.47	7.21	6.53
	h_2	5.29	5.98	6.16	6.01	5.63	4.93
1/2	h_1	10.8	12.8	12.8	8.16	4.88	2.62
	h_2	7.66	8.33	8.13	5.33	3.20	1.65
3/4	h_1	23.1	13.1	5.87	1.90	1.23	0.883
	h_2	12.1	7.88	3.84	1.01	0.454	0.240

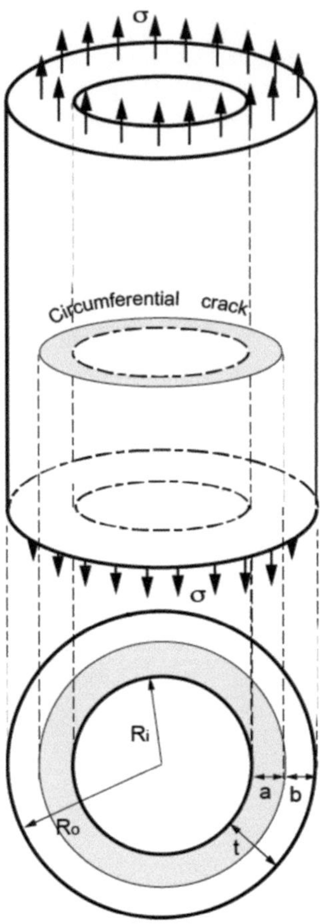

Fig. 16.12 Cylinder in tension with an internal circumferential crack

Table 16.17 Values of the functions h_1, h_2 and h_3 for a cylinder in tension carrying an internal circumferential crack for several values of the ratio, t/R_i and a/t ratio, and for several values of the strain hardening exponent, n [2]

a/t		n = 1	n = 2	n = 3	n = 5	n = 7	n = 10
$t/R_i = 0.2$							
0.125	h_1	3.78	5.00	5.94	7.54	8.99	11.1
	h_2	4.56	5.55	6.37	7.79	9.10	11.0
	h_3	0.369	0.700	1.07	1.96	3.04	4.94
0.25	h_1	3.88	4.95	5.64	6.49	6.94	7.22
	h_2	4.40	5.12	5.57	6.07	6.28	6.30
	h_3	0.673	1.25	1.79	2.79	3.61	4.52
0.5	h_1	4.40	4.78	4.59	3.79	3.07	2.34
	h_2	4.36	4.30	3.91	3.00	2.26	1.55
	h_3	1.33	1.93	2.21	2.23	1.94	1.46
0.75	h_1	4.12	3.03	2.23	1.546	1.30	1.11
	h_2	3.46	2.19	1.36	0.638	0.436	0.325
	h_3	1.54	1.39	1.04	0.686	0.508	0.366
$t/R_i = 0.1$							
0.125	h_1	4.00	5.13	6.09	7.69	9.09	11.1
	h_2	4.71	5.63	6.45	7.85	9.09	10.9
	h_3	0.548	0.733	1.13	2.07	3.16	5.07
0.25	h_1	4.17	5.35	6.09	6.93	7.30	7.41
	h_2	4.58	5.36	5.84	6.31	6.44	6.31
	h_3	0.757	1.35	1.93	2.96	3.78	4.60
0.50	h_1	5.40	5.90	5.63	4.51	3.49	2.47
	h_2	4.99	5.01	4.59	3.48	2.56	1.67
	h_3	1.555	2.26	2.59	2.57	2.18	1.56
0.75	h_1	5.18	3.78	2.57	1.59	1.31	1.10
	h_2	4.22	2.79	1.67	0.725	0.48	0.300
	h_3	1.86	1.73	1.26	0.775	0.561	0.360
$t/R_i = 0.05$							
0.125	h_1	4.04	5.23	6.22	7.82	9.19	11.1
	h_2	4.82	5.69	6.52	7.90	9.11	10.8
	h_3	0.680	0.759	1.17	2.13	3.23	5.12
0.25	h_1	4.38	5.68	6.45	7.29	7.62	7.65
	h_2	4.71	5.56	6.05	6.51	6.59	6.39
	h_3	0.818	1.43	2.03	3.10	3.91	4.69
0.50	h_1	6.55	7.17	6.89	5.46	4.13	2.77
	h_2	5.67	5.77	5.36	4.08	2.97	1.88

(continued)

Table 16.17 (continued)

a/t		n = 1	n = 2	n = 3	n = 5	n = 7	n = 10
	h_3	1.80	2.59	2.99	2.98	2.50	1.74
0.75	h_1	6.64	4.87	3.08	1.68	1.30	1.07
	h_2	5.18	3.57	2.07	0.808	0.472	0.316
	h_3	2.36	2.18	1.53	0.772	0.494	0.330

References

1. Newman. J.C., Raju. I.S.: Stress intensity factor equations for cracks in three-dimensional finite bodies subjected to tension and bending loads, NASA Technical Memorandum 85793, NASA Langley Research Center, Hampton, VA (1984)
2. Kumar, V., German, M.D., Shih, C.F.: An engineering approach for elastic-plastic fracture analysis, EPRI Report NP-1931, Electric Power Research Institute, Palo Alto, California (1981)

Index

P. P. Milella, *Fundamentals of Fracture Mechanics*,
https://doi.org/10.1007/978-3-032-23081-2

MIX
Papier aus verantwortungsvollen Quellen
Paper from responsible sources
FSC® C105338

If you have any concerns about our products,
you can contact us on
ProductSafety@springernature.com

In case Publisher is established outside the EU,
the EU authorized representative is:
Springer Nature Customer Service Center GmbH
Europaplatz 3, 69115 Heidelberg, Germany

Printed by Libri Plureos GmbH
in Hamburg, Germany